Materials Horizons: From Nature to Nanomaterials

Series Editor

Vijay Kumar Thakur, School of Aerospace, Transport and Manufacturing, Cranfield University, Cranfield, UK

Materials are an indispensable part of human civilization since the inception of life on earth. With the passage of time, innumerable new materials have been explored as well as developed and the search for new innovative materials continues briskly. Keeping in mind the immense perspectives of various classes of materials, this series aims at providing a comprehensive collection of works across the breadth of materials research at cutting-edge interface of materials science with physics, chemistry, biology and engineering.

This series covers a galaxy of materials ranging from natural materials to nanomaterials. Some of the topics include but not limited to: biological materials, biomimetic materials, ceramics, composites, coatings, functional materials, glasses, inorganic materials, inorganic-organic hybrids, metals, membranes, magnetic materials, manufacturing of materials, nanomaterials, organic materials and pigments to name a few. The series provides most timely and comprehensive information on advanced synthesis, processing, characterization, manufacturing and applications in a broad range of interdisciplinary fields in science, engineering and technology.

This series accepts both authored and edited works, including textbooks, monographs, reference works, and professional books. The books in this series will provide a deep insight into the state-of-art of Materials Horizons and serve students, academic, government and industrial scientists involved in all aspects of materials research.

More information about this series at http://www.springer.com/series/16122

Sabu Thomas · Preetha Balakrishnan
Editors

Green Composites

Springer

Editors
Sabu Thomas
School of Chemical Sciences
Mahatma Gandhi University
Kottayam, Kerala, India

Preetha Balakrishnan
IIUCNN
Mahatma Gandhi University
Kottayam, Kerala, India

ISSN 2524-5384 ISSN 2524-5392 (electronic)
Materials Horizons: From Nature to Nanomaterials
ISBN 978-981-15-9642-1 ISBN 978-981-15-9643-8 (eBook)
https://doi.org/10.1007/978-981-15-9643-8

This Springer imprint is published by the registered company Springer Nature Singapore Pte Ltd.
The registered company address is: 152 Beach Road, #21-01/04 Gateway East, Singapore 189721, Singapore

Preface

This book comprehensively addresses eco-friendly green composites, their applications and modifications. The introductory chapters examine why green composites should be considered. The book then discusses properties of cellulose and wood. The book is essential for government agricultural departments, automotive companies, and others devoted to eco-friendly materials and production. Global awareness of environmental issues has resulted in the emergence of economically and environmentally friendly bio-based materials free from the traditional side effects of synthetics. This book delivers an overview of the advancements made in the development of natural bio-renewable resources-based materials, including processing methods and potential applications in green composites. Bio-renewable polymers are a special class of natural material found in nature, such as natural fibers, wheat straw, rice husk, and saw dust. In addition to offering renewable feedstocks, natural bio-renewable materials are compostable, recyclable, edible, and more energy efficient to process than plastic.

Preetha Balakrishnan
ADSO Naturals Private Limited
KINFRA park
Koratty, India

Curesupport B. V.
The Netherlands

Contents

About the Editors

Prof. Sabu Thomas is the Director of the International and Interuniversity Centre for Nanoscience and Nanotechnology and full professor of Polymer Science and Engineering at the School of Chemical Sciences of Mahatma Gandhi University, Kottayam, Kerala, India. He is an outstanding leader with sustained international acclaims for his work in polymer science and engineering, polymer nanocomposites, elastomers, polymer blends, interpenetrating polymer networks, polymer membranes, green composites and nanocomposites, nanomedicine and green nanotechnology. Dr. Thomas's ground breaking inventions in polymer nanocomposites, polymer blends, green bionanotechnological and nano-biomedical sciences, have made transformative differences in the development of new materials for automotive, space, housing and biomedical fields. Professor Thomas has received a number of national and international awards which include: Fellowship of the Royal Society of Chemistry, London, Distinguished Professorship from Josef Stefan Institute, Slovenia, MRSI medal, Nano Tech Medal, CRSI medal, Distinguished Faculty Award, and Sukumar Maithy Award for the best polymer researcher in the country. He is in the list of most productive researchers in India and holds a position of No.5. Very recently, because of the outstanding contributions to the field of Polymer Science and Engineering, Prof. Thomas has been conferred Honoris Causa (DSc) by the University of South Brittany, Lorient, France. Professor Thomas has published over 600 peer reviewed research papers, reviews and book chapters. He has co-edited 50 books published by Royal Society, Wiley, Wood head, Elsevier, CRC Press, Springer, Nova etc. He is the inventor of 6 patents. The h-index of Prof. Thomas is 72 and has more than 21,000 citations. Prof. Thomas has delivered over 300 Plenary/Inaugural and Invited lectures in national/international meetings over 30 countries. He has established a state of the art laboratory at Mahatma Gandhi University in the area of Polymer Science and Engineering and Nanoscience and Nanotechnology through external funding form DST, CSIR, TWAS, UGC, DBT, DRDO, AICTE, ISRO, DIT, TWAS, KSCSTE, BRNS, UGC-DAE, Du Pont, USA, General Cables, USA, Surface Treat

Czech Republic, MRF Tyres and Apollo Tyres. Professor Thomas has several international collaborative projects with a large number of countries abroad. He has already supervised 70 doctoral theses.

Dr. Preetha Balakrishnan is a Principal Scientist, QA/QC ADSO Naturals India and Curesupport Netherlands. She Completed her Ph. D in Chemistry from Mahatma Gandhi University Kottayam. Her area of expertise is polymer science and engineering, polymer nanocomposites, elastomers, polymer blends, interpenetrating polymer networks, polymer membranes, green composites and nanocomposites, nanomedicine and green nanotechnology. Her area of research includes preparation of starch-based films, blends and nanocomposites and nutraceuticals. She has published over 20 peer-reviewed research papers, reviews and book chapters. She has also co-edited 3 books published by Wood head, Elsevier. She has more than 5-year experience in polymer synthesis and composite science, and has presented several papers in national and international conferences.

Chapter 1
Green Composites: Introductory Overview

M. Roy Choudhury and K. Debnath

1 Introduction

The fiber-reinforced polymer composites have grabbed the commercial market vary rapidly in past few years. The reinforced polymer composite offers several advantages including conservation of dwindling reserves of conventional materials like metals and their alloys. However, the disposal of this kind of composite is a huge concern due to the non-biodegradable nature of the petroleum-based polymer and synthetic fiber. Looking into the environmental concern, engineers have focused on the sustainability of materials and shifted to the development of biomaterials. The first biomaterial composed of cotton fiber and phenol or melamine–formaldehyde was developed in the year 1908 [1]. All biomaterials are not fully biodegradable if the constituent materials are not derived from renewable sources. Research efforts have been given to develop fully degradable and environmentally friendly biomaterial called green composites [2]. These materials are consisted of biodegradable polymer and lignocellulosic bio-fiber derived from natural sources. These materials can degrade into water and carbon dioxide by the enzymatic action of a living organism. The green composites can be used for outdoor consumer products having short life cycles of one to two years (nondurable) or one-time products and indoor products with a useful life of several years [2]. The applications of green composites have been extensively increased in the various fields of engineering by replacing many engineering components made of synthetic fiber-reinforced composites. The topic of green composites has earned massive impetus in both the educational and manufacturing units due to their attractive properties such as low density, high specific strength, recyclability, economic, and eco-friendly. A variety of natural and synthetic biodegradable resins are available for use in green composites. Many reports and research articles have

M. R. Choudhury · K. Debnath (✉)
Department of Mechanical Engineering, NIT Meghalaya, Shillong 793003, India
e-mail: debnath.iitr@gmail.com

S. Thomas and P. Balakrishnan (eds.), *Green Composites*,
Materials Horizons: From Nature to Nanomaterials,
https://doi.org/10.1007/978-981-15-9643-8_1

been reported to lighten up various aspects of green composites. Zini and Scandola [3] discussed the environmental benefits deriving from the use of green composites. Fazita et al. [4] and Oyama et al. [5] discussed several functional properties related to packaging applications to explore the potential of green composites. Koronis et al. [6] and Ashori [7] reviewed the green composites adequate for automotive applications. Georgios et al. [8] and Dicker et al. [9] reported the applications of green composites in consumer products. Netravali and Chabba [2], Manita and Morreale [10], and Khalil et al. [11] reported the available biopolymer and natural fibers to develop green composites and their properties. An extensive review of properties, modification, characterization, processing techniques, applications of green composites was reported by Gholampour and Ozbakkaloglu [12] and Potluri [13].

From the above discussion, it can be seen that many researchers have tried to bring important information regarding green composites in one place in the review articles. This present chapter includes an introductory overview of green composites. The available green composites for appropriate applications, properties, processing, machinability, and joining behavior have been discussed.

2 Types of Green Composites

2.1 Biodegradable Polymer

The biodegradable polymer has a natural origin and can be converted to simpler compounds such as the carbon, nitrogen, and sulfur, mineralized, and redistributed through elemental cycles upon enzymatic reaction. Biodegradable polymers are mainly classified into two categories such as agro-polymers and biopolyesters. The agro-polymers are biomass products, while the biopolysters are obtained by fermentation of biomass or from genetically modified plants. A large number of these biodegradable polymers are commercially available nowadays. The different types of biodegradable polymers with their classifications are presented in Fig. 1. Among the agro-polymers such as starch, lignocellulose, casein, collagen, zein, soya, gluten, the soy protein polymers are proved to be an appropriate candidate for matrix material in green composites [14–19]. The soy proteins are manly available in the form of flour, isolate, and concentrate. Some researchers also studied starch and its modified blends as a matrix material for green composites [20–23]. Polylactic acid (PLA) which is a thermoplastic biopolyester has gained huge attention from researchers as it is derived from natural sources like corn, starch, etc. [24]. PLA is obtained by condensation of lactic acid or ring-opening polymerization of lactide. The thermoplastic and thermosetting polymers are classified based upon their response to heat. The thermoplastic polymer can re-melt to the liquid state upon heating, while thermosetting polymer remains in the permanent solid-state. Almost all biodegradable polymers obtained so far are thermoplastic polymer. The main problem associate with thermoplastic polymer is the low heat resistance capacity. In the year 2017, the

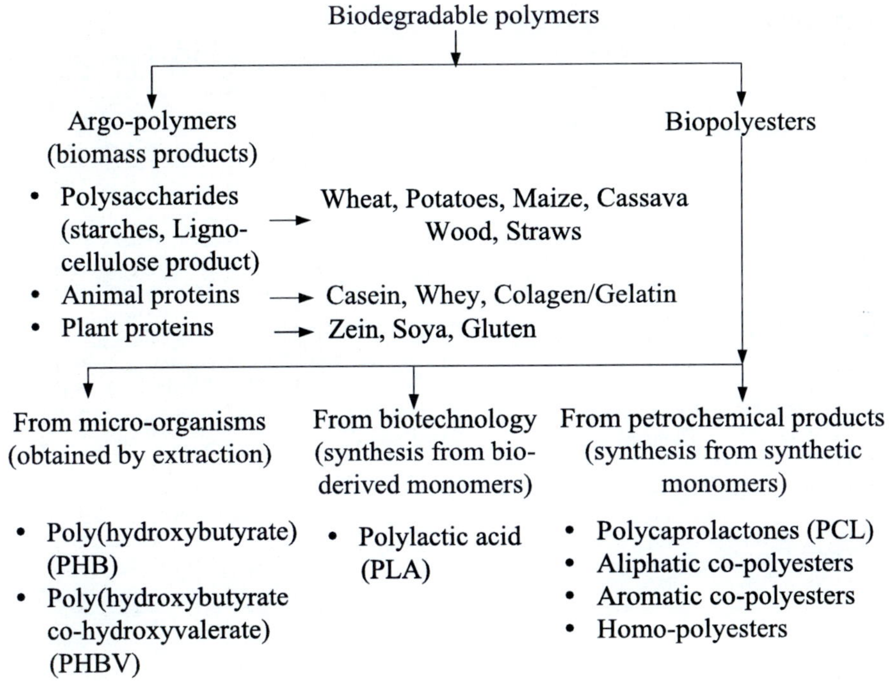

Fig. 1 Classification of biodegradable polymers. *Source* Author

researchers at the University of Amsterdam have developed a biodegradable thermosetting polymer called Glycix [25]. This was a major step forward in the battle against the waste disposal problem. Glycix can be converted into two completely natural compounds called glycerol and citric acid, as it comes in contact with water. The decomposition rate of Glycix depends on the degree to which the plastic has been hardened. Cai et al. [26] developed a biodegradable thermosetting polymer called poly(glycerol-succinate) in the year 2019. The properties of various biodegradable polymers are presented in Table 1.

2.2 *Bio-Fiber*

Natural fibers are emerged as potential reinforcing materials and have utilized as a replacement of many synthetic reinforcing materials. They have gained valuable attention from the researchers looking into the view of recyclability, biodegradability, low cost, high specific strength and stiffness, and low density. Natural fibers are classified based on their origins, such as plants, animals, or minerals. The classification of natural fiber can easily be understood from Fig. 2. All plant fibers are composed of cellulose, hemicelluloses, lignin, pectin, and waxes. Thousands of cellulose fibrils

Table 1 Properties of biodegradable polymers [27–36]

Polymers	P (g/cm^3)	T_m (°C)	T_g (°C)	TS (MPa)	EB (%)	E (GPa)	FM (GPa)
PLA	1.24	173–178	60–65	28	6	27	1.2
PGA	1.53	225–230	35–440	14.03	16.9	7	3
PLGA	1.30	140	57	4	10	2.2	$\sigma = 54.6$ MPa
PCL	1.11–1.14	58–65	−60	20.7–42	–	0.43	0.411
PEA	1.09	76–87.8	−34 to 3	8.5–19.6	1–14.3	0.25	2.9
Starch	1.3	110	−20 to 43	25	3	1.02	0.12
PBS	1.23	90–120	−40 to −10	62	–	0.05	0.47
PHB	1.25	160–165	2–5	25	4	3	3.20

PGA Polyglycolic acid; *PLGA* Poly(lactic-co-glycolic acid); *PEA* Polyesteramides; *PBS* Polybutylene succinate; ρ Density; T_m Melting temperature; T_g Glass transition temperatur; *TS* Tensile strength; *EB* Elongation at break; *E* Elastic modulus; *FM* Flexural modulus; σ Flexural strength

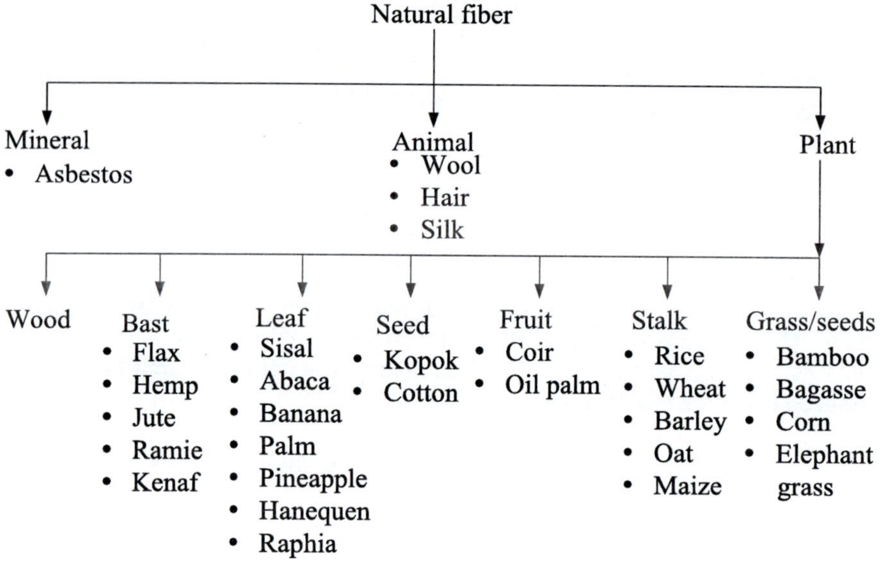

Fig. 2 Classification of bio-fibres. *Source* Author

are bind together by lignin and hemicelluloses naturally [37]. The diameter of cellulose fiber is usually in the range of 10–30 nm [1]. Moreover, these fibers have a higher aspect ratio that can transfer a higher load due to the flattened oval cross section. Animal fibers are the second most natural fibers after plant fibers. They are consisting of proteins and have an elliptical cross-section. The overall properties of the fibers can be determined by the sequence of the types of amino acid forming polypeptide chains. Generally, animal fibers are costlier, and the availability is lower than the plant fibers. That makes them expensive for normal day-to-day applications. Asbestos fibers are naturally occurring silicate minerals composed of long thin fibrous crystals. Asbestos is a unique mineral combining unusual physical and chemical properties, such as fibrous nature, thermal stability, high tensile strength, flexibility, and incombustibility, which make it highly suitable as a reinforcing material for polymers [38]. However, due to the fact of hazards to the health of human and animal life, its application is limited to a very small area. The properties of some extensively used natural fibers are mentioned in Table 2. Researchers have also developed some innovative ways to preserve the land by extracting the fibers from agricultural waste. Agricultural waste such as sunflower stalk [43], rice husk [44], wheat straw [45, 46], soy stalk [46], and cornhusk [47] was studied by researchers as reinforcement in green composites. The strength of natural fibers is relatively lower

Table 2 Properties of bio-fibers [3, 4, 9, 12, 39–42]

Fiber	*L* (mm)	*D* (μm)	*ρ* (g/cm³)	TS (MPa)	*E* (GPa)	EB (%)	WA (%)
Bamboo	~90	10–20	0.6–0.91	193–600	20–46	1.4	7
Flax	~900	12–16	1.5	345–2000	15–80	1.2–3.2	12
Hemp	~4000	16–50	1.48	550–900	26–80	1.6	8
Jute	~4000	17–20	1.3	393–800	13–55	1.16–1.5	12
Kenaf	~6	20–30	1.45	157–930	22–60	1.6	17
Ramie	~1900	25–30	1.5	400–938	61–128	1.2–3.8	12–17
Banana	~5000	240–260	0.72.-0.88	161–789	7–9	2.0–3.34	60
Pineapple	~200	274	1.07	126	4.4	2.2	11.8
Sisal	~1000	200–400	1.5	468–700	9–22	3.0–7.0	11
Coir	350	12–25	1.2	175	4–6	17–47	10
Oil palm	0.59–142	150–500	0.7–1.55	50–400	0.57–9.0	4–18	23.63
Softwood Kraft	2.8–4.1	28–47	1.5	1000	18–40	-	8
Hardwood Kraft	2.67–3.98	31	1.2	-	37.9	-	8–25
Cotton	100–650	11–22	1.5–1.6	287–800	5.5–12.6	3–10	8–25
Chicken feather	0.005–1.7	10–40	0.89	100–220	3–10	-	11
Silk worm silk	~1500	10–13	1.3–1.4	293	4.5–57.1	0.03–25	98
Kapok	20–32	20	0.31–0.38	93.3	4	1.2	10.9
Abaca	~3000	151	1.5	12	41	3.4	14
Bagasse	1.2	15	1.2	20–290	19–27	3–4.7	8.8
Aloe Vera	~700	8–24	-	116.7	3.91	–	8.98
Curaua	~1500	100	1.4	87–1150	11.8–96	1.3–4.9	–
Wool	40–75	11.5–24	1.31	50–315	2.3–5	13–335	–
Grewia optiva	730	30	0.45	25.43	–	10.28	–

L Length; *D* Diameter; *ρ* Density; *TS* Tensile strength; *EB* Elongation at break; *E* Elastic modulus; *WA* Water absorption

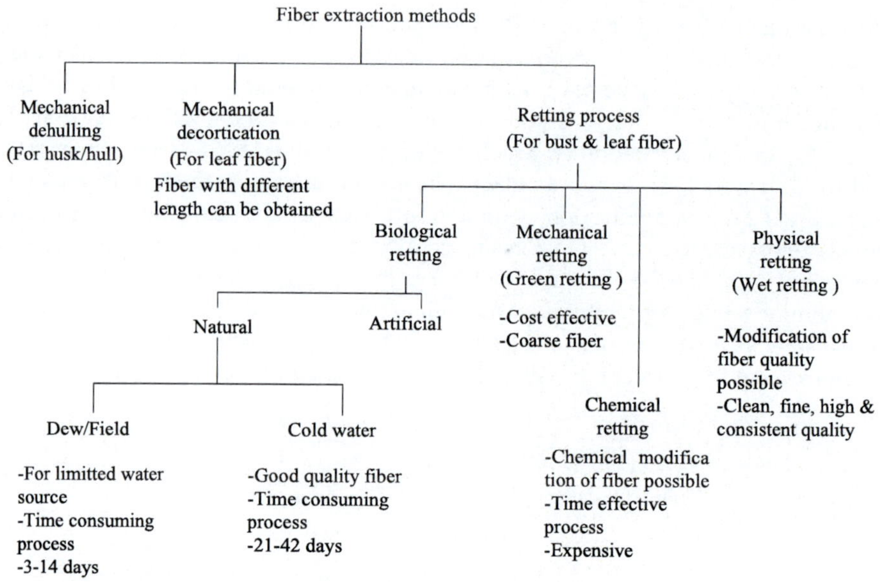

Fig. 3 Fiber extraction methods. *Source* Author

than that of synthetic fiber. However, the specific mechanical properties of natural fibers are pretty nearer to the properties of synthetic fibers.

Separation or extraction of plant-based fibers bundles from the leaf, bust, seed, fruit, stalk, grass, or wood usually carried out either manually, using the machine, or combining both. Retting process and decortications are the two commonly used process of fiber extraction. In retting process, the plant sources are chemically or biologically treated. In the decortications process, some kind of machine is used. In this process, the plant sources are crushed and beaten by a rotating wheelset with blunt knives to separate the fibers from the source. The extracted fibers are then washed and dried under the sun. Mechanical dehulling is another extraction process mainly used for the separation of hull/husk fibers. The various fiber extraction methods are presented in Fig. 3.

2.3 *Green Composites*

Various possible combinations of biodegradable polymer and natural fiber have been attempted by many engineers and scientists to develop green composites. The combinations of biodegradable polymer and natural fiber studied by researchers are given in Table 3. Different fiber orientation can be adopted for developing green composites of different properties. The fiber orientation may be one-directional at an angle

Table 3 Different combinations of biodegradable polymer and natural fiber [48–87]

Matrix	PLA
Fiber	Kenaf, Chicken feather, Ramie, Jute, Cotton, Hemp, Hemp + Kenaf, Hemp + Lyocell, Lyocell, Cordenka rayon, Flax, Triacetin + Flax, Cellulose, Triacetin + Flax, Abaca, Denim fabric, Chopped recycled newspapers, Rice starch + Epoxidised natural rubber, Sisal, Sugar beet pulp, Silk, Wood flour, Coconut, Bamboo flour, Chitin, Fique Fibers, Bamboo fiber, Basalt, Coir, Banana, Aloe Vera
Matrix	PCL
Fiber	Flax, Rice husk, Abaca, Oil palm, Wheat gluten, Sisal, Chitin, Gelatin, Ramie, Fruit bunch, Bagasse, Chitin, Banana, Wood floor, Jute, Silk-fibroin, Hemp, Fique fibers, Macaiba
Matrix	PHBV
Fiber	Abaca, Luffa, Cellulose, Flax, Bamboo, Pineapple, Wheat straw, Coir, Kenaf
Matrix	PBS
Fiber	Abaca, Caraua, Sisal, Basalt, Coir, Jute, Cotton stalk bast, Bamboo, Straw, Sugarcane rind/bagasse, Silk, Hemp, Flax seed, Kenaf, Coconut

of 0°, 90°, or 45° or bidirectional by combining two or more orientations. The bidirectional fiber orientation is also called woven fabric. Woven fabric is produced by interlocking the warp (0°) and weft fibers (90°). Different style of weave is available to maintain the fabric's integrity such as plain, twill, satin, basket, leno, and mock leno [48]. In plain weave style, the warp and weft fibers are alternate to each other, and the warp fibers are weaving over and under one weft fiber. While the warp fibers are weaving over and under two or more weft fiber in twill fabric. When twill fabric is modified to have fewer intersections, it is called satin fabric. The plain weave can also be modified to basket fabric with interlocking two or more warp fibers alternately with two or more weft fibers. The spiral pair obtained by twisting the warp fiber around the consecutive weft fibers is called leno fabric. Mock leno is a type of plain weave where fibers are warped at several fibers apart at regular intervals.

3 Applications

Green composites have the potential to create a positive environmental effect if they are adopted by aerospace, automotive, construction, consumer product, and medical product industries. In the past decades, many of these industries have shifted their materials of product from conventional to green. At early stage, green composites find their application only in aerospace industries. The materials used for indoor structures such as seat cushions, cabin linings, parcel shelves, doors, and door shutters of aircraft have been replaced by green composites [88]. With the development of high-performance or 'advanced' composites, they are now admired in automotive industries, electronics industries, building materials, and sporting goods [6, 9, 89]. In 2007, a 90% biodegradable vehicle named Agri-Car was developed as a joint

Table 4 Applications of green composites

Sectors	Applications
Short life product	Packaging applications [10, 94]; Rope, bag, broom [39]
Electronics industry	PLA/kenaf dummy cards in personal computer [95]; Mobile phone [96]; Printed circuit board [97]; Radio [12]
Sports industry	Toys [97]; Flax reinforced snowboards [98]; Tennis rackets [99]; Bicycle frames [100]
Biomedical industry	Articular cartilage [101]; Trachea [102]; Sutures, drug delivery systems and grafts [39]; Scaffolds for tissue engineering and bone fixators [3]
Automobile industry	Door trim panels[103]; Spare tire cover [104]; Seat backs, dashboard, headliners [105]; Seat foams [106]; Lower door panel [107]; Luggage compartment [108]; Engine and transmission enclosures for sound insulation [109]; Spare tire well covers [110]; Center console and trim [48]; Damping and insulation parts [48]; C-pillar trim [48]
Household components	Ceiling, floor, window, wall partition, table chair, kitchen cabinet, decks, suitcases [48]; Roofing [1]
Structural applications	Concrete elements, fencing, decking, siding, bridge, fiber cement [48, 111]; Composite soil [1]

effort of Ohio State University and the University of Akron (Goodyear Polymer Center) [90]. World First Formula 3 racing car or 'green car' was developed by University of Warwick in 2009 [91]. PLA/flax green composite was used to fabricate a small pleasure ships/boats by collaborative work of companies and research centers at PoleMer Bretagne, France in 2009 [92, 93]. Some of the applications of green composites in different sectors are listed in Table 4.

4 Properties

4.1 Mechanical Properties

Mechanical properties are important to predict how the green composites would behave under subjected loads when applied to practical applications during their service life. Important mechanical properties which to be evaluated before introducing the green composites to the practice field are tensile strength (stress in tension), Young's modulus (stiffness), compressive strength (stress in compression), flexural strength (stress at bending), flexural modulus, impact strength (stress at impact loading), and shear strength (stress at shearing). To evaluate the tensile and compressive strength of the green composites, uniaxial tensile and compressive tests are carried out in a universal testing machine (UTM). The flexural strength is obtained in terms of a three-point or four-point bending test. The flexural test of green composites

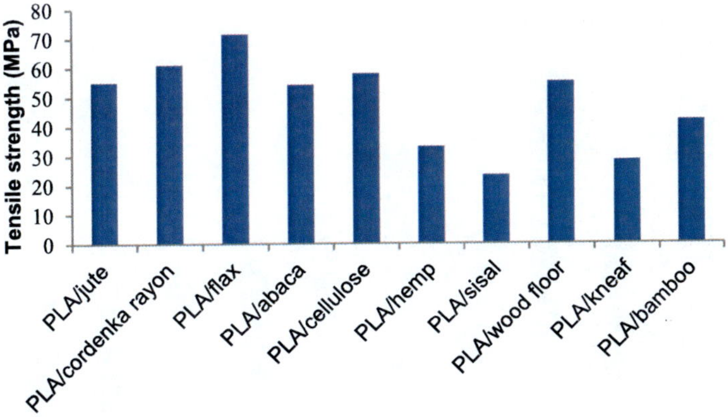

Fig. 4 Tensile strength of different green composites. *Source* Author

can also perform in UTM. Traditionally, the impact strength of green composites is obtained by Charpy and Izod impact tests. However, these traditional testing methods do not give accurate impact strength data for green composites. Researchers have developed other impact testing methods particularly to evaluate the impact strength of composites, such as, instrumented impact test and drop weight impact test [4].

The mechanical properties of green composites depend on many factors such as fiber orientation, fiber content (%), length of the fiber, and bonding between the matrix and the reinforcement. Many researchers have studied the tensile, flexural, and impact strength of green composites [48]. The tensile strength of the green composites can be enhanced by improving the bonding between the matrix and fibers. Ochi [112] reported that the tensile strength of the green composites can be improved by adapting accurate fabrication methods. The tensile strength, elastic modulus, flexural strength, and impact strength of different PLA-based green composites at 30% (weight) fiber content are given in Figs. 4, 5, 6 and 7. Data related to compressive strength and shear strength are handful as researchers have not given focus in this area yet. Roy Choudhury et al. [28] studied the tensile and compressive failure load of PLA/bamboo green composites. The authors have found that the compressive failure load of green composites is more than that of tensile failure load.

4.2 *Thermal Properties*

The thermal properties of green composites are necessary to understand to know the heat resistance capacity in different environmental conditions. The melting temperature and glass transition temperature of the green composites are the critical factors while preparing the composites for particular applications to prevent premature failure. The relative service temperature of the composites is determined by evaluating heat deflection temperature (the temperature up to which the composite

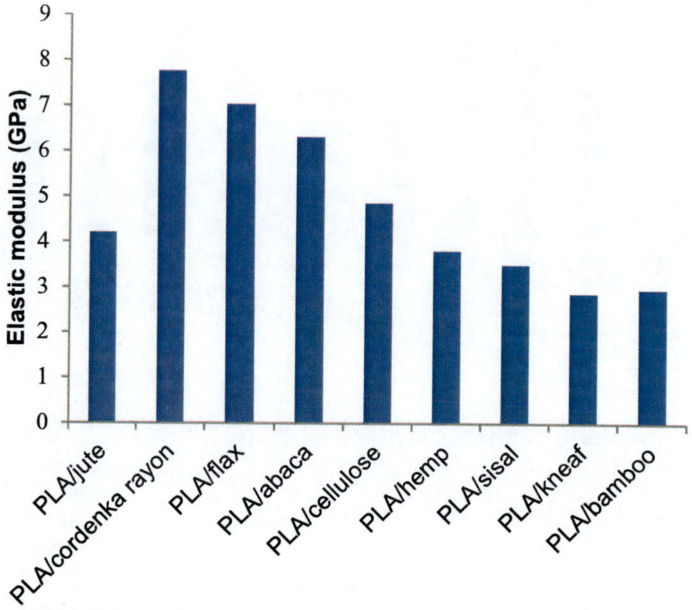

Fig. 5 Elastic modulus of different green composites. *Source* Author

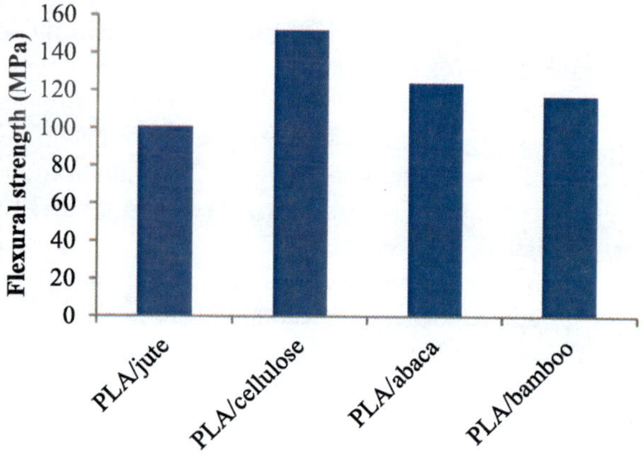

Fig. 6 Flexural strength of different green composites. *Source* Author

is stable). The PLA-based green composites are found to have lower heat deflection temperature. For this reason, their application is limited to a low temperature environment [113]. However, the heat deflection temperature can be improved by adding fillers [114]. Heat resistance capacity of green composites can also be significantly increased by decreasing the polymerization of the matrix. Lower cellulose

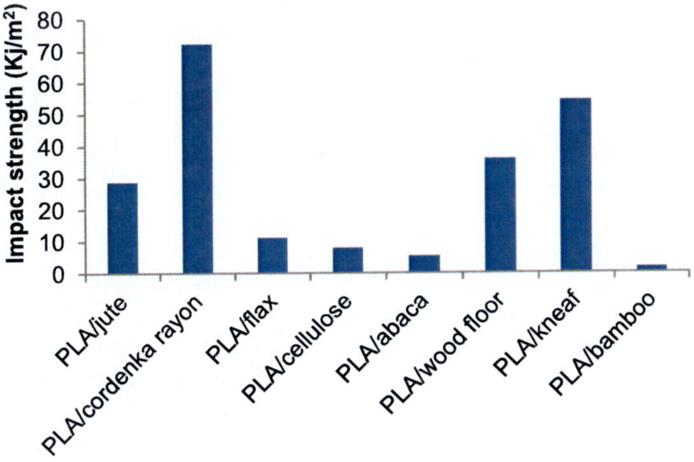

Fig. 7 Impact strength of different green composites. *Source* Author

percentage of natural fiber can also help in increasing the heat deflection temperature of the composites. Lower crystalline composites are more prone to degrade in higher temperatures [115]. There are many techniques available to evaluate the thermal properties of the green composites, such as, thermogravimetry analysis (TGA), differential scanning calorimetry (DSC), and differential thermal analysis (DTA).

4.3 Biodegradability

Green composites can be degraded to CO_2 and water under enzymatic action of a living organism. Degradation initiates through an arbitrary chain breakdown due to changes in environmental temperature. The rate of degradation of the green composites is affected by many factors such as chemical and crystal structure, molecular weight, glass transition and melting temperature, crystallinity, surface area, and condition of green composites [4]. The degradation of green composites is faster under higher temperatures and humid conditions. The amorphous part of the matrix of the green composites degrades more rapidly compared to the crystalline part. This indicates that the rate of degradation decreases with an increase in the crystallinity of the composites. Again, when the molecular weight, glass transition temperature, and melting temperature of green composites are higher, the degradation rate is slower [116]. When degradation temperature is higher than the glass transition temperature, the anaerobic biodegradation of the green composite is found to be rapid [117]. The degradation rate of green composites without enzymatic action is lower compared to the enzymatic action [118]. The reinforced fibers in green composites can also slow down the rate of degradation of the matrix material [4].

5 Processing

The processing of green composites is influenced by many factors such as moisture content, types of fiber, fiber orientations, and the melting point of the polymer matrix. The moisture present in the fiber and the polymer can lead to debonding of the fiber–matrix interface. Preheating of the fiber and the polymer is an essential step before processing of green composites to remove the moisture. Researchers have developed many processing technologies to obtain high quality of green composites. Compression molding is one of the oldest and commonly used processing technologies to obtain lightweight and high-performance green composites. In the compression molding technique, the fiber and polymer matrix are together compressed between two hot molds. The main controllable parameters in compression molding are heating time, applied pressure, cooling time, de-molding, etc. Sheet molding is a type of compression molding in which the concept of film stacking method is used. These two types of fabrication methods are mainly used for woven fabric and low volume fraction of fiber. The short fiber-reinforced green composites can be fabricated by using injection molding and extrusion molding techniques. In these techniques, the chopped fiber and polymer matrix are delivered to the heated barrel through the hopper. The polymer gets melted in the barrel and molded together with fiber as required shape. The composite components obtained from extrusion molding are of uniform cross section. While many complex shapes of green composite components can be fabricated by an injection molding technique. Chaitanya et al. [53] used the single screw extruder to pelletize the PLA pellets and short aloe vera into 5 mm long pellets. The extruded pellets were then transferred to the injection molding machine to fabricate the composites. A similar procedure was applied by Liu et al. [119]. The extruded fiber-reinforced pellets can also be used in the compression molding to fabricate short fiber-reinforced green composites [120]. Researchers have also used some other methods like wet impregnation, solution casting, and melt blending to fabricate green composites [121–123]. Murariu et al. [124] developed green composites by combining melt blending and compression molding methods. Hand layup, resin infusion molding, etc., are some of the methods used to fabricate thermosetting-based green composites.

6 Machining and Joining Behavior

To enhance the areas of application of green composites and to produce more complicated products, some kind of secondary manufacturing processes such as machining and joining are vital. However, secondary manufacturing of green composites is not an easy task due to the anisotropic and inhomogeneous nature of the natural fiber. Researchers have found that delamination, fiber pullout, matrix burning, chipping, spalling, micro-cracks, and reduction of strength are the common defects associated with the secondary processing of green composites [28]. This damage is due to the higher forces and temperature induced during machining operations. The drilling behavior of the green composites has been extensively studied by many researchers. However, very few studies are found on milling and turning of green composites. The stress concentration that arises during the machining operation can be reduced to eliminate the immature failure of composites during the service life. The stress concentration can be eliminated by reducing the damage produced during machining. This can be achieved by selecting a proper parametric setting by optimizing the process parameters. The machining behavior of green composites is influenced by many factors, such as, feed rate, spindle speed, tool geometry, depth of cut, and machining conditions. In drilling operation, good quality of hole can be produced in green composites by controlling the higher induced forces and temperature. It is obtained from the research study that by selecting lower feed rate and higher spindle speed, the induced forces and temperature can be lower down. The geometry of the tool also significantly influences the machining behavior of green composites [125]. Drilling of green composites with traditional drill bit produced higher machining force resulting in severe damage in and around the drilled hole. The drilling forces are largely dependent on the indentation effect induced by the quasi-stationary tip of the drill bit. To overcome this difficulty, various drill bits have been specifically designed for drilling polymer composite laminate. The commonly used drill bits for making a hole in green composites are step drill, parabolic drill, 8-facet drill, dagger drill, core drill, step-core drill, straight flute, and Jo drill [28, 126]. Similarly, the milling cutter used during milling operation of green composites has a significant effect on the quality of the milled surface. The six-flute end mill cutter is found to produce more damage compared to the two-flute end mill [127]. The fiber orientation of the green composites also significantly affects the surface quality during machining. A good surface finish can be obtained at 0° orientation of fiber. The surface roughness value increases as the orientation angle increases [128].

Along with the machining operations, some kind of joining operations are also essential for assembling various small green composites parts to make a complicated design. Different joining methods available for green composites are mechanical joining, adhesive joining, and fusion bonding. The joining of green composites is a very challenging task due to different load transfer paths. The improper joint design may lead to stress concentration and cause the failure of the joint during the service life of the product. Therefore, joining behavior of green composites is necessary to understand. Mechanical joining and adhesive bonding are the two conventional methods of joining of composites. In mechanical joining, the composites are assembled by a mechanical fastener. The mechanical fasteners available are nut and bolt, rivets, pin, etc. In adhesive joints, adhesive materials such as epoxy, polyurethane, and acrylic are introduced between the interfaces of the joint to obtain the bonding. The mechanical and adhesive joining behavior of PLA/bamboo green composites was studied by Roy Choudhury et al. [28, 49]. The authors found that bolt fastening torque of mechanical fastener and the adhesive materials used for joining are important parameters while determining the joint strength. In recent times, the hybrid joining (bonded/bolted) technique is emerging as a potential joining technique as it prevents premature failure of the composite joint. In hybrid joining, the mechanical and adhesive joining techniques are combined to produce stronger and stiffer joint than the individual joining methods. The defects associated with mechanical and adhesive joining such as stress concentration, deboning, and kissing bonds can be eliminated in hybrid joining. The hybrid joint can offer both the jigging and sealing action in a composite joint. The fusion bonding technique of joining thermoplastic composites appears to be a promising method that has replaced many of the traditional joining methods. In fusion bonding, the bonding at the interface of the joining takes place by the generation of heat utilizing thermal, friction, and electromagnetic energy. In thermal welding, the heat at the interface is generated by the infrared, laser, or hot gas. In friction welding method, ultrasonic vibration or stirring mechanism is used to generate heat at the interface. Similarly, induction, microwave, and resistance heating mechanism are used to induce heat in electromagnetic welding technique. The main advantage of fusion bonding techniques is the time saving, eco-friendly, cost-effective, elimination of stress concentration, and additional weight. The fusion bonding behavior of green composites is studied by Roy Choudhury et al. [129]. The authors have used high-frequency ultrasonic vibration to generate heat in the interface of PLA/bamboo composites parts. The welding parameters such as applied pressure, welding time, and cooling time are found to have a significant effect on the joining strength. The selection of an individual joining method depends on the application of the final product. The potentiality of green composite materials can be completely utilized by developing an efficient joint design.

7 Conclusions

In this chapter, an attempt has been made to present a brief introductory overview of green composites. The green composites are biodegradable material, and they have replaced many synthetic fiber-reinforced polymer composites in various fields of applications. By exploitation of more biodegradable polymer and fiber, the potential of green composites as lightweight sustainable material can be increased. Economic fabrication methods and improved properties are the keys to implement green composites in more complicated applications. More attention should be given to the problems associated with green composites such as poor interfacial adhesion between the natural fibers and the matrix, moisture absorption, poor fire resistance, low impact strength, and low durability. The research on machining and joining behavior of the green composites is still in the initial stage to meet the demands of the commercial appeal.

8 Future Scope

The processing temperature of most of the composites is restricted to 200 °C. Beyond this temperature, the matrix as well the natural fibers undergoes degradation. Also, the high moisture absorption nature of natural fibers results in poor mechanical properties of green composites consequently reduces dimensional stability. There is a huge opportunity for researchers to improve the thermal and mechanical properties of green composites and to make it long durable and long-lasting.

References

1. John MJ, Thomas S (2008) Biofibres and biocomposites. Carbohydr Polym 71(3):343–364
2. Netravali AN, Chabba S (2003) Composites get greener. Mater Today 4(6):22–29
3. Zini E, Scandola M (2011) Green composites: an overview. Polym Compos 32(12):1905–1915
4. Nurul Fazita MR, Jayaraman K, Bhattacharyya D, Mohamad Haafiz MK, Saurabh CK, Hussin MH, Abdul Khalil HPS (2016) Green composites made of bamboo fabric and poly (lactic) acid for packaging applications—a review. Materials 9(6):435
5. Oyama IC, de Souza GP, Rezende MC, Montagna LS, Passador FR (2020) A new eco-friendly green composite for antistatic packaging: green low-density polyethylene/glassy carbon. Polym Compos. https://doi.org/10.1002/pc.25572
6. Koronis G, Silva A, Fontul M (2013) Green composites: a review of adequate materials for automotive applications. Compos B 44(1):120–127
7. Ashori A (2008) Wood–plastic composites as promising green-composites for automotive industries! Bioresour Technol 99(11):4661–4667
8. Georgios K, Silva A, Furtado S (2016) Applications of green composite materials. Biodegrad Green Compos 16:312
9. Dicker MP, Duckworth PF, Baker AB, Francois G, Hazzard MK, Weaver PM (2014) Green composites: a review of material attributes and complementary applications. Compos A 56:280–289
10. La Mantia FP, Morreale M (2011) Green composites: a brief review. Compos A 42(6):579–588
11. Khalil HA, Bhat AH, Yusra AI (2012) Green composites from sustainable cellulose nanofibrils: a review. Carbohydr Polym 87(2):963–979
12. Gholampour A, Ozbakkaloglu T (2020) A review of natural fiber composites: properties, modification and processing techniques, characterization, applications. J Mater Sci 55:829–892
13. Potluri R (2019) Natural fiber-based hybrid bio-composites: processing, characterization, and applications. Green composites. Springer, Singapore, pp 1–46
14. Thames SF, Zhou L (1998) Effect of preparation and processing on mechanical properties and water absorption of soy protein based biocomposite. In: 5th international conference on composites engineering, ICCE, Las Vegas, Nevada, 5–11 July 1998
15. Li K, Peshkova S, Geng X (2004) Investigation of soy protein-Kymene® adhesive systems for wood composites. J Am Oil Chem Soc 81(5):487–491
16. Drzal LT (2002) Environmentally friendly bio-composites from soy-based bio-plastic and natural fiber. Polym Mat Sci Eng 87:117
17. Lodha P, Netravali AN (2002) Characterization of interfacial and mechanical properties of "green" composites with soy protein isolate and ramie fiber. J Mater Sci 37(17):3657–3665
18. Chabba S, Netravali AN (2004) 'Green' composites using modified soy protein concentrate resin and flax fabrics and yarns. JSME 47(4):556–560
19. Nam S, Netravali AN (2004) Characterization of ramie fiber/soy protein concentrate (SPC) resin interface. J Adhes Sci Technol 18(9):1063–1076
20. Takagi H et al (2002) International workshop on 'Green' Composites, 4
21. Ochi S et al (2002) International workshop on 'Green' Composites, 22
22. Ichihara Y, Takagi H (2002) International workshop on 'Green' Composites, 26
23. Goda K et al. (2002) International workshop on 'Green' Composites, 8

24. Murariu M, Dubois P (2016) PLA composites: from production to properties. Adv Drug Deliv Rev 107:17–46
25. Alberts AH, Rothenberg G (2017) Plantics-GX: a biodegradable and cost-effective thermoset plastic that is 100% plant-based. Faraday Discuss 202:111–120
26. Cai M, Liu H, Jiang Y, Wang J, Zhang S (2019) A high-strength biodegradable thermoset polymer for internal fixation bone screws: Preparation, in vitro and in vivo evaluation. Colloid Surfaces B 183:110445
27. Manvi PK, Beckers M, Mohr B, Seide G, Gries T, Bunge CA (2019) Polymer fiber-based biocomposites for medical sensing applications. Mater Biomed Eng. https://doi.org/10.1016/B978-0-12-816872-1.00003-0
28. Choudhury MR, Debnath K (2019) Experimental analysis of tensile and compressive failure load in single-lap bolted joint of green composites. Compos Struct 225:111180
29. de Oca HM, Ward IM (2006) Structure and mechanical properties of PGA crystals and fibres. Polymer 47(20):7070–7077
30. Takayama T, Daigaku Y, Ito H, Takamori H (2014) Mechanical properties of bio-absorbable PLA/PGA fiber-reinforced composites. J Mech Sci Technol 28(10):4151–4154
31. Gentile P, Chiono V, Carmagnola I, Hatton PV (2014) An overview of poly (lactic-co-glycolic) acid (PLGA)-based biomaterials for bone tissue engineering. Int J Mol Sci 15(3):3640–3659
32. Melo LPD, Salmoria GV, Fancello EA, Roesler CRDM (2017) Effect of injection molding melt temperatures on PLGA craniofacial plate properties during in vitro degradation. Int J Biomater. https://doi.org/10.1155/2017/1256537
33. Rudnik E (2008) Compostable polymer properties and packaging applications. In: Rudnik E (ed) Compostable polymer materials. Elsevier, Józefów, Poland. https://doi.org/10.1016/B978-1-4557-3112-1.00013-2
34. Serrano PJM, Thüss E, Gaymans RJ (1997) Alternating polyesteramides based on 1, 4-butylene terephthalamide: 2. Alternating polyesteramides based on a single, linear diol (4NTm). Polymer 38(15):3893–3902
35. Jamaluddin N, Razaina MT, Ishak ZM (2016) Mechanical and morphology behaviours of polybutylene (succinate)/thermoplastic polyurethaneblend. Procedia Chem 19:426–432
36. https://omnexus.specialchem.com/polymer-properties/properties/stiffness. Last viewed: 15 May 2020
37. Jayaraman K (2003) Manufacturing sisal-polypropylene composites with minimum fibre degradation. Compos Sci Technol 63:367–374
38. Wang Q, Kaliaguine S, Ait-Kadi A (1992) Catalytic grafting: a new technique for polymer–fiber composites I. Polyethylene–asbestos composites. J Appl Polym Sci 44(6):1107–1119
39. Shekar HS, Ramachandra M (2018) Green composites: a review. Mater Today Proc 5(1):2518–2526
40. Jha K, Kataria R, Verma J, Pradhan S (2019) Potential biodegradable matrices and fiber treatment for green composites: a review. AIMS Mater Sci 6(1):119–138
41. Ramamoorthy SK, Skrifvars M, Persson A (2015) A review of natural fibers used in biocomposites: plant, animal and regenerated cellulose fibers. Polym Rev 55(1):107–162
42. Sailesh A, Arunkumar R, Saravanan S (2018) Mechanical properties and wear properties of Kenaf–aloe vera–jute fiber reinforced natural fiber composites. Mater Today Proc 5(2):7184–7190
43. Ashori A, Nourbakhsh A (2010) Bio-based composites from waste agricultural residues. J Waste Manag 30(4):680–684
44. Fávaro SL, Lopes MS, de Carvalho Neto AGV, de Santana RR, Radovanovic E (2010) Chemical, morphological, and mechanical analysis of rice husk/post-consumer polyethylene composites. Compos A 41(1):154–160
45. Pfister DP, Larock RC (2010) Green composites from a conjugated linseed oil-based resin and wheat straw. Compos A 41(9):1279–1288
46. Ahankari SS, Mohanty AK, Misra M (2011) Mechanical behaviour of agro-residue reinforced poly (3-hydroxybutyrate-co-3-hydroxyvalerate),(PHBV) green composites: A comparison with traditional polypropylene composites. Compos Sci Technol 71(5):653–657

47. Reddy N, Yang Y (2005) Properties and potential applications of natural cellulose fibers from cornhusks. Green Chem 7(4):190–195
48. Bajpai PK, Singh I, Madaan J (2014) Development and characterization of PLA-based green composites: a review. J Thermoplast Compos Mater 27(1):52–81
49. Choudhury MR, Debnath K (2020) Experimental analysis of tensile and compressive failure load in single-lap adhesive joint of green composites. Int J Adhes Adhes 99:102557
50. Yu S, Hwang YH, Hwang JY, Hong SH (2019) Analytical study on the 3D-printed structure and mechanical properties of basalt fiber-reinforced PLA composites using X-ray microscopy. Compos Sci Technol 175:18–27
51. Rigolin TR, Takahashi MC, Kondo DL, Bettini SHP (2019) Compatibilizer acidity in coir-reinforced PLA composites: matrix degradation and Composite properties. J Polym Environ 27(5):1096–1104
52. Komal UK, Lila MK, Singh I (2020) PLA/banana fiber based sustainable biocomposites: a manufacturing perspective. Compos B 180:107535
53. Chaitanya S, Singh I (2016) Novel Aloe Vera fiber reinforced biodegradable composites—development and characterization. J Reinf Plast Compo 35(19):1411–1423
54. Wahit MU, Akos NI, Laftah WA (2012) Influence of natural fibers on the mechanical properties and biodegradation of poly (lactic acid) and poly (ε-caprolactone) composites: a review. Polym Compos 33(7):1045–1053
55. Cyras VP, Iannace S, Kenny JM, Vázquez A (2001) Relationship between processing and properties of biodegradable composites based on PCL/starch matrix and sisal fibers. Polym Compos 22(1):104–110
56. Chen B, Sun K, Ren T (2005) Mechanical and viscoelastic properties of chitin fiber reinforced poly (ε-caprolactone). Eur Polym J 41(3):453–457
57. Xu H, Wang L, Teng C, Yu M (2008) Biodegradable composites: Ramie fibre reinforced PLLA-PCL composite prepared by in situ polymerization process. Polym Bulletin 61(5):663–670
58. Hamid MZA, Ibrahim NA, Yunus WMZW, Zaman K, Dahlan M (2010) Effect of grafting on properties of oil palm empty fruit bunch fiber reinforced polycaprolactone biocomposites. J Reinf Plast Comp 29(18):2723–2731
59. Cao Y, Goda K, Shibata S (2007) Development and mechanical properties of bagasse fiber reinforced composites. Adv Compos Mater 16(4):283–298
60. Yang A, Wu R (2002) Enhancement of the mechanical properties and interfacial interaction of a novel chitin-fiber-reinforced poly (ϵ-caprolactone) composite by irradiation treatment. J Appl Poly Sci 84(3):486–492
61. Misra R, Kumar S, Sandeep K, Misra A (2008) Dynamic analysis of banana fiber reinforced high-density polyethylene/poly (ε-caprolactone) composites. J Mech Mater Struct 3(1):107–125
62. Lee SH, Ohkita T (2003) Mechanical and thermal flow properties of wood flour–biodegradable polymer composites. J Appl Poly Sci 90(7):1900–1905
63. Goriparthi BK, Suman KNS, Nalluri MR (2012) Processing and characterization of jute fiber reinforced hybrid biocomposites based on polylactide/polycaprolactone blends. Polym Compos 33(2):237–244
64. Qiao X, Li W, Sun K, Xu S, Chen X (2009) Nonisothermal crystallization behaviors of silk-fibroin-fiber-reinforced poly (ϵ-caprolactone) biocomposites. J Appl Poly Sci 111(6):2908–2916
65. Dhakal HN, Ismail SO, Beaugrand J, Zhang Z, Zekonyte J (2020) Characterization of nano-mechanical, surface and thermal properties of hemp fiber-reinforced polycaprolactone (HF/PCL) biocomposites. Appl Sci 10(7):2636
66. Mina JH, González AV, Muñoz-Vélez MF (2020) Micro-and macromechanical properties of a composite with a ternary PLA–PCL–TPS matrix reinforced with short Fique fibers. Polymers 12(1):58
67. Siqueira DD, Luna CBB, Araújo EM, Ferreira ESB, Wellen RMR (2019) Biocomposites based on PCL and macaiba fiber. Detailed characterization of main properties. Mater Res Express 6(9):095335

68. Guo Y, Wang L, Chen Y, Luo P, Chen T (2019) Properties of Luffa fiber reinforced PHBV biodegradable composites. Polymers 11(11):1765
69. Panaitescu DM, Nicolae CA, Gabor AR, Trusca R (2020) Thermal and mechanical properties of poly (3-hydroxybutyrate) reinforced with cellulose fibers from wood waste. Ind Crops Prod 145:112071
70. Muniyasamy S, Ofosu O, Thulasinathan B, Rajan AST, Ramu SM, Soorangkattan et al (2019) Thermal-chemical and biodegradation behaviour of alginic acid treated flax fibres/poly (hydroxybutyrate-co-valerate) PHBV green composites in compost medium. Biocatal Agric Biotechnol 22:101394
71. Jiang L, Huang J, Qian J, Chen F, Zhang J, Wolcott MP, Zhu Y (2008) Study of poly (3-hydroxybutyrate-co-3-hydroxyvalerate)(PHBV)/bamboo pulp fiber composites: Effects of nucleation agent and compatibilizer. J Polym Environ 16(2):83–93
72. Luo S, Netravali AN (1999) Interfacial and mechanical properties of environment-friendly "green" composites made from pineapple fibers and poly (hydroxybutyrate-co-valerate) resin. J Mater Sci 34(15):3709–3719
73. Berthet MA, Angellier-Coussy H, Chea V, Guillard V, Gastaldi E, Gontard N (2015) Sustainable food packaging: valorising wheat straw fibres for tuning PHBV-based composites properties. Compos a 72:139–147
74. Javadi A, Srithep Y, Pilla S, Lee J, Gong S, Turng LS (2010) Processing and characterization of solid and microcellular PHBV/coir fiber composites. Mater Sci Eng C 30(5):749–757
75. Buzarovska A, Bogoeva-Gaceva G, Grozdanov A, Avella M, Gentile G, Errico M (2007) Crystallization behavior of poly (hydroxybytyrate-co-valerate) in model and bulk PHBV/kenaf fiber composites. J Mater Sci 42(16):6501–6509
76. Zhang Y, Yu C, Chu PK, Lv F, Zhang C, Ji J et al (2012) Mechanical and thermal properties of basalt fiber reinforced poly (butylene succinate) composites. Mater Chem Phy 133(2–3):845–849
77. Nam TH, Ogihara S, Tung NH, Kobayashi S (2011) Effect of alkali treatment on interfacial and mechanical properties of coir fiber reinforced poly (butylene succinate) biodegradable composites. Compos B 42(6):1648–1656
78. Bin T, Qu JP, Liu LM, Feng YH, Hu SX, Yin XC (2011) Non-isothermal crystallization kinetics and dynamic mechanical thermal properties of poly (butylene succinate) composites reinforced with cotton stalk bast fibers. Thermochim Acta 525(1–2):141–149
79. Ohkita K, Takagi H (2010) Flexural properties of injection-molded bamboo/PBS composites. Int J Mod Phys B 24:2838–2843
80. Zhang M, Ding F, Li C, Ge Z, Tian Y (2011) Effect of different treatment and modifiers on the straw fiber/PBS composites property. Acta Materiae Compositae Sinica 28(1):56–60
81. Huang Z, Qian L, Yin Q, Yu N, Liu T, Tian D (2018) Biodegradability studies of poly (butylene succinate) composites filled with sugarcane rind fiber. Polym Test 66:319–326
82. Frollini E, Bartolucci N, Sisti L, Celli A (2013) Poly (butylene succinate) reinforced with different lignocellulosic fibers. Ind Crop Prod 45:160–169
83. Nam TH, Ogihara S, Kobayashi S, Goto K (2015) Effects of surface treatment on mechanical and thermal properties of jute fabric-reinforced poly (butylene succinate) biodegradable composites. Adv Compos Mater 24(2):161–178
84. Song R, Kimura T (2011) Mechanical properties of silk/bamboo hybrid paper reinforced PBS green composite. J Tex Eng 57(1):1–7
85. Li J, Ben G, Yang J (2014) Fabrication of hemp fiber-reinforced green composites with organoclay-filled poly (butylene succinate) matrix by pultrusion process. Sci Eng Compos Mater 21(2):289–294
86. Azhar SW, Xu F, Zhang Y, Qiu Y (2019) Fabrication and mechanical properties of flaxseed fiber bundle-reinforced polybutylene succinate composites. J Ind Text, 1–16
87. Soatthiyanon N, Aumnate C, Srikulkit K (2020) Rheological, tensile, and thermal properties of poly (butylene succinate) composites filled with two types of cellulose (kenaf cellulose fiber and commercial cellulose). Polym Compos. https://doi.org/10.1002/pc.25575

88. Subash T, Pillai SN (2015) Bast fibers reinforced green composites for aircraft indoor structures applications: Review. J Chem Pharm Sci 7:305–307
89. Dittenber DB, GangaRao HV (2012) Critical review of recent publications on use of natural composites in infrastructure. Compos A 43(8):1419–1429
90. Irwin L (2007) Biomass Magazine, BBI International, July. Available from: https://issuu.com/bbiinternational/docs/bmm-june.07_print. Last viewed: 15 May 2020
91. www.warwick.ac.uk/newsandevents/pressreleases/racing_car (2011). Last viewed: 15 May 2020
92. www.pole-mer-bretagne.com/navecomat_0.php. Last viewed: 15 May 2020
93. Le Duigou A, Davies P, Baley C (2009) Seawater ageing of flax/poly (lactic acid) biocomposites. Polym Degrad Stabil 94(7):1151–1162
94. Mohanty AK, Misra MA, Hinrichsen GI (2000) Biofibres, biodegradable polymers and biocomposites: an overview. Macromole Mater Eng 276(1):1–24
95. Iji M, Kiuchi Y. ITR (2012) Highly functional PLA composites used for electronic products. Green Innovation Reseach, Laboratories NEC Corporation. Available from: https://www.innovationtakesroot.com/~/media/ITR2012/2012/presentations/durables/01_Highly-Functional-PLA-Composites_Iji_pdf.pdf. Last viewed: 15 May 2020
96. Herrera T (2012) Sprint to require green certification for all cell phones. <http:// www.greenbiz.com/blog/2012/01/09/sprint-require-green-certification-allcell-phones. [last viewed: May 15, 2020].
97. Lucintel (2012) XBoards introduces snowboard made from flax fibre composite. Available from: https://www.lucintel.com/news/xboards_introduces_snowboard_made_from_flax_fibre_composite.aspx. Last viewed: 15 May 2020
98. JEC Composites (2012) Biocomposite snowboard using Biotex flax fabric. https://www.jeccomposites.com/news/features/biocomposites/biocompositesnowboard-using-biotex-flax-fabric. Last viewed: 15 May 2020
99. Artengo (2011). Artengo Flaxfiber. Available from: https://www.artengo.com/EN/tennis-178543753/. Last viewed: 15 May 2020
100. Museeuw (2012) MF-5. Available from: https://en.museeuwbikes.be/bikes/race/mf-5. Last viewed: 15 May 2020
101. Dai W, Kawazoe N, Lin X, Dong J et al (2010) The influence of structural design of PLGA/collagen hybrid scaffolds in cartilage tissue engineering. Biomaterials 31(8):2141–2152
102. Macchiarini P, Jungebluth P, Go T, Asnaghi MA, Rees LE, Cogan TA et al (2008) Clinical transplantation of a tissue-engineered airway. The Lancet 372(9655):2023–2030
103. Mohanty AK, Misra M, Drzal TL, Selke SE, Harte BR, Hinrichsen G (2005) Natural fibers, biopolymers, and biocomposites: an introduction. In: Natural fibers, biopolymers, and biocomposites. CRC Press-Taylor & Francis Group, Boca Raton, USA, pp 1–36
104. Anonymous (2007) Bioplastics in automotive applications. Bioplastics Magazine. Available from: https://bioplastics-cms.de/bioplastics/. Last viewed: 15 May 2020
105. Mitsubishi Motors develops plant-based green plastic floor mat. Tokyo: Mitsubishi Motors Co. mitsubishi-motors.com. Available from: https://www.mitsubishimotors.com/en/corporate/pressrelease/corporate/detail1475.html. Last viewed: 15 May 2020
106. North America enviromental report, report recycling use. Available from: https://www.toyota.com/about/enviroreport2008/pdfs/2008Report_Recycling_Use.pdf. Last viewed: 15 May 2020
107. Stewart R (2010) Automotive composites offer lighter solutions. Reif Plast 54(2):22–28
108. Automotive news world congress, Detroit, toyota.com. Available from: https://www.toyota.com/about/news/corporate/2011/01/13-1-Automotive.html. Last viewed: 15 May 2020
109. Anonymous (2000) Daimler Chrysler turns to natural fibres. Reinf Plast 44:21
110. Daimler Chrysler Uses a Natural-fiber Component in the Exterior of the Mercedes-Benz A-Class, Stuttgart, 2005. Available from: https://media.daimler.com/dcmedia/0-921-657582-1-815396-1-0-0-0-0-1-11701-854934-0-1-0-0-0-0-0.html. Last viewed: 15 May 2020

111. Ghavami K (2005) Bamboo as reinforcement in structural concrete elements. Cement Concrete Compos 27:637–649
112. Ochi S (2006) Development of high strength biodegradable composites using Manila hemp fiber and starch-based biodegradable resin. Compos Part a 37:1879–1883
113. Lee JT, Kim MW, Song YS, Kang TJ, Youn JR (2010) Mechanical properties of denim fabric reinforced poly(lactic acid). Fibers Polym 11:60–66
114. Shi QF, Mou HY, Li QY, Wang JK, Guo WH (2012) Influence of heat treatment on the heat distortion temperature of poly(lactic acid)/bamboo fiber/talc hybrid biocomposites. J Appl Polym Sci 123:2828–2836
115. Annicchiarico D, Alcock JR (2014) Review of factors that affect shrinkage of molded part in injection molding. Mater Manuf Process 29(6):662–682
116. Tokiwa Y, Calabia B (2006) Biodegradability and biodegradation of poly(lactide). Appl Microbiol Biotechnol 72:244–251
117. Shi B, Palfery D (2012) Temperature-dependent polylactic acid (PLA) anaerobic biodegradablity. Int J Environ Waste Manag 10:297–306
118. Yussuf A, Massoumi I, Hassan A (2010) Comparison of polylactic acid/kenaf and polylactic acid/rise husk composites: the influence of the natural fibers on the mechanical, thermal and biodegradability properties. J Polym Environ 18:422–429
119. Liu W, Misra M, Askeland P, Drzal LT, Mohanty AK (2005) ‘Green’ composites from soy based plastic and pineapple leaf fiber: fabrication and properties evaluation. Polymer 46(8):2710–2721
120. Oksman K, Skrifvars M, Selin JF (2003) Natural fibers as reinforcement in polylacticacid (PLA) composites. Compos Sci Technol 63:1317–1324
121. Aluigi A, Vineis C, Ceria A, Tonin C (2008) Composite biomaterials from fibre wastes: characterization of wool–cellulose acetate blends. Composite Part A 39:126–132
122. Morreale M, Scaffaro R, Maio A, La MFP (2008) Effect of adding wood flour to the physical properties of a biodegradable polymer. Composite Part A 39(3):503–513
123. Nishino T, Hirao K, Kotera M, Nakamae K, Inagaki H (2003) Kenaf reinforced biodegradable composite. Compos Sci Technol 63:1281–1286
124. Murariu M, Dechief AL, Bonnaud L, Paint Y, Gallos A, Fontaine G et al (2010) The production and properties of polylactide composites filled with expanded graphite. Polym Degrad Stabil 95:889–900
125. Choudhury MR, Srinivas MS, Debnath K (2018) Experimental investigations on drilling of lignocellulosic fiber reinforced composite laminates. J Manuf Process 34:51–61
126. Piquet R, Ferret B, Lachaud F, Swider P (2000) Experimental analysis of drilling damage in thin carbon/epoxy plate using special drills. Compos A 31:1107–1115
127. Davim JP, Reis P (2005) Damage and dimensional precision on milling carbon fiber-reinforced plastics using design experiments. J Mater Process Tech 160(2):160–167
128. Palanikumar K, Karunamoorthy L, Karthikeyan R (2006) Assessment of factors influencing surface roughness on the machining of glass fiber-reinforced polymer composites. Mater Des 27(10):862–871
129. Choudhury MR, Debnath K (2020) Analysis of tensile failure load of single-lap green composite specimen welded by high-frequency ultrasonic vibration. Mater Today Proc

Chapter 2
Green Composite Using Agricultural Waste Reinforcement

M. Ramesh, L. Rajeshkumar, D. Balaji, and V. Bhuvaneswari

1 Introduction

Present decade is the era of using the organic wastes as reinforcements or fillers in natural fibre reinforced composite materials (NFCs) which became the research interest in the field of novel material development. Utilization of organic wastes as a part of NFCs results in numerous merits including the minimization of non-renewable resources used to manufacture NFCs like synthetic fillers or polymers. This may in turn reduce the overall cost of the raw material used to prepare the composites, as majority of the resources used are from renewable sources particularly from organic or agricultural wastes. Many researchers stated the various possible ways of manufacturing NFCs by adding organic wastes. In few cases, NFCs were manufactured from animal manures along with the addition of high density polyethylene (HDPE) or high density polypropylene (HDPP) [1]. Even though the mechanical strength of the above-stated composite material cannot be compared with pine flour filled thermo composites, they can be used in some applications whose load ranges are low and medium.

Few researches were performed in order to compare the properties of NFCs made up of agro-wastes such as sugarcane bagasse, sunflower and corn stalks. Among them, composites made of sugarcane bagasse possessed better mechanical characteristics owing to its rich cellulose content when compared with other agro-wastes. It was also stated that in agro-waste reinforced NFCs instead of fibre morphology, like aspect

M. Ramesh (✉)
Department of Mechanical Engineering, KIT-Kalaignarkarunanidhi Institute of Technology, Coimbatore, Tamil Nadu 641402, India
e-mail: mramesh97@gmail.com

L. Rajeshkumar · D. Balaji · V. Bhuvaneswari
Department of Mechanical Engineering, KPR Institute of Engineering and Technology, Coimbatore, Tamil Nadu 641407, India

S. Thomas and P. Balakrishnan (eds.), *Green Composites*,
Materials Horizons: From Nature to Nanomaterials,
https://doi.org/10.1007/978-981-15-9643-8_2

ratio, governing the mechanical characteristics, microscopic chemical contents of the fibres like cellulose and lignin governed them. This was purely due to the direct governance of microscopic contents over the interfacial interaction between matrix and the reinforcements [2]. Thermo-chemical conversion of biomass results in biochars and liquid residues follow an uphill trend in being used as reinforcements for manufacture of NFCs. Char can potentially be used as reinforcement or filler in wood flour composites (WFC) and is investigated by many researchers. Further, when WFC was filled with charcoal powder, results were reported to render numerous advantages like minimal water absorption [3–5], reduced rate of heat release [6], better tensile, impact and flexural properties [3–7] and increased thermal stability and conductivity [3, 4, 6, 7]. Apart from WFCs, char can be used as filler or additive for other NFCs with jute or hemp as their primary ingredient.

Very few or no studies has reported the preparation of NFCs with char as an additive but without the use of coupling agents and lubricants which are yet to be revealed. In spite of many positive results, the possibility of using the organic waste and residue materials as an alternative for non-renewable derivatives like plastics has to be studied deeply. Currently, char has been investigated as a substitute for wood flour in equal quantity as the fraction of PP remains the same. Though the wood-based composites filled with biochar reinforced in PP matrix has proven to be economically sound, pure biochar filled NFCs have not yet been discussed in economic point of view. Meanwhile, the possibilities of using various pyrolysis liquids for the synthesis of biochar can also be studied. Vaisanen et al. [8] extensively discussed the effectiveness of biochar prepared by slow pyrolysis liquid and its utilization in WFCs. Results exposed that the addition of pyrolysis liquids into WFC greatly influenced the volatile organic compounds (VOC) of the composites. Safety of material usage and the properties of VOC should be taken into account meticulously when such organic wastes are added to the composites. Complete studies on the pyrolysis liquids for the preparation of biochar and its association with polymers and natural fibres can be extensively conducted.

Similarly, when waste paper sludge was added to WFCs, a positive change in dimensional stability, mechanical properties and hydrophobicity were noted to occur. Hamzeh et al. [9] noted that even without the use of coupling agent while manufacturing WFCs filled with paper waste, a notable increase in flexural properties could be noted. Addition of paper mill sludge into HDPE composites was also noted to bring up significant changes in composites characteristics [10]. Constituents of the paper sludge like cellulose and ash influenced the mechanical characteristics of the composites much along with various types of sludge used. Researchers also tried to prepare hybrid NFCs made of waste newsprint fibre and wood flour reinforced in polypropylene (PP) and maleic anhydride grafted polypropylene (MAPP). During the preparation, the aspect ratio of newspaper fibres were at least nine times greater than the wood flour and it resulted in enhanced strength of composites [11]. These composites exhibited better strengths and thermal stability [measured using thermogravimetric analysis (TGA)] than unreinforced PP in absence of coupling agents and these observations were similar to Hamzeh et al. [9].

Yet, the mechanical characteristics of hybrid NFCs reinforced in PP performed better after the incorporation of MAPP into them [2]. When the content of wood flour was increased in the hybrid composites, thermal stability has gotten better which could be due to the presence of large amount lignin content in wood flour. Such hybridization, therefore, renders better results which could possibly be explored further more. These hybridizations would be feasible in any of the following ways: (i) unlike single fibre reinforced composite materials, usage of chemically and physically distinct fibres ends up in lot of advantages, (ii) usage of many organic wastes in manufacturing of a single composite material offers the irrepressible use of waste materials for producing NFCs.

Researchers are over-concerned in using organic industrial wastes to manufacture NFCs as the natural and renewable resources are currently over-used and exploited more [9, 12]. Hybridization of sand dust from medium-density fibreboards (MDF) with nano-clay reinforced in PP has been found to improve the strength characteristics and hydrophobicity of the composites. Since the usage of residues is a novel in material usage, few authors could not explain the results broadly even they contribute to the economical point of view and applications greatly. The importance of resin elements and fillers used was masked by the MDF itself while manufacturing MDF-based composite materials. This could be understood from the influence of nano-clay particles and matrix interaction over the experimental results obtained. Slow pyrolysis method of liquid addition was also found to enhance the hydrophobicity and mechanical characteristics of the viable WFC [8, 13–15].

Lignin that was the residue out of paper and pulp industries was considered to be a surplus ingredient that can be used as additive while manufacturing NFCs. Few experimenters used lignin as a filler material while preparing coir-PP composite materials. TGA results of the composite materials revealed that the initial decomposition temperature during the oxidation induction has increased solely because of lignin addition but upon compromising the enhancement of tensile strength of the composites. Water absorption of the coir-PP composites was reduced due to the addition of lignin fillers [16]. Norway spruce pulp derived tall oil fractions were utilized to prepare the ethane and propene monomers which are very effective constituents that can be used for the preparation of NFCs [17]. 35 wt% of ethene and 18 wt% of propene can be obtained from the tall oil fatty acids by means of pulp hydrodeoxygenation in the presence of nickel-molybdenum hydrotreating catalysts. Hence, the adoption of organic waste as additive or reinforcement in manufacturing of NFCs can be clearly witnessed from the above facts. Studies and experimental gist prove the swift increase of researches in the use of agricultural wastes and residues. Many articles were published during the recent times by exploring various manufacturing methods for the waste and residue materials usage in NFC preparations and this minimizes the use of fossil fuels that in turn reduces the environmental effects. This has also been a trendsetter as far as composite industries are concerned.

Nevertheless, improvement the interaction between the key ingredients of NFCs and accomplishing desired properties pose a great challenge for which the usage of coupling agents of other additive materials becomes inevitable for preparing a full-fledged viable NFCs. Current volume of NFC markets and industries are not sufficient to minimize the fast-growing annual wastes by utilizing them as additives in NFCs [18]. Hamzeh et al. [9] estimated that the annual organic wastes may shoot up to 19 billion tons by the end of 2025. Demands for a novel technology to treat wastes and use them in NFCs are currently required for the waste management and overcoming the hurdles in manufacturing of NFCs. The studies reveal versatile ways of using the organic waste as reinforcement. In this review, the focus is all about two-specific hybrid application of the agricultural waste as a reinforcement that is in construction industry along with metal matrix composites.

2 Agricultural Waste Composites

2.1 Oil Palm Shell Reinforced Composites

Teo et al. [19] conducted the most primitive research primarily on the oil palm shell (OPS) reinforced concrete composite beam and the flexural characteristics of concrete beam under static loading were investigated. Reinforcement ratios of 046%, 0.67% and 0.98% were maintained for the while preparing the composites. Authors stated that the failure of the concrete beam followed conventional flexural failure mode preceded by concrete cover crushing with steel yielding. It was also noted that the ultimate moment values of the beam were 19–35% more than that of theoretical moment predicted by BS8110 code. As the results were traditional, it was concluded that no negative effects were observed in reinforcing OPS with concrete beam.

Muda et al. [20] adopted a new method of using geogrid concrete slab reinforced with OPS and investigated the flexural strength of the slab by static tests. Geogrid is a polymer material primarily used as reinforcement which has openings and net-like surface texture characterized by high modulus values. Authors reported the increase of flexural strength values by almost 57% with the increase in geogrid reinforcement layers and decrease of flexural strength with the increase on content of OPS. This can be attributed to poor OPS properties as the material is inherently possess large amount of porosity. Bond breakage between OPS and concrete paste was found to be predominant failure mode in OPS filled geogrid concrete slabs. Yet it was observed that the OPS incorporated geogrid concrete slabs can be readily used in residential building construction because the flexural strength of these slabs satisfies the ultimate strength specified by BS8110 code for residential slab specification. In the meantime, OPS added geogrid concrete slab satisfies the serviceability deflection which also falls within the limit for residential construction as stipulated by BS8110 code. Above-experimental results showcase that OPS concrete slabs are safe while preparing them for construction and they were found to conform the design code norms and serves as better alternative for traditional concrete slab. This could be possible due to the minimization of dead load of a structural member covering a larger area at the site.

2.2 Palm Oil Clinker Reinforced Composites

Mohammed et al. [21] investigated the palm oil clinker (POC) reinforced concrete composite beam for its flexural strength under various loading conditions and various combinations of tension reinforcement ratios ranging between 0.35 and 2.23% were discussed. Results revealed that there was no significant change brought by reinforcing POC into concrete in terms of failure mode as all the reinforced beams failed in typical modes. It is analysed and compared the flexural behaviour of short and long shear POC reinforced concrete slab contoured by steel sheets with traditional normal weight concrete (NWC) slab [22]. It was noticed from the experiments that both the slabs failed by shear-bond mode of failure apart from showcasing better ductile behaviour. Due to low elasticity modulus of the POC material, the POC reinforced concrete slab exhibited high deflection values when compared with traditional NWC slab. Nevertheless, POC material can be potentially used for composite slab construction as the characteristics of the POC slabs were observed to be fitting. Composite deck slab that is used as in building construction is inherently characterized by a lightweight structure and reinforcing POC with that would render better applicability due to the fact that POC slab was proven to have 18% lesser weight compared to NWC slabs.

2.3 Coconut Shell Reinforced Composites

Gunasekaran et al. [23] analysed the single and double layer of coconut shell (CS) reinforced concrete beam for its flexural strength under static loading. Failure mode of single layer CS reinforced concrete beam (nomenclated as under reinforced) was by yield mode trailed by the crushing of concrete in compression zone while the failure of double layer CS reinforced concrete (nomenclated as over reinforced) was before yield that occurred in tension zone. Above results state that CS reinforced concrete beam has likely behaviour with traditional NWC beam. It was also stated by the authors that the general characteristics of the CS beams were mostly in line with other types of concrete beams such as lightweight aggregate (LWA) concrete beams. Authors concluded that IS 456 and BS8110 codes can be readily used to create a conventional design method for finding the ultimate moment of CS reinforced concrete beam and this fact was proven by the experimental ultimate moment values from the experimental evaluations.

2.4 Cellulosic Fibres Composites

Vegetable fibres are available in various forms such as staple, strands and pulp and in different morphologies such as length, diameter, surface roughness and aspect ratio as shown in Fig. 1. Fibre surfaces can also be readily altered for making the fibres additionally hydrophobic or hydrophilic and can be made ready for attaching functional groups with it [24].

Soroushian et al. [26] used a slurry vacuum de-watering method in Hatschek technique to manufacture pulp fibre reinforced cement composites in a weight fraction of 8% of reinforcement. Here, the slurry is formed with 20% solid materials by mixing the matrix and corresponding weight of reinforcements which are water dispersed thus forming a slurry. Then, the slurry is poured on to a mould with drilled holes and was applied with vacuum. Then, the board placed over the slurry was applied with compressed load till the thickness of the slurry falls down to 15 mm (Fig. 2).

Experiments were also carried out by reinforcing around 4 wt% of pinus and cotton pulp reinforced cellulose composites by using a de-watering under pressure methods [27]. Cement, fluidizer, water and sand are the constituents of cement paste with water to cement mixture ratio of 1:1. This paste is added with the water-dispersed cellulose fibres. In order to have a least amount of loss of sand and cement and to allow most of the water out of the mixture, the specimens were placed in a micro-grilled mould and the load was applied to it. Mould was then transferred to a plate, compacted using a vibrating table and compressed for about 24 h to obtain an ultimate pressure of

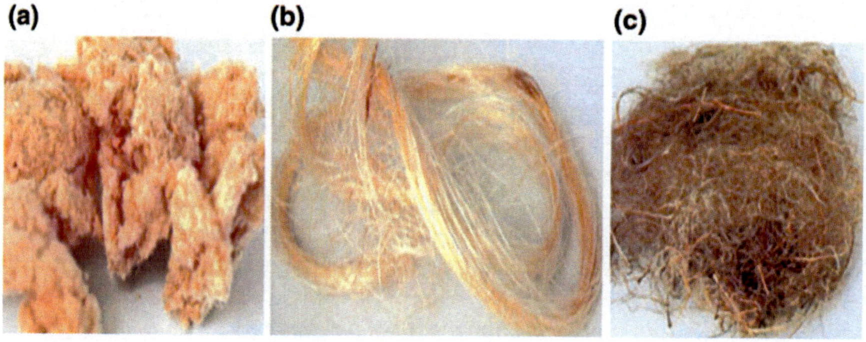

Fig. 1 Vegetable fibres as **a** pulp, **b** strands and **c** staple [25]

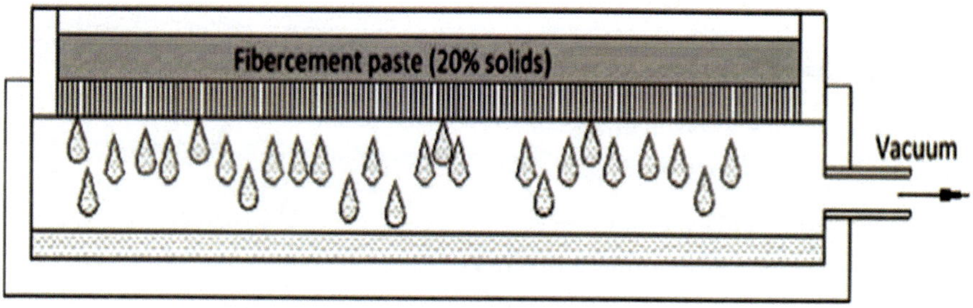

Fig. 2 Processing of composite [25]

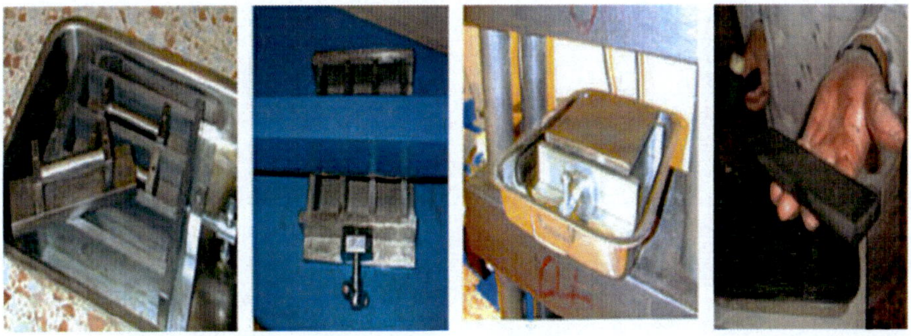

Fig. 3 Mould and press system to prepare the specimens [25]

4 MPa. Figure 3 shows the mould and press used to prepare the specimen and Fig. 4 shows the scanning electron microscopy (SEM) images of the specimen prepared.

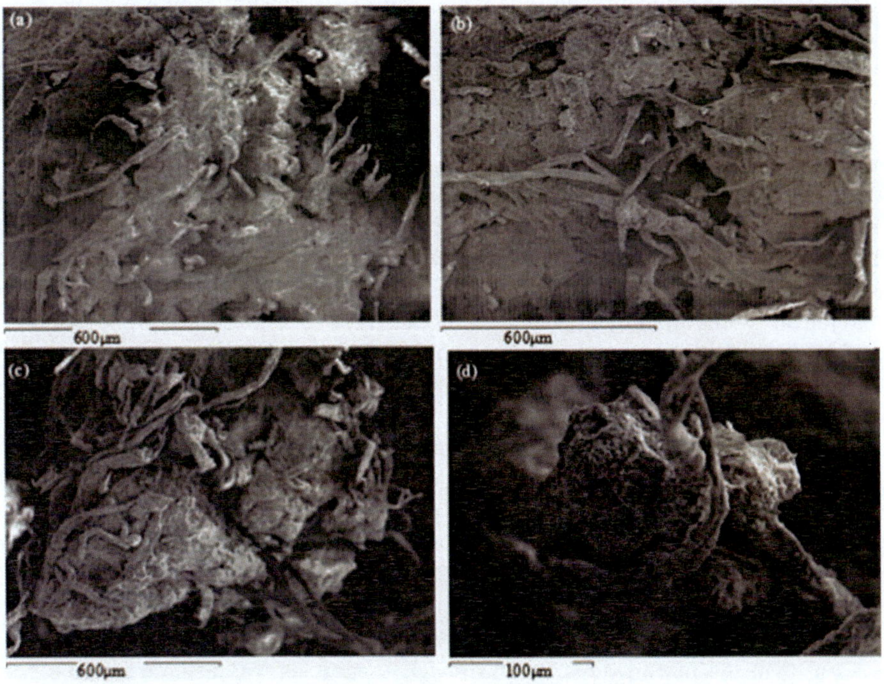

Fig. 4 SEM images of cement cellulose fibre reinforced composites [25]

3 Agricultural Waste-Based Metal Matrix Composites

Materials created in view for sustainable development paved way for most number of current day applications. Many years of research in the field of metal matrix composites (MMCs) were spent by utilizing different chemical constituents as reinforcements for them. Particularly, aluminium MMCs resulted in positive change of characteristics in terms of wear, corrosion, thermal, microstructural and mechanical properties when such chemicals were used as reinforcements. Search for potential alternative materials that can be used as reinforcements for MMCs replacing traditional reinforcements like titanium nitride, graphite, zirconium, silicon nitride and so on. Many literature reviews revealed that the likeliness of using agro-waste materials as substitutes for chemical reinforcements like magnesium oxide or silica can be least expensive and easily manufactural. Few waste materials researched to be a potential alternative for ceramic or particulate reinforcements are cow horn, groundnut shell, bean pod ash, aloe vera, rice husk, coconut shell, breadfruit seed hull ash and many other materials. Many authors determined that the agro-waste reinforcements in powder form are having potential abilities for using as reinforcements in MMCs. They improve the mechanical characteristics equally when they are used as reinforcements when compared with unreinforced cast materials.

Atuanya and Aigbodion [28] prepared the Al-Cu-Mn MMCs reinforced with bean pod ash through double layer feed stir casting technique and the mechanical and microstructural characteristics of bean pod ash reinforced aluminium alloy (BPAAC). BPA was reinforced in aluminium alloys in 1–4% by weight. Hardness, tensile and impact strength of BPAAC were evaluated and they are characterized through XRD and SEM. Experimental results revealed that the interfacial bond of the resulting composites was better than unreinforced composites and the mechanical properties like tensile strength and hardness were better for 4 wt% and were higher by 44.1 and 35% than other counterparts. Even though the organic wastes were reinforced in lower weight percentages, the mechanical characteristics of the AMMCs were enhanced.

Saravanan and Kumar [29] done the investigation on aluminium–silicon-magnesium alloy for the betterment of mechanical properties. Here, the base metal is easily available with low cost.

Rice husk ash is used as reinforcing element. Different weight fraction of RHA was used for fabrication. Those fractions are 3%, 6%, 9% and 12% for the improvement of mechanical properties with liquid metallurgy. Analysis of particle distribution of RHA was done from scanning electron microscope (SEM). From SEM, it was evident that, the uniform distribution of RHA particles was achieved which was useful for achieving the improvement of hardness and tensile. Interfacial surface was increased among the reinforcement and matrix. Ochieze et al. [30] done the tribological property analysis on A356 alloy with the reinforcement of cow horn. Production of cow horn particles was achieved by sintering of the spark plasma. Design of experiments was used for the experimental investigation. Here, Tahuchi's (L9) technique was implemented. Tribometer was effectively utilized for wear test.

SEM experiment analysis was done for the surface morphology of the composite. By comparing unreinforced alloy A356, reinforced alloy shown an improved sliding resistance to the wear. From this investigation, it is evident that, reinforcing corn horn particulates with A356 alloy enhancing the composite wear resistance by reducing wear rate.

Mishra et al. [31] done the research on the aluminium matrix composite. In that aluminium alloy, LM6 as a base metal and rice husk is a reinforcing agent. Here, the weight percentage of rice husk was utilized as 6. Hardening of the composites was conducted at different temperature conditions like 135, 175 and 225 °C. Fabrication of metal matrix composites was done by using stir casting technique. Hardness of the composite was increased from 54.8 to 78.4. The high increment was observed in the composite with the temperature of 175 °C. Atuanya et al. [32] focused their research on aluminium silicon ferrous alloy for the betterment of mechanical property and microstructural behaviour. In this regard, breadfruit seed hull is added as a reinforcing agent with the Al–Si–Fe alloy. Here, the seed hull is added as a particulate with a size of 500 nm which was examined and analysed from six different fractional weights. As expected, the mechanical property of a composite was good and enhanced but with small decrement of the energy absorbed by the composite. The reason behind, this is, compatibility between the base metal and reinforcement was excellent.

Hima Gireesh et al. [33] done the fabrication of aluminium metal matrix composite with the reinforcement of aloe vera powder in order to enhance the mechanical property and characterization behaviour. They also observed there was a tremendous change of AMMC reinforced with aloe vera powder than reinforced with fly ash done by many researchers. Significant improvements were observed in the composites in case of elongation, energy observed and hardness. Alaneme et al. [34] carried a research on fabrication of AMMC with hybrid synthetic and natural reinforcement. Here, hybrid reinforcement indicates silicon carbide and groundnut shell ash particle. The experiment has been done for characterizing fracture, SEM and mechanical properties. The mixing ratio of both the particles varied in the range of 0:10, 2.5:7.5, 5:5, 7.5:2.5, and 10:0 with 6 and 10 wt%. Continuous observation shown that with increased amount of groundnut shell ash (GSA) the percentage elongation was constant, but the fracture toughness was increased by increasing the content of the groundnut shell ash.

Nwobi-Okoye and Ochieze [35] carried out the research work of aluminium alloy A356 reinforced with cow horn powder. This work was applied on brake drum. In this case, several methods have been used for the betterment. Those methods include artificial neural network (ANN), response surface methodology (RSM) and simulated annealing. The main aim of this work concentrated the modelling of the age hardness process with the help of A356 reinforced by cow horn particulates using ANN and RSM. Final result shows the improvement of age hardening by 0.9921 predictions in ANN and 0.9583 predictions in RSM. Alaneme and Olubambi [36] examined the tribological property of aluminium alloy with hybrid reinforcement. In this work for wear and corrosion behaviour, Al–Mg–Si alloy is taken and reinforced by rice husk and ash particulate. Reinforcing elements were used with a weight percentage of 2, 3 and 4 for preparing 10 wt% of strengthening element with aluminium metal matrix composites. Corrosion and wear behaviour of composites were examined by potentiodynamic polarization measurement, open circuit corrosion potential (OCP) and coefficient of the friction parameter, respectively. It was observed that drastic improvement for corrosion resistance in the hybrid reinforced composite with 10 wt% of alumina of 3.5% NaCl solution.

In 2017, the research on hybrid aluminium metal matrix composite done with two-step compo-casting method which is a novel technique. In this, fly ash chemosphere with the weight percentage of 10 and graphite with a weight percentage of 2, 4 and 6 added with the base matrix. Relating to the study of structure reveals the formation of dendrite of chemosphere and graphite in Al 7075 matrix. Moreover, mechanical properties like hardness, tensile of the composites were enhanced by the addition of chemosphere and graphite. It was observed that tensile strength of the composite was improved as 213 N/mm from 178 N/mm. Tribological behaviour like wear resistance was improved by graphite because of its lubricating nature. Bear resistance was enhanced by chemosphere. It was also observed that machinability of the composite enhanced and optimization of machining parameters employed with the help of artificial neural network (ANN) technique. It was evident that from ANOVA, the surface roughness of the composite was decreased in which percentage of graphite increased [37]. Sharma et al. [38] concentrated their research on tribiological properties by fabricating the composites from aluminium with the reinforcement of fly ash. Various proportions of base metal and reinforcement were used for fabrication in metal matrix composite. For this fabrication, stir casting technique was chosen with 2, 4 and 6 wt% of fly ash. Pin on disc tester was used for wear, frictional force and bear calculation of the composite and there was a need of keen examination of tribe pairs which are connecting the soft surfaces of cast iron disc and flat metal matrix composite. Fly ash with weight percentage of 6 is shown a better outcome in the bear resistance of 0.32 g and 4 wt% of fly ash gives short coefficient of resistance of 0.12.

Rice husk ash is used as a reinforcing agent with aluminium metal matrix composites which is extracted from milling paddy. Demand of conventional aluminium material can be reinforced with RHA. Industrial economy can be improved from agricultural waste by using RHA as a reinforcing agent. Since it is an agricultural waste, the inventory and discarding are easy. Figure 5 shows the images of rice husk, its carbonized form and the ground form.

Hardness and tensile strength of the composite enhanced because of better uniformity exist between RHA and aluminium matrix. It was observed that, appreciable increment in case of strength due to the better interfacial bond between both matrix and reinforcement. It is also observed that hardness, tensile and compressive strength of the composite were improved and ductility was decreased by increasing the weight percentage of RHA. Figure 6 shows the process of stir casting of aluminium alloys reinforced with rice husk ash.

It was concluded that the enhancement in the mechanical properties can be well attributed to the high dislocation density. Even though the tensile strength of the composite was reduced by adding the RHA with wt% of 12 [39]. Figure 7 shows

Fig. 5 **a** Raw rice husk, **b** carbonize rice husk ash, **c** grounded carbonized rice husk ash and **d** combusted rice husk ash (grayish white) [39]

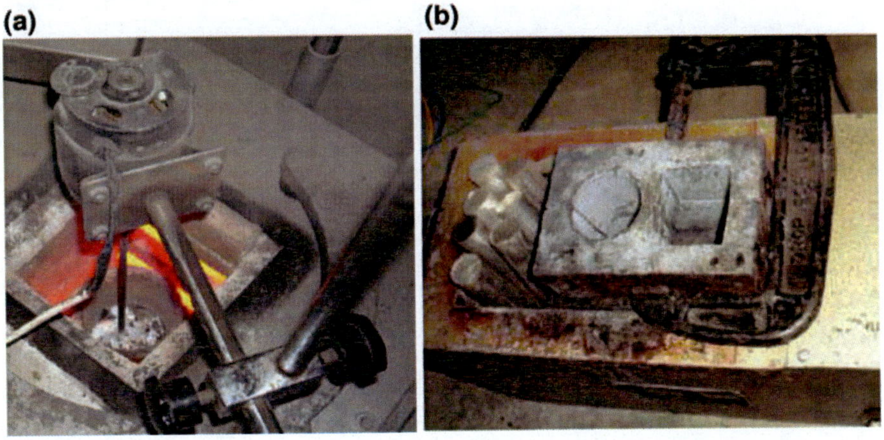

Fig. 6 Stir casting equipment: **a** electrical furnace and **b** preheated permanent mould [39]

the scanning electron microscopic images of aluminium–silicon-magnesium metal matrix composites incorporated with rice husk ash.

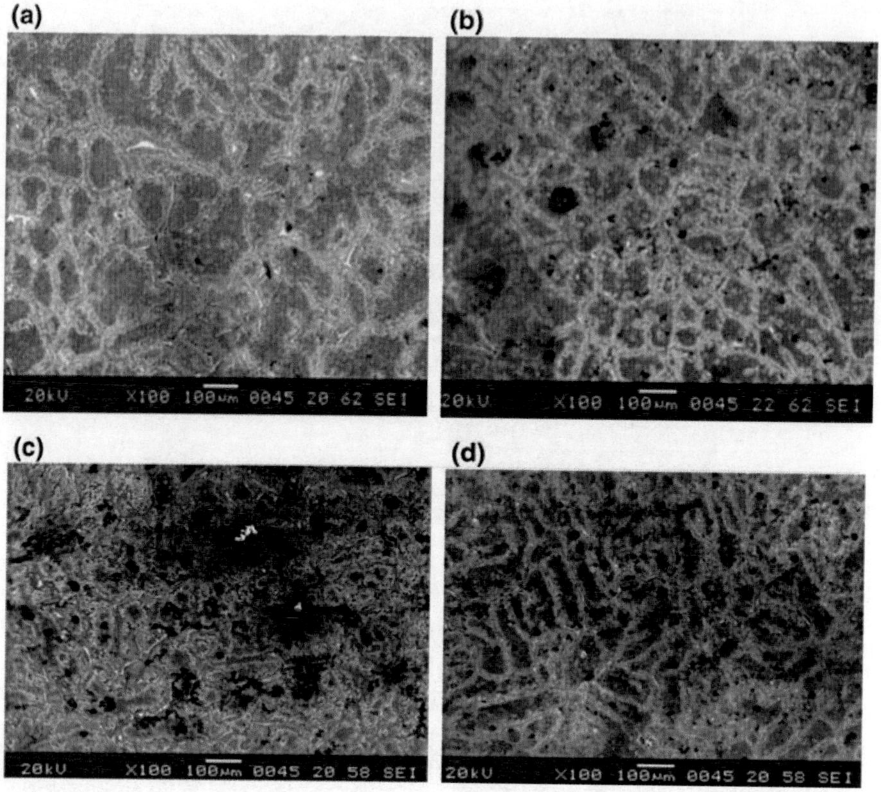

Fig. 7 SEM images of MMC with RHA. **a** $AlSi_{10}Mg$; **b** $AlSi_{10}Mg$ + 6% RHA; **c** $AlSi_{10}Mg$ + 9% RHA; **d** $AlSi_{10}Mg$ + 12% RHA [39]

4 Conclusion

The thrust area in the research generally varies with the advancements in the technology. Only few areas always hold a position in the list, one among of them is waste reusability (waste to wealth). Agriculture waste has special attention owing to environmentally friendly in nature. This chapter focused on the agricultural waste reusability as reinforcement in the wired combination as in the construction industry and along with the metal matrix composite are discussed. In construction industry, the agricultural waste is added with concrete materials as the reinforcement like slab, beam and cement mortar. The waste materials from agri-like oil palm shell, palm oil clinker, coconut shell and cellulosic fibre for cement mortar. It is replaced in the concrete to form the variant of lightweight concrete (LWC), as a known fact that these waste materials is much less in weight. These waste materials inclusion as a reinforcement in the LWC is while comparing with the normal weight concrete (NWC), the possibility of increasing the bond strength is observed. The biomaterials are added in the metal matrix composite which enhanced the material property.

Further, in all the cases, fibre treatment improves the strength additionally along with specific property can be added to the composite materials. Hence, the application of such NFCs reinforced with organic wastes is wide spread such as in construction industries and in fabrication of non-structural elements also. Further, explorations are possible if these pre-fabricate elements are analysed experimentally for their special curing conditions.

References

1. Rowell RM, O'Neill E, Krzysik A, Bossman D, Gallaway DF, Hemenover M (2007) Incorporation of animal manures as reinforcing fillers in high-density polyethylene and high-density polypropylene composites. In: Stark NM (ed) Proceedings of ninth international conference on wood & biofiber plastic composites, 2007 May 21–23, Forest Products Society, Madison, WI, 7224, pp 371–374. ISBN: 1892529505: 9781892529503
2. Ashori A, Nourbakhsh A (2010) Bio-based composites from waste agricultural residues. Waste Manage 30:680–684
3. Li X, Lei B, Lin Z, Huang L, Tan S, Cai X (2014) The utilization of bamboo charcoal enhances wood plastic composites with excellent mechanical and thermal properties. Mater Des 53:419–424
4. DeVallance DB, Oporto GS, Quigley P (2015) Investigation of hardwood biochar as a replacement for wood flour in wood–polypropylene composites. J Elastom Plast 1–13
5. Ayrilmis N, Kwon JH, Han TH, Durmus A (2015) Effect of wood-derived charcoal content on properties of wood plastic composites. Mater Res 18:654–659
6. Das O, Sarmah AK, Bhattacharyya D (2015) Biocomposites from waste derived biochars: mechanical, thermal, chemical, and morphological properties. Waste Manage 49:560–570
7. Das O, Sarmah AK, Bhattacharyya D (2015) A novel approach in organic waste utilization through biochar addition in wood/polypropylene composites. Waste Manage 38:132–140
8. Vaisanen T, Heikkinen J, Tomppo L, Lappalainen R (2016) Improving the properties of wood–plastic composite through addition of hardwood pyrolysis liquid. J Thermoplast Compos Mater. https://doi.org/10.1177/0892705716632862
9. Hamzeh Y, Ashori A, Mirzaei B (2011) Effects of waste paper sludge on the physico-mechanical properties of high density polyethylene/wood flour composites. J Polym Environ 19:120–124
10. Soucy J, Koubaa A, Migneault S, Riedl B (2015) Chemical composition and surface properties of paper mill sludge and their impact on high density polyethylene (HDPE) composites. J Wood Chem Technol 36:77–93
11. Stark NM, Rowlands RE (2003) Effects of wood fiber characteristics on mechanical properties of wood/polypropylene composites. Wood Fiber Sci 35:167–174
12. Viksne A, Berzina R, Andersone I, Belkova L (2010) Study of plastic compounds containing polypropylene and wood derived fillers from waste of different origin. J Appl Polym Sci 117:368–377
13. Ramesh M, Rajeshkumar L (2018) Wood flour filled thermoset composites. In: Thermoset composites: preparation, properties and applications, materials research forum, vol 38, pp 33–65. https://doi.org/10.21741/9781945291876-2
14. Vaisanen T, Heikkinen J, Tomppo L, Lappalainen R (2016) Softwood distillate as a bio-based additive in wood-plastic composites. J Wood Chem Technol 36(4):278–287. https://doi.org/10.1080/02773813.2015.1137947
15. Vaisanen T, Tomppo L, Selenius M, Lappalainen R (2016) Effects of slow pyrolysis derived birch distillate on the properties of wood-plastic composites. Eur J Wood Prod 74:131–133
16. Morandim-Giannetti AA, Agnelli JAM, Lanças BZ, Magnabosco R, Casarin SA, Bettini SH (2012) Lignin as additive in polypropylene/coir composites: thermal, mechanical and morphological properties. Carbohydr Polym 87:2563–2568

17. Pyl SP, Dijkmans T, Antonykutty JM, Reyniers M, Harlin A, Van Geem KM, Marin GB (2012) Wood-derived olefins by steam cracking of hydrodeoxygenated tall oils. Bioresour Technol 126:48–55
18. Mohammed L, Ansari M, Pua G, Jawaid M, Islam MS (2015) A review on natural fiber reinforced polymer composite and its applications. Int J Polym Sci. https://doi.org/10.1155/2015/243947
19. Teo DCL, Mannan MA, Kurian VJ (2006) Structural concrete using oil palm shell (OPS) as lightweight aggregate. Turk J Eng Environ Sci 30(4):251–257
20. Muda ZC, Sharif SFA, Hong NJ (2012) Flexural behavior of lightweight oil palm shells concrete slab reinforced with geogrid. Int J Sci Eng Res 3(11)
21. Mohammed BS, Foo WL, Abdullahi M (2014) Flexural strength of palm oil clinker concrete beams. Mater Des 53:325–331
22. Mohammed BS, Al-Ganad MA, Abdullahi M (2011) Analytical and experimental studies on composite slabs utilising palm oil clinker concrete. Constr Build Mater 25(8):3550–3560
23. Gunasekaran K, Annadurai R, Kumar PS (2013) Study on reinforced lightweight coconut shell concrete beam behavior under flexure. Mater Des 46:157–167
24. Faruk O, Bledzki AK, Fink H-P, Sain M (2012) Biocomposites reinforced with natural fibers: 2000–2010. Prog Polym Sci 37:1552–1596
25. Monica A, Claramunt J, Filho RDT (2015) Cellulosic fiber reinforced cement-based composites: a review of recent research. Constr Build Mater 79:115–128
26. Soroushian P, Won J-P, Hassan M (2012) Durability characteristics of CO_2-cured cellulose fiber reinforced cement composites. Constr Build Mater 34:44–53
27. Claramunt J, Ardanuy M, García-Hortal JA, Filho RDT (2011) The hornification of vegetable fibers to improve the durability of cement mortar composites. Cem Concr Compos 33:586–595
28. Atuanya CU, Aigbodion VS (2014) Evaluation of Al-Cu-Mg alloy/bean pod ash nanoparticles synthesis by double layer feeding-stir casting method. J Alloys Compd 601:251–259
29. Saravanan SD, Kumar MS (2013) Effect of mechanical properties on rice husk ash reinforced aluminium alloy (AlSi10Mg) matrix composites. Proc Eng 64:1505–1513
30. Ochieze BQ, Nwobi-Okoye CC, Atamuo PN (2018) Experimental study of the effect of wear parameters on the wear behavior of A356 alloy/cow horn particulate composites. Def Technol 14:77–82
31. Mishra P, Mishra P, Rana RS (2018) Effect of rice husk ash reinforcements on mechanical properties of aluminium alloy (LM6) matrix composites. Mater Today 5:6018–6022
32. Atuanya CU, Ibhadode AOA, Dagwa IM (2012) Effects of breadfruit seed hull ash on the microstructures and properties of Al-Si-Fe alloy/breadfruit seed hull ash particulate composites. Res Phys 2:142–149
33. Hima Gireesh C, Durga Prasad KG, Ramji K, Vinay PV (2018) Mechanical characterization of aluminium metal matrix composite reinforced with aloe vera powder. Mater Today 5:3289–3297
34. Alaneme KK, Bodunrin MO, Awe AA (2018) Microstructure, mechanical and fracture properties of groundnut shell ash and silicon carbide dispersion strengthened aluminium matrix composites. J King Saud Univ Eng Sci 30:96–103
35. Nwobi-Okoye CC, Ochieze BQ (2018) Age hardening process modeling and optimization of aluminium alloy A356/Cow horn particulate composite for brake drum application using RSM, ANN and simulated annealing. Def Technol 14:336–345
36. Alaneme KK, Olubambi PA (2013) Corrosion and wear behavior of rice husk ash—alumina reinforced Al-Mg-Si alloy matrix hybrid composites. J Mater Res Technol 2:188–194
37. Kumarasamy SP, Vijayananth K, Thankachan T, Muthukutti GP (2017) Investigations on mechanical and machinability behavior of aluminum/flyashcenosphere/Gr hybrid composites processed through compo casting. J Appl Res Technol 15:430–441
38. Sharma VK, Singh RC, Chaudhary R (2017) Effect of fly ash particles with aluminium melt on the wear of aluminium metal matrix composites. Eng Sci Technol Int J 20:1318–1323
39. Saravanan SD, Kumar MS (2013) Effect of mechanical properties on rice husk ash reinforced aluminum alloy ($AlSi_{10}Mg$) matrix composites. Procedia Eng 64:1505–1513

Chapter 3
Green Fiber Thermoplastic Composites

Gulcihan Guzel Kaya and Huseyin Deveci

1 Composite Materials

Composites as functional advanced materials take place near the top among required materials worldwide. There is a growing interest in utilization of composite materials in many engineering fields such as automotive, construction, aerospace, electronics, and biomedical industry [1]. Composite materials compose of at least two components whose promising features combine to form a material with better properties [2]. Regarding to matrix type, composite materials are subclassified as polymer-matrix composites, ceramic-matrix composites, and metal-matrix composites [3].

Polymer composites are commonly used materials with the advantages of low density, high mechanical strength, excellent resistance to wear and corrosion and flexible operability [4, 5]. Polymer matrix as continuous phase in the composite structure is divided into two groups: thermoplastics and thermosets [6]. Compared to thermosets, thermoplastic polymers have been mostly focused on due to their remoldability as well as recyclability [7]. To enhance properties of different types of polymer matrices, fillers as dispersed phase are incorporated into the matrix. Various fillers including carbon-based materials (carbon fiber, carbon black, graphene, carbon nanotube, etc.), glass fiber, silica, clays, metal oxides (TiO_2, Al_2O_3, Fe_2O_3, ZnO, MgO, etc.) and calcium carbonate have been used in polymer composites [8–14]. However, green fibers play an important role as filler due to increasing environmental awareness and economic issues. Moreover, legislative policies have been applied for using renewable sources like green fibers instead of conventional sources in many countries [15, 16].

In this chapter, the authors centered upon green fiber-reinforced thermoplastic polymers. Firstly, different types of thermoplastic polymers and green fibers were

G. Guzel Kaya (✉) · H. Deveci
Department of Chemical Engineering, Konya Technical University, Konya, Turkey
e-mail: ggkaya@ktun.edu.tr

S. Thomas and P. Balakrishnan (eds.), *Green Composites*,
Materials Horizons: From Nature to Nanomaterials,
https://doi.org/10.1007/978-981-15-9643-8_3

investigated in general. Production processes of green fiber thermoplastic composites were expressed briefly. Mechanical, thermal, and water absorption properties of green fiber–polymer matrix composites were examined. Lastly, wide range of applications of green fiber thermoplastic composites was summarized.

1.1 Thermoplastic Polymers

Thermoplastics consist of linear polymer chains which are connected each other by intermolecular or van der Waals interactions [17]. Softening of thermoplastic polymers under heating makes their formability easy, and thermoplastics in viscous state can be converted into solid phase by cooling without chemical changes. Therefore, repeatable heating and cooling process provide them recycling ability [18]. In case of effective drying before using thermoplastic polymers, no gas or vapor release is observed which is significant in terms of environmental policy [19].

Thermoplastic polymers can be classified as semi-crystalline and amorphous according to molecular arrangement. Polyethylene (PE), polypropylene (PP), polyamide-6 (Nylon-6), and polyamide-6,6 (Nylon-6,6) are most known semi-crystalline polymers that have mixture of well-ordered and random arrangement [20]. Amorphous polymers composes of randomly arranged molecules including polystyrene (PS), polyvinyl chloride (PVC), polycarbonate (PC), polymethylmethacrylate (PMMA), etc. [21].

Commonly used thermoplastic polymers called as engineering plastics are given with typical properties in Table 1 [20–27]. Low-density polyethylene (LDPE) produced with the ethylene polymerization under high pressure has a density as low as 0.91 g/cm^3 due to many braches in its structure originated from intermolecular chain transfer in polymerization. Formation of branches causes low mechanical strength that can restrict utilization of LDPE. However, high-density polyethylene (HDPE) is prepared at atmospheric pressure in the presence of catalyst leading no or fewer branching structure. When compared to LDPE, HDPE possesses better mechanical properties depending on branching degree [28, 29]. PP is one of the commodity polymers which is highly resistant to many chemicals [30]. PS has attracted great attention owing to ability of chemical modification in various fields [31]. Nylon-6 and Nylon-6,6 exhibit higher thermal stability than many thermoplastic polymers. These types of polymers are preferred especially in high-temperature required processes [32, 33]. PVC is the second most widely used thermoplastic polymer following PE with the advantages of low-cost and optical properties [34].

1.2 Green Fibers

Natural fibers are sustainable and environmental-friendly green materials that have an edge over synthetic fibers [35]. Based on sources, natural fibers can be classified

Table 1 Properties of commonly used thermoplastic polymers [20–27]

Property	LDPE	HDPE	PP	PS	Nylon-6	Nylon-6,6	PVC
Tensile strength (MPa)	12	26	26–41	46	80	60–80	48
Young modulus (GPa)	0.2	1.4	1.3	2.8	2.5	2.8	2.2
Elongation at break (%)	90–800	840	15–200	1.0–2.5	20–150	0.7–19	10–100
Izod impact strength (J/m)	854	27–1068	21–267	17	60	100	32
T_g (°C)	−125	−120	−10	100	48	80	80
T_m (°C)	100–110	130	175	110–135	215–216	250–269	180
T_d (°C)	32–50	43–60	50–63	83	44–80	75–90	67
α_T (mm/mm/°C × 10^5)	10	10–20	6.8–13.5	6–8	8.0–8.6	7.2–9.0	6–8
Density (g/cm^3)	0.91	0.96	0.92	1.04	1.14	1.14	0.70–1.35
Water absorption (%)	0.01	0.01–0.20	0.01–0.02	0.03–0.10	1.3–1.8	1.0–1.6	0.1–0.4

T_g: glass transition temperature, T_m: melting temperature, T_d: heat deflection temperature at 1.8 MPa, and α_T: linear thermal expansion coefficient

as animal, mineral, and plant fibers. Plant fibers are generally derived from bast (flax, jute, kenaf, hemp and ramie), leaf (abaca, pina, sisal, raffia and banana), seed (coir, cotton and kapok), grass (bamboo, bagasse and reed), stalk (corn, rice, wheat and barley), and wood (soft wood and hard wood) [36]. Wood is an abundant and cheap source of fiber with the annual production of ~1750,000 × 10^3 tons (0.3–0.6 US$/kg). Wheat (720,000 × 10^3 tons/year) and rice husk (120,000 × 10^3 tons/year) are most produced fibers among stalk-based natural fibers. Bamboo production (30,000 × 10^3 tons, 0.5 US$/kg) is considerably higher than that of the grass-based plant fibers. Moreover, production of green fibers derived from bast such as jute (2300 × 10^3 tons, 0.4–1.5 US$/kg), kenaf (970 × 10^3 tons, 0.3–0.5 US$/kg), and flax (830 × 10^3 tons, 2.1–4.2 US$/kg) is increasing day by day [7, 25, 37].

Typical structure of green fibers consisting of primary and secondary wall is shown in Figure 1. Primary cell wall composes of randomly arranged cellulose microfibrils in addition to pectin, lignin and hemicellulose [38]. Secondary wall of green fiber forms through combination of helically arranged crystalline cellulose microfibrils and amorphous phase mostly formed by lignin and hemicellulose. Due to the crystalline parts, secondary wall especially middle wall affects the properties of green fiber [39].

Chemical composition of commonly used green fibers is given Table 2 [38, 40–45]. It is obvious that cellulose, hemicellulose, lignin, and pectin are primary components of green fibers. Depending on content of components, physical and mechanical properties can change as shown in Table 3 [2, 37, 46–52]. Most of the green fibers consist

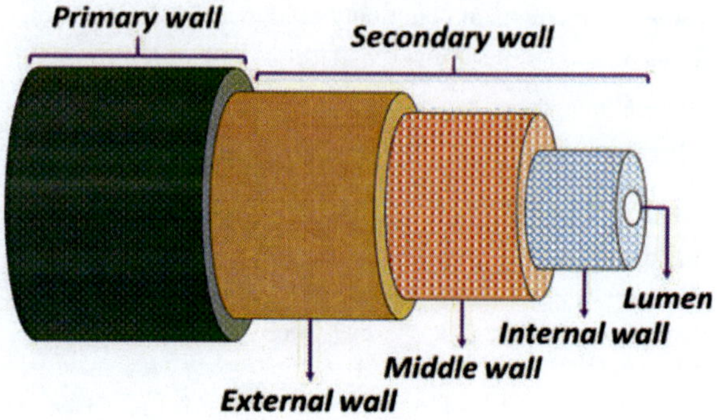

Fig. 1 Schematic representation of green fiber structure

Table 2 Chemical composition of commonly used green fibers [38, 40–45]

Fiber	Cellulose (wt%)	Hemicellulose (wt%)	Lignin (wt%)	Pectin (wt%)	Wax (wt%)	Ash (wt%)
Wood	40–50	30–40	20–34	0–1	0.4–0.5	0.2–0.8
Flax	71–81	18–20	2–3	2.2–2.3	1.5–1.7	1.5
Jute	61–71	14–20	12–13	0.2	0.5	0.5–2
Kenaf	45–57	12–22	8–13	3–5	0.8	2–5
Hemp	70–77	18–22	3.7–5.7	0.9	0.8	0.8
Ramie	68–76	13–16	0.6–0.7	1.9	0.3	–
Abaca	56–63	20–25	7–9	1	–	3
Pina	81	7.1	12	–	–	2
Sisal	66–78	10–14	8–14	10	2	0.6–1
Banana	63–64	12.1	5	–	–	2.2
Coir	32–43	0.15–0.25	40–45	3–4	1–2	2.7
Cotton	85–90	5.7	0.7–1.6	–	0.6	–
Bamboo	26–43	30	21–31	0–0.2	1–2	1.7–5
Bagasse	33–55	17	18–25	–	–	1.7–1.8
Rice	38–57	19–33	8–20	10–15	14–17	10–20
Wheat	28–45	15–31	12–20	0–1	0.5–1	6–8

of cellulose more than 60 wt%. Cellulose is a polysaccharide in which D-glucopyranose units are connected with each other through β-(1 $\rightarrow$ 4)-glycosidic links [53]. Long-chain cellulose microfibrils have synergistic effects on the mechanical strength of green fibers as well as alkali and oxidation resistance [54]. Hemicellulose is responsible for biodegradation, thermal degradation, and moisture absorption [49]. Lignin provides thermal stability and rigidity to green fibers, and flexibility of green fibers is originated from pectin [55, 56].

In spite of their advantages, green fibers have some drawbacks that adversely affect their compatibility with many polymer matrices. Due to the hydroxyl groups

Table 3 Physical and mechanical properties of commonly used green fibers [2, 37, 46–52]

Fiber	Density (g/cm^3)	Moisture content (%)	Tensile strength (MPa)	Elongation at break (%)	Young modulus (GPa)
Wood	1.4	5–8.5	90–180	8–14	18–40
Flax	1.5	7–12	350–1100	2–4	28–70
Jute	1.3	9–12	345–800	1.5–1.8	20–55
Kenaf	1.4	9–12	223–930	1.5–2.7	14–53
Hemp	1.5	6–9	270–900	1.5–4	23–70
Ramie	1.5	7.5–17	400–1000	1.2–3.8	44–128
Abaca	1.5	7–15	418–813	3–10	31–33
Pina	1.5	9–13	400–1627	1–3	60–82
Sisal	1.5	9–11	400–700	3–7	9–38
Banana	1.4	8–12	529–759	5–6	27–32
Coir	1.2	8–10	108–252	15–30	4–6
Cotton	1.6	8–8.5	287–800	3–8	5–13
Bamboo	1.3	8.8–8.9	140–230	4–7	11–17
Bagasse	1.2	8.8–10	222–290	1.1–4	20–27
Rice	0.6–0.8	8–9.1	10–200	2.7	1–12
Wheat	0.7–1	5.1–8.3	55	2–5	22

in the structure of green fibers, interfacial adhesion between fiber and hydrophobic matrix weakens. High hydrophilic fibers are tending to absorb moisture leading to reduce wettability of fiber by a polymer. Poor interfacial adhesion between green fiber and polymer matrix negatively influences properties of final composite materials [57–59]. To overcome this problem, many physical and chemical treatments of green fibers have been developed. Physical treatments such as corona, laser, and plasma treatment alter surface energy of green fibers without chemical changes [60]. There are many chemical treatment methods to control adhesion in fiber–polymer matrix interface. Alkali treatment is an effective method for the removal of free hydroxyl groups in cellulose and dissolving hemicellulose that restrict moisture absorption [58, 61]. Additionally, silane treatment, acetylation, benzoylation, acrylation, permanganate treatment, peroxide treatment, maleated coupling, isocyanate treatment, triazine treatment are commonly used chemical modification methods to make green fibers suitable as filler [62–64]. And also, enzyme treatment as a new biological method is used to remove non-cellulosic components of green fibers [36].

2 Green Fiber Reinforced Thermoplastic Composites

Combination of a green fiber and thermoplastic polymer results in production of composite materials with low process cost, high performance, eco-friendly, and sustainability. Many factors including fiber type, size and amount of fiber, aspect ratio, fiber orientation, fiber treatment, manufacturing process, and operation conditions play a significant role on the properties of green fiber thermoplastic composites [65, 66]. Due to the degradation of a lot of green fibers above 200 °C, process temperature is one of the most important factors to avoid of deteriorated performance of the materials [67]. Manufacturing technique is another factor affecting feature of final materials.

To prepare thermoplastic composites including green fibers, various techniques such as compression molding, extrusion, injection molding, long fiber thermoplastic direct (LFT-D) method, thermoforming, foaming and film casting are used [24, 68–70]. Schematic representation of commonly used manufacturing techniques is shown in Fig. 2. Compression molding is utilized for the production of thermoplastic composites including randomly dispersed or oriented long or short fiber mat or chopped fiber in general. It is a close molding method in which fiber and thermoplastic

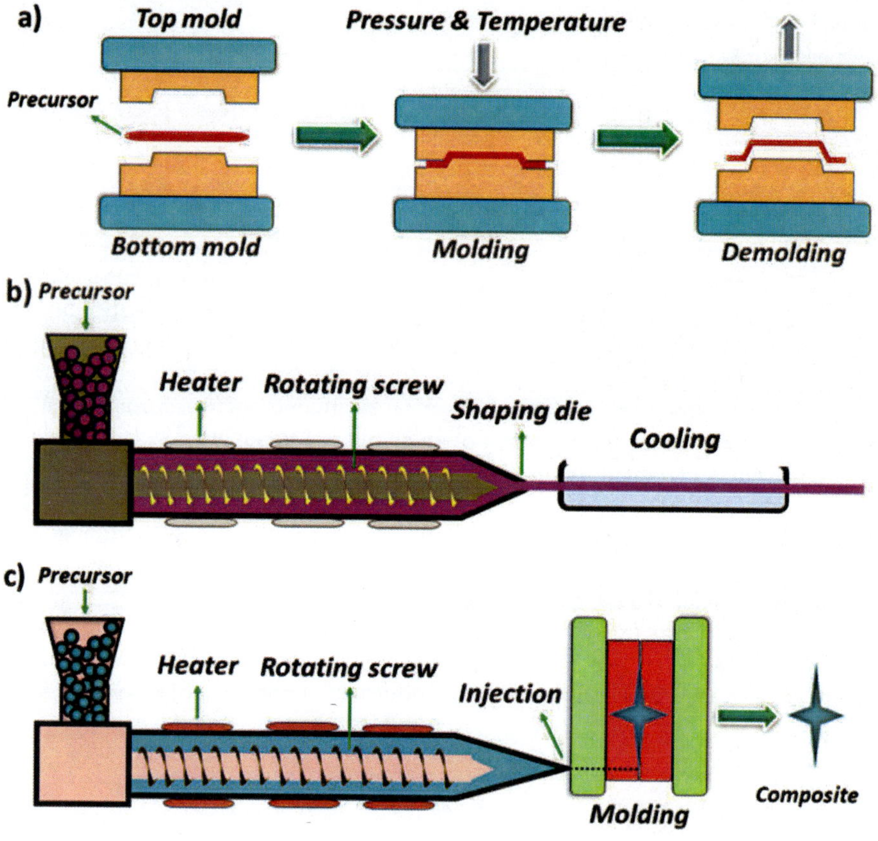

Fig. 2 Schematic representation of commonly used manufacturing techniques. **a** Compression molding, **b** extrusion, and **c** injection molding

are compressed between two molds under the optimized temperature and pressure. Easy controlling viscosity, holding time, temperature, and pressure is advantage of the technique as well as ability to produce complex components in low cycle time [71, 72]. Extrusion is widely used manufacturing technique owing to low-cost process, high-quality product, and the absence of effluent. In this technique, granule and pellet thermoplastic and fiber are mixed for homogeneous blends through single or twin screw and transferred to compression zone to be melted. Shaped composite materials with high strength and stiffness are obtained after cooling with water [73, 74]. For elaborated thermoplastic composites, injection molding is a promising technique. Following injection of melted precursors into molds with hopper, different shaped composite materials can be obtained for especially large-scale applications [75].

2.1 Mechanical Properties

It is well known that effective load transfer from thermoplastic polymer to green fiber enhances mechanical properties of composite materials related with interfacial adhesion between polymer and filler. In Table 4, improved mechanical properties of various green fiber thermoplastic composites such as tensile, flexural and impact properties, dynamic mechanical, and creep behavior are remarked depending on amount of green fiber, fiber length and orientation, modification of fiber, manufacturing method, etc.

Certain amount of green fiber in thermoplastic matrix generally increases mechanical strength of the composite materials. Ng et al. studied on mechanical properties of pineapple leaf fiber/PP composites consisting of different ratios of fiber (10, 20, 30, 40, and 50 wt%) [85]. Maximum tensile strength was belonged to PP composite including 30 wt% fiber (17.1 MPa). More fiber content resulted in decreasing tensile strength because of insufficient load transfer by PP matrix. Tensile modulus of PP composites slightly increased depending on increased fiber content. Flexural strength of PP composite including 30 wt% fiber was higher than that of the other composites. Flexural modulus was increased up to 2.75 GPa with fiber loading. Kaewkuk et al. investigated mechanical properties of PP composites including 10, 20, and 30 wt% alkali-treated sisal fiber [83]. An increase in fiber content had synergistic effects on tensile strength and Young's modulus of the PP composites. However, elongation at break and impact strength of the composites showed a decrease due to agglomerations of sisal fiber and voids. Yuan et al. revealed effects of wood fiber content (10, 20, 30, 40, and 50 wt%) on the mechanical properties of HDPE composites [78]. With incorporation of 50 wt% wood fiber, maximum tensile modulus (4064 MPa) and flexural modulus (3036 MPa) were achieved, respectively. Tensile strength of neat HDPE reduced from 24 to 14.2 MPa with increasing amount of wood fiber. In spite of an increase in impact strength with 20 wt% wood fiber, more fiber content caused to decrease impact strength of HDPE composite because of increasing viscosity of the composite. Tajvidi et al. compared dynamic mechanical properties of PP composites including 25 and 40 wt% hemp fiber [84]. It was indicated that storage modulus of

Table 4 Mechanical properties of green fiber reinforced thermoplastic composites

Matrix	Fiber	Fiber treatment	Manufacturing method	Properties	References
LDPE	Doum	Alkali	Extrusion	Higher tensile and flexural modulus	[76]
LDPE	Jute	Alkali silane	Compression molding	Higher tensile and flexural strength	[77]
HDPE	Wood	–	Extrusion	Higher tensile and flexural modulus	[78]
HDPE	Wood	–	Injection molding extrusion	Higher tensile and flexural modulus with injection molding	[79]
HDPE	Bark with different aspect ratio	–	Extrusion	Higher tensile strength and flexural modulus at higher fiber aspect ratio	[80]
HDPE	Poplar wood Radiate pine Rice husk	–	Extrusion	Higher flexural strength and modulus, higher impact strength and lower creep strain at 0°	[81]
HDPE	Kenaf	–	Extrusion at low and high temperatures	Higher storage and loss modulus, lower tan δ and higher tensile strength at high temperature	[82]
PP	Sisal	Alkali	Injection molding	Higher tensile strength and Young's modulus	[83]
PP	Hemp	–	Injection molding	Higher storage modulus and lower tan δ	[84]

(continued)

Table 4 (continued)

Matrix	Fiber	Fiber treatment	Manufacturing method	Properties	References
PP	Pineapple leaf (10, 20, 30, 40 and 50 wt%)	–	Compression molding	Higher tensile and flexural strength with 30 wt% fiber, and higher tensile and flexural modulus with 50 wt% fiber	[85]
PP	Flax Hemp	–	Compression molding	Higher flexural modulus with wrap yarn structure than twisted yarn structure	[86]
PP	Flax	Acetylation with different degree (3.6, 12, 18 and 34%)	Injection molding	Higher tensile strength and modulus, higher flexural strength and modulus at 18% acetylation	[87]
PP	Flax	–	Injection molding with kneading process, Henschel kinetic mixer, extrusion, and LFT	Higher tensile strength and stiffness with kneading process	[88]
PP	Coir	Chemical treatment with diazonium salt in pH 3, 7 and 10.5	Injection molding	Higher tensile strength and modulus, higher flexural strength and modulus, and higher charpy impact strength in pH 10.5	[89]
PP	Kenaf	Alkali treatment with different NaOH content (1–8 wt%)	Compression molding	Higher tensile and flexural strength, and higher tensile modulus with 5 wt% NaOH–treated fiber	[90]

(continued)

Table 4 (continued)

Matrix	Fiber	Fiber treatment	Manufacturing method	Properties	References
PP	Banana	$KMnO_4$ MAPP NaOH VTMO	Compression molding	Higher tensile strength with VTMO treatment	[91]
PP	Kenaf	–	Compression molding at different temp. (190–240 °C) with different barrel speed (12–24 Hz)	Higher tensile strength at 230 °C with barrel speed of 16 Hz	[92]
PP	Sisal Banana Jute Flax	–	Extrusion	Higher flexural stiffness with jute fiber and higher impact energy with sisal fiber	[93]
PS	Chopped agave Short agave Long agave	–	Compression molding	Higher tensile, flexural and compressive strength with chopped fiber, higher impact strength with long fiber	[94]
PS	Banana Hemp Sisal	Maleic anhydride	Compression molding	Higher tensile and flexural modulus, higher izod impact strength with sisal fiber	[95]
PS	Sugar palm	Alkali treatment with 4 and 6 wt% NaOH	Compression molding	Higher flexural strength and modulus, and higher impact strength	[96]
Nylon-6	Wheat straw Maize straw Rice straw	Alkali	Injection molding	Higher tensile and impact strength with wheat straw fiber	[97]

(continued)

Table 4 (continued)

Matrix	Fiber	Fiber treatment	Manufacturing method	Properties	References
Nylon-6	Pineapple leaf (20, 30 and 40 wt%)	Alkali Silane	Injection molding	Higher tensile strength with silane treated 30 wt% fiber and tensile modulus with silane treated 40 wt% fiber	[98]
Nylon 6/66	Pineapple leaf	Alkali Silane	Injection molding	Higher tensile and flexural strength, and higher flexural modulus with silane-treated fiber, higher tensile modulus with alkali-treated fiber	[99]
PVC	Bagasse Rice straw Rice husk Pine	–	Compression molding	Higher tensile and impact strength with rice straw	[100]
PVC	Bagasse (15, 25 and 35 wt%)	Benzoic acid (3, 5 and 10 wt%) at different temperatures (140–160 °C)	Compression molding	Higher tensile strength with 10 wt% benzoic acid treated 35 wt% bagasse fiber at 160 °C	[100]
PVC	Bamboo	Sodium silicate (0.5, 1, 2, 5 and 10 wt%)	Compression molding	Higher tensile strength and modulus with 5 wt% sodium silicate-treated bamboo fiber	[101]

40 wt% hemp fiber/PP composite was higher than that of the 25 wt% hemp fiber/PP composite. Additionally, 40 wt% hemp fiber/PP composite exhibited lower mechanical damping factor (tan δ) which was evidence of better fiber-matrix adhesion related with restriction of chain mobility.

Fiber dimension, aspect ratio (length/diameter), and orientation are critical factors for mechanical properties of green fiber thermoplastic composites. Singha and Rana prepared PS composites with the addition of raw chopped (90 μm), short (3 mm), and long (8 mm) Agave fiber to examine effects on fiber dimensions on mechanical properties of PS composites [94]. At optimum fiber loading (20 wt%), tensile strength of PS composite materials including chopped, short, and long Agave fiber was specified as 34.2 MPa, 24.7 MPa, and 34.2 MPa, respectively. In case of chopped fiber loading, maximum flexural and compressive strength were observed with the help of homogeneous dispersion of fiber. However, PS composite showed highest impact factor in the presence of long Agave fiber. It was attributed interfacial debonding following cracks that enable more energy absorption. In other words, impact resistance of the composites was increased with more energy dissipation. As well as fiber length, fiber aspect ratio influences especially flexural and tensile properties of composite materials. Synergistic effects on mechanical properties of composites are observed at higher aspect ratio in contrast to lower aspect ratio. Yemele et al. studied mechanical properties of HDPE composites reinforced by bark fiber with different aspect ratio (8.85, 9.45, and 13.83) [80]. Lowest flexural modulus (22.7 MPa) and tensile strength (13.7 MPa) were determined for HDPE composite reinforced by bark fiber with the aspect ratio of 8.85 due to stress concentration that limits homogeneous fiber dispersion in matrix. Hao et al. used poplar wood fiber (PW), radiate pine fiber (RP), and rice husk fiber (RH) as reinforcement for HDPE composites [81]. HDPE composites were prepared with various fiber orientations (0°, 30°, 45°, 60°, and 90°). Flexural strength and modulus of all type of HDPE composites decreased with increasing angles as shown in Fig. 3a. Highest unnotched impact strength was observed at 0° in general which was explained higher longitudinal fracture strength than transverse one leading a decrease in energy dissipation in Fig. 3b. However, creep strain of HDPE composites increased from 0° to 90° which was resulted from that better load carrying ability through on axis orientation than off axis direction. Zhang and Miao investigated commingled yarn structure of PP and hemp or flax fiber on the mechanical performance of the composites [86]. It was revealed that wrap yarn structure of composites had better flexural properties when compared to twisted yarn structure. Better fiber alignment in wrap yarn provided a significant increase in flexural modulus of the composites.

To enhance mechanical performance of the green fiber thermoplastic composites, compatibility of green fiber with thermoplastic matrix is improved with treatment of green fibers. Effective fiber treatment generally has synergistic effects on mechanical properties of composites with the help of easy load transfer from matrix to green fiber. Nopparut and Amornsakchai evaluated untreated pineapple leaf fiber (UPALF), alkali-treated pineapple leaf fiber (TPALF), and silane-treated pineapple leaf fiber (SiPALF) following alkali treatment as filler for nylon 6/66 copolymer [99]. Changes in hydrophilic structure of PALF were investigated by Fourier transform

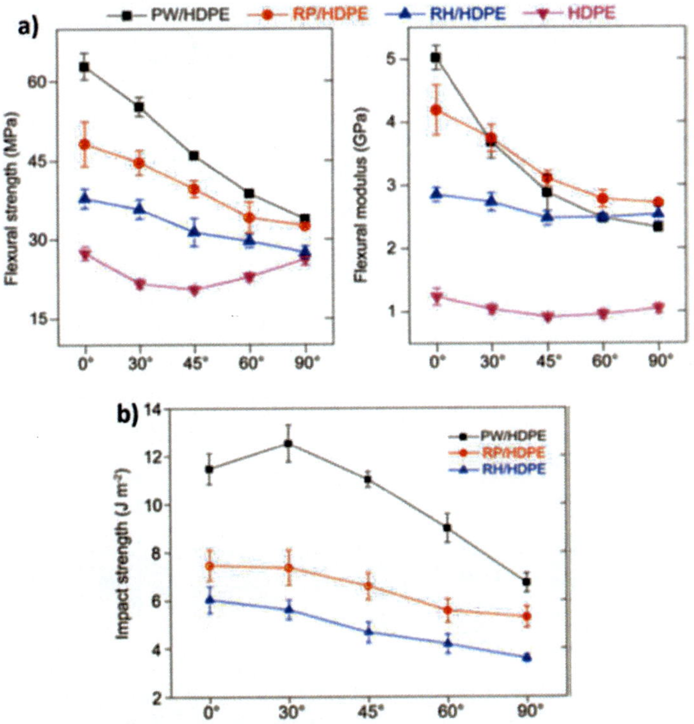

Fig. 3 **a** Flexural strength and modulus, and **b** unnotched impact strength of HDPE composites as a function of different fiber orientations (Reprinted from Ref. [81] with permission from Elsevier)

infrared spectroscopy (FTIR) after treatments. After treatments, the peak attributed –OH bending vibrations on hemicellulose disappeared at 1640 cm^{-1} as can be clearly seen in Fig. 4a. Additionally, the absence of peaks at 1737 cm^{-1} and 1240 cm^{-1} assigned to hemicellulose C=O and C–O stretching vibrations, respectively, was indication of successful fiber treatment. In case of silane treatment, no new peak was observed due to low concentration of siloxane. After treatments, mechanical performance of neat nylon was enhanced with PALF addition. However, SiPALF/nylon composites had highest tensile and flexural strength in addition to flexural modulus as shown in Fig. 4b–c. In the presence of UPALF, impact energy of neat nylon considerably decreased because of notch sensitivity of nylon and phase discontinuity. Fiber treatments also exhibited no effects on the initial and propagation impact energy of nylon composites. Haque et al. prepared coir fiber reinforced PP composites in pursuit of chemical treatment of coir fiber with benzene diazonium salt in pH 10.5, 7, and 3 [89]. Treatment in alkali medium enabled to dissolve lignin in coir fiber and more hydroxyl groups belonged to cellulose reacted with diazonium salt to convert hydrophilic –OH groups to hydrophobic –O–Na groups. In the presence of alkali-treated coir fiber, improved interfacial adhesion between coir fiber and PP increased tensile strength and modulus, flexural strength and modulus, and charpy

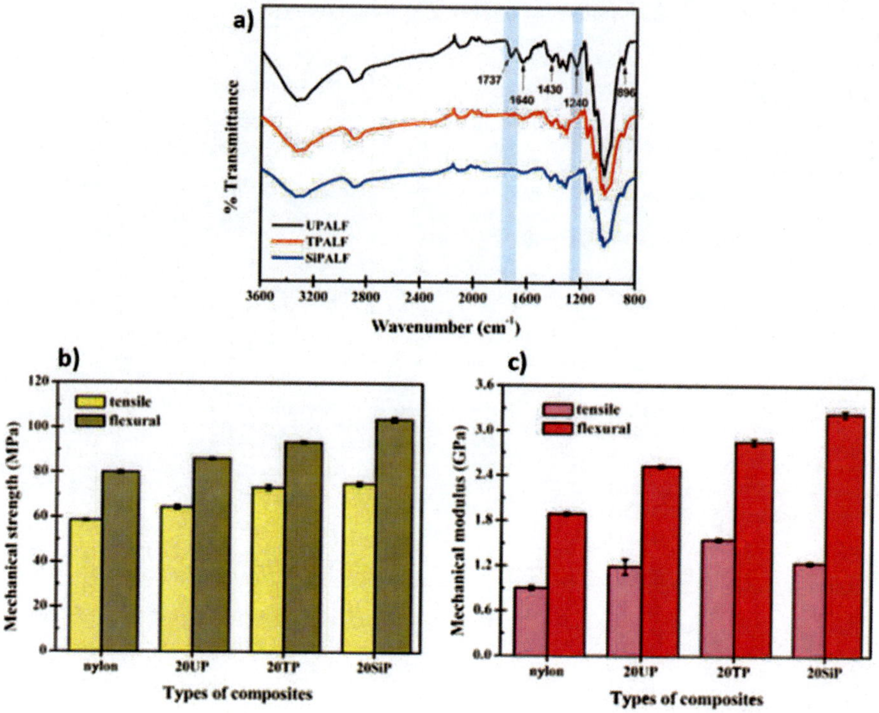

Fig. 4 **a** FTIR spectra of UPALF, TPALF and SiPALF, **b** tensile and flexural strength, and **c** tensile and flexural modulus of nylon composites (20UP: 20UPALF/nylon, 20TP: 20TPALF/nylon and 20SiP: 20SiPALF/nylon; 20 indicates fiber loading by the weight of %20) (Reprinted from Ref. [99], with permission from Elsevier)

impact strength at optimized coir fiber loading. Asumani et al. used kenaf fiber which was alkali treated with different NaOH concentrations (1, 2, 3, 4, 5, 6, 7, and 8 wt%) as filler for PP matrix [90]. Effect of NaOH concentration on the mechanical properties of kenaf fiber/PP composites was investigated. PP composite including 5 wt% NaOH-treated kenaf fiber exhibited higher tensile strength and modulus in addition to flexural strength than that of the other PP composites. More concentrated NaOH solution caused a degradation of fiber structure leading to decrease mechanical strength of the PP composites slightly.

Manufacturing process type and parameters such as process temperature, barrel speed, compression pressure and time, and cooling properties significantly affect residual stress distribution in green fiber thermoplastic composites. Depending on especially uniform temperature and effective pressure gradient, composite materials with high mechanical performance can be produced related with homogeneous stress distribution [102]. Barkoula et al. prepared flax fiber/PP composites by injection molding through four different compounding method: (1) kneading process, (2) Henschel kinetic mixer, (3) extrusion, and (4) LFT [88]. Kneading process provided to obtain flax fiber/PP composites which had maximum tensile strength and stiffness due to better retention of fiber length. Salleh et al. examined effect of extrusion temperature on tensile and dynamical mechanical properties of kenaf fiber/HDPE composites [82]. Extrusion was performed on low process temperature (160–180 °C between hopper and die) and high process temperature (165–185 °C). In case of high temperature, enhanced storage and loss modulus were observed as well as lower tan δ. Moreover, kenaf fiber/HDPE composites showed higher tensile strength owing to homogeneous mixing of fiber and matrix at high temperature. At low temperatures, an increase in shear between fiber and matrix caused poor fiber-matrix adhesion that had adverse effects on the mechanical properties of kenaf fiber/HDPE composites. Bernard et al. investigated mechanical properties of kenaf fiber/PP composites prepared by compression molding at different temperatures (190, 200, 210, 220, 230, and 240 °C) with different barrel speed (12, 16, 20, and 24 Hz) [92]. Maximum tensile strength of kenaf fiber/PP composite was determined as 31.5 MPa at optimum conditions (230 °C and 16 Hz) that can be contributed to well dispersion of fiber in the PP matrix.

2.2 *Water Absorption Properties*

Water absorption of green fiber thermoplastic composites highly originated from hydrophilic structure of green fibers adversely affect physicomechanical and thermal properties. Water penetrates into the composites through three ways: (1) water diffusion along micro gaps which in polymer chain connections, (2) water transportation into gaps in addition to flaws at interface between green fiber and thermoplastic by capillary forces, and (3) water flow along micro cracks and voids [103, 104]. Moreover, water absorption behavior of composite materials is expressed with diffusion or percolation mechanism according to amount of fiber. Diffusion theory is commonly used mechanism for high fiber-loaded composite materials. In this theory, it is assumed that water absorption is governed with randomly flow of water molecules from high concentration to low concentration [105].

There are many parameters such as chemical composition of green fiber, fiber size and orientation, treatment of fiber, and amount of fiber that have important influences on water absorption behavior of green fiber thermoplastic composites. In Table 5, water absorption of green fiber thermoplastic composites is given. Sultana Mir et al. prepared coir fiber/PP and coir fiber/PE composites and examined their water absorption behavior [115]. It was obvious that PP composites had lower water absorption than PE composites in the presence of alkali-treated coir fiber. It was explained with better compatibility between coir fiber and PP matrix. Ibrahim et al. compared water absorption of PP composites including 30 wt% untreated and alkali-treated sisal fiber [114]. While water absorption of untreated sisal fiber/PP was about 1.1%, water absorption of PP composite decreased to 0.41% with the addition of alkali-treated sisal fiber. Sahai and Pardeshi studied water absorption of wheat straw fiber/PS composites after treatment of fiber with alkali, maleic anhydride, silane and alkali

Table 5 Water absorption of green fiber thermoplastic composites

Matrix	Fiber	Fiber treatment	Manufacturing method	Water absorption (%)	References
LDPE	Hemp Flax Wood pulp Wood chips Wheat straw Triticale pulp (20 wt%)	–	Injection molding	4.8 4.1 2.3 4.0 5.0 3.2 for 6000 h	[106]
LDPE	Banana (25 wt%)	Untreated Alkali Alkali + acrylation	Compression molding	19 13 10 for 480 h	[107]
HDPE	Wood (60 wt%)	–	Compression molding	15 for 720 h	[108]
HDPE	Short birch (40 wt%)	–	Compression molding	9.9 for 1000 h at 65 °C	[109]
HDPE	Hemp Banana Sisal (45 wt%)	Maleic anhydride	Compression	14 12 9 for 30 h	[110]
PP	Jute (50 wt%)	–	Compression molding	17.5 for 96 h	[111]
PP	Hard wood Soft wood Long wood Wood chips (50 wt%)	–	Injection molding	0.42 0.73 0.08 0.11 for 50 h	[112]

(continued)

Table 5 (continued)

Matrix	Fiber	Fiber treatment	Manufacturing method	Water absorption (%)	References
PP	Wood Rice hull Kenaf (25 wt%)	–	Injection molding	1.5 1.4 1.9 for 840 h	[113]
PP	Sisal (30 wt%)	Untreated Alkali	Compression molding	1.1 0.41 for 7 h	[114]
PP	Coir (20 wt%)	Untreated Alkali	Hot-press	6 2 for 24 h	[115]
PS	Rice straw (30 wt%)	–	Hot-press	2.2 for 24 h	[116]
PS	Wheat straw (20 wt%)	Alkali Maleic anhydride Silane Alkali + silane	Compression molding	2.02 0.90 1.03 0.37 for 24 h	[117]
PVC	Banana (24 wt%)	–	Compression molding	1.8 for 24 h	[118]
PVC	Bamboo (25 wt%)	–	Extrusion	3.9 (75 μm fiber size) 4.1 (1 mm fiber size) for 1500 h	[119]
Nylon-6	Curauá (20 wt%)	Alkali	Injection molding	12 for 960 h	[120]
Nylon-6	Coconut shell Empty fruit bunch (15 wt%)	–	Injection molding	2.4 1.2	[121]

+ silane [117]. It was observed that water absorption of PS composites increased with increasing treated fiber loading. In case of 20 wt% alkali, maleic anhydride and silane-treated fiber, water sorption of PS composites was determined as 2.02%, 0.90%, and 1.03%, respectively. Lowest water sorption (0.37%) was belonged to PS composite including 20 wt% alkali + silane-treated fiber due to better elimination of fiber hydroxyl groups. Bahari and Krause investigated effect of fiber loading (25 and 50 wt%) and size (75 μm and 1 mm) on the water absorption of bamboo/PVC composites [119]. As expected, an increase in amount of fiber caused to increase water absorption of the composites. With the incorporation of 25 wt% 75 μm and 1 mm bamboo fiber, water absorption of the PVC composites was specified as approximately 3.9 and 4.1%. In contrast to high size fibers, small size fibers dispersed homogeneously in PVC matrix providing less voids and good interfacial fiber–matrix adhesion that restricted water penetration.

2.3 *Thermal Properties*

Thermal properties of green fiber thermoplastic composites are mainly associated with chemical composition of green fibers. Thermal decomposition of green fiber thermoplastic composites is generally divided into five steps: (1) hemicellulose decomposition (100–200 °C), (2) cellulose decomposition (200–270 °C), (3) lignin decomposition (280–500 °C), (4) decomposition of polymer structure (200–500 °C) and combustion of polymer composite (400–650 °C) [7]. Polymer matrix crystallinity, fiber treatments, and adhesion degree between fiber and matrix considerably influence thermal stability, flammability and thermal conductivity of composite materials in addition to composition of the green fiber [67]. Especially, great green fiber compatibility with thermoplastic improves thermal properties of green fiber thermoplastic composites due to requirement of high energy for separating components [25].

In many times, using green fiber as filler for thermoplastic matrices deteriorates thermal resistance of the composite materials without flame retardancy, inorganic additives, or coupling agents. In spite of this, some improvements such as an increase in thermal stability and a decrease in combustion rate with the help of fiber treatments or a decrease in thermal stability with increasing fiber content have been reported. Lu and Oza studied effect of alkali and silane treatments on thermal stability of hemp fiber/HDPE composites [122]. Decomposition temperatures of treated hemp fiber/HDPE composites were higher than that of the untreated hemp fiber/HDPE composites. Maximum initial decomposition temperature was determined as 362 °C in the presence of silane-treated hemp fiber. In case of silane treatment, formation of covalent bonds through reaction of silanol molecules with hydroxyl groups enhanced thermal stability of HDPE composites. Subasinghe et al. investigated flammability of kenaf fiber/PP composites [123]. According to UL-94 V flammability test, dripping time of PP was reduced with the addition of kenaf due to lignin in the structure of fiber. Moreover, a decrease in combustion rate was assigned to heat flux barrier effect of the kenaf fiber. Cone calorimetry tests showed that heat release rate of neat PP decreased by 37% with incorporation of 30 wt% kenaf fiber. And also, formation of char layer increased fire resistance of the PP composite materials. Annie Paul et al. examined effect of different fiber treatments on thermophysical properties of banana fiber/PP composites [124]. With increasing fiber loading, thermal conductivity and thermal diffusivity of PP composites reduced owing to cellulosic structure of banana fiber which acts as an insulator. Thermal conductivity and diffusivity of PP composites exhibited an increase with the following treatment order: benzoylation > alkali (10 wt%) > $KMnO_4$ > alkali (2 wt%) > silane. It was explained with that benzoylation significantly decreased hydrophilicity of banana fiber resulting in high interfacial adhesion between fiber and matrix. The compatible composite structure made heat transfer easy leading to increase thermal conductivity and diffusivity.

3 Applications

In consideration of economic issues and environmental legislations, sustainable green fiber thermoplastic composites are produced. Due to their low cost, low energy consumption for production, lightness and promising physicomechanical properties, green fiber thermoplastic composites have gained great attention in many applications such as automotive, construction, aerospace, marine, electronics and consumer goods, recently [125, 126].

Automotive industry is a major market place for utilization of these types of composites. Especially, European Guideline 2000/53/EG implemented by the European Commission was important force to recycle 85% of the vehicle weight in 2005 that was increased to 95% by 2015. And also, this legislation mandated that there should be recyclable components by the weight of 95% in the composition of vehicles which has significantly increased consumption of green fiber-based composites [127]. Moreover, the European Commission and the European Automobile Manufacturers Association collaborated to reduce CO_2 emission to 147 and 95 g/km for vehicles and passenger cars by 2020, respectively [128]. Green fiber thermoplastic composites can be used in door panel, seat back, floor tray, insulation panel, and so on. Wood, cotton, flax, kenaf, hemp, sisal, coir, and jute are commonly preferred fibers as filler instead of glass fibers for PP and PE matrices in automotive industry [129].

Building and construction compose 27% of application fields of green fiber-based polymer composites. Thermoplastic composites reinforced with hemp, oil palm, rice husk, bagasse, sisal, kenaf, ramie, and jute fiber can be utilized as window and door frame, furniture coating, insulation panel, floor, roof, and pipeline. Jute fiber/PP composites have been widely used for especially indoor housing [67]. In marine industry, green fiber thermoplastic composites are promising materials to decrease weight of components and prevent corrosion and biological agent attacks [130, 131]. In the production of consumer goods including packaging, tennis racket, snowboarding, laptop cases, filter clothes, fishing tools, etc., hemp, flax, coir, cotton, and kenaf-based thermoplastic composites play an important role [132].

4 Conclusion

Thermoplastic composites reinforced with green fibers have attracted great attention due to the advantages of green fibers such as low density, their renewable sources, and biodegradability. In spite of these advantages, hydrophilic structure of green fibers generally decreases compatibility between fiber and hydrophobic thermoplastic matrix. So, different physical and chemical treatments are utilized to improve fiber-matrix adhesion. Strong adhesion between fiber and matrix is significant factor for effective load transfer that enhances mechanical performance of green fiber thermoplastic composites as well as fiber size and orientation, aspect ratio, surface treatment of fiber, amount of fiber, and manufacturing process conditions. Depending on chemical composition of green fiber, water absorption of green fiber thermoplastic composites changes. Removal of hydroxyl groups from green fibers increases fiber-matrix compatibility that decreases water absorption of the composite materials. Due to the decomposition of green fibers at low temperatures, thermal stability of the composites can be deteriorated. However, fiber treatments can provide to increase decomposition temperatures of the composites. And also, a decrease in flammability rate is observed due to heat flux barrier effect of green fibers. With increasing green fiber loading, thermal conductivity of the composite materials decreases that can be explained with insulation behavior of green fibers. Enhanced properties of green fiber thermoplastic composites make them suitable materials for many industrial and engineering applications such as automotive, construction, marine, packaging, and consumer goods.

References

1. Thiruchitrambalam M, Athijayamani A, Sathiyamurthy S, Thaheer ASA (2010) A review on the natural fiber-reinforced polymer composites for the development of roselle fiber-reinforced polyester composite. J Nat Fibers 7(4):307–323. https://doi.org/10.1080/15440478.2010.529299
2. Mohd Nurazzi N, Khalina A, Sapuan SM, Dayang Laila AHAM, Rahman M, Hanafee Z (2017) A review: fibres, polymer matrices and composites. Sci Technol 25(4):1085–1102
3. Mann GS, Singh LP, Kumar P, Singh S (2018) Green composites: a review of processing technologies and recent applications. J Thermoplast Compos Mater 089270571881635. https://doi.org/10.1177/0892705718816354
4. Huang Y, Meng X, Xie Y, Wan L, Lv Z, Cao J, Feng J (2018) Friction stir welding/processing of polymers and polymer matrix composites. Compos A Appl Sci Manuf 105:235–257. https://doi.org/10.1016/j.compositesa.2017.12.005

5. Guzel G, Deveci H (2017) Physico-mechanical, thermal, and coating properties of composite materials prepared with epoxy resin/steel slag. Polym Compos 38:1974–1981. https://doi.org/10.1002/pc.23768
6. Yao S-S, Jin F-L, Rhee KY, Hui D, Park S-J (2018) Recent advances in carbon-fiber-reinforced thermoplastic composites: a review. Compos B Eng 142:241–250. https://doi.org/10.1016/j.compositesb.2017.12.007
7. Vaisanen T, Haapala A, Lappalainen R, Tomppo L (2016) Utilization of agricultural and forest industry waste and residues in natural fiber-polymer composites: a review. Waste Manag 54:62–73. https://doi.org/10.1016/j.wasman.2016.04.037
8. Thomassin J-M, Jérôme C, Pardoen T, Bailly C, Huynen I, Detrembleur C (2013) Polymer/carbon based composites as electromagnetic interference (EMI) shielding materials. Mater Sci Eng: R: Rep 74(7):211–232. https://doi.org/10.1016/j.mser.2013.06.001
9. Du JH, Bai J, Cheng HM (2007) The present status and key problems of carbon nanotube based polymer composites. Express Polym Lett 1(5):253–273. https://doi.org/10.3144/expresspolymlett.2007.39
10. Kuilla T, Bhadra S, Yao D, Kim NH, Bose S, Lee JH (2010) Recent advances in graphene based polymer composites. Prog Polym Sci 35(11):1350–1375. https://doi.org/10.1016/j.progpolymsci.2010.07.005
11. Smitthipong W, Tantatherdtam R, Chollakup R (2013) Effect of pineapple leaf fiber-reinforced thermoplastic starch/poly(lactic acid) green composite. J Thermoplast Compos Mater 28(5):717–729. https://doi.org/10.1177/0892705713489701
12. Mohanty AK, Vivekanandhan S, Pin J-M, Misra M (2018) Composites from renewable and sustainable resources: Challenges and innovations. Science 362:536–542. https://doi.org/10.1126/science.aat9072
13. Bouclé J, Ravirajan P, Nelson J (2007) Hybrid polymer–metal oxide thin films for photovoltaic applications. J Mater Chem 17(30):3141. https://doi.org/10.1039/b706547g
14. Ng LY, Mohammad AW, Leo CP, Hilal N (2013) Polymeric membranes incorporated with metal/metal oxide nanoparticles: a comprehensive review. Desalination 308:15–33. https://doi.org/10.1016/j.desal.2010.11.033
15. Monteiro SN, Calado V, Rodriguez RJS, Margem FM (2012) Thermogravimetric behavior of natural fibers reinforced polymer composites-an overview. Mater Sci Eng, A 557:17–28. https://doi.org/10.1016/j.msea.2012.05.109
16. Kaya GG, Yilmaz E, Deveci H (2018) Sustainable bean pod/calcined kaolin reinforced epoxy hybrid composites with enhanced mechanical, water sorption and corrosion resistance properties. Constr Build Mater 162:272–279. https://doi.org/10.1016/j.conbuildmat.2017.12.046
17. Papanicolaou GC, Zaoutsos SP (2011) 1-viscoelastic constitutive modeling of creep and stress relaxation in polymers and polymer matrix composites. In: Guedes RM (ed) Creep and fatigue in polymer matrix composites. Woodhead Publishing, pp 3–47. https://doi.org/10.1533/9780857090430.1.3
18. Taj S, Munawar MA, Khan S (2007) Natural fiber-reinforced polymer composites. Proc Pak Acad Sci 44(2):129–144
19. Biron M (2013) 1-Outline of the actual situation of plastics compared to conventional materials. In: Biron M (ed) Thermoplastics and thermoplastic composites. William Andrew Publishing, pp 1–29. https://doi.org/10.1016/b978-1-4557-7898-0.00001-9
20. Mallick PK (2010) 5-Thermoplastics and thermoplastic–matrix composites for lightweight automotive structures. In: Mallick PK (ed) Materials, design and manufacturing for lightweight vehicles. Woodhead Publishing, pp 174–207. https://doi.org/10.1533/9781845697822.1.174
21. Grigore M (2017) Methods of recycling, properties and applications of recycled thermoplastic polymers. Recycling 2(4):24. https://doi.org/10.3390/recycling2040024
22. Yashas Gowda TG, Sanjay MR, Subrahmanya Bhat K, Madhu P, Senthamaraikannan P, Yogesha B, Pham D (2018) Polymer matrix-natural fiber composites: an overview. Cogent Eng 5(1). https://doi.org/10.1080/23311916.2018.1446667

23. Gupta MK, Srivastava RK (2015) Mechanical properties of hybrid fibers-reinforced polymer composite: a review. Polym-Plast Technol Eng 55(6):626–642. https://doi.org/10.1080/03602559.2015.1098694
24. Sain M, Panthapulakkal S (2004) 9-Green fibre thermoplastic composites. In: Baillie C (ed) Green composites polymer composites and the environment. Woodhead Publishing, pp 181–203
25. Väisänen T, Das O, Tomppo L (2017) A review on new bio-based constituents for natural fiber-polymer composites. J Clean Prod 149:582–596. https://doi.org/10.1016/j.jclepro.2017.02.132
26. Patrick SG (2005) Practical guide to polyvinyl chloride. In: Patrick SG (ed) Rapra Technology Limited, pp 59–60
27. Sobkowicz-Kline M, Budhlall BM, Mead JL (2017) Synthetic resins and plastics. In: Kent JA, Bommaraju T, Barnicki, SD (eds) Handbook of industrial chemistry and biotechnology. Springer International Publishing, pp 1397–1462. https://doi.org/10.1007/978-3-319-52287-6_25
28. Kumar Sen S, Raut S (2015) Microbial degradation of low density polyethylene (LDPE): a review. J Env Chem Eng 3(1):462–473. https://doi.org/10.1016/j.jece.2015.01.003
29. Kumar S, Panda AK, Singh RK (2011) A review on tertiary recycling of high-density polyethylene to fuel. Res Conserv Recycl 55(11):893–910. https://doi.org/10.1016/j.resconrec.2011.05.005
30. Vaidya UK, Chawla KK (2013) Processing of fibre reinforced thermoplastic composites. Int Mater Rev 53(4):185–218. https://doi.org/10.1179/174328008x325223
31. Sarı A, Alkan C, Biçer A (2012) Synthesis and thermal properties of polystyrene-graft-PEG copolymers as new kinds of solid–solid phase change materials for thermal energy storage. Mater Chem Phys 133(1):87–94. https://doi.org/10.1016/j.matchemphys.2011.12.056
32. Pant HR, Pandeya DR, Nam KT, Baek WI, Hong ST, Kim HY (2011) Photocatalytic and antibacterial properties of a TiO_2/nylon-6 electrospun nanocomposite mat containing silver nanoparticles. J Hazard Mater 189(1–2):465–471. https://doi.org/10.1016/j.jhazmat.2011.02.062
33. Yousefian H, Rodrigue D (2016) Effect of nanocrystalline cellulose on morphological, thermal, and mechanical properties of Nylon 6 composites. Polym Compos 37(5):1473–1479. https://doi.org/10.1002/pc.23316
34. Zhang L, Luo M, Sun S, Ma J, Li C (2010) Effect of surface structure of nano-$CaCO_3$ particles on mechanical and rheological properties of PVC composites. J Macromol Sci, Part B 49(5):970–982. https://doi.org/10.1080/00222341003609336
35. Xu X, Jayaraman K, Morin C, Pecqueux N (2008) Life cycle assessment of wood-fibre-reinforced polypropylene composites. J Mater Process Technol 198(1–3):168–177. https://doi.org/10.1016/j.jmatprotec.2007.06.087
36. Faruk O, Bledzki AK, Fink H-P, Sain M (2012) Biocomposites reinforced with natural fibers: 2000–2010. Prog Polym Sci 37(11):1552–1596. https://doi.org/10.1016/j.progpolymsci.2012.04.003
37. Sanjay MR, Madhu P, Jawaid M, Senthamaraikannan P, Senthil S, Pradeep S (2018) Characterization and properties of natural fiber polymer composites: a comprehensive review. J Clean Prod 172:566–581. https://doi.org/10.1016/j.jclepro.2017.10.101
38. Pereira PHF, Rosa MdF, Cioffi MOH, Benini KCCdC, Milanese AC, Voorwald HJC, Mulinari DR (2015) Vegetal fibers in polymeric composites: a review. Polímeros 25(1):9–22. https://doi.org/10.1590/0104-1428.1722
39. Rohit K, Dixit S (2016) A review—future aspect of natural fiber reinforced composite. Polym Renew Res 7(2):43–59
40. Asim M, Abdan K, Jawaid M, Nasir M, Dashtizadeh Z, Ishak MR, Hoque ME (2015) A review on pineapple leaves fibre and its composites. Int J Polym Sci 2015:1–16. https://doi.org/10.1155/2015/950567
41. Kiaei M, Samariha A, Kasmani JE (2011) Characterization of biometry and the chemical and morphological properties of fibers from bagasse, corn, sunflower, rice and rapeseed residues in Iran. Afr J Agric Res 6(16):3762–3767. https://doi.org/10.5897/ajar10.752

42. Sun L, Chen JY, Negulescu II, Moore MA, Collier BJ (2011) Kinetics modeling of dynamic pyrolysis of bagasse fibers. Biores Technol 102(2):1951–1958. https://doi.org/10.1016/j.biortech.2010.08.109
43. Abdul Khalil HPS, Bhat AH, Ireana Yusra AF (2012) Green composites from sustainable cellulose nanofibrils: a review. Carbohyd Polym 87(2):963–979. https://doi.org/10.1016/j.carbpol.2011.08.078
44. Malkapuram R, Kumar V, Yuvraj Singh N (2008) Recent development in natural fiber reinforced polypropylene composites. J Reinf Plast Compos 28(10):1169–1189. https://doi.org/10.1177/0731684407087759
45. Pai AR, Jagtap RN (2015) Surface morphology & mechanical properties of some unique natural fiber reinforced polymer composites a review. J Mater Env Sci 6(4):902–917
46. Yeh S-K, Hsieh C-C, Chang H-C, Yen CCC, Chang Y-C (2015) Synergistic effect of coupling agents and fiber treatments on mechanical properties and moisture absorption of polypropylene–rice husk composites and their foam. Compos A Appl Sci Manuf 68:313–322. https://doi.org/10.1016/j.compositesa.2014.10.019
47. Mani S, Tabil LG, Sokhansanj S (2006) Effects of compressive force, particle size and moisture content on mechanical properties of biomass pellets from grasses. Biomass Bioenerg 30(7):648–654. https://doi.org/10.1016/j.biombioe.2005.01.004
48. Satpathy SK, Tabil LG, Meda V, Naik SN, Prasad R (2014) Torrefaction of wheat and barley straw after microwave heating. Fuel 124:269–278. https://doi.org/10.1016/j.fuel.2014.01.102
49. Mahir FI, Keya KN, Sarker B, Nahiun KM, Khan RA (2019) A brief review on natural fiber used as a replacement of synthetic fiber in polymer composites. Mater Eng Res 1(2):88–99. https://doi.org/10.25082/mer.2019.02.007
50. Ku H, Wang H, Pattarachaiyakoop N, Trada M (2011) A review on the tensile properties of natural fiber reinforced polymer composites. Compos B Eng 42(4):856–873. https://doi.org/10.1016/j.compositesb.2011.01.010
51. Bajpai PK, Singh I, Madaan J (2012) Development and characterization of PLA-based green composites. J Thermoplast Compos Mater 27(1):52–81. https://doi.org/10.1177/0892705712439571
52. Lotfi A, Li H, Dao DV, Prusty G (2019) Natural fiber–reinforced composites: a review on material, manufacturing, and machinability. J Thermoplast Compos Mater. https://doi.org/10.1177/0892705719844546
53. Akil HM, Omar MF, Mazuki AAM, Safiee S, Ishak ZAM, Abu Bakar A (2011) Kenaf fiber reinforced composites: a Review. Mater Des 32(8–9):4107–4121. https://doi.org/10.1016/j.matdes.2011.04.008
54. Sood M, Dwivedi G (2018) Effect of fiber treatment on flexural properties of natural fiber reinforced composites: a review. Egyptian J Petrol 27(4):775–783. https://doi.org/10.1016/j.ejpe.2017.11.005
55. John MJ, Anandjiwala RD (2008) Recent developments in chemical modification and characterization of natural fiber-reinforced composites. Polym Compos 29(2):187–207. https://doi.org/10.1002/pc.20461
56. Koohestani B, Darban AK, Mokhtari P, Yilmaz E, Darezereshki E (2018) Comparison of different natural fiber treatments: a literature review. Int J Environ Sci Technol 16(1):629–642. https://doi.org/10.1007/s13762-018-1890-9
57. Goriparthi BK, Suman KNS, Mohan Rao N (2012) Effect of fiber surface treatments on mechanical and abrasive wear performance of polylactide/jute composites. Compos A Appl Sci Manuf 43(10):1800–1808. https://doi.org/10.1016/j.compositesa.2012.05.007
58. Dittenber DB, GangaRao HVS (2012) Critical review of recent publications on use of natural composites in infrastructure. Compos A Appl Sci Manuf 43(8):1419–1429. https://doi.org/10.1016/j.compositesa.2011.11.019
59. Kalia S, Kaith BS, Kaur I (2009) Pretreatments of natural fibers and their application as reinforcing material in polymer composites-a review. Polym Eng Sci 49(7):1253–1272. https://doi.org/10.1002/pen.21328

60. Ragoubi M, Bienaimé D, Molina S, George B, Merlin A (2010) Impact of corona treated hemp fibres onto mechanical properties of polypropylene composites made thereof. Ind Crops Prod 31(2):344–349. https://doi.org/10.1016/j.indcrop.2009.12.004
61. Obi Reddy K, Uma Maheswari C, Shukla M, Song JI, Varada Rajulu A (2013) Tensile and structural characterization of alkali treated Borassus fruit fine fibers. Compos B Eng 44(1):433–438. https://doi.org/10.1016/j.compositesb.2012.04.075
62. Nunna S, Chandra PR, Shrivastava S, Jalan AK (2012) A review on mechanical behavior of natural fiber based hybrid composites. J Reinf Plast Compos 31(11):759–769. https://doi.org/10.1177/0731684412444325
63. Xie Y, Hill CAS, Xiao Z, Militz H, Mai C (2010) Silane coupling agents used for natural fiber/polymer composites: a review. Compos A Appl Sci Manuf 41(7):806–819. https://doi.org/10.1016/j.compositesa.2010.03.005
64. Hassan MM, Wagner MH (2016) Surface modification of natural fibers for reinforced polymer composites: a critical review. Rev Adhes Adhes 4(1):1–46. https://doi.org/10.7569/raa.2016.097302
65. Pickering KL, Efendy MGA, Le TM (2016) A review of recent developments in natural fibre composites and their mechanical performance. Compos A Appl Sci Manuf 83:98–112. https://doi.org/10.1016/j.compositesa.2015.08.038
66. Al-Oqla FM, Sapuan SM (2014) Natural fiber reinforced polymer composites in industrial applications: feasibility of date palm fibers for sustainable automotive industry. J Clean Prod 66:347–354. https://doi.org/10.1016/j.jclepro.2013.10.050
67. Gholampour A, Ozbakkaloglu T (2019) A review of natural fiber composites: properties, modification and processing techniques, characterization, applications. J Mater Sci 55(3):829–892. https://doi.org/10.1007/s10853-019-03990-y
68. Saba N, Tahir P, Jawaid M (2014) A review on potentiality of nano filler/natural fiber filled polymer hybrid composites. Polymers 6(8):2247–2273. https://doi.org/10.3390/polym6082247
69. Panthapulakkal S, Raghunanan L, Sain M, Kc B, Tjong J (2017) 4-Natural fiber and hybrid fiber thermoplastic composites: Advancements in lightweighting applications. In: Baillie C, Jayasinghe R (eds) Green composites. Woodhead Publishing, pp 39–72. https://doi.org/10.1016/b978-0-08-100783-9.00003-4
70. Zini E, Scandola M (2011) Green composites: an overview. Polym Compos 32(12):1905–1915. https://doi.org/10.1002/pc.21224
71. Abilash N, Sivapragash M (2016) Optimizing the delamination failure in bamboo fiber reinforced polyester composite. J King Saud University—Eng Sci 28 (1):92–102. https://doi.org/10.1016/j.jksues.2013.09.004
72. Maghsoudi K, Momen G, Jafari R, Farzaneh M (2018) Direct replication of micro-nanostructures in the fabrication of superhydrophobic silicone rubber surfaces by compression molding. Appl Surf Sci 458:619–628. https://doi.org/10.1016/j.apsusc.2018.07.099
73. Crawford DE, Casaban J (2016) Recent developments in mechanochemical materials synthesis by extrusion. Adv Mater 28(27):5747–5754. https://doi.org/10.1002/adma.201505352
74. Moad G (2011) Chemical modification of starch by reactive extrusion. Prog Polym Sci 36(2):218–237. https://doi.org/10.1016/j.progpolymsci.2010.11.002
75. Srinivas K, Lakshumu Naidu A, Raju Bahubalendruni MVA (2017) A review on chemical and mechanical properties of natural fiber reinforced polymer composites. Int J Performability Eng 13(2):189–200. https://doi.org/10.23940/ijpe.17.02.p8.189200
76. Arrakhiz FZ, El Achaby M, Malha M, Bensalah MO, Fassi-Fehri O, Bouhfid R, Benmoussa K, Qaiss A (2013) Mechanical and thermal properties of natural fibers reinforced polymer composites: doum/low density polyethylene. Mater Des 43:200–205. https://doi.org/10.1016/j.matdes.2012.06.056
77. Sever K (2016) The improvement of mechanical properties of jute fiber/LDPE composites by fiber surface treatment. J Reinf Plast Compos 29(13):1921–1929. https://doi.org/10.1177/0731684409339078

78. Qiang Y, Donglyang W, Gotama J, Bateman S (2008) Wood fiber reinforced polyethylene and polypropylene composites with high modulus and impact strength. J Thermoplast Compos Mater 21(3):195–208. https://doi.org/10.1177/0892705708089472
79. Migneault S, Koubaa A, Erchiqui F, Chaala A, Englund K, Wolcott MP (2009) Effects of processing method and fiber size on the structure and properties of wood–plastic composites. Compos A Appl Sci Manuf 40(1):80–85. https://doi.org/10.1016/j.compositesa.2008.10.004
80. Yemele MCN, Koubaa A, Cloutier A, Soulounganga P, Wolcott M (2010) Effect of bark fiber content and size on the mechanical properties of bark/HDPE composites. Compos A Appl Sci Manuf 41(1):131–137. https://doi.org/10.1016/j.compositesa.2009.06.005
81. Hao X, Zhou H, Mu B, Chen L, Guo Q, Yi X, Sun L, Wang Q, Ou R (2020) Effects of fiber geometry and orientation distribution on the anisotropy of mechanical properties, creep behavior, and thermal expansion of natural fiber/HDPE composites. Compos B Eng 185:107778. https://doi.org/10.1016/j.compositesb.2020.107778
82. Salleh FM, Hassan A, Yahya R, Azzahari AD (2014) Effects of extrusion temperature on the rheological, dynamic mechanical and tensile properties of kenaf fiber/HDPE composites. Compos B Eng 58:259–266. https://doi.org/10.1016/j.compositesb.2013.10.068
83. Kaewkuk S, Sutapun W, Jarukumjorn K (2013) Effects of interfacial modification and fiber content on physical properties of sisal fiber/polypropylene composites. Compos B Eng 45(1):544–549. https://doi.org/10.1016/j.compositesb.2012.07.036
84. Tajvidi M, Motie N, Rassam G, Falk RH, Felton C (2009) Mechanical performance of hemp fiber polypropylene composites at different operating temperatures. J Reinf Plast Compos 29(5):664–674. https://doi.org/10.1177/0731684408100266
85. Ng LF, Dhar Malingam S, Selamat MZ, Mustafa Z, Bapokutty O (2019) A comparison study on the mechanical properties of composites based on kenaf and pineapple leaf fibres. Polym Bull 77(3):1449–1463. https://doi.org/10.1007/s00289-019-02812-0
86. Zhang L, Miao M (2010) Commingled natural fibre/polypropylene wrap spun yarns for structured thermoplastic composites. Compos Sci Technol 70(1):130–135. https://doi.org/10.1016/j.compscitech.2009.09.016
87. Bledzki AK, Mamun AA, Lucka-Gabor M, Gutowski VS (2008) The effects of acetylation on properties of flax fibre and its polypropylene composites. Express Polymer Lett 2(6):413–422. https://doi.org/10.3144/expresspolymlett.2008.50
88. Barkoula NM, Garkhail SK, Peijs T (2009) Effect of compounding and injection molding on the mechanical properties of flax fiber polypropylene composites. J Reinf Plast Compos 29(9):1366–1385. https://doi.org/10.1177/0731684409104465
89. Haque MM, Ali ME, Hasan M, Islam MN, Kim H (2012) Chemical treatment of coir fiber reinforced polypropylene composites. Ind Eng Chem Res 51(10):3958–3965. https://doi.org/10.1021/ie200693v
90. Asumani OML, Reid RG, Paskaramoorthy R (2012) The effects of alkali–silane treatment on the tensile and flexural properties of short fibre non-woven kenaf reinforced polypropylene composites. Compos A Appl Sci Manuf 43(9):1431–1440. https://doi.org/10.1016/j.compositesa.2012.04.007
91. Anil A, Tomlal JE, George G, Mathew JT, Manilal VB (2016) Novel eco-friendly commingled polypropylene/banana fiber composite: studies on thermal and mechanical properties. Polym Bull 73(11):2987–3005. https://doi.org/10.1007/s00289-016-1636-0
92. Bernard M, Khalina A, Ali A, Janius R, Faizal M, Hasnah KS, Sanuddin AB (2011) The effect of processing parameters on the mechanical properties of kenaf fibre plastic composite. Mater Des 32(2):1039–1043. https://doi.org/10.1016/j.matdes.2010.07.014
93. Oksman K, Mathew AP, Långström R, Nyström B, Joseph K (2009) The influence of fibre microstructure on fibre breakage and mechanical properties of natural fibre reinforced polypropylene. Compos Sci Technol 69(11–12):1847–1853. https://doi.org/10.1016/j.compscitech.2009.03.020
94. Singha AS, Rana RK (2012) Natural fiber reinforced polystyrene composites: effect of fiber loading, fiber dimensions and surface modification on mechanical properties. Mater Des 41:289–297. https://doi.org/10.1016/j.matdes.2012.05.001

95. Mishra S, Naik JB (2005) Effect of treatment of maleic anhydride on mechanical properties of natural fiber: polystyrene composites. Poly-Plast Technol Eng 44(4):663–675. https://doi.org/10.1081/pte-200057814
96. Bachtiar D, Sapuan SM, Khalina A, Zainudin ES, Dahlan KZM (2012) Flexural and impact properties of chemically treated sugar palm fiber reinforced high impact polystyrene composites. Fibers Polym 13(7):894–898. https://doi.org/10.1007/s12221-012-0894-1
97. Huang Z, Yin Q, Wang Q, Wang P, Liu T, Qian L (2017) Mechanical properties and crystallization behavior of three kinds of straws/nylon 6 composites. Int J Biol Macromol 103:663–668. https://doi.org/10.1016/j.ijbiomac.2017.05.121
98. Panyasart K, Chaiyut N, Amornsakchai T, Santawitee O (2014) Effect of surface treatment on the properties of pineapple leaf fibers reinforced polyamide 6 composites. Energy Procedia 56:406–413. https://doi.org/10.1016/j.egypro.2014.07.173
99. Nopparut A, Amornsakchai T (2016) Influence of pineapple leaf fiber and it's surface treatment on molecular orientation in, and mechanical properties of, injection molded nylon composites. Polym Test 52:141–149. https://doi.org/10.1016/j.polymertesting.2016.04.012
100. Xu Y, Wu Q, Lei Y, Yao F, Zhang Q (2008) Natural fiber reinforced poly(vinyl chloride) composites: effect of fiber type and impact modifier. J Polym Environ 16(4):250–257. https://doi.org/10.1007/s10924-008-0113-8
101. Wang H, Sheng K, Chen J, Mao H, Qian X (2010) Mechanical and thermal properties of sodium silicate treated moso bamboo particles reinforced PVC composites. Sci China Technol Sci 53(11):2932–2935. https://doi.org/10.1007/s11431-010-4123-0
102. Ho M-p, Wang H, Lee J-H, Ho C-k, Lau K-t, Leng J, Hui D (2012) Critical factors on manufacturing processes of natural fibre composites. Compos B Eng 43(8):3549–3562. https://doi.org/10.1016/j.compositesb.2011.10.001
103. Azwa ZN, Yousif BF, Manalo AC, Karunasena W (2013) A review on the degradability of polymeric composites based on natural fibres. Mater Des 47:424–442. https://doi.org/10.1016/j.matdes.2012.11.025
104. Akil HM, Santulli C, Sarasini F, Tirillò J, Valente T (2014) Environmental effects on the mechanical behaviour of pultruded jute/glass fibre-reinforced polyester hybrid composites. Compos Sci Technol 94:62–70. https://doi.org/10.1016/j.compscitech.2014.01.017
105. Akil HM, Cheng LW, Mohd Ishak ZA, Abu Bakar A, Abd Rahman MA (2009) Water absorption study on pultruded jute fibre reinforced unsaturated polyester composites. Compos Sci Technol 69(11):1942–1948. https://doi.org/10.1016/j.compscitech.2009.04.014
106. Robertson N-LM, Nychka JA, Alemaskin K, Wolodko JD (2013) Mechanical performance and moisture absorption of various natural fiber reinforced thermoplastic composites. J Appl Polym Sci 130(2):969–980. https://doi.org/10.1002/app.39237
107. Prasad N, Agarwal VK, Sinha S (2016) Banana fiber reinforced low-density polyethylene composites: effect of chemical treatment and compatibilizer addition. Iran Polym J 25(3):229–241. https://doi.org/10.1007/s13726-016-0416-x
108. Saeed Kazemi N, Sharifnia H, Tajvidi M (2008) Effects of water absorption on creep behavior of wood—plastic composites. J Compos Mater 42(10):993–1002. https://doi.org/10.1177/0021998307088608
109. Mejri M, Toubal L, Cuillière JC, François V (2018) Hygrothermal aging effects on mechanical and fatigue behaviors of a short- natural-fiber-reinforced composite. Int J Fatigue 108:96–108. https://doi.org/10.1016/j.ijfatigue.2017.11.004
110. Naik JB, Mishra S (2007) Esterification effect of maleic anhydride on swelling properties of natural fiber/high density polyethylene composites. J Appl Polym Sci 106(4):2571–2574. https://doi.org/10.1002/app.25329
111. Das SC, Paul D, Fahad MM, Das MK, Rahman GMS, Khan MA (2018) Effect of fiber loading on the dynamic mechanical properties of jute fiber reinforced polypropylene composites. Adv Chem Eng Sci 08(04):215–224. https://doi.org/10.4236/aces.2018.84015
112. Bledzki AK, Faruk O (2003) Wood fibre reinforced polypropylene composites: effect of fibre geometry and coupling agent on physico-mechanical properties. Appl Compos Mater 10:365–379. https://doi.org/10.1023/a:1025741100628

113. Tajvidi M, Najafi SK, Moteei N (2006) Long-term water uptake behavior of natural fiber/polypropylene composites. J Appl Polym Sci 99(5):2199–2203. https://doi.org/10.1002/app.21892
114. Ibrahim ID, Jamiru T, Sadiku RE, Kupolati WK, Agwuncha SC (2016) Dependency of the mechanical properties of sisal fiber reinforced recycled polypropylene composites on fiber surface treatment, fiber content and nanoclay. J Polym Environ 25(2):427–434. https://doi.org/10.1007/s10924-016-0823-2
115. Sultana Mir S, Hasan M, Hasan SMN, Hossain MJ, Nafsin N (2017) Effect of chemical treatment on the properties of coir fiber reinforced polypropylene and polyethylene composites. Polym Compos 38(7):1259–1265. https://doi.org/10.1002/pc.23690
116. Tawfik ME, Eskander SB, Nawwar GA (2017) Hard wood-composites made of rice straw and recycled polystyrene foam wastes. J Appl Polym Sci 134(18):44770. https://doi.org/10.1002/app.44770
117. Sahai RSN, Pardeshi RA (2019) Comparative study of effect of different coupling agent on mechanical properties and water absorption on wheat straw-reinforced polystyrene composites. J Thermoplast Compos Mater: 089270571984397. https://doi.org/10.1177/0892705719843975
118. Dan-asabe B (2018) Thermo-mechanical characterization of banana particulate reinforced PVC composite as piping material. J King Saud Univ-Eng Sci 30(4):296–304. https://doi.org/10.1016/j.jksues.2016.11.001
119. Bahari SA, Krause A (2016) Utilizing Malaysian bamboo for use in thermoplastic composites. J Clean Prod 110:16–24. https://doi.org/10.1016/j.jclepro.2015.03.052
120. Santos PA, Spinacé MAS, Fermoselli KKG, De Paoli M-A (2007) Polyamide-6/vegetal fiber composite prepared by extrusion and injection molding. Compos A Appl Sci Manuf 38(12):2404–2411. https://doi.org/10.1016/j.compositesa.2007.08.011
121. Savetlana S, Mulvaney-Johnson L, Gough T, Kelly A (2017) Properties of nylon-6-based composite reinforced with coconut shell particles and empty fruit bunch fibres. Plast, Rubber Compos 47(2):77–86. https://doi.org/10.1080/14658011.2017.1418711
122. Lu N, Oza S (2013) Thermal stability and thermo-mechanical properties of hemp-high density polyethylene composites: effect of two different chemical modifications. Compos B Eng 44(1):484–490. https://doi.org/10.1016/j.compositesb.2012.03.024
123. Subasinghe ADL, Das R, Bhattacharyya D (2015) Fiber dispersion during compounding/injection molding of PP/kenaf composites: flammability and mechanical properties. Mater Des 86:500–507. https://doi.org/10.1016/j.matdes.2015.07.126
124. Annie Paul S, Boudenne A, Ibos L, Candau Y, Joseph K, Thomas S (2008) Effect of fiber loading and chemical treatments on thermophysical properties of banana fiber/polypropylene commingled composite materials. Compos A Appl Sci Manuf 39(9):1582–1588. https://doi.org/10.1016/j.compositesa.2008.06.004
125. Chegdani F, Mansori ME (2018) Mechanics of material removal when cutting natural fiber reinforced thermoplastic composites. Polym Testing 67:275–283. https://doi.org/10.1016/j.polymertesting.2018.03.016
126. Sun Z (2018) Progress in the research and applications of natural fiber-reinforced polymer matrix composites. Sci Eng Compos Mater 25(5):835–846. https://doi.org/10.1515/secm-2016-0072
127. Zorpas AA, Inglezakis VJ (2012) Automotive industry challenges in meeting EU 2015 environmental standard. Technol Soc 34(1):55–83. https://doi.org/10.1016/j.techsoc.2011.12.006
128. Chauhan V, Kärki T, Varis J (2019) Review of natural fiber-reinforced engineering plastic composites, their applications in the transportation sector and processing techniques. J Thermoplast Compos Mater: 089270571988909. https://doi.org/10.1177/0892705719889095
129. Peças P, Carvalho H, Salman H, Leite M (2018) Natural fibre composites and their applications: a review. J Compos Sci 2(4):66. https://doi.org/10.3390/jcs2040066

130. Verma D, Goh KL (2019) 3-Natural fiber-reinforced polymer composites: application in marine environments. In: Verma D, Fortunati E, Jain S, Zhang X (eds) Biomass, biopolymer-based materials, and bioenergy. Woodhead Publishing, pp 51–73. https://doi.org/10.1016/b978-0-08-102426-3.00003-5
131. Neşer G (2017) Polymer based composites in marine use: history and future trends. Procedia Eng 194:19–24. https://doi.org/10.1016/j.proeng.2017.08.111
132. Mohammed L, Ansari MNM, Pua G, Jawaid M, Islam MS (2015) A review on natural fiber reinforced polymer composite and its applications. Int JPolym Sci 2015:1–15. https://doi.org/10.1155/2015/243947

Chapter 4
Processing and Properties of Starch-Based Thermoplastic Matrix for Green Composites

Laura Ribba, Maria Cecilia Lorenzo, Maribel Tupa, Mariana Melaj, Patricia Eisenberg, and Silvia Goyanes

L. Ribba
Dirección de Materiales Avanzados, Áreas del Conocimiento, INTI, CONICET, San Martín, Buenos Aires, Argentina
e-mail: ribba@df.uba.ar

L. Ribba · S. Goyanes (✉)
Departamento de Física, Facultad de Ciencias Exactas y Naturales, Universidad de Buenos Aires, Buenos Aires, Argentina
e-mail: goyanes@df.uba.ar

M. C. Lorenzo
Dirección de Materiales Avanzados, INTI-Áreas del Conocimiento, San Martín, Buenos Aires, Argentina
e-mail: clorenzo@inti.gob.ar

M. C. Lorenzo · P. Eisenberg
Instituto de Investigación e Ingeniería Ambiental, Universidad Nacional de San Martín, San Martín, Buenos Aires, Argentina
e-mail: peisenberg@unsam.edu.ar

M. Tupa · M. Melaj
Facultad de Ingeniería, ITPN. Instituto de Tecnología en Polímeros y Nanotecnología, CONICET—UBA, Buenos Aires, Argentina
e-mail: mtupa@fi.uba.ar

M. Melaj
e-mail: mmelaj@fi.uba.ar

S. Goyanes
Instituto de Física de Buenos Aires (IFIBA), CONICET—Universidad de Buenos Aires, Buenos Aires, Argentina

M. Tupa
Departamento de Ingeniería Química, Facultad de Ingeniería, Universidad de Buenos Aires, Buenos Aires, Argentina

S. Thomas and P. Balakrishnan (eds.), *Green Composites*,
Materials Horizons: From Nature to Nanomaterials,
https://doi.org/10.1007/978-981-15-9643-8_4

Abbreviations

ACR	Acrylonitrile-chlorinated polyethylene styrene
a_w	Water activity
bio-PE	Bio-polyethylene
bio-PEF	Bio-polyethylene furanoate
bio-PET	Bio-polyethylene terephthalate
BP	Benzoyl peroxide
BT	Bentonite
C/OS	Corn/octenylsuccinated starch
C30B	Organically modified montmorillonite
CA	Contact angle
CAc	Citric acid
CF	Cellulose fiber
ChCl	Choline chloride
CNF	Cellulose nanofiber
CNFs	Cellulose nanofibrils
CS	Corn starch
1D	One-dimensional
2D	Two-dimensional
DES	Deep eutectic solvent
DMSO	Dimethyl sulfoxide
DMTA	Dynamic mechanical thermal analysis
DSC	Differential scanning calorimetry
EB	Elongation at break
EMA	Ethylene–methyl acrylate
EVA	Ethylene-vinyl acetate
EVOH	Poly(ethylene-co-vinyl alcohol)
Gl	Glycerol
GMA	Glycidyl methacrylate
HPDSP	Hydroxypropyl distarch phosphate
HSM	High-speed mixer
HV	3-Hydroxyvalerate
Im	Imidazole
M0	Monolayer moisture content
MA	Maleic anhydride
MCC	Microcrystalline cellulose
MFI	Melt flow index
MMT	Montmorillonite
MMTDA	Modified MMT
NCC	Nanocrystalline cellulose
nCOM	Nanoclay organically modified
O/W	Oil-in-water
OMMT	Derivate of montmorillonite

OP	Oxygen permeability
OS	Octenylsuccinated starch
OSA starch	Octenyl succinic anhydride-modified starch
PBAT	Polybutylene adipate-co-terephthalate
PBAT-g-MA	Maleate PBAT
PBT	Polybutylene terephthalate
PCF	Plasma-treated cellulose fiber
PCL	Polycaprolactone
PDA	Polydopamine
PE	Polyethylene
PEG	Polyethylene glycol
PET	Polyethylene terephthalate
pEVOH	Plasticized EVOH
PHA	Polyhydroxyalkanoate
PHB	Polyhydroxybutyrate
PHB-g-AA	Acrylic-acid-grafted PHB
PHBV	Poly-(3-hydroxybutyrate-co-3-hydroxyvalerate)
PLA	Polylactic acid
POE-g-GMA	Glycidyl methacrylate-functionalized polyolefin elastomer
POE-g-MAH	Maleic anhydride-grafted ethylene–octene copolymer
PPG	Poly propylene glycol
PS	Polystyrene
PVA	Polyvinyl alcohol
PVC	Polyvinyl alcohol
SEM	Scanning electron microcopy
SF6	Sulfur hexafluoride
SME	Specific mechanical energy
SNC	Starch nanocrystal
SO	Soybean oil
SSE	Single-screw extruder
TA	Tartaric acid
TBC	Tributyl citrate
TBoAC	Tributyl o-acetylcitrate
T_d	Decomposition temperature
TEC	Triethyl citrate
T_g	Glass transition temperature
T_m	Melting temperature
TOMC	Oxidized MCC
TPS	Thermoplastic starch
TS	Tensile strength
TSE	Twin-screw extruder
U	Urea
UM	Urea-intercalated montmorillonite
VAc	Vinyl acetate
WC	Water content

WF	Wood fiber
WS	Water sorption
WSNC	Waxy starch nanocrystal
WVP	Water vapor permeability

1 Introduction

The market for plastic materials and products faces the challenge of increasing demand for materials with higher performance and new functionalities with focus on both, biobased and biodegradable polymers as alternatives to the non-biodegradable plastics produced from fossil resources. Bioplastics can be classified into three main groups: (1) biobased, non-biodegradable materials (bio-PE, bio-PET and bio-PEF), (2) biobased and biodegradable materials (TPS, PHA, PHB, PLA), and (3) fossil-based and biodegradable materials (PBAT, PCL, PVA, mostly blended with group 2). Their input to sustainability contributes to lower carbon footprint, high recycling value and complete biodegradability/compostability [1]. The development of sustainable bioplastics made of either biobased or biodegradable polymers and the principles of the circular economy opens new opportunities to alleviate the intensive use of non-renewable sources, innovation, competitiveness, new more qualified jobs and plastic pollution [1]. In this context, global bioplastics production capacity is set to increase from around 2.11 million tonnes in 2019 to approximately 2.43 million tonnes in 2024, according to the latest market data compiled by European Bioplastics and the research institute nova-Institute [2].

Among bioplastics, starch has been widely studied for a longtime since it is an inexpensive, renewable, versatile, fully biodegradable and widely available raw material. Starch and its blends account for over 38% of the global biodegradable plastic production capacities (Fig. 1). Today, bioplastics can be used in almost all market segments and applications, but each polymer has main application fields. The fields of starch-based materials' greatest application are flexible packaging and agri- and horticulture [2]. Since starch sources, availability and structure have been discussed in many reviews as well as chapters, particularly during the last two decades, they are not going to be detailed in this chapter [3, 4].

Before being thermally processable as thermoplastic polymers, starch must be transformed into thermoplastic *starch* (TPS) by the addition of plasticizers combined with the application of elevated temperatures and shear forces. Although the clear advantages of using starch-based plastics for a sustainable development, their application is still restricted by many factors. TPS great sensitivity to water, low moisture barrier capacity and poor mechanical properties are considered as major drawbacks when compared to conventional plastics.

The addition of organic and inorganic fillers of a different nature is an efficient way to improve the performance of TPS, and some may even grant unique properties to be used in new and demanding applications. Moreover, blending TPS with other

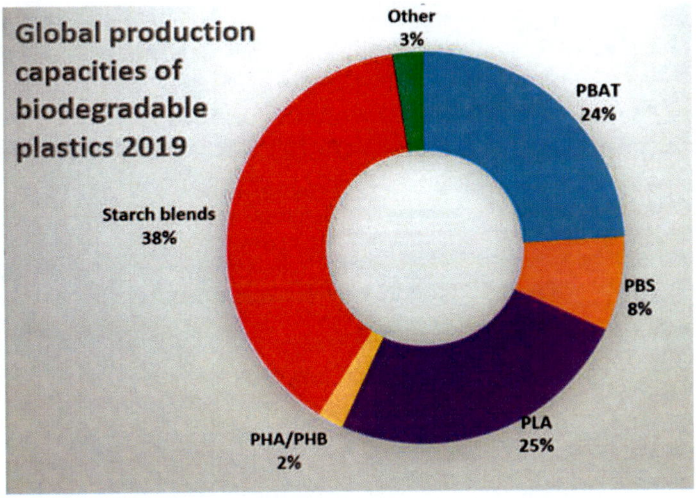

Fig. 1 Global production capacities of biodegradable plastics in 2019. *Source* European Bioplastics, nova-Institute (2019). Figure drawn by the authors

biopolymers to develop starch blend composites could bring their properties closer to those of conventional non-biodegradable plastics. A careful choice of blending polymer should be made according to the pretended application. Among biodegradable polymers, polybutylene adipate-co-terephthalate (PBAT) and polyhydroxybutyrate (PHB) have the important advantage of being biodegradable in soil. Both polymers are very stable in water due to their hydrophobic character. PHB has higher modulus and stress at break than starch but a very low elongation at break, while PBAT has higher stress at break values than that of starch and deformations at break as large as 700%, which makes it an ideal material for different applications, particularly in agriculture as mulching. One of the main drawbacks of both polyesters is their high production cost, being higher in the case of PHB. On the other hand, PBAT has the extra disadvantage of coming from fossil resources. Blending them with starch is an effective way to lower the final cost of the PHB or PBAT blends and increase the biobased character of PBAT. Moreover, the inclusion of fillers to develop starch/PHB and starch/PBAT composites could lead to superior properties. However, this poses a series of challenges to be solved mainly due to the incompatibility between polymers as a consequence of the strong difference in their polarity. While several researches that focus on the study of starch/PBAT blends can be found in the literature, there are far fewer that deal with TPS/PHB blends, possibly due to the higher cost and lower world production of the latter.

Expectations of achieving a green starch-based composite using blends with other polymers and the addition of nano-microfillers grow day by day and have already materialized in the form of commercial products, such as bags, disposable tableware and mulching. However, the costs of these products are still high compared to those of traditional plastics.

In the next sections, processing strategies for filler incorporation into starch-based composites will be described. Starch and starch blend matrices will be studied separately since different challenges are faced in each case. Then, these composites will be described from a point of view of their thermal, barrier and mechanical properties, again distinguishing between TPS and starch blend matrices. Finally, the main conclusions obtained are listed, as well as future perspectives of these materials.

2 Processing of Starch-Based Green Composites

2.1 Processing of Starch Matrix Composites

2.1.1 Usual Strategies for Processing Starch

Like most materials, polymers must be melted or solubilized, with the objective to transform them into different kinds of products with specific properties and shape. The most common ways to process polymers include extrusion, blowing, meltblown, spunbond, thermoforming and injection. The method chosen for the processing of a material will be decisive in its final properties.

Processing starch by the techniques usually used in the plastic industry is extremely complicated. Conventional equipment was designed to work with thermoplastic materials such as PE, PET, PVC or PS. Starch is not a thermoplastic, but with favorable processing conditions and proper formulation development, it can become this type of material [5, 6]. However, thermoplastic starch (TPS) processing has several additional problems, as a consequence, for example, of its high molecular weight or poor melt tenacity [7].

The melting temperature (T_m) of dry starch is generally higher than its decomposition temperature (T_d), which makes the incorporation of plasticizing agents necessary to convert it into a thermoplastic material and thus allow its processing. Although the incorporation of water is a usual strategy, for many industrial processing techniques high water content (WC) is a problem. The solution casting technique involves large amounts of water, and for that reason, it is called wet method. It is one of the most widely used techniques at laboratory scale since it is extremely simple, but as it involves drying times that are too long to permit large-scale manufacturing, it cannot be carried out at industrial scale. In this chapter, we will focus on the so-called dry methods that involve shear conditions and therefore require a lower WC in comparison with the wet process techniques, making them applicable on an industrial scale. In particular, to produce TPS from starch and plasticizers, the most widely used industrial technique is extrusion. Other usual techniques for polymer processing, such as injection, blowing or compression molding, can only be performed on thermoplastic materials. Therefore, in order to process starch in these ways, it must be first extruded together with plasticizers to become TPS.

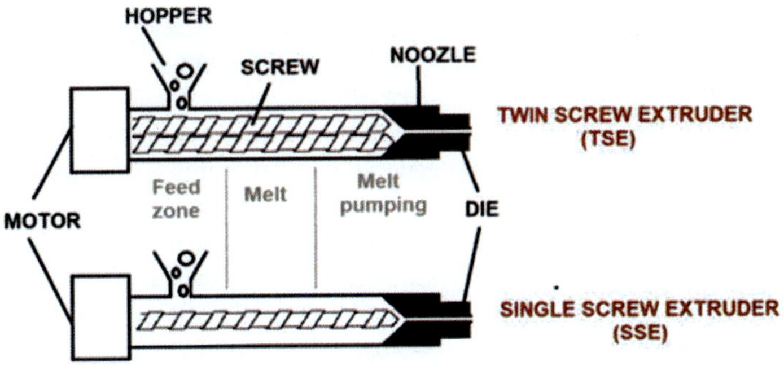

Fig. 2 Cross-sectional diagrams of a single- and a twin-screw extruder. Figure drawn by the authors

Extrusion combines high shear, temperature and pressure to break down starch granule structure, facilitating the diffusion of the plasticizers and the subsequent melting of the new blend, generating TPS. During this process, multiphase transitions occur such as gelatinization, granule expansion, melting and decomposition [8, 9].

In general form, an extruder consists of a hopper, barrel, feed screw, thermocouples and dies. There are two different extrusion equipment designs: the single-screw extruder (SSE) and the twin-screw extruder (TSE). Cross-sectional diagrams of a single- and a twin-screw extruder are shown in Fig. 2. While single-screw extrusion is ideal to produce basic and simple compounds, corotating twin-screw extrusion is better for composites which require a relatively high level of mixing in the extruder and flexibility during the process [10, 11]. Moreover, TSEs have a large operational flexibility (individual barrel zone temperature control, multiple feeding/injection ports and diverse screw configurations for different degrees of mixing/kneading). In both SSE and TSE, residence times and specific mechanical energy (SME) inputs can be controlled, achieving highly efficient production. It is worth noting that, as TSE's screws have a self-wiping ability, they are more suitable to process raw starch powder than SSEs, which may encounter the problem of conveying the starch powder at the feeding port [6].

Both the equipment configuration and its operating conditions, such as temperature profile and screw speed, will be decisive in achieving mechanical disruption and starch transformation [8, 12]. SME is the amount of mechanical energy (work) dissipated as heat inside the material, expressed per unit mass of the material. It can be calculated as

$$\mathrm{SME} = \frac{P \times \tau \times \mathrm{RPM_{act}}/\mathrm{RPM_{rated}}}{m}$$

where P is the motor power, expressed in kW, τ is the difference between the running torque and the torque when the extruder is running empty divided by the maximum allowable torque, $\mathrm{RPM_{act}}$ is the actual screw rpm, $\mathrm{RPM_{rated}}$ is the maximum allowable screw rpm and m is the mass flow rate of the system (kg/s) [13]. During TPS extrusion, mechanical energy must be enough to break starch grains and efficiently complete its extrusion process. Different authors studied how operating parameters influence on SME [8, 14].

Ahead of the extruder's screw barrel, a die system produces shaped products to facilitate further product processing and/or to determine the properties of end products.

Sheet products can be obtained through continuous extrusion of the TPS melt through a horizontal or plate die. The uniformity of the final sample thickness and of the exit velocity distributions across the width of the die exit must be guaranteed by the design of the die [10]. Different authors used this type of die to obtain band-shaped sheets of starch and starch-based composites after extrusion [14, 15]. However, the obtained materials are usually too thick for many applications (more than 0.5 mm).

Other possibility is that the die extrudes single or multiple strands which can be cut into pellets once dried. A schematic drawing of a strand pelletizer is shown in Fig. 3. Strand extruding and pelletizing is a simple and straightforward process, but an extra step to obtain sheets or films is necessary. Generally, strands obtained from the extruder are pelletized and these pellets are converted into films through other processes such as thermocompression or single-screw extrusion followed by blowing. Many authors choose to manufacture TPS pellets using a twin-screw extruder, pelletize TPS strands and subsequently obtain films by thermocompression. For example, González-Seligra et al. [8] obtained different threads of TPS by TSE varying the screw speed used. Then, pellets from each system were obtained by the pelletization of the threads using an automatic pelletizer. TPS films were finally prepared by thermocompression using a thermostatized hydraulic press in two stages: First, pellets were heated to 140 °C for 15 min, then, pressure was

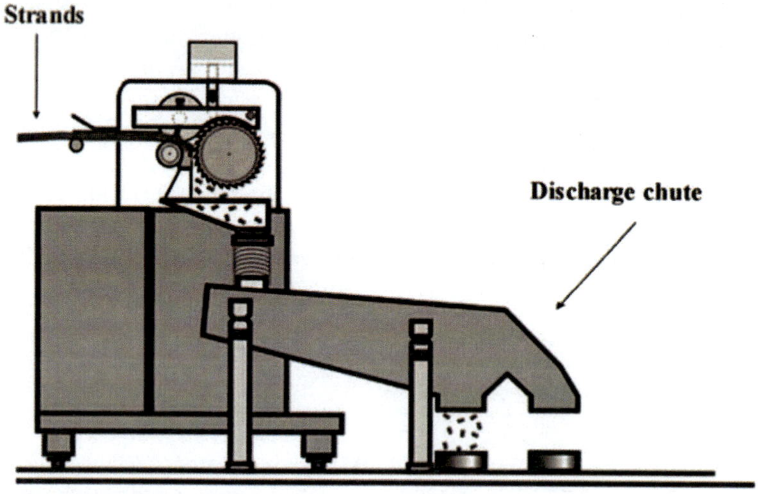

Fig. 3 Schematic drawing of a strand pelletizer. Reprinted from Drobny [18], "Processing Methods Applicable to Thermoplastic Elastomers", 33–173, Copyright (2014), with permission from Elsevier

increased to 56 kPa and temperature decreased up to 40 °C. The resultant film thickness was approximately 0.3 mm. Similarly, Estevez-Areco et al. [16] used extrusion followed by thermocompression, but in their case, they developed bioactive starch composites with antioxidant activity. They used a thermo-stated hydraulic press to convert the thermoplastic threads into films. Basically, they placed a piece of thread between Teflon sheets and heated to 130 °C for 15 min. Pressure was then increased to 45 kPa and maintained for 15 min. Finally, temperature was decreased to room temperature while keeping the pressure constant. Gutiérrez et al. [17] also prepared bio-nanocomposite films by twin-screw extrusion followed by thermo-molding, in this case from corn starch and pH-sensitive nanoclays packaged with Jamaica flower extract. After extrusion, the strands were pelletized using an automatic pelletizer and pellets were hot-pressed using a hydraulic press at 120 °C and 1×10^4 kPa for 20 min, after which a cooling cycle was applied until they reached a temperature of 30 °C.

The most common method for making plastic films, particularly for the packaging industry, is blown-film extrusion. In this process, a SSE coupled to a blower is used (Fig. 4). At the exit of the extruder, the blower has a first stage in charge of transforming the melted TPS into a hollow tube. As the tube is extruded, it expands due to a pressure increase produced by the blower. The ability to process a system by this technique will depend on the tensile properties of the melt. The melt tenacity of TPS, understanding this as the ability of the melt to deform without rupture, is not usually enough to be processed by this methodology. Thunwall et al. [19] studied possible routes for film-blowing TPS on a laboratory scale by a suitable choice of processing conditions, amount of glycerol and moisture content, finding that the

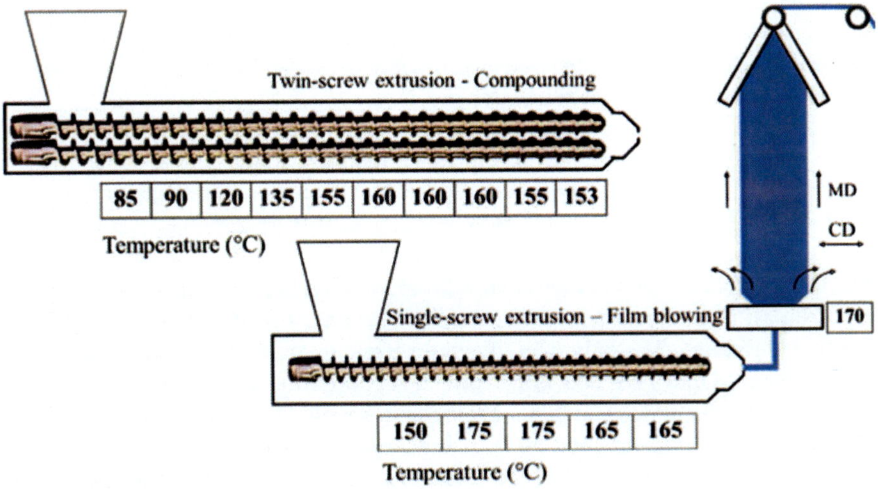

Fig. 4 Diagram of the process to obtain acetylated cassava starch composite films with pea protein isolate incorporation through blow molding, including temperature profiles. Reprinted from Huntrakul et al. [25], Copyright (2020)

possibility of obtaining films was given in a too narrow processing window. The difficulties encountered were mainly related to a sticky surface of the film, insufficient tenacity and foaming. Even in the cases in which they managed to obtain films, their final properties were poor. Several studies have explored the film blowing of starch-containing materials, introducing moderate quantities of starch (between 8 and 30 wt%) in a regular synthetic polymeric system [20, 21, 22] or developing blends with biodegradable polymers, such as poly(vinyl alcohol), poly(caprolactone) or other polyester [23, 24]. In the case of regular synthetic polymers, starch gives them the enormous benefit of its biodegradability, although it worsens its mechanical and water barrier properties. In the case of blends with other biodegradable polymers, the main benefit is economical, thanks to the low cost of starch. Other strategies to improve the melt tenacity of thermoplastic starch and allow its processing through blow molding are the development of composites. Huntrakul et al. [25] obtained acetylated cassava starch composites with pea protein isolate incorporation using a twin-screw extruder. Pellets obtained from the extruder were dried in a hot air oven at 60 °C overnight, and films were produced using a single-screw blown-film extruder connected to an annular ring-shaped film-blowing die. Temperature profiles used are shown in Fig. 4. The incorporation of pea protein isolate helped to avoid stickiness and improve processability during the blown extrusion. Moreover, the high structural deformation shown by acetylated starch films during storage was effectively prevented by pea protein isolate incorporation. Dang and Yoksan [26] improved blown-film extrusion processability of TPS films by incorporating low contents of plasticized chitosan (0.37–1.45%). The composite was extruded and pelletized, obtaining pellets which were blown into films using a single-screw extruder with an L/D 30:1, a screw diameter of 25 mm and four controlled temperature zones, connected to a film-blowing attachment with a ring-shaped die. Screw speed and nip roll speed were adjusted to 35–45 rpm and 3 rpm, respectively, while the barrel temperature profile was maintained at 130–140–140–140 °C (from feed inlet to die) and the die temperature was set at 150 °C.

Injection molding is another typical technique for thermoplastics processing, where the polymer is fed into a heated barrel, mixed (using a helical-shaped screw) and injected into a mold cavity, where it cools and hardens to the shape of the cavity. The behavior of starch during injection molding and the processing parameters have been studied by different authors during the last decades [27, 28, 29]. TPS high viscosity and poor flow properties make the application of this methodology complex [6]. Besides, TPS injection molded parts have problems related to its serious distortion and shrinkage. The only way to avoid shrinkage in hydrophilic polymers is to process them so that the products are formed at approximately the equilibrium in-use water content. For potato starch, for example, this would be water contents of around 14% used under ambient conditions. If higher contents are involved as plasticizer, the material will deform, as the equilibrium is naturally achieved after processing [29]. There are commercial products based on starch that can be injected without problems. The first commercial products made of injected molded TPS were starch capsules designed for drug delivery (Capill®). Other developments include blends of starch with eco-friendly polymers that allow to improve not only the processability but also the final properties of the materials, such as the case of the Mater-Bi®, which is offered in molding pellet presentations.

Besides processing starch to produce TPS, extrusion can also be used for its chemical modification, addition of cross-link agents or copolymer creation in a continuous process with a more consistent product quality. Inside the extruder, very efficient mixing processes of highly viscous liquids are produced, which allows starch modification to be performed in a homogeneous medium. This methodology, called reactive extrusion, is one of the typical methods in the industrial modification of starches due to the aforementioned advantages, added to its low cost. It should be mentioned that for some particular processes special equipment designs are needed.

Ye et al. [30] synthesized citric acid-esterified rice starch using a one-step reactive extrusion method. They dissolved citric acid in distilled water and slowly added the solution to the rice starch. After equilibration in sealed bags, reactive extrusion was performed using a twin-screw extruder. The four sections of the extruder were set to 80–100–90 and 75 °C, the feed rate was 18 kg/h, and the screw (30 mm in diameter, L/D 16:1) speed was 250 rpm. The starch citrate extrudates were collected as powder, dried in an oven until constant weight at 45 °C and sieved through an 80 mesh sieve. Finally, it was washed with absolute ethanol to remove unreacted citric acid. This method to modify starch chemical structure has advantages over conventional methods because it can be carried out rapidly using a continuous process. On the other hand, Siyamak et al. [31] developed a method of synthesis based on reactive extrusion combining the benefits of continuous manufacturing with the use of green chemistry principles. They grafted four different types of starches with acrylamide monomers via free radical copolymerization using a 16-mm corotating twin-screw extruder, with a horizontally split barrel of 40:1 L/D comprising 10 barrel sections with independent temperature control zones. They designed an experimental set consisting of three different phases. The initial phase was finding optimum conditions for the gelatinization of starch using extrusion processing. The second phase involved the transfer of these optimized formulations and conditions to the development of the copolymers. Nitrogen gas was used in all experiments to eliminate oxygen from the extruder, while deionized water and acrylamide (50 wt%) solutions were sequentially injected into the barrel sections 1 and 2, followed by ammonium persulfate (5 wt%) solution, which was injected into the barrel sections 5 and 7 using peristaltic pumps. Figure 5 shows the screw profile used. Finally, the third phase consisted in

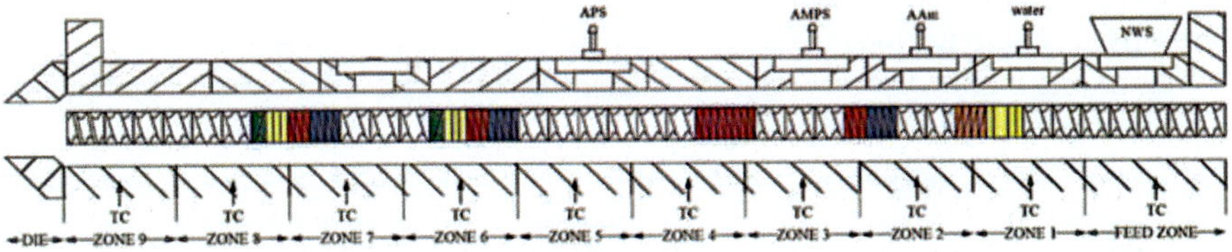

Fig. 5 Screw profile for reactive extrusion of starch copolymers. Reprinted from Siyamak et al. [31], Copyright (2020)

preparing the graft copolymers of other types of starches using the optimized conditions identified in the previous phases. The botanical origin of the starch determines the relationship between the amount of amylose and amylopectin, and the type of crystal structure. This establishes the temperature profile necessary for extrusion, as well as the efficiency and type of grafting that can be performed. Since extrusion is currently applied for large-scale manufacturing of starch-based commodities, the modification method they propose could lead to the successful production of complex smart systems based on starch-grafted copolymers.

2.1.2 Filler Incorporation During Starch Matrix Composite Processing

There are different strategies to incorporate fillers throughout the processing of composites. The final objective is to obtain a proper dispersion of the filler in the matrix, what will depend not only on the processing but also on their chemical similarity, filler's morphology and surface area [32].

One of the main factors to consider when deciding how to introduce a filler during composite processing is the chemical affinity between it and the matrix, and what interactions are expected to occur. Depending on the filler–matrix compatibility, different effects will be produced in the matrices. When the filler and the matrix are compatible and interactions between them enhance adhesion, several improvements can be achieved. For example, a good stress transfer between the matrix and the filler will occur, so that, if the filler is more rigid than the matrix, the composite will have greater Young's modulus and stress at break than the matrix material. Thermal stability, water resistance and degradation behavior can also be improved due to high compatibility and strong bonds between filler and matrix. Ilyas et al. [33] demonstrated these effects by incorporating nanocrystalline cellulose (NCC) as a reinforcing filler into sugar palm starch matrices and attributed them to the excellent compatibility achieved thanks to the presence of abundant hydroxyl groups on the NCC surface which can interact through hydrogen bonding with the hydrophilic polymer matrix.

However, on occasions fillers that are not naturally compatible with the starch matrix are used as reinforcements. In these cases, different processes can be carried out to improve filler dispersion. One possibility to incorporate a hydrophobic filler into a starch matrix is to encapsulate it. Moreover, this methodology can also provide the filler protection from temperature and stress generated during processing, as well as modulate its release kinetics in the case of active compounds [34]. Different techniques have been proposed for micro- and nano-capsule preparation and incorporation into starch matrices.

The most commonly used technique at industrial level for microencapsulation is spray-drying. In this technology, a dry powder is produced from a liquid or slurry by rapidly drying with a hot gas. The choice of the shell materials used for the encapsulation of actives is determinant in the active load of spray-dried microparticles. Talón et al. [35] employed whey protein and soy lecithin as wall materials together with maltodextrin as drying coadjuvant and oleic acid as a carrying agent, for the production of microencapsulated eugenol through spray-drying. Different encapsulated eugenol powders were obtained by using a spray dryer with a rotary atomizer at an inlet air temperature of 180 °C. These microcapsules exhibited a mean particle diameter of around 15 μm, regardless of the wall material, and a very small percentage of finest particles (0.5 μm). To develop composite films, a premixture of the starch, microcapsules and glycerol using a starch:microencapsulate powder:glycerol mass ratio of 1:0.35:0.3 was obtained in a two-roll mill at 160 °C and 12 rpm for 10 min. The obtained pellets were process through compression molding using a hot-plate press in order to obtain films. The functional properties and the release kinetics of the final materials were evaluated. They found that the incorporation of the carrying agent (oleic acid) was essential for the protective effect of microcapsules during film processing. Other possibility is to use modified starches as the shell material. Octenyl succinic anhydride-modified starch (OSA starch) is widely used in the microencapsulation of oil-soluble nutrients, flavors, agrichemicals, fragrances and pharmaceutical actives, thanks to its hydrophobic character [36]. In spray-drying microencapsulation, modified starches offer high oil retention, long shelf life and high manufacturing efficiency [37]. Figure 6 illustrates the morphological evolution of an emulsion droplet in a highly dynamic spray-drying process, together with the SEM images of spray-dried microcapsules and the cross section of a microcapsule [37].

Extrusion process can also be used for encapsulation of different agents. As it was mentioned before, during TPS extrusion starch and plasticizers are converted into a homogenous molten state, involving various structural changes, such as granule disruption, crystal melting and molecule entanglement. This molecule entanglement can be exploited for the incorporation and sustained release of the filler to be encapsulated. Several authors have used starch in extrusion encapsulation due to its ability to form stable inclusion complexes with the encapsulated agents [37–40]. Chen et al. [39], for example, presented a simple preparation of microparticles with extruded starch as the shell material, resveratrol as the core material and the thermostable α-amylase as the release-improvement reagent of resveratrol. They extruded a mixture including 10 kg of corn starch (water content was adjusted to 24%), 10 g of resveratrol and 5 g of α-amylase with 65 °C barrel temperature (50–55–60–65 °C for the four parts starting from the feed part) and 110 rpm screw speed. The extrudate was smashed by using ultra-micro-pulverizer, and microparticles with 0.15–1-mm diameter were selected. Modified starches have also been proposed for extrusion encapsulation, being CAPSUL® (product of Ingredion) a preferred choice due to its low viscosity [37].

Other possibility to encapsulate agents is through the formation of complexes with amphiphilic molecules, such as cyclodextrin. This molecule is an enzymatically modified starch which has a truncated cone shape with a hydrophobic interior, allowing the formation of an inclusion complex with oils that protects them against harsh temperature and shear conditions during extrusion [41].

The way in which the filler is incorporated into the starch matrix during processing will be highly influential in the filler's degree of dispersion, filler–matrix interaction and specific filler arrangement in the starch matrix for both, hydrophilic

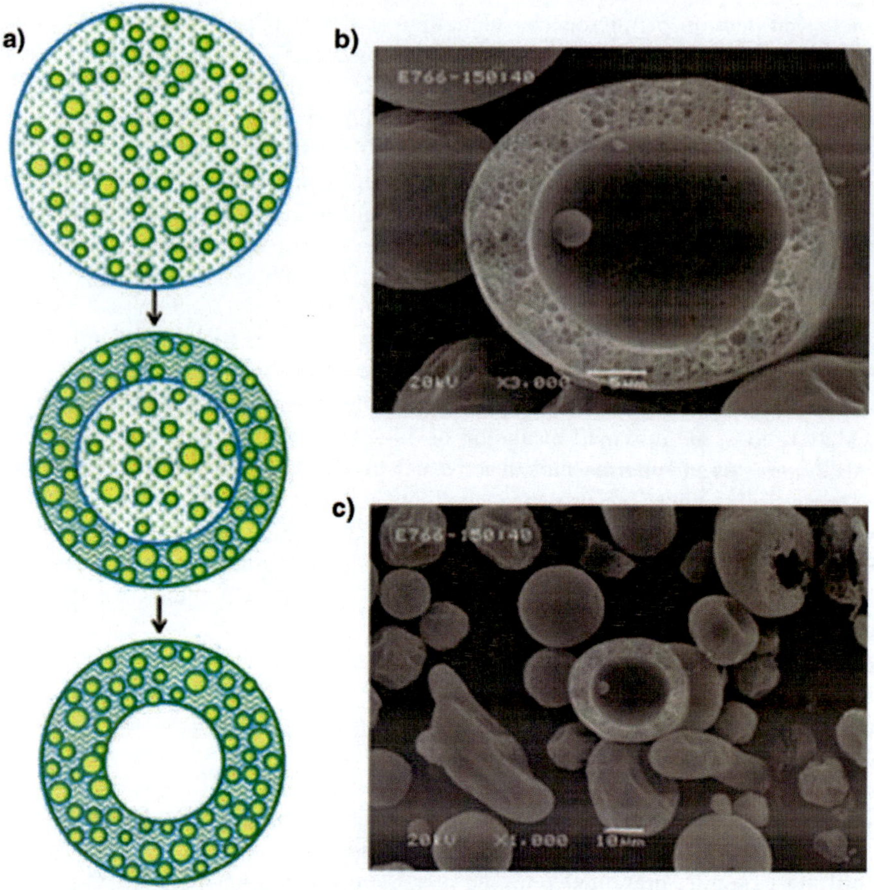

Fig. 6 **a** Schematic illustration of spray-drying dynamics. The big circles represent atomized water droplets, small circles represent oil droplets, and dots represent soluble carbohydrate molecules. **b** Scanning electron microcopy (SEM) images of spray-dried microcapsules and **c** the cross section of a microcapsule. Reprinted from Jin et al. [37], Copyright (2018)

and hydrophobic fillers. One possibility when producing starch-based composites through extrusion is to introduce the filler in the extruder barrel together with starch, plasticizers and/or other components. As the active compound is added at the beginning of the extrusion process, in this option, temperature and wear sensitive molecules, such as flavors, are more likely to be degraded being necessary to previously encapsulate them [38]. In the case that fillers support the temperature profile and shear stresses inside the extruder, a usual procedure is to carry out a premixed step to guarantee a proper dispersion. This is the case of Ochoa-Yepes et al. [42] who mixed cassava starch, glycerol, water and a desired amount of lentil proteins at 20 rpm for 15 min in a horizontal mixer. Mixtures sieved with a number 10 mesh were stored for 24 h in sealed containers and then extruded in a corotating twin-screw extruder at a screw speed of 80 rpm (with feeding rate of 12 g/min) and temperature profile (from the feeder to the die) of 90–100–110–120–130–130–140–140–130–120 °C. Ghanbari et al. [43] also performed a premix, but in their case it was made manually. They mixed corn starch and glycerol in polyethylene bags for 5 min, and subsequently, the resulting blend was further mixed with the corresponding amount of cellulose nanofibers for more than 10 min. Dean et al. [44] explored three different dispersion methods to develop clay–starch mixtures before the extrusion process. Nanoclays were either dry blended with starch in a high-speed mixer (HSM); dispersed in water using conventional mixers prior to blending with starch in a HSM; or dispersed in water using ultrasonics prior to blending with starch in a HSM. All mixtures were later processed through a corotating twin-screw extruder, with diameter 30 mm and L/D 40:1, using a profile producing a melt temperature of 110 °C. They demonstrated that when the level of clay, water and starch was optimized an exfoliated structure was produced via standard mixing which exhibited comparable improvements in mechanical properties to ultrasonically treated samples.

Some authors propose to perform a double extrusion in order to enhance the filler dispersion, although this is not the most common methodology. For example, Chaves da Silva et al. [45] obtained starch films with different percentages of gelatin using a single-screw extruder in two stages. In the first stage, the starch, gelatin and a plasticizer were mixed manually and extruded to produce pellets; in the second step, the pellets were extruded to produce the films. The screw (diameter of 25 mm) speed was 30 rpm, and the zone temperature was set between 90 and 120 °C in the four heating zones for the two stages. Huang et al. [46] proceeded similarly by first extruding glycerol and corn starch to obtain small particles of TPS, which were later mixed with montmorillonite (MMT) and fed into the single-screw plastic extruder again. The temperature profile along the extruder barrel was 110–115–120–120 °C (from feed zone to die).

In some cases, for TPS extruded films a preliminary gelatinization is produced by heating a mixture of starch, distilled water and the desired plasticizer. One option for filler incorporation is to add them to the obtained gel and homogenize it. This is the case of González et al. [47], who added polysaccharide nanocrystals dispersed in distilled water by ultrasonication to a gel composed of corn starch, glycerol and water. The material was freeze-dried, then extruded using a MiniLab extruder at 120 °C and 50 rpm and pelletized. The obtained pellets were compressed at 120 °C for 5 min without pressure and 120 °C under a pressure of 2.5 t for 5 min. This work is an example of the importance of a previous mix step with high shear stresses, necessary to process this type of composites with a single-screw extruder.

Other possibility, which is recommended by some extruder's manufacturers, is the direct injection of fillers into some middle barrel section of the extruder. This approach is usually used for the addition of volatiles that can be lost due to high temperatures [38]. Basically, in the feeding section on the extruder, starch and plasticizers are metered and conveyed into the first mixing zone, where thanks to the heat and shearing, the mixture is transformed into TPS melt. In a secondary feed zone, the filler is added by means of a liquid feeding pump, reaching a further mixing zone where it is dispersed and evenly distributed into the matrix.

In most cases, the fillers are introduced at some stage of the extrusion process with the sole objective of being mixed under processing conditions that allow them to maintain their structure and properties throughout the entire process. In other cases, the extrusion process also serves as a treatment for tailoring the final properties of the filler. Such is the case of Fourati et al. [48] who produced nanocomposites based on TPS filled with cellulose nanofibrils (CNFs) in a single step by twin-screw extrusion of corn starch granules, glycerol and oxidized cellulose fibers. CNFs were produced in situ during the processing of the nanocomposite. For the extrusion process, starch granules, glycerol and fibers containing 50 wt% water were premixed manually. The mix was continuously extruded using a corotating conical twin-screw extruder at 100 rpm during 15 min at a temperature of 25 °C. Temperature was progressively increased during 10 min up to 110 °C, and the extrusion was completed at 200 rpm during 15 min to complete the gelatinization of starch and the breakdown of fibers. These nanocomposites displayed a higher strength and similar transparency degree than those produced by incorporating readily prepared CNFs.

Other processing techniques such as compression molding also offer the possibility to simultaneously process the TPS composite and modify the filler. Grylewicz et al. [49] manufactured TPS composites from potato starch, wood fiber (WF) and deep eutectic solvent (DES) based on choline chloride with urea or glycerol (Gl) as well as imidazole (Im) with Gl. The processing method included a first stage of component mechanical mixing followed by thermocompression with simultaneous WF modification. They found that DES played a triple role as starch plasticizer, as WF surface modifier and as composite component interfacial adhesion improver.

Beyond good compatibility and good dispersion, the way in which fillers are oriented within the polymeric structure will also be decisive on its final properties [50]. In the case of layered silicates, the clay platelets can be either intercalated by macromolecules and/or exfoliated, depending on the processing conditions and on the matrix–filler affinity. The best performances are commonly observed with exfoliated structures in which the clay platelets are individually delaminated and fully dispersed into the polymer matrix [51]. Different strategies have been performed to achieve these structures. For example, Adamus et al. [52] explored the possibility to modify montmorillonite through intercalation with urea. They mixed urea and montmorillonite (1:1 weight ratio) with different amounts of distilled water (0.15, 0.20 or 0.30 g H_2O/1 g M) using a laboratory mixer to obtain a homogenous material, which was subsequently extruded with an Eurolab digital laboratory extruder (L/D 40:1, screw diameter 16 mm). After extrusion, the material was dried for 2 h at 105 °C and then turned into a fine powder with mortar and pestle. The obtained powder was mixed together with starch, water and deep eutectic solvents that acted as plasticizers using laboratory mixer (25 °C, 350 rpm). After mixing, the compositions were extruded using a twin-screw extruder. The obtained extrudate was turned into 3-mm diameter pellets which were finally pressed between Teflon sheets using a hydraulic press for 5 min, at 140 °C, under pressure of 6 tonnes to obtain films. Results from the film characterization showed that an intercalated structure of MMT sheets was obtained. It must be mentioned that the inclusion of these fillers had an important effect in the processing of the material. The authors found that MMT presence helps polymer flow through the extruder die as it decreases melt viscosities of TPS. These could be explained in terms of a limitation of polymer chain entanglement caused by intercalated clay particles. This leads to lower energy dissipation in composite materials compared with neat polymer melt and, therefore, easier movement of polymer chains. This approach results promising in the development of new starch composites, which could be easier processed by melt processing (extrusion, extrusion with film blowing, injection molding or thermocompression).

2.2 Processing of Starch Blend Matrix Composites

2.2.1 Usual Strategies for Processing Starch Blend Matrix Composites

Blending different polymers constitutes a simple and effective approach and gives the possibility to improve material or product performance, with the desired properties at a lower cost, compared to the synthesis of a new polymer. For example, through mixing technology, it is possible to improve polymer modulus and dimensional stability by blending with a more rigid and more heat-resistant polymer. In addition, it is possible to design a compounding strategy to rebuild molecular weights of partially degraded polymers, thus to produce articles from scrap or post consumer plastics. Another important advantage of polymer blending is the improvement in the processability of plastics. For example, it is possible to process a high glass transition temperature (T_g) polymer at a temperature below the thermal degradation limit by blending with a miscible polymer of lower T_g. A reduction of the pressure drop can be achieved through the incorporation of an immiscible polymer of low viscosity, thus increasing productivity. Blending allows scrap reduction and rapid change in formulations giving rise to greater adaptability and flexibility in production and market demands [53].

The performance of polymer blends depends strongly on the final morphology achieved. The morphology of immiscible polymer blends is defined by the concentration of the blended components, phase identity, viscosity ratio, compatibility and interfacial tension between components, and finally, by the processing conditions [54]. That is, in the case of polymer blends, studying the behaviour at the interface as well as the rheological response results of great importance to understand the final material'The thermal properties of starch/PHB performance [55].

As it was previously mentioned, TPS is a biobased, biodegradable and cost-effective polymer, although it cannot meet all the application requirements in terms of processability, mechanical properties and durability [56]. Its strong inherent sensitivity to moisture, high water absorption capacity and low water resistance lead to a decrease in the barrier and mechanical properties that limits its applications [57]. Many studies have been conducted to evaluate the feasibility of combining TPS with hydrophobic biodegradable polyesters, with the aim of improving its physical properties and processability [58–60]. According to Muthuraj et al. [54], the main challenges with polymer melt blending are the improvement of adhesion between components, the reduction of interfacial tension between them and the generation of limited inclusion phase size. In this context, the addition of a reactive coupling agent or a compatibilizer is a prerequisite to improve interfacial adhesion and therefore the final properties of the blend.

The key to optimizing polymer blend performance is to control the morphology achieved after processing, as the development of a specific morphology will give rise to the desired properties [53]. For example, to optimize the material impact strength, a matrix/dispersed phase morphology is desirable, while co-continuity is expected to yield a better stiffness/ductility balance. Optimized properties, therefore, result from the proper equipment design selection, the strict control over processing conditions and the methods of generating and stabilizing the morphology, such as reactive compatibilization and the addition of coupling agents [53]. The material characteristics, such as bulk, rheological and thermal properties, have a significant impact on processing [53]. Biobased and biodegradable polymers can be processed using conventional techniques such as injection molding, extrusion and compression molding. Special attention must be paid due to their hygroscopic characteristics. Unlike starch, PHB and PBAT need to be dried before processing to avoid a drop in molecular weight and melt viscosity due to hydrolytic degradation. Furthermore, processing should be performed under controlled humidity to reduce the potential for flashing and brittle products [61]. Drying temperature and time conditions must be optimized for each material, especially when blending different polymers. The processing window of biobased and biodegradable polymers is usually narrower than most of the commodity plastics. Hence, the processing temperature profile must be designed and optimized for each system. High temperatures, the development of high shear stresses and high residence time during the process, are not desirable and can lead to polymer degradation [53, 61]. The forming stage often has a dramatic influence on the final morphology. Thus, the type of processing technology used, the equipment design and the processing conditions will determine the final morphology. During processing, the material undergoes complex deformation (e.g., elongational flow, uniaxial orientation, contraction) that affects the morphology and, thus, the material's final properties. Therefore, tailoring polymer blends for specific applications requires control of the morphology through tighter control over processing parameters [53].

Other useful strategy to improve polymer compatibility is the addition of a suitable filler which can improve interfacial adhesion between polymer phases due to preferable localization at the interface. Moreover, the addition of natural fillers can enhance the final performance of polymer blends, reducing the overall material cost and ensuring the total biodegradability of the material [62]. As in the case of neat TPS matrices, improvements in mechanical, thermal and barrier properties can be obtained when uniform filler distribution in the blend-polymer matrix is achieved. The challenge is to create a favorable interaction between the polymers and the filler, and thus avoid phase separation and agglomeration of the filler particles [63]. Polymer composites can be obtained by three different methods, in situ polymerization, melt mixing and solvent casting. Melt mixing, using conventional extrusion or injection molding technologies, is the most common method for preparing thermoplastic-based composites at large scale [62]. When adding a filler to polymer blends with a different molar mass and viscosity, selective localization of the filler can be achieved. The preferred localization is mainly driven by thermodynamic (viz., enthalpic interaction between each polymer and particles) and kinetic factors (i.e., viscosity ratios of the two polymers). Hence, micro- and nanofillers may result in localization of the filler within different polymeric phases, specific polymeric phases and/or at the interphase [64–66]. Thus, the final properties of biocomposites can be adjusted by controlling the filler location, its distribution and the adhesion between both polymers and with the filler [66].

Melt mixing provides a high shear force method to promote dispersion and distribution of the micro- and nanofillers, providing large-scale mass production. Adjusting processing conditions such as order of mixing, mixing time and shearing forces may result in an adequate strategy to promote the desired filler location and distribution in the polymer blend. Enough shearing force and residence time must be chosen to facilitate the migration of the particle from the preferent polymeric component to the less compatible phase. If processing conditions are not enough to allow the particle to move from one polymeric phase to the other, particles will be located as a function of the shear stress and time in any of the polymers in the mixture or the interface [66]. Different mixing strategies have been described. They depend on the polymers and filler characteristics, as well as on the versatility that the processing method can offer. Simultaneous mixing of all the components and its subsequent incorporation to the processing equipment, and the addition of particles to the molten blend where polymer pairs have already been mixing for some time are the most common strategies used. In more complex systems, premixing fillers with a component that is thermodynamically preferred (or not) is the recommended choice to promote further interaction between them [66].

In addition, chemical structure and viscosity of blends may negatively affect the dispersion of fillers (usually hydrophilic and with a great tendency to agglomerate), decreasing the ultimate mechanical properties of the composite. Different methods have been proposed to overcome these issues, such as the chemical modification of fillers to decrease their inherent hydrophilicity, chemical modification of polymers and the incorporation of a suitable compatibilizer [62, 63].

There are many studies focusing on the development of green composites based on starch combined with other biobased/biodegradable polymers reported in the literature. The most common are polycaprolactone (PCL), which is incorporated in some commercial Mater-Bi® (Italy) product, and polybutylene adipate terephthalate (PBAT), which is incorporated, together with polylactic acid (PLA). In the commercial product Ecowill FS0330® (China), PLA that is easy to process is biobased and is incorporated in the commercial product INZEA® (Spain), and polyhydroxybutyrate (PHB) like starch is biobased and does not require a synthesis process in addition to being biodegradable in soil and is incorporated in the commercial product Arboblend® (Germany). In the next sections, we will focus on processing strategies of two specific blends with potential applications in agriculture: starch/PHB and starch/PBAT.

2.2.2 Processing of Starch/PHB Composites

PHB is the most common type of polyhydroxyalkanoate (PHA). PHAs are linear polyesters with good mechanical strength and similar modulus to polypropylene. They are semicrystalline, hydrophobic and biocompatible. PHAs are degraded in the environment by soil microorganisms which are able to secrete PHA depolymerases, enzymes responsible for the hydrolysis of polymer ester bonds.

PHB is a linear polyester of D(-)-3-hydroxybutyric acid (Fig. 7) that was first discovered in bacteria by Lemoigne in 1925 [67]. It is synthesized biochemically by microorganisms in response to conditions of physiological stress and then accumulated in granular form in their body as energy storage material. PHB can be fermented from a variety of sources, such as sugars, molasses or hydrogen and carbon dioxide depending on the bacteria used. Commercial PHB is synthesized by several bacterial strains such as *Alcaligenes* sp. (PHB Industrial SA, Brazil); *Cupriavidus necator* (Bio-Oil SRL, Italy); and *Ralstonia eutropha* (Tianan Biologic, China; Telles, USA; and Kaneka Corporation, Japan) [68]. Due to the high cost of production, several studies have been carried out to develop better bacterial strains, more efficient fermentation/recovery processes and the use of inexpensive carbon resources as substrates, such as cyanobacteria [69]. Bhatia et al. [70] obtained an *Escherichia coli* strain which can accumulate intracellular PHB up to 57.4% of cell dry mass. The molecular weight of PHB can vary from about 50,000 to over a million, depending on the organism, growing conditions and the extraction methodology used.

PHB is water insoluble and relatively resistant to hydrolytic degradation; it has good ultraviolet resistance and poor resistance to acids and bases. It is soluble in chlorinated solvents such as chloroform and dichloromethane, and is insoluble in non-chlorinated solvents such as hexane [68]. Besides, PHB is less "sticky" than traditional polymers when it is melted [71].

In addition, PHB can be used in food packaging applications [72], controlled drug delivery carriers [73], and wound dressing and tissue engineering [74, 75]. However, moderate mechanical, thermal and barrier properties of PHB limit its applicability in these fields. Hence, blends of PHB with other polymers are prepared in order to balance these drawbacks [76].

As previously discussed, plasticized starch films or composite starch films have low mechanical resistance and a poor barrier property toward water vapor compared

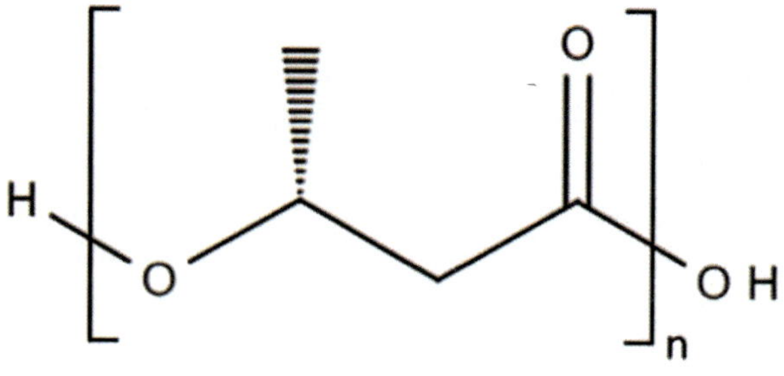

Fig. 7 Chemical structure of PHB. Figure drawn by the authors

to conventional plastics. Given the background, blending starch with other biopolymers like poly-3-hydroxybutyrate, which has low deformation at break, is one approach to produce starch-based materials keeping its biodegradability and renewability while improving its Young's modulus, tensile strength and even its barrier properties and water sensitivity.

Starch/PHB blends can be produced via compression molding, extrusion and injection molding. As PHB contains bound water, it is necessary to dry pellets before processing the material. Bugnicourt et al. [71] suggest drying PHB for over 2 h at 80 °C in dry air dryers. It is important to take into account that pellets recover the original humidity within 30 min after they are removed from the dryer. On the other hand, some producers of commercial PHB recommend a maximum of 0.005 wt% of humidity before processing the material, drying the PHB at temperatures not exceeding 100 °C. Garrido-Miranda et al. [77] dried PHB at 80 °C for 12 h before processing it with a previously dried TPS.

The major inconvenience during processing starch/PHB blends is the narrow temperature range of working without degrading the polymers, owing to the poor thermal properties of PHB. This polymer has low resistance to thermal degradation that involves chain scission and a rapid drop in viscosity and molecular weight [78, 79]. It decomposes at temperatures just above its melting point (T_m = 170–175 °C). If PHB is exposed to temperatures near 180 °C, it could suffer severe degradation generating products like olefinic and carboxylic acid compounds, e.g., crotonic acid and various oligomers [71]. When PHB is processed for industrial applications, melt-processing techniques such as extrusion and/or injection expose the polymeric chains to high temperatures (above 170 °C for PHB) and shearing tension. Therefore, as degradation may occur rapidly, the acceptable residence time in the processing equipment is only a few minutes. Pachekoski et al. [80] studied the PHB degradation after being extruded and/or injected. Melt flow index (MFI) value before processing was (17 $\pm$ 2) g/10 min, and it increased up to (21 $\pm$ 2) and (26 $\pm$ 2) g/10 min after the materials were extruded, and extruded and injected, respectively. The greater fluidity was related to the increase of the polymeric chain mobility due to the reduction in molecular mass, not only as a consequence of thermal degradation but also by the shearing of the polymeric chains caused by the mechanical forces involved at extrusion and injection processes. One strategy to decrease MFI is the addition of secondary antioxidants, which increases molecular weight since they act like chain extender agents. Correa [81] found that the addition of antioxidants stabilizes the PHB during the melt mixing, increases the processing window and generates chain extension reactions that increase the melt viscosity of the matrix, allowing for better processing conditions.

After processing, the viscosity of PHB decreases due to polymer degradation. In extrusion, the molecular weight of obtained PHB is related to the residence time, and it decreases when temperature and screw speed are higher [82]. Higher temperatures also increase the biodegradation rate of PHB due to a decrease in crystallinity, aiding extracellular enzymes to attack PHB chains [83].

Some strategies have been applied to overcome the thermal drawbacks during processing of starch/PHB blends. For example, the addition of plasticizers decreases

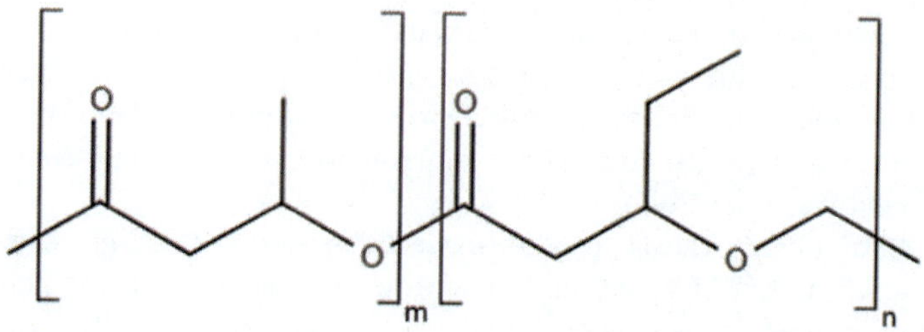

Fig. 8 Chemical structure of PHBV. Figure drawn by the authors

the glass transition and melting temperature of the mixture. However, this route is accompanied with a decrease in tensile properties; the extent depends on the amount and type of the plasticizer added [84]. Likewise, poly-(3-hydroxybutyrate-co-3-hydroxyvalerate) (PHBV) (Fig. 8) is a copolymer of PHB that is widely used to be blended to starch because it has a lesser melting temperature (T_m = 150–155 °C) than its homopolymer [85–87]. PHBV is obtained by the incorporation of 3-hydroxyvalerate (HV) during the fermentation process (the content can reach up to 95 mol%). It is produced by a fermentative process similar to that of PHB, only differing in the use of propionic acid (responsible for the concentration of HV) with glucose, as a carbon source [88]. Higher HV contents contribute to decrease the melting point of the copolymer, expanding the processing window. Furthermore, PHBV is less crystalline, easy to mold and more resistant than its homopolymer [89–91].

Another possibility is to improve the interfacial adhesion of starch and PHB in order to increase thermal stability. For example, Don et al. [92] observed that pure PHB decreased its molecular weight in 70% when it was processed at 175 °C for 5 min, while for blends prepared with potato starch grafted with vinyl acetate the reduction was lower, reaching 46% when the composition was 20:80 of PHB:modified starch.

In general, compression molding is the most used method for obtaining starch/PHB-based blends/composites. Before this, a compounding step is carried out to obtain a homogeneous mixture of all the components. The first step is often done in a conventional mixer or in a twin-screw extruder at temperatures around to that of PHB melting. Then, the melted mixtures (frequently grounded) are processed to obtain films with desirable thickness. By compression molding, the assemblies are often processed between 160 and 180 °C, at 8–69 MPa, and for 5–30 min when PHB is used [73, 93, 94].

The resulting starch/PHB blends are often heterogeneous due to a lack of affinity between both components. Discontinuities at the interface generally lead to cracks and thus low deformation at break and poor barrier properties. Over the years, several strategies have been addressed to get better interaction between both polymers and thus obtain completely biodegradable materials with desirable properties. Among them, the adjustment of processing conditions, the extent of disruption/plasticization of starch, the botanical source and type of starch (different amylose contents), the employment of starch/PHB derivates (physical or chemical modified starches and/or copolymers of PHB) and the use of compatibilizers have been taken into account.

The addition of fillers has also been proposed to increase the polymer compatibility. Depending on the type of filler or the simultaneous use with other additives, differences are obtained in the compatibility of the composite's components. For example, Garrido-Miranda et al. [95] investigated the influence of clay's incorporation on the properties of the resultant starch/PHB-based composite. The researchers obtained a bio-nanocomposite of PHB, TPS and clay by melt mixing with different concentrations of clay (1 and 5%). They worked with a derivate of montmorillonite (OMMT) modified by surfactants which has the same affinity for hydrophilic and hydrophobic polymers. TEM and DRX analysis indicated that OMMT structures were exfoliated and intercalated. The results showed that when concentrations of OMMT were higher (5%), only layers of clay and no starch granules were observed in the matrix. That could be attributed to the behavior of the clay as compatibilizing and reinforcing agent in PHB/TPS blend, improving the interfacial adhesion between immiscible polymers. This better compatibility of the polymers resulted in higher hardness and elastic modulus of TPS/PHB/OMMT composites with respect to that of TPS/PHB.

2.2.3 Processing of Starch/PBAT Composites

PBAT is a 100% biodegradable aliphatic–aromatic copolyester based on fossil resources. It is obtained by a polycondensation reaction between butanediol, adipic acid and terephthalic acid, using conventional polyester manufacturing technologies. Its chemical structure, shown in Fig. 9, combines the biodegradability and processability characteristic of aliphatic polyesters with the thermal and mechanical properties provided from aromatic polymers [96]. PBAT offers higher physical properties, including flexibility (elongation at break close to 700%), Young's modulus (20–35 MPa), tear resistance and tensile strength (32–36 MPa) than most biodegradable polyesters, such as poly(lactic acid) and polybutylene succinate [62]. Furthermore, its mechanical properties and processing conditions are comparable to those of low-density polyethylene. Therefore, it has become a promising biodegradable material for a wide range of potential applications [62, 97]. PBAT is a semicrystalline

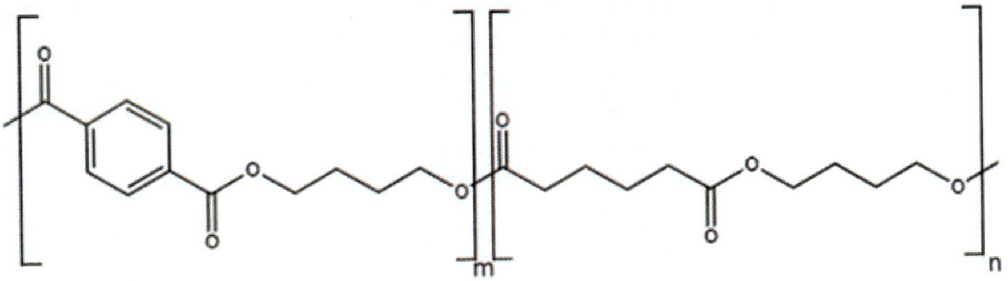

Fig. 9 Chemical structure of PBAT. Figure drawn by the authors

polymer with a thermal behavior well characterized. A crystallization exotherm is usually observed around 55–60 °C, and the melting temperature ranges from 110 to 140 °C. PBAT glass transition temperature is reported to be between −30 and −15 °C, depending on the testing conditions [64, 96]. The moderate crystallinity and good thermal stability allow PBAT to be melt-processed by conventional techniques such as injection molding, extrusion and compression molding [61]. As reported by Dammak et al. [56], PBAT presents short molding cycles and good processability at high extrusion speeds. Typical processing temperature profile ranges from 110 to 160 °C according to the formulation, the processing technique and the characteristic of the neat PBAT. Special attention must be given to moisture removal before processing to prevent the hydrolytic degradation of PBAT. When processing biopolymers at high temperature and shear stress, the presence of moisture can accelerate polymer chain scission and promote downstream processing issues [61, 98]. The degradation routes of biodegradable polymers are generally related to a depolymerization step due to the action of external driving forces (temperature, mechanical stress, radiations, etc.) and the subsequent reactions of the radicals generated with the environment. As reported by Ferreira et al. [62], two main degradation routes have been proposed for PBAT. One considers the enzymatic action of microorganisms such as bacteria, fungi and algae in the disposal environment. The other one is based on a depolymerization process promoted by a no-enzymatic reaction (e.g., chemical hydrolysis and thermal degradation) and the subsequent assimilation and metabolization of the generated intermediates by the present microorganisms. The downsides of PBAT are its poor tensile strength, low modulus, low barrier properties and especially its relatively high production cost when compared with most conventional plastics. Hence, its biodegradability alone is not enough for consumer acceptance and makes it difficult for its large-scale application [54, 62, 65, 99]. These limitations might be addressed by blending with other biodegradable, more cost-effective polymers, and/or adding biobased fillers to design biocomposites, without compromising its biodegradability [54].

PBAT is presented as a proper candidate to combine with TPS, improving TPS toughness and water resistance while maintaining its biodegradability performance [96]. According to Manepalli and Alavi [50], blending starch (0.20–0.40 $/lb) with high-cost polymers like PLA and PBAT (1.50–3.00 $/lb) can be competitive in cost with respect to commodity plastics. Biodegradability degrees greater than 90% for PBAT and TPS systems were obtained by various authors. However, the rate of CO_2 production varies depending on the composition of the mixture. This variation is related to the components added to the TPS/PBAT mixture, the developed morphology and the strength of the interphase between the polymers [56].

Simple physical blending leads to the deterioration of mechanical properties caused by the incompatibility between phases, limiting the maximum TPS content in the blend to approximately 25–30 wt% [100, 101]. Hence, the major problem of the TPS/PBAT blends is the poor interfacial adhesion between the hydrophobic PBAT and the hydrophilic TPS [102, 103]. Extensive efforts have been made to lower the interfacial adhesion between TPS and PBAT, and a variety of compatibilizers, including low molecular and macromolecular compatibilizers, have been studied

[64, 102] The reactive extrusion of TPS with polyester in the presence of maleic anhydride, citric acid and tartaric acid, as well as low molecular coupling agents, has been investigated. They have been used to improve the plasticization process and to increase the compatibility between TPS and other polymers [100, 102, 104, 105]. Maleated PBAT, maleated TPS and epoxy additives have also been used to improve PBAT/TPS blend performance [106]. In all of these compatibilizers, the anhydride group of maleic anhydride, the epoxy group of glycidyl methacrylate (GMA) and the carboxyl group of acrylic acid, tartaric acid and citric acid are efficient to improve the mechanical properties of PBAT/TPS blends by enhancing the interfacial adhesion between the two phases [56].

Citric acid (CAc) is an organic tricarboxylic acid present in most fruits, especially citrus fruits such as lemon, orange and tangerine. Its molecular formula is $C_6H_8O_7$. The multi-carboxylic structure and low molecular weight make possible its use as plasticizer, cross-linking agent, hydrolytic agent and compatibilizer in polymer formulations. Several authors reported the effect of different contents of citric acid on TPS/PBAT blend properties [56, 100, 97].

Maleic anhydride (MA) is an organic compound, the acid anhydride of the bifunctional maleic acid. The anhydride group has been extensively used as an in situ or ex situ coupling agent for immiscible polymer blends due to its high reactivity with water, alcohols and amines. Maleic anhydride can react with amine and hydroxyl end groups (e.g., of polyesters) to yield the desired graft copolymer necessary for compatibilization. The MA-modified polymers can then be employed to compatibilize polyesters (e.g., PET and PBT) with other polymers due to the potential of anhydride reaction with terminal hydroxyl groups. Methods for grafting MA into polymer backbone have been studied as a technique to promote hydrophilicity, adhesion, dyeability, functionality for cross-linking and other chemical reactions in polymer development. In addition, the MA grafting has been utilized to promote compatibility between polymers and polymer/fillers [107]. Fourati et al. [100] studied the incorporation of citric acid, maleic anhydride and maleated PBAT, in PBAT/starch formulations and their effect on the mechanical properties, rheological behavior at a solid and molten state, the co-continuity and the morphology of the phases. In the work of Dammak et al. [56], the authors continued their investigation concerning the effect of MA and maleate PBAT (PBAT-g-MA) on the mechanical, rheological and biodegradability of PBAT/TPS blend with a TPS content ranging from 40 to 60%.

Tartaric acid (TA) is a dicarboxylic acid extracted from plants and fruitages, widely used as an edible acid derivative in pharmaceutical, food and general industrial chemical fields [108]. As for citric acid, it is reported that TA acts as a coupling agent by promoting esterification (grafting) and transesterification reactions (cross-linking) with polymers, improving the compatibility between them. Its effect on promoting the acid hydrolysis of starch chains has also been studied. For example, Zhang et al. [108] proposed a new facile strategy to reduce the TPS shear viscosity and improve compatibility with PBAT by adding increasing contents of tartaric acid. In their work, Olivato et al. [109] explored the use of TA as a compatibilizer of starch/PBAT blends obtained by a one-step reactive extrusion process based on a constrained mixture design.

A study of the torque recovery and polymer degradation for a TPS/PBAT blend with up to 30% of TPS, processed in a laboratory mixer with and without a commercial chain extender additive as a compatibilizer, was conducted by Marinho et al. [110]. They proposed that the compatibilizer action of the oligomer used as a chain extender is due to the presence of epoxy and methacrylate residues that can interact with the polymeric system. As an alternative for conventional macromolecular compatibilizers, Garcia et al. [111] studied the effect of using sericin (at low concentrations) in the performance of starch–PBAT blown films. Sericin protein is obtained as a by-product during silk production from the *Bombyx mori* silkworm. Its molecular weight ranges from 24 to 400 kDa, and the predominant amino acids are serine (40%), glycine (16%), glutamic acid, aspartic acid, threonine and tyrosine. Due to its chemical characteristics, sericin is reported to be used as an additive or adjuvant in polymer blends [111].

Different strategies have been proposed to process TPS/PBAT blends with the inclusion of different compatibilizers and coupling agents. The preparations of TPS/PBAT/compatibilizer (CA, MA and PBAT-g-MA) blends studied by Fourati et al. [100] first involved obtaining TPS from potato starch mixed with 20 wt% glycerol using a dough mixer (40 rpm, 10 min). The starch–glycerol blend was then extruded twice on a single-screw extruder (L/D 28:1; 200 rpm, temperature profile 110–110–120–120–130–130–120 °C from hopper to die). Before mixing with PBAT, TPS pellets were dried overnight at 70 °C. TPS and PBAT granules (Ecoflex F Blend C1200; TPS60/PBAT40) and citric acid (2 and 4%) were manually fed and extruded twice to improve the dispersion of the blend components at a screw speed of 120 rpm, with a temperature profile 120–130–145–150–150–140–130–120 °C. A flat die was used to obtain films with a thickness between 0.3 and 0.2 mm. Blends with MA (2, 4 and 6 wt%; based on the whole TPS/PBAT blend) were processed as described above for CAc. PBAT-g-MA was prepared by radical grafting in the melt of PBAT with 2% MA and benzoyl peroxide (BP) (1%) as a radical initiator, using a batch mixer (140 °C, screw speed of 45 rpm, 20 min). The PBAT-g-MA was pelletized and used without further purification. The extent of maleation was around 0.55%. Using rheological measurements, they found that the rheological behavior of the TPS60/PBAT60 blend showed a typical viscous-like behavior (G'' higher than G') and followed that of PBAT, where PBAT was the continuous phase and the TPS the dispersed one. When TPS/PBAT was processed in the presence of CA, the authors found a rheological behavior typical of a gel-like material with G' higher than G'' and nearly frequency-independent. The authors investigated the rheological properties of TPS-CA 2% blend and found that both the viscosity and G' decreased by one order of magnitude in comparison with the neat TPS. They explained the drop in viscosity, as other authors did, due to the capacity of CAc to disrupt starch inter- and intramolecular interaction and also a certain degree of starch acidolysis. The rheological behavior of TPS/PBAT/MA, at 150 °C, was also typical of a gel-like material. As the content of MA increases, a decrease of both G' (G'TPS-PBAT-2MA > G'TPS-PBAT-4MA > G'TPS-PBAT-6MA) and complex viscosity (η*) was observed. The SEM and co-continuity analysis showed a direct correlation with the rheology results. TPS/PBAT/MA 2% presented a co-continuous structure, while at 4% MA, a phase inversion occurred (as was evidenced by SEM) with TPS being the matrix, and PBAT as droplets, the dispersed phase. In accordance with SEM results, the authors observed that the rheological behavior of TPS/PBAT-g-MA/PBAT was similar to that of the TPS/PBAT blend, dominated by a liquid-like character, where PBAT was the continuous phase and TPS the dispersed one, although PBAT represented the 40 wt% of the total weight of the blend.

Dammak et al. [56] prepared PBAT-g-MA using the same procedure and processing conditions as Fourati et al. [100]. However, the MA content grafted was 1.5%. The materials used in the work of Dammak et al. [56] were similar to the ones employed in the previously published paper [100]. However, the method of preparing the blends differs from that presented in the previous study. In their work, TPS was prepared by mixing starch with glycerol at a ratio of 80/20 wt% and 10% of water was added. Water acts as a plasticizer for starch, so the plasticizer content in this blend is higher than in the previous work [100]. Then, the starch–glycerol–water mixture was extruded in a twin-screw extruder (L/D 38:1, screw speed 100 rpm, temperature profile 130–130–130–130–135 °C) instead of the dough mixer and the two-extrusion process on a single-screw extruder previously used. Films from TPS (content 40, 50 and 60%), PBAT granules and the compatibilizer (MA and PBAT-g-MA, 2%) were extruded using the same twin-screw extruder. Films were obtained with a flat die set at the end of the extruder (thickness range 0.2 and 0.3 mm).

Zhang et al. [108] obtained TPS modified with TA by premixing dried corn starch (Langfang Starch Company) and glycerol (70:30 w/w) for 5 min with the incorporation of 0.5, 1, 2 and 4% TA (weight based on corn starch). All components were mixed for 10 more minutes. Mixtures were kept for 30 min at 80 °C, and afterward melt compounded using a twin-screw extruder (L/D 40:1) at a screw speed of 80 rpm. The temperatures of the extruder, from the feed zone to die end, were 90–120–130–140–145–145–140–135 °C. The effect of TA contents on the properties of PBAT/TPS-TA blends was studied by melt compounding TPS-TA mixtures with PBAT (TH-801, Xinjiang Blue Ridge Tunhe Chemical Industry Joint Stock Co.). TPS-TA mixtures were dried at 80 °C for 24 h and extruded with PBAT to obtain PBAT30/TPS-TA70 blends using the same equipment. The temperatures of the extruder from the feed zone to die were 140–140–150–150–155–155–160 and 150 °C, and screw speed was kept at 70 rpm. TPS and PBAT/TPS were performed at the same procedure and were considered as control materials. All PBAT/TPS-TA blends displayed a shear thinning behavior at high frequency, with complex viscosity (η^*) and storage modulus (G') of PBAT/TPS-TA higher than those of PBAT in the whole frequency region. However, η^* and G' of PBAT/TPS-TA decreased as the TA content increases from 1 to 4%, confirming that the compatibility of PBAT/TPS-TA at levels up to 1% is better than at higher contents and thus might act as a coupling agent at low content. When TA's content was higher, η^* and G' of PBAT/TPS-TA decreased, suggesting its lubricant action.

The effect of TA on the properties of a starch/PBAT blend was studied by Olivato et al. [109] using a constrained mixture design. TA proportions used ranged from 0 to 1.1%. The maximum proportion of glycerol was set to 12.0%, and the third component was a mixture of starch and PBAT with a 55:45 proportion between the phases. Native cassava starch and PBAT (BASF) were used. The formulations were processed using a laboratory single-screw extruder (L/D 28:1), with a barrel temperature profile of 100–120–120–120 °C and the screw speed set to 40 rpm. Pellets were obtained afterward. In a second step, blown films were obtained using the same equipment with a barrel temperature profile of 100–120–120–130 °C and 130 °C for the 50-mm film-blowing die, with a screw speed of 40 rpm. The film thickness was maintained between 80 and 100 μm.

The incorporation of an oligomer with epoxy and methacrylate residues (commercial name Joncryl PR010) as chain extender and compatibilizer, in blends of PBAT with 10%, 20% and 30 wt% of TPS, was studied by Marinho et al. [110]. PBAT (Ecoflex® F Blend C1200) and thermoplastic starch (TPS Beneform 4180, Ingredion) were compounded in a laboratory internal mixer at rotor speed of 60 rpm, and chamber wall temperature kept constant at 140, 170 and 200 °C (total processing time of 15 min). PBAT and TPS were also subjected to the same process, and 1% Joncryl additive was added after 10 min without interrupting the process. The chain extender is recommended to compensate for degradation during processing in polyesters and polyamides.

Films of starch/PBAT/glycerol with increasing contents of sericin (0.5, 1.0 and 1.5 wt%) were obtained by Garcia et al. [111] using a single mixing step followed by blown-film extrusion. Cassava starch (Indemil, Brazil), PBAT (Ecoflex® S BX 7025, BASF), glycerol and sericin (extracted from silkworm cocoons, *Bombyx mori*) were manually mixed and extruded in a twin-screw extruder (L/D 35:1) to obtain pellets. The screw speed was 100 rpm, and the temperature profile used was 90–120–120–120-120 °C. In a second step, pellets were extruded using a mono-screw extruder (L/D 26:1) with a 50-mm film-blowing die. The temperature profile was 90–120–120–130 °C and 130 °C at the die, and the screw speed was 40 rpm. The control formulation contained 61, 26 and 13 wt% starch, PBAT and glycerol, respectively. As the authors' preliminary results showed that sericin could perform as a plasticizer, its addition replaced glycerol content in every formulation.

As it was mentioned before, in some cases adding fillers to polymer blends results in an improvement of the interfacial adhesion between the polymer phases caused by preferable filler localization at the interface [65]. As it was previously mentioned, the addition of low-cost materials (organic and inorganic fillers) as natural reinforcing agents is an effective way to improve the starch/PBAT properties and decrease their final cost, while maintaining (but preferably accelerating) the inherent biodegradability of the matrix. However, some concerns related to common problems of filler dispersion, the interaction between filler/matrix and reduction filler content must be overcome [62, 112]. Several factors, such as processing conditions, the miscibility and viscosity of the polymer phases, and the composition of the final mixture, will define the composite's performance.

The enhanced mechanical, thermal and barrier properties of a composite are related to the formation of strong interactions between the filler and matrix. Thus, when some external stress is applied, part of the energy can be absorbed by the filler, and part of it can be dissipated by frictions between particle–particle and particle–polymer interaction through the interphase. The uniform filler dispersion within the starch/PBAT and the improved interactions between the filler and polymer matrix better dissipates the energy throughout the matrix.

The final properties of starch/PBAT and TPS/PBAT blends depend on several factors. The most important ones are associated with the structure of the starch in blends, influenced by the origin of the native starch, the presence of plasticizers, processing aids, use of compatibilizer agents, among others. The parameters used to obtain the thermoplastic starch and the way in which it is mixed with PBAT will influence the morphology and properties of the compound obtained [97, 101]. Different microstructures and component distributions in the starch/PBAT or TPS/PBAT blend can be achieved according to the process and processing conditions.

In order to improve the interaction between fillers and PBAT/starch (or PBAT/TPS) matrices and the filler dispersion, different strategies were proposed [62, 63]. For example, Liu et al. [103] proposed the incorporation of MA and poly(ethylene-co-vinyl alcohol) (EVOH) to improve the properties of starch-based nanocomposites/PBAT blends using nano-SiO_2 as a reinforcing agent. In their work, Zhai et al. [97] proposed a single-step process to obtain starch–glycerol/PBAT nanocomposite containing an organically modified montmorillonite as a filler and citric acid as a compatibilizer. The effect of the starch content on the morphology, mechanical properties and hydrophobicity of blown films were investigated.

The impact of different compatibilizers, including fillers, on the morphology of the TPS/PBAT blends was evaluated by several authors. Fourati et al. [100] studied by SEM the morphology of cryo-fractured cross section of TPS/PBAT (60/40 wt%) films, processed with 2 and 4% of citric acid as coupling agent, on samples with and without extraction with HCl solution (TPS extracted) and CHCl3 (PBAT soluble). Also, the continuity index was calculated as

$$\%\text{continuity} = \frac{W_{\text{init}} - W_{\text{end}}}{W_{\text{init}}} \times 100,$$

where W_{init} corresponds to the weight of PBAT or TPS before the solvent extraction step and W_{end} was the weight remaining after extraction. For TPS-PBAT (60/40), in the absence of a compatibilizer, the continuity index was about 100 and 20% for PBAT and TPS, respectively. This means that the PBAT formed the continuous matrix phase with TPS being the dispersed phase, despite its volume fraction exceeding that of PBAT. SEM images confirmed a dispersed morphology in which TPS appears as dispersed elongated droplets in the PBAT matrix (TPS droplet size between 1 and 5 μm). The presence of holes and gaps in the interfacial area between the matrix and the dispersed phase, observed in the image after the cryofracture breaking of the film, indicated a weak interfacial adhesion between TPS particles and PBAT matrix. For TPS/PBAT/2% CAc, the SEM images showed a co-continuous morphology, consisting of interconnected TPS structures with fiber-like morphology (observed in SEM image after film treatment with HCl solution). SEM micrograph for blend processed in the presence of 4% CAc showed a continuous phase of TPS and PBAT dispersed (spherical particles with a size between 2 and 4 μm). In the presence of CAc as a compatibilizer, TPS/PBAT 60/40 films totally disintegrated in water within several minutes, especially for 4% CAc. For TPS/PBAT/2% MA, SEM images showed a co-continuous morphology, where TPS structures were interconnecting with a fiber-like morphology. This effect is highlighted in these mixture morphologies more than in those prepared with CAc. The continuity test and SEM images of TPS/PBAT/4 and 6% MA showed a dispersed morphology in which the PBAT is dispersed (droplets) in a TPS matrix. This was confirmed by the appearance of voids in the SEM micrograph of the cross section of the film after the extraction with chloroform. In addition, after immersion in the HCl solution, the film fully disintegrated, and PBAT particles (size between 1 and 4 μm) were observed. As for CA, TPS/PBAT 60/40 MA 6% totally disintegrate in water within several minutes. For the blend TPS/PBAT-g-MA/PBAT (TPS/PBAT 60/40) obtained by Fourati et al. [100], in the presence of 2% PBAT-g-MA, the morphology resembles that of TPS/PBAT (PBAT formed the continuous matrix phase with TPS being the dispersed phase). However, they observed that during cryogenic breaking for SEM evaluation a lesser number of TPS particles were pulled out and were less visible in the micrograph, compared to TPS/PBAT. They explained these results as an indication of improved interactions between TPS and PBAT when PBAT-g-MA was used as a compatibilizer.

Low interfacial adhesion and poor compatibility between TPS and PBAT were reported by several authors [108]. TPS-TA/PBAT blends presented small particles of TPS-TA phases dispersed in the PBAT matrix without agglomeration and good wettability between phases, for TA content range from 0.5 to 2%. However, the increase in the TA content from 2 to 4% leads to the phase morphology of TPS-TA/PBAT 4% with TPS-TA particles agglomerated and showing low interfacial adhesion and poor compatibility. As reported by Zhang et al. [108], increasing contents of TA in TPS mixtures improve its processability. An important reduction of the melt viscosity was reported as the melt flow index increased from 0.68 to 15.7 g/10 min when the TA contents increased from 0.5 to 4%. Melt viscosity decrease was associated with the acid hydrolysis of starch macromolecular chains observed from the viscosity-average molecular weight measurements. TA can promote the hydrolysis of starch during processing in the presence of glycerol and water vapor. FTIR and titration measurement analysis of TPS-TA presented by Zhang et al. [108] evidenced that TA does not esterify TPS or esterifying reaction degrees were very low during the extrusion process (without an external catalytic agent) and proposed that TA could be acting as a lubricant agent.

Contrary to Zhang et al. [108] results, Olivato et al. [109] demonstrated that TA was able to promote esterification reactions with the OH groups of starch (characterized by FTIR and solid-state 13C cross-polarization/magic angle spinning nuclear magnetic resonance). These esterification and transesterification reactions were responsible for the compatibilizer effect observed in starch/PBAT-TA blends. The addition of TA changes the morphology of starch–PBAT blends showing a reduction in the interfacial tension between starch and PBAT and promoting more homogeneous structures.

Olivato et al. [109] and Olivato et al. [113] published SEM images that showed a good dispersion of sepiolite in TPS80/PBAT20-sepiolite 5%, and they found no effect of nanofiller content on the morphology of the blends. TPS50/PBAT50 with and without sepiolite showed a round TPS dispersed phase. A phase inversion was observed for TPS80/PBAT20 samples with and without nanofiller, where PBAT was the dispersion phase.

Morphology of starch/PBAT/glycerol films obtained by Garcia et al. [111] presented a fibrillar structure with effects that indicate chain orientation promoted by the blown-film process. They also reported the absence of starch granules, demonstrating that the process was adequate to enable the disruption of the starch granular structure. Sericin addition at 1.0 and 1.5 wt% resulted in more homogenous and compact structures, suggesting a cohesive effect when incorporated into polymer matrices.

Lendvai et al. [114] obtained nano-biocomposites using native maize starch, PBAT (Ecoflex® F Blend C1200, BASF), natural bentonite (BT) and an organically modified montmorillonite (OMMT, Cloisite 30B). They reported the presence of a quasi-continuous phase when 40 wt% of TPS was added to PBAT, with some less plasticized starch domains as a second phase (size range of 5–15 μm). EDS mapping for Si in 40 wt% TPS nanocomposites/PBAT blend revealed a continuous phase for TPS, with a uniform distribution of bentonite (BT). As expected from the compounding method described by the authors, BT nanoparticles were predominantly located in the TPS phase and showed an intercalated morphology when 40 wt% of TPS nanocomposite was incorporated into PBAT. The intercalated morphology progressed further with the increasing content of PBAT. These results indicated that a decrease of the melt viscosity during compounding of blends with higher PBAT content could promote the disintegration of the BT agglomerates and facilitate their dispersion on the polymer matrix. SEM micrograph for TPS-OMMT/PBAT was not presented by the authors in the discussion.

In the work of Nunes et al. [55], the effect of processing temperature and TPS contents on the rheological and mechanical performance of TPS/PBAT blends and TPS/PBAT biocomposite reinforced with babassu mesocarp was reported. Regardless of TPS content in the blends, they described that immiscible morphologies were obtained, with starch granules of heterogeneous sizes dispersed in the PBAT matrix. TPS/PBAT SEM micrograph showed poor adhesion between phases. When TPS content was increased from 10 to 30%, an increase in the agglomeration of TPS particles was observed. Biocomposite SEM micrographs also showed partial adhesion between matrix and fibers. These results are consistent with the mechanical properties reported. Blend morphology did not present significant differences as a function of the processing temperature.

In the work of Liu et al. [103], SEM micrographs showed low compatibility for the system TPS/PBAT (TPS particle-like could be found). In their work, only the influence of the compatibilizer in the TPS/PBAT blend was evaluated. When MA or plasticized EVOH (pEVOH) was used as compatibilizer, a co-continuous phase was formed. The continuous phase showed a more homogeneous structure when a combination of MA + pEVOH was added to TPS-nSiO_2/PBAT composites. The XRD results for the samples confirmed the absence of significant changes in the crystalline structures of starch and PBAT with the addition of nano-SiO_2 and the compatibilizers. The structure and morphology of TPS-nSiO_2 nanocomposite were not presented in their results.

X-ray diffraction analysis was used by Zhai et al. [97] to evaluate the intercalated or exfoliated structures of hydroxypropyl distarch phosphate HPDSP/PBAT/nCOM nanocomposite. The nanocomposite XRD pattern showed diffraction peaks that indicated an intercalated clay nanostructure. As the PBAT content increases, polymer chains were intercalated into nanoclay organically modified (nCOM), and the d-spacing values increase from 3.46, 3.61, 3.95, 4.16 and 4.27 nm, for PBAT contents 10, 20, 30, 40 and 50%, respectively (nanoclay d-spacing 2.76 nm). The authors proposed that the PBAT has better affinity with the hydrophobic nCOM, promoting an improvement in the fluidity of the blends and facilitating the polymer chains to enter the nanoclay interlayer during extrusion processing. He et al. [115] found that nCOM has better dispersion in PLA/PBAT two-phase blends than pure PLA or PBAT as d-spacing and relative intercalation was higher than in PLA or PBAT nanocomposites. In addition, they found that d-spacing was greater when PBAT was the continuous phase in the composite blend. They explained this result due to better compatibility of the clay organic modifier with PBAT. SEM micrographs showed smoother surfaces in HPDSP/PBAT/nCOM nanocomposite as PBAT content increased. For blends with 10% and 30% PBAT, the film's surfaces became less rough and exhibited better structural integrity than the film from 100% modified starch HPDSP. When the PBAT content was equal to 50 wt%, the surface structure of the film exhibited a smooth, dense and uniform appearance. No cracks, wrinkles or irregularities were observed on the surface of the HPDSP/PBAT 50/50, and also, the films exhibited a coarser and ductile fracture surface, probably due to a better interaction between the two polymers. The role of citric acid as a compatibilizer for HPDSP and PBAT was reflected and referenced in the work of Zhai et al. [97] through the FTIR spectra analysis, although as the peak at approximately 1713–1716 cm^{-1} (carbonyl stretch of the ester group) originally present in PBAT did not allow to find evidence of grafting and/or cross-linking reaction of the HPDSP chains, either the esterification or interesterification reaction promoted by the compatibilizer. They mentioned that these reactions are feasible to take place due to the temperatures used in the extrusion process, as was referenced by the authors.

3 Properties of Starch-Based Green Composites

3.1 Properties of Starch Matrix Composites

3.1.1 Thermal Properties of Starch Matrix Composites

Three characteristic thermal transitions may exist for such semicrystalline polymers as starch: a glass transition for the amorphous fraction (T_g); a thermal transition for the melting of crystallites (T_m); and a transition due to crystallization (T_c). Starch is partially miscible with its most used plasticizer: glycerol. For that reason, TPS films plasticized with glycerol generally have two glass transition temperatures, one associated with the plasticizer-rich phase ($T_{\alpha 1}$) and the other with the starch-rich one ($T_{\alpha 2}$). The presence of nano-/microfillers can change this behavior through interactions with the matrix and/or the plasticizer. It has been demonstrated, for example, that waxy starch nanocrystals affect preferably the mobility of the starch-rich phase, due to strong hydrogen bonding interactions between them and the amylose/amylopectin chains of the matrix, and therefore a more remarkable shift in $T_{\alpha 2}$ is obtained [47]. It should be noted that other plasticizers could lead to different behaviors. In the case of sorbitol, for example, as it is miscible with starch, only one T_g is presented. Other possible plasticizers are currently being studied, like the use of D-isosorbide, 1,3-propanediol and deep eutectic solvents [47, 52]. In both cases, two glass transition temperatures are observed.

The effective attachment of TPS to fillers can therefore constrain the segmental motion of TPS chains by strong hydrogen bonding interactions, increasing the material's T_g. Ghanbari et al. [43], for example, found that the T_g of neat TPS film (37 °C) increased 11.5 °C with the addition of 1.5 wt% of cellulose nanofibers and ascribed this difference to the strong matrix–filler adhesion. A similar effect was found by Nessi et al. [15] for cellulose nanocrystals (CNCs)–starch nanocomposites, who proposed that CNCs may act like a junction and promote the intermolecular interaction of starch chains reducing their relaxation. In this case, the T_g of the starch-rich phase shifted from 68 °C for the TPS matrix to 75 °C for the composites with the inclusion of 2.5, 5 and 10 wt% of CNCs, regardless of the filler content.

On the other hand, the thermal degradation of TPS occurs as a three-step process. In the first stage, the initial mass decrease is in the range 60–120 °C and is related to the evaporation of water. In the second one, between 180 and 270 °C the plasticizer thermal degradation of thermoplastic starch usually occurs. Finally, the third step in thermal curves, from 280 to 350 °C, corresponds to starch degradation. All these stages may be modified due to filler addition. The first stage will depend on the possibility of fillers to modify films' moisture content. Some fillers could interact with the plasticizer, retaining it or allowing easy evaporation [16, 116, 117]. In those cases, changes in the second step are expected. Ochoa-Yepes et al. [42] showed that protein addition into TPS led to a slightly lower mass loss of the composites in the second step with respect to the matrix. When fillers with higher thermal stability than starch are incorporated into TPS matrices and many interactions occur between the polymer and the filler, a slight increase in the composite thermal stability is usually observed, due to the increasing dissociation energy. This means a delay in the decomposition of the material and has been demonstrated for gelatin [45], cellulose nanofibers [43], cassava and ahipa peels and bagasse [118], SiO_2 [103] composites, among others.

The importance of the matrix–filler interaction was demonstrated by Fazeli et al. [119], who studied TPS and TPS composite films reinforced with cellulose fibers (CFs) and plasma-treated cellulose fibers (PCFs). They found that the main decomposition temperature of the neat TPS and TPS/CF 6 wt% composite was about 260 °C, whereas the one obtained for the TPS/PCF 6 wt% composite raised to approximately 284 °C. The achieved improvements are attributed to the fact that the PCF compounded into the TPS matrix strongly, making decomposition of the composite arduous. Hence, better thermal stability was obtained with PCF as the reinforcement.

3.1.2 Gas Barrier Properties and Water Sensitivity of Starch Matrix Green Composites

Many potential applications of starch-based materials need specific gas barrier properties, especially in the packaging field, where a product needs to be protected from the outside. Moreover, food packaging materials need the presence of specific atmospheric conditions to maintain the freshness and quality of food during storage [120]. In particular, water vapor permeability (WVP) and oxygen permeability (OP) are the two most studied gas barrier properties in the literature. In general, starch-based films show high WVP values, whereas they present better behavior against nonpolar molecules such as O_2 and CO_2, because the transport of gas molecules depends on both diffusion and solubility coefficients. The WVP of starch-based films is mainly governed by the high ability of water molecules to interact and penetrate through them because of the strong starch/water affinity and is little influenced by the type of plasticizer used. González et al. [47] showed that WVP values of starch films plasticized with glycerol and D-isosorbide had no significant differences. However, the OP was found to be strongly influenced by the type of plasticizer used. An OP value near twenty times lower for the D-isosorbide plasticized film comparing to that obtained for glycerol-plasticized one was observed. They attribute this significant decrease to the reduction of oxygen solubility due to the hydrogen bonding interactions between starch and D-isosorbide, saturating the sorption sites. On the other hand, Battegazzore et al. [121] studied the influence of ambient humidity in oxygen transmission of TPS films prepared with isosorbide as a plasticizer. They found no significant variations of oxygen permeability till 75% RH. However, over this RH, an exponential increase was found. From these results, it can be inferred the need to take into account both the plasticizer used for the development of TPS and the ambient humidity at which the material will be tested, especially for packaging applications.

The incorporation of micro- and nanofillers has proven to be highly effective in improving the barrier properties of starch films. In many cases, its success is associated with the introduction of a tortuous pathway for gas molecules to pass through. In this context, the filler shape plays a very important role in the improvement of barrier properties. Most effective fillers are usually those with platelet shape, which can create a sort of winding road hindering and delaying the passage of gases and water [13]. Most common nanofillers with this shape are clays and some starch nanocrystals. González et al. [47] compared the OP of TPS films developed from maize starch, glycerol, waxy starch nanocrystals (WSNCs) and CNC. A reduction in the OP value was found for the TPS nanocomposite containing 1 wt% of WSNC, from (108 ± 35) to (43 ± 10) $cm^3\ \mu m\ m^{-2}\ day^{-1}\ kPa^{-1}$ for the matrix and the nanocomposite, respectively. The incorporation of 1 wt% of CNC also decreased the permeability value, but the improvement was less notorious, being the OP value (70 ± 6) $cm^3\ \mu m\ m^{-2}\ day^{-1}\ kPa^{-1}$. The greater effectiveness in the decrease of permeability against O_2 molecules achieved by the WSNC was attributed to their platelet-shaped morphology, which gave them the capacity to generate a tortuous path for the O_2 molecules, decreasing the diffusivity. Significant improvements in permeability to both oxygen and water vapor thanks to the incorporation of platelet-shaped fillers were found by other authors [122, 123].

As it has been known for many years, the morphology of the fillers is not the only factor to consider. The way in which they are oriented within the matrix is also determinant for the produced effect in barrier properties. If the largest surface of each nanoparticle is oriented perpendicular to the direction of the gas diffusion or permeation, an optimal decreasing effect on permeability will be achieved. However, it is an extremely challenging task to achieve fillers' regular arrangements within TPS matrices, specially through industrially scalable methods. In general, the obtained composite materials have randomly oriented fillers, diminishing the impact on the decrease of permeability [124].

Possible interactions between filler and matrix components could also highly influence the composite's permeability due to the conformation of different interface morphologies. A defect-free TPS–filler interface is difficult to achieve, as the incorporation of filler particles usually modifies the properties of the neighboring polymer phase. Therefore, depending on the TPS–filler adhesion, various structures can be observed at the interface. If voids are formed in the filler–matrix interface, an increase in gas permeability can be observed, as the formation of these defects allows the gases to pass. This can occur due to poor adhesion, polymer packing disruption in the vicinity of the dispersed particles, repulsive force between the two phases or different thermal expansion coefficients, among other possible reasons. On the contrary, when the adhesion between the two phases is good, polymer chains located near the filler surface are rigidified, due to a reduction of the polymer-free volume in this region. This interfacial defect usually leads to a decrease in permeability [125, 126]. For example, while platelet-shaped starch nanocrystals can strongly reduce the permeability of cassava starch–glycerol matrices [127], the effect of these fillers is the opposite in waxy corn starch matrices. García et al. [128] proposed that this important difference was due to unlike nanocrystal distribution and interface composition. While in the case of mandioca starch matrices the nanocrystals were well distributed with excellent matrix–filler adhesion, for waxy maize starch matrices the nanocrystals were glue–glycerol bonded, forming threads with high concentration of OH groups and thus forming a preferential path for water vapor diffusion through the nanothreads.

The internal microstructure of the final composite is therefore a key aspect in its permeability properties. Whether generating impediments to the diffusion of molecules or generating pathways through which the material can penetrate thanks to chemical affinities, this property will be modified by the inclusion of fillers. Besides filler–matrix interactions, the used processing methodology can be highly influential, as due to the filler's inclusion voids and structural defects in the matrix can be created during processing. Florencia et al. [118] studied cassava and ahipa peels and bagasse as potential fillers of TPS films. They found that all TPS composite films presented WVP values similar to those of the TPS matrix. This behavior was attributed to two offset factors. On the one hand, the filler presence increases tortuosity within the film matrix and therefore reduces water vapor diffusion. On the other hand, the processing method could promote the generation structural defects in the matrix facilitating the water molecule transport. The processing of the composite could also influence in the final water content of the composite and therefore its WVP. Ochoa-Yepes et al. [42] compared WVP of starch/lentil protein composite films obtained by extrusion/thermocompression and casting techniques. Extruded samples had lower WVP values tan that obtained by casting. This is consequence of a difference in the materials with water content. The extrusion/thermocompression process involved less water and higher temperatures than casting methodology, leading to materials with lower water contents. As water plasticizes TPS, it reduces internal hydrogen bonding between polymer chains, increasing molecular volume. It is therefore expected for materials with higher water content to have greater diffusion coefficient of water vapor and therefore to experiment increased in WVP.

Another possible strategy to improve the water vapor permeability of starch-based materials is the inclusion of hydrophobic agents, such as lipids. In general, the addition of this type of compounds to hydrophilic film matrices decreases the WVP due to the promotion of hydrophobicity and increases the OP values due to the greater oxygen solubility in the hydrophobic regions of the matrices [129]. In these cases, the addition of emulsifying or compatibilizing agents or a previous encapsulation process is mandatory to improve compatibility. Talón et al. [35] studied the effect of incorporating free or spray-dried encapsulated eugenol on the barrier properties of compression-molded corn starch composite films. Films containing non-encapsulated eugenol showed higher WVP and OP than both pure starch films and films incorporating microencapsulated eugenol. When non-encapsulated eugenol is incorporated into the starch matrix, starch–eugenol complexes are formed, in which the hydrophobic cavity of the helical conformations of amylose and amylopectin chains are involved. These complexes could limit eugenol's active role at reducing the matrix's water affinity, reaching a higher equilibrium water content which enhanced all the diffusion-dependent processes, such as water vapor or gas permeation. When microencapsulated eugenol was added to the starch films, the WVP values and water content decreased as expected for the incorporation of more hydrophobic components. Octenylsuccinated starch (OS) can act as an emulsifier for oil-in-water (O/W) system, and therefore it can be used to prepare starch–lipid composite films. The process of molecular self-assembly of film-forming components is shown in Fig. 10. Gao et al. [130] prepared corn/octenylsuccinated starch (C/OS) composite films incorporating soybean oil (SO) at 0, 0.5, 1.0, 1.5 and 2.0 wt%. The WVP of control film without SO was 2.93×10^{-12} g cm/cm^2 s Pa, while for the composite films it gradually decreased from 2.72×10^{-12} g cm/cm^2 s Pa to 2.46×10^{-12} g cm/cm^2 s Pa, when the SO concentration increased from 0.5 to 1.5%. A similar strategy was used by Kang et al. [131], who investigated the physicochemical properties of amylose–lipid inclusion complexes and their film-forming capacities. The authors used ultrasonication to promote the complexation between amylose and liquid and solid fatty acids. The WVP values of the films formed from starch–lipid composites were significantly lower than that of the native starch film. This decrease was ascribed to the presence

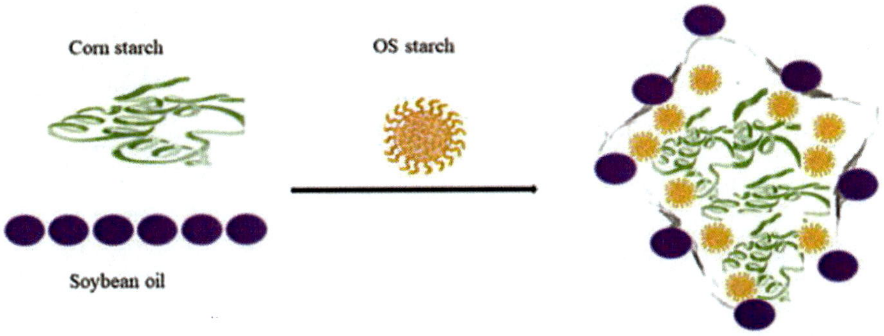

Fig. 10 Process of molecular self-assembly of corn starch, corn/octenylsuccinated starch (OS) and soybean oil. Reprinted from Gao et al. [130], Copyright (2020)

of hydrophobic carbon chains and amylose–lipid complexes in the polymer matrix. When comparing the results obtained with the incorporation of liquid or solid lipids, it was found that the former led to lower WVP values. This result is attributed to the fact that fatty acids in liquid state have shorter carbon chains than solids, so they can be more easily incorporated between the polymeric chains of the matrix, leading to a more compact structure and restricting the water permeability. Parallel, lower WVP values were found for films prepared from ultrasonicated starch–lipid complexes than those formed from untreated ones, demonstrating that the ultrasonic treatment improved the moisture resistance of films. The authors suggested that the ultrasonication process was favorable to construct a more compact and uniform structure in the polymer matrix, thus decreasing the WVP values.

The incorporated amount of filler is also of utmost importance when evaluating its effect on barrier properties of the composite. In the case of oleo-fillers, while very low contents may not be enough to hinder the passage of vapor molecules, high contents could interrupt the hydrogen bonds between the starch molecules, destroy the network structure and expand the molecular interstice of the films, thus leading to higher WVP value [130].

There are different ways to evaluate the water sensitivity of TPS composites. Due to their hydrophilic nature, it is clear that they will be highly sensitive to moisture, and this can be a problem for many applications. Depending on the aspect to be evaluated, it is possible to study the absorption isotherms, the water uptake from the environment, the wettability or directly the solubility of starch-based films. In each case, there are strategies to improve the material stability. Although the inclusion of fillers into TPS matrices cannot completely change its hydrophilic nature, if the right combination is chosen substantial improvements can be achieved.

As it was mentioned before, TPS absorbs water from the environment when the environmental RH increases or loses water while the RH decreases. This change in moisture content of TPS materials leads to modifications in their structure and, therefore in many of their properties, including mechanical, thermal and barrier ones. Moisture sorption isotherms show the relationship between water content of

TPS materials and the environmental RH, at a constant temperature. Although there are many models proposed for the behavior of these materials, no single equation can describe accurately the relationship of equilibrium moisture content and environmental RH for various TPSs over a broad range of RH and temperature, and even less in the case of composites. Therefore, for each TPS-based material, it is necessary to investigate its behavior at different RHs. Several authors used the GAB model to fit starch-based material water sorption isotherms, indicating the monolayer moisture content (M_0) as the most representative parameter [132, 133, 134]. Typically, starch and protein-rich product behavior corresponds to Type II isotherms, according to BET classification. At low RH values, film humidity content increases gradually up to a limit water activity (a_w), after which an strong increment is obtained. This significant increase of equilibrium water content is attributed to a phenomenon called "water clustering" [132].

López et al. [132] studied the influence of talc nanoparticle inclusion in corn TPS films on their sorption isotherms. Films containing talc presented a similar behavior than TPS ones, but particle incorporation reduced water sorption for $a_w > 0.4$. Strong polar interactions among starch, glycerol and mineral edge surface establish a competition mechanism which could explain the reduction of water sorption capacity of composites based on TPS by talc presence. The M_0 parameter was reduced with talc incorporation, but a net tendency with particle concentration was not evidenced. The authors explain this effect as the consequence of the low filler contents used to develop starch-based composites. Similar effects were observed by other authors, indicating that generally polymer matrix governs the sorption mechanism of TPS composites [135].

Other possibility is to study the material's water uptake while it is in a constant RH over time. Samples with the same size and equally stabilized are exposed to constant temperature and relative humidity for a time period. Water sorption (W_S) is calculated as

$$W_S(t) = \frac{W_1 - W_0}{W_0} \times 100\%$$

where W_1 is the measured moisture content at time t and W_o is the initial moisture content.

These tests can be carried out in the extreme case of 100% RH or in a particular RH value where the material is supposed to be applied, comparing how different materials (matrix and composites) absorb humidity from the environment. Fillers with lower hydrophilic character than TPS could reduce composites' moisture absorption. On the other hand, hydrophilic fillers could also reduce water uptake, as functional groups on their surface could interact with starch OHs resulting in good interfacial adhesion between the two phases and fewer sites capable of interacting with absorbed water. Cellulosic fillers have shown the ability to reduce starch-based composite water uptake thanks to this phenomenon [43, 48, 136]. However, the improvements are usually moderate. Fourati et al. [48] studied the moisture absorption maximum of TPS and TPS/CNF nanocomposites at 50% RH. They found that it was slightly reduced by the addition of CNF decreasing from about 10% for unfilled TPS to about 8.5% for TPS/CNF. Ghanbari et al. [43] also included CNF in a TPS matrix and studied the moisture absorption, but at a RH level of 98%. They found that the water uptake of nanocomposites filled with 1.0 and 1.5 wt% of CNFs compared to the neat TPS was reduced to 6.4% and a 10% after 24 h, respectively. Other authors also found moderate improvements with the addition of sugarcane bagasse [137] and clays [138] into TPS matrices.

The wettability of TPS films can be evaluated through water contact angle (CA) measurements. Usual contact angle of glycerol-plasticized TPS films is around 40°. This value is of course related to the sample's surface polarity and can be modified thanks to the addition of fillers. One of the most hydrophobic compounds that can be introduced into TPS matrices is those of the lipid type. As it was mentioned before, the use of emulsifiers is a usual strategy for the preparation of starch–lipid composite films. Gao et al. [130] developed corn/octenylsuccinated starch (C/OS) composite films with soybean oil at 0, 0.5, 1.0, 1.5 and 2.0 wt%. OS acted as an emulsifier leading to the formation of molecular structures represented in the diagram of Fig. 10. Those conformations where a lipid-rich out layer surrounds a starch-rich phase can create a lipid-rich phase at the surface increasing the contact angle due to its higher hydrophobicity. In fact, the CA values were 39.9°, 45.4°, 58.6° and 76.1° for SO additions of 0, 0.5, 1.0 and 1.5%, respectively. However, in the presence of higher concentrations of SO, the aggregation of oil droplets in the film occurs, and therefore the conformations of Fig. 10 can no longer be formed. In this case, a lower CA value is obtained (37.0° for 2%).

Other lipophilic possible fillers are modified clays. The incorporation of MMT has turned out to be an adequate strategy to increase the contact angle of TPS films in many cases [138, 139]. In the cases of glycerol-plasticized TPS composites, a fraction of the glycerol plasticizer can migrate to the clay phase, reducing the glycerol content of the polymeric matrix and therefore its hydrophilicity [140]. Other plasticizers have proven to be successful in the reduction of TPS hydrophilicity, regardless of the incorporation of fillers. It is the case, for example, of deep eutectic solvents (DESs), which led to TPS films with contact angles in a range 80–90°. The filler–plasticizer interaction can be determinant in the final composite wettability. Adamus et al. [52] showed that the introduction of urea-intercalated montmorillonite (UM) to TPS films plasticized with two different deep eutectic solvent systems, choline chloride (ChCl) with urea (U) or with imidazole (IM), caused unlike influence on wettability depending on the particular DES. Contact angle decreased from 89° to a range 62–74° for starch/ChCl/U-UM and almost constant values 82–85° for starch/ChCl/IM-UM. The authors attribute the wettability reduction caused by the incorporation of UM to the fact that intercalated UM seems to exhibit higher polarity and then easier wettability than sodium montmorillonite itself. Differences between wettability of the films plasticized with ChCl/U and ChCl/IM are probably caused by various tendencies of particular DES to withdraw U from interlayer clay space at equilibrium state of ChCl/hydrogen donor compound into starch bulk matrix, as well as a result of specific interaction between IM and U molecules. Besides urea intercalation, there are other strategies usually applied to enhance filler compatibility with the starch matrix. These modifications usually include the incorporation of more hydrophilic groups, which could lead to an increase in the composite's polarity. It is the case of Abreu et al. [141] who found that the incorporation of MMT modified with a quaternary ammonium salt C30B into a TPS matrix caused a significant decrease in the water contact angle value of the composite with respect to the matrix. Chen et al. [142] studied the wettability of TPS composites with the incorporation of microcrystalline cellulose (MCC) and oxidized MCC (TOMC), finding that water contact angle of TPS films (38.72°) grows with the incorporation of MCC to values between 46° and 68°, but in the case of TOMC composites its contact angle value and wettability were close to native TPS films.

Water solubility is another important aspect to be evaluated. Depending on the final application of the material, different behaviors will be needed when submerged in water. A clear example is that of coating materials for laundry condensate beads, in which different sections are expected to dissolve at different times [142]. The presence of discontinuities (microparticles) in the polymer matrix usually promotes a greater water disintegration of the films usually related to a greater solubility. Talón et al. [35] showed that the incorporation of microencapsulated eugenol into a starch matrix promoted its water disintegration. Moreover, they showed that the interfacial adhesion between the filler and the matrix plays a key role. Azevedo et al. [143] also showed this for corn starch–whey protein blend films obtained by extrusion.

3.1.3 Mechanical Properties of Starch Matrix Composites

Depending on the materials' final application, different mechanical properties will be needed. Packaging films, for example, are expected to possess high tensile strength, which helps them to endure the regular stress met while handling food, but not very high elongation at break [144]. On the other hand, films for agricultural mulches must withstand high elongations so that they can be placed on the ground without deterioration [145]. Mechanical properties of starch films are usually not enough for many potential applications. Furthermore, the great variability in these properties, due to the high water sensitivity of films, introduces an additional problem. In 0% relative humidity, starch-based composites' mechanical strength can reach values as high as 20 MPa, while when the moisture contents are high, the tensile strength may be as low as 1 MPa [144].

The improvement in mechanical properties and handling of TPS materials thanks to the incorporation of micro- and nanofillers has been known for a long time [13, 47, 136, 146]. As it has been mentioned before, interactions in the filler–matrix interface will greatly influence the composite mechanical properties. Strong interaction at the interface leads to the transmission of stress from one component to the other, so if the added filler has greater Young's modulus and tensile strength than the matrix material, these two properties will grow in the composite. On the other hand, the elongation at break depends on failure propagation. In general terms, many well-distributed nanofillers with well interaction at the interface can anchor cracks, increasing the elongation at break. Simultaneous increases in elongation and stress at break lead to an increase in tensile toughness. In the case of microfillers, generally it is not possible to increase the elongation at break because although the achieved dispersion is excellent, there will be large domains in the matrix that do not have any filler capable of anchoring cracks. Figure 11 illustrates the difference in the distribution of nano- and microfillers in a matrix.

Beigmohammadi et al. [136], for example, worked with microfillers, investigating the effect of rye flour and cellulose as reinforcement agents in starch-based composites on mechanical properties. They found that filler incorporation had positive effect on tensile strength (TS) and Young's modulus, increasing from about 7 MPa for the starch matrix to more than 65 MPa for starch composites, reflecting the chemical and structural compatibility between starch and fillers. Unlike TS, the elongation at break decreased from 31.96 to 5.02%. On the other hand, Li et al. [139] incorporated montmorillonite (MMT) into a corn starch (CS) matrix to generate CS/MMT nanocomposite. They proposed a facile biomimetic method to enhance the interfacial adhesion between layered clay and polymer matrix based on the coating of clay nanofiller with a thin layer of polydopamine (PDA). The obtained results demonstrated that PDA coating benefited not only the intercalation and dispersion of the modified MMT (MMTDA) in the starch matrix but also the strong interfacial adhesion between filler and matrix. Thanks to this strong adhesion and the nano-character of the filler, simultaneous improvements of strain and stress at break were achieved.

Among the most widely used reinforcements for starch matrices, starch nanocrystals (SNCs) allow to create self-reinforcing starch materials. It has been demonstrated that the reinforcing effect of SNC is more significant in TPS than in other matrices, probably because of the strong interactions between the filler and amylopectin chains and a possible crystallization at the filler/matrix interface [148]. The same reinforcing effect has been reported by other authors [128, 149, 150, 151].

Other polysaccharides can also be transformed into nanofillers and introduced in starch matrices. That is the case of chitosan, which can be introduced into starch matrices as nanoparticles and thanks to their similar polysaccharide structures and

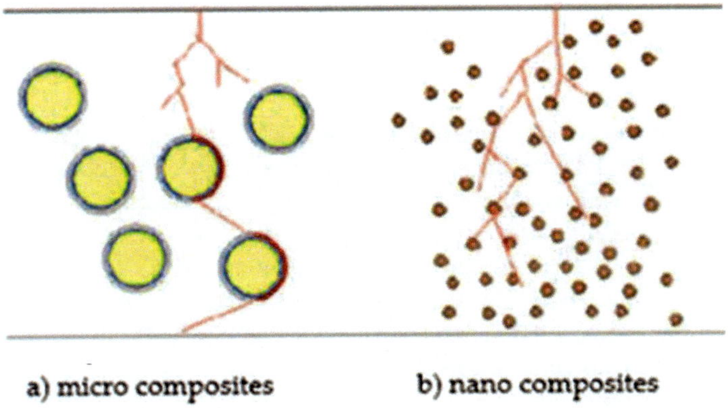

Fig. 11 Distribution of micro- and nanofillers in a matrix. Adapted from Cheng and Yu [147]

their great interaction can greatly improve the matrices' mechanical behavior [152, 153]. Moreover, chitosan has the great advantage of having antibacterial activity, which makes it very promising for applications in medicine, agriculture, drug release and edible film packaging [154].

Cellulosic reinforcements also have high affinity with starch matrices. Different authors have reported important mechanical properties' improvements thanks to this type of fillers [142], [155]. The formation of a rigid nanofiller network, the mutual entanglement between the nanofiller and the matrix, the efficient stress transfer from the matrix to the nanofiller, and the increase in the overall crystallinity of the system resulting from the nucleating effect of the fillers are the main reasons of these improvements [142], [155]. One important variable to be considered when analyzing the quality of the reinforcement is the filler morphology. It has been demonstrated that cellulose nanofibers (CNFs) have better reinforcing ability and stress transfer character than nanocrystals. This is attributed to a stronger intermolecular attraction between starch and nanofibers, consequence of a higher entanglement degree [156]. However, it is important to clarify that the degree of improvement achieved by CNF into starch matrices is considerably less than that achieved in other polymers, such as acrylics [157]. This lower reinforcing effect in TPS matrix could be due to two possible causes. On the one hand, the plasticizing agents (glycerol and water) could accumulate at the CNF/matrix interface which might reduce the possibility of interaction among neighboring fibrils through hydrogen bonding. On the other hand, the chemical similarity of CNF and starch will favor a high interfacial interaction between the two phases at the expense of CNF–CNF interaction, reducing the strength of the nanocellulose network which is known to play a key role in the mechanism of nanocellulose reinforcement [48, 158].

Besides the fillers' morphology, the way in which they accommodate in the matrix will also define the composite's mechanical properties, specially in the case of 2D fillers, such as clays and graphene sheets. Three different polymer nanocomposites can be prepared from 2D clays: (a) intercalated structure, (b) exfoliated structure and (c) flocculated structure-based nanocomposites. In general terms, exfoliated and intercalated structures will lead to greater mechanical property enhancement [159].

A useful strategy to obtain a stable dispersion of two-dimensional (2D) MMT plates in aqueous systems is through the incorporation of one-dimensional (1D) cellulose nanofibers. These systems could be used to develop starch/MMT/CNFs ternary nanocomposites. CNF can interact with MMT sheets via hydrophobic interaction between MMT sheets and specific crystalline faces (hydrophobic (200) planes) of CNF, and also with starch molecular chains from numerous hydrogen bonding through the hydroxyl groups on CNF surface. It is the case of Li et al. [138] who studied the synergistic reinforcing mechanisms of the MMT-CNFs system. They found that the tensile strength of starch/MMT/CNF ternary nanocomposites showed a remarkable improvement, which was much higher than that of the binary starch/MMT and starch/CNF nanocomposites. They ascribed this to the homogeneous dispersion of MMT, strong interfacial interaction between the fillers and matrix, and the lamellar structure with alternate stacking of 2D MMT platelet and 1D CNF fibril network layers in the ternary system. A schematic illustration for preparation

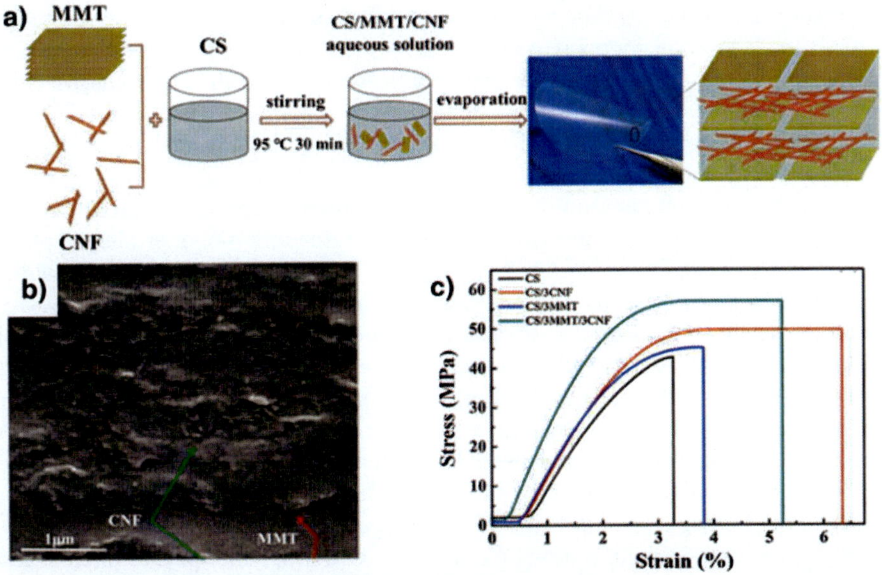

Fig. 12 **a** Schematic illustration for preparation of CS/MMT/CNF composites, **b** SEM image of CS/3MMT/5CNF × 80,000, **c** tensile stress–strain curves of CS film, binary CS/3CNF and CS/3MMT films, ternary CS/3MMT/3CNF film. Reprinted from Li et al. [138], Copyright (2019)

of starch/MMT/CNF composites, a SEM image of CS/3MMT/5CNF and tensile stress–strain curves are presented in Fig. 12.

The filler content is also determinant in the improvement effect. For each system, there will be a formulation with an optimal filler concentration that will lead to the greatest improvements in mechanical properties. Fazeli et al. [119], for example, showed that the maximum improvement in tensile strength for starch-based composite films reinforced by cellulose nanofibers was obtained for a concentration of 0.4 wt%. Tensile strength and elastic modulus increased by up to 80% and 170%, respectively. However, above 0.5 wt% CNF, tensile strength started to deteriorate. The authors proved that an excess of CNF content provokes particle agglomeration or phase separation in the starch matrix.

3.2 Properties of Starch Blend Matrix Composites

3.2.1 Introduction to Starch Blend Properties

Blend's properties can be manipulated according to their end use by correct selection of component polymers, additives and processing conditions. The properties of the polymer blend will depend on the final morphology, making the components' miscibility and phase behavior a key issue. In the specific case of heterogeneous polymer blends (obtained from immiscible polymers), the compatibility between polymer phases decides the final achieved properties.

A very important aspect when studying the properties of a blend is knowing how they behave with temperature, since this will influence their future processing to give it a certain shape, by either injection, blowing or thermoforming. In PHB/starch or PBAT/starch blends, as in the vast majority of engineering polymer blends, phase separation occurs as a consequence of polymers' immiscibility. The glass transition behavior associated with these systems demonstrates an elevated level of complexity, showing multiple transitions, ascribed to pure component phases and regions of partial mixing [160]. Properties such as crystallization and melting behavior of blends are also strongly influenced by the miscibility of involved polymers [161]. On the other hand, thanks to the characteristic hydrophobicity of both polymers, the water sensitivity is expected to be improved in the blend compared to starch-based materials. This will highly depend on the compatibility achieved between phases, as cracked morphologies will not allow the improvement of starch's moisture stability. In the same way, poor adhesion at the interface of the polymer phases in the blend results in diminished mechanical properties such as lower impact resistance and elongation at break [161].

In the next sections, the properties previously described for TPS composites are detailed for the case of composites obtained from starch blends with PHB and PBAT. Taking into account the importance of the affinity degree achieved between the blended polymers, the results obtained when using different compatibilizers as well as modified polymers are detailed.

3.2.2 Properties of Starch/PHB Composites

Thermal Properties of Starch/PHB Composites

The thermal properties of starch/PHB blend strongly influence their processability. As discussed earlier, melting point of PHB limits the temperature range for processing it with starch. PHB is a thermoplastic polymer with relatively high melting temperature which is in the range of 170 and 176 °C [73, 77, 92, 94, 162].

In the case of starch/PHB blends, the melting temperature of the binary mixtures is slightly reduced by the increment of starch amount in the mixture as consequence of the lower melt viscosity of starch contrary to that of PHB [162, 163]. For example, Liao and Wu [163] prepared PHB/starch blends by compression molding method using the acrylic-acid-grafted PHB (PHB-g-AA) as an alternative to PHB. With both, PHB and the grafted polymer, the melting temperature of the mixtures gradually decreased up to 4 °C from 0 to 30% of starch added.

On the other hand, the incorporation of plasticizers into the formulation has been applied to considerably reduce the melting and glass transition temperatures of the starch/PHB mixtures, thus enhancing the blend processability. For example,

Innocentini-Mei et al. [164] blended native and modified starches (starch adipate and grafted starch–urethane) with PHB by injection molding. Triacetin was added as plasticizer of PHB resulting in blends with lower melting temperatures (T_m = 141.5–159.1 °C) and glass transition temperature (T_g = −19.2 °C and −38.9 °C) than that of neat PHB (T_m = 171 °C and T_g = 0.7 °C). Likewise, the utilization of glycerol as plasticizer of starch led to reduce the melting temperature range of the binary blends (TPS/PHB, 70:30 wt%) by 10 °C compared to that of pure PHB [165].

As it was mentioned before, the blend glass transitions will depend on the components' miscibility. While TPS plasticized with glycerol typically shows two T_gs, PHB exhibits just one T_g between −2.5 and 0.7 °C [92], [164]. In the case of TPS/PHB blends, the three T_gs are maintained due to the inherent immiscibility of both components. However, the Tgs of starch on the blends cannot always be observed using the differential scanning calorimetry (DSC) technique because of the chain rigidity and the strong hydrogen bonding of starch molecule [92]. In addition, the temperature at which transitions occur can vary. For example, Innocentini-Mei et al. [164] prepared three types of blends based on PHB (plasticized with triacetin) and TPS from native starch, starch adipate and grafted starch–urethane by injection molding. The amount of starch added to the plasticized PHB matrix varied from 10 to 30%. The plasticized PHB matrix had a T_g at −25.7 °C. However, this T_g value shifted to a lower (−19.2 °C) or higher (−38.9 °C) temperature in the three kinds of TPS/PHB blends depending on the amount of PHB and the type and amount of starch used. For blends of PHB with native and adipate starches, no other T_g values were observed. However, the blends prepared with grafted starch–urethane at concentrations over 20% showed an extra T_g in an interval of −42.2 °C and −58.8 °C attributed to the glass transition of these modified starches.

The thermal decomposition of PHB-TPS blend has also been studied. Generally, it occurs in three stages that are associated with the decompositions of the PHB and the TPS. As it was mentioned before, TPS has three-step degradation. The first step corresponds to the loss of water, the second to the degradation of other plasticizers, such as glycerol, and the third to starch degradation (T_d = 323 °C). On the other hand, degradation of PHB occurs in two steps [77, 166]. The first is associated with the rupture of the ester groups, and the second is attributed to the plasticizer in the commercial PHB [166]. In the PHB-TPS blend, three stages of decomposition have been found and associated with the decompositions of PHB and TPS [77]. PHB stability is expected to be improved by blending it with starch due to the formation of hydrogen bonding between hydroxyl groups of starch and carbonyl groups of PHB that may inhibit early PHB degradation [94]. Different authors have reported that with the addition of PHB, the beginning of thermal decomposition of starch/PHB blends remains as that of the starch and the maximum temperature at which it occurs is shifted to higher temperatures as PHB amount increases [93, 165].

The inclusion of fillers into starch/PHB blends can influence its thermal stability. The effect of clay and eugenol inclusion was studied by Garrido-Miranda et al. [77], who observed temperatures for 5 wt% weight loss (T95%) and 10 wt% weight loss (T90%). The T95%, corresponding to the first step of starch degradation (water loss), was displaced to higher temperatures. This could be attributed to a reorganization caused by the clay's inclusion that restricted starch chains, and therefore, higher energy for the degradation process is needed.

Gas Barrier Properties and Water Sensitivity of Starch/PHB Composites

One of the main causes that motivates the formulation of starch/PHB blends is the necessity of reducing the water uptake of starch and therefore enhances and extends the range of application of starch-based materials. An inherent characteristic of PHB is its hydrophobic character as a result of the large number of carbonyl groups in its structure, which makes it a suitable material for packaging applications. In this context, blending starch with PHB may improve the starch WVP and water sensitivity if both polymers are compatibilized, obtaining a co-continuous phase without the presence of cracks at the interface. It should be noted that PHB barrier properties against other gases, such as oxygen, are poorer than starch ones, being the OP 13.4 mL * 700 μm/(m^2 * 24 h * 0.21 atm O_2) and 9.1 mL * 700 μm/(m^2 * 24 h * 0.21 atm O_2) for PHB and starch, respectively [167]. Taking into account that the property to improve in starch matrices is their moisture barrier capacity, in this section we will focus on WVP along with other aspects of materials' water sensitivity.

Thiré et al. [93] explored the water absorption of PHB/starch blends by contact angle assay studying the kinetics of spreading a drop of water on the materials' surface every 15 s. To prepare the blends, previously melted PHB and different amounts of starch (0–50%) were processed by injection molding. PHB film showed the highest value of contact angle owing to its hydrophobic character and intrinsic rigidity, contrary to that in TPS films in which the drop of water rapidly spread on their surface due to its high water adsorption tendency. These values were intermediate for all blends with starch content from 20 to 40%. Over time, an increase in the water absorption rate was observed with the increment in starch content. The contact angle for PHB80/starch20 was 76° and diminished to 69° for PHB60/starch40. Likewise, Liao and Wu [163] showed similar findings on the study of water absorption of PHB/starch blends produced by compression molding for six weeks. The study included different contents of starch (10–50 wt%), and the possibility to include acrylic-acid-grafted PHB was also studied. Better water resistance was obtained with PHB-g-AA/starch than with PHB/starch at the same amount of starch. These results were attributed to the formation of ester carbonyl function groups by the reaction between starch's hydroxyl groups and anhydride carbonyl groups of PHB-g-AA. In addition, the increment of starch content in the blends led to higher water uptake again which was assigned to lower miscibility of both polymers. The positive effect of PHA on the increasing water resistance was showed also in TPS matrix combined with other polymer. Tingwei [168] prepared TPS/PBAT/modified-PHA biodegradable films with enhanced water repellency to be applied in a variety of industries such as sanitary products.

The addition of less hygroscopic fillers into TPS/PHB blends could improve the water sensitivity and barrier properties even more. Magalhães and Andrade [169] showed that the incorporation of PHBV and an organically modified montmorillonite (C30B) into a starch matrix was able to reduce the moisture absorption of the extruded materials in more than a 50%. While the maximum water absorption of TPS was about 14% of its dry weight, this value was reduced to 6.3% for the PHBV50/TPS50 neat blend, decreasing up to 3% when C30B was added. This result is in agreement with the property of organoclays as a barrier against humidity.

Mechanical Properties of Starch/PHB Composites

At room temperature, TPS films are soft and ductile while PHB films are rigid and brittle with high Young's modulus and tensile strength at break due to its high crystallinity (about 80%). PHB brittleness is attributed to the formation of large volume-filling spherulites from few nuclei which are accompanied by interspherulitic cracks, and to the secondary crystallization of the amorphous phase which takes place during storage at room temperature [170].

The crystallinity level of PHB is usually tuned by crystallization rate that affects the nucleation rate and spherulite size [171]. Generally, PHB presents slow crystallization rates which will depend on the production method. At temperatures between 80 and 100 °C, the crystallization rate is fast, but it is slow below 60 °C and above 130 °C resulting in a higher crystallinity. In addition, PHB suffers recrystallization with aging at room temperature, and a progressive reduction of the amorphous content in the partially crystalline polymer, changing its mechanical properties with time, for example, an increase in yield stress and modulus, and decrease in elongation at break and fracture toughness. The incorporation of some additives such as plasticizers and nucleating agents contributes to the reduction of crystallization process and to impart flexibility and toughness to the material. Râpă et al. [172] investigated the incorporation of three plasticizers, triethyl citrate (TEC), tributyl o-acetylcitrate (TBoAC) and tributyl citrate (TBC) in PHB via melt-mixing procedure and studied the effect of these additives on thermal and mechanical properties. Then, the incorporation of plasticizers into PHB decreased tensile strength and Young's modulus, but increased elongation at break; PHB/plasticizer blends exhibited a wider melt-processing window with a lower melting temperature.

Mechanical properties of starch/PHB blends are mainly affected by their degree of compatibility and miscibility to a greater or lesser extent, as well as processing conditions. As morphological characteristics of blends are strictly related to the mechanical properties, they will be also described along this section. Poor interfacial interaction between starch and PHB often leads to heterogeneous blends. Zhang and Thomas [94] investigated the effect of blending two types of starch (high amylose corn starch and regular corn starch) with PHB in a ratio of 30:70. No plasticizer or other additives were added in the formulation. Starch and PHB were melted at 175 °C for 10 min at 50 rpm, and then the assembly was hot-pressed at 180 °C. According to dynamic mechanical thermal analysis (DMTA), high amylose starch had better interfacial bonding with PHB due to its linear structure compared to regular starch. On the other hand, SEM micrographs of cryogenic fracture blends' surfaces showed that starch granules still kept their structure which negatively affected the blends' mechanical properties according to the researchers. As the major component of the blend was PHB and starch granules were not disintegrated through processing, the researchers inferred that starch acted as a filler of PHB matrix. Similar results were obtained by Thiré and collaborators [93]. The authors prepared PHB/starch blends with different contents of maize starch (0–50%) by compression molding. Herein, PHB was melted in a mixer at 165 °C for 15 min, then a mixture of starch, glycerol and water was added into the melted PHB, and this resulting mixture was continuously mixed for 30 min at 60 rpm. Then, the mixes were hot-pressed at 160 °C and 69 MPa for 30 min to form sheets. TPS for comparative purposes was prepared at 60 °C and 20 rpm for 20 min, and then films were obtained by hot-pressing at 90 °C. SEM micrographs of fractured surface samples obtained after tensile tests revealed that starch granules were not disintegrated during processing. These granules kept their original size and were dispersed throughout the matrix, forming agglomerates when starch content was higher. Starch incorporation of up to 30 wt% into PHB matrix resulted in materials as hard as pure PHB ones. Young's modulus of blends with those starch contents remained practically constant in a value around 1300 MPa. If the amount of starch was increased, the module's value decreased until reaching 850 MPa for a starch content of 50%. These values are much higher than that reported by the authors for TPS (300 MPa). However, all the binary blends were much less flexible than TPS films and PHB since they presented lower values for elongation at break. The authors explained these results by the lack of interfacial adhesion between starch and PHB and by the heterogeneous dispersion of starch granules over PHB-rich matrix, and they concluded that both polymers were processed at very light processing conditions which did not allow starch degradation.

In addition to the polarity differences of starch and PHB, the difficulty of an adequate process of starch is added when it is blended with PHB. A suitable gelatinization/destructurization and/or plasticization of starch is required to promote the compatibilization between starch and PHB to produce blends with higher performance of their mechanical properties. Lai et al. [173] prepared TPS/PHB blends from potato, corn and soluble potato starches. The starches were gelatinized at different degrees, adding glycerol (25 and 33%) and water (25 and 17%) at 90 °C for corn starch and 70 °C for both potato starch and soluble potato starch, during 30 min at 50 rpm. TPSs and PHB (1–7%) were mixed for 10 min at 180 °C. Then, the prepared batch was compression-molded to obtain 1-mm-thin sheet. In all cases, the dimension of starch granules in the sheets was in the order of 1 μm according to SEM micrographs. Besides, as PHB content increased, tensile strength of blends generally grew from 0.6 to 5.8 MPa, which was attributed to the high strength of PHB and a reasonable compatibility between TPS and PHB. In addition, the higher molecular weight and the more compact structure of potato starch with respect to soluble potato starch and corn starch conferred it better tensile properties. The authors ascribed the better performance of the blends to the suitable gelatinization of starch.

A processing method that could lead to the development of materials with better properties is reactive extrusion, in which interfacial adhesion between components can be enhanced. Avella et al. [174] developed PHBV/starch blends using between 20 and 30% of a high amylose content starch by both traditional and reactive extrusions. They found that in the case of traditional extrusion, as the starch content increases, impact resistance decreases, and attribute this behavior to the poor interfacial addition between polymers and failure propagation. In the case of reactive extrusion in the presence of peroxides, a better interaction between the two components is achieved, which is responsible for good stress transfer. In particular, with a 20% of starch the blend's impact resistance improves, while with a 30% of starch it is in the same order, in both cases comparing with PHBV.

Researchers have been exploring other alternatives for enhancing mechanical and physical properties of starch/PHB blends through upgrading their compatibility. Recently, Florez et al. [165] modified the surface of PHB with an atmospheric plasma treatment to compatibilize its particles with the matrix of TPS. The selected gases for the mentioned process were atmosphere air and sulfur hexafluoride (SF6) gas which caused surface ablation of the particles. Starch (35–75%), glycerol (25–35%) and PHB (10–30%) were melted at high rotation for short times (1800 RPM for 20 s followed by 3600 RPM for 5 s). Disks of ≈2 mm of thickness were obtained after hot-pressing the melted mass at 90 °C for 15 s with 8 tonne force. Pristine TPSs had homogeneous surface suggesting that processing was capable of promoting disruption of the starch granules; however, the blends exhibited bimodal morphology of the equiaxed and elongated PHB grains dispersed in the starch matrix. Besides, agglomerations of PHB particles were also found on the surface of blends. These agglomerations could be generated during plasma treatment leading to a fragile starch/PHB blends that induced the rupture of the materials, especially when PHB content was higher.

Knowing the miscibility degree between polymers is critical to choose compatibilizing strategies. Concerning this, binary blends of starch and PHB were produced using modified starches such as acetate starch, adipate starch and grafted starches [164]. These starches are characterized by their less hydrophilicity that might result in better affinity and interaction with the hydrophobic PHB. For example, Innocentini-Mei et al. [164] investigated the behavior of starch derivatives (10, 20, 30 wt%) blended with PHB by injection molding. Triacetin (30 wt%) was used as plasticizer of PHB. The assayed starches were natural starch, starch adipate and grafted starch–urethane. Grafted starches were obtained with 0.25 mol of PEG, PPG or Rucoflex polyester as grafting agents and the addition of 0.25 mol of diisocyanate. Before the injection molding process, natural starch and starch adipate were gelatinized in dimethyl sulfoxide (DMSO) (50 wt%) in a mechanical mixer. However, starches retrograded as a result of a presumable migration of DMSO to PHB, due to their affinity. Therefore, stiffer and brittle materials with poor mechanical properties were obtained as a consequence of the insufficient amount of DMSO employed and the poor affinity of these starches with PHB, even for the adipate starch as its modification level was very low according to its FTIR spectrum. On the other hand, for grafted starches/PHB, Young's modulus and tensile strength showed slightly higher performance than those obtained with the natural and adipate starch due to a greater affinity of its structure with PHB. However, elongation at break of all modified starches remained as that of natural starch.

Other proposed modifications include grafting EVA [175] or poly(vinyl acetate) [92, 162] to starch chains or acrylic acid to PHB [163]. Ma et al. [175] showed that the presence of EVA chemically bounded with starch inhibited the starch agglomeration and enhanced the toughness of starch–EVA/PHB materials as shown by an increment of elongation at break and tensile-fractured energy. Similarly, the increased compatibility between PHB and soluble potato starch grafted with vinyl acetate (VAc) resulted in blends with higher toughness than pure PHB [92]. However, all these strategies do not achieve enough improvements for many of the applications proposed for these materials. Going one step further, Yingxin et al. [176] used compatibilizers and at the same time included a new polymer in the blend. They used different combinations of compatibilizers, including glycidyl methacrylate (GMA)-functionalized polyolefin elastomer (POE) (POE-g-GMA), maleic anhydride-grafted ethylene–octene copolymer (POE-g-MAH), ethylene–methyl acrylate (EMA), acrylonitrile-chlorinated polyethylene styrene (ACR), ethylene–vinyl acetate (EVA), and different combinations of coupling agents, including silane, titanate and aluminate, to develop starch-based degradable polypropylene/PHB materials by extrusion process with good mechanical properties (high tensile strength, elongation at break and impact strength).

Filler incorporation into PHB/TPS matrices has been a successful strategy to improve the blends' mechanical properties. Garrido-Miranda et al. [77] analyzed the influence of eugenol and clay incorporation into a PHB/TPS matrix on mechanical properties. They worked with a derivate of montmorillonite (OMMT) modified by surfactants which has the same affinity for hydrophilic and hydrophobic polymers. The authors found that the OMMT presence improves the interaction between PHB and TPS by the formation of hydrogen bonds between eugenol and the available hydroxyl groups of the components. Then, the elastic modulus of the nanocomposites was higher than that of PHB-TPS. These results were explained by the reinforcing effect of the OMMT which caused partial immobilization of the polymer chains and the increment of the composites' rigidity. Another derivate of MMT was used by Magalhaes and Andrade [169] to produce bio-nanocomposite with TPS and PHBV. The researchers investigated the properties of 1:1 PHBV (with 3.4 mol% hydroxyvalerate units) and TPS nanocomposites, prepared by melt extrusion in the presence of the different contents of an organically modified montmorillonite (C30B, 2.5, 5.0, 7.5 and 10.0 wt%). Due to the reinforcement effect of the organoclay and the rigidity of PHBV, the nanocomposites were stiffer as the organoclay content increased resulting in a lower elongation at break and an increase in the Young modulus and tensile strength.

On the other hand, a recent alternative proposed to include starch in PHB materials is through the use of starch nanocrystals. Zhang et al. [177] used SNC to make matrices of immiscible PHB and PBS mixtures compatible. More degraded and nearly disappeared banded spherulite morphology was obtained with the addition of SNC to the matrix of PHB as SNC loading increased, which evidences the SNCs compatibilizer role. Consequently, improved phase adhesion and reinforcement were obtained resulting in a rigid material with a deformation at break 4 times greater than PHB.

3.2.3 Properties of Starch/PBAT Composites

Thermal Properties of Starch/PBAT Composites

Following the principles detailed in Sect. 3.2.1, the final blend's processability of starch/PBAT blend differs from that of the neat polymers. Generally, there is a new T_m of the blend, as well as a new viscosity and rheological behavior. PBAT has good crystallization and thermal stability, and it melts at about 123 °C and crystallizes at about 60 °C. As a result, it has good processing stability to be used alone or blended with other material [96].

When PBAT is blended with TPS, its melting point can slightly increase due to the stabilizing effect of TPS caused by the hydrogen bond interaction between the polymeric chains after mixing with PBAT. Lendavi et al. [114] reported an increase in the T_m of PBAT, from 123 °C to 125.7 °C for the blend PBAT50/TPS50.

Filler inclusion can increase the blends' T_m. Olivato et al. [113] showed a slight T_m increase of PBAT phase in TPS50/PBAT50 (124 °C) and TPS80/PBAT20 (124 °C) with the incorporation of 3 wt% of sepiolite nanoclay (126–127 °C). An increase in the nanoclay content from 3 to 5 wt% did not significantly affect composites' T_m. On the other hand, both blends exhibited a crystallization peak at 79 °C for TPS50/PBAT50 and 63 °C for TPS80/PBAT20. A significant increase in the T_c (84–86 °C) and decrease in the crystallinity content were obtained with the addition of sepiolite. The authors proposed that the clay was acting in the crystallization process mainly during the nucleation step, but could reduce the crystallinity degree, possibly due to confinement effect and steric hindrance, which restrain the crystal growth. Liu et al. [103] studied the T_m and T_c of TPS-SiO_2 composite blended with PBAT in a 60:40 ratio. They found that Tm of the PBAT phase in the composite did not differ from that of PBAT. Moreover, they found that the addition of two compatibilizers, MA and EVOH, to the blend caused no influence in the melting point. In the case of crystallization temperature of the PBAT phase, an important shift occurred for the TPS-nSiO_2/PBAT composite (83 °C) compared to neat PBAT (45 °C). The authors explained that this sharp increase could be due to the TPS-nSiO_2 particles acting as a nucleating agent. When MA and pEVOH were incorporated into the TPS-nSiO_2/PBAT blend, a higher T_c was observed for PBAT phase (88–89 °C).

These results suggest that, unlike the case of PHB, when blending PBAT with TPS, the melting temperature of the PBAT phase does not undergo significant changes. Moreover, the addition of the additives, such as compatibilizers or fillers, had no significant influence on the melting transitions of the composites. However, this is not the case of crystallization temperatures, which could be highly affected, modifying the properties of the final material.

As it was mentioned before, TPS plasticized with glycerol has two T_gs, one related to the plasticizer's rich phase and the other to the starch's one. In the case of PBAT, the principal T_g is found in the range −16 to −35 °C and is associated with its soft segments (aliphatic). In some cases, other T_g is observed around 60 °C, which is associated with the T_g of the hard segment (aromatics). In the case of starch/PBAT blends, as they are immiscible polymers, all the T_gs of the components are expected to be present. Generally, three distinct transitions (from low to high temperature) are reported: the secondary relaxation of the glycerol-rich domain, the principal glass transition of the PBAT phase and the glass transition of the starch-rich phase which is overlapped with the one of PBAT's hard segments. This behavior was reported by several authors [100, 103, 108].

The addition of compatibilizing agents can have different effects. Better compatibility between phases tends to generate a shift in their T_gs toward each other. This effect has been reported as a consequence of the inclusion of different compatibilizers [97, 178]. In particular, Zhai et al. [97] found that the T_gs of the phases rich in PBAT and starch shifted toward each other when the content of the two polymers became close. They proposed that during the reactive extrusion process of a TPS/PBAT blend in the presence of CAc, some interactions may have occurred between starch and PBAT at the molecular level. CAc could graft single ester groups in the starch chains, decreasing their hydrophilicity and increasing their compatibility with PBAT. Simultaneously, if the interfacial adhesion is strong, T_g shifts to higher temperatures can occur, as more energy will be needed for molecules to gain mobility. For example, it has been demonstrated that the inclusion of PBAT-g-MA into TPS/PBAT blends leads to an upward shift of the relaxation temperatures of PBAT (+5 °C) and TPS (+20 °C) [100]. Zhai et al. [97] deduced that in HPDSP/PBAT 50/50 film, using citric acid as a compatibilizer, hydroxypropyl distarch phosphate interacts with PBAT, from changes in the alpha phase relaxation temperatures of both polymers. They found a slight shift in the T_g of the PBAT phase from −24.9 to −22.6 °C for HPDSP/PBAT 10 to 50 wt%. The glass transition for the HPDSP phase decreased from 84.4 to 81.7 °C. The authors suggested that during extrusion, some interactions between PBAT and HPDSP may have occurred and also a graft reaction between citric acid and modified starch HPDSP. Zhang et al. [108] blended PBAT with a previously extruded starch with glycerol as plasticizer and tartaric acid as modifier (TPS-TA). Using the blend PBAT30/TPS-TA70, they demonstrated that the compatibility between starch and the polyester matrix could be improved when starch had been pretreated. The incorporations of TPS-TA into PBAT shift the T_g associated with the polyester's soft segment (aliphatic) from −16.4 to −13.8 °C and one of the primary glass transitions of the starch-rich phase from 28 to 11.8 °C. These variations in glass transition temperatures demonstrate the compatibility of PBAT/TPS-TA composites. Starch modification with TA serves as a coupling reagent and reduces the phase separation.

Some compatibilizers can also act as plasticizers, and therefore its effect can be varied. DMA analysis of starch/PBAT/glycerol blends with different amounts of sericin performed by Garcia et al. [111] showed both sericin's plasticizing and compatibilizing effects depending on its concentration in the blends. For contents up to 1 wt%, T_g of the starch-rich phase was shifted to a higher temperature indicating that the molecular mobility of the starch chains was reduced and that sericin could be acting as a compatibilizer. The highest content of sericin (1.5 wt%) showed the opposite behavior, indicating a plasticizing effect. The authors proposed that the short-chain segments of sericin could act as plasticizers improving the mobility of the starch short-chain segments, while the larger sericin segments could act as compatibilizers between polymers, increasing stiffness.

The addition of fillers can, of course, also have effects on material transitions. Depending on whether the filler is located in the interface or if it tends to be interacting only in one phase, its effect will be different. For example, Olivato et al. [113] studied the effect of sepiolite inclusion in TPS50/PBAT50 blend. They found that strong interaction between starch molecules and sepiolite limited the mobility of the starch chains, producing a shift in the T_g of starch-rich phase to higher temperatures. This gives account of the high interfacial interactions between sepiolite and starch. When the filler is highly interacting with one of the polymers, then part of the plasticizer can migrate to the other phase. Lendvai et al. [114] blended PBAT with TPS nanocomposites (reinforced with BT and OMMT). They found that for blended nanocomposites both T_gs of TPS increased, probably due to the depletion of the plasticizer in the TPS phases that could have migrated into the PBAT phase. Similar results were found by Liu et al. [103], who blended PBAT with TPS-nSiO_2 nanocomposite. Compared with that for the pure PBAT ($T_g = -30$ °C), the glass transition of the PBAT phase in each of the composites was observed at a lower temperature by about 5–6 °C. They explained this as a consequence of the increased plasticizing effect on the PBAT phase by the migration of glycerol from TPS. The glass transition of the starch-rich phase for the TPS-nSiO_2 (60 °C) shifted to a higher temperature (70 °C) in TPS-nSiO_2/PBAT sample. The authors explained this result probably due to the poor compatibility between TPS and PBAT and the weakened plasticizing of the starch-rich phase from the migration of glycerol, which reduced the mobility of the starch chain segment. On the contrary, after the addition of the compatibilizer of MA, EVOH and MA + EVOH, the glass transition of the starch-rich phase in the nanocomposites shifted to a lower temperature (about 60 °C for MA, 62 °C for EVOH and 55 °C for MA + EVOH) than that for the sample without compatibilizer (about 70 °C). The lowered transition temperatures indicated that the introduction of both MA or/and EVOH could improve the compatibility between TPS and PBAT, resulting in the increased mobility of the starch chain segment. When the combination of MA and EVOH was used, the T_g of the starch-rich phase was the lowest (about 55 °C).

While neat PBAT decomposes in one step at 400 °C, thermal degradation of TPS/PBAT blends occurs as a three-step process, including a first stage attributed to the elimination of water and plasticizers (50–250 °C), a second much sharper transition at $\approx$320 °C, which represents the degradation of TPS components amylose and amylopectin, and a final third stage around 400 °C corresponding to the PBAT degradation [114].

The inclusion of fillers could cause changes in the different stages, depending on their interactions. Lendvai et al. [114] blended TPS-OMMT and TPS-BT nanocomposites with PBAT. They found a lower temperature and faster plasticizer degradation step for PBAT60/TPS-OMMT40 than for PBAT60/TPS-BT40 composite and PBAT60/TPS40 neat blend. This phenomenon was also observed in torque measurements, which revealed that OMMT-filled composites presented a higher degree of dehydration than the others. TGA results revealed that BT and OMMT incorporation on TPS affected its dehydration and degradation, suggesting that the clays remained predominantly in the TPS phase, where they were initially introduced. Liu et al. [103]

also blended a TPS nanocomposite with PBAT, but in their case it was TPS-SiO_2. The thermal stability of starch and PBAT in the PBAT40/TPS-$SiO_2$60 sample was not affected by the blending of PBAT with the starch nanocomposite. Moreover, when MA or/and plasticized EVOH were added as compatibilizers, the two main degradation temperatures were similar to those for the PBAT40/TPS-$SiO_2$60 nanocomposite, which indicated that the compatibilizers had no significant influence on the thermal stability of the composites.

Gas Barrier Properties and Water Sensitivity of Starch/PBAT Composites

While PBAT presents a good balance between biodegradability and thermal–mechanical performance, its barrier properties are lower comparing to traditional petrochemical non-biodegradable polymers. As it is well known, compared to starch, PBAT has lower WVP but its barrier properties against oxygen are poorer. Therefore, depending on the needed property different combinations can be performed. Generally, increasing the PBAT content in PBAT/TPS blends decreases the WVP and simultaneously increases the OP [97, 179, 180]. In this section, we will focus on water vapor permeability, as it is the property to improve in starch matrices.

When additives with compatibilizing action are added to starch/PBAT blends, stronger intramolecular bonds are expected, which could lead to the formation of a tighter structure that could restrict water vapor diffusion. It is the case of Garcia et al. [111], who incorporated sericin protein to improve the compatibility of a starch/PBAT/glycerol blend and found a reduction of almost 20% in the WVP of sericin containing films with respect to the blend without compatibilizer.

As in the cases of starch/PHB blends, blending starch with PBAT together with fillers using an appropriate processing strategy and conditions to promote an intercalated nanostructure, where better polymer–nanofiller interaction is achieved, will result in better barrier properties of the TPS/PBAT composite [181].

Manepalli and Alavi [50] used the modified Nielsen equation to model the WVP of the ternary blend of PLA/PBAT/TPS with the inclusion of nanocrystalline cellulose (NCC), as a function of filler morphology, filler content and the degree of filler–polymer interaction. As expected, more TPS contents lead to larger WVP values, since both PLA and PBAT have a more hydrophobic response than TPS. In the same way, as it was predictable due to the tortuousness generated by the addition of nanofillers, the incorporation of NCC to the PBAT6/TPS40/PLA54 matrix decreased the WVP to 15 and 32%, for a filler content of 1 and 2 wt%, respectively. Similarly, nanocomposite films obtained by Zhai et al. [97] showed a significant reduction in WVP when PBAT content increased in PBAT/HPDSP/nCOM composite (PBAT10/HPDSP90 24%; PBAT20/HPDSP80 38%; PBAT30/HPDSP70 40%; PBAT40/HPDSP60 57%; and PBAT60/HPDSP50 66%). Moreover, they showed that the PBAT content increases the *d*-spacing values (evaluated by XDR), evidencing an increase in the intercalated nanostructure in the composite, due to better polymer interaction with the nanoclay. The increase in the PBAT content (less water sensitivity) and the appropriate melt extrusion condition that promote the formation of an intercalated nanostructure, due to better PBAT interaction with the organic modified clay, results in better water vapor barrier properties of the TPS/PBAT composite.

In agreement with the WVP results, the surface hydrophobicity of the films increased with increasing PBAT content, especially beyond 30 wt%. In contrast to the WVP of the HPDSP/PBAT/nCOM, the oxygen and carbon dioxide permeability values increased with increasing PBAT content. Although the addition of PBAT promoted the formation of a filler-intercalated nanostructure, which could prevent oxygen and carbon dioxide molecules from passing through the film matrix, the clay's dispersion was not enough to generate a tortuous path homogeneously distributed to reduce the permeability of oxygen and carbon dioxide efficiently.

As in the case of PHB, due to its hydrophobic character, it is expected that the incorporation of PBAT in starch matrices will lead to more stable blends in water, compared to TPS. The composition and morphology of PBAT/TPS blends will be decisive in the amount of water they can absorb from the environment. Depending on whether the predominant phase is TPS or PBAT, the behavior will be very different. Dammak et al. [56] showed that when TPS was the continuous phase in the blends of TPS/PBAT, this phase was accessible to water by diffusion from the surface increasing film's water absorption. This will be influenced by compatibilizer's presence. Fourati et al. [100] found that the moisture absorption for PBAT40/TPS60 was affected by the compatibilizer, reaching the highest level around 6% in the presence of MA and CAc, and the minimum was observed in the absence of the compatibilizer and in the presence of PBAT-g-MA. This evolution is connected with the continuity index for PBAT, as the lowest moisture absorption was attained when the PBAT is the continuous phase. For example, PBAT40/TPS60 formulated with CAc 2, 4 and 6% and presented a 33% increase regardless of CAc content, where the TPS formed a co-continuous (2% CAc) or continuous phase (4 and 6%) in the composite.

Changes in moisture content usually lead to differences in surface properties. Zhang et al. [108] showed that the incorporation of tartaric acid influenced the water absorption and contact angle of PBAT/TPS-TA blends. The authors reported that when 0.5% of TA was added, the water contact angle of PBAT/TPS-TA-0.5 increased from that neat PBAT and TPS/PBAT blend. These results confirmed the TA coupling effect in PBAT/TPS-TA-0.5. In these materials, TPS-TA-0.5 particles dispersed uniformly in the PBAT matrix to achieve a homogeneous phase, where the interface intensity was stronger than that of PBAT/TPS and prevented water molecules from permeating the boundary. On the contrary, the introduction of TPS and TPS-TA with higher TA content enhanced the hydrophilicity of PBAT, resulting in a decrease in the contact angle and an increase in water absorption. This is probably due to a new phase morphology reverse from the homogeneous phase of PBAT/TPS-TA-0.5.

Investigations of Spiridon et al. [182] revealed that the addition of ground and sieved particles of grape pomace, celery, treated waste of *Asclepias syriaca* floss and poplar seed hair fibers used as biomass fillers in starch/PBAT blends increased the water uptake capacity, compared to starch/PBAT. This result was awarded by the authors due to the strong hydrophilic character of the biomass fillers, related to the presence of hydroxyl groups present in the fibers. However, slight differences were found depending on the biomass incorporated into the mixture, due to a different cellulose and lignin content in the biomass fiber composition [183]. When lignin was added to the formulation, a reduction in the water uptake capacity was observed. Vapor sorption capacity of biocomposites in the dynamic regime and isothermal conditions was also studied. The authors reported that the addition of biomass fillers to the starch/PBAT system did not change the shape of the isotherms, though the water uptake capacity was higher than that of the reference sample in the whole range of the curve. Other fillers, such as clays, have also been successful in reducing the blend water absorption. For example, sepiolite nanoclay's addition, studied by Olivato et al. [113], in PBAT80/TPS20 and PBAT20/TPS80 decreased the water adsorption rate and the water absorption capacity of nano-biocomposites, due to intermolecular links between starch and sepiolite silanol groups.

Mechanical Properties of Starch/PBAT Composites

The mechanical properties of the final blend will strongly depend on the compatibilization degree achieved. The role of various compatibilizers on TPS/PBAT mechanical properties was evaluated by several authors. The effect achieved by a compatibilizer will depend on the way in which it is added. For example, both Dammak et al. [56] and Fourati et al. [100] studied the case of maleic anhydride inclusion, placed as a compatibilizer in the blend or through a graft in PBAT (PBATg-MA). Dammak et al. [56] presented stress–strain curves for PBAT40/TPS60, PBAT50/TPS50 and PBAT60/TPS40 with and without compatibilizer. All blends exhibited three regions involving elastic, plastic deformation and strain hardening behavior. The tensile strength and the elongation at break of the samples improved as the TPS content decreased in the presence and absence of MA. For PBAT50/TPS50 blends, the tensile strength and elongation at break attained 12.8 MPa and 205% in the absence of any additive, reaching, respectively, 15.1 MPa and 614% when PBATg-MA was used as a compatibilizer, but decreasing to 6.5 MPa and 36%, respectively, when the blend was processed in the presence of 2% MA. In the blend PBAT50/TPS50 without compatibilizer, starch was in the form of dispersed droplets of about 2–10-μm diameter, with a low connection between them, while in the presence of MA, a full continuous morphology was observed, where the TPS formed an interconnected structure. This continuous morphology of the TPS phase was not observed when the blend was processed in the presence of PBATg-MA, where a dispersed morphology was obtained. These important morphological changes could be the reason for the marked differences in the materials' mechanical properties. Similarly, Fourati et al. [100] studied the mechanical properties of PBAT40/TPS60 processed in the presence of MA or PBATg-MA. They also found that the important improvements of strain at break achieved by blending TPS with PBAT (185%) reversed when MA was added (to values between 12 and 30%). The tensile strength of films also decreased in the presence of MA from 10 MPa for the blend without compatibilizer to around 8 MPa. By including PBATg-MA in the blends, the negative effect of MA was no longer observed, but on the contrary, strength and strain at break values around 12 MPa and 380%, respectively, were obtained. The effective interfacial adhesion brought by the PBAT-g-MA coupling agent explains the improvement in the performance of the TPS/PBAT blends. The negative effect of MA on mechanical properties of PBAT50/TPS50 and PBAT40/TPS60 seems to be in disagreement with previous works highlighting the beneficial effect of MA in PBAT-TPS blends [105, 184]. Nevertheless, it is worth noting that all these results are being compared for blends with compositions close to that for which phase inversion occurs (PBAT40/TPS60 and PBAT45/TPS55). It is possible then that the difference in behavior obtained for the MA is due to morphological changes that produce different physical–chemical responses.

In summary, the effect of compatibilizer's inclusion in blend's mechanical properties is strongly dependent on the PBAT/TPS ratio and on the way in which it is included. In all cases, more important improvements are obtained when the compatibilizer is modified with one of the blend's components to maximize the interaction between them. The enhancement of tensile strength and elongation at break when PBATg-MA was included in PBAT/TPS blends confirmed the efficiency of this reactive compatibilizer to promote the interfacial adhesion between both phases. A presumably reason for this effect is the ester linkage between grafted MA onto PBAT and TPS.

Zhang et al. [108] show a case where the compatibilizer (tartaric acid) is initially mixed with starch and glycerol for the manufacture of TPS-TA, which is later blended with PBAT. Tensile test analysis of PBAT30/TPS-TA70 blends revealed that the use of TA at 0.5% increased the tensile strength value of the blend from 15 to 16 MPa and the elongation at break from 942 to 1311%. However, it was observed that increasing the TA content leads to a continuous decrease in these values, with a tensile strength of 9 MPa and elongation at break of 942% for a TA addition of 4%. A combination of different factors was proposed by the authors to explain this behavior. They observed a drastic reduction in the molecular weight of TPS-TA as the TA content increased. In addition, the interface interaction between phases decreased in the presence of excessive TA concentration. These facts lead to a reverse in the morphology and the presence of TPS-TA particle agglomerations, all of which promoted a decrease in the tensile strength for TA contents higher than 0.5%.

Other additives, such as proteins, have been used as compatibilizers. It is the case of Garcia et al. [111] who reported that in PBAT/starch/glycerol blends (61, 26 and 13 wt%) the addition of sericin increased the Young modulus and tensile strength but decreased the strain at break. These results were observed for all studied sericin contents (0.5–1.5%). As the sericin content increased from 0.5 to 1%, the tensile strength was slightly improved, from 5.7 to 6.4 MPa, Young's modulus increased from about 66–91 MPa and the deformation remained unchanged. By increasing more the sericin concentration, to 1.5%, the authors found no differences in the tensile strength values, but a strong increment in the Young modulus (to approximately 130 MPa) and an important decrease in elongation at break (of about a 50%) were obtained, compared with the blend containing 1% of sericin. These results were attributed to the fact that the presence of sericin reduced the molecular mobility of starch and PBAT chains, and due to its amphoteric character, it could act similarly to a surfactant agent. The side chains of amino acid residues with hydrophobic character (such as glycine) and hydrophilic character (such as serine, aspartate and asparagine) could interact with PBAT and starch, respectively, leading to molecular interactions when located in the interface. The compatibilizer effect explains the mechanical performance and the more homogeneous and compact morphology observed from sericin containing PBAT/starch/glycerol films.

The influence of compatibilizers' inclusion on the material's mechanical properties was also studied in the case of composites. In these cases, the compatibilizers' effect needs to be reevaluated, as new effects can appear. Liu et al. [103] blended a TPS composite reinforced with $nSiO_2$ (TPS-$nSiO_2$) and PBAT, and studied the influence of adding compatibilizers in mechanical properties. They worked with MA, one of the typical compatibilizers of these systems, and plasticized EVOH, which meets the requirement of a starch–PBAT compatibilizer, as it has a lipophilic long chain on one end and many hydroxyl groups on the other. The authors showed that the addition of MA reduces dramatically the impact strength and elongation at break of PBAT40/TPS-$nSiO_2$60/MA compared to PBAT40/TPS-$nSiO_2$60. This decrease was explained as a consequence of a plasticization effect of MA or its participation in hydrolysis reactions of the polymer chains. The impact strength of PBAT40/TPS-$nSiO_2$60/pEVOH was higher than for PBAT40/TPS-$nSiO_2$60/MA, and close to the PBAT40/TPS-$nSiO_2$60hg, without compatibilizer. CA is other possible compatibilizer usually used in these systems. Garcia et al. [102] stated that it might promote not only cross-linking reactions, which could reduce film flexibility but also starch hydrolysis or plasticization, reducing hydrogen bonds between starch chains that weaken the blend structure. De Campos et al. [57] found that the addition of 0.75 wt% of curcumin into a PBAT30/starch49/glycerol21 blend with CAc caused a significant increase in the evaluated mechanical properties (Young's modulus increases 32%, tensile strength 7% and 39% elongation at break), compared to the film without the addition of curcumin. They explained that the presence of curcumin could hinder or inhibit the interaction between citric acid and starch, so better mechanical properties were observed. Furthermore, the change in mechanical properties was not associated with a plasticizing effect since this behavior was undetected by DSC, and they did not find evidence of improvement in compatibilization when comparing the glass transition temperature of the individual materials and the produced blends.

The degree of improvement achieved by filler inclusion usually depends on their dispersion in the polymer matrix as well as their properties. Manepalli and Alavi [50] not only studied the effect of NCC inclusion in PLA/PBAT/TPS blends, but also made a detailed description of the filler's morphology effect on mechanical properties and predicted them by applying the well-known Halpin–Tsai model. They found that the addition of NCC increased the mechanical properties of PLA/PBAT/TPS films and predicted that the increase in aspect ratio (L/d of the particle) and modulus of nanofiller lead to a nanocomposites' modulus increase.

The affinity of the filler to each component of the blend will define its reinforcement effect. Olivato et al. [113] studied the effect of sepiolite addition on PBAT/TPS blends' mechanical properties, varying PBAT/TPS ratio and nanofiller content. Their results suggested that sepiolite was more compatible with the TPS phase, probably due to its more hydrophilic character. For blends with higher TPS content, the dispersion of sepiolite was facilitated and so its relative contribution to blend final mechanical properties. In some cases, significant improvements are achieved when the filler is incorporated into one of the polymers, but in the case of the blend this positive result is no longer observed. Lendvai et al. [114] reported a different mechanical behavior as a function of clay characteristics, when studying the effect of increasing contents of TPS nanocomposites in PBAT. The incorporation of BT and OMMT particles into the starch formulation resulted in the typical reinforcement effect, showing a significant increase in TPS composite strength and modulus without affecting elongation at break with respect to TPS. The incorporation of increasing contents of TPSnBT to PBAT, however, had no significant effect on the strength and modulus data when compared to neat TPS/PBAT blend with the same TPS content. Elongation at break for blends with contents of TPSnBT up to 60% exceeded that of the neat TPS/PBAT blends suggesting a compatibilizing effect of BT particles. When TPSnOMMT was incorporated into PBAT at contents of 40 and 50%, mechanical properties (tensile strength, modulus and elongation at break) were similar to blends without clay.

The costs of industrial processes and new environmental legislation promoted development of composites containing directly natural fibers, in particular from biomass residues. Its cellulose and lignin content strongly depend on the chosen waste, and these values greatly affect the composite's mechanical properties. Spiridon et al. [182] explored the effect of incorporating waste biomass fibers into the PBAT/starch blends in mechanical properties. Young's modulus of biocomposites registered an increase in the rage of 141–191% with respect to starch/PBAT blend as the content of cellulose in fibers increased. Tensile strength values obtained for the biocomposites were similar to those of the starch/PBAT blend, demonstrating that low interactions between fibers and the polymer matrix were achieved. At the same time, the ductility of biocomposites was significantly reduced with the addition of waste fibers. The presence of rigid fibers could restrain the matrix deformation and therefore reduce the elongation of biocomposites. Nunes et al. [55] found similar effects studying the reinforcing effect of babassu mesocarp fiber into PBAT70/TPS30 blends. Their biocomposite presented higher elastic modulus and lower elongation at break with regard to neat PBAT and PBAT70/TPS30. When lignocellulosic fibers are dispersed in a polymer matrix with low affinity (as with PBAT), and the system is stretched, they tend to hinder chain mobility and lower the elongation at break despite maintaining or improving tensile strength and Young's modulus. Although morphological analysis showed poor adhesion between the fiber and the polymer matrix, the increase reported in the elastic modulus for the biocomposite could be due to some interactions of the hydroxyl groups present in TPS and cellulose fiber. The composition of the employed fiber resulted influential in the achieved effect. Spiridon et al. [182] found that higher lignin contents of fiber could result in an improved fiber distribution promoting the interaction with the matrix. Furthermore, lignin addition to the biocomposites resulted in more homogeneous structures, with improved interfacial adhesion due to partial miscibility between cornstarch and lignin. A slight increase in plastic deformation, higher elastic modulus, impact performance and tensile strength was achieved when compared to the neat PBAT/starch blend and PBAT/starch/waste biomass fibers without lignin. The improvement observed could be due to the partial miscibility related to the formation of interactions such as hydrogen bonding and other physical forces between lignin and the polymer matrix.

Zhai et al. [97] simultaneously used two strategies to improve the performance of the material: using a modified starch and adding a filler. They developed PBAT/TPS blends using hydroxypropyl distarch phosphate (HPDSP), citric acid as compatibilizer and a nanoclay as filler (an organic montmorillonite modified with octadecyl dimethyl benzyl ammonium chloride). The authors found that at fixed amounts of nanofiller and compatibilizer content, an increase in the amount of PBAT improved mechanical strength and flexibility, especially for nanocomposite films with PBAT content higher than 30%. The maximum tensile strength of the PBAT50/HPDSP50 nanocomposite film was approximately five times higher than that of the HPDSP, and the elongation at break (EB) was seven times higher than that of the starch nanocomposite film. The maximum tensile strength and elongation at break of the PBAT50/HPDSP50 nanocomposite films were 7.4 MPa and 614%, respectively.

4 Conclusions and Future Perspectives

Although the proposal to replace conventional plastics with starch-based materials is a very attractive idea from an ecological and economical point of view, the properties of TPS are not enough to meet the requirements of many applications. The development of starch-based composites with the incorporation of natural fillers helps improve their performance, but it is still a long way from commodity plastics.

The best strategy found so far is the use of starch blends with biodegradable polyesters that improve the properties of TPS together with the inclusion of fillers. The results obtained by different researchers show that a critical key to achieving the desired properties is the proper design of compounding–processing strategies. However, desirable properties such as good processability, thermal stability, mechanical strength and improved barrier properties still must be achieved to increase the market share among commodity materials. Hence, further studies, including the introduction of new biobased plasticizers, cross-linkers, additives and/or different types of micro-nanofillers, are required, to fulfill the needs of each particular plastic sector. Rationally designing these materials, including higher biomass contents as a valuable resource, and the use of green methodologies to obtain them, would establish a sustainable production value.

One of the most promising fields of application for these materials is agri-horticulture. In much of the recent scientific works and published patents related to agricultural materials, modified starches blended with other biopolymers are used, being today the critical points to solve the compatibility and processing problems of the mixtures [185, 186]. In relation to plastics destined for agriculture, a strong growth in mulching and covering materials is expected, being PBAT one of the most attractive candidates to be used since it gives the mixture high toughness and elongation at break while maintaining its biodegradability [99, 108]. PHB is other excellent candidate that has the extra advantage of being biobased, but which production cost is much higher and can be used in applications where hydrophobicity improvements are required, as well as higher Young's modulus and stress at break [165].

Although PBAT is still fully fossil derived, sustainable alternatives have been proposed looking for obtaining a whole biobased PBAT. In this sense, the production of biobased 1,4 butanediol (BDO) has been achieved through industrial biological fermentation; sebacic acid (as a substitute of adipic acid) coming from castor oil has been used as a monomer to prepare poly(butylene sebacinate-co-butylene terephthalate) (PBSeT) co-polyesters; 2, 5-furandicarboxylic acid (FDCA) has been regarded as a biobased alternative to the petroleum-based terephthalic acid.

PHB, on the other hand, has the great advantage of being fully biobased and biodegradable in soil. However, the challenge of obtaining good compatibility with TPS remains unsolved. At this moment, the studies carried out were orientated to develop different strategies to improve compatibility and therefore the physical and mechanical properties of the blends. These strategies were focused on using TPS from various botanical sources, modified polymers and the addition of compatibilizing agents as well as micro-nanofillers.

The amount of filler incorporated into the composites is usually very small, most of the times much lower than 5%. When choosing the reinforcement to use, a cost–benefit analysis must be carried out in all cases. The only truly green fillers are renewable, biobased and biodegradable, such as cellulose. However, other fillers, such as clays, are much more economical, do not generate a high environmental impact and can generate strong improvements if a good dispersion is achieved. For that reasons, they were and are one of the most studied fillers as reinforcement of starch blend. In some cases, it is preferred to choose fillers that do not generate the most important improvements, but that are completely green. However, this can be tricky, since the method of obtaining/purifying them can include the use of non-environmentally friendly processes, as happens for example with some cellulosic reinforcements. To date, there is not enough experimental evidence to show that 100% green starch-based composites with suitable properties to replace commodity plastics can be produced, including the use of green chemistry to obtain the modified starches or to make the grafts in the bioplastics with which it is usually combined.

Nowadays, trend aims to generate plastics with the least environmental impact. Phrases like "a better bioplastic for a better world", "plastic more sustainable, compostable, and eco-friendly" or "zero waste to make a difference" are the head of many of today's developments and even appear on the presentation pages of many companies. In this framework, starch composites such as those presented in this chapter have great potential to be inserted into the agricultural and packaging industry. It is expected that the problems that still persist in these materials' properties will be solved in the near future, responding to the demands of society for a sustainable world.

Acknowledgements This work was supported by Agencia Nacional de Promoción científica y Tecnológica (ANPCyT PICT 2017-2362), Universidad de Buenos Aires (UBACyT 20020170100381BA).

References

1. Rameshkumar S, Shaiju P, O'Connor KE (2020) Bio-based and biodegradable polymers—state of the art, challenges and emerging trends. Curr Opin Green Sustain Chem 21:75–81
2. Skoczinski P, Chinthapalli R, Carus M, Baltus W, de Guzman D, Käb H, Raschka A, Ravenstijn J (2020) Global markets and trends of bio-based building blocks and polymers 2019–2024. Carus M (Publ) Nova-Institut GmbH, Germany
3. Villar MA, Barbosa SE, García MA, Castillo LA, López OV (eds) (2017) Starch-based materials in food packaging. Processing, characterization and applications. Academic Press, United Kingdom and United States
4. Visakh PM (2015) Starch: state-of-the-art, new challenges and opportunities. In: Visakh PM, Yu L (eds) Starch-based blends, composites and nanocomposites. RSC Green Chemistry, United Kigndom, pp 1–16
5. Ribba L, Garcia NL, D'Accorso N, Goyanes S (2017) Disadvantages of starch-based materials, feasible alternatives in order to overcome these limitations. In: Villa MA, Barbosa SE, Garcia MA, Castillo LA, Lopez OV (eds) Starch-based materials in food packaging, 1st edn. Academic Press, Cambridge, pp 37–76
6. Xie F, Zhang B, Wang DK (2017) Starch thermal processing: technologies at laboratory and semi-industrial scales. In: Villa MA, Barbosa SE, Garcia MA, Castillo LA, Lopez OV (eds) Starch-based materials in food packaging, 1st edn. Academic Press, Cambridge, pp 187–227
7. Thunwall M, Boldizar A, Rigdahl M, Kuthanová V (2006) On the stress-strain behavior of thermoplastic starch melts. Int J Polym Anal Charact 11(6):419–428
8. González-Seligra P, Guz L, Ochoa-Yepes O, Goyanes S, Famá L (2017) Influence of extrusion process conditions on starch film morphology. LWT 84:520–528
9. von Borries-Medrano E, Jaime-Fonseca MR, Aguilar-Mendez MA (2016) Starch–guar gum extrudates: microstructure, physicochemical properties and in-vitro digestion. Food Chem 194:891–899
10. Bouvier JM, Campanella OH (2014) Extrusion processing technology: food and non-food biomaterials. Wiley, Hoboken
11. Lawal A, Kalyon DM (1995) Mechanisms of mixing in single and co-rotating twin screw extruders. Polym Eng Sci 35(17):1325–1338
12. Lai LS, Kokini JL (1991) Physicochemical changes and rheological properties of starch during extrusion (a review). Biotechnol Prog 7(3):251–266
13. García NL, Famá L, D'Accorso NB, Goyanes S (2015) Biodegradable starch nanocomposites. Adv Struct Mater 75:17–77
14. Jebalia I, Maigret JE, Réguerre AL, Novales B, Guessasma S, Lourdin D, Della Valle G, Kristiawan M (2019) Morphology and mechanical behaviour of pea-based starch-protein composites obtained by extrusion. Carbohydr Polym 223:115086
15. Nessi V, Falourd X, Maigret JE, Cahier K, D'orlando A, Descamps N, Gaucher V, Chevigny C, Lourdin D (2019) Cellulose nanocrystals-starch nanocomposites produced by extrusion: structure and behavior in physiological conditions. Carbohydr Polym 225:115123
16. Estevez-Areco S, Guz L, Famá L, Candal R, Goyanes S (2019) Bioactive starch nanocomposite films with antioxidant activity and enhanced mechanical properties obtained by extrusion followed by thermo-compression. Food Hydrocolloids 96:518–528
17. Gutiérrez TJ, Toro-Márquez LA, Merino D, Mendieta JR (2019) Hydrogen-bonding interactions and compostability of bionanocomposite films prepared from corn starch and nano-fillers with and without added Jamaica flower extract. Food Hydrocolloids 89:283–293
18. Drobny JG, (2014) Handbook of thermoplastic elastomers. PDL (Plastics Design Library)/William Andrew, Pensilvania
19. Thunwall M, Kuthanova V, Boldizar A, Rigdahl M (2008) Film blowing of thermoplastic starch. Carbohydr Polym 71(4):583–590
20. Fishman ML, Coffin DR, Onwulata CI, Willett JL (2006) Two stage extrusion of plasticized pectin/poly (vinyl alcohol) blends. Carbohydr Polym 65(4):421–429

21. Jana T, Maiti S (1999) A biodegradable film, 1. Pilot plant investigation for production of the biodegradable film. Angew Makromolek Chem 267(1):16–19
22. Otey FH, Westhoff RP, Doane WM (1987) Starch-based blown films. 2. Ind Eng Chem Res 26(8):1659–1663
23. Halley P, Rutgers R, Coombs S, Kettels J, Gralton J, Christie G, Jenkins M, Beh H, Griffin K, Jayasekara R, Lonergan G (2001) Developing biodegradable mulch films from starch-based polymers. Starch-Stärke 53(8):362–367
24. Matzinos P, Tserki V, Kontoyiannis A, Panayiotou C (2002) Processing and characterization of starch/polycaprolactone products. Polym Degrad Stabil 77(1):17–24
25. Huntrakul K, Yoksan R, Sane A, Harnkarnsujarit N (2020) Effects of pea protein on properties of cassava starch edible films produced by blown-film extrusion for oil packaging. Food Packaging Shelf 24:100480
26. Dang KM, Yoksan R (2016) Morphological characteristics and barrier properties of thermoplastic starch/chitosan blown film. Carbohyd Polym 150:40–47
27. Abbégs B, Ayad R, Prudhomme JC, Onteniente JP (1998) Numerical simulation of thermoplastic wheat starch injection molding process. Polym Eng Sci 38(12):2029–2038
28. Onteniente JP, Abbès B, Safa LH (2000) Fully biodegradable lubricated thermoplastic wheat starch: mechanical and rheological properties of an injection grade. Starch-Stärke 52(4):112–117
29. Stepto RFT (2009) Thermoplastic starch. Macromol Symp 279:163–168
30. Ye J, Luo S, Huang A, Chen J, Liu C, McClements DJ (2019) Synthesis and characterization of citric acid esterified rice starch by reactive extrusion: a new method of producing resistant starch. Food Hydrocolloids 92:135–142
31. Siyamak S, Luckman P, Laycock B (2020) Rapid and solvent-free synthesis of pH-responsive graft-copolymers based on wheat starch and their properties as potential ammonium sorbents. Int J Biol Macromol 149:477–486
32. Okonkwoa EG, Aigbodiona VS, Akinlabib ET, Daniel-Mkpumea CC (2020) Starch-based composites and their applications. In: Al-Ahmed A, Inamuddin (eds) Advanced applications of polysaccharides and their composites, 1st edn. Materials Research Forum LLC, Millersville, pp 73–198
33. Ilyas RA, Sapuan SM, Ibrahim R, Abral H, Ishak MR, Zainudin ES, Atiqah A, Atikah MSN, Syafri E, Jumaidin R (2020) Thermal, biodegradability and water barrier properties of bio-nanocomposites based on plasticised sugar palm starch and nanofibrillated celluloses from sugar palm fibres. J Biobased Mater Bio 14(2):234–248
34. Valencia-Sullca C, Jiménez M, Jiménez A, Atarés L, Vargas M, Chiralt A (2016) Influence of liposome encapsulated essential oils on properties of chitosan films. Polym Int 65(8):979–987
35. Talón E, Vargas M, Chiralt A, González-Martínez C (2019) Eugenol incorporation into thermoprocessed starch films using different encapsulating materials. Food Packaging Shelf 21:100326
36. Wandrey C, Bartkowiak A, Harding SE (2010) Materials for encapsulation. In: Zuidam NJ, Nedovic V (eds) Encapsulation technologies for active food ingredients and food processing, 1st edn. Springer Science & Business Media, New York, pp 31–100
37. Jin Y, Li JZ, Nik AM (2018) Starch-based microencapsulation. In: Sjöö M, Nilsson L (eds) Starch in food: structure, function and applications, 1st edn. Woodhead Publishing, Cambridge, pp 661–690
38. Castro N, Durrieu V, Raynaud C, Rouilly A, Rigal L, Quellet C (2016) Melt extrusion encapsulation of flavors: a review. Polym Rev 56(1):137–186
39. Chen S, Zong J, Jiang L, Ma C, Li H, Zhang D (2020) Improvement of resveratrol release performance and stability in extruded microparticle by the α-amylase incorporation. J Food Eng 274:109842
40. Zhu F (2017) Encapsulation and delivery of food ingredients using starch based systems. Food Chem 229:542–552
41. Yuliani S, Torley PJ, D'Arcy B, Nicholson T, Bhandari B (2006) Extrusion of mixtures of starch and d-limonene encapsulated with β-cyclodextrin: flavour retention and physical properties. Food Res Int 39(3):318–331

42. Ochoa-Yepes O, Di Giogio L, Goyanes S, Mauri A, Famá L (2019) Influence of process (extrusion/thermo-compression, casting) and lentil protein content on physicochemical properties of starch films. Carbohydr Polym 208:221–231
43. Ghanbari A, Tabarsa T, Ashori A, Shakeri A, Mashkour M (2018) Preparation and characterization of thermoplastic starch and cellulose nanofibers as green nanocomposites: extrusion processing. Int J Biol Macromol 112:442–447
44. Dean K, Yu L, Wu DY (2007) Preparation and characterization of melt-extruded thermoplastic starch/clay nanocomposites. Compos Sci Technol 67(3–4):413–421
45. Chaves da Silva NM, Fakhouri FM, Fialho RLL, Cabral Albuquerque ECDM (2018) Starch–recycled gelatin composite films produced by extrusion: physical and mechanical properties. J Appl Polym Sci 135(19):46254
46. Huang MF, Yu JG, Ma XF (2004) Studies on the properties of montmorillonite-reinforced thermoplastic starch composites. Polymer 45(20):7017–7023
47. González K, Iturriaga L, González A, Eceiza A, Gabilondo N (2020) Improving mechanical and barrier properties of thermoplastic starch and polysaccharide nanocrystals nanocomposites. Eur Polym J 123:109415
48. Fourati Y, Magnin A, Putaux JL, Boufi S (2020) One-step processing of plasticized starch/cellulose nanofibrils nanocomposites via twin-screw extrusion of starch and cellulose fibers. Carbohydr Polym 229:115554
49. Grylewicz A, Spychaj T, Zdanowicz M (2019) Thermoplastic starch/wood biocomposites processed with deep eutectic solvents. Compos Part A Appl Sci Manuf 121:517–524
50. Manepalli PH, Alavi S (2019) Mathematical modeling of mechanical and barrier properties of poly(lactic acid)/poly(butylene adipate-co-terephthalate)/thermoplastic starch based nanocomposites. J Food Eng 261:60–65
51. Chivrac F, Pollet E, Dole P, Avérous L (2010) Starch-based nano-biocomposites: plasticizer impact on the montmorillonite exfoliation process. Carbohydr Polym 79(4):941–947
52. Adamus J, Spychaj T, Zdanowicz M, Jędrzejewski R (2018) Thermoplastic starch with deep eutectic solvents and montmorillonite as a base for composite materials. Ind Crop Prod 123:278–284
53. Utracki LA (ed) (2002) Polymer blends handbook. Kluwer Academic Publishers, Dordrecht
54. Muthuraj R, Misra M, Mohanty AK (2018) Biodegradable compatibilized polymer blends for packaging applications: a literature review. J Appl Polym Sci 135(24):45726
55. Nunes MABS, Marinho VAD, Falcão GAM, Canedo EL, Bardi MAG, Carvalho LH (2018) Rheological, mechanical and morphological properties of poly(butylene adipate-co-terephthalate)/thermoplastic starch blends and its biocomposite with babassu mesocarp. Polym Test 70:281–288
56. Dammak M, Fourati Y, Tarrés Q, Delgado-Aguilar M, Mutjé P, Boufi S (2020) Blends of PBAT with plasticized starch for packaging applications: mechanical properties, rheological behaviour and biodegradability. Ind Crop Prod 144:112061
57. de Campos SS, de Oliveira A, Moreira TFM, da Silva TBV, da Silva MV, Pinto JA, Bilck AP, Gonçalves OH, Fernandes IP, Barreiro M-F, Yamashita F, Valderrama P, Shirai MA, Leimann FV (2019) TPCS/PBAT blown extruded films added with curcumin as a technological approach for active packaging materials. Food Packaging Shelf 22:100424
58. Nesic A, Castillo C, Castaño P, Cabrera-Barjas G, Serrano J (2020) Bio-based packaging materials. In: Galanakis CM (ed) Biobased products and industries, 1st edn. Elsevier, Amsterdam, pp 279–309
59. Platt DK (2006) Biodegradable polymers—market report. Smithers Rapra Limited, United Kingdom
60. Shogren R, Wood D, Orts W, Glenn G (2019) Plant-based materials and transitioning to a circular economy. Sustain Prod Consum 19:194–215
61. Ashter SA (2016) Introduction to bioplastics engineering. William Andrew Publishing, Oxford
62. Ferreira FV, Cividanes LS, Gouveia RF, Lona LMF (2019) An overview on properties and applications of poly(butylene adipate-co-terephthalate)–PBAT based composites. Polym Eng Sci 59:E7–E15

63. Markovic G, Visakh PM (2017) Polymer blends: state of art. In: Visakh PM, Markovic G, Pasquini D (eds) Recent developments in polymer macro, micro and nano blends, 1st edn. Woodhead Publishing, UK, pp 1–15
64. Liu W, Liu S, Wang Z, Liu J, Dai B, Chen Y, Zeng G (2020) Preparation and characterization of compatibilized composites of poly(butylene adipate-co-terephthalate) and thermoplastic starch by two-stage extrusion. Eur Polym J 122:109369
65. Mochane MJ, Sefadi JS, Motsoeneng TS, Mokoena TE, Mofokeng TG, Mokhena TC (2020) The effect of filler localization on the properties of biopolymer blends, recent advances: a review. Polym Compos 1–22. https://doi.org/10.1002/pc.25590
66. Ray SS, Salehiyan R (2019) Nanostructured immiscible polymer blends: migration and interface. Elsevier, Amsterdam
67. Lemoigne M (1926) Produit de deshydratation et de polymerisation de l'acide b-oxybutyrique. Bull Soc Chim Biol 8:770–782
68. Kumar M, Rathour R, Singh R, Sun Y, Pandey A, Gnansounou E, Lin KYA, Tsang DCW, Thakur IS (2020) Bacterial polyhydroxyalkanoates: opportunities, challenges, and prospects. J Clean Prod 121500:1–20
69. Balaji S, Gopi K, Muthuvelan B (2013) A review on production of poly β hydroxybutyrates from cyanobacteria for the production of bio plastics. Algal Res 2(3):278–285
70. Bhatia SK, Shim YH, Jeon JM, Brigham CJ, Kim YH, Kim HJ, Seo HM, Lee JH, Kim JH, Yi DH, Lee YK, Yang YH (2015) Starch based polyhydroxybutyrate production in engineered *Escherichia coli*. Bioproc Biosyst Eng 38(8):1479–1484
71. Bugnicourt E, Cinelli P, Lazzeri A, Alvarez VA (2014) Polyhydroxyalkanoate (PHA): review of synthesis, characteristics, processing and potential applications in packaging. Express Polym Lett 8(11):791–808
72. Manikandan NA, Pakshirajan K, Pugazhenthi G (2020) Preparation and characterization of environmentally safe and highly biodegradable microbial polyhydroxybutyrate (PHB) based graphene nanocomposites for potential food packaging applications. Int J Biol Macromol 154:866–877
73. Souza Jd, Chiaregato CG, Faez R (2018) Green composite based on PHB and montmorillonite for KNO_3 and NPK delivery system. J Polym Environ 26:670–679
74. Coltelli MB, Danti S, Trombi L, Morganti P, Donnarumma G, Baroni A, Fusco A, Lazzeri A (2018) Preparation of innovative skin compatible films to release polysaccharides for biobased beauty masks. Cosmetics 5(4):70
75. Coltelli MB, Panariello L, Morganti P, Danti S, Baroni A, Lazzeri A, Danti S, Baroni A, Lazzeri A, Fusco A, Donnarumma G (2020) Skin-compatible biobased beauty masks prepared by extrusion. J Funct Biomater 11(2):23
76. Din MI, Ghaffar T, Najeeb J, Hussain Z, Khalid R, Zahid H (2020) Potential perspectives of biodegradable plastics for food packaging application—review of properties and recent developments. Food Addit Contam A 37(4):665–680
77. Garrido-Miranda KA, Rivas BL, Pérez-Rivera MA, Sanfuentes EA, Peña-Farfal C (2018) Antioxidant and antifungal effects of eugenol incorporated in bionanocomposites of poly (3-hydroxybutyrate)-thermoplastic starch. LWT-Food Sci Technol 98:260–267
78. Weinmann S, Bonten C (2019) Thermal and rheological properties of modified polyhydroxybutyrate (PHB). Polym Eng Sci 59(5):1057–1064
79. Corre YM, Bruzaud S, Audic JL, Grohens Y (2012) Morphology and functional properties of commercial polyhydroxyalkanoates: a comprehensive and comparative study. Polym Test 31(2):226–235
80. Pachekoski WM, Dalmolin C, Agnelli JAM (2013) The influence of the industrial processing on the degradation of poly (hidroxybutyrate)-PHB. Mater Res 16(2):237–332
81. Correa JP (2018) Nanocompuestos poliméricos: estudio de la interacción de mezclas de polímeros biodegradables con refuerzos químicamente modificados. Ph. D. thesis, Universidad de San Martín, Buenos Aires, Argentina
82. Yamaguchi M, Arakawa K (2006) Effect of thermal degradation on rheological properties for poly (3-hydroxybutyrate). Eur Polym J 42(7):1479–1486

83. Renstad R, Karlsson S, Albertsson AC (1999) The influence of processing induced differences in molecular structure on the biological and non-biological degradation of poly (3-hydroxybutyrate-co-3-hydroxyvalerate), P (3-HB-co-3-HV). Polym Degrad Stabil 63(2):201–211
84. Janigová I, Lacík I, Chodák I (2002) Thermal degradation of plasticized poly (3-hydroxybutyrate) investigated by DSC. Polym Degrad Stabil 77(1):35–41
85. Koller I, Owen AJ (1996) Starch-filled PHB and PHB/HV copolymer. Polym Int 39(3):175–181
86. Kotnis MA, O'Brien GS, Willett JL (1995) Processing and mechanical properties of biodegradable poly (hydroxybutyrate-co-valerate)-starch compositions. J Polym Environ 3(2):97–105
87. Rosa DDS, Rodrigues TC, Gracas Fassina Guedes CD, Calil MR (2003) Effect of thermal aging on the biodegradation of PCL, PHB-V, and their blends with starch in soil compost. J Appl Polym Sci 89(13):3539–3546
88. Gomes Brunel D, Pachekoski WM, Dalmolin C, Marcondes Agnelli JA (2014) Natural additives for poly (hydroxybutyrate-CO-hydroxyvalerate)-PHBV: effect on mechanical properties and biodegradation. Mater Res 17(5):1145–1156
89. Chen L, Zhu M, Song L, Yu H, Zhang Y, Chen Y, Adler H (2004) Crystallization behavior and thermal properties of blends of poly-(3-hydroxybutyrate-co-3-hydroxyvalerate) and poly(1,2-propandiolcarbonate). Macromol Symp 210:241–250
90. Modi SJ (2010) Assessing the feasibility of poly-(3-hydroxybutyrate-co-3-hydroxyvalerate) (PHBV) and poly-(lactic acid) for potential food packaging applications. M.Sc. thesis, The Ohio State University, EEUU
91. Yu L (2009) Biodegradable polymer blends and composites from renewable resources. Wiley, Hoboken
92. Don TM, Chung ChY, Lai SM, Chiu HJ (2010) Preparation and properties of blends from poly(3-hydroxybutyrate) with poly(vinyl acetate)-modified starch. Polym Eng Sci 50(4):709–718
93. Thiré RM, Ribeiro TA, Andrade CT (2006) Effect of starch addition on compression-molded poly (3-hydroxybutyrate)/starch blends. J Appl Polym Sci 100(6):4338–4347
94. Zhang M, Thomas NL (2010) Preparation and properties of polyhydroxybutyrate blended with different types of starch. J Appl Polym Sci 116(2):688–694
95. Garrido-Miranda KA, Rivas BL, Pérez MA (2017) Poly (3-hydroxybutyrate)–thermoplastic starch–organoclay bionanocomposites: surface properties. J Appl Polym Sci 134(34):45217(1–8)
96. Jian J, Xiangbin Z, Xianbo H (2020) An overview on synthesis, properties and applications of poly(butylene-adipate-co-terephthalate)–PBAT. Adv Ind Eng Polym Res 3(1):19–26
97. Zhai X, Wang W, Zhang H, Dai Y, Dong H, Hou H (2020) Effects of high starch content on the physicochemical properties of starch/PBAT nanocomposite films prepared by extrusion blowing. Carbohydr Polym 239:116231
98. Meraldo A (2016) Introduction to bio-based polymers. In: Wagner JR (ed) Multilayer flexible packaging, 2nd edn. Plastics design library. William Andrew Publishing, New York, pp 47–52
99. Nunes MABS, Castro-Aguirre E, Auras RA, Bardi MAG, Carvalho LH (2019) Effect of babassu mesocarp incorporation on the biodegradation of a PBAT/TPS blend. Macromol Symp 383:1800043
100. Fourati Y, Tarrés Q, Mutjé P, Boufi S (2018) PBAT/thermoplastic starch blends: effect of compatibilizers on the rheological, mechanical and morphological properties. Carbohydr Polym 199:51–57
101. Ivanič F, Kováčová M, Chodák I (2019) The effect of plasticizer selection on properties of blends poly(butylene adipate-co-terephthalate) with thermoplastic starch. Eur Polym J 116:99–105
102. Garcia PS, Grossmann MVE, Shirai MA, Lazaretti MM, Yamashita F, Muller CMO, Mali S (2014) Improving action of citric acid as compatibiliser in starch/polyester blown films. Ind Crop Prod 52:305–312

103. Liu W, Liu S, Wang Z, Dai B, Liu J, Chen Y, Zeng G, He Y, Liu Y, Liu R (2019) Preparation and characterization of reinforced starch-based composites with compatibilizer by simple extrusion. Carbohydr Polym 223:115122
104. Da Roz AL, Carvalho AJF, Gandini A, Curvelo AAS (2006) The effect of platicizers on thermoplastic starch compositions obtained by melt processing. Carbohydr Polym 63:417–424
105. Olivato JB, Grossmann MVE, Yamashita F, Eiras D, Pessan LA (2012) Citric acid and maleic anhydride as compatibilizers in starch/poly(butylene adipate-co-terephthalate) blends by one-step reactive extrusion. Carbohydr Polym 87(4):2614–2618
106. Wei D, Wang H, Xiao H, Zheng A, Yang Y (2015) Morphology and mechanical properties of poly(butylene adipate-co-terephthalate)/potato starch blends in the presence of synthesized reactive compatibilizer or modified poly(butylene adipate-co-terephthalate). Carbohydr Polym 123:275–282
107. Robeson LM (2007) Polymer blends: a comprehensive review. Hanser, Munich
108. Zhang S, He Y, Lin Z, Li J, Jiang G (2019) Effects of tartaric acid contents on phase homogeneity, morphology and properties of poly (butyleneadipate-co-terephthalate)/ thermoplastic starch bio-composities. Polym Test 76:385–395
109. Olivato JB, Müller CMO, Carvalho GM, Yamashita F, Grossmann MVE (2014) Physical and structural characterisation of starch/polyester blends with tartaric acid. Mater Sci Eng C 39:35–39
110. Marinho V, Pereira CAB, Vitorino MBC, Silva AS, Carvalho LH, Canedo EL (2017) Degradation and recovery in poly(butylene adipate-co-terephthalate)/thermoplastic starch blends. Polym Test 58:166–172
111. Garcia PS, Turbiani FRB, Baron AM, Yamashita F, Eiras D, Grossmann VE (2018) Sericin as compatibilizer in starch/polyester blown films. Polímeros (São Carlos, Online) 28(5):389–394
112. Xiong SJ, Pang B, Zhou SJ, Li MK, Yang S, Wang YY, Shi Q, Wang SF, Yuan TQ, Sun RC (2020) Economically competitive biodegradable PBAT/lignin composites: effect of lignin methylation and compatibilizer. ACS Sustain Chem Eng 8(13):5338–5346
113. Olivato JB, Marini J, Yamashita F, Pollet E, Grossmann MVE, Avérous L (2017) Sepiolite as a promising nanoclay for nano-biocomposites based on starch and biodegradable polyester. Mater Sci Eng C 70:296–302
114. Lendvai L, Apostolov A, Karger-Kocsis J (2017) Characterization of layered silicate-reinforced blends of thermoplastic starch (TPS) and poly(butylene adipate-co-terephthalate). Carbohydr Polym 173:566–572
115. He H, Liu B, Xue B, Zhang H (2019) Study on structure and properties of biodegradable PLA/PBAT/organic-modified MMT nanocomposites. J Thermoplast Compos Mater 1–18. https://doi.org/10.1177/0892705719890907
116. Guz L, Famá L, Candal R, Goyanes S (2017) Size effect of ZnO nanorods on physicochemical properties of plasticized starch composites. Carbohydr Polym 157:1611–1619
117. Morales NJ, Candal R, Famá L, Goyanes S, Rubiolo GH (2015) Improving the physical properties of starch using a new kind of water dispersible nano-hybrid reinforcement. Carbohydr Polym 127:291–299
118. Florencia V, López OV, García MA (2020) Exploitation of by-products from cassava and ahipa starch extraction as filler of thermoplastic corn starch. Compos Part B Eng 182:107653
119. Fazeli M, Keley M, Biazar E (2018) Preparation and characterization of starch-based composite films reinforced by cellulose nanofibers. Int J Biol Macromol 116:272–280
120. Pandit P, Nadathur GT, Maiti S, Regubalan B (2018) Functionality and properties of bio-based materials. In: Ahmed S (ed) Bio-based materials for food packaging: green and sustainable advanced packaging materials, 1st edn. Springer, Singapore, pp 81–103
121. Battegazzore D, Bocchini S, Nicola G, Martini E, Frache A (2015) Isosorbide, a green plasticizer for thermoplastic starch that does not retrogradate. Carbohydr Polym 119:78–84
122. Preetha B, Sreerag G, Geethamma VG, Sabu T (2019) Mechanical and permeability properties of thermoplastic starch composites reinforced with cellulose nanofiber for packaging applications. J Siberian Federal Univ Biol 12(3):287–301

123. Sadegh-Hassani F, Nafchi AM (2014) Preparation and characterization of bionanocomposite films based on potato starch/halloysite nanoclay. Int J Biol Macromol 67:458–462
124. Wolf C, Angellier-Coussy H, Gontard N, Doghieri F, Guillard V (2018) How the shape of fillers affects the barrier properties of polymer/non-porous particles nanocomposites: a review. J Membr Sci 556:393–418
125. Rybak A, Rybak A, Sysel P (2018) Modeling of gas permeation through mixed-matrix membranes using novel computer application MOT. Appl Sci 8(7):1166
126. Vinh-Thang H, Kaliaguine S (2013) Predictive models for mixed-matrix membrane performance: a review. Chem Rev 113(7):4980–5028
127. García NL, Ribba L, Dufresne A, Aranguren MI, Goyanes S (2009) Physico-mechanical properties of biodegradable starch nanocomposites. Macromol Mater Eng 294(3):169–177
128. García NL, Ribba L, Dufresne A, Aranguren MI, Goyanes S (2011) Effect of glycerol on the morphology of nanocomposites made from thermoplastic starch and starch nanocrystals. Carbohydr Polym 84(1):203–210
129. Jiménez A, Fabra MJ, Talens P, Chiralt A (2012) Effect of re-crystallization on tensile, optical and water vapour barrier properties of corn starch films containing fatty acids. Food Hydrocolloids 26(1):302–310
130. Gao W, Wu W, Liu P, Hou H, Li X, Cui B (2020) Preparation and evaluation of hydrophobic biodegradable films made from corn/octenylsuccinated starch incorporated with different concentrations of soybean oil. Int J Biol Macromol 142:376–383
131. Kang X, Liu P, Gao W, Wu Z, Yu B, Wang R, Cui B, Qiu L, Sun C (2020) Preparation of starch-lipid complex by ultrasonication and its film forming capacity. Food Hydrocolloids 99:105340
132. López OV, Castillo LA, Barbosa SE, Villar MA, Alejandra García M (2018) Processing–properties–applications relationship of nanocomposites based on thermoplastic corn starch and talc. Polym Compos 39(4):1331–1338
133. Moghaddam MRA, Razavi SMA, Jahani Y (2018) Effects of compatibilizer and thermoplastic starch (TPS) concentration on morphological, rheological, tensile, thermal and moisture sorption properties of plasticized polylactic acid/TPS blends. J Polym Environ 26(8):3202–3215
134. Müller CM, Laurindo JB, Yamashita F (2012) Composites of thermoplastic starch and nanoclays produced by extrusion and thermopressing. Carbohydr Polym 89(2):504–510
135. Masclaux C, Gouanve F, Espuche E (2010) Experimental and modelling studies of transport in starch nanocomposite films as affected by relative humidity. J Membr Sci 363(1–2):221–231
136. Beigmohammadi F, Barzoki ZM, Shabanian M (2020) Rye flour and cellulose reinforced starch biocomposite: a green approach to improve water vapor permeability and mechanical properties. Starch-Stärke 72(5–6):1900169
137. dos Santos BH, de Souza do Prado K, Jacinto AA, da Silva S, Aparecida M (2018) Influence of sugarcane bagasse fiber size on biodegradable composites of thermoplastic starch. J Renew Mater 6(2):176–182
138. Li J, Zhou M, Cheng G, Cheng F, Lin Y, Zhu PX (2019) Fabrication and characterization of starch-based nanocomposites reinforced with montmorillonite and cellulose nanofibers. Carbohydr Polym 210:429–436
139. Li J, Zhou M, Cheng F, Lin Y, Zhu P (2019) Bioinspired approach to enhance mechanical properties of starch based nacre-mimetic nanocomposite. Carbohydr Polym 221:113–119
140. Gutiérrez TJ, Ollier R, Alvarez VA (2018) Surface properties of thermoplastic starch materials reinforced with natural fillers. In: Thakur VK, Thakur MK (eds) Functional biopolymers, 1st edn. Springer International Publishing, Cham, pp 131–158
141. Abreu AS, Oliveira M, de Sá A, Rodrigues RM, Cerqueira MA, Vicente AA, Machado AV (2015) Antimicrobial nanostructured starch-based films for packaging. Carbohydr Polym 129:127–134
142. Chen J, Chen F, Meng Y, Wang S, Long Z (2019) Oxidized microcrystalline cellulose improve thermoplastic starch-based composite films: thermal, mechanical and water-solubility properties. Polymer 168:228–235

143. Azevedo VM, Borges SV, Marconcini JM, Yoshida MI, Neto ARS, Pereira TC, Pereira CFG (2017) Effect of replacement of corn starch by whey protein isolate in biodegradable film blends obtained by extrusion. Carbohydr Polym 157:971–980
144. Niranjana Prabhu T, Prashantha K (2018) A review on present status and future challenges of starch based polymer films and their composites in food packaging applications. Polym Compos 39(7):2499–2522
145. Merino D, Gutiérrez TJ, Alvarez VA (2019) Potential agricultural mulch films based on native and phosphorylated corn starch with and without surface functionalization with chitosan. J Polym Environ 27(1):97–105
146. Famá, LM, Goyanes, S, Pettarin, V, Bernal, CR (2015) Mechanical behavior of starch–carbon nanotubes composites. In: Kar KK, Pandey JK, Rana S (eds) Handbook of polymer nanocomposites. Processing, performance and application. Springer, Berlin, pp 141–171
147. Cheng Y, Yu G (2020) The research of interface microdomain and corona-resistance characteristics of micro-nano-ZnO/LDPE. Polymers 12(3):563
148. Angellier H, Molina-Boisseau S, Dole P, Dufresne A (2006) Thermoplastic starch−waxy maize starch nanocrystals nanocomposites. Biomacromolecules 7(2):531–539
149. Dai L, Zhang J, Cheng F (2020) Cross-linked starch-based edible coating reinforced by starch nanocrystals and its preservation effect on graded Huangguan pears. Food Chem 311:125891
150. Li X, Qiu C, Ji N, Sun C, Xiong L, Sun Q (2015) Mechanical, barrier and morphological properties of starch nanocrystals-reinforced pea starch films. Carbohydr Polym 121:155–162
151. Mukurumbira AR, Mellem JJ, Amonsou EO (2017) Effects of amadumbe starch nanocrystals on the physicochemical properties of starch biocomposite films. Carbohydr Polym 165:142–148
152. Chang PR, Jian R, Yu J, Ma X (2010) Fabrication and characterisation of chitosan nanoparticles/plasticized-starch composites. Food Chem 120(3):736–740
153. Othman SH, Kechik NR, Shapi'i RA, Talib RA, Tawakkal IS (2019) Water sorption and mechanical properties of starch/chitosan nanoparticle films. J Nanomater. https://doi.org/10.1155/2019/3843949
154. Shapi'i RA, Othman SH, Nordin N, Basha RK, Naim MN (2020) Antimicrobial properties of starch films incorporated with chitosan nanoparticles: in vitro and in vivo evaluation. Carbohydr Polym 230:115602
155. Siqueira G, Bras J, Dufresne A (2010) Cellulosic bionanocomposites: a review of preparation, properties and applications. Polymers 2(4):728–765
156. Balakrishnan P, Gopi S, Geethamma VG, Kalarikkal N, Thomas S (2018) Cellulose nanofiber vs nanocrystals from pineapple leaf fiber: a comparative studies on reinforcing efficiency on starch nanocomposites. Macromol Symp 380:1800102
157. Boufi S, Gandini A (2015) Triticale crop residue: a cheap material for high performance nanofibrillated cellulose. RSC Adv 5(5):3141–3151
158. Anglès MN, Dufresne A (2000) Plasticized starch/tunicin whiskers nanocomposites. 1. Structural analysis. Macromolecules 33(22):8344–8353
159. Iqbal MZ (2016) Structure-property relationships in graphene/polymer nanocomposites. Ph. D. thesis, Colorado School of Mines, Arthur Lakes Library
160. Kalogeras IM (2016) Glass-transition phenomena in polymer blends. In: Isayev AI (ed) Encyclopedia of polymer blends, vol 3: structure. Wiley-VCH, Weinheim, pp 1–134
161. Jabarin SA, Majdzadeh-Ardakani K, Lofgren EA (2016) Crystallization and melting behavior in polymer blends. In: Isayev AI (ed) Encyclopedia of polymer blends, vol 3: structure. Wiley-VCH, Weinheim, pp 135–190
162. Lai SM, Sun WW, Don TM (2015) Preparation and characterization of biodegradable polymer blends from poly (3-hydroxybutyrate)/poly (vinyl acetate)-modified corn starch. Polym Eng Sci 55(6):1321–1329
163. Liao HT, Wu ChS (2007) Performance of an acrylic-acid-grafted poly(3-hydroxybutyric acid)/starch bio-blend: characterization and physical properties. Des Monomers Polym 10(1):1–18

164. Innocentini-Mei LH, Bartoli JR, Baltieri RC (2003) Mechanical and thermal properties of poly(3-hydroxybutyrate) blends with starch and starch derivatives. Macromol Symp 197:77–88
165. Florez JP, Fazeli M, Simão RA (2019) Preparation and characterization of thermoplastic starch composite reinforced by plasma-treated poly (hydroxybutyrate) PHB. Int J Biol Macromol 123:609–621
166. Panaitescu DM, Nicolae CA, Frone AN, Chiulan I, Stanescu PO, Draghici C, Iorga M, Mihailescu M (2017) Plasticized poly(3-hydroxybutyrate) with improved melt processing and balanced properties. J Appl Pol Sci 134(19):1–14
167. Petersen K, Nielsen PV, Olsen MB (2001) Physical and mechanical properties of biobased materials starch, polylactate and polyhydroxybutyrate. Starch-Stärke 53(8):356–361
168. Tingwei W (2020) PHA-modified TPS/PBAT biodegradable resin and preparation method therefor. WO2020088214 (A1), 7 May 2020
169. Magalhães NF, Andrade CT (2013) Properties of melt-processed poly (hydroxybutyrate-co-hydroxyvalerate)/starch 1: 1 blend nanocomposites. Polímeros 23(3):366–372
170. Bhattacharyya B, Behera HT, Mojumdar A, Raina V, Ray L (2019) Polyhydroxyalkanoates: resources, demands and sustainability. In: Jamil N, Kumar P, Batool R (eds) Soil microenvironment for bioremediation and polymer production, 1st edn. Wiley, Hoboken, pp 253–270
171. Kushwah BS, Kushwah AVS, Singh V (2016) Towards understanding polyhydroxyalkanoates and their use. J Polym Res 23:153
172. Râpă M, Darie-Niţă RN, Grosu E, Tănase EE, Trifoi AR, Pap T, Vasile C (2015) Effect of plasticizers on melt processability and properties of PHB. J Optoelectron Adv Mater 17:1778–1784
173. Lai SM, Don TM, Huang YC (2006) Preparation and properties of biodegradable thermoplastic starch/poly (hydroxy butyrate) blends. J Appl Polym Sci 100(3):2371–2379
174. Avella M, Errico ME, Rimedio R, Sadocco P (2002) Preparation of biodegradable polyesters/high-amylose-starch composites by reactive blending and their characterization. Appl Polym Sci 83(7):1432–1442
175. Ma P, Xu P, Chen M, Dong W, Cai X, Schmit P, Spoelstra AB, Lemstra PJ (2014) Structure–property relationships of reactively compatibilizedPHB/EVA/starch blends. Carbohydr Polym 108:299–306
176. Yingxin H, Tianci H, Peng W (2019) Starch-based degradable PP/PHB composite material and preparation method thereof. CN109679305 (A), 26 Apr 2019
177. Zhang G, Xie W, Wu D (2020) Selective localization of starch nanocrystals in the biodegradable nanocomposites probed by crystallization temperatures. Carbohydr Polym 227:115341
178. Pan HW, Ju DD, Zhao Y, Wang Z, Yang HL, Zhang HL, Dong LS (2016) Mechanical properties, hydrophobic properties and thermal stability of the biodegradable poly(butylene adipate-co-terephthalate)/maleated thermoplastic starch blown films. Fiber Polym 17:1540–1549
179. Hablot E, Dewasthale S, Zhao Y, Zhiguan Y, Shi X, Graiver D, Narayan R (2013) Reactive extrusion of glycerylated starch and starch–polyester graft copolymers. Eur Polym J 49(4):873–881
180. Stagner JA, Alves VD, Narayan R (2012) Application and performance of maleated thermoplastic starch–poly (butylene adipate-co-terephthalate) blends for films. J Appl Polym Sci 126(S1):E135–E142
181. Rhim J-W, Park H-M, Ha C-S (2013) Bio-nanocomposites for food packaging applications. Prog Polym Sci 38(10):1629–1652
182. Spiridon I, Anghel NC, Darie-Nita RN, Iwańczuk A, Ursu RG, Spiridon IA (2019) New composites based on starch/Ecoflex®/biomass wastes: mechanical, thermal, morphological and antimicrobial properties. Int J Biol Macromol. https://doi.org/10.1016/j.ijbiomac.2019.11.185
183. Spiridon I, Darie-Nita RN, Hitruc GE, Ludwiczak J, Cianga Spiridon IA, Niculaua M (2016) New opportunities to valorize biomass wastes into green materials. J Clean Prod 133:235–242

184. Olivato JB, Grossmann MVE, Yamashita F, Nobrega MM, Scapin MRS, Eiras D, Pessan LA (2011) Compatibilisation of starch/poly(butylene adipate co-terephthalate) blends in blown films. Int J Food Sci Technol 46:1934–1939
185. Lekube BM, Fahrngruber B, Kozich M, Wastyn M, Burgstaller C (2019) Influence of processing on the mechanical properties and morphology of starch-based blends for film applications. J Appl Polym Sci 136(39):47990
186. Sun S, Liu P, Ji N, Hou H, Dong H (2018) Effects of various cross-linking agents on the physicochemical properties of starch/PHA composite films produced by extrusion blowing. Food Hydrocolloids 77:964–975

Chapter 5
Green Composites from Sustainable Cellulose Nanofibrils

Gonzalo Martínez-Barrera, Irna Zukeyt Garduño-Jaimes, Enrique Vigueras-Santiago, Julián Cruz-Olivares, Nelly González-Rivas, and Osman Gencel

1 Cellulose and Nanocellulose

Cellulose is considered to be the most abundant renewable polymer on earth. It is obtained from diverse sources as plants and algae, as well as by bacterial, enzymatic and chemical processes. Natural fibers are essentially made of cellulose, hemicellulose and lignin. Pectin, pigments and extractables can be found in low amounts. For this reason, natural fibers are also referred to as cellulosic fibers or lignocellulosic fibers [1]. The chemical structures of natural fibers are sophisticated. Each fiber is a compound in which the rigid cellulose microfibrils are embedded in a soft matrix mainly composed of hemicellulose and lignin. The cellulose fiber properties depend on the chemical composition, microfibril angle, cell dimensions and defects; they differ either from different sections of the plant or from different plants. The

G. Martínez-Barrera (✉) · E. Vigueras-Santiago
Laboratorio de Investigación y Desarrollo de Materiales Avanzados (LIDMA), Facultad de Química, Universidad Autónoma del Estado de México, Km.12 de la carretera Toluca-Atlacomulco, 50200 San Cayetano, México
e-mail: gonzomartinez02@yahoo.com.mx

I. Z. Garduño-Jaimes
Posgrado en Materiales, Facultad de Química, Universidad Autónoma del Estado de México, Paseo Colon Esquina Paseo Tollocan S/N, 50180 Toluca, México

J. Cruz-Olivares
Facultad de Química, Universidad Autónoma del Estado de México, Paseo Colón Esq. Paseo Tollocan S/N, 50120 Toluca, Estado de México, México

N. González-Rivas
Centro Conjunto de Investigación en Química Sustentable UAEM-UNAM, Km. 14.5 Carretera Toluca-Atlacomulco, Campus UAEMex "El Rosedal" San Cayetano-Toluca, 50200 Toluca de Lerdo, Estado de México, México

O. Gencel
Department of Civil Engineering, Bartin University, Bartin 74100, Turkey

S. Thomas and P. Balakrishnan (eds.), *Green Composites*, Materials Horizons: From Nature to Nanomaterials,
https://doi.org/10.1007/978-981-15-9643-8_5

Table 1 Types of nanocellulose and their properties [author]

Parameter	Cellulose nanocrystals (CNCs)	Cellulose nanofibrils (CNFs)	Bacterial cellulose (BC)
Diameter (nm)	3–50	2–10	1.5–4
Length (μm)	0.1–1	>2	1–9
Crystallinity index (CI)	85–100%	40–78%	84–90%
Degree of polymerization (DP)	140–6,000	200–10,000	300–10,000
Tensile strength (MPa)	7,500–7,700	2–2,000	200–2,000
Young's module (GPa)	130–250	13–180	15–138

mechanical qualities of natural fibers also depend on the cellulose type, because each one has its own degree of crystallinity [2].

In recent years, cellulose has been studied at nanostructural level. Cellulose nanoparticles can be obtained from lignocellulosic fibers, which are located around the world, mainly in tropical countries. They can be obtained from sisal fibers (from Agave sisalana leaves), which are easily cultivable in India, Brazil and Tanzania [3]. Three types of structures are known for nanocellulose: (a) cellulose nanocrystals (CNCs), also referred to as nanocrystalline cellulose (NCC) and cellulose nanowhiskers (CNWs); (b) cellulose nanofibrils (CNFs), also referred to as nanofibrillated cellulose (NFC); and (c) bacterial cellulose (BC) [4–6], as it is shown in Table 1.

The relevance of the nanocellulose investigations is related to its green nature, physical and chemical properties and applications. Moreover, its crystallinity, surface area and mechanical properties depend on the extraction and processing methods [7].

The cellulose nanocrystals (CNCs) are obtained mainly by acid hydrolysis. They have high resistance, large surface area, 3–50 nm diameters and 0.1–1 μm length [8]. Their properties make it an excellent material for manufacturing biopolymers, antimicrobial films, medical implants and automotive components [9]. Cellulose nanocrystals have been mixed with different polyvinyl alcohol (PVA) concentrations, which reducing the fibers diameter [10].

Cellulose is also biosynthesized by some bacteria, which is called bacterial cellulose (BC), the bacteria that generate it are mainly those from the genders Gluconacetobacter, Sarcina and Agrobacterium [11]. By Gluconacetobacter (formerly called Acetobacter) is possibly to produce cellulose at commercial levels [12]. The main applications of BC are in biomedicine, explosives, membranes, magnetic materials and even in the food industry [13, 14].

Nanocellulose-based materials are non-toxic, carbon neutral, sustainable and recyclable. Investigations of novel, efficient and environmentally safe treatments continue as main objective. Despite all difficulties for nanocellulose production, it is easily found on the market. However, many established and proposed methods have emerged with respect to its large-scale production [15].

2 Cellulose Nanofibrils (CNFs)

Cellulose nanofibrils (CNFs), also referred to as nanofibrillated cellulose (NFC), consist of long, flexible, entangled nanometer-sized cellulose fibrils (2–10 nm diameter and >2 μm length). It exhibits amorphous and crystalline domains, with high aspect ratio and specific surface area. Currently, cellulose nanofibrils are considered high-performance, abundant, renewable, biodegradable and biocompatible materials, as well as with great potential for several industrial applications [6, 16, 17].

The first CNFs were obtained in the early eighties by Herrick and Turbak, who disintegrated cellulose by mechanical treatments in a high-pressure homogenizer. The cellulose was brought to a low aqueous consistency several times during the mechanical process. The final product had highly viscous gel consistency [18].

Various pretreatments and processes for obtaining CNFs have been developed. The objective is to eliminate some disadvantages for its production as improving the quality and reducing the large amounts of energy required; thus, the production cost is noticeably affected. Another important disadvantage after producing CNFs is the produced chemical wastes; so many efforts are focused for obtaining sustainable processes.

The most common mechanical procedures for producing CNFs are high-pressure homogenization (HPH), microfluidization, grinding, high-intensity ultrasound and ball grinding (Table 2). However, the scaling of mechanical processes becomes quite complex due to its high energy consumption. In the case of homogenizer treatment,

Table 2 Mechanical processes for producing CNFs [author]

High-pressure homogenization (HPH)	The velocity of the fibrous suspension is stopped by a homogenization valve, which causes low pressure, but high turbulence, temperature, impact and shear forces. Pressures of 30–150 MPa and energy ranging from 12,000–70,000 kWh/ton are used. The fibrillation degree depends on the applied pressure and the number of the passes
Microfluidization	The fibers are placed in an inlet tank and are propelled to interaction chamber by an intensifier pump, which provides high pressure. Inside, the pulp circulates through microchannels. High pressures (>100 MPa) are required and 500–2,550 kWh/ton energy; the fibrillation degree depends on the chamber size and the number of the passes
Grinding or ball process	It is based on impact and friction forces. Pressures of 30–50 MPa are used. However, it consumes high energy, due to the large number of passes, namely 16–30 times
High-intensity ultrasound	Hydrodynamic forces produce vibrations in the cellulose fibers, generating millions of microbubbles, which break the cell structure forming cellulose fibrils. It is a process that does not allow large-scale production

the electrical consumption reaches 70,000 kWh/ton; however, pilot-scale plants have been developed in some companies and research centers for such purposes [17].

The mechanical processes are normally accompanied by an enzyme or chemical pretreatment; otherwise, the energy costs rise to a great extent to achieve the cellulose defibrillation. The most widely used pretreatment process involves enzymatic hydrolysis, alkaline acids and ionic liquids.

In the case of enzymes, those with selective hydrolysis capacity as laccase can degrade or modify the lignin and hemicellulose content without altering the cellulose content [19]. A single enzyme is not capable for fiber degradation, as cellulose contains many organic compounds; thus, a set of cellobiohydrolase-type enzymes is required for the breakdown of crystalline cellulose and endoglucanases for disordered cellulose [20].

Pretreatment with alkaline acids is the most widely used method for removal of lignin, hemicellulose and pectin. In a study, sodium hydroxide and hydrochloric acid were used to remove cellulose, and then, it was submitted to mechanical treatment to obtain CNFs, and the results showed improvement on the cellulose yielding from 43 to 84% [21].

In the case of ionic liquids, they are organic salts, which have a high melting point and work at very low vapor pressure, prior to a mechanical treatment. Moreover, they have been used as solvents to dissolve cellulose and exhibit good results because an equivalent number of cations accompany anions and the electro-neutrality is maintained. The interactions of the ions (cation or anion) differ from each other but lead to self-organizing behavior [22].

Anions with high hydrogen bond basicity can solvate cellulose by binding to their hydroxyl groups, when the total hydrogen bond basicity of the solvent or the solvent mix has certain values [23]. The ionic liquids are salts with melting point less than 100 °C and better physicochemical properties compared to conventional organic solvents. They can dissolve polar, non-polar, organic, inorganic compounds and polymer materials. Recently, alternative solvents have been used for polymer production, such as supercritical carbon dioxide and water. In the CNFs production, high propagation rates and decreased termination rates have been observed in free radical polymerizations. Moreover, in the production of electrospinning CNFs can happen moderate reaction conditions, reuse of the catalytic system without decreasing of its activity and higher yields [24].

Catalytic oxidation by TEMPO is a process to obtain CNFs with a basic pH, very small diameters and transparent gels. Other characteristics also studied are quality, microfibrils angles and residual chemical composition (lignin and hemicellulose content). In this process, the production costs are reduced due to an energy efficient methodology, based first on the TEMPO process with catalytic oxidation and then on subsequent mechanical destructing. Moreover, this process is widely used to obtain nanopaper as a final product, with a high degree of transparency [25].

In recent years, a lot of investigations are concerning to use agro-industrial wastes as raw material for obtaining NFCs. The products are rich in natural fibers with structural arrangements and have high tensile strength. The chemical structure of such wastes are essentials for the nanofibers production, mainly the lignin and hemicellulose quantities, for example, the lignin content in the pulp can produce nanocellulose with diameters up to six times larger, and the hemicellulose content can influence on the diameter distribution [17].

The CNFs' properties depend on the chemical structure, surface chemistry, crystallinity and polymerization degree. The cellulose type and the used process are determinant on the morphological characteristics, which are studied by electron scanning microscopy (SEM), electron transmission (TEM) and atomic force (AFM) [26].

3 Green Composites

Research in sustainable technologies and highly resistant smart materials is an actual and future topic, supporting the need to balance economic growth with social and environmental concerns. One of the most actual research is concerning to nanocomposites, mainly those based on nanocellulose, which is an environmentally friendly material, and used in medical, automotive, electronics, packaging and construction industries as well as in the wastewater treatment, besides being an excellent substitute of synthetic materials.

Nanocomposites' production involves a lot of parameters, for example, process improvements, industrial scale production, energy consumption and matrix blending. Green composites include nanofibril-based polymer nanocomposites, as well as bacterial cellulose, which have been used in biomedical and technological applications. They show improvements on the mechanical, optical and barrier properties.

The global demand for textile fibers is supposed to increase as the population increases, and the quality of life improves. It is a fact that the CNFs' market is constantly increasing, due to the lack of cotton production and its high demand around the world. Cotton production cannot increase significantly; therefore, it is expected that, in few years, the CNFs' production will cover from 33 to 37% of the total market. Nevertheless, it would cause an increase in demand by 2030 year and would represent the production of 15 million tons CNFs by year. Such forecasts are due to some properties of the CNFs as absorption and humidity, among others [27]. The large-scale production of nanocellulose is already a fact. Both research laboratories and industries around the world promote its research; however, there are many challenges to overcome for its manufacture.

In the investigations about CNFs have been studied natural and synthetic polymer matrices. However, the scope of the polymers is very broad. Thus, it is essential to conduct more investigations. The objective is to obtain materials with good optical, mechanical and barrier qualities for their use in biomedical and technological applications. The nanocomposites based on CNFs show great advances, but it is necessary to develop properties, such as high tensile resistance, low density, high barrier properties (for sound, oxygen or other gases), optical transparency, biodegradability, renewability, among others [28].

Table 3 Applications of green composites based on CNFs [author]

Biomedical industry	• In moisturizing creams with ingredients capable of penetrating the skin and carrying bioactive agents • As matrix for tissue engineering (bone, cartilage, vascular, nerves and skin) • In the development of multifunctional matrices that allow the controlled release of therapeutic agents
Automotive and space industry	• As substitute of steel in the manufacture of bonnets and other auto parts. For reducing fuel consumption and carbon dioxide emissions • In aircraft fuselage manufacturing
Food industry	• Used as films in food wraps that produce strong barrier effect against gases penetration (mainly oxygen) are effective for maintaining food freshness • As food additive to reinforce the body feeling or food chewiness
Paper industry	• Filters that collect dust, small particles or deodorant substances that absorb odor-bearing microparticles • Manufacture of sheets or foils with deodorizing and antibacterial functions • Gel ink pens in which cellulose nanofibers are used as a thickener

Nowadays, there is a great deal of information related to the processing and properties of nanocellulose, which include several topics such as processing strategies, chemical modification of surfaces, biocompatibility, toxicity, characterization and possible applications. In addition, obtaining cellulose nanofibrils has been studied too; the reports are focused on optimizing chemical–mechanical treatments for the extraction of nanofibrillar cellulose and the final properties. Moreover, in recent years CNFs have been used in different sectors (Table 3).

Production of CNFs hydrogels is a novel topic; they have been recently used for the segregation of pollutants in water, non-polar hydrocarbons, solvents and oils as well as in chloroform adsorption.

In the coming years, the use of CNFs is expected to grow considerably, especially in tissue engineering, in the development of implants where surgeons replace the damaged tissue resulting from trauma, as well as in orthopedic reconstruction, where congenital deformity or pathological deterioration is replaced with autogenous grafts, made with CNF-reinforced composites.

Cellulose nanofibers are extensively used as reinforcing agents (up to 2.5 wt%) for manufacturing strong, highly flexible polymer gels, which are obtained by in situ polymerization, or heat treatment. They are designed for biomedical applications [29].

CNFs obtained from vegetables as well as synthetic nanofibers have been used as reinforcements of polymer materials. In a study, water soluble polymer matrices were used, due to their polar hydrogen bridges, which can bond with cellulose surface by hydroxyl groups, and thus generate a single and more stable phase. Polyvinyl alcohol

(PVA) matrices were dissolved in water at low percentages of CNFs. According to the results, an increase of approximately 100% in tensile strength was obtained when CNFs are added [30].

Another application of CNFs hydrogels is for elimination of heavy metals in water treatments to obtain purified water, due to their elevated specific surface area, nanoscale size, low toxicity, hydrophilicity and bioadsorption ability. Hydrogels produced with chemically cross-linked CNFs, alginate and polyvinyl alcohol (PVA) exhibited high swelling and adsorption capacity, as well as high-storage modulus [31]. The main objective is to produce CNFs with polymer networks or with nanomaterials nets. For example, aerogels are novel materials that are being applied in catalysis, filtration, graft, damping and liquid storage [29].

High-performance biomaterials are produced adding nanocellulose fibers, which produce high uniformity and few defects [32]. Addition of nanocellulose to biodegradable polymers produces improvement on the mechanical properties and accelerates the biodegradation rate. Moreover, nanocellulose is used in water soluble polymer solutions to modify their viscosity and mechanical properties.

In the biomedical area, the biodegradable CNFs are not toxic to humans and biocompatible. It is used for personal hygiene products, cosmetics, biomedicines, burned skin treatment. They hold the stabilization of medical suspensions and decrease the sedimentation and phase separation of heavy ingredients. Moreover, after its chemical modification, it serves as a scaffold for enzymes and other medications [33].

Currently, the care and preservation of natural resources require the use of technologies based on biodegradable materials. For example, an ecological sport vehicle, called nanocellulose vehicle (NCV), was made with cellulose nanofibers from plants and agricultural waste, which are a fifth lighter than steel but five times more resistant. Various automotive parts were developed, such as doors, roof and hood. The cellulose nanofibers reduced the NCV body weight up to 50% compared to a traditional car. In addition, this novel technology helps to reduce the carbon emissions related to automobile manufacturing [34].

The paper industry has developed novel researches and applications based on the use of CNFs, becoming one of their main activities. As it is known, fibers decrease their properties during recycling processes, mainly due to drying. To resolve such problems, mechanical refining techniques have been used for their rehydration, but they cause other irreversible structural damages [35].

Production of electrospinning CNFs depend on the polymer concentration, tip-collector distance, viscosity, solution flow and applied voltage. Moreover, require of environmentally friendly solvents. Several investigations are related to the collector shape, which directly affects the diameter and resistance of the CNFs.

One of the advantages of the electrospinning technique is its ability to mix CNFs with synthetic polymer matrices. In this case, prior chemical modification is often necessary to ensure good compatibility with the matrix. Poly (vinyl alcohol), poly (lactic acid), polyurethanes, conductive polymers and acrylic resins are a few examples of used synthetic polymers. Electrospun CNFs have small pore sizes and a large surface area compared to commercial fabrics. They have controllable dissolution properties and are used as tissue engineering scaffolds and drug delivery systems [32].

Inside the production, 3D printing manufacture has been added. Now, it is possible to create three-dimensional scaffold-like structures. These nanomaterials can respond to the temperature and color changes and are known as intelligent materials. Due to their swelling, they become soft and elastic. In the case of nanocomposites produced with CNFs and impenetrable network (IPN) systems, the 3D structures show the matrix and an interwoven network [36].

In the future, it is planned to program the shape and behavior of 3D-printed objects, but adding a fourth dimension related to their temporal changes. This new scientific revolution is called 4D printing and is based on novel printing materials called "intelligent," they undergo a controlled structural change when they are submitted to external stimulus. Their shape and properties change with temperature, humidity, light and time. For example, it could be possible design smart fabrics that react to our thermal sensation, that is to say, if we feel heat they can release this energy and on the contrary, when we feel cold they can retain body heat. Thus, 4D printing is a new technological paradigm in areas such as textile, membranes, medicine, robotics and energy production.

4 Experimental Studies

In the investigation group belonging to Laboratory of Research and Development of Advanced Materials (LIDMA) at Autonomous University of the State of Mexico (UAEM), some experimental studies related to the production of CNFs, involving raw materials, production methods, characterization and applications are conducted.

In the first experimental stage, electrodynamic methods were used for to obtain bacterial cellulose (BC) by using environmentally friendly solvents. According to the literature, BC has been applied as a skin transitory substitute in the treatment of wounds, burns and ulcers, as well as in dental implants or acoustic transducers, among others, due to its high mechanical resistance acquired after chemical treatments. With elasticity modulus of 16–18 GPa, tensile stress of 260 MPa, 2.1% deformation and high purity and degree of crystallinity [37, 38].

The BC synthesis was carried out at 28–32 °C and using the *Gluconacetobacter xylinus*. The produced BC was crystalline, free of lignin and hemicelluloses, with 4–6 pH. According to the literature, highest cellulose yield can achieve with controlled quantities of glucose; high concentration of this can inhibit the cell growth and production as well as decrease pH, due to accumulation of (keto)gluconic acids [39]. The carbon source is a fundamental factor in the production of BC, whereby two types of carbon sources were used in the static bioreactors: sugar and beet molasses.

Gluconacetobacter xylinum was obtained from apple cider vinegar. Hestrin and Schramm were used as culture medium, whose composition was glucose (20 g/L), peptone (5 g/L), yeast extract (15 g/L), disodium phosphate (2.7 g/L) and acid citric (1.15 g/L). The pH was adjusted to 5.5 using 0.1 N sodium hydroxide, while the culture medium was autoclaved at 15 psi and 121 °C by 15 min. Then, 9 mL of culture medium and 1 mL of apple cider vinegar were placed in test tubes. The mixtures were homogenized with a shaker and incubated by 6 days at 32 °C. The variables were the chemical structure, pH, volume loss, the cellulose sizes, cellulose production and the bioreactor size.

Bacterial cellulose (BC) growth was performed in bioreactors, which were sterilized in an autoclave at 15 psi and 121 °C by 15 min. The bioreactors contained the test tube solution (1 mL), potassium sorbate (0.13%), sucrose (12.66%), yeast extract (1.26%), calcium chloride (0.76%), potassium phosphate (0.37%), distilled water (84.9%) and the corresponding percentage of each substrate. The bioreactors were covered with a cotton cloth to allow aeration in the culture medium, in order to promote contact with oxygen and achieve the bacteria growth. Then, the bioreactors were placed in an incubator with circulation at 30 °C. Produced BC is shown in Fig. 1.

The electrospinning process was used, due to its advantages to produce nanofibers. In recent years, numerous types of materials, including synthetic and natural polymers have been electrospun to obtain continuous fibers. They produce a highly porous spun bond nonwoven membrane, with fibers size ranging from nanometers to few microns [36]. Moreover, cellulose derivatives with greater solubility have been synthesized, as cellulose acetate, cellulose triacetate and methyl cellulose. Cellulose acetate nanofibers (247–265 nm diameters) have been used for protection of vitamins A and E, while cellulose acetate is mixed with water/acetic acid for producing pyranose 2-oxidase fibers, with 200–400 nm diameters. In the case of cellulose triacetate (4.7 k dielectric constant), this is used for produce CNFs with low surface area and without lumps or beads [40, 41].

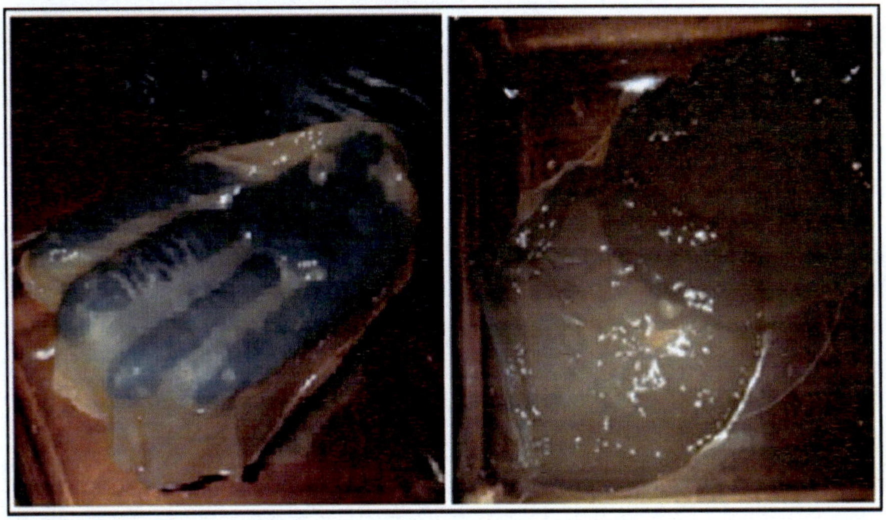

Fig. 1 Bacterial cellulose (BC) obtained after 28 days of culture [author]

The parameters involved in the CNFs production are determinant on the applications; some of them are shown in Table 4.

The electrospinning equipment (shown in Fig. 2) consist of an infusion pump, which controls the flow rate, a high-voltage source (10,000 V as minimum) and the collecting system.

The configuration can include more variants, for example, type of XY movements, rotary nozzles, rotocollectors, arrangements with auxiliary electrodes, cryogenic collectors and coaxial nozzles. The electrospinning equipment was designed with CAD software and has movement on the three axes. Moreover, a collector with

Table 4 Types of CNFs produced and its applications [author]

Fiber type	Parameter	Applications
Flattened or tapes	It is attributed to the emergence of a polymer layer on the surface of the fiber, due to the uneven solvent evaporation. The atmospheric pressure tends to collapse the round shape of the fiber. It can be related to the solvent type and the addition of salts to the solution	• Biosensors systems, due to its electrochemical activity and the ability to transfer electrons
Helical	It occurs due to the jet deformation during the impact with the collecting plate. Solution concentration promotes this behavior. The incidence angle of the jet influences on its maintenance	• Drug delivery systems • Electromechanical and electromagnetic microsystems • Advanced optical components
Ramified	It is related to the small jets on the surface, produced by the initial jet. The instability between electrical forces and surface tension produces instability in the jet	• Drug delivery systems
Hollow	It is obtained by coaxial electrospinning processes or by chemical processes applied to electrospun fibers	• Nanoelectronic and optoelectronic devices • Energy conversion • Drug release • Environmental protection • Sensors
Fibers with beads	They are due to the surface tension and the viscoelastic properties of the solution. They depend on the outlet flow, distance between capillary and collector and voltage as well as from molecular weight or viscosity of the solution	• Tissue engineering
Fibers with pores	It is the consequence of the relative humidity and vapor pressure of the solvent	• Tissue engineering • Catalysis • Sensors

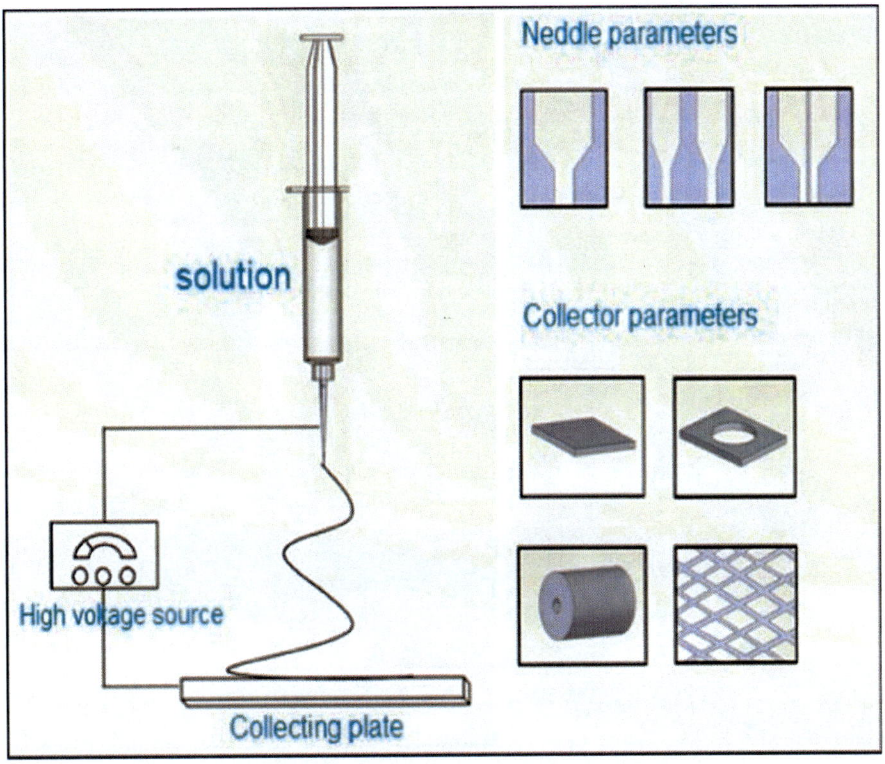

Fig. 2 Electrospinning process and their parameters for producing CNFs [author]

speed was also designed to determine the accuracy deposition of CNFs. The objective was to have several needles connected to the dosing pump and a collector in motion, as it is shown in Fig. 3.

The electrospinning design gives the possibility of using a single injector with specific movement, in order to reproduce the patterns designed in the CAD software. The parameters shown in Table 5 were taken into account for production of electrospinning CNFs.

In the second experimental stage of the electrospinning process, the cellulose triacetate was replaced with eutectic mixtures having low melting point. In the process, sustainable and biodegradable solvents with low toxicity and ecological footprint were used. As it is known, molecular weight and solvent type are the most important factors for producing electrospinning cellulose. A homogeneous distribution of solvents produces a homogeneous material; otherwise, the needle capillary will be obstructed.

A possible application of the produced CNFs was as an interchangeable mask filter, as it is shown in Fig. 4. The mask was designed with CAD software and 3D printing in a molten deposition printer, by using a flexible and low-density filament.

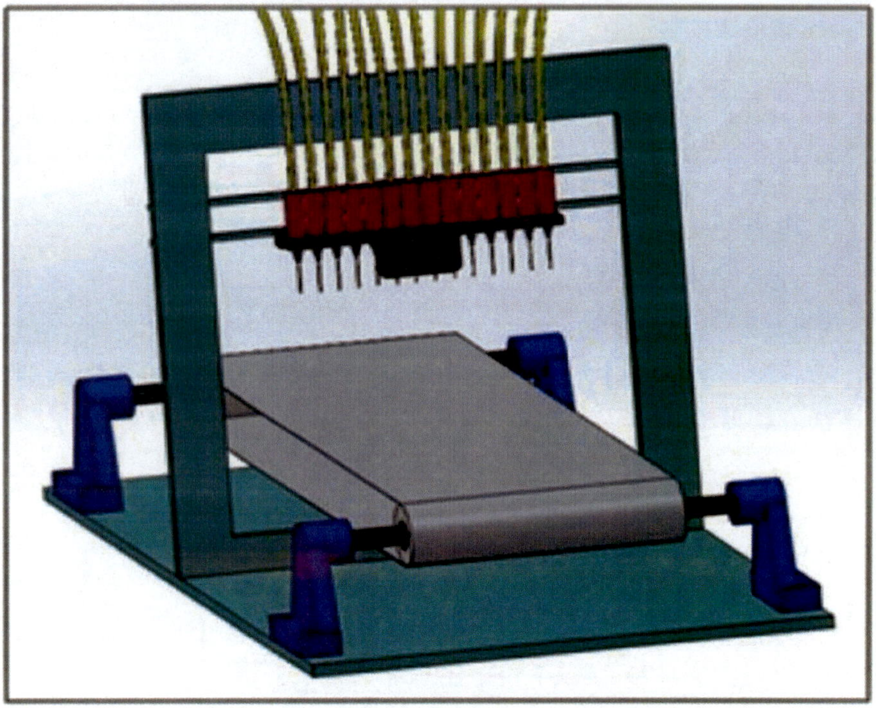

Fig. 3 Solidworks design of a multiple electrospinning machine with dynamic collector [author]

Table 5 Parameters in the electrospinning process [author]

	Parameters
Dissolution	• Concentration of the polymer solution • Surface tension • Solution conductivity • Dielectric effect of the solvent
Process	• Voltage • Outflow • Collector distance
Environmental	• Room temperature • Humidity

Undoubtedly, research and developments of CNFs will continue. The future nanocomposites are innumerable, and their success will depend on many factors including their novel methods and applications.

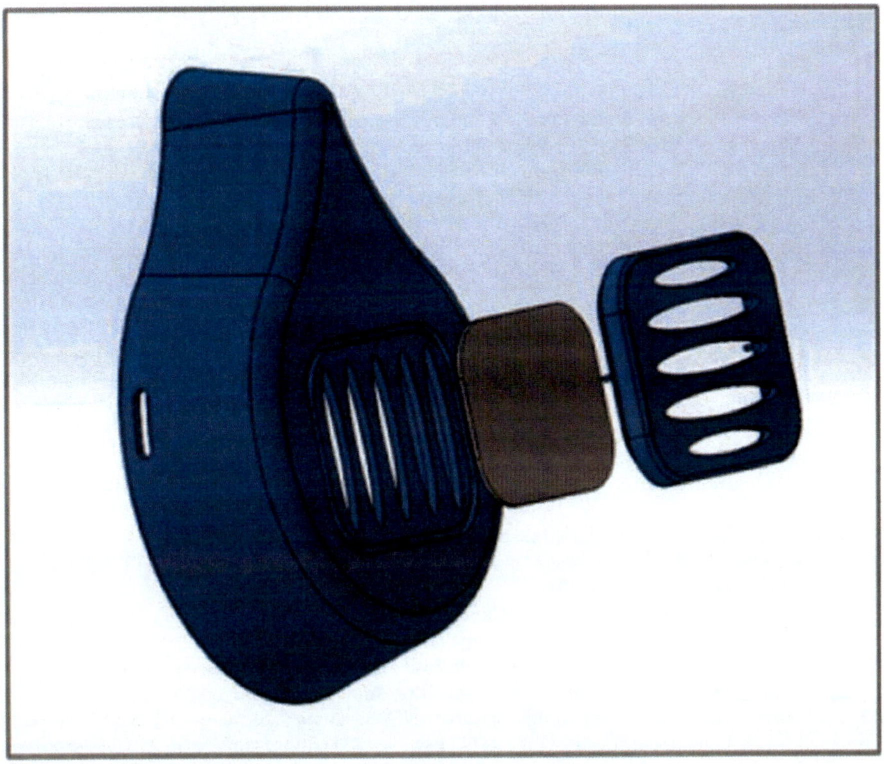

Fig. 4 Prototype mask including filter produced with NFC [author]

5 Conclusions

The CNFs' era continues to evolve; however, there are still many activities to do respect to the current production processes, to the use of available resources and mainly to the related to environment care. The electrospinning technique gives a very encouraging image for the improvement of the CNFs' properties as well as for producing green composites, which have a wide range of applications in many industrial sectors. Such materials will allow in the near future improving some needs of the human being.

Acknowledgements The authors thank to Council of Science and Technology of the State of Mexico (COMECYT), for the scholarship granted to one of the authors (I.Z. Garduño-Jaimes).

References

1. Satyanarayana KG, Guilmaraes JL, Wypych F (2007) Studies on lignocellulosic fibers of Brazil. Part I: Source, production, morphology, properties and applications. Compos A Appl Sci Manufact (7)38:1694–1709. https://doi.org/10.1016/j.compositesa.2007.02.006
2. Dufresne A (2008) Cellulose-based composites and nanocomposites. In: Gandini A, Belgacem MN (eds) Monomers, polymers and composites from renewable resources. Elsevier, Oxford, UK, pp 401–418. https://doi.org/10.1016/B978-0-08-045316-3.00019-3
3. Sreekumar PA, Joseph K, Unnikrishnan G, Thomas S (2007) A comparative study on mechanical properties of sisal-reinforced polyester composites prepared by resin transfer and compression moulding techniques. Compos Sci Technol 3–4(67):453–461. https://doi.org/10.1016/j.compscitech.2006.08.025
4. Nasir M, Hashim R, Sulaiman O, Nordin NA, Lamaming J, Laccase AM (2015) An emerging tool to fabricate green composites: a review. Bioresources 10: 6262–6284. https://doi.org/10.15376/biores.10.3.Nasir
5. Lamaming J, Hashim R, Leh CP, Sulaiman O, Sugimoto T, Nasir M (2015) Isolation and characterization of cellulose nanocrystals from parenchyma and vascular bundle of oil palm trunk (*Elaeisguineensis*). Carbohydr Polym 134:534–540. https://doi.org/10.1016/j.carbpol.2015.08.017
6. Klemm D, Kramer F, Moritz S, Lindstrom T, Ankerfors M, Gray D, Dorris A (2011) Nanocelluloses: a new family of nature-based materials. Angew Chem Int Ed 50(24):5438–5466. https://doi.org/10.1002/anie.201001273
7. Abitbol T, Rivkin A, Cao Y (2016) Nanocellulose, a tiny fiber with huge applications. Curr Opin Biotechnol 39:76–88. https://doi.org/10.1016/j.copbio.2016.01.002
8. Habibi Y, Lucia LA, Rojas OJ (2010) Cellulose Nanocrystals: Chemistry, self-assembly, and applications. Chem Rev 110(6):3479–3500. https://doi.org/10.1021/cr900339w
9. Kaboorani A (2012) Nanocrystalline cellulose (NCC): A renewable nano-material for polyvinyl acetate (PVA) adhesive. Eur Polym J 48(11):1829–1837. https://doi.org/10.1016/j.eurpolymj.2012.08.008
10. Siqi H, Long B, Wanli C, Guangping H (2016) Manufacture of electrospun allaqueous poly(vinyl alcohol)/cellulose nanocrystal composite nanofibrous mats with enhanced properties through controlling fibers arrangement and microstructure. Polymer 92:25–35. https://doi.org/10.1016/j.polymer.2016.03.082
11. Campano C (2016) Enhancement of the fermentation process and properties of bacterial cellulose: a review. Cellulose 23(1):57–91. https://doi.org/10.1007/s10570-015-0802-0
12. Shi Z, Zhang Y, Phillips GO, Yang G (2014) Utilization of bacterial cellulose in food. Food Hydrocoll 35:539–545. https://doi.org/10.1016/j.foodhyd.2013.07.012
13. Mohite BV, Patil SV (2014) A novel biomaterial: bacterial cellulose and its new era applications. Biotechnol Appl Bioc 61(2):101–110. https://doi.org/10.1002/bab.1148
14. Huang Y (2014) Recent advances in bacterial cellulose. Cellulose 21(1):1–30. https://doi.org/10.1007/s10570-013-0088-z
15. Asim M, Abdan K, Jawaid M (2015) A review on pineapple leaves fibre and its composites. Int J Polym Sci 2015:16. https://doi.org/10.1155/2013/950567
16. Plackett DV, Letchford K, Jackson JK, Burt HM (2014) A review of nanocellulose as a novel vehicle for drug delivery. Nord Pulp Pap Res J 29(1):105–118. https://doi.org/10.3183/npprj-2014-29-01-p105-118
17. Lavoine N, Desloges I, Dufresne A, Bras J (2012) Microfibrillated cellulose—its barrier properties and applications in cellulosic materials: a review. Carbohydr Polym 90(2):735–764. https://doi.org/10.1016/j.carbpol.2012.05.026
18. Siró I, Plackett D (2010) Microfibrillated cellulose and new nanocomposite materials: a review. Cellulose 17(3):459–494. https://doi.org/10.1007/s10570-010-9405-y
19. Nasir M, Gupta A, Beg MDH, Chua GK, Asim M (2014) Laccase application in medium density fibreboard to prepare a bio-composite. RSC Adv 4:11520–11527. https://doi.org/10.15376/biores.10.3.Nasir

20. Henriksson M, Berglund LA (2007) Structure and properties of cellulose nanocomposite films containing melamine formaldehyde. J Appl Polym Sci 106:2817–2824. https://doi.org/10.1002/app.26946
21. Bhatnagar A, Sain M (2005) Processing of cellulose nanofiber-reinforced composites. J Reinf Plast Compos 24:1259–1268. https://doi.org/10.1177/0731684405049864
22. Kuzina SI, Shilova IA, Mikhailov (2011) Chemical and radiation-chemical radical reactions in lignocellulose materials. Radiat Phys Chem 80:937–946. https://doi.org/10.1515/hf-2013-0119
23. Hauru LK, Hummel M, Nieminen K, Michud A, Sixta H (2016) Cellulose regeneration and spinnability from ionic liquids. Soft Matter 12(5):1487–1495. https://doi.org/10.1039/c5sm02618k
24. Erdmenger T, Guerrero-Sanchez C, Vitz J, Hoogenboom R, Schubert US (2010) Recent developments in the utilization of green solvents in polymer chemistry. Chem Soc Rev 39(8):3317–3333. https://doi.org/10.1039/b909964f
25. González I, Alcalà M, Chinga-Carrasco G, Vilaseca F, Boufi S, Mutjé P (2014) From paper to nanopaper: evolution of mechanical and physical properties. Cellulose 21(4):2599–2609. https://doi.org/10.1007/s10570-014-0341-0
26. Nechyporchuk O, Belgacem MN, Bras J (2016) Production of cellulose nanofibrils: a review of recent advances. Ind Crop Prod 93:2–25. https://doi.org/10.1016/j.indcrop.2016.02.016
27. Haemmerle FM (2011) The cellulose gap (the future of cellulose fibers). Lenzinger Ber 89(1):12–21. https://pdfs.semanticscholar.org/e0fd/9707cc4b03187e5adc3ec53857912f3d968c.pdf?_ga=2.215517983.966850805.1589074223-94483054.1583769934
28. Shatkin JOA, Wegner TH, Bilek EM, Cowie J (2014) Market projections of cellulose nanomaterial-enabled products—Part 1: Applications. TAPPI J 13(5):9–16. https://www.fpl.fs.fed.us/documnts/pdf2014/fpl_2014_shatkin001.pdf
29. Tarrés Q, Oliver-Ortega H, Llop M, Pèlach M, Delgado-Aguilar M, Mutjé P (2016) Effective and simple methodology to produce nanocellulose-based aerogels for selective oil removal. Cellulose 23:3077–3088. https://doi.org/10.1007/s10570-016-1017-8
30. Xiao S, Gao R, Lu Y, Li J, Sun Q (2015) Fabrication and characterization of nanofibrillated cellulose and its aerogels from natural pine needles. Carbohydr Polym 119:202–209. https://doi.org/10.1016/j.carbol.2014.11.041
31. Yue Y, Han J, Han G, French AD, Qi Y, Wu Q (2016) Cellulose nanofibers reinforced sodium alginate-polyvinyl alcohol hydrogels: Core-shell structure formation and property characterization. Carbohydr Polym 147:155–164. https://doi.org/10.1016/j.carbpol.2016.04.005
32. Spence K, Habibi Y, Dufresne A (2011) Nanocellulose-based composites cellulose fibers: bio-and nano-polymer composites. Springer, Berlin, Heidelberg and New York, pp 179–213. https://doi.org/10.1007/978-3-642-17370-7_7
33. Sabino M, Loiaza M, Dernowsek J, Rezende R, Silva J (2017) Techniques for manufacturing polymer scaffolds with potential applications in tissue engineering. Revista Latinoamericana de Metalurgia y Materiales 37:1–27. https://www.researchgate.net/publication/317003912
34. Dufresne A (2013) Nanocellulose: a new ageless bionanomaterial. Mater Today 16:220–227. https://doi.org/10.1016/j.mattod.2013.06.004
35. Hubbe MA (2014) Prospects for maintaining strength of paper and paperboard products while using less forest resources: a review. BioResources 1(9):1787–1823. https://www.lib.ncsu.edu/resolver/1840.2/2613
36. Lin T (2011) Nanofibers production, properties and functional applications. InTech, Croatia. ISBN 978-953-51-4404-5
37. Credou J, Berthelot (2014) Cellulose: from biocompatible to bioactive material. J Mater Chem B 2(30):4767–4788. https://doi.org/10.1039/C4TB00431K
38. Chawla PR, Bajaj IB, Survase SA, Singhal RS (2009) Microbial cellulose: Fermentative production and applications. Food Technol Biotech 47(2):107–124. https://public.carnet.hr/ftbrfd/

39. Keshk S, Sameshima K (2005) Evaluation of different carbon sources for bacterial cellulose production. Afr J Biotechnol 4(6):478–482. https://doi.org/10.5897/AJB2005.000-3087
40. Mendes AC, Stephansen K, ChronakisIoannis S (2016) Electrospinning of food proteins and polysaccharides. Food Hydrocolloids 6(4):62–67. https://doi.org/10.1016/j.foodhyd.2016.10.022
41. Weerapha P, Thammasit V, Jeerus S, Pimchai C, Pramuan T (2015) Functionalized electrospun regenerated cellulose fibers for immobilizing pyranose 2-oxidase. React Funct Polym 85:45–51. https://doi.org/10.1016/j.reactfunctpolym.2014.11.008

Chapter 6
Green Composite as an Adequate Material for Automotive Applications

Magdi El Messiry

1 Introduction

The natural fiber composite (NFC) is one of the oldest technologies used by old civilizations in building technology by the production of bricks from mixing the mud with stacks of the papyrus. The NFC during the 7000 years has penetrated in different areas such as aviation, automotive, marine, building, and other areas of modern industries. Natural fibers from renewable natural resources offer the potential to act as a biodegradable reinforcing material substitute for the utilization of glass or carbon fibers and inorganic fillers. The specific properties of these natural fibers, specifically low cost, lightweight, biodegradability, noncorrosive properties, high specific strength, and Young's modulus make it more and more worthwhile. All types of natural fiber–polymer composites are used in automotive and aerospace, defense, transportation, and other sectors of industrial textile applications, and their consumption is continuously growing by the average triple growth rate of the average growth rate of composite material. The natural fiber biocomposites are growing on average by 7.5% annually.

Fibers like cotton, flax, kenaf, hemp, jute, agricultural waste as well as recycle waste, and wood flour are used with several types of polymers, either synthetic or natural polymers, such as epoxy polyethylene, polypropylene, polyvinyl chloride, polylactide acid, polyester, phenolic, vinyl ester, and others for the manufacturing of NFC. The natural fiber–polymer composites can be processed by using one of the following techniques: extrusion, compression molding, injection molding, etc.

This chapter is classified into three subjects:

M. El Messiry (✉)
Textile Engineering Department, Faculty of Engineering, Alexandria University, Alexandria, Egypt
e-mail: mmessiry@yahoo.com

S. Thomas and P. Balakrishnan (eds.), *Green Composites*,
Materials Horizons: From Nature to Nanomaterials,
https://doi.org/10.1007/978-981-15-9643-8_6

Natural fibers as an acceptable material for the automotive industry

The analysis of different natural fibers, their global production, physical and mechanical properties are investigated. Assessment of the specific strength versus the specific elastic modulus of different natural fibers, agriculture waste, and textile recycling waste shows their accessibility to be used in NFC for certain applications in the automotive industry. Meanwhile, the physical and mechanical properties of the different polymers, used as a matrix of NFC, are discussed.

Application of natural fiber composites

Since the noise level represents one of the quality factors that evaluate the satisfaction of the passenger in automotive engineering, there is a growing trend and prospects for natural fiber composite applications in the automotive industry as sound absorption for different types of vehicles. The assessment of the noise level attributable to different parts of the vehicle and the method of reducing it through the use of textile fibers, either in free form or as an NFC, was presented. The coefficient of the sound absorption of the different natural fibers and the panels of agro-waste that manufactured as a composite and shaped are considered as well as the factors affecting the sound absorption of the natural fibers or NFC.

Natural fiber–polymer composites design for the automotive industry

Natural fiber composites utilized in the automotive, passenger rail and aviation have an increasing opportunity in the production of interior and exterior applications due to the growing demand for lightweight materials with enhanced mechanical performance, low cost, and low environmental impact, especially for the high-speed vehicle. The demand for several parts of the car is contemplated NFC, for instance, in automobile bumper, passenger car's side door impact beam, and door panel. The design parameters of such elements were discussed.

The main objective of this chapter is an attempt to outline the state-of-the-art of the NFC applications in different the automotive interior and exterior parts due to growing concern for reducing the car weight, cost, and environmental concern, keeping in mind a passenger safety.

2 Natural Fibers Material for the Automotive Industry

2.1 Introduction

In the last decades, there has been an increasing environmental awareness of the need for reducing the CO_2 emission and raised the interest in using natural fibers as reinforcement in the polymer composites to replace synthetic fibers [1, 2]. Also, weight-saving opportunities could be obtained by replacing traditional fiber composites with natural fiber composites. Biocomposite materials are defined as composite materials in which at least one of the constituents is derived from natural resources. Generally, the term biocomposites covers composite materials made from the combination of biopolymer-reinforced synthetic fibers, or natural fiber-reinforced petroleum-derived polymers [2], while green composite in which natural fibers are reinforced by biopolymers [1].

2.2 *Natural Fibers for a Green Composite*

Natural fibers are the most important components in various applications such as textile manufacturing, textile polymer composites, smart and technical textiles of different applications, Fig. 1. The technical textiles are designed to fulfill a certain function in various industries, for instance the automotive, aviation, sport, medical, etc. The statistics indicate that the global technical textiles market is expected to reach $334,938 million by 2025 [3] at an average growth rate of 4.5%. Global market demand was 26.58 million tons in 2014 and is expected to reach 35.47 million tons by 2022, growing at an average growth rate of 3.7%. The automotive components were the leading application segment and accounted for 15.2% of the total market volume in 2014. Growing demand for high-performance materials for the automotive industry is expected to remain a key driving factor for this segment over the forecast period [4]. The major reasons for the exponential growth in this market are the increasing demand for lightweight noncorrosive materials in the aerospace and the automotive industry, with growing awareness regarding green products and environmental protection. For comparison, the energy consumption to produce a flax-fiber mat (9.55 MJ/kg), including cultivation, harvesting, and fiber separation, amounts to approximately

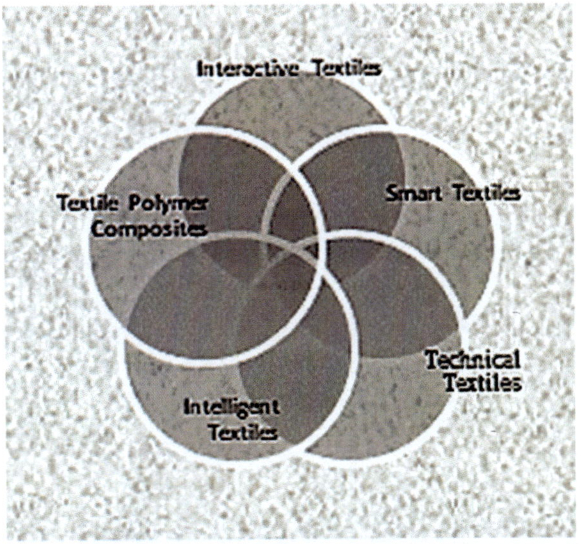

Fig. 1 Different applications of technical textiles in the market

17% of the energy to produce a glass fiber mat (54.7 MJ/kg) [5]. The global natural fiber composites market size was valued at $4.46 billion in 2016, at an average growth rate of 11.8% [6].

The percentage of the automotive technical textiles is the second after the home textiles and represents 14.5%, Fig. 2 [7]. The global technical textiles market is expected to reach USD 193.16 billion by 2022 [7]. The market has grown exponentially in the last few years looking for ecological lightweight materials with a low-carbon footprint. Composites made from natural fibers can help in the reduction of the component mass and lowering the production cost.

Natural fiber composites are biobased materials manufactured using materials such as wood, cotton, flax, kenaf, and hemp. Some natural fiber composites (NFC) specific strength is 30.0% stronger than glass fibers. The consumption of the natural fiber composite in the market is moving fast [8] as illustrated in Fig. 3.

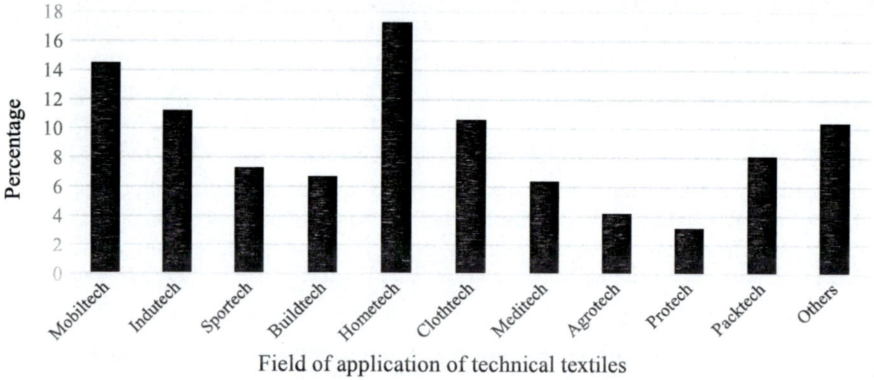

Fig. 2 Percentage of technical textiles market [7]

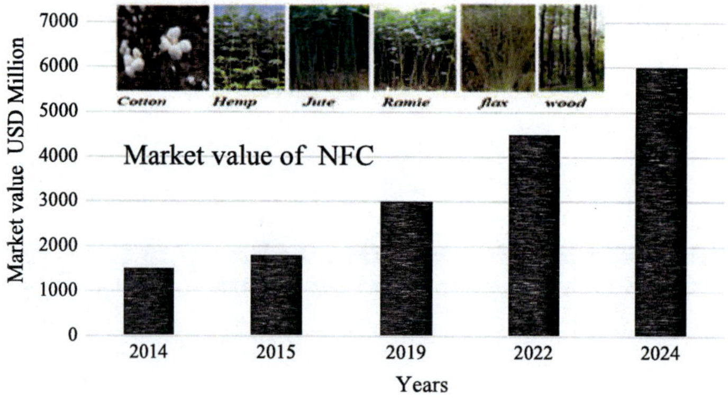

Fig. 3 Market value of the natural fiber composite [8]

Table 1 Percentage of different global natural fibers production [1, 9]

Fiber type	%	Fiber type	%
Cotton lint	78.97	Sisal	0.53
Flax	0.72	Other bast fibers	0.88
Hemp	0.16	Coir	3.22
Jute	10.19	Silk, raw	0.48
Kapok	0.29	Wool, clean	3.27
Ramie	0.45	Other	0.17

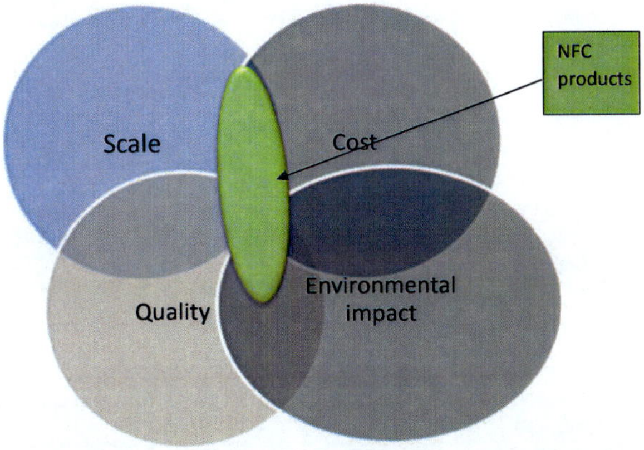

Fig. 4 their significant utilization in fibers for the manufacturing of NFC (by the author)

Table 1 gives the percentage of the different natural fibers production, which indicates that all bast fibers represent about 17.11% of the total natural fibers produced each year [9].

The components that define the use of certain material for the production chain, Fig. 4, are the scale of availability, the quality of the product, the cost of the part, and finally, the environmental impact.

NFC properties, such as recyclability and biodegradability, have also shown their significant utilization in the automotive industry.

2.3 *Classification of Cellulosic Textile Fibers*

The vegetable fibers exist in about 250,000 species of plants, but less than 0.1% of these are commercially important as fiber sources [1]. Some of the fibers are suitable to be used for the production of the garment but most can be used to produce the composite materials. The fibers have been defined as units of matter characterized by their length, flexibility, fineness, and a high ratio of length to thickness. Natural fibers can be divided into two classes, according to the diameter size: microfibers and

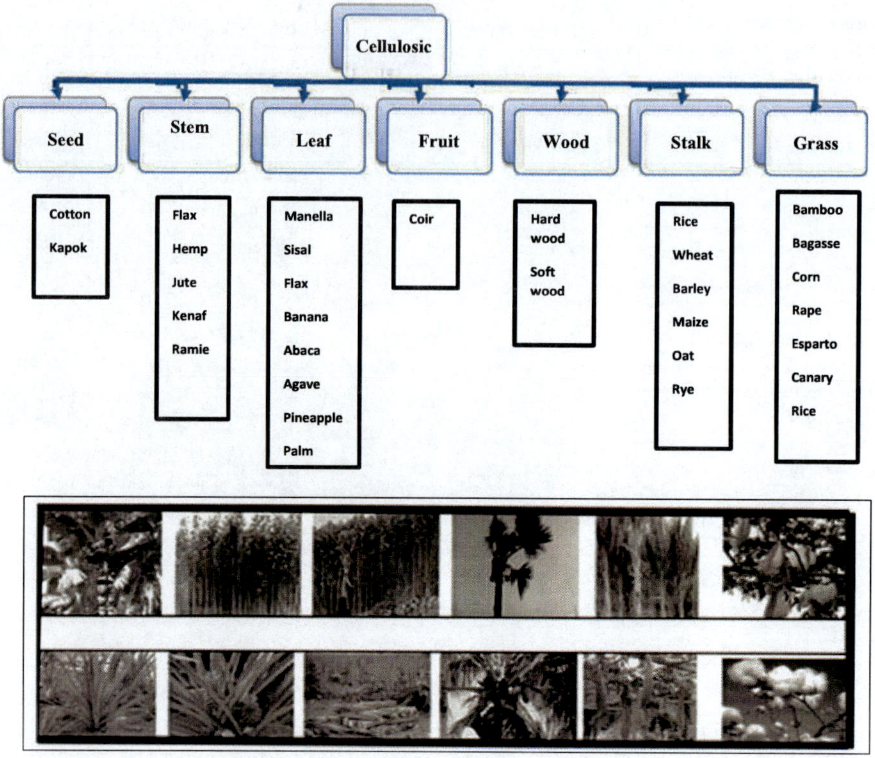

Fig. 5 Natural cellulosic fiber classification (by the author)

nanofibers. Usually, the fibers of the microdiameter are used for composite materials. Natural fibers can be categorized according to its source too, Fig. 5:

- Vegetable fibers: Cotton, Jute, Flax, Hemp, Ramie, Kenaf, Sisal, Coconut fibers, etc.
- Animal fibers: Wool, Silk, Alpaca, Camel Alpaca, Lama wool, Mohair, Cashmere, Angora, etc.
- Mineral fibers: Asbestos.
- Inorganic fibers: Glass, Carbon.

The fibers are classified according to their lengths:

1. Short fibers (1–5 mm), (wood)
2. Long fibers (5–50 mm), (cotton, flax, hemp, jute)
3. Very long fibers mm (flax, hemp, jute, ramie, coir).

For the same type of natural fibers, the variation of the properties depends greatly on the cultivation condition of the plant, soil properties, irrigation, environmental conditions, and the methods of fiber extraction, which are various [1]. The range of variations in some types of fibers was found very high, as illustrated in Table 2, for

Table 2 Properties of natural fibers for composites applications [1, 19–30]

Origin	Fiber name	World production (10^3 tons)	Density (gm/cm^3)	Strength (MPa)	Elongation (%)	Young's modulus (GPa)	Cellulose content (%)
Seed	Cotton	25,000	1.55	300–700	6–10	6–10	85–90
	Kapok	93.612	0.45	93.2	–	4	38
Stem	Flax	810	1.5	500–910	1.5–4.0	50–70	65–85
	Hemp	215	1.47	300–800	2–4	30–60	30–77
	Jute	3530	1.4	200–500	2–3	20–55	45–63
	Kenaf	770	1.2–1.3	284–1200	1.5–2.7	22–60	45–57
	Ramie	100	1.55	220–938	2.0–3.8	24.5–128	68.6–76.2
	Nettle	Abundant	0.72	650–1500	1.7	38	53.0–86.0
	Sisal	380	1.45	100–800	3–14	9–22	50–73
	Banana	200	1.4	500–700	1–4	7–29	63
	Abaca	70	1.5	400–980	1–10	6.2–20	56–63
	Pineapple	Abundant	1.44	400–1600	0.8–1.6	35–80	81
	Palm	Abundant	0.7–1.55	150–500	17–25	3.24	30
Fruit	Coir	100	1.2	175–593	15.0–30.0	4.0–6.0	15–51.4
	Hard wood		0.3–0.88	51–120.7	–	5.2–15.6	40–50
	Soft wood		1.5	1000	4.4	40	45–50
Stalk	Rice	Abundant					
	Wheat	Abundant					
	Barley	Abundant					
	Maize	Abundant					
	Oat	Abundant					
	Rye	Abundant					
Grass	Bamboo	10,000	0.6–1.1	140–800	2.5–3.7	11–32	74
	Bagasse	75,000	1.25	222–290	1.1	17–27.1	45–55
Hair	Wool	1705	1.3	50–315	13.5–35	2.3–5	–

example the strength of kenaf fiber recorded to vary between 284 and 1200 MPa, sisal varies between 100 and 800 MPa, and ramie 220 and 938 MPa. The designer of the natural fiber composite must take into consideration the variability of the fiber reinforcement properties. The global production [9, 10] of natural fibers is illustrated in Fig. 6. Total fiber production expected to grow 3.7% per annum to 2025. The global production of cellulosic fiber has grown by up to 5.8% [1, 9].

Another source of textile fibers is the textile waste produced by the fashion industry and textile mills as well as the recycling of textile products. The Environmental Protection Agency reports that 15.1 million tons of textile waste were generated in 2013, of which 12.8 million tons were discarded. About 15% of fabric intended

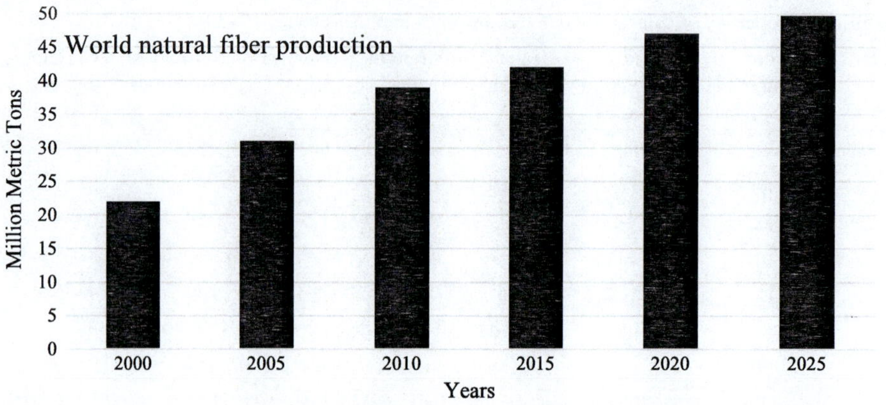

Fig. 6 Production of the natural fibers and their projection to the year 2025 [7, 9]

for clothing ends up on the cutting room floor. This waste rate has been tolerated industry-wide for decades [11, 12]. The recycled fiber waste can be used in the productions of some parts for the vehicle, for instance as a sound absorption sheet, in-car interior—as seatbacks, parcel shelves, front and rear door liners [13, 14].

2.4 *Properties of the Textile Fibers*

The natural fibers derived from legionellosis material are considered to be the most suitable for the formation of the composites used in the different the automotive parts. The choice of the proper natural fibers depends on the mechanical properties of the fibers to withstand the forces acting on the NFC during the use and its life-cycle analysis. For life-cycle accountability, a manufacturer is responsible not only for direct production impacts, but also for impacts associated with product inputs, use, transport, and disposal. There is a large variety of properties that mainly dependent upon plant species, growth conditions, and methods of fiber extraction as well as the structure of the layout of the febrile arrangement and the degree of polymerization. Moreover, the physical properties depend on the fiber cell geometry of each type of cellulose and its degree of polymerization [1, 15]. The percentage of cellulose, hemicellulose, and lignin, which are linked by hydrogen bonds that not only hold fibers together but also the cellulose within the fiber cell wall, is closely associated with fiber stiffness [1, 16]. The fiber mechanical properties depend on the ratio of crystallin and amorphous potion in the fiber structure [17, 18]. Based on the latest data, the use of natural fibers is expected to increase significantly in the future as they are starting to enter other markets than just the automotive sector, in 2020 it may reach 650,000 tons. Cotton, flax, kenaf fibers are mostly used for the manufacturing of different parts. The mechanical properties of the main cellulosic natural fibers are given in Table 2.

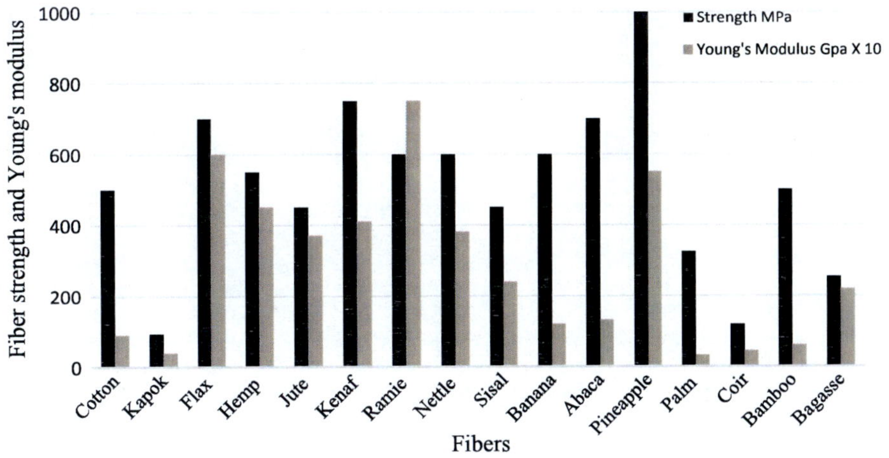

Fig. 7 Comparison between different natural fiber properties (by the author)

Figure 7 shows the comparison between the tensile strength and Young's modulus for the most used natural fibers. For the automotive industry NFC, the cotton, jute, flax, hemp, ramie, kenaf, and sisal are the most candidates. Due to the different values of the fiber densities, the relation of the specific strength and specific elastic modulus should be considered. Figure 8 shows the relation between them, and it was found to be almost linear, especially for all bast fibers.

2.5 *Natural Fibers Treatments*

To enhance the properties of natural fibers, in most cases the natural fibers are pretreated before it can be used in forming a composite material. The objective of this treatment is to improve both the different fiber morphology and properties as well as the interfacial adhesion between the fiber surface and polymer matrix, thereby the mechanical properties of the composites [1, 31]. The natural fiber consists of lignin, pectin, waxy materials, and natural oils that cover the outside layer of the fiber cell wall [32, 33].

Some chemical treatments are applied to improve the fiber properties such as alkaline treatment, silane treatment, acetylation treatment, peroxide treatment, benzoylation treatment, potassium permanganate ($KMnO_4$) treatment, stearic acid treatment. The chemical treatments of the natural fibers mainly enhance the properties of the fiber by modifying their microstructure along with improvement in wettability, surface morphology, chemical groups, and tensile strength of the fibers [1, 35]. The best result for a specific fiber type is reached by applying one of the above treatments. Physical methods involve surface fibrillation, electric discharge (corona, cold

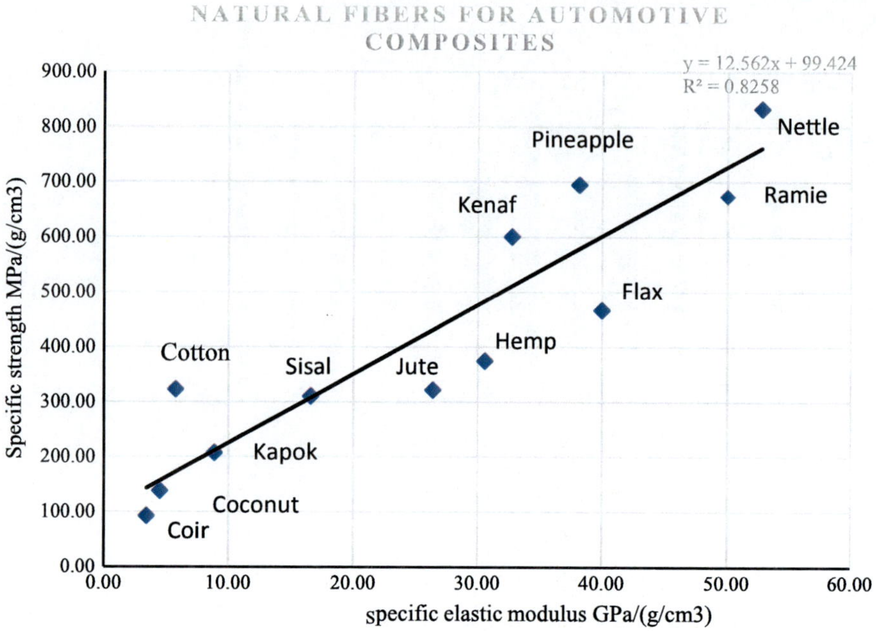

Fig. 8 Specific strength versus the specific elastic modulus of different natural fibers (by the author)

plasma); this changes the structural and surface properties of the fibers and thereby influences the mechanical bonding with the matrix.

2.6 Agriculture Waste

Another source of fiber material is massive of the unused quantities of agricultural waste, Fig. 9. The world statistics of the wheat and rice straw records that about 710 million metric tons of wheat straw and 670 million tons of rice straw are produced each year as the agricultural waste.

Agriculture waste is one of the problems of modern intensive agriculture which needs innovative methods for its useful industrial applications. The development of composites from renewable raw materials has increased considerably during the last years as they are considered to be environmentally friendly materials [34]. Wheat and rice straw both are of the same constitution with about 39% of cellulose. Generally, agriculture waste can be grouped into two broad categories:

- Hard straw: cereal, sugarcane bagasse, bamboo, reeds, and grasses cotton, etc.
- Soft waste: cotton staple and linters, flax, hemp, and kenaf bast fibers, sisal, abaca, bamboo, etc.

Fig. 9 Wheat straw bale

Agricultural residues account for 73% of the world's nonwood pulp capacity, natural plants such as reed and bamboo account for 18%, and the remainder consists mainly of industrial crops. Table 3 gives the estimated global availability of the most common agricultural waste.

The fiber length of the agriculture waste varied between 0.5 and 1.5 mm, while the fiber diameter is between 8 and 23 μ.

The idea of using biobased materials in vehicle parts was first pondered by the founder of the Ford Motor Company in the early 1930s [40, 41]. Henry Ford was fond of using plants like hemp and straw to reinforce plastic components for his cars. This had a body featuring mostly soy resin-reinforced hemp, sisal, and wheat straw composites; this had its weight two-thirds that of a regular car [41].

Table 3 Estimated global agricultural residues [1, 36–39]

Crop	Plant component	Availability	Crop	Plant component	Availability
		1000 tons			1000 tons
Barley	Straw	218.5	Bast fibrous plants	Straw	25.0
Oats	Straw	50.8	Seed grass	Straw	3.0
Rice	Straw	465.2	Oil flax	Straw	3.0
Rye	Straw	41.9	Sorghum	Stalks	104.7
Wheat	Straw	739.7	Sugarcane	Bagasse	100.2
Corn	Stalks	750.3	Bamboo	Bagasse	7.2
Cotton	Linters	2.3	**Total**		**3447.7**
	Stalks	35.9			
	Mote	900			

Table 4 Analysis of the percentage of cellulose and lignin for different agro-residuals [1, 36]

Type of agro-residual	Cellulose	Lignin	Type of agro-residual	Cellulose	Lignin
Rice straw	28–48	12–16	Bamboo	26–34	21–31
Wheat straw	29–51	16–21	Sugarcane bagasse	42	20
Barley straw	31–45	14–15	Grasses	25–40	10–30
Oat straw	31–48	16–19	Palm	40	24
Rye straw	33–50	16–19	Softwood	40–45	26–34
Cereal straw	31–45	16–19	Hardwood	38–49	23–33
Corn straw	32–35	16–27		43–47	16–24
Cotton straw	45–49	11–27			

The 2010 Ford Flex car will be the first car to feature a plastic part that contains wheat straw. Ford claims it is "advancing a strategy to migrate this biobased material to numerous other interiors, exterior, and under-hood applications for multiple product lines." According to the company, the wheat straw/plastic performs better than conventional resins, having better "dimensional integrity," and weighing 10% less than plastics reinforced with glass fibers [42]. Several types of agricultural waste can be used in forming NFC. They variate according to their source. Table 4 gives the analysis of the percentage of cellulose and lignin for different agro-residuals.

From Table 4, it is revealed that the percentage of cellulose in the agricultural waste is much less than in the natural fibers (cotton 85–96%) with an increase of the lignin, which is 11–34% compared to 0.7–1.6% for cotton. This directly reflects on the mechanical properties of agricultural waste, for example, the properties of the wheat straw are linear density 35–100 den, strength 2.2 g/den, elongation 2.8%, modulus of elasticity 11.2 g/den, while for rice straw: linear density 41 den, strength 4 g/den, elongation 2.35%, modulus of elasticity 22.6 g/den. The properties of palm fibers are linear density 35.7 den, strength 1.6 g/den, elongation 6.2%, modulus of elasticity 24.8 g/den. Banana stem properties are strength 5 g/den, elongation 2.25%, modulus of elasticity 110 g/den. Coconut husk properties are diameter 90 μm, strength 1 g/den, elongation 51.15%, modulus of elasticity 38.5 g/den [43].

The agricultural waste is not suitable to be used directly for the composite production, so a pretreatment process is required to transfer it to high-density short fibers appropriate to be manufactured into the thermoplastic composite. For this purpose, several methods can be applied that classified as physical, physicochemical, chemical, mechanical, and biological techniques [1, 44]. It will convert the fiber straw into a loose fiber which can be easily mixed with the polymer to design the composite forms. Each type of agriculture waste has a suitable method to be reformed into particles, flour, fibers, and in some cases nanoparticles. The steam exposition process is an efficient method to convert the agro-residual to nanofibers [1, 45].

3 Application of Natural Fiber Composites

3.1 Application of Textile Material for Sound Absorption in Automobile

Noise represents a quality factor in the automotive engineering [46, 47]. A variety of sources contribute to the interior noise of a vehicle which can be structure-borne or airborne sound. The main sources of noise and vibration in vehicles include those related to the power system (engine, cooling fans, gearboxes and transmissions, brakes and inlet and exhaust systems) and those nonpower system sources generated by the vehicle motion (tire/road interaction noise and aerodynamic noise caused by flow over the vehicles) [48].

In general, the structure-borne sound is the dominant source of interior total vehicle noise below 400 Hz, and airborne noise is dominant above 400 Hz. It was revealed that the acoustical coupling between the power train component excitation and the vehicle cabin cavity is also a dominant source of up to 50 Hz of vehicle interior low-frequency boom noise [49]. The frequency response of different vehicle noise emitted parts of the vehicle is at normal highway speed, tire–pavement interaction noise is the most important noise contributor (up to 80–90% at a speed of over 70–80 km/h [50]) and it may reach 78 dB at vehicle velocity 130 km/h, even more than the aerodynamic noise. Generally, the overall sound level is linearly elevated as the vehicle's speed increases, raising the discomfort for the passenger that determines his satisfaction. The source of the noise can be divided into interior noise and the exterior noise of the vehicle. The level of the external noise is different as illustrated in Fig. 10.

It was found that the interior noise of the vehicle is less than the exterior noise, around low frequencies (50 Hz). The control of noise, using noise-absorbing

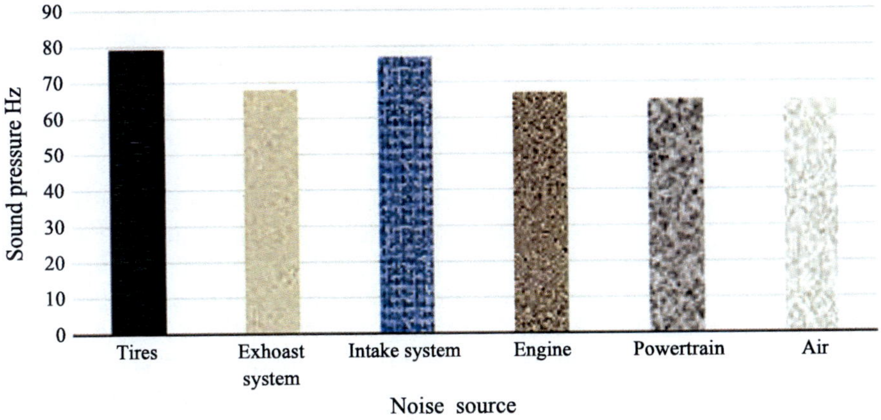

Fig. 10 Sound pressure (Hz) response of different vehicle noise-emitted parts [50]

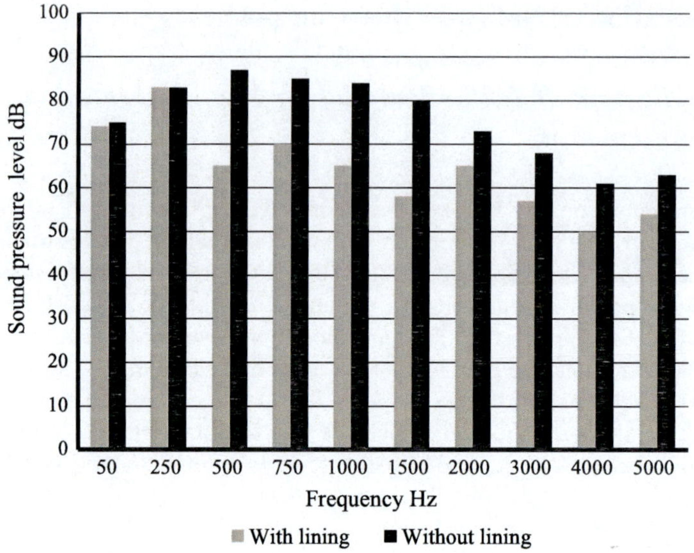

Fig. 11 Frequency response of a cabin enclosure with and without an absorbing lining [49]

shielding, has several methods, especially for modern the automotive designs, which have higher engine performance and lightweight construction, using thinner body panels. For this reason, the contribution of the sound absorption materials such as textile materials in achieving a low level of interior noise is highly appreciated [51, 52]. Figure 11 illustrates the effect of using sound absorber at different frequencies which is more effective for frequencies above 500 Hz [49]. Excessive noise can lead to mental and physical health problems, especially when driving for a long time. In-vehicle noise is overall higher than the maximum safe level of 70 dB for 24 h exposure without harmful effects. It was revealed that drivers are more likely to be exposed to the noise level between 75 and 85 dB [53]. At the driving speed of above 70 km/h, the in-vehicle noise level exceeds the public health and welfare marginal safe level of 8 h 75 dB [53, 54].

Typically, the dominant noise in the vehicle interior has a frequency varied from 30 Hz to 8.5 kHz. The nature of the interior noise can be classified into structure-borne noise, below 500 Hz, and above due to airborne-noise elements [55]. It was revealed that the interior noise resonances caused by modes of tires and suspensions were mainly below 100 Hz. Therefore, the acoustic textiles used in vehicles must have a damping effect for the different frequencies, additionally to other serviceability properties such as abrasion, stain, ultraviolet resistance, formability, recyclability [56]. In modern vehicles, the sound absorption materials are added in the different parts of the body to reduce the metal plates that vibrate and increase the noise. The sound-absorbing material is used in floor carpets, wheelhouse, doors, roof, front of the dash, and any other body panels subjected to high levels of vibration to reduce the sound pressure inside the vehicle (in the frequency range up to 400 Hz). Consequently, the sound absorption materials differ for the diverse parts of the vehicle body: engine soundproofing, truck soundproofing, etc., depending on the range of the frequency of sound exited and coefficient of sound absorption at different frequencies. There are various types of absorbers classified according to their structure, like woven, knitted, nonwoven, felt as well as the type of the fibers used. In some cases, the absorber is a composite material from different structures and different fiber blends. Figure 12 shows some types of absorber; felt acoustic, mechanically glue-free pressed fibers, bitumen pad bonded, the basis of which is a felt absorber, needle-punched nonwoven material.

Made with the highest proportion of recycled cotton and synthetic fibers .	Jute Felt Soundproofing Compressed 'Jute' felt.

Under bonnet, behind door panels, car sound systems, behind side paneling of motor homes / vans, next to engine block	Bitumen pad bonded to a thick layer of felt allows for extra sound protection. Product is flexible.
	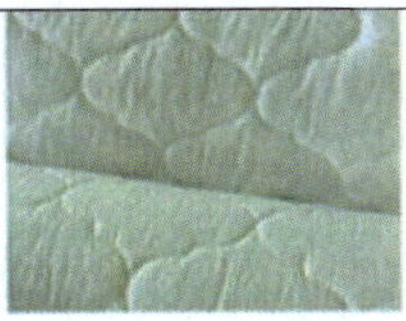
self-adhesive needle-punched synthetic non-woven material, which is made by interlacing and subsequent mechanical glue-free pressing of fibers, the basis of which is a felt silencer.	Sound Deadening Cotton Panels Car Soundproof Materials

Jute felt is made from natural plant fibers compressed into a flexible mat with a black latex finish on the back.	Compressed Jute Felt Soundproofing Material

Fig. 12 Some types of sound absorber used in vehicles (by the author) [57–61]

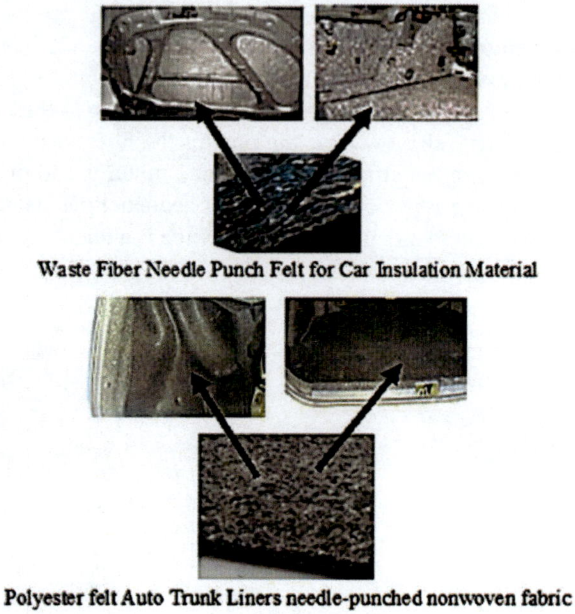

Fig. 13 The lining of the internal parts of the vehicle with sound absorber material (by the author)

The sound absorber for the road noise is fixed in the different parts of the vehicle, Fig. 13, such as front fender, dashboard, floorboard, doper panel, rear fender, trunk engine cover, tailgate [62].

Needle-punched nonwoven material is usually used as a second layer in the processing of the car body, cabs of specialty machines, bonnets, insulation panels, and transport facility bodies. It may serve as a principal layer, which includes insulating and absorbing properties [63]. The sound absorber under the driver and passenger floor will be a responsible damper for the road and tire noise to be transmitted to the passenger cabinet. An example of the application of the absorber in 2012 Buick Verano car is given in Fig. 14, where different types of absorbers are used according to their location for the noise level reduction inside the passenger's cabinet [63].

In Buick's car, extensive types of noise-reducing and noise-canceling materials and constructions were used, for instance, steel front-of-dash panel is sandwiched between two damping mats, the headliner comprises five layers of thermal fiber acoustic material, including a premium woven fabric on the visible outer layer, sound insulation material between rear-body structural components is made from recycled denim, triple-sealed doors feature fiberglass "blankets," windshield and side glass consist of 5.4 and 4.85-mm-thick acoustic laminated glass, respectively [63]. As a whole, natural fibers sound absorber materials can be purposely used to reduce the noise level, to decrease reverberation time, or to eliminate echoes as well as to prevent sound from being trapped by concave surfaces [64]. Several types of natural fibers have been used for many years as a source of raw material to produce porous sound-absorbing and sound-isolating materials [65, 66], including stalk or wood fiber

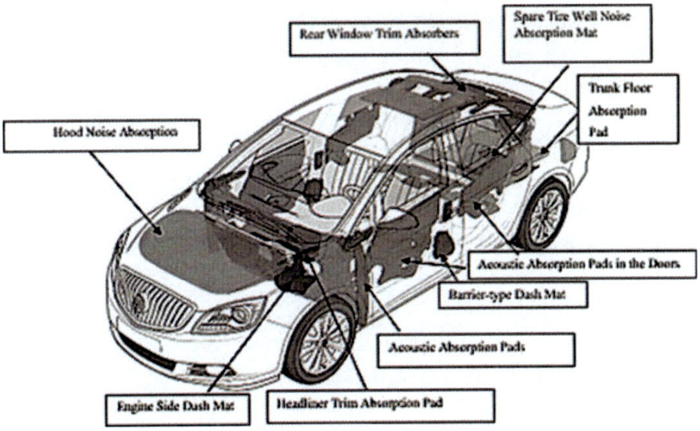

Fig. 14 Buick Verano noise-reducing elements [63]

(straw of wheat or rice, softwood, or hardwood), bast fibers (flax, jute, kenaf, hemp, ramie), leaf fibers (sisal, palm, and agave), seed fibers (cotton and kapok), and fruit fiber (coconut).

Analysis of the mechanism of fibrous materials to absorb sound energy, formed by the nonhomogeneous pores between the fibers, indicates that the sound waves are transmitted through the cross section of the fibers, which have lumina at its center, and attenuate part of sound energy, the other part of the sound energy will dissipate due to the friction between the air and fibers when the air moves inside the specimen's pores as well as the vibration of the air inside the bulk of the specimen [67]. The multi-scale and hollow lumen structures of natural fibers contributed to the high sound absorption performance. It was established that multi-functional composite materials can be made from the natural fibers so that both the mechanical and acoustical functions can be achieved [68].

From the analysis of Table 5, it can be revealed that the kapok fibers have the best sound absorption at sound frequency 500 Hz characterized by having a large lumina

Table 5 Coefficient of sound absorption of some natural fibers at 500 Hz [46–52]

Fiber	Coefficient of sound absorption at 500 Hz	NRC	Fiber	Coefficient of sound absorption at 500 Hz	NRC
Hemp	0.6	0.54	Wool	0.66	0.45
Ramie	0.25	0.6	Degreasing cotton	0.3	0.4
Jute	0.51	0.65	Cotton	0.50	0.35
Flax	0.58	0.65			
Kenaf	0.74	0.7			
Coconut	0.42	0.42			
Kapok	0.9	0.627			

which will absorb a substantial part of the sound energy, followed by kenaf fibers. The value of the coefficient of sound absorption varies according to the thickness, bulk density as well as the porosity of the absorber. Therefore, different values were recorded in the literature [67, 68]. The sound absorption of the different natural fibers is considerably high in the low and medium frequencies, either in fibrous form or fiber/epoxy composite plates [69]. It is clearly evident that increasing layer thickness of any porous sound absorption material promotes the sound absorption coefficient at the low-frequency region [70]. The high values of absorber material's density and thickness increase the coefficient of sound absorption at medium and high frequencies. While the fiber length has no significant effect on the sound absorption coefficient. The random fiber arrangement in the absorber usually provides higher values of the sound absorption coefficient, especially at the high frequencies. The effective absorption of the incident sound wave occurs when the thickness of the material is one-tenth of its wavelength [48]. Most of the absorbers are a nonwoven of a structure, manufactured in one way or another; consequently, the structure will be different. The relationship of the nonwoven structure and the pore size, the pore size diameter distribution in the nonwoven web and the pore shape, which are usually irregular in both shape size and continuity through the web thickness, interconnective pore or closed pore [71], depends on the method of manufacturing the web, fiber properties, and crimp. There are several parameters that influence the sound absorption of textile nonwoven material. Mechanism of sound absorption of nonwoven structures can be explained in the following domain: When sound enters porous materials, owing to sound pressure, air molecules oscillate in the interstices of the porous material with the frequency of the exciting sound wave. This oscillation results in frictional losses. A change in the flow direction of sound waves, together with expansion and contraction phenomenon of flow through irregular pores, results in a loss of momentum. The porous nonwoven material is a space structure of fibers of different length that contains cavities, channels, or interstices so that sound waves can enter through them. Open pores have a continuous channel of communication with the external surface of the fiber structure, influencing the absorption of sound, such as the needle hole in needle punched nonwoven. Closed pores are substantially less efficient in absorbing sound energy [72, 73]. Another category of pores is blind, which is opened from one end only. The increase in the number of fiber layers in the nonwoven structure will lead to a change in the percentage of the different types of pores. This explains the variation in the air permeability of the multilayer structures. The sound absorption depends on the pore shapes and sizes in the nonwoven fabric, the spaces between the pores, the pore size, and also the flexibility of the fibers, their orientation, concerning the direction of the sound wave, as their resonance frequency will affect the absorbed energy.

The sound absorption occurs if the area of absorber separating two media dissipates the acoustic energy (energy that is not reflected).

The coefficient of sound absorption is given by

$$\alpha = (A_a/A_i) \quad (1)$$

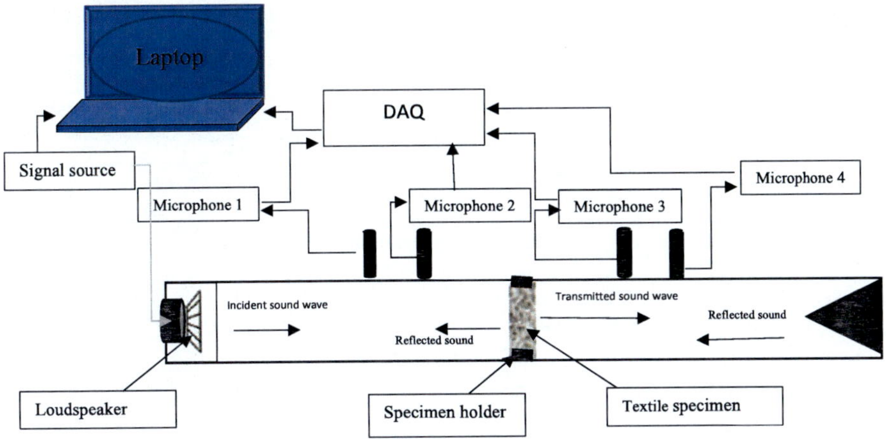

Fig. 15 Sketch of the principles of four-microphone impedance tube (by the author)

$$\alpha = 1 - (A_r/A_i) \tag{2}$$

where α is the sound absorption coefficient, A_r is reflected acoustic energy flow, A_i is the energy flux incident, A_a is the sound energy absorbed.

There are several methods that can be used to measure the sound absorption coefficient: impedance tube and reverberation room. The procedure to measure the sound absorption coefficient of the material was carried out using the impedance tube, Fig. 15, according to ASTM E2611-17 or ASTM E1050-19. Figure 15 shows the impedance tube and a sketch of the principle for measuring the coefficient of sound absorption of the textile material.

In almost the entire frequency range, the textile material absorption coefficient first increases to the maximum at lower frequencies and then fluctuates at higher frequencies, Fig. 16, similarly to the sound-absorbing property of the porous sound-absorbing materials. The analysis of the factors affecting the sound absorption of the textile structures is summarized [72] and presented in Table 6. Many studies attempt to optimize the use of natural fibers as sound insulation materials replacing readily available synthetic products in the market, such as rice straw, coconut coir, palm oil, tea-leaf, kenaf, hemp, bamboo, cotton, wood particles, and wool [74].

The noise reduction coefficient (NRC) of a product will tell us how much sound absorber absorbs as well as how much it reflects. The NRC rating is an average of how absorptive material is at only four frequencies (250, 500, 1000, and 2000 Hz). These industry-standard ranges are from zero (perfectly reflective) to 1 (perfectly absorptive). Noise reduction coefficient (NRC) can be calculated by the following equation for the range of the sound frequency considered:

$$\text{NCR} = (\alpha_{250} + \alpha_{500} + \alpha_{1000} + \alpha_{2000})/4 \tag{3}$$

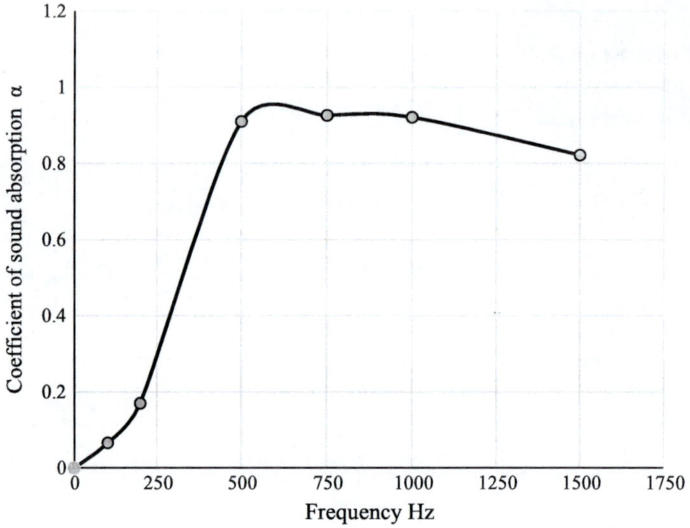

Fig. 16 Coefficient of sound absorption versus frequency (by the author)

where α_p is the coefficient of sound absorption at sound pressure p Hz.

The value of the NCR is affected by several factors such as the material areal density, air permeability, thickness, and fiber diameter as illustrated in Figs. 17, 18, 19 and 20 [72].

The use of several layers of the absorptive material usually affects the value of the coefficient of sound absorption at the low frequencies as shown in Fig. 21.

Nevertheless, as the absorption performance (coefficient of sound absorption) increased, the peak frequencies shifted from high frequency to lower frequencies region. Furthermore, thick sound absorber absorbs more sound energy at lower-frequency regions, while thin absorptive materials are more suitable for higher frequency applications. In terms of porosity, materials with fewer pores normally contain more fiber elements per unit volume, which significantly increases the resistance between sound energy and fiber elements. By increasing resistance, more heat will be dissipated resulting in higher sound absorption. However, porosity and density are closely related. High-density materials that low in porosity normally absorb less sound energy [75]. The influence of the thickness of nonwoven material on the sound-absorption capability shows that the sound-absorption coefficient increases with material thickness, and this relationship is more distinct for low-mid frequencies than for high frequencies [76].

The acoustics of vehicle trims and in-cabin noise reduction is becoming more and more important for consumer satisfaction and comfort. Some perceptions into various aspects of analyzing and designing a new vehicle carpet system improved acoustic performance. The simulation results of the vehicle carpet show that floor trims can significantly contribute (12 dB range) in reducing engine noise, compared to tire/road noise (4 dB range) [77].

Table 6 Factors affecting the textile absorber sound absorption coefficient (by the author)

Sound absorption coefficient	Fiber specifications	Fiber diameter
		Fiber aspect ratio
		Fiber cross-sectional shape
		Fiber sonic modulus
		Fiber surface properties
		Lumen size
		Fiber material
		Fiber morphology
		Surface friction between the fiber and air molecules
	Fabric specifications	Areal density
		Porosity
		Air permeability
		Fabric thickness
		Fiber volume fraction
		Method of manufacturing
		Surface properties
		Yarn morphology
	NFC structural parameters	Fiber arrangements
		Fiber blending ratio
		Pore size
		Pores distribution
		Pores structure
		Composite thickness
		Polymer sound absorption coefficient

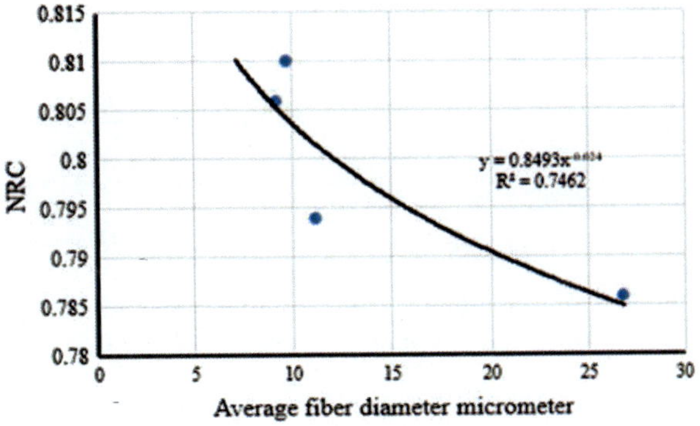

Fig. 17 NRC versus fiber diameter

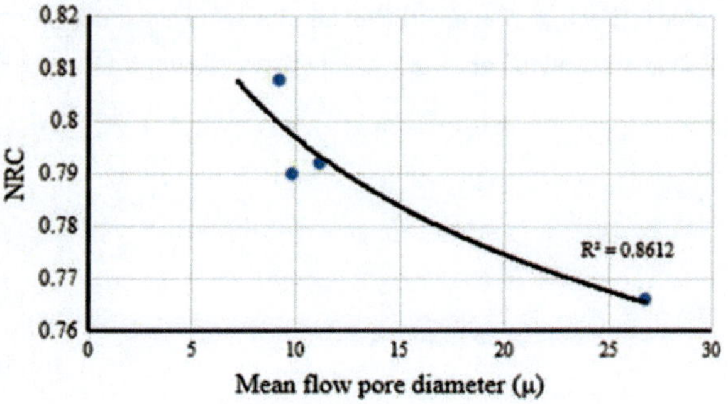

Fig. 18 NRC versus mean flow pore diameter

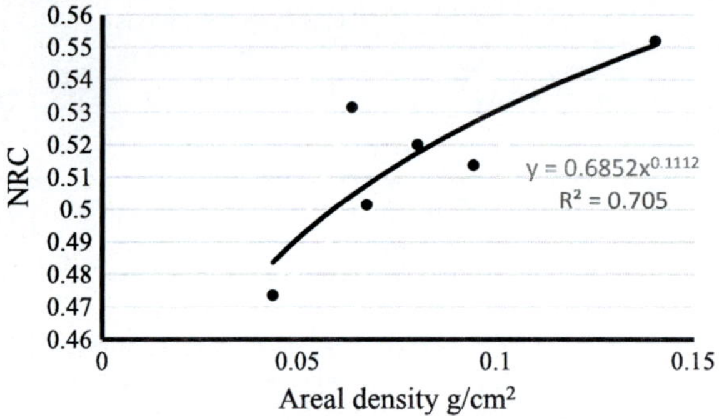

Fig. 19 NRC versus areal density (by the author)

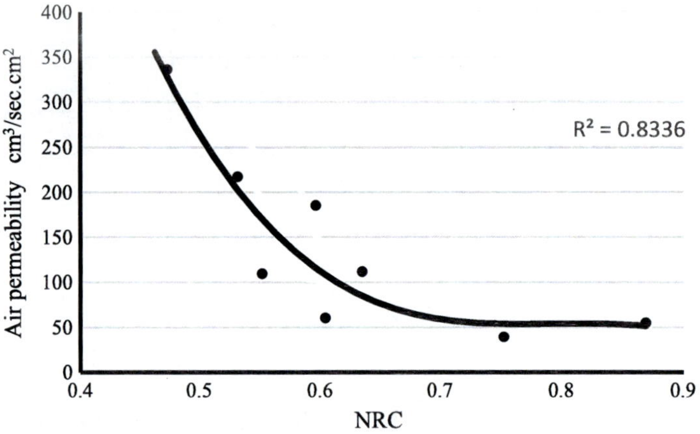

Fig. 20 Air permeability versus NRC (by the author)

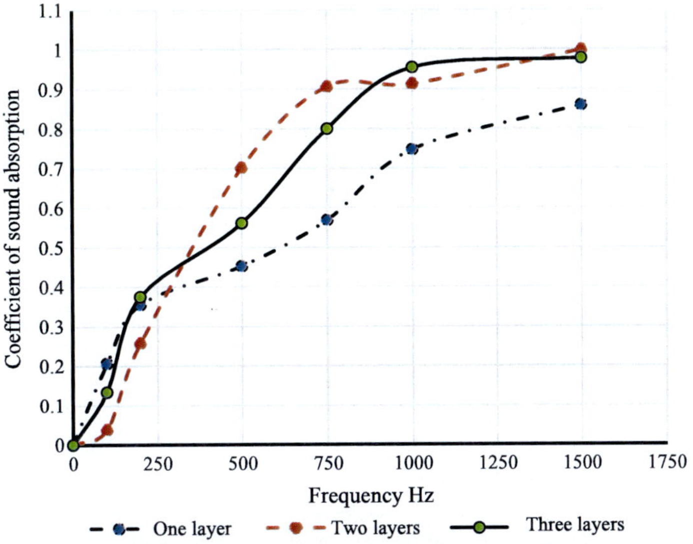

Fig. 21 Coefficient of sound absorption versus sound frequency (by the author)

3.2 Acoustic Absorption of Natural Fiber Composites (NFC)

The sound absorption of the different natural fibers is considerably high at low and medium frequencies either in fibrous form or fiber/epoxy composite plates [78]. In recent years, natural fiber-reinforced polymer composites are gaining more attention due to their cheap production cost, eco-friendly composition, and their relevant properties related to the application of interest. The microscopic structure and the surface morphology of these lignocellulosic materials make them favorable to be used as acoustic absorbers noise insulation requirements in automobiles, a porous laminated composite material manufactured by lamination exhibits a very high sound absorption property at the frequency range 0.5–2 kHz. Sound absorption panels, produced from particle composite boards using agricultural wastes or natural fibers, have challenged researchers to develop novel enhanced soundproofing material. The panel is a combination of natural cellulose fiber (rice straw, bamboo, sawdust, coconut coir fiber, kenaf fiber) with the matrix as biopolymers, resins, or binders. The sound absorption coefficient of the composite increases as the frequency increased. However, it reduces somewhat at a frequency of 1200 Hz and then increases again at a higher frequency [79]. The combination of natural fiber, the ratio of the fiber to the polymer may reach 55% [80], and rubber granular materials exhibit an encouraging sound absorption performance at a low-frequency region when compared with either pure natural fiber or granular composites (fibrogranular composites). In fibrogranular composite materials, there is a good potential in filling the small pores by the granular component and the formation of bridges between the fibers as well [81].

3.3 *Acoustic Characterization of Natural Fibers Agro-residuals*

The waste materials have shown better acoustical performance, especially at mixed levels, than the pure levels due to the improvement in the porosity [82]. Panels of agro-waste can be processed as a composite with polymer and shaped. For example, when using the wood fibers, the noise reduction coefficient (NRC) is 0.55 but 0.72 for the hemp fiber. For coconut fibrous husks, it was revealed that fresh coir fiber of 20 mm thickness has an average absorption coefficient of 0.8 at frequency >1360 Hz. Increasing the thickness improves the sound absorption at a lower frequency but maintaining the same average at frequency >578 Hz for 45 mm thickness [83].

Fire sawdust, beech sawdust, and particles of recycled rubber are also having a good NRC in the range of 1000–2000 Hz. Sawdust-filled recycled-PET composites are also a good candidate for the manufacturing of the sound absorber. Multilayers absorber of different materials can be designed to obtain the highest coefficient of sound absorption at the range of the targeted sound level [84], since it was found that the maximum coefficient of sound absorption at a particular frequency depends on the material, its thickness, and sequence in the different layers. The fibers from coir, corn, sisal, kenaf, hemp, corn, date palm, bagasse, jute, and banana are some examples, these materials are cheaper and environmentally superior to glass fiber-reinforced composites. The agro-waste, like rice husk or rice hull, wheat husk, in fibers, powder, or dust form is used as a reinforcing element in making composites for low-frequency sound absorber, Fig. 22 [1].

Also, other fiber stacks like jute, hemp, kenaf proved to be good absorber at low-frequency range, below 500 Hz [85, 86]. Wood residuals and wood husk can be processed with polyethylene, polypropylene, or polyvinyl-chloride for the production of acoustic absorbers sheets, and wood flour is utilized in the manufacturing of wood plastics composites.

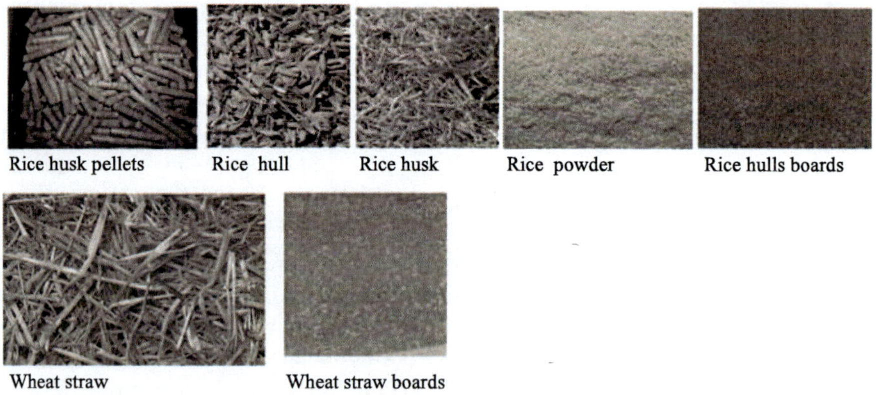

Fig. 22 Rice and wheat straw different products (by the author)

3.4 Recycled Textile Materials

In the last few years, research on the use of eco-materials that come from residues, either from industrial plants or processes, has received much concern [1, 2, 48]. It is well known that recovered materials and the use of environmentally friendly materials for noise control will be increased in the future. There are many examples of recycled eco-materials. Textile recycling strategies are based on the knowledge of their classification: (a) post-production materials, such as yarns, textiles, fibers which come from the loss of the production process, and (b) post-consumption materials, such as clothing, carpets. It was found that up to 95% of all discarded clothing ends up in the landfill. The recycling of textile waste, Fig. 23, can serve as a means of providing solutions to many economics and can be used as recycled composites. Some studies, concerned with the composite sound-absorbing materials made of recycled textile waste, exhibited that the NRC for given thickness was generally higher than 0.5 from 500 Hz on and higher than 0.9 from 1 kHz on [87]. Recycled fabric waste can supply reprocessed fibers for various acoustical applications, including the automotive industry and acoustical panels.

Several authors presented studies on the sound absorption and sound insulation properties of this type of the materials [88]. In some trials, composite materials are manufactured from wastes generated from the textile, maize, and newspapers wastes. These raw materials were bonded using polyvinyl acetate (PVA) adhesives. The textile waste material can be also mixed with newspaper or maize waste in different ratios. It was proved that NRC value is the highest when mixing maize and textile fiber waste equally [89]. The biodegradable composite materials from textile waste, wood (flakes or fibers), and textiles (cotton, wool, or jute) and binders as wheat flour and/or ecological acrylic copolymers represent the new structures with a very good absorption capacity [90]. This has prompted an urge to substitute nearly 40 the automotive interior components, that currently contain traditional materials, such as glass fibers and other synthetic fibers and foams, that are difficult to recycle,

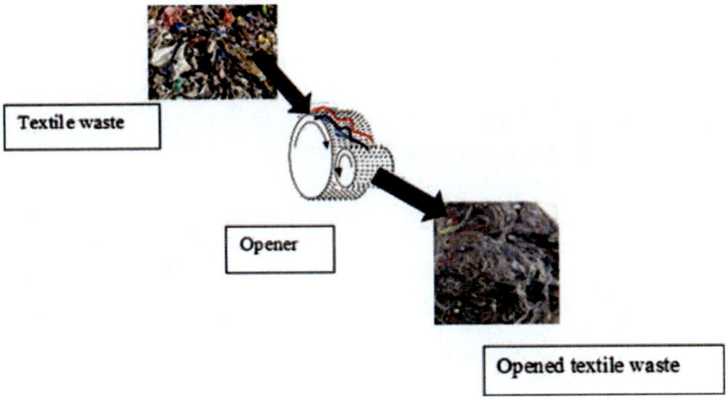

Fig. 23 Textile waste-recycled fibers for a sound absorber (by the author)

for biodegradable sound absorptive materials [91]. It was revealed that the ability to reduce noise inside the vehicle enhances the recognized value of the vehicle to the customer and offers a competitive advantage to the manufacturer [92].

Numerous methods are presently employed to reduce noise and its sources, one of which uses sound-absorbing materials attached to various components, such as floor-coverings, package trays, door panels, headliners, and trunk liners [93]. Nonwoven selvages were recycled to produce sound absorption composites with acceptable sound absorption efficiency under a proper compression process [94]. Most of the fibers in NFC need to be treated with coupling agents to modify the natural fiber–polymer matrix interface by increasing the interfacial strength, such as the silane coupling agent [72]. Industrial prepared coir fiber is obtained from coconut husk and combined with latex and other additives to enhance its structural characteristics. A perforated plate was added to the multilayer structure to further enhance the sound absorption. Moreover, when the perforated plate was backed by coir fiber and air gap, the porosity of the plate had superior influence in adjusting the amount of low-frequency sound absorption. The coefficient of absorption of perforated plate backed with 50-mm-coir fiber has a value of 0.9 at frequency 1500 Hz and fluctuated till it reached 0.95 at 3000 Hz. In such a design, the presence of air gap between the coir fiber and the perforated plate affects the values of the coefficient of absorption [13, 95, 96]. The investigation of raw paddy fibers from the panicles has a good acoustic performance with a normal incidence absorption coefficient greater than 0.5 from 1 kHz and can reach the average value of 0.8 above 2.5 kHz [97].

It was revealed that the sugarcane bagasse has a good coefficient of sound absorption at a frequency above 1000 Hz and NRC 0.65 at high frequencies 1.2–4.5 kHz [98]. Wasted ramie fiber-treated and nontreated with alkalization can also produce promising results with an average absorption coefficient of 0.6 at the frequency range of 315–3200 Hz. The fibers waste either is used as it is to form the sound absorber or it should be pretreated to be suitable to be utilized as a sound absorber or used with a suitable polymer to form a sheet to be used as a composite sound absorber, Fig. 24. The design of such composite may contain several types of fibers and, in some cases, fabrics are used to cover the natural fiber or perforated plate to improve the value of the coefficient of the sound absorption. The design of the absorber maybe with an air gap or without an air gap to improve the NRC of the absorber. The sound absorption coefficient results measured in an impedance tube show that a significant improvement in the sound absorption performance of the bulk materials can be achieved by incorporating the nanofibers layer on them [96].

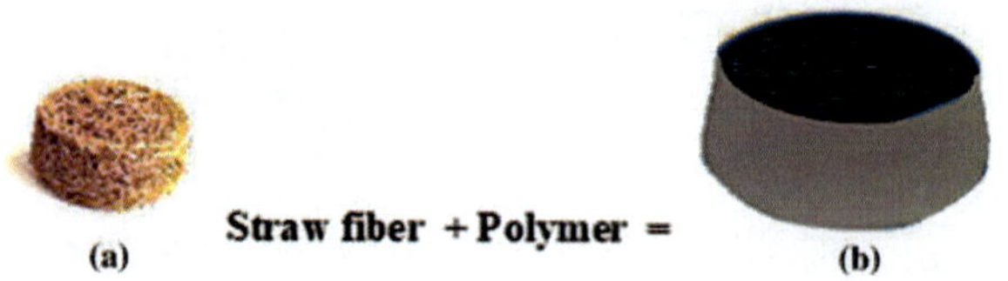

Fig. 24 Straw fiber–polymer composite, **a** straw fiber laminate and **b** straw fiber (by the author)

Recycled denim fibers were used to successfully produce insulation materials with good performance properties, 60% Denim/40% biodegradable fiber (Sorona). Web formation of blended fibers was done using nonwoven techniques. Heat and pressure were applied to the blended fiber webs to produce composites. Several layers of fiber webs produced from the card were combined. These layers of fiber webs were placed between two plates of a carver hot press and then formed into a composite panel (temperature 235°, pressing force 750 kg, for 10 min) of thickness 1.85 mm. The transmission coefficient is the fraction of the incident airborne sound power that is transmitted through the material and was found to be 0.4. The transmission loss of 10.45 dB was observed at different frequencies. Composite panel made of 60% recycled cotton/40% PLA has an average of 15.05 dB [99].

4 Natural Fiber–Polymer Composites Design for the Automotive Industry

4.1 Introduction

The automotive textiles are a branch that deals with the textile parts used in the automotive industry, either cars, trains, buses, airplanes, using textile fabrics, or textile polymer composites. The textiles may be in the form of compressed fiber padding, nonwoven fabrics, woven or knitted fabrics, or fiber–polymer composite (FPC). The textiles materials are used for different objectives: seat cover, carpets, and roof and door liners. The rest is utilized to reinforce tires, hoses, safety belts, airbags, door kick panels, parcel shelves, and heat insulation [100]. Besides, the textile fabrics are used as reinforcement of the main element of the automobile tires. This increases the number of textile fibers and FPC materials used in modern vehicles. The NFC can be biodegradable when using biodegradable polymers. One of the main advantages of using textile polymer composite is to reduce the weight of the vehicle that makes it possible to cut the fuel consumption, the noise level at high speed, and the cost. The natural fibers have a low cost, the estimated cost of various plant fibers in its loose form is given in Table 7 [15, 101], and low density in the range of 0.9–1.5 kg/cm^3, not mention the least impact on the environment.

Table 7 Cost of the different types of fibers [15, 101]

Fibers	Price (US$/kg)	Fibers	Price (US$/kg)
Flax	2.1–4.2	Coir	0.3–0.5
Hemp	0.5–2.1	Cotton linter	1.5–4.5
Kenaf	0.4–0.6	Banana	0.89
Jute	0.3–1.5	Ramie	1.5–2.5
Sisal	0.6–0.7	Wood	0.5–1.5
Bamboo	0.5	Abaca	0.35

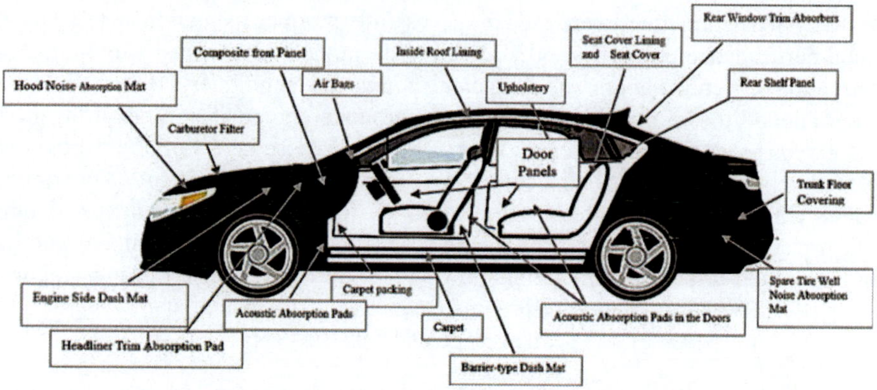

Fig. 25 Natural fiber textile composites in the modern automotive industry (by the author)

The car manufacturers Ford, Mercedes Benz, Audi, Audi, Toyota, BMW, Mazda, Fiat, GM, Chrysler and many other companies used mostly the bast fibers to form the different parts of their new models [41], such as package trays, internal engine cover, engine insulation, sun visor, interior insulation, bumper, wheel box, roof cover, front-rear door panels, parcel shelves, boot linings, interior door paneling, pillar cover panel, head-liner panel, boot-lid finish panel, spare tire-lining, side back door panel, hat rack, car windshield/car dashboard, noise insulation panels, molded foot linings, instrument panel support, insulation, molding rod/apertures, trunk panel, seat surface/backrest, floor mats, indoor cladding, seat-back linings, cargo area floor, body panels, spoiler, cargo floor tray, and ceilings, door inserts. Figure 25 illustrates some application of the textile material in a modern car [102]. Different automotive makers used different types of fibers and matrix polymers.

4.2 Natural Fiber Composites for the Automotive Industry

The natural fibers used for the manufacturing of the various parts of the automotive differ depending on the type of vehicle (passenger, transport, truck, etc.). Table 8 shows the different fibers used by machine makers. The amount of the fibers used by the car manufacturers for each vehicle varies, for instance Ford uses from 5 to 13 kg, BMW up to 24 kg.

BMW introduced the combination of 80% flax with a 20% sisal blend for increased strength and impact resistance. The main application is in interior door linings and paneling. Wood fibers are also used to enclose the rear side of seat backrests, and cotton fibers are utilized as a sound absorber material; the exterior skirting panels are flax-based. The 2010 Ford Flex was the first car to feature plastic parts that contained wheat straw. All the machine makers claimed to reduce the weight of their

Table 8 Natural fibers used for vehicle manufacturing [13, 104]

Natural fibers used for NFC	Origin	Type of fibers	Machine maker[a]
	Bast	Flax, hemp, jute, kenaf, ramie	1, 2, 3, 4, 6
	Leaf	Abaca, banana, pineapple, sisal	1, 2
	Seed	Cotton, kapok	2, 7
	Fruit	Coir	1
	Wood	Hardwood, softwood	4, 5, 6, 7
	Stalk	Wheat, maize, oat, rice	7
	Grass/Reed	Bamboo	3

[a] 1—Mercedes Benz, 2—BMW, 3—Toyota, 4—GM, 5—Chrysler, 6—Fiat, 7—Ford

vehicle by more than 20% using different processing techniques (extrusion, compression molding, injection molding, and others), resin (polyethylene, polypropylene, polyvinyl chloride, polylactide acid, and others), and fibers (wood flour, flax, kenaf, hemp, jute, and others) [103]. The global composite market trends, opportunities, and forecast in the biocomposites toward the end of the first quarter of the twenty-first century by end-use industry, such as the automotive, industrial and others, is expected to reach an estimated $6.4 billion by 2024 with an average annual increase rate of growth 4.6%. The biocomposites market is expected to reach an estimated $8.2 billion by 2024 and is forecast to annual growth of 7.5% [103, 104]. These aspects will increase the use of natural fibers and their agricultural waste.

The application of the different natural fibers polymer composite in the automotive parts depends on the properties of the fibers and the stresses applied to it during the use finally the cost. For example, jute fiber is used for the manufacturing of the roofing panel, soft armrests, coir fiber for floor mats, seat upholstery, kenaf for trays, door panels, flax for seatbacks, oil palm for decking, roofing panels, hemp for hard arm seats, ramie for sound absorber, rice husk for roof panels, while sisal for door panels and roofing sheets [104].

4.3 Natural Fiber–Polymer Composites (NFPC)

The eco-friendliness of the natural fibrous materials can be applied for the automotive composites. The current trends in the application of composites have 25% of the market share [105]. With natural fiber composites, car weight reduction up to 35% is possible [106]. This can be translated into lower fuel consumption and lower environmental impact. Natural fiber-based composites also offer good mechanical performance, good formability, high sound absorption, and cost savings due to low material costs. For such a composite, the different properties of the fibers have different importance. The manufacturing of natural fiber composite for the automotive applications (NFPC) includes the process of interfering with the reinforcement textile structure into the polymer matrix to produce the final composite shapes. The reinforcement may be in form of fibers, wood flour, short fibers in bulk form (two-dimensional fiber mat or random fiber mat) or entangled forms (non-woven), yarns, 2D fabric (woven, knitted, biaxial braided, triaxial, multidirectional fabric, or 3D fabric). The basic requirements for the automotive textile materials are illustrated in Fig. 26.

The configuration of the textile preforms, which define the final fiber architect in the polymer matrix, has a great influence on the final mechanical properties of the composites. The reinforcements will take several shapes, depending on the final targeted composite shape and the properties of the raw materials. Thus, fiber reinforcement may be in the form of short staple fibers, long-staple fibers, continuous filament, chopped fibers, microparticles, whiskers, fabric reinforcement, nonwoven reinforcement [72]. Figure 27 illustrates the ranking of the different main requirements for the fibers according to their end-use, either in apparel and domestic or industrial applications, ranked from 0 to 10.

Table 9 gives the requirement of the NFC to be used in manufacturing different parts in the automotive industry. Of course, not every part should have all the

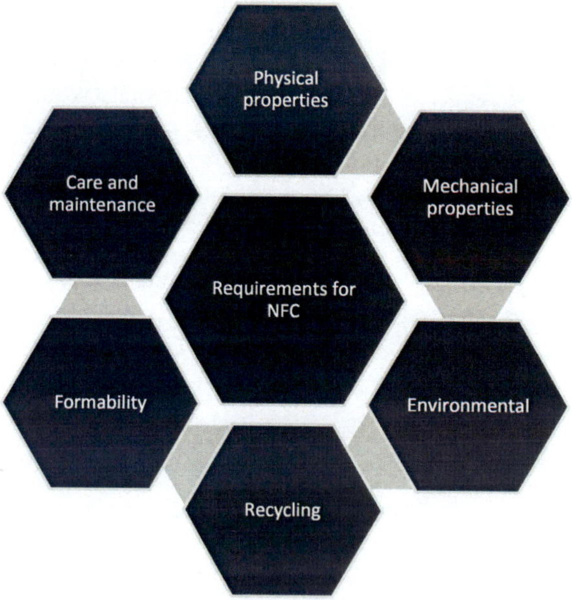

Fig. 26 The basic requirement of the textile material for the automotive applications (by the author)

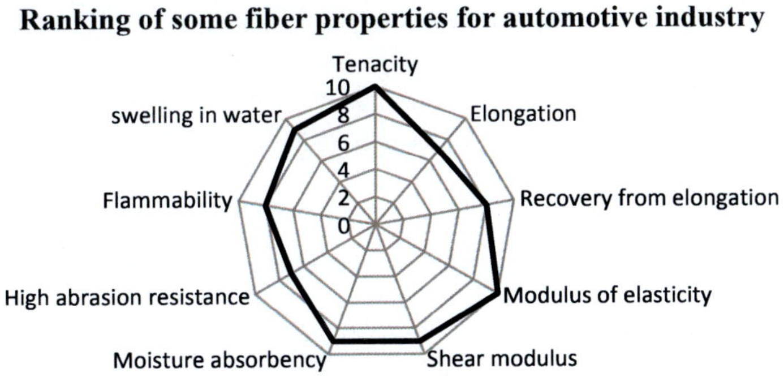

Fig. 27 Ranking of some fiber properties requirements for the automotive (by the author)

Table 9 NFC requirements for the automotive industry application (by the author)

NFC requirement	NFC requirement
High strength	Low moisture sensitivity
High elongation at break	Low thermal sensitivity
High modulus of elasticity	High acoustic absorption coefficient
High shear modulus	Suitability during processing under temperature and pressure
Moderate flexural properties	Good recyclability
High impact strength	High interfacial shear stress with the matrix polymer
Low swelling in water	Dimensional stability under working conditions
Low flammability	Behavior of the material during crash (shattering behaviour)

mentioned properties, but it depends on the stresses applied on it during the manufacturing as well as during its service life. The pioneer in this area was Henry Ford, he built the first Ford plastic car in 1941 containing hemp, sisal, wheat straw [1]. He claimed that the bast fibers are the candidates to be used in the automotive industry to form NFC for different applications. In the last decades, focusing on sustainable designs and environmentally friendly products has attracted the interest of the researchers and engineers in the context of replacing metals and synthetic fibers with natural-based fibers, especially in the automotive industry [107]. Natural fiber-based composites are finding increasing use in products like the interior and exterior parts of the automotive [69, 108]. One might already say that the era of biopolymers is just beginning, as they can make up as much as 60 kg of mass in some car models. Some leaders in the automotive industry predict a decrease in the use of glass fibers for the sake of natural ones such as wood, flax, hemp, bamboo, jute, and kenaf [109].

This will result in higher safety due to the mechanical parameters of the fibers, which absorb energy very efficiently and do not break easily or shatter in a crash.

Natural fiber–polymer composites can be classified in several ways:

1. According to the constitutes of the textile element used
   a. Randomly Dispersed

      Fibers
      Particles
      Shavings
      Flour

   b. Continuously Aligned

      Yarns
      Fabric

2. Structural

   Laminate
   Sandwich panels

3. Matrix polymer components

   Natural base polymer
   Biodegradable petroleum base
   Partial biodegradable base
   Non-biodegradable petroleum base

4.4 Polymers for NF in the Automotive Composite

Natural fiber–polymer composites consist of a polymer matrix embedded with high-strength natural fibers. There are several types of matrix used in the manufacturing of NFC for the automotive parts that can be categorized into thermoplastic (polypropylene, nylon, acrylic, polyethylene, polystyrene, polycarbonate), thermoset (epoxy, polyester, polyimide, etc.), and rubbers (polybutadiene, styrene-butadiene, nitrile-butadiene). Table 10 gives the constituent of NFPC, while Table 11 gives the mechanical properties of polymers used as a matrix in the automotive composite formation [1, 110].

Thermoplastic polymer becomes soft on heating and solidified on cooling, while the thermoset polymer, which is a highly cross-linked polymer, is cured using heat and pressure [1, 13]. Mixing biopolymers or biodegradable polymers with each other can improve their intrinsic properties, for example, starch-based blends, starch-poly (ethylene-co-vinyl alcohol), starch-polyvinyl alcohol, starch/PLA, PHB/starch, PHBV/starch or PHBV/polylactic acid [1, 13].

Table 10 Natural fiber/polymer composite (NFPC) constituents [1, 110]

Reinforcement	Matrix	Reinforcement (agricultural waste)	Matrix
Natural fibers, nonwoven mats, granulated natural fibers, wood flour	Polyester, thermoplastic, high-density, or low-density polyethylene	Natural granulated fibers	Thermoplastic resin
Bagasse mixed with other agricultural fibers	Thermosetting resin, phenol formaldehyde resin, methylene diphenyl diisocyanate, urea formaldehyde	Agriculture residues (straw, bark, coir fiber, bagasse, etc.)	Thermoplastic resin
Natural fibers (jute, sisal, ramie, etc.) in mats, fabrics, or hybrids	Thermosetting liquid resin, polyurethanes, epoxy resin, polyimides, polyester resins, urea-formaldehyde		
Nonwoven mat	Unsaturated polyester resin		

4.4.1 Automobile Bumper

Automobile bumper is a structural component of the automotive vehicle, which contributes to vehicle crashworthiness or occupant protection during the front or rear collisions. The bumper systems also protect the hood, trunk, fuel, exhaust, and cooling system as well as safety-related equipment [110]. There are several factors to be considered when selecting a bumper system. The most important consideration is the ability of the bumper system to absorb enough energy to meet the original equipment manufacturers (OEM's) internal bumper standard [111]. The shape of the front and rare bumpers varied in the last years, Fig. 28; however, its function became more important, especially with the increase in the car speed, when the wind resistance consumes more power and depends on the bumper shape.

The application of natural fibers as reinforcement in polymer matrix focused on the environmental consciousness. NFPC may be manufactured from one type of fiber or formed as a hybrid composite as a combination of two or more different types of fiber to reach the final requirement of the designed part [112]. There are attempts to design the passenger car bumper beam from a composite of kenaf/glass epoxy material [113]. Oil palm empty fruit bunch (EFB) fibers also were suggested to reinforce epoxy resin for bumper beam in cars to replace epoxy/glass fiber composite. EFB fibers were extracted by two methods: chemical method by treating with 10–30% sodium hydroxide (% by weight of fiber) and the mechanical method by steam explosion process at 12–20 kg/cm^2 for 5 min. Then the obtained fibers were bleached by hydrogen peroxide. It was revealed that EFB fiber-reinforced epoxy composite

Table 11 Mechanical properties of polymers used as a matrix in the automotive composite [1, 110]

Polymer	Density (g/cm^3)	Tensile strength (MPa)	Tensile elongation to break (%)	Young's modulus (GPa)	Melting temperature T_m (°C)	Poisson's ratio "ν"
Polypropylene (PP)	0.9–0.94	27–33	200–700	1.3–4.4	134–174	0.45
Low-density polyethylene (LDPE)	0.95	20–30	20–100	0.700	124–136	0.4
High-density polyethylene (HDPE)	0.93–0.97	10–60	4–18	0.60–2.90	115–125	0.40–0.46
Polystyrene (PS)	1.050	48	1–60	1.4–4	240	0.35
Polyvinyl chloride (PVC)	1.330	48	40–450	0.014–4	212	0.40
Acrylic	1.19	74	6	3	265	0.38
Nylon	1.15	55–83	60–200	1.4–2.8	265	0.32–0.40
General purpose epoxy	1.1–1.4	35–100	1–6	3–6		
Polyester	1.2–1.5	40–90	5–300	2–4.5	200–265	0.38
Polylactic acid (PLA)[a]	1.21–1.43	58	4–8.7	3.5	145–186	0.36
Poly-L-lactide (PLLA)[a]	1.25–1.29	15.5–65.5	3–4	0.83–2.7	170–190	
Polyhydroxybutyrate (PHB)[a]	1.18–1.26	24–40	5–8	3.5–4	168–182	
Copolymer poly (3-hydroxybutyrate-co-3-hydroxyvalerate), (PHBV)[a]	1.23–1.25	20–25	0.5–1.5	17.5–25	144–172	
Starch[a]	1–1.39	35–80	31	0.125	110–115	

[a]Biopolymers

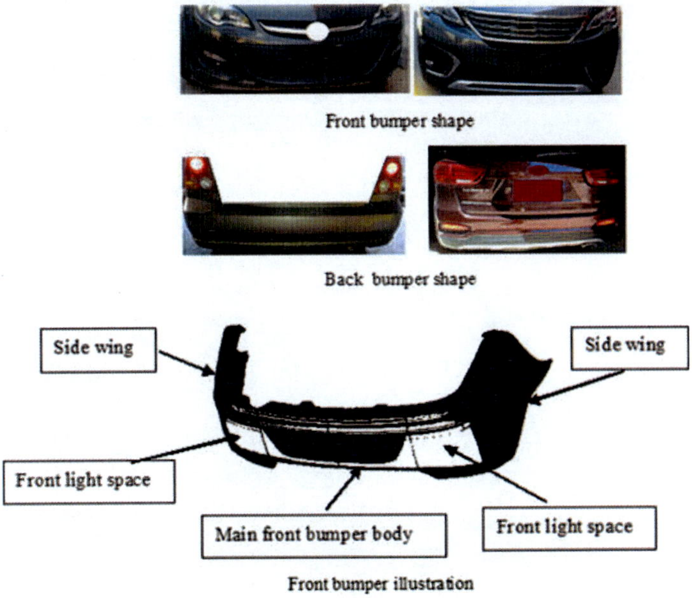

Fig. 28 Front bumper shape (by the author)

could be an alternative green material for bumper beam in automobiles [114]. Okra and banana fibers were mixed at different blending ratios before chopping them to lengths of 10, 30, and 50 mm. The fibers were treated with 2% of NaOH for an hour under constant stirring and allowed for 24 h at room temperature. The matrix material was vinyl ester resin [115]. Another trial to manufacture car bumper was from jute fiber-based composite with the matrix of polyvinyl acetate (PVA). It was found that the weight and cost of the jute fiber composite bumper were found to be reduced by 56.1% and 58%, respectively, and the impact strength increased by 54.5% in comparison with the steel bumper [116]. The results of various trials to design bumper with sisal/epoxy and sisal/polyester concluded that not only the tensile strength of the bumper but also the flexural strength of the material can compete with steel [117].

4.4.2 Passenger Car Side Door Impact Beam

Natural fiber composites are the subject of enormous interest for its use in the manufacturing of different parts of the interior and exterior components of the automotive. One of these components is the side door impact beam. On a demand to reduce the weight of the body of the vehicle, the material used became thinner consequently, side door impact beam was implemented to ensure that the maximum energy absorption by the beam during the side door impact collision can prevent serious injuries and reduce the deformation of the side door components. The side door impact beam's (SDIB)

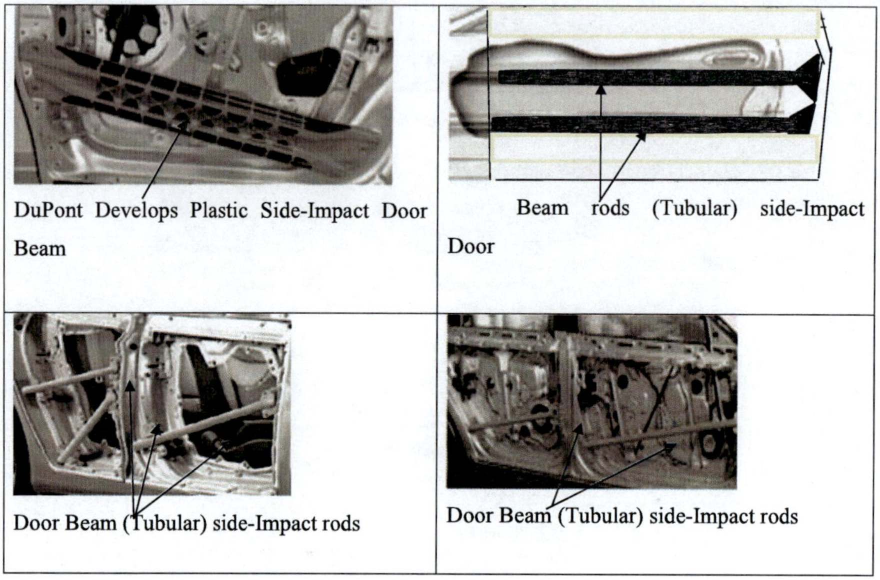

Fig. 29 Some designs of side door impact beams [119–122]

shape may be: beam (tubular) or panel of different designs [107]; therefore, they can withstand the oblique side pole impact and side impact mobile deformable barrier (Federal Motor Vehicle Safety Standard FMVSS 214) [118]. Figure 29 illustrates some designs of (SDIB).

Several designs using a composite material instead of steel to reduce the total weight of the car [123] were studied; moreover, the synthetic fiber–polymer composites, the biocomposites with natural fibers, have been developing rapidly for the manufacturing of the SDIB [124].

It was suggested that kenaf fiber is the best natural fiber material to be used as the reinforcement material in polymer composites for the side-door impact beam [101]. In another study, when different composites were compared, PLA-flax, PLA-kenaf, PLLA-hemp, and PHB-ramie [125], PLA/flax ranked first in both average specific strength and cost/volume, while PHB-ramie was the stiffest of all composites, and therefore, these two dominated all the others. Usually, Polyamide 6.6 thermoplastic composites reinforced with continuous fiberglass strands are used for the manufacturing of the SDIB [126], but the specific modulus of the glass fiber is 29 while flax is 45 and for hemp 40 GPa/cm^3, making them a better candidate composite for SDIB.

4.4.3 Door Panel

The automotive companies, Ford, Opel, Volkswagen, Fiat, Saturn, Saab, Toyota, Mitsubishi, Mercedes Benz, and other carmakers, use the natural fiber–polymer composites for manufacturing of the door panel. Daimler-Benz is working with a range of natural fibers, like sisal, jute, coconut, hemp, and flax, as reinforcing fibers in high-quality polypropylene components to replace glass fibers polymer composite [13]. By using flax/sisal thermoset composite in the door panels, a reduction of 20% weight saving was achieved. Recently, banana fiber-reinforced composites are coming into in interest due to the innovative application of banana fiber in under-floor protection for passenger cars [31]. Automobile parts, such as rearview mirror, visor in two-wheelers, seat cover, indicator cover, cover L-side, the nameplate, are fabricated using sisal and roselle fibers hybrid composites [32]. The mechanical properties of some natural fiber–polymer composites depend on several factors: fiber properties, interfacial shear strength, polymer mechanical properties, fiber to matrix ratio, the fiber length, fiber diameter, fiber cross-sectional shape, structure of the composite and the method of composite manufacturing (molding under hydraulic pressing, extrusion, compression molding, pultrusion or resin transfer molding) [1, 72]. Many works have been reported on the surface-treated cellulosic fiber-reinforced biocomposites, such as sisal/PP, flax/epoxy, jute/epoxy, sisal/polyester by mercerization process (usually using NaOH with concentration varied between 1 and 5%, while in the case of banana, 10% was recommended) [110]. To improve the final composite properties, such as tensile, flexural, and impact strengths, through increase the adhesion force between the fibers and the matrix, the NaOH concentration can go up to 20%. Silane of concentration 0.5–10% may be used with some fibers to improve the interfacial force between the fiber and the matrix. It was revealed that mechanical properties of the poly(lactic acid) (PLA)/30% jute composite strength reaches 81.9 MPa and Young's modulus 9.6 GPa, while the polypropylene polymer (PP)/30% glass fiber composite has the same values but the density for jute is 1.4 g/cm^3 against 2.55 g/cm^3 for glass fiber. This results in a reduction of the weight by about 40%. The poly-L-lactide (PLLA) polymer/30% flax composite strength is 99.9 MPa and Young's modulus 9.5 GPa [121]. This explains why the use of NFC has ascended dramatically in recent years, gaining: comparative weight reduction; good mechanical and impact performance; recycling possibilities; environmentally friendly at low cost [1, 39, 121]. The percentage of natural fibers in some composites in the automotive industry may reach 26–90% in the case of wood fibers composites and 20–65% when using other types of fibers [122–124].

4.5 *Some Aspects of Natural Fibers Composite Design*

Composites began to be used more and more in the car industry, started initially for some of the parts and later on for the whole structure of the vehicle, as the composites exhibited better fabrication performance. Among these, the textile composites filled their own niche, and at the same time, shared the production problems [1]. Textile composite defects often are due to the faults in the fiber discontinuity, shear stress concentration at the end of fibers, fiber de-bonding defects, fabrics defects, presence of voids or variation of the volume fraction at the different parts of the composite in *xy*-coordinate plane or the three coordinate axes (*xyz*). Any flaws in these parameters might lead to early failure of the composite. In the case of laminated composite, structure interlayer delamination can cause a microcrack which accelerates the rapture under the applied load. Fiber–matrix interface is problematic due to the hydrophilic nature of a polymer; therefore, it can lead to a heterogenous structure whose properties are inferior because of the lack of adhesion between fibers and matrix. The necessity for the fiber's treatment to improve interfacial strength of fiber–matrix is a critical step in the development of such composite [129–132]. The tensile strength is more sensitive to matrix properties, whereas its modulus depends on the fiber modulus. To improve the tensile strength, a composite with high fiber volume fraction and the strong interface is required.

Theoretical values that express the volume fraction were carried out for different textile preforms [132]. In the case of spun yarns, natural fibers that are nonhomogeneous fibers, the yarns are usually twisted. The yarns have a variable diameter and variable packing density [133]. The fiber volume fraction is the most decisive factor in determining the mechanical properties of the NFC; hence, it plays an important role in the behavior of a composite under different loading. Composite tensile stress (σ_c) of NFC materials often follows the rule of mixtures:

$$\sigma_c = V_f\sigma_f + (1 - V_f)\sigma_{um} \tag{4}$$

where σ_f is the fiber tensile strength, σ_{um} is the matrix strength at the fiber failure strain, V_f is the fiber volume fraction.

The elastic modulus of a composite can be expressed as:

$$E_{comp} = (1 - V_f)E_m + V_f E_f \tag{5}$$

E_m, E_f are the elastic modulus of matrix and fiber, respectively.

The value of the volume fraction varies depending on the structure of reinforcement: unidirectional, woven, or random mat. There is an optimal space between fibers that will fully exploit the uniform load transfer between fibers [134]. With minimum or critical fiber volume fraction, if there are very few fibers present ($0 < V_f < V_{f\,min}$), the stress on a composite may be high enough to break the fibers. The broken fibers, which carry no load, can be then regarded as an array of aligned holes ($0 < V_f < V_{f\,min}$) [135]. The reinforcing action of the fibers is only observed once the fiber volume fraction exceeds the critical fiber volume fraction ($V_f > V_{f\,crit}$). For instance, for short banana fiber-reinforced vinyl-ester composite, the value of $V_{f\,min} \cong 15\%$ and $V_{f\,crit} \cong 25\%$ [136].

NFC may consist of loose fiber aggregate compressed in the thin layer and, in the case of short fibers, it will be randomly laid. However, for long fibers, they may be laid unidirectionally or multi-directionally in several layers. Figure 30 illustrates the different structures of some NF prepregs. Consequently, the fiber volume fraction and fiber orientation play an important role in the final composite's characteristics.

When a composite is formed of random orientation fibers, the value of volume fraction is reduced depending on the spacing ratio [132]. For the ring spun yarn, the

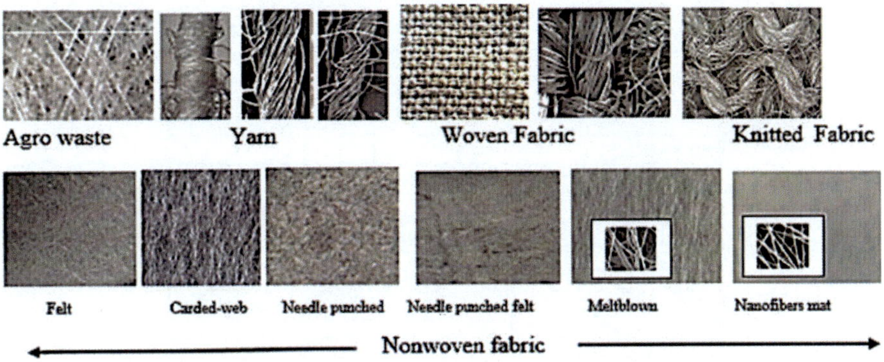

Fig. 30 Some textile product structures for automotive industry (by the author)

variation of diameter along its length indicates the high variability of the number of fibers in the yarn cross section along its length, as well as its packing density [137–139]. Figure 31 shows the distribution of the fibers inside the yarn cross section.

Experimental evaluation of radial packing density trace is based on the division of yarn cross section to the radial system of rings and calculation of relative portion of fiber areas to the ring area [132, 137–140]. If the yarn in the fabric was covered with polymer, the packing density of the polymer will vary from the surface of the fabric to its center, and it will be 1 at the surface to 0.26 at the fabric centerline. Consequently, the failure of the yarns will start at its surface and propagate towards its center. Figure 32 illustrates the variation of the packing density across the yarn cross section and polymer distribution in the composite cross section, which is the

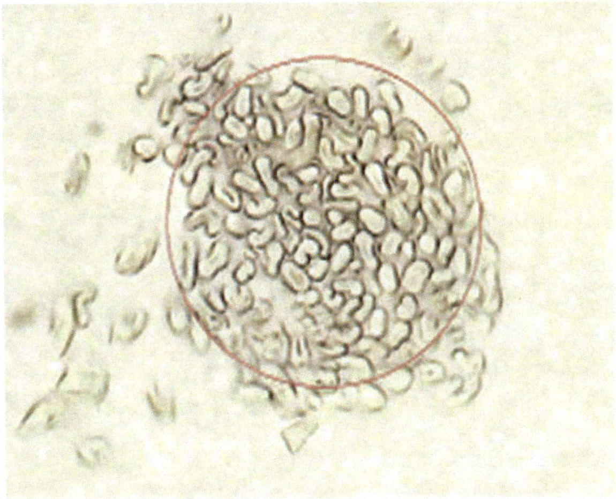

Fig. 31 Distribution of fibers inside the yarn cross section (by the author)

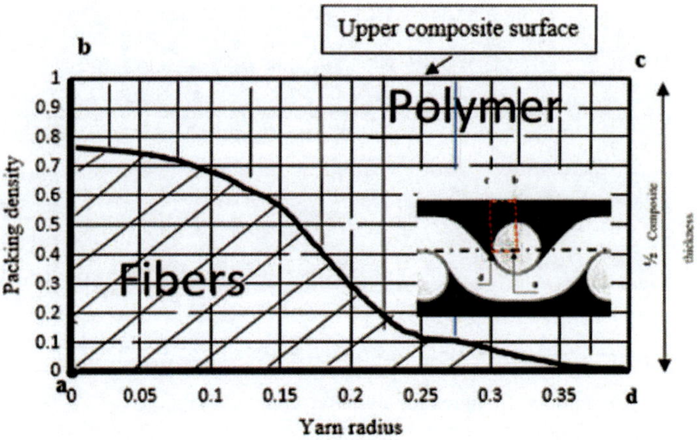

Fig. 32 Yarn packing density and polymer distribution in the part of composite's cross section (*abcd*) (by the author)

characteristic being generally a function of fibers and yarn fineness, twist, yarn count, method of spinning and the thickness of the composite [138].

Normally, the yarn diameter (D_y) varies along its length as illustrated in Fig. 33. Consequently, there will be another source of fiber volume fraction variation along the length and the width of the fabric.

Due to the variation of the fiber volume fraction of the yarn along its length, it will directly reflect on the fiber volume fraction of the formed composite (FRPCY) across its cross section. Figure 34 illustrates the shape of the yarns molded in polymer matrix [139]. The relation between the yarn fiber volume fraction and the composite fiber volume fraction was studied, taken into consideration fiber volume fraction variation along the yarn length. The effect of the composite ratio (β) (rod (D_c)/yarn diameter (D_y)) and yarn fiber volume fraction variation was investigated in the light of the variability of yarn diameter [139–141]. The relation between the average composite

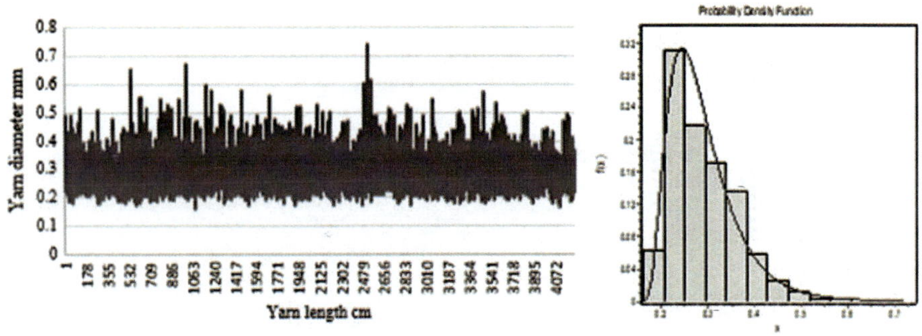

Fig. 33 Yarn diameter variation along its length and yarn diameter histogram (by the author)

Fig. 34 Yarn in composite (FRPCY) [139]

fiber volume fraction V_{fc} and that of the yarn V_{fy} for different ratios of β indicated that V_{fc} is inversely proportional to β^2 and the range of V_{fc} variation reduces as β increased. The value of average V_{fc} in the case of $V_{fy} = 0.7$ reaches 0.37 for $\beta = 1.5$ and is 0.048 for $\beta = 4$, these values are a function of yarn diameter variability [139–141]. Moreover, for the value of β, the yarns compression during the formation of the rod should be considered. The change in the average value of fiber volume fraction of the yarn will result in changing the final fiber rod polymer composite yarn (FRPCY) average fiber packing density, as shown in Fig. 35.

Figure 36 illustrated the relationship between the average composite fiber volume fraction (V_{fc}) and that of the yarn (V_{fy}) for different values of (β).

Spun yarns have a certain fiber volume fraction and, when they are architected in a weave structure, the total fiber volume fraction of the weave will depend upon the spaces between the yarns in the fabric. The value of the volume fraction varies, depending on the structure of reinforcement: unidirectional, woven, or random mat. Therefore, there is an optimal space between fibers that will fully exploit the uniform

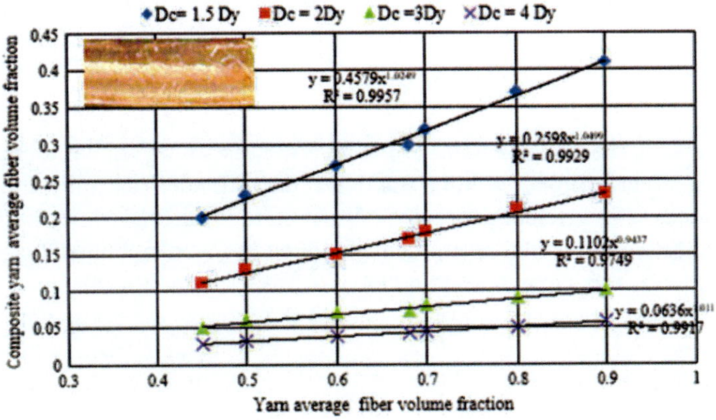

Fig. 35 Yarn composite V_{fc} versus $V_{f\,yarn}$ [139]

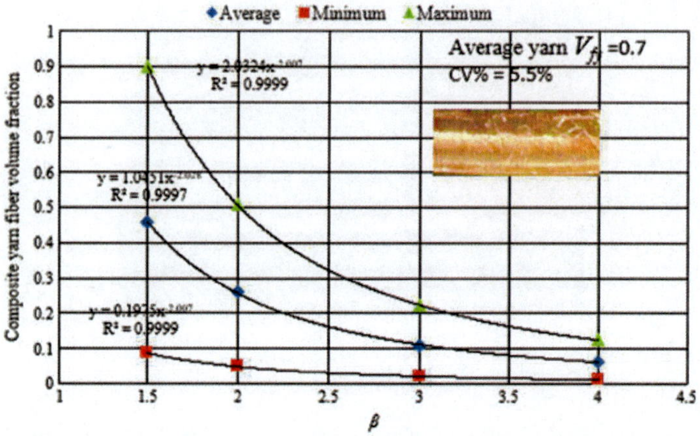

Fig. 36 Yarn composite V_{fc} versus rod/yarn diameter ratio (β) [139]

load transfer between fibers [140–142]. When there are very few fibers present ($0 < V_f < V_{f\,min}$), the stress on a composite may be high enough to break the fibers. To cover the surface of the yarn by the matrix for the benefit of the utilizing all the fibers in the yarn cross section, the polymer has to penetrate inside the yarn as well as form a sheath of diameter D_c, which is greater than the average yarn diameter [133–139]. That will lead to a reduction in the composite fiber volume fraction value. As the thickness of the composite is reduced, a certain percentage of the fibers will not share the load applied to the composite. In all cases, the yarns with the different diameters and fiber packing density result in the lower composite fiber volume fraction. The distribution shape of the yarn diameter is expressed by its kurtosis, which is a measure that describes the shape of a distribution's tails concerning its overall shape and reflects on the value of $V_{f\,comp}$. It is recommended that the yarns of low kurtosis value should be used for the production of the NFC.

4.5.1 Plain Weave Composite Model

For plain weave, the yarn diameter of the natural fiber spun yarn has a great heterogeneity that affects the fiber volume fraction at the different basic units of the fabric. That, in turn, creates a rich fiber or rich polymer point in the final composite, resulting in the composite with varied reinforcement which has a crucial impact on the stress transfer between yarns and matrix and the composite fracture mechanism. The fiber volume fraction of the weave unit is equal to the ratio of the volume of the fibers to the volume of the unit weave [132, 143]. The fiber volume fraction of each basic weave unit, Fig. 37, was targeted to calculate the minimum, average, and maximum values of the fabric's fiber volume fraction distribution.

Spun yarns have a certain fiber volume fraction and, when they are architected in a weave structure, the total fiber volume fraction of the weave will depend upon

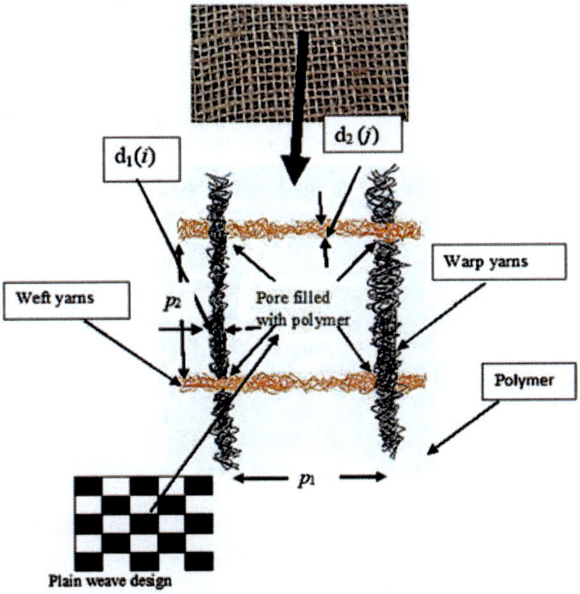

Fig. 37 Repetitive unit cell of a plain weave (by the author)

the spaces between the yarns in the fabric. The fiber volume fraction of the weave unit is associated with the level of the dependency determined by the yarn diameter variations along with the warp and weft yarns. Consequently, the variable diameter of the yarns in different unit cells of the fabric preform weave results in different fiber volume fraction $V_{\mathrm{f\,fabric\,i}}$. From the analysis of fiber volume fraction of the unit cell distribution, critical regions with low values of fiber volume fraction could be marked as shown in Fig. 38.

Since the distribution of the fiber volume fraction is randomly distributed all over the fabric preform area, it is expected to have a failure in the cross section of the higher percentage of weave cells with least values of fiber volume fraction. The strain distribution in the fabric preform at these areas will be more irregular. The results of calculation for the plain weave fabric specimen give the distribution of the fiber volume fraction illustrated in Fig. 39. One of the important conclusions of this analysis is that the yarn fiber volume fraction, the variation of yarn diameter, and the warp and weft densities of the fabric affect the distribution of the fiber volume fraction all over the fabric area [132].

The average fiber volume fraction of the fabric preform is lower than that of the used weft and warp yarns. Figure 40 illustrates the fabric unit cell fiber volume fraction versus yarn fiber volume fraction for different ends/cm.

Furthermore, the analysis of the distribution of fiber volume fraction indicates the presence of rich polymer spots in the composite that affects the strain distribution under loading. The compaction behavior of the preform should be considered to secure the designed value of the composite fiber volume fraction. It is recommended

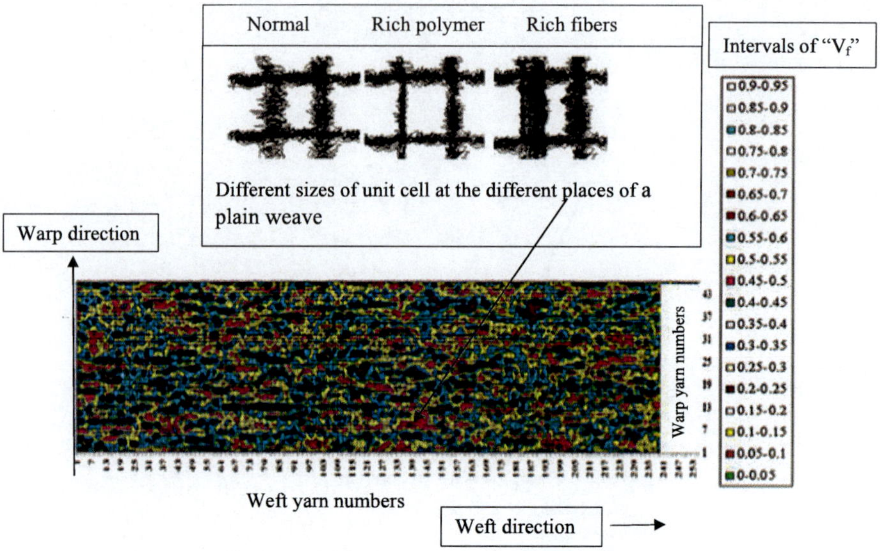

Fig. 38 Fiber volume fraction distribution map of the fabric preform area (by the author)

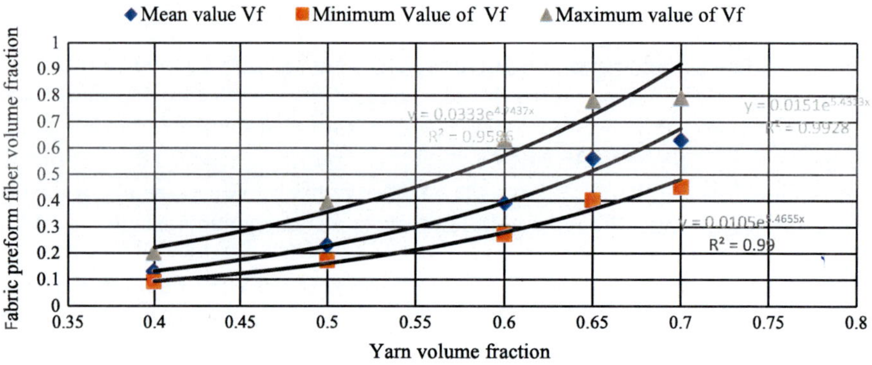

Fig. 39 Mean, maximum and minimum values of fabric preform fiber volume fraction versus yarn fiber volume fraction [132]

to use yarns of high packing density and evenness for the reliable fabric enforcement polymer composite. Average value of fabric unit cell fiber volume fraction, in the case equal ends and picks per cm (E), is given by [132]:

$$V_{\mathrm{f\ fabric\ average}} = 0:00045E\,\mathrm{e}^{(5:34\phi)} \tag{6}$$

where ϕ is the yarn packing density.

During the consolidation phase of liquid composite molding of 2D woven fabric, the composite is subjected to pressure to get the final form. In this phase, the final

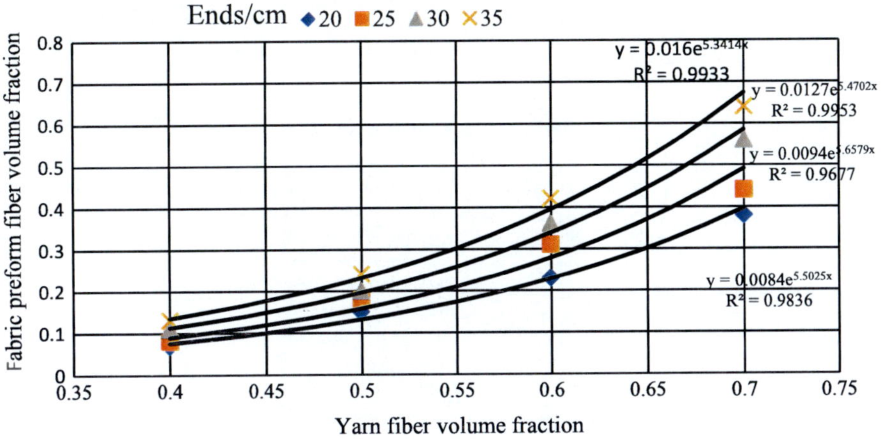

Fig. 40 Fabric unit cell fiber volume fraction versus yarn fiber volume fraction for different ends/cm [132]

fiber volume fraction of the composite can be attained. The yarns are stressed to reduce their volume. The final thickness depends on the compactness of the yarns and the fabric structure [144]. Overall thickness and hence the fiber volume fraction of the composite are directly affected by the external mechanical pressure applied during molding, targeting to increase the fiber volume fraction in the final formed composite, ϕ_{comp} [145]. In this case, the value of the ϕ will be modified to be ϕ_{comp}, Eq. (6).

4.6 *Some Mechanical Properties of NFC*

The composite structure consists of several materials with different properties: bulk matrix, fibers/matrix on the fabric surface, and fabric impregnated with the matrix polymer, forming one coherent composite, as illustrated in Fig. 41. The effect of the matrix or fiber surface treatment on the performance of the composite depends on the type of matrix. For a nonpolar polymer matrix, the adhesion occurs through van der Waals interaction only; therefore, a matrix modification is necessary to improve the adhesion strength. For a polar polymer matrix, adhesion is sensitive to fiber surface treatments [144].

4.6.1 Mechanism of Failure of Fabric/Polymer Composite

During NFC manufacturing, the penetration of the polymer through the porous yarn will bind the fibers together more firmly. Moreover, the resin might penetrate inside the fiber itself, increasing the yarn strength and reducing the breaking elongation. The

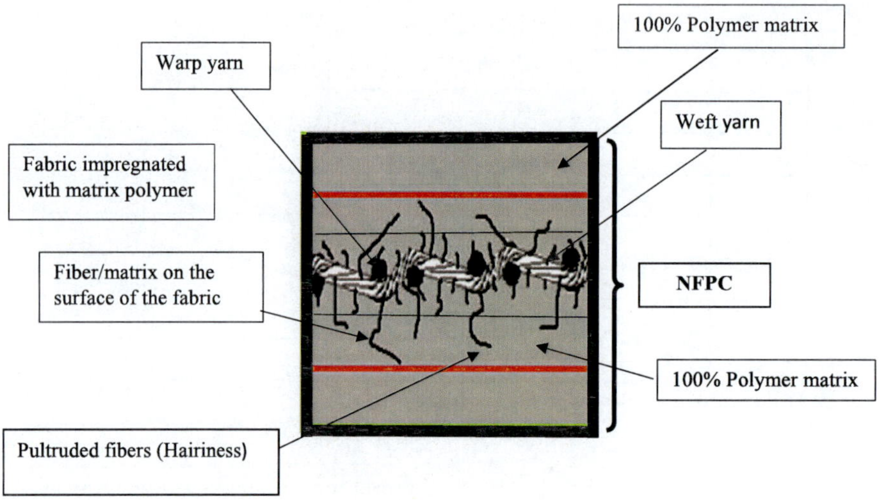

Fig. 41 Model of fabric-reinforced polymer matrix [144]

mechanism of the composite failure is rather complicated, depending on the properties and structure of fabric, the matrix properties, and the adhesive force between the fabric and the matrix. When the fabric and the polymer become a part of the composite, the final mechanism of composite failure under tension will depend on the following models:

- The applied strain of the matrix is greater than that of the fibers causing the material to shear at the interface between matrix and fibers. The polymer matrix will support the stress.
- Applied stress exceeds the tolerance of the fabric impregnated with matrix polymer causing the fibers themselves to fracture. This will lead to strain the matrix till its failure, and the total structure of the composite will fail, too. The microstructural parameters, that control the properties of the composite, are the properties of reinforcement impregnated fabric, the properties of the matrix, and the fiber volume fraction ratio [146]. The strength of the composite can be expressed in the case of fiber failed first or matrix failed earlier. If the matrix failed first, then the failure stress will be:

$$\sigma_c = [V_f E_f + (1 - V_f) E_m] \varepsilon_m \tag{7}$$

But, if the fiber failure occurs before the matrix, then:

$$\sigma_c = [V_f E_f + (1 - V_f) E_m] \varepsilon_f \tag{8}$$

where σ_c is composite stress, E_c is Young's modulus of composite, E_f is Young's modulus of NFC fabric, E_m is Young's modulus of matrix, ε_c is composite strain, ε_f is fabric strain, ε_m is matrix strain, V_f is fiber volume fraction.

The epoxy mechanical properties can be changed by altering the percentage of the hardener; the ratio between the hardener to the polymer by weight denoted by α. Figure 42 shows a comparison between the mechanical properties of the matrix as a function of the α. Figure 43 illustrates the mechanical properties of the original fabric, matrix, and composite for different values of α.

The mechanical properties of the final composite in all cases were higher than its components, since the penetration of the polymer inside the yarns increases the bonding of the fiber leading to the increase in the utility of fiber strength. This effect depends on the fabric/resin properties; hence, the increase of blending ratio α_2 infuses low strength resin reducing the benefit of strengthening the fabric through pultrusion. The mechanical properties of the matrix play a determining role in defining which model of failure will take place, as shown in Fig. 44 [146].

Another approach in the preparation of the natural fiber composite from fabric and matrix is two-step technique [145–148]. The pultruded fabric preforms (jute in this case) were processed firstly and cured separately using a polymer with different components blending ratio "α_1" (resin/solvent ratio) and the two-step where the composite was prepared with a different combination of polymer blending ratio α_2 consuming the pultruding jute fabric to get different mechanical properties of the final NFC. This two-step technique has contributed positively to final NFC tensile stress, breaking strain, toughness, and Young's modulus. Figure 44 gives the matrix mechanical properties formed by different ratio of α_2. Figure 45a–c represents a comparison between the normal technique of preparing the composite and the two-step technique where better values of the mechanical properties were achieved.

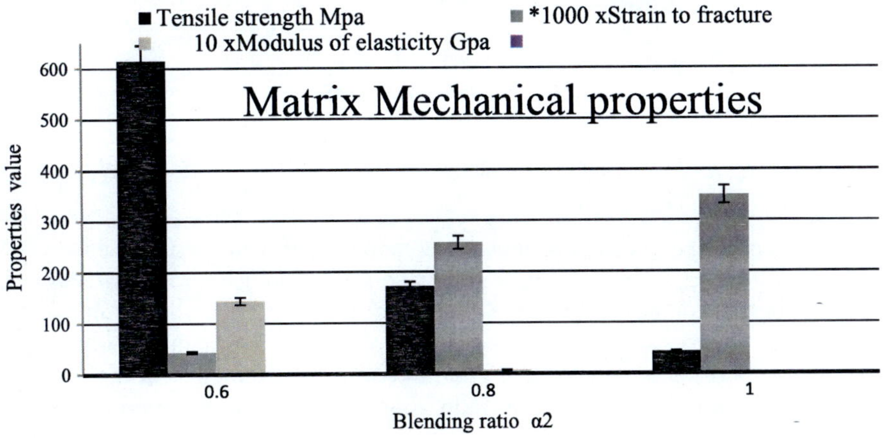

Fig. 42 Mechanical properties of NFC versus matrix versus blending ratio α

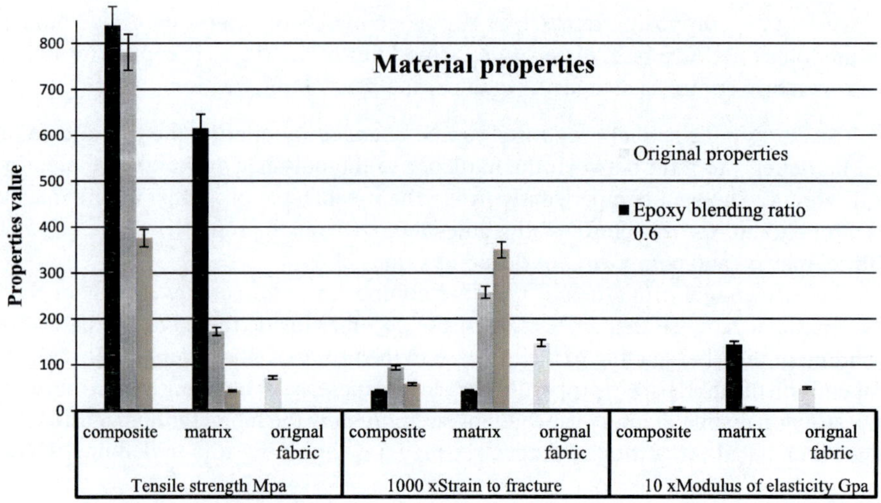

Fig. 43 Comparison between the mechanical properties of original fabric, pultruded fabric, and composite for different values of α

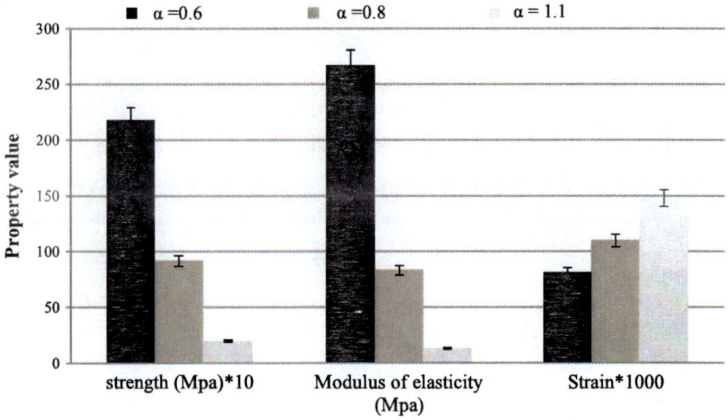

Fig. 44 Effect of the matrix properties on the mechanical properties of the NFC [136]

The effect of epoxy component blending ratio in pultruded jute fabric and matrix on composite mechanical properties in the two-step technique for fabric-reinforced polymer matrix composite gives the designer of the NFC the ability to change the mechanical properties of the pultrusion polymer and/or use another polymer for the final composite manufacturing. The investigation of the effect using different polymer constituent by changing the ratio α_1 and α_2 results in the different mechanical properties, as illustrated in Fig. 46. It is obvious, that the ratio of α_1 and α_2 affect the mechanical properties of the composite. The strength of the composite will reduce as the value of α_2 increases, while the modulus of elasticity decreases, but

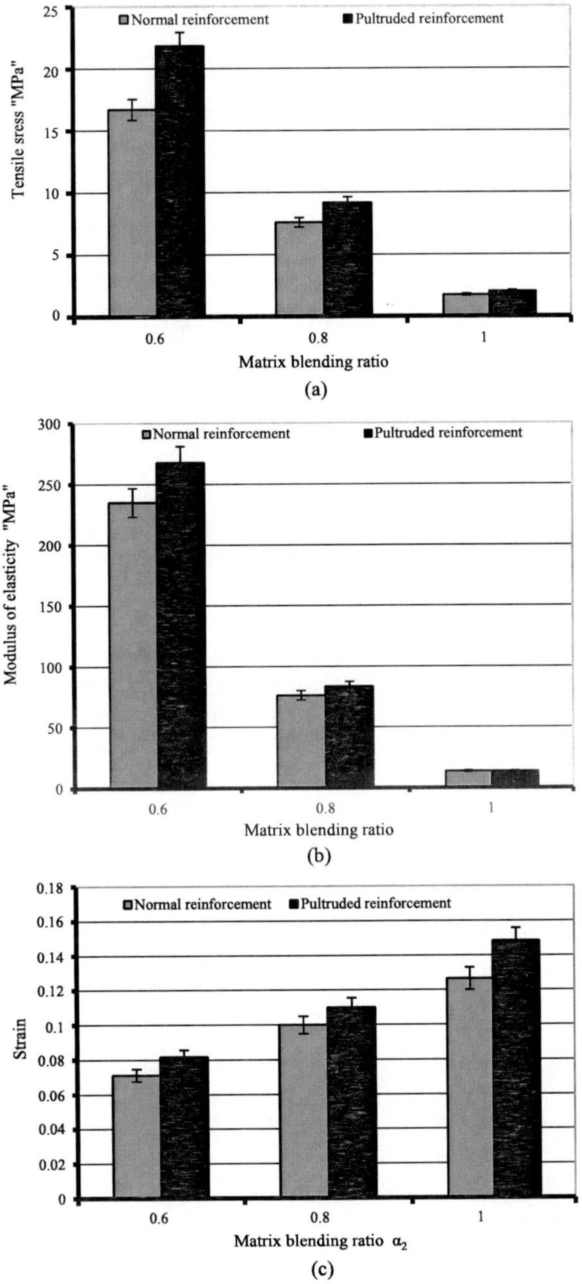

Fig. 45 **a–c** Comparison between the mechanical properties of the normal and two-step technique of manufacturing NFC (by the author)

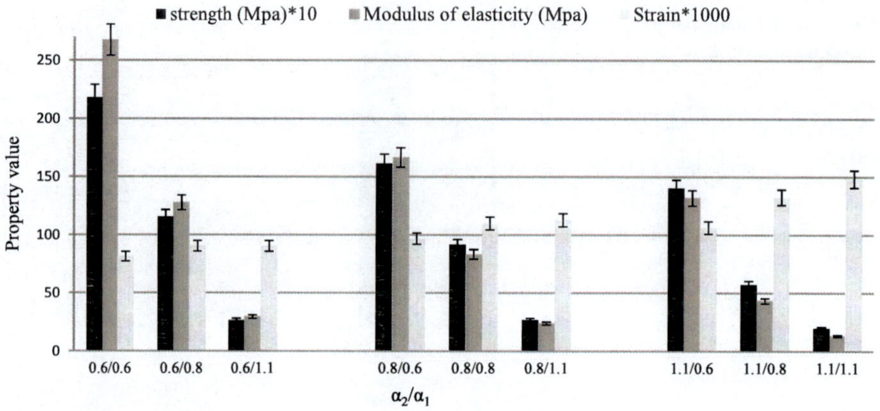

Fig. 46 Mechanical properties of the composite using different ratios of α_1 and α_2 (by the author)

the strain increases as values of α_1 surges. Applying the two-step technique, it is recommended to manufacture the automotive composites sheets from uniform and sandwich 2-D or 3-D composite structures through the use of polymer pultruded fabric and then manufacture the final composite form by adding another polymer with different properties.

The agriculture waste is one of the problems of the modern intensive agriculture which needs innovative methods for its useful industrial utilization. A round cross-sectional shaft has been suggested [149]. Round shaft hybrid composite was designed using wheat straw/flax fibers and covered by a layer of flax fabric. The construction of shaft is made of several layers of wheat straw; first layer (core) consists of three straws winded around by a bundle of flax fibers. Figure 47a illustrates the forming process. The rotation of the laid layer of flax fibers and the stacks around the core structure continues till the final diameter is reached. Flax fabric was wrapped on the formed structure as the outside layer. During the whole process, the resin (PVA) is continuously fed at the rate to insure the required fiber volume fraction of the final composite. In this case, the weight fraction for the straw, flax fibers, and fabric was 51%, and composite density was 0.1125 g/cm^3. The shaft formation and cross section are shown in Fig. 47a, b, respectively. This method can be used for the manufacturing of a hybrid round shafts with different diameters. The bending stiffness and impact stress were found to be suitable for several applications.

The natural fiber composites are the future trends in the automotive industry, utilizing the original fibers, woven, knitted, and nonwoven fabrics, agro-waste or recycling textile materials, to construct the different parts of the vehicles from the eco-friendly materials, achieving the low cost of production, lower environmental impact, lighter in weight, improved fuel efficiency and low noise levels. The reasons for the application of natural fibers in the automotive industry include their low density, good acoustic properties, acceptable mechanical properties, satisfactory accident

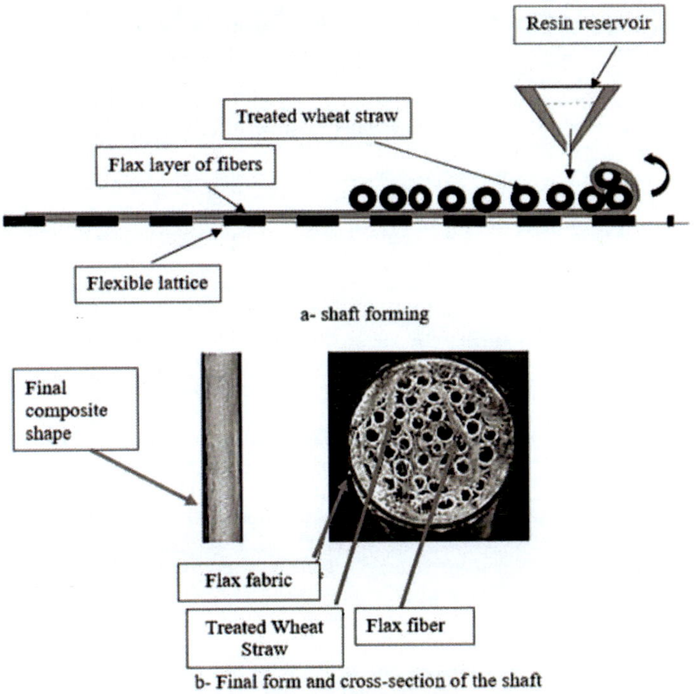

Fig. 47 **a**, **b** Wheat straw/flax fiber/flax fabric reinforced resin composite [139]

performance, high stability, less splintering, favorable eco-balance for part production, during vehicle operation due to weight savings, occupational health benefits compared to the glass fibers. At the same time, they provide the machine maker with the required parts that can withstand the stresses acting on it during the lifespan of the vehicle.

The natural fiber composite provides a high degree of flexibility to produce lightweight shell parts that opens another advantage for design flexibility and manufacturing processes. Besides, vehicle recycling at the end of its service life will be more optimal. One might already say that the era of biocomposite is just beginning.

References

1. El Messiry M (2017) Natural fiber textile composite engineering. Apple academic press, USA
2. Ngo T-D (2017) Natural fibers for sustainable bio-composites. In: Natural and artificial fiber-reinforced composites as renewable sources, Chap. 7. https://dx.doi.org/10.5772/intechopen.71012
3. Allied Market Research (Online). https://www.alliedmarketresearch.com/technical-textile-market. Accessed 20 May 2020

4. Grand view research (Online). Technical textiles market size, share & trends analysis report by technology (thermoforming, 3D weaving, 3D knitting, nanofibers), by end use (mobiltech, indutech, hometech, clothtech), and segment forecasts, 2012–2022. https://www.grandviewresearch.com/industry-analysis/technical-textiles-market. Accessed 20 May 2020
5. Holbery J, Houston D (2006) Natural-fiber-reinforced polymer composites in automotive applications. JOM 58(11):80–86
6. Natural Fiber Composites (NFC). Market size, share & trends analysis report by raw material, by matrix, by technology, by application, and segment forecasts, 2018–2024 (Online). https://www.grandviewresearch.com/industry-analysis/natural-fiber-composites-market. Accessed 20 May 2020
7. Market Analysis Report (Online). https://www.grandviewresearch.com/industry-analysis/technical-textiles-market. Accessed 13 Aug 2020
8. Grand view research. Natural fiber composites (NFC) market size, share & trends analysis report, by raw material, by matrix, by technology, by application, and segment forecasts, 2018–2024. https://www.grandviewresearch.com/industry-analysis/natural-fiber-composites-market. Accessed 18 May 2020
9. Discover Natural Fibres Initiative (2019) Natural fibres and world economy, July 2019 (Online). https://dnfi.org/coir/natural-fibres-and-the-world-economy-july-2019_18043/. Accessed 13 Aug 2020
10. Dhanabalan V, Laga K. Kapok fibre: a perspective fibre (Online). https://pdfs.semanticscholar.org/285a/59dfc301047bb66b6c086dc0e2b425b5ec82.pdf. Accessed 14 May 2020
11. Koszewska M (2018) Circular economy—challenges for the textile and clothing industry. AUTEX Res J 18(4):337–347. https://doi.org/10.1515/aut-2018-0023 © AUTEX
12. Claudio L (2007) Waste couture: environmental impact of the clothing industry. Environ Health Perspect 115(9):A448–A454
13. Abd Ali L, Ansari M, Pua G et al (2015) A review on natural fiber reinforced polymer composite and its applications. Hindawi Publishing Corporation, Int J Polym Sci 2:1–15
14. Gowda Y, Girijappa T, Rangappa S, Parameswaranpillai J, Siengchin S (2019) Natural fibers as sustainable and renewable resource for development of eco-friendly composites: a comprehensive review. Front Mater Sci 6:1–14, Article 226
15. Peças P, Carvalho H, Salman H et al (2018) Natural fibre composites and their applications: a review. J Compos Sci 2:66. https://doi.org/10.3390/jcs2040066, 1–20
16. Chen H (2014) Biotechnology of lignocellulose. Springer, Dordrecht, The Netherlands. ISBN 978-94-007-6897-0
17. Carus M, Eder A, Dammer L et al (2015) Wood-plastic composites (WPC) and natural fibre composites (NFC): European and global markets 2012 and future trends in automotive and construction. Nova-Institute, Hürth, Germany
18. Pickering K, Efendy A, Le T (2016) A review of recent developments in natural fibre composites and their mechanical performance. Compos Part A 83:98–112
19. Asim M, Saba N, Jawaid M, Nasir M (2018) Potential of natural fiber/biomass filler-reinforced polymer composites in aerospace applications. In: Jawaid M, Thariq M (eds) Sustainable composites for aerospace applications. Woodhead Publishing Series in Composites science and engineering, USA, pp 253–268
20. Gohil P, Chaudhary V, Shaikh A (2015) Natural fiber-reinforced composites: potential, applications, and properties. In: Hakeem KR, Jawaid M, Alothman O (eds) Agricultural biomass based potential materials. Springer, Cham, USA
21. Ranakoti L, Pokhriyal M, Kumar A (2018) Natural fibers and biopolymers characterization: a future potential composite material. J Mech Eng–Strojnícky Časopis 68(1):33–50
22. Ali A, Shaker K, Nawab Y, Jabba M et al (2018) Hydrophobic treatment of natural fibers and their composites—a review. J Ind Text 47(8):2153–2183
23. Hernandez C, Rosa D (2016) Extraction of cellulose nanowhiskers: natural fibers source, methodology and application. In: Polymer science: research advances, practical applications, and educational aspects. https://www.semanticscholar.org/paper/Extraction-of-cellulose-nanowhiskers-%3A-natural-%2C-Hernandez-Rosa/f0f78c9a9628c4a1bdc17151b8caf8b8156f0113

24. Pradeep P, Dhas J (2015) Characterization of chemical and physical properties of palm fibers. Adv Mater Sci Eng Int J (MSEJ) 2(4):1–6
25. Asim M, Abdan K, Nasir M et al (2015) A review on pineapple leaves fibre and its composites. Int J Polym Sci 2015:1–16, Article ID 950567. https://dx.doi.org/10.1155/2015/950567
26. Taj S, Munawar M, Khan S (2007) Natural fiber-reinforced polymer composites. Proc Pak Acad Sci 44(2):129–144
27. Faruk O, Bledzki A, Fink H, Sain M (2012) Biocomposites reinforced with natural fibers: 2000–2010. Prog Polym Sci 37(11):1552–1596
28. John M, Thomas S (2008) Biofibres and biocomposites. Carbohydr Polym 71(3):343–364
29. Purnawati R, Febrianto F, Wistara N et al (2018) Physical and chemical properties of kapok (Ceiba pentandra) and balsa (Ochroma pyramidale) fibers. J Korean Wood Sci Technol 46(4):393–401
30. Sathishkumar T et al (2013) Characterization of natural fiber and composites – a review. J Reinf Plast Comp 32(19):1457–1476
31. El Messiry M, Fadel N (2019) Tailoring the mechanical properties of jute woven/cement composite for innovation in the architectural constructions. J Nat Fibers 1–13. https://doi.org/10.1080/15440478.2019.1688748
32. Li X, Tabil L, Panigrahi S (2007) Chemical treatments of natural fiber for use in natural fiber-reinforced composites: a review. J Polym Environ 15:25–33
33. Edeerozey A, Akil H, Azha A et al (2007) Chemical modification of kenaf fibers. Mater Lett 61:2023–2025. https://doi.org/10.1016/j.matlet.2006.08.006
34. Saba N, Tahir P, Jawaid M (2014) A review on potentiality of nano filler/natural fiber filled polymer hybrid composites. Polym Basel 6:2247–2273. https://doi.org/10.3390/polym6082247
35. Bakker R, Elbersen W, Poppens R, Lesschen JP (2013) Rice straw and wheat straw, potential feedstocks for the biobased economy. Wageningen UR, Food & Biobased Research
36. Havarsson S (1992) Manufacture of straw MDF and fiberboard. PhD theses, Mid Sweden University, Sweden
37. Kozlowski R. Green fibres and their potential in diversified applications (Online). https://www.fao.org/3/y1873e/y1873e0b.htm. Accessed 14 May 2020
38. Fahmy T, Mobarak F, El-Sakhawy M et al (2017) Agricultural residues (wastes) for manufacture of paper, board, and miscellaneous products: background overview and future prospects. Int J Chemtech Res 10(2):425–448
39. El Messiry M, El Deeb R (2013) Study wheat straw composite properties reinforced by animal glues as eco-composite materials. In: The 2nd conference for industrial textile research "Manpower development, manufacturing technologies and management in textile industries". National Research Centre (NRC), Cairo, Egypt
40. Hurley LG (1942) U.S. Patent No. 277, 426
41. Akampumuza O, Wambua P, Ahmed A et al (2016) Review of the applications of biocomposites in the automotive industry. Polym Compos 1–17. https://doi.org/10.1002/pc.23847
42. Barrett A (2019) History of bioplastics in the automotive industry, 26 Nov 2019 (Online). https://bioplasticsnews.com/2019/11/26/history-bioplastics-automotive-car-industry/. Accessed 22 May 2020
43. Reddy N, Yang Y (2015) Wheat and rice straw fibers. In: Innovative biofibers from renewable resources. Springer, Berlin, Heidelberg
44. Yadav C, Saini A, Maji P (2017) Energy efficient facile extraction process of cellulose nanofibres and their dimensional characterization using light scattering techniques. Carbohydr Polym 165:276–284. https://doi.org/10.1016/j.carbpol.2017.02.049
45. Brinchi I, Contnaa F, Fortumati E et al (2013) Production of monocrystalline cellulose from lignocellulosic agro residuals. Technol Appl Carbohydr Polym 94:154–169
46. Mohamed Z, Wang X, Jazar R (2013) A survey of wheel tire cavity resonance noise. IJVNV 9(3–4):276–293

47. Braun M, Walsh S, Horner J et al (2013) Noise source characteristics in the ISO 362 vehicle pass-by noise test: literature review. Appl Acoust 74(11):1241–1265
48. Crocker M (2007) Introduction to transportation noise and vibration sources. In: Crocker MJ (ed) Handbook of noise and vibration control. Wiley, New York, pp 1013–1023
49. Arenas J (2016) Applications of acoustic textiles in automotive/transportation. In: Padhye R, Nayak R (eds) Acoustic textiles, textile science and clothing technology. Springer Science, Business Media Singapore. https://doi.org/10.1007/978-981-10-1476-5-7
50. Li T (2018) Literature review of tire-pavement interaction noise and reduction approaches. J Micromech Microeng 20(6):2424–2452
51. Roy N, Germès S, Lefebvre B et al (2006) Damping allocation in automotive structures using reduced models. In: Conference paper, SDTools, Leuven, Belgium
52. Morello L, Rossini L, Pia G et al (2011) A noise, vibration, harshness in the automotive body, vol II, Chap 11. Springer, Italy, pp 239–263
53. Environmental Protection Agency (1974) Information on levels of environmental noise requisite to protect public health and welfare with an adequate margin of safety. EPA, Washington D.C., USA
54. Li Q, Qiao F, Yu L (2017) Risk assessment of in-vehicle noise pollution from highways. Environ Pollut Climate Change 1(107):1–5
55. Pellettieri J (2018) Noise, vibration & harshness challenges in vehicle light weighting (Online). https://lightweightingworld.com/noise-vibration-harshness-challenges-in-vehicle-lightweighting/. Accessed 14 May 2020
56. Fung W, Hardcastle M (2001) Textiles in automotive engineering. Woodhead Publishing Limited in association with the Textile Institute Cambridge, UK
57. Qatu M, Abdelhamid M, Pang J et al (2009) Overview of automotive noise and vibration. IJVNV 5(1–2):150–170
58. https://www.premiergroup.com.au/pg-pad-uf72-felt-carpet-underlay-pad-with-cerex-backing.html
59. http://www.automobiletrim.com/sound-proofing.html
60. https://www.automotiveinteriors.com/auto-carpet-kits-and-accessories-s/17774.htm
61. http://www.vibrofiltr.eu/en
62. https://sgm-techno.com/catalog/sound-absorber-materials-cars/bb-ton-bb-ton-ultra
63. Bose Breaks the Silence in Buick Verano. Thousands of sound measurements taken for musical clarity in luxury sedan's interior (Online). https://media.gm.com/media/us/en/gm/news.detail.html/content/Pages/news/us/en/2011/Nov/1128_BuickVerano.html. Accessed 1 May 2020
64. Samsudin E, Ismail L, Kadir A (2016) A review on physical factors influencing absorption performance of fibrous sound absorption material from natural fibers. J Eng Appl 11(6):3703–3711
65. Arenas J, Crocker M (2010) Recent trends in porous sound absorbing materials for noise control. Sound Vib 44:12–17
66. Arenas J, Asdrubali F (2018) Eco-materials with noise reduction properties. In: Martínez LMT et al (eds) Handbook of ecomaterials. Springer International Publishing AG. https://doi.org/10.1007/978-3-319-48281-1_137-1
67. Berardi U, Iannace G (2015) Acoustic characterization of natural fibers for sound absorption applications. Build Environ 94:840–852, Part 2
68. Wei Dong Y, Yan L (2012) Sound absorption performance of natural fibers and their composites. Sci China Technol Sci 55(8):2278–2283
69. Lee H, Pun Ng B, Rammohan A et al (2017) An investigation of the sound absorption properties of flax/epoxy composites compared with glass/epoxy composites. J Nat Fibers 14(1):71–77. https://doi.org/10.1080/15440478.2016.1146643
70. Mamtaz H, Fouladi M, Al-Atabi M et al (2016) Acoustic absorption of natural fiber composites. J Eng 1–11, Article ID 5836107
71. Michalski B (2009) The validation of test methods for defining structural characteristics of nonwovens for tissue engineering applications. MSc theses, ITA, Germany

72. Bhat G, El Messiry M (2019) Effect of microfiber layers on acoustical absorptive properties of nonwoven fabrics. J Ind Text 2020 50(3):312–332. https://doi.org/10.1177/1528083719830146
73. Neithalath N, Marolf A, Weiss J et al (2005) Modeling the influence of pore structure on the acoustic absorption of enhanced porosity concrete. https://citeseerx.ist.psu.edu/viewdoc/download?doi=10.1.1.485.5395&rep=rep1&type=pdf. Accessed 1 Jan 2014
74. Dunne R, Desai D, Sadiku R (2017) A review of the factors that influence sound absorption and the available empirical models for fibrous materials. Acoust Aust 45:453–469. https://doi.org/10.1007/s40857-017-0097-4
75. Ramis J, Alba J, del Rey R et al (2010) New absorbent material acoustic based on kenaf's fibre. Mater Constr 60(299):133–143
76. Wen S, Xin-yu L, Shu-sen L (2013) Influence of thickness and density on sound-absorption capability of nonwoven sound-absorbing material. Appl Mech Mater 393:102–107
77. John T, Wang X (2009) Vehicle floor carpet acoustic optimization using statistical energy analysis. IJVNV 5(1/2):141–157
78. Zhang J, Shen Y, Jiang B et al (2018) Sound absorption characterization of natural materials and sandwich structure composites. Aerospace 5(75):1–13. https://doi.org/10.3390/aerospace5030075
79. Jayamani E, Hamdan S, Heng S et al (2014) Sound absorption property of agricultural ligno-cellulsic residue fiber reinforced polymer matrix composites. Appl Mech Mater 663:464–468. https://doi.org/10.4028/www.scientific.net/amm.663.464
80. Yuhazri Y, Kamarul A, Sihambing H et al (2010) The potential of agriculture waste material for noise insulator application toward green design and material. Int J Civ Environ Eng IJCEE IJENS 10(05):13–17
81. Yahya M, Chin D (2017) A review on the potential of natural fibre for sound absorption application. IOP Conf Ser Mater Sci Eng 226, 012014. International research and innovation summit (IRIS2017). https://doi.org/10.1088/1757-899X/226/1/012014
82. Balan A, Shivasankaran N (2019) Noise control using waste materials reinforced composites. LJRS 19(2):47–52, Compilation 1.0
83. Fouladi M, Ayub M, Nor M (2011) Analysis of coir fiber acoustical characteristics. Appl Acoust 72:35–42
84. Ancuţa E, Ovidiu V, Horaţiu V et al (2018) New multilayered composite for sound absorbing applications. RJAV 15(2):115–121
85. Yahya M, Sambu M, Abdul Latif H et al (2017) A study of acoustics performance on natural fibre composite. IOP Conf Ser Mater Sci Eng 226, 012013. https://doi.org/10.1088/1757-899X/226/1/012013
86. Asdrubali F, Schiavoni S, Horshenkov K (2012) A review of sustainable materials for acoustic applications. Build Acoust 19:283–312
87. Rubino C, Aracil M, Gisbert-Payá J et al (2019) Composite eco-friendly sound absorbing materials made of recycled textile waste and biopolymers. Mater Basel 12(23):4020
88. Seddeq H, Aly N, Marwa A et al (2013) Investigation on sound absorption properties for recycled fibrous materials. J Ind Text 43:56–73
89. Sandin G, Peters GM (2018) Environmental impact of textile reuse and recycling—a review. J Cleaner Prod 184:353–365
90. Curtu I, Stanciu M, Coªereanu C et al (2012) Assessment of acoustic properties of biodegradable composite materials with textile inserts. Mater Plast 49(1):68–72
91. Parikh D, Calamri T, Swahney A et al (2002) Thermoformable automotive composites containing kenaf and other cellulosic fiber. Text Res J 72(8):668–672
92. Thilagavathi G, Pradeep E, Kannaian T et al (2010) Development of natural fiber nonwovens for application as car interiors for noise control. J Ind Text 39(3):267–277
93. Parikh D, Chen Y, Sun L (2006) Reducing automotive interior noise with natural fiber nonwoven floor covering systems. Text Res J 76(11):813–820
94. Lou C, Lin J, Su K (2005) Recycling polyester and polypropylene nonwoven selvages to produce functional sound absorption composites. Text Res J 75(5):390–394

95. Fouladi M, Nor M, Ayob R, Leman Z (2010) Utilization of coir fiber in multilayer acoustic absorption panel. Appl Acoust 71:241–249
96. Ulrich T, Arenas J (2020) Sound absorption of sustainable polymer nanofibrous thin membranes bonded to a bulk porous material. Sustainability 12:2361. https://doi.org/10.3390/su12062361
97. Putra A, Abdullah Y, Efendy H et al (2013) Utilizing sugarcane wasted fibers as a sustainable acoustic absorber. Procedia Eng 53:632–638
98. Carvalho S, Cesar A, Flórez J et al (2015) Acoustic characterization of sugarcane bagasse particleboard panels (*Saccharum officinarum* L). Mater Res 18(4):1–11
99. Islam S, El Messiry M, Seylar J, Sikdar P, Bhat G, (2020) Microstructure and performance characteristics of acoustic insulation materials from post-consumer recycled denim fabrics. J Ind Text 2020. https://doi.org/10.1177/1528083720940746
100. Shivendra P, Tanveer M (2018) Application of textiles in automobile (Online). https://www.researchgate.net/publication/326508226_Application_of_Textiles_In_Automobile_Application_of_Textiles_In_Automobile. Accessed 14 May 2020
101. Fortea-Verdejoa M, Bumbarisa E, Burgstallerb C et al (2017) Plant fibre-reinforced polymers: where do we stand in terms of tensile properties? Int Mater Rev 62(84):41–464
102. Ilyas R, Sapuan S, Norizan M et al (2019) A potential of natural fibre composites for transport industry: a review. In: Proceedings seminar Enau Kebangsaan, Malaysia
103. Lucintel (2019) Bio-composites market report: trends, forecast and competitive analysis (Online). https://www.lucintel.com/bio-composites-market.aspx. Accessed 29 May 2020
104. Kurien RA, Philip Selvaraj D, Sekar M, Koshy CP (2020) Green composite materials for green technology in the automotive industry. IOP Conf Ser Mater Sci Eng 872:012064 https://doi.org/10.1088/1757-899X/872/1/012064.
105. Zimniewska M, Mankowski W (2011) Cellulosic bast fibers, their structure and properties suitable for composite applications. In: Kalia S, Kaith B, Kaur I (eds) Cellulose fibers: bio- and nano-polymer composites. Springer, Berlin, Heidelberg
106. Shanmugasundaram L (2009) Green composites: manufacturing techniques & applications. Indian J Fibre Text. https://indiantextilejournal.com/articles/FAdetails.asp?id=2489. Accessed 14 May 2020
107. Shaharuzaman M, Sapuan S, Ridzuan M et al (2019) Decision support strategy in selecting natural fiber materials for automotive side-door impact beam composites. https://www.researchgate.net/publication/336761257_Decision_Support_Strategy_in_Selecting_Natural_Fiber_Materials_for_Automotive_Side-Door_Impact_Beam_Composites. Accessed 14 May 2020
108. Dunne R, Desai D, Sadiku R et al (2016) A review of natural fibres, their sustainability and automotive applications. J Reinf Plast Compos 35(13):1041–1050
109. Kalia S, Kaith B, Kaur I (2011) Cellulose fibers: bio- and nano-polymer composites green chemistry and technology. Springer, Berlin, Heidelberg
110. Gurunathan T, Mohanty S, Sanjay K et al (2015) A review of the recent developments in biocomposites based on natural fibres and their application perspectives. Compos Part A 77:1–25
111. Sudin N, Din A, Abdullah M et al (2007) Conceptual design of automotive bumper beam. In: Conference on design, simulation, product development and optimization (product & design). https://www.researchgate.net/publication/216465827_Conceptual_design_of_automotive_bumper_beam
112. Cheon SS, Lee DG, Jeong KS (1997) Composite side door impact beam for passenger cars. Compos Struct 38:229–239
113. Sanjay M, Arpitha G, Yogesha B (2015) Study on mechanical properties of natural-glass fibre reinforced polymer hybrid composites a review. Mater Today Proc 2:2959–2967. https://doi.org/10.1016/j.matpr.2015.07.264
114. Davoodi M, Sapuan M, Ahmad D et al (2010) Mechanical properties of hybrid kenaf/glass reinforced epoxy composite for passenger car bumper beam. Mater Des 31:4927–4932. https://doi.org/10.1016/j.matdes.2010.05.021

115. Witayakran S, Kongtud W, Boonyarit J et al (2017) Development of oil palm empty fruit bunch fiber reinforced epoxy composites for bumper beam in automobile. Key Eng Mater 751:779–784
116. Onyedum O, Aduloju C, Sheidu S et al (2015) Comparative mechanical analysis of okra fiber and banana fiber composite used in manufacturing automotive car bumpers. AJETS 2(6):193–199
117. Ragupathi P, Sivaram N, Vigneshm G et al (2018) Enhancement of impact strength of a car bumper using natural fiber composite made of jute. i-manager's J Mech Eng 81(31)
118. Lakshminarayana V (2016) Effect and analysis of natural fibre polymer composite plates used for passenger vehicles bumper. IJCEM 2(12)
119. Setiawan R, Salim M (2017) Crashworthiness design for an electric city car against side pole impact. J Eng Technol Sci 49(5):587–603
120. WAEDAAUTO (Online) https://www.wardsauto.com/industry/dupont-develops-plastic-side-impact-door-beam. Accessed 1 May 2020
121. Nichit Y, Battu A (2017) Development of side door intrusion beam of passenger car for maximum bending load
122. Li M, Chiang T, Tseng J et al (2014) Hot stamping of door impact beam. In: 11th International conference on technology of plasticity, ICTP 2014, 19-24 October 2014, Nagoya Congress Center, Nagoya, Japan, Procedia Engineering, vol 81, pp 1786–1791
123. Saraf S, Bajaj P (2017) Design & experimentation of side impact beam for Hyundai Verna. IRJET 04(06):62–69
124. Veeraswamy K, SudheerBabu V (2016) Design and analysis of a composite beam for side impact protection of a car door. IRJET 03(02):464–469
125. Bledzki A, Faruk O, Sperber V (2006) Cars from bio-fibres. Macromol Mater Eng 291(5):449–457
126. Koronis G, Silva A, Fontul M (2013) Green composites: a review of adequate materials for automotive applications. Compos Part B-Eng 44(1):120–127
127. Jahan A, Ismail M, Sapuan S et al (2010) Material screening and choosing methods—a review. Mater Des 31:696–705
128. Silva A, Koronis G (2018) Green composites for automotive applications. Woodhead Publishing, USA, Published 12th Nov 2018
129. Saheb D, Jog J (1999) Natural fiber polymer composites: a review. Adv Polym Technol 18(4):351–363, Winter 1999; Khan G et al (2013) Studies on the mechanical properties of woven jute fabric reinforced poly(l-lactic acid) composites. J King Saud Univ Eng Sci, 10 Dec 2013. https://www.sciencedirect.com/science/article/pii/S1018363913000548
130. Lomov S, Willems A, Verpoes I et al (2006) Picture frame test of woven composite reinforcements with a full-field strain registration. Text Res J 76(3):243–252
131. Sapuan S, Leenie A, Harimi M et al (2006) Mechanical properties of woven banana fiber reinforced epoxy composites. Mater Des 27(8):689–693
132. El Messiry M (2018) Theoretical determination of the fiber volume fraction distribution for natural fiber fabric reinforced polymer composite. J Ind Text 48(5):904–925
133. El Messiry M, El Deeb R (2016) Engineering fiber volume fraction of natural fiber staple - spun yarn reinforced composite. J Text Sci Eng 6(5):1–6
134. Madsen B (2004) Properties of plant fiber yarn polymer composites: an experimental study. Technical University of Denmark
135. Shah DU, Schubel PJ, Licence P et al (2012) Determining the minimum, critical and maximum fiber content for twisted yarn reinforced plant fiber composites. Compos Sci Technol 72:1909–1917
136. Ghosh R, Reena G, Krishna A et al (2011) Effect of fiber volume fraction on the tensile strength of Banana fibre reinforced vinyl ester resin composites. Int J Adv Eng Sci Technol 4:89–91
137. Shao Ming Z, Zhuan Yong Z, Wei S et al (2008) Computer aided textile design 'LibTex'. Indian J Fibre Text 400–404

138. Křemenáková D, Militký J (2003) Internal structure and strength of cotton yarns. In: World textile conference—4th AUTEX conference Roubaix, CZ, 22–24 June 2003
139. El Messiry M, El Deeb R (2018) Analysis of the natural fiber reinforced polymer: composite fiber rods (FRPCY). IOP Conf Ser Mater Sci Eng 406:1–10, 012039
140. Thygesen A (2006) Properties of hemp fiber polymer composites. PhD thesis, The Royal Agricultural and Veterinary University of Denmark
141. Behera BK, Militky J, Mishra R et al (2012) In: Jeon H-Y (ed) Modeling of woven fabrics geometry and properties, woven fabrics. InTech, Rijeka, Croatia. ISBN: 978-953-51-0607-4. Available at: https://www.intechopen.com/books/woven-fabrics/modelingof-woven-fabrics-geometry-and-properties. Accessed 15 Dec 2017
142. Thompson A, El Said B, Ivanov D et al (2017) High fidelity modelling of the compression behaviour of 2D woven fabrics. Int J Solids Struct 1–10
143. Ibrahim S, Militky J, Kremenakova D et al (2012) Characterization of yarn diameter measured on different. In: RMUTP International conference: textiles & fashion, Bangkok, Thailand, 3–4 July 2012
144. Maldas D, Kokta B (1990) Improving adhesion of wood fiber with polystyrene by the chemical treatment of fiber with a coupling agent and the influence on mechanical properties of composite. J Adhes Sci Technol 3(7):529–539
145. El Messiry M, El Deeb R (2018) Investigation of 2-step technique for jute fabric reinforced polymer matrix composite. J Text Inst 109(10):1293–1303. https://doi.org/10.1080/00405000.2018.1423882
146. El Deeb R, El-Tarfawy S, El Messiry M (2016) Engineered pultruded fabric cementitious thin composites of low fiber volume fraction. In: The twenty-fourth annual international conference on composites/nano engineering (ICCE-24) near Sanya, Hainan Island, China, 17–23 July 2016
147. El Messiry M, El-Tarfawy S, El Deeb R (2017) Enhanced impact properties of cementitious composites reinforced with pultruded flax/polymeric matrix fabric. Alexandria Eng J 56(3):297–307
148. El Messiry M, El Deeb R (2016) Mechanical properties of pultruded jute-polymer composite. In: The twenty-fourth annual international conference on composites/nano engineering (ICCE-24) near Sanya, Hainan Island, China, 17–23 July 2016
149. El Messiry M, El Deeb R (2016) Analysis of the wheat straw/flax fiber reinforced polymer hybrid composites. J Appl Mech Eng 5(7):1–5

Chapter 7
Development and Characterization of PLA-Based Green Composites: Experimental and Simulation Studies

G. Surya Rao, K. Debnath, and R. N. Mahapatra

1 Introduction

In the present day, public attention has shifted toward the development of materials by utilizing natural resources. The natural fibers are widely used as reinforcement in partially and fully biodegradable green composites. Natural fibers are the potential alternative to synthetic fiber for the manufacturing of cheap, renewable, and environmentally friendly composites. The strong bonding between the fiber and the matrix results in good mechanical properties of composites which lead to high load-bearing performance. The composites are widely used in boat hulls, bulkheads, inner and outer surfaces of different components due to their high corrosion resistance and lightweight characteristics. The high specific stiffness and toughness, easy fabrication, high creep, and fatigue strength are the key characteristics of the composites for which these materials are extensively used in aerospace, automobile, construction, and electrical applications, to name a few. The behavior of composite mainly depends on the part design and the manufacturing process. Simulation is the best way to create an adequate relationship between these two stages for attaining the quality product. The initial focus should be on the statistically derived values like strength properties while designing the components of airspace, aircraft, automobile, etc. It is difficult to design job parts with multiple materials, complex layups, and huge number of plies. Thus, simulating the composite part is a challenging task when compared with homogeneous materials like steel and plastic. Numerous experimental investigations have been performed to study the behavior of several types of polymer and its composites. However, the present work is focused on the development and characterization of PLA-based green composites.

G. Surya Rao · K. Debnath (✉) · R. N. Mahapatra
Department of Mechanical Engineering, NIT Meghalaya, Shillong 793003, India
e-mail: debnath.iitr@gmail.com

S. Thomas and P. Balakrishnan (eds.), *Green Composites*,
Materials Horizons: From Nature to Nanomaterials,
https://doi.org/10.1007/978-981-15-9643-8_7

2 Characteristics of PLA and Its Composites

PLA and its composites are emerging as a potential alternative to petroleum-based plastics and composites. Many attempts have been made to improve the performance of PLA and PLA-based composites. Abdulkhani et al. [1] prepared PLA nanocomposite embedded with acetylate cellulose nano-fiber (CNF-Ac) and microcrystalline cellulose (MCC) through solvent casting. PLA-CNF-Ac composite exhibited better mechanical properties when compared with PLA-MCC composite. The tensile properties of the composite with different loadings of MCC and 1 wt% CNF-Ac did not show any significant change. However, the tensile strength, strain, and elastic modulus of PLA-CNF-Ac composite were improved with an increase in the weight percentage of CNF-Ac to 3 and 5%. Nurnadia et al. [2] studied the mechanical behavior of PLA-bamboo composite prepared by the twin screw extruder and compression molding. The effect of fiber content and fiber length on the mechanical and flexural properties of the composite was evaluated. It was revealed that the mechanical performance of the developed composite was significantly affected by the fiber aspect ratio and fiber content. Sun et al. [3] studied the mechanical and thermal properties of chemically treated PLA-coir fiber composite fabricated by injection molding. The effect of treated coir fiber was evaluated in terms of tensile, impact, and thermal properties of the composite. The increasing content of treated coir fiber resulted in improved tensile modulus of the developed composite which was also confirmed through morphological analysis. Eng et al. [4] studied the enhancement in mechanical and thermal properties of PLA-polycaprolactone (PCL) blend with an addition of hydrophilic clay nanomer PGV. PLA-PCL blend with nanomer PGV showed significant enhancement in mechanical properties. Lay et al. [5] compared the physical and mechanical properties of PLA, acrylonitrile butadiene styrene (ABS), and nylon 6 fabricated by fuse deposition modeling (FDM) and injection molding techniques. The tensile strength, Young's modulus, elongation at break, and impact strength of the fabricated samples were compared which showed that the mechanical performance of the injection molded samples was better than the FDM samples. Asadi et al. [6] analyzed the effect of different weight proportions of nanographene (NG) on the mechanical properties of PLA-wood composites. The tensile and the bending properties of the PLA-wood composite were improved by incorporating the NG. It was found that the incorporation of 1.5% NG in PLA-wood fiber composite resulted in significant improvement in tensile and bending properties. Aliotta et al. [7] investigated the mechanical and interfacial properties of biocomposite composed of natural cellulose fiber and PLA. The analytical models were developed to predict the stiffness of the material and to estimate the adhesion between matrix and fiber. Chen et al. [8] studied the performance of PLA-sugar beet pulp (SBP) composite using polymeric diphenylmethane diisocyanate (pMDI) as a coupling agent. The tensile strength of PLA-SBP composite with an addition of 0.5 and 2% pMDI resulted in significant improvement in tensile strength. Parida et al. [9] investigated the tensile, flexural, and impact properties of PLA-based composites reinforced with cellulose nanofibers extracted from luffa cylindrical (LC). The composite exhibited better mechanical properties due to the incorporation of a low weight percentage of LC fiber. The mechanical properties deteriorated as the fiber content was increased due to the agglomeration of the cellulose fiber. Fujiura et al. [10] investigated the mechanical properties of PLA-long jute fiber (LJF) composite fabricated by injection molding. It was revealed that the tensile and flexural strength of the developed composite are dependent on the molecular weight of the PLA. It was concluded that fiber dispersion and the mechanical properties of the PLA-chopped jute fiber (chopped-JF) were better when compared with PLA-LJF composite. Huda et al. [11] compared the PLA-wood fiber composite with PP-wood fiber composite. It was reported that the PLA-wood fiber composite had better tensile and flexural properties when compared with virgin resin. The flexural modulus of PLA-wood fiber composite was comparable to that of PP-wood fiber composite. Musyarofah et al. [12] investigated a hybrid composite of PLA-coir-fleece fiber composite fabricated by hot press molding. The author concluded that the tensile properties of the composite were affected by the addition of fiber. Kuciel et al. [13] studied the mechanical properties of PLA-basalt fiber (BF) and PLA-wood fiber (WF) composite fabricated by injection molding. The addition of WF led to a decrease in the tensile strength of the composite. PLA-BF exhibited better mechanical properties. Righetti et al. [14] studied the mechanical, thermal, and rheological properties of a biocomposite based on PLA and potato pulp powder. It was concluded that increasing the percentage of potato pulp powder resulted in a decrease in the tensile strength, elastic modulus, and elongation at break. Rawi et al. [15] investigated the performance of PLA-bamboo composite fabricated by compression molding. The impact strength of the composite in warp direction was enhanced by 20% when compared one-on-one with pure PLA. Whereas the tensile and flexural properties of the composite were found to be lower than expected. Sujaritjun et al. [16] investigated the tensile properties of PLA-natural fiber (untreated and epoxy treated bamboo, vetiver grass fiber, and coconut fiber) biocomposite. It was found that the stiffness increased and tensile strength decreased while increasing the fiber content in untreated biocomposite. PLA-vetiver grass biocomposite showed less improvement in the tensile properties when compared with other biocomposites. Ramesh et al. [17] investigated the thermal, mechanical, and barrier properties of PLA bio-composite and PLA hybrid bio-composite. It was reported that the addition of montmorillonite (MMT) clay to PLA hybrid bio-composite resulted in improved properties while comparing with PLA bio-composite. Sujito et al. [18] investigated the effect of bamboo fiber content on the mechanical properties of PLA-bamboo composite fabricated by hot press molding. Zhao et al. [19] studied the mechanical behavior of PLA-bamboo composite prepared by biaxial weft knitted fabrics. The composite demonstrated excellent tensile and flexural properties. Hu et al. [20] investigated the effect of fiber treatment, namely alkali, alkali-saline, and alkali-titanate treatment on the properties of PLA-bamboo composites. The author concluded that the alkali-titanate treated fiber composite exhibited better tensile, flexural, and impact strength when compared with other treated fiber composites. Huda et al. [21] compared the properties of chopped glass fiber (CGF) and recycled newspaper cellulose fiber (RNCF) reinforced PLA composite using twin screw extruder and injection molding. The tensile and flexural properties of PLA-RNCF composite were significantly higher as compared to virgin resin. Iwatake et al. [22] studied a sustainable green composite of PLA-micro-fibrillated cellulose (MFC). The author concluded that the incorporation of MFC resulted in increased tensile strength and Young's modulus of PLA by 40% and 25%, respectively. Bledzki et al. [23] investigated tensile, flexural, and impact strength of biocomposite based on PLA-abaca and PLA-man-made cellulose fiber. The performance of both PLA-abaca and PLA-cellulose fiber composite was superior to the native PLA. Lee et al. [24] studied the impact, tensile, and dynamic mechanical properties of PLA-denim composite. The author concluded that the impact, tensile, and dynamic mechanical properties of the composite were improved by piling denim fiber to PLA. Huda et al. [25] studied a biocomposite of PLA-recycled cellulose fiber processed by extrusion and injection molding. The mechanical and morphological properties were compared by varying the content of cellulose fiber to composite. The tensile and flexural properties of the composite were significantly higher than the neat resin. Tayommai and Aht-Ong [26] investigated the mechanical and biodegradability characteristics of PLA-coconut green composite by changing the fiber content. The addition of maximum fiber content resulted in improved impact strength but no significant change in the tensile strength of the green composite. Suryanegara et al. [27] studied the mechanical and thermal properties of PLA-micro-fibrillated cellulose nanocomposite (MFC). The various contents of MFC and PLA were mixed in an organic solvent followed by drying, kneading, and hot pressing into sheets. Increasing the MFC content in both amorphous and crystallized state resulted in improved tensile strength and modulus of neat PLA.

From the above discussion, it is clear that the researchers have majorly concentrated on improving the performance of PLA and its composites by various means through experimentally. The modeling and simulation performed to understand the mechanical behavior of green composites are rarely reported. Thus, in this study, the mechanical behavior of PLA-bamboo green composite is evaluated both experimentally and analytically. Both shell and solid green composites are modeled through ANSYS Composite PrepPost module and analyzed by ANSYS Static Structural. PLA-bamboo green composite is fabricated by hot compression, and the tensile properties are evaluated experimentally. The experimental results are then compared with the simulated results.

3 Modeling of Solid and Shell Composites

A rectangular plate of 150 mm × 150 mm × 1.5 mm is created as shown in Fig. 1. ANSYS Composite PrepPost is used to enter the material data. The properties of some materials already exist in ANSYS. In this study, the shell and solid composites are created by entering the properties of PLA and bamboo. The properties of PLA and bamboo considered are shown in Tables 1 and 2, respectively.

The mesh for the geometry is created after entering the materials data, as shown in Fig. 2. The size of the mesh element and the mesh thickness is considered as

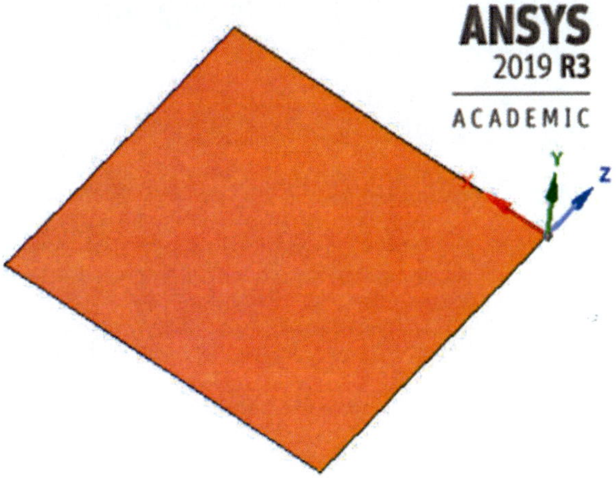

Fig. 1 Rectangular composite plate (*Source* Author)

Table 1 Properties of PLA

Properties	Values
Density	1250 kg/m^3
Melting temperature	180 °C
Young's modulus	3.45E+09 Pa
Poisson's ratio	0.33
Bulk modulus	5.2273E+09 Pa
Shear modulus	1.241E+08 Pa
Tensile yield strength	5.41E+07 Pa
Tensile ultimate strength	5.92E+07 Pa
Thermal conductivity	0.144 J m^{-1} s^{-1} c^{-1}

Source Author

Table 2 Properties of bamboo

Properties	Values
Density	693 kg/m^3
Young's modulus	1.73E+10 Pa
Poisson's ratio	0.384
Bulk modulus	2.4856E+10 Pa
Shear modulus	6.255E+09 Pa
Tensile yield strength	0.17 J m^{-1} s^{-1} c^{-1}

Source Author

5 and 1 mm, respectively. Both shell and solid models are created and converted into composites by defining the fabric and stackup properties. The thickness of the bamboo fabric and polymer film is considered as 0.187 and 1.5 mm, respectively.

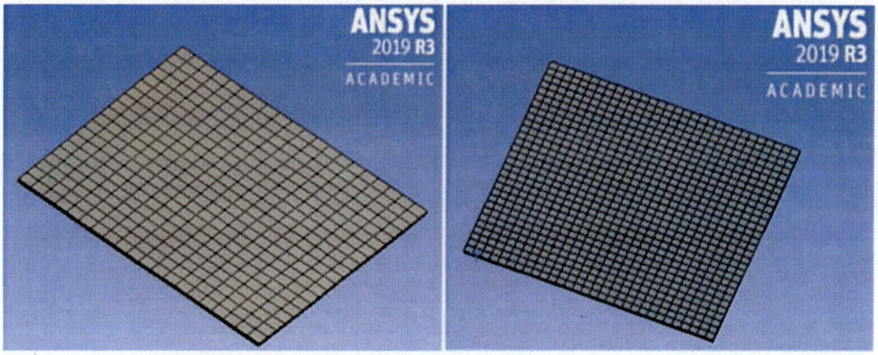

Fig. 2 Models of shell and solid composites (*Source* Author)

ANSYS Composite PrepPost follows both top-down and down-top stackup. In the top-down sequence, the first defined ply is placed first which is on the bottom of the stackup, and the other plies are placed over it. In this study, stackup is defined as top-down for unidirectional fabric.

After defining the material and the rosettes, the orientations and offset directions are defined by oriented selection sets. Rosettes are coordinate systems that used to set the reference direction of oriented selection sets. The offset direction is specified by the orientation point and direction, as shown in Fig. 3. The fiber alignment is specified by the reference direction and the relative angle of the modeling plies. The

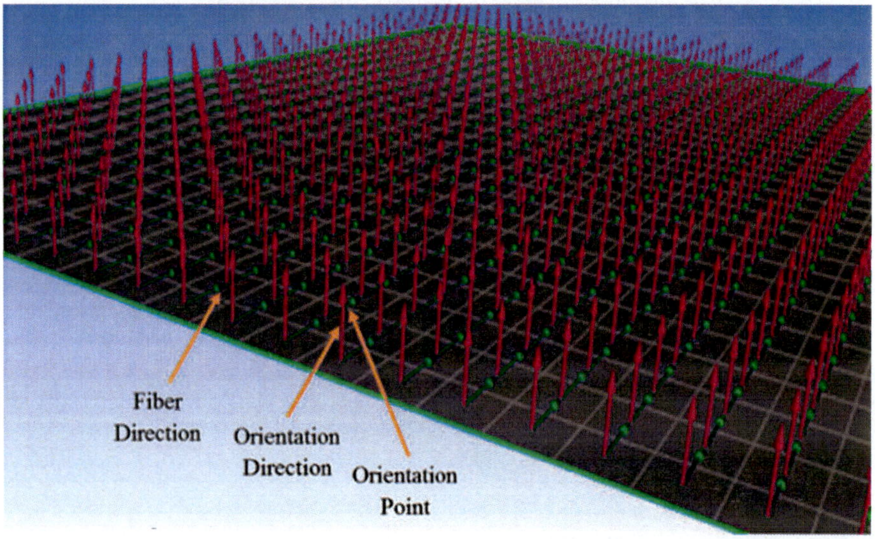

Fig. 3 Orientation point, orientation direction, and fiber direction (*Source* Author)

stacking sequence is executed by the order of modeling plies to the offset of the oriented selection sets. The sequential order of ply generation is shown in Fig. 4.

The composite models are created in ANSYS Composite PrepPost module by updating the properties of solid models. The ply sequence and the thickness of each ply are considered while generating the composite models. Each ply is bonded to each other in solid composite model. The ply creation, stackup, and properties play a vital role in the creation of composite models. The different views of the solid and shell composite models are shown in Figs. 5 and 6, respectively.

ANSYS Composite PrepPost is connected to the ANSYS Static Structural for the purpose of analysis after generating the shell and solid models of composites. The stress, strain, displacement, deformation, and forces on the components can be determined using ANSYS Static Structural. The effect of static and time-varying loads under different conditions can be executed using ANSYS Static Structural. The

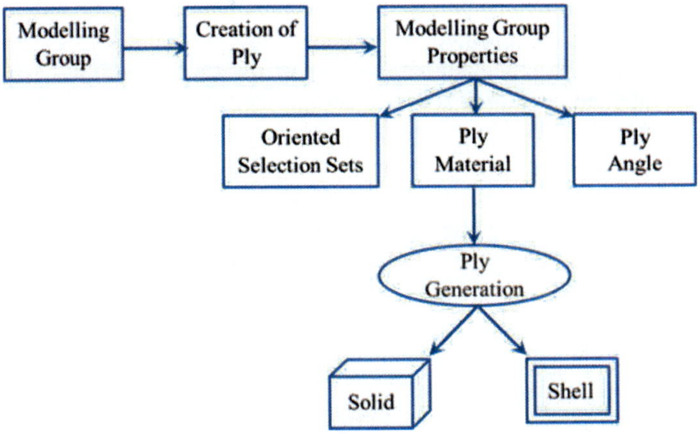

Fig. 4 Flowchart of ply generation (*Source* Author)

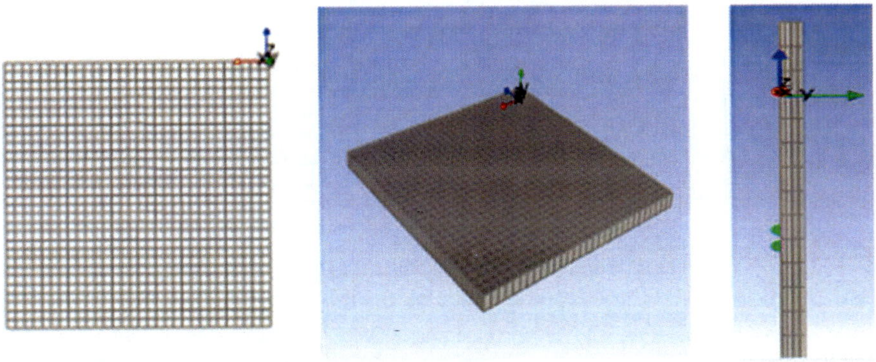

Fig. 5 Different views of solid composite model (*Source* Author)

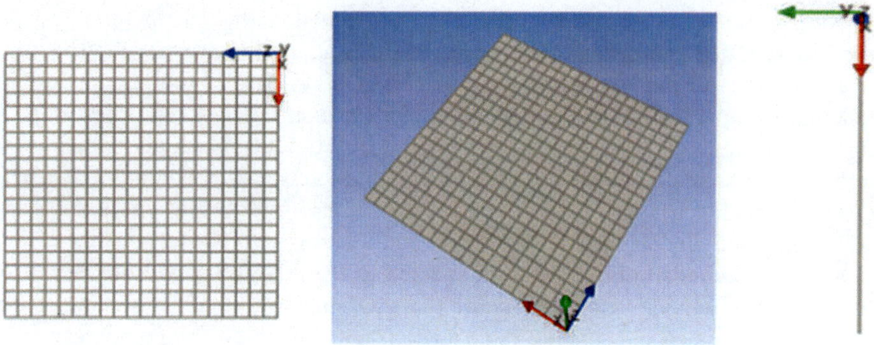

Fig. 6 Different views of shell composite model (*Source* Author)

general procedure to perform the analysis is started with the creation of component in workbench. The proper meshing of the component generates a model for analysis. The solution taskbar is updated after applying the boundary conditions. Finally, the results are evaluated according to the loading conditions of the component. After updating the ANSYS Composite PrepPost setup, the setup is directly connected to the model of Static Structural, as shown in Fig. 7. The model consists of two options: (i) transfer shell composite data for analyzing shell composite and (ii) transfer solid composite data for analyzing solid composite, as shown in Fig. 8.

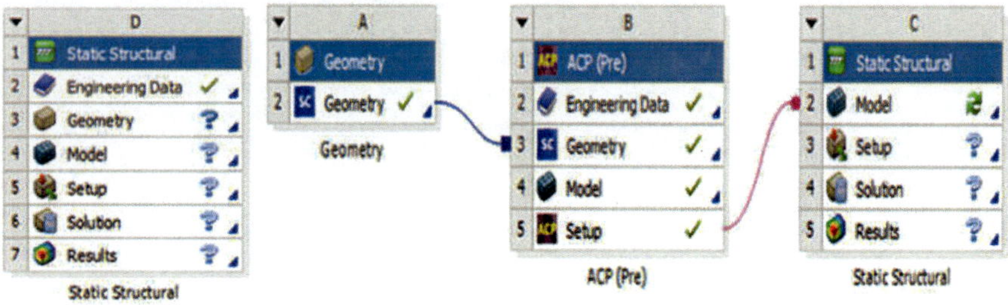

Fig. 7 ANSYS composite PrepPost module connection with static structural (*Source* Author)

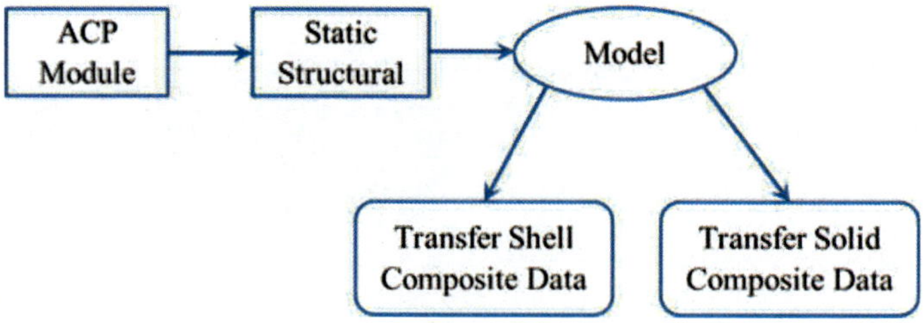

Fig. 8 Flowchart of model connection with shell and solid composite (*Source* Author)

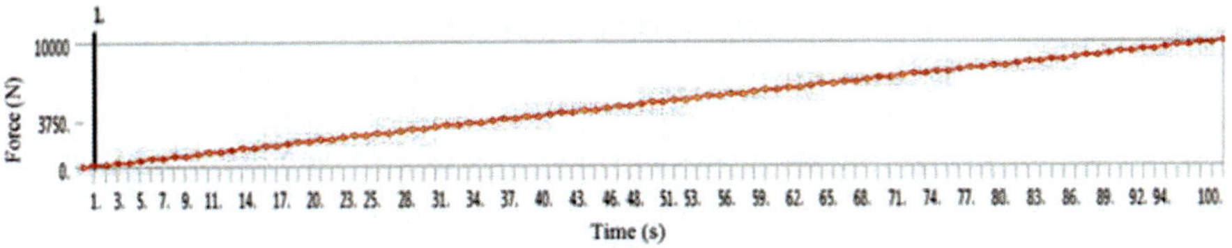

Fig. 9 Time-variant force graph for the application of force in stepwise (*Source* Author)

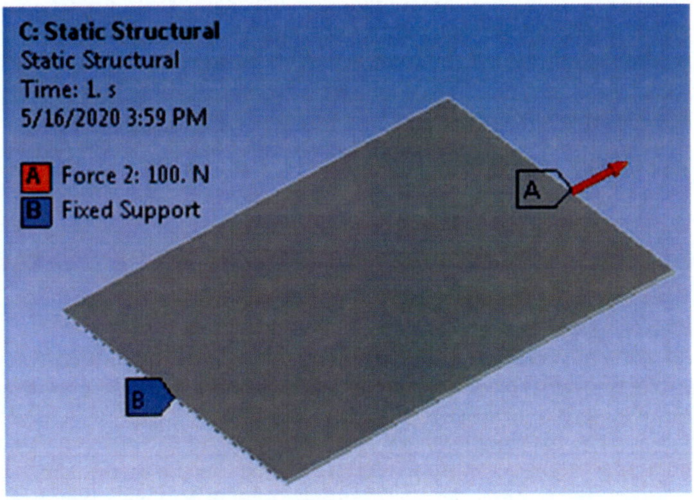

Fig. 10 Application of force in the fiber direction (*Source* Author)

ANSYS Static Structural setup for analysis blog is open once the data of shell and solid composite is updated. In the analysis, the minimum time for every step is fixed according to the force variant. The force can be applied either in a static or time-variant form. In this study, the number of steps and minimum elapsed step time for every step is considered as 1000 and 1 s, as shown in Fig. 9. For every step, a force of 100 N is applied on the plate of shell and solid composites incrementally up to 100 steps. The one end of the plate is fixed by using a fixed support option in the Static Structural. The other end of the plate is selected and applied a force of 100 N in vector form. The force is applied in the direction of the fiber as shown in Fig. 10.

4 Composite Preparation and Testing

The green composite is developed using bamboo fiber and PLA. PLA pellet used as binding material is purchased from Natur Tec India Pvt. Ltd, India. The crystalline melting and glass transition temperature of PLA are 150–170 and 55–60 °C. The density of PLA is 1.25 g/cm^3. The bamboo fiber is supplied by Sri Lakshmi Group, Andhra Pradesh, India. The fabrication of the green composite is performed using the compression molding setup. A compression molding setup is developed in house by retrofitting the upper and lower plates of the mold in the universal testing machine (UTM). The required pressure for compression is applied using the UTM. Two heating elements are placed above and below the top and bottom mold plates. The temperature of the mold plates is measured using *K*-type thermocouples. Initially, the moisture from the fiber and PLA pellets is removed by keeping them in an oven at 80 °C for 4 h. Then PLA film of 1.5 mm in thickness is fabricated by maintaining the temperature and pressure of 170 °C and 3 MPa. The shaped film is allowed to cool for 3 h under the same pressure in the mold. A total of three layers of bamboo fiber are stacked alternatively between the PLA films and compressed at a temperature of 170 °C. The fabricated composite plate is removed from the mold once the consolidation is completed under the pressure of 6 MPa. The tensile testing of the composite is performed at a crosshead speed of 2 mm/min using the UTM as per ASTM 3039.

5 Results and Discussion

The effect of tensile load on the developed composite is analyzed both experimentally and analytically. The maximum tensile stress of the composite obtained experimentally is 73.23 MPa; whereas, the maximum tensile stress of the shell composite obtained analytically is 72.53 MPa, as shown in Fig. 11. This indicates that the analytical value is quite close to the experimental value. The deviation between the experimental and analytical tensile stress is 0.009%. Figure 12 shows the maximum tensile stress of the solid composites is 65.20 MPa. This indicates that there is a 10.9% deviation in the tensile stress while comparing the analytical tensile stress with experimental tensile stress. In shell composite, tensile stress is the maximum

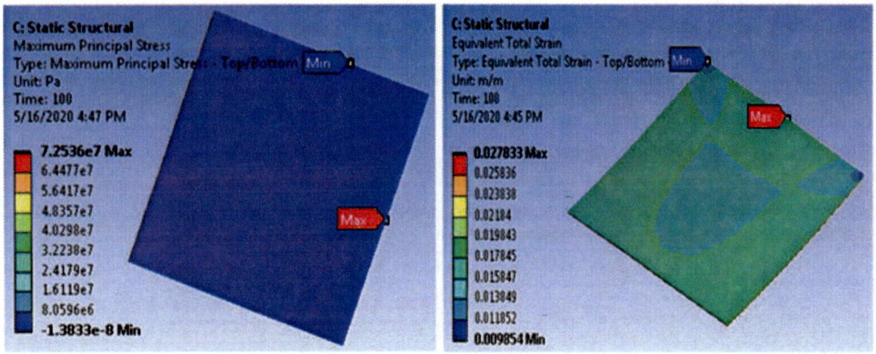

Fig. 11 Maximum stress and strain of shell composite (*Source* Author)

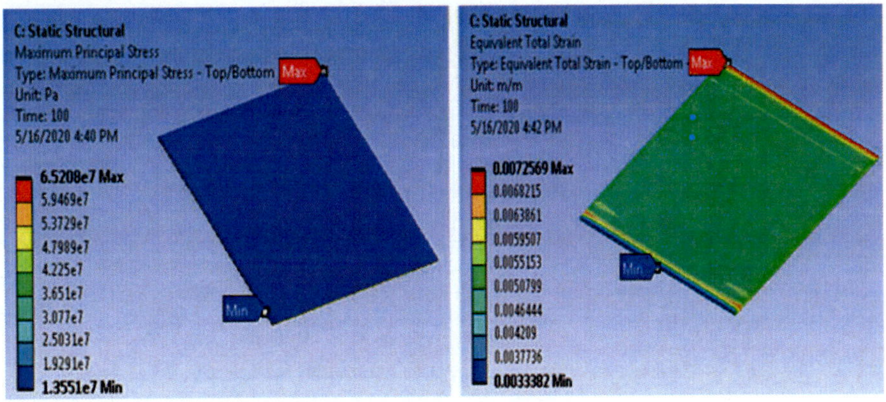

Fig. 12 Maximum stress and strain of solid composite (*Source* Author)

at the fixed end, and the stress is the minimum at the side end. Similarly, the tensile stress is maximum at the fixed end and minimum at the force applying end for the solid composite. The maximum tensile stress deviation between the shell and solid composites is 7.33 MPa. This shows that the performance of shell composite is better than the solid composite. The maximum strain in terms of displacement is 7.2 mm for solid composites. This exhibit almost the same maximum tensile strain obtained both analytically and experimentally tested specimens. The strain is more in the shell composite when compared one-on-one with solid composites. Figure 13 shows stress versus strain plots for both solid and shell composites which clearly indicate a slight variation in the maximum tensile stress between the solid and shell composite.

Fig. 13 Stress versus strain of shell and solid composites (*Source* Author)

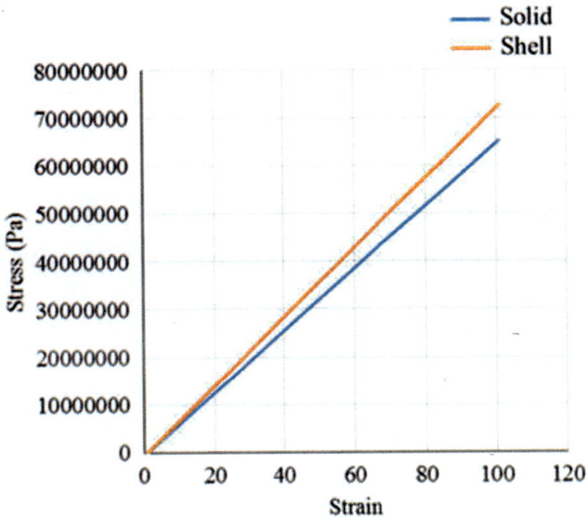

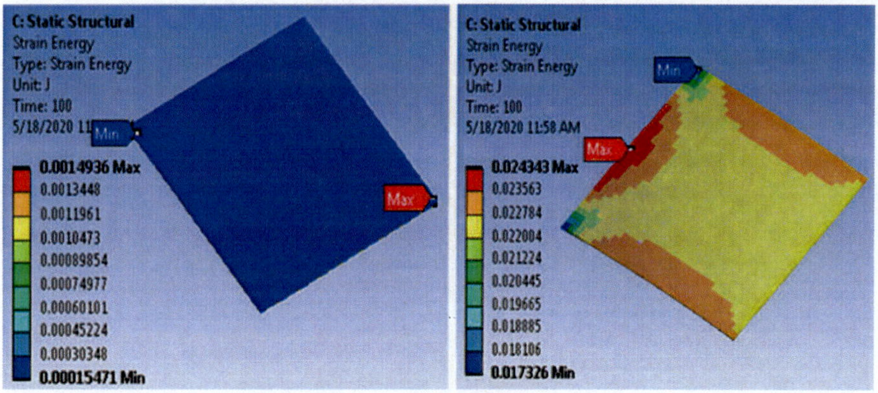

Fig. 14 Strain energy of solid and shell composite (*Source* Author)

The strain energy of the solid and shell composite is shown in Fig. 14. The strain is the maximum at the force applying end and minimum at the fixed end in solid composite; whereas, the scenario is completely different in shell composite. The maximum strain energy is in the middle of the fixed end region and minimum at the corner of the fixed end of the shell composite. In solid composite, all the particles are closely packed, and the applied force acts on every node of the composite. The strain energy is more due to the presence of shell elements in the shell composite. Both strain and strain energy at the corner of the fixed end are the minimum in shell composite. Figures 15 and 16 show the behavior of strain energy with respect to the strain. The figures clearly indicate that the strain energy increases with an increase in the strain. The figures also reveal that the strain energy of solid composite is less when compared with the strain energy of shell composite.

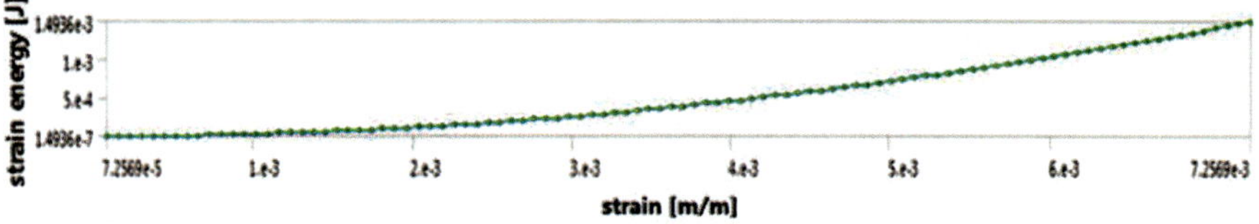

Fig. 15 Variation of strain energy with strain in solid composite (*Source* Author)

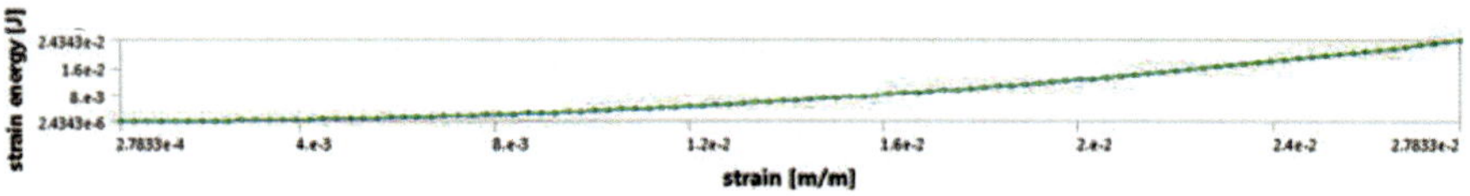

Fig. 16 Variation of strain energy with strain in shell composite (*Source* Author)

6 Comparative Analysis of the Mechanical Properties

It was reported that the PLA-untreated bamboo fiber composites had the maximum tensile strength of 60.6 MPa [20]. Another study reported that compatibilization between polymer and fibers is necessary for obtaining composites with improved tensile and impact properties with respect to raw PLA which is required in the automotive or electronic sectors [7]. Ramesh et al. [17] showed that the PLA-treated aloe vera fiber (TAF) hybrid biocomposites containing 1% montmorillonite (MMT) clay resulted in improvement in the impact, flexural, and tensile strength by 10.43, 6.08, and 5.72% when compared with PLA biocomposite. Sujito et al. [18] studied the effect of bamboo fiber loading (15–75%) on the tensile and flexural properties of PLA-short single bamboo fiber green composites. The investigation showed that the tensile and flexural properties were significantly affected by the fiber content. The best tensile and flexural properties of the developed green composites were obtained at the bamboo fiber loading of 40%. The green composite at this fiber loading had the maximum tensile strength of 110.47 ± 5.35 MPa, Young's modulus of 4.33 ± 0.27 GPa, flexural strength of 164.47 ± 43.34 MPa, and flexural modulus of 9.93 ± 2.67 GPa; whereas, the neat PLA considered for the purpose of investigation had the maximum tensile strength of 11.5 MPa. The improvement in the interfacial adhesion between the reinforcement and matrix due to the addition of polymeric diphenylmethane diisocyanate (pMDI) coupling agent resulted in significant improvement in the tensile strength of PLA-sugar beet pulp (SBP) composite (64 MPa) as compared to the tensile strength of the composites without the addition of coupling agent (37.5 MPa) [8]. PLA-recycled newspaper cellulose fiber (RNCF) composite containing 30 wt% RNCF exhibited the tensile strength of 67.9 MPa which was quite higher than the tensile strength of virgin PLA (62.9 MPa) [21]. The uniform dispersion of 10 wt% micro-fibrillated cellulose (MFC) fiber in PLA resulted in improvement in Young's modulus (4.7 GPa) and tensile strength (75 MPa) by 40% and 25%, respectively [22]. PLA composite reinforced with man-made cellulose achieved the highest tensile and flexural strengths of around 92 and 152 MPa, respectively [23]. Composite of PLA and three layers of denim fabric showed tensile strength and modulus of 75.76 MPa and 4.65 GPa, respectively [24]. The above discussion shows that many attempts have been made to improve the mechanical properties of PLA-based green composites. Chemical treatment of fibers, use of compatibilizer, varying the fiber loading, etc., are some of the attempts made to improve the mechanical performance of PLA-based green composites. In this work, the maximum tensile strength of 73.23 MPa is obtained for the developed PLA-bamboo composite which is comparable to the other type of PLA-based green composites.

7 Conclusions

In this study, both experimental and simulation works are performed to investigate the characteristics of PLA-bamboo green composites. The PLA-bamboo green composite is fabricated by compression molding and then mechanical testing is performed. The simulation is performed by developing shell and solid composite models using ANSYS Composite PrepPost module and then analyzed through ANSYS Static Structural. The simulation results show that the maximum tensile stress of shell and solid composites are close to the experimental tensile stress (73.23 MPa) of PLA-bamboo composite. PLA-based green composites reinforced with different fibers like sugar beet pulp (SBP), recycled newspaper cellulose fiber (RNCF), micro-fibrillated cellulose (MFC), and abaca fiber exhibiting almost the same tensile strength. The maximum tensile stress of the shell composite is more (72.53 MPa) than the solid composite (65.20 MPa). The experiential strain and the strain obtained through simulation are almost same. However, the strain in shell composite is more as compared to the solid composite. The strain energy of the shell composite is also more than the solid composite.

References

1. Abdulkhani A, Hosseinzadeh J, Dadashi S, Mousavi M (2015) A study of morphological, thermal, mechanical and barrier properties of PLA based biocomposites prepared with micro and nano sized cellulosic fibers. Cellul Chem Technol 49(7–8):597–605
2. Nurnadia MJ, Nurul Fazita MR, Abdul Khalil HPS, Mohamad Haafiz MK (2017) Optimization of mechanical properties of bamboo fiber reinforced-PLA bio composites. AIP Conf Proc 1901:030019
3. Sun Z, Zhang L, Liang D, Xiao W, Lin J (2017) Mechanical and thermal properties of PLA biocomposites reinforced by coir fibers. Int J Polym Sci 2017:2178329
4. Eng CC, Ibrahim NA, Zainuddin N, Ariffin H, Yunus WZW, Then YY, Teh CC (2013) Enhancement of mechanical and thermal properties of polylactic acid/polycaprolactone blends by hydrophilic nanoclay. Indian J Mater Sci 2013:816503
5. Lay M, Thajudin NLN, Hamid ZAA, Rusli A, Abdullah MK, Shuib RK (2019) Comparison of physical and mechanical properties of PLA, ABS and nylon 6 fabricated using fused deposition modeling and injection molding. Compos B Eng 176:107341
6. Asadi M, Bazyar B, Hooman Hemmasi A, Ghsemi I, Talaeipoor M (2018) Assessment of mechanical and morphological properties of new poly lactic acid (PLA)/wood fibers/nanographene composite. Drvna Industrija 69(2):127–134
7. Aliotta L, Gigante V, Coltelli MB, Cinelli P, Lazzeri A (2019) Evaluation of mechanical and interfacial properties of bio-composites based on poly (lactic acid) with natural cellulose fibers. Int J Mol Sci 20(4):960
8. Chen F, Liu L, Cooke PH, Hicks KB, Zhang J (2008) Performance enhancement of poly (lactic acid) and sugar beet pulp composites by improving interfacial adhesion and penetration. Ind Eng Chem Res 47(22):8667–8675
9. Parida C, Dash SK, Chaterjee P (2015) Mechanical properties of injection molded poly (lactic) acid-luffa fiber composites. Soft Nanosci Lett 5(04):65
10. Fujiura T, Okamoto T, Tanaka T, Imaida Y (2010) Improvement of mechanical properties of long jute fiber reinforced polylactide prepared by injection molding process. WIT Trans Ecol Environ 138:181–188

11. Huda MS, Drzal LT, Misra M, Mohanty AK (2006) Wood-fiber-reinforced poly (lactic acid) composites: evaluation of the physicomechanical and morphological properties. J Appl Polym Sci 102(5):4856–4869
12. Musyarofah L, Puspita D, Hidayah E (2019) Tensile properties of coir and fleece fibers reinforced poly-lactic acid hybrid green composites. J Phys Conf Ser 1217(1):012008
13. Kuciel S, Mazur K, Hebda M (2020) The influence of wood and basalt fibres on mechanical, thermal and hydrothermal properties of PLA composites. J Polym Environ 28:1204–1215
14. Righetti MC, Cinelli P, Mallegni N, Massa CA, Bronco S, Stäbler A, Lazzeri A (2019) Thermal, mechanical, and rheological properties of biocomposites made of poly (lactic acid) and potato pulp powder. Int J Mol Sci 20(3):675
15. Rawi NFM, Jayaraman K, Bhattacharyya D (2013) A performance study on composites made from bamboo fabric and poly (lactic acid). J Reinf Plast Compos 32(20):1513–1525
16. Sujaritjun W, Uawongsuwan P, Pivsa-Art W, Hamada H (2013) Mechanical property of surface modified natural fiber reinforced PLA biocomposites. Energy Procedia 34:664–672
17. Ramesh P, Prasad BD, Narayana KL (2019) Effect of MMT clay on mechanical, thermal and barrier properties of treated aloevera fiber/PLA-hybrid biocomposites. Silicon. https://doi.org/10.1007/s12633-019-00275-6
18. Sujito, Pandey JK, Takagi H (2011) Mechanical properties of green composite based on poly-lactic acid resin and short single bamboo fibers. In: Proceedings of 18th international conference on composite materials, Jeju Island, Korea
19. Zhao C, Song GL, Liu HW (2011) The mechanical properties of bamboo/PLA composites made by biaxial weft knitted fabrics. Adv Mater Res 331:89–92
20. Hu G, Cai S, Zhou Y, Zhang N, Ren J (2018) Enhanced mechanical and thermal properties of poly (lactic acid)/bamboo fiber composites via surface modification. J Reinf Plast Compos 37(12):841–852
21. Huda MS, Drzal LT, Mohanty AK, Misra M (2006) Chopped glass and recycled newspaper as reinforcement fibers in injection molded poly (lactic acid) (PLA) composites: a comparative study. Compos Sci Technol 66(11–12):1813–1824
22. Iwatake A, Nogi M, Yano H (2008) Cellulose nanofiber-reinforced polylactic acid. Compos Sci Technol 68(9):2103–2106
23. Bledzki AK, Jaszkiewicz A, Scherzer D (2009) Mechanical properties of PLA composites with man-made cellulose and abaca fibres. Compos A Appl Sci Manuf 40(4):404–412
24. Lee JT, Kim MW, Song YS, Kang TJ, Youn JR (2010) Mechanical properties of denim fabric reinforced poly (lactic acid). Fibers Polym 11(1):60–66
25. Huda MS, Mohanty AK, Drzal LT, Schut E, Misra M (2005) "Green" composites from recycled cellulose and poly (lactic acid): physico-mechanical and morphological properties evaluation. J Mater Sci 40(16):4221–4229
26. Tayommai T, Aht-Ong D (2010) Natural fiber/PLA composites: mechanical properties and biodegradability by gravimetric measurement respirometric (GMR) system. Adv Mater Res 93:223–226
27. Suryanegara L, Nakagaito AN, Yano H (2009) The effect of crystallization of PLA on the thermal and mechanical properties of microfibrillated cellulose-reinforced PLA composites. Compos Sci Technol 69(7–8):1187–1192

Chapter 8
Green Hydrogels

K. Viswanath Allamraju

1 Introduction

Hydrophilic polymeric frameworks that are prepared for absorbing goliath volumes of water and encountering developing and shrinkage suitably to energize controlled medicine release are called hydrogels. Their porosity and comparability with liquid circumstances make them significantly engaging biocompatible prescription transport vehicles. Their applications are intricate and for a couple of biomedical needs as they are adaptable into moved physical structures, for instance, nanoparticles, microparticles, areas, films, and coatings. USA has been the greatest creator of hydrogels and is required to remain so for a few additional years [1]. Hydrogels are promising, stylish, astute, and 'wise' sedate movement vehicles that consider the specific necessities for concentrating on drugs to the specific regions and controlling prescription release. Enzymatic, hydrolytic, or environmental enhancements much of the time take care of business to control the hydrogels for the medicine release at the alluring site [2]. Like the various sides of a coin, there are furthermore the shortcomings related with their use. The fundamental weight in sedate movement would be the hydrophobicity of most drugs. The water-appreciating polymeric focus is likely not very immaculate to hold opposite hydrophobic drugs, which is a test since various that are correct presently used and fruitful in ailment treatment are hydrophobic. The flexibility of these hydrogels is frail and this at times causes unexpected appearance of the prescription before arriving in the goal site. The going with review discusses on how hydrogels are being constrained eventually for improved concentrated on steady movement. The propelled design wherein the hydrogels are manhandled for sedate transport is made sure about. The engaging physical properties of hydrogels,

K. V. Allamraju (✉)
Department of Mechanical Engineering, Institute of Aeronautical Engineering, Dundigal, Hyderabad, India
e-mail: akvn87@gmail.com

S. Thomas and P. Balakrishnan (eds.), *Green Composites*,
Materials Horizons: From Nature to Nanomaterials,
https://doi.org/10.1007/978-981-15-9643-8_8

especially their porosity, offer gigantic focal points in sedate transport applications, for instance, upheld appearance of the stacked drug. A high close by centralization of the dynamic pharmaceutical fixing is held over a huge stretch of time by methods for a suitable release framework compelled by scattering, developing, creation, or subject to some regular overhauls.

Scattering-controlled drug movement with hydrogels uses storehouse or cross section contraptions that license dispersal-based prescription release through a hydrogel work or pores stacked up with water. In the store transport structure, the hydrogel layer is secured on a drug containing focus making cases, circles, or pieces that have a high medicine center in the especially point of convergence of the system to energize a predictable prescription release rate. While the store transport system produces time-self-ruling and predictable drug release, the system structure works by methods for the macromolecular pores or work. This kind of release is time-subordinate drug release wherein the basic release rate is comparing to the square base of time, rather than being consistent.

The developing controlled medicine release from hydrogels uses drugs dissipated inside a smooth polymer which when in contact with a bio-fluid beginning extending. The augmentation during developing occurs past its breaking point empowering the prescription dispersal close by the polymer chain loosening up. The system, regardless suggested as Case II transport, supports without time, consistent drug release vitality. Since the slant between the dissipated prescription in the hydrogel and its general condition allows the dynamic fixing spread from a territory of higher obsession inside the hydrogel to a lower one, the methodology is in like manner suggested as odd vehicle as it solidifies both the strategies of scattering and developing for engaging medication release. Visual drug transport bearers have been made using hydrogels that are covalently crosslinked. These fragile, biodegradable hydrogels with high developing cutoff remain in-situ in the lacrimal channel offering increasingly critical comfort for the patient. Collagen or silicone that may be used picks if the punctal-plug structure could be used by chance or always, separately. Poly(ethylene glycol) hydrogels are typically used for making ophthalmic drug movement systems. Prescription release considering common changes would be an ideal transport system as the release ends up being controlled, and ambiguous indications at off-target regions are facilitated. Along these lines, sensitive prescription movement contraptions open to changes in pH, temperature, ionic quality, or glucose center have been developed that are gainful in the treatment of sicknesses, for instance, dangerous development, and diabetes, depicted by close by physiological changes express to the diverse infirmity stages. The polymer game plan of the hydrogel open to the biological upgrades is controlled to make it responsive to nature [3].

Hydrogels broadly improve the medicinal consequence of prescription movement and have found enormous clinical use. The transient and spatial transport of macromolecular meds, little particles, and cells have exceptionally improved through hydrogel use for quiet movement [2]. Medicine transport using hydrogels, regardless, has not been freed from troubles, and anyway consistent upgrades are being made to separate the hydrogel arrangement generally fitting for express drug movement purposes. Likewise, this paper discusses the continuous examples in steady movement applications using hydrogels, including their translation to the middle and their applications to viably pass on hydrophobic drugs.

Hydrogels can be characterized into two particular classes, the characteristic and the manufactured hydrogels. Common hydrogels incorporate collagen, fibrin, hyaluronic corrosive, matrigel, and derivatives of normal materials, for example, chitosan, alginate, and aptitude filaments. They remain the most physiological hydrogels as they are segments of the extracellular framework (ECM) in-vivo. Two fundamental disadvantages of characteristic hydrogels, be that as it may, make their last microstructures and properties hard to control reproducibly between tests. To begin with, the fine subtleties of their mechanical properties and their reliance on polymerization or gelation conditions are regularly ineffectively comprehended. Second, because of their common root (ox-like fibrogen, rodent tail collagen …), their piece may fluctuate starting with one bunch then onto the next. Conversely, engineered hydrogels, for example, poly ethylene glycol diacrylate, poly acryl amide, poly vinyl liquor are progressively reproducible, despite the fact that their last structure can likewise rely upon polymerization conditions in an unpretentious manner, with the goal that a thorough control of the planning convention, including temperature and condition control, might be fundamental. As a rule, manufactured hydrogels offer greater adaptability for tuning synthetic arrangement and mechanical properties; clients can, for instance fluctuate the fixation or sub-atomic load of the forerunner, or adjust the level of cross-linkers. They can likewise be chosen or tuned to be hydrolysable or biodegradable over factor timeframes.

Hydrogels might be artificially steady or they may debase and in the long run break down and disintegrate. They are called reversible or physical gels when the systems are held together by atomic traps and/or auxiliary powers including ionic, H-holding r hydrophobic powers [4].

Physical hydrogels are not homogeneous, since groups of sub-atomic snares, or hydrophobically or ionically related spaces, can make in homogeneities. Free chain finishes or chain circles additionally present transient system abandons in physical gels. At the point when a poly electrolyte is joined with a multivalent particle of the contrary charge, it might shape a physical hydrogel known as an ionotropic hydrogel, and calcium alginate is a case of this kind of hydrogel. Hydrogels are called changeless or concoction gels when they are covalently cross-connected systems [5].

The manufactured hydrogels of Wichterle and Lim depended on copolymerizaton of HEMA with the cross-linker EGDMA. Substance hydrogels may likewise be produced by cross-connecting of water-dissolvable polymers, or by transformation of hydrophobic polymers to hydrophilic polymers in addition to cross-connecting which is not fundamental. Now and again, contingent upon the dissolvable arrangement, temperature, and solids focus during gel development, stage detachment can happen, and water-filled voids or macropores can shape. I compound gels, free chains closes speak to gel arrange "deserts" which do not add to the flexibility of the system. Other system surrenders are chain circles and entrapments, which likewise do not add to the lasting system versatility.

In numerous applications, a utilitarian-added substance is mixed into a polymer framework to improve its properties. Notwithstanding, when the polymer and utilitarian-added substance are applied to a surface, the useful atom might be effectively lost. In ideal cases, it might be conceivable to fuse the added substance legitimately into the polymer as a comonomer. In ongoing investigation, a functionalized polymer has been acquired through the blend of connecting a photodynamics, antimicrobial color, rose Bengal, to vinyl benzyl chloride by means of etherification and afterward polymerizing this into an igure water-dissolvable polymer using chain development copolymerization [6]. The measure of unincorporated color was resolved utilizing dialysis I support arrangement at pH = 7.4. At the point when rose Bengal was first responded with vinyl benzyl chloride before polymerization, it was discovered that it was about 100% joined into the copolymer.

2 History

The word "hydrogel," as per Lee, Kwon, and Parl, goes back to an article distributed in 1894. Anyway, the material portrayed there was not a hydrogel as we depict it today; it was without a doubt a colloidal gel made with inorganic salts. Is yet wonderful to see how the historical backdrop of the term itself is reliably long [7]. At any rate, the primary cross-connected system material that showed up in writing and has been portrayed by its common hydrogel properties, one for all the high water proclivity, was a poly hydroxyl ethyl meth acrylate (PHEMA) hydrogel that grew a lot later, in 1960, and with the driven objective of utilizing them in changeless contact applications with human tissues, hydrogels are in actuality the principal materials created for utilizing inside the patient [8, 9]. From that point forward, the quantity of learnings about hydrogels for biomedical applications started to rise, particularly from the time of 70s [7]. The points and objectives and the quantity of materials changed and expanded continually throughout the years. As recommended by Buwalda et al. [8], the historical backdrop of hydrogels can be isolated in three fundamental squares. A hydrogel's original that includes a wide scope of cross-connecting methods including the concoction changes of a monomer or polymer with an initiator. The general point is to create material with high expanding, great mechanical properties, and moderately straightforward basis. At that point, beginning in the seventies, an alternate idea of hydrogel developed in importance: a second era of materials equipped for a reaction to explicit upgrades, for example, varieties in temperature, pH, or grouping of explicit particles in arrangement. These particular upgrades can be misused to trigger in like manner explicit occasions, for instance the polymerization of the material, a medication conveyance or an in-situ pore development [10].

At long last, a third era of hydrogels concentrating on the examination and improvement of stereo-complexed materials (e.g., PEG-PLA association) [11, 12] hydrogels cross-connected by other physical cooperations such as cyclodextrines) [13, 14]. This advancement in hydrogels science is rapidly prompting an expanding enthusiasm for the improvement of the supposed "keen hydrogels," polymeric grids with a wide range of tunable properties and trigger upgrades. The subject is hypothetically boundless, and the potential applications, the designing, and clinical gadgets that can be gotten from it ate over any creative mind. Since the spearheading work of Wichterle and Lim in 1960 on cross-connected hydrogels [8], and in view of their hydrophilic character and potential to be biocompatible, hydrogels have been of extraordinary enthusiasm to biomaterial researchers for a long time [15–17]. The significant and persuasive work of Lim and Sun in 1980 [18] exhibited the fruitful use of calcium alginate microcapsules for cell embodiment. Later during the 1980s, Yannas and associates [19] fused regular polymers, for example, collagen and shark ligament into hydrogels for use as counterfeit consume dressings. Hydrogels dependent on both common and manufactured polymers have kept on being of enthusiasm for epitome of cells [20], and most as of late such hydrogels have gotten particularly alluring to the new field of "tissue designing" as frameworks for fixing and recovering a wide assortment of tissues and organs [21].

3 Physical and Chemical Properties

Despite a great deal progress, a fundamental understanding of gel properties is not yet sufficient for a rational design of novel gel systems. For such designs, it is miles important to understand how solute molecules interact with the gel, in particular, how they partition between the gel phase and the surrounding liquid section. Partitioning relies upon on two predominant effects: size exclusion and molecular attraction/shock. Growing: Hydrogels are crosslinked polymer networks swollen in a liquid medium. The imbibed liquid serves as a selective clear out to allow unfastened diffusion of a few solute molecules, while the polymer network serves as a matrix to keep the liquid together. Hydrogels may absorb from 10 to 20% (an arbitrary lower limit) up to lots of times their dry weight in water. The person of the water in a hydrogel can decide the overall permeation of vitamins into and mobile merchandise out of the gel. When a dry hydrogel starts to soak up water, the primary water molecules getting into the matrix will hydrate the maximum polar, hydrophilic groups, leading to number one sure water. As the polar corporations are hydrated, the network swells and exposes hydrophobic companies, which also engage with water molecules, leading to hydrophobically sure water, or secondary sure water. Primary and secondary certain water are regularly combined and simply called the whole sure water. After the polar and hydrophobic Web sites have interacted with and bound water molecules, the community will imbibe additional water, because of the osmotic driving pressure of the network chains in the direction of limitless dilution. This extra swelling is opposed by way of the covalent or physical crosslink, leading to an elastic network retraction pressure.

Thus, the hydrogel will attain an equilibrium swelling level. The additional swelling water is imbibed after the ionic, polar, and hydrophobic companies that turn out to be saturated with bound water which is referred to as loose water or bulk water, and is believed to fill the distance between the community chains, and/or the center of larger pores, macropores, or voids. As the network swells, if the network chains or crosslink are degradable, the gel will start to collapse and dissolve, at a rate relying on its composition. There are a number of methods utilized by researchers to estimate the relative quantities of free and bound water, as fractions of the overall water content. All of them are controversial, since there may be proton NMR proof that the interchange of water molecules between the so-referred to as certain and loose states is extraordinarily rapid, possibly as rapid as one H_2O molecule every 10–9 s. The three essential strategies used to symbolize water in hydrogels are primarily based on the use of small molecular probes, DSC and NMR. When probe molecules are used, the labeled probe solution is equilibrated with the hydrogel, and the attention of the probe molecule within the gel at equilibrium is measured. Assuming that most effective unfastened water inside the gel can dissolve the probe solute, it is easy to calculate the free water content material from the amount of the imbibed probe molecule and the known (measured) probe molecule concentration inside the outside answer. Then, the sure water is obtained through difference of the measured total water content of the hydrogel and the calculated free water content. The use of DSC is based totally on the belief that only the free water may be frozen, so it is assumed that the endotherm measured while warming the frozen gel represents the melting of the loose water, and that value will yield the quantity of unfastened water inside the HG pattern being tested. Then, the bound water is acquired by way of difference of the measured general water content of the HG test specimen and the calculated unfastened water content material. In another formulation, swelling is the belongings to absorb water and maintain it for a relative lengthy time. It may be evaluated by using measuring the dry weight and the swollen-state weight and computing both a ponderal variation (water uptake) or a volume of adsorbed solvent (both the quantities are considered as percentages).

$$\text{W.U.} = \frac{\text{swollen weight} - \text{dry weight}}{\text{dry weight}} \times 100 \tag{1}$$

$$\text{V.A.S.} = \frac{\text{swollen weight} - \text{dry weight}}{\text{water density}} \times 100 \tag{2}$$

As simple as it may seem, the assessment of swelling is the important assay to be performed on hydrogel samples, as it is able to be a degree for lots of their properties: crosslinking degree, mechanical properties, degradation charge, and so on. For many gels, the evaluation of swelling and swollen-state stability is the simplest, cheapest, and surest way to discriminate between crosslinked gels and the now not crosslinked authentic polymer [21].

4 Mechanical Properties

The mechanical properties can be varied and also be compromised depending on the rationale of the material. The possibility of obtaining the higher stiffness gel is by increasing the cross-linking degree via heating the material. The variation in mechanical properties are connected to a broad range of dependent and independent variables and reasons, and therefore, different analysis should be implemented according to the material, the surroundings, and objective of study. Suppose, while gelatin shows a recognizable improvement in modulus of elasticity by cross-linking [22], silk fibroin has a very high modulus of elasticity, but it will be reduced after the revitalization [23]. The mechanical properties such as modulus of elasticity, Poisson's ratio, storage capacity, and loss moduli can be measured with the help of dynamic mechanical analysis device or rheometer [24, 25]. In the case of hydrogel, the modulus of elasticity is the results of the bond between water and gel matrix. In the case of tissue engineering, high stiffness material is required for seeding osteoblast than the culture adipocyte. The same method is valid for the growth of a varied prosthetic mechanism. This type of mechanism can be observed in the case of substitute for intervertebral disk.

5 Porosity and Permeation

The formation of pores in hydrogels is occurred by phase separation in synthesis process, or pores can be existed in small size within the network. The important factors of a hydrogel matrix are difficult to quantify such as pore size average, pore size distribution, and the pore interconnections. The parameter called "tortuosity" which includes sum of pore size average, pore size distribution, and the pore interconnections. This parameter helps to determine the effective diffusion path length across a HG film barrier.

$$\text{The effective diffusion path length} = \text{Film thickness} \times \frac{\text{Pore volume fraction}}{\text{Tortuosity}}$$

The composition and cross-link density of hydrogel polymer network effects the film thickness, pore volume fraction and tortuosity. The geometric properties of molecules are used to explore pore sizes in hydrogels [26]. Pore length distributions of hydrogels are strongly suffering from three factors: concentration of the chemical cross-links of the polymer strands. That attention is determined through the initial ratio of cross-linker to monomer. Concentration of the bodily entanglements of the polymer strands. And also which is determined by means of the preliminary attention of all polymerizable monomers inside the aqueous solution. Net price of the polyelectrolyte hydrogel is determined by the initial attention of the cationic and/or anionic monomer. These three elements can be quantified the use of the composition of the hydrogel, that is, with the aid of the nominal concentrations of monomer and cross-linker. The porous structure of a hydrogel is influenced by the properties of surrounding solution and Donnan and Osmotic effects. The dissolved ionic solutes are called as Donnan effects. Uneven partition between gel phase and the solution phase caused by dissolved uncharged solutes is called as Osmotic effects. Porosity describes the presence of void cavity within the bulk material. Porosity is a morphological aspect of a material.

In a sample, pores can show extraordinary morphologies: They can be closed, opened as a blind give up or interconnected, again divided in cavities and throats. These porosities were studied and evaluated in papers in the beyond decades the usage of a numerous spectrum of techniques. First of all, porosity can be evaluated with the aid of theoretical methods, which includes unit dice analysis, mass method, Archimedes approach, and liquid-displacement approach. These evaluations are generally coupled with optical and digital microscopy. Other thrilling strategies are the mercury porosimetry, based totally on Washburns equation, with the inconvenience of being a detrimental assay, the gasoline pycnometry, the gasoline adsorption (that can be issued the usage of one of a kind strategies which includes small quantity adsorption, monolayer, and multilayer adsorption), liquid extrusion porosity, and capillary flow porosity, X-ray microtomography, microscopy techniques (optical microscopy, stereo microscopy, scanning electron microscopy, tunneling microscopy, and atomic force microscopy).

6 Applications of Hydrogels

6.1 Diapers

An exciting application of hydrogels thermodynamical affinity for water, as not fancy as it can be, is the manufacturing of super-adsorbent diapers with the property of being dry even after a great adsorption of fluids. This is due to, as previously said, the character of hydrogels water adsorption: these materials do not act as sponges, unstably trapping drinks into their pores; however, alternatively they hold water (or, now and again, other solvents) if they are carrying good sized portions of water at the identical time. The development of hydrogel-containing diapers, most of them loaded with extraordinary formulations of sodium polyacrylate [27–29], within the past decades reduce down on a massive wide variety of dermatological conditions related to a prolonged contact with wet tissues. Nevertheless many health challenge arose on the big usage of disposable diapers: $\approx$95% of nappies in western international locations are disposable ones and critiques approximately fabric nappies are still conflicting [30]. Indeed, many chemical substances used inside the production of such products like scents, leak-proof substances, and super-adsorbent polymers, appear to be key elements for the improvement of many conditions, from chronic diaper rash and asthma, to greater serious problems which include male infertility or even testicular cancer. Furthermore, disposable diapers, considering that they are used in large quantity, create a first-rate environmental issue on account that it is no longer clean to take away them [31]. However, this is a subject that, as some distance as interesting and important, goes beyond the reason of our review. Watering beads for plants: Another easy utility of hydrogels consists in hard powders of polyacrylamide or potassium polyacrylate matrix bought with a massive variety of names (plant gel, super crystals, and water gel crystals) and used as long-term reservoir of water for plant boom in gardening, domestic, and every now and then industrial horticulture. On the opposite aspect as the only of diaper's hydrogel, these substances are optimized for their capability of freeing water, in place of the ability of keeping it. The sustained release of many diverse species is, indeed, one of the main electricity of hydrogels on the market, from gardening to genetic engineering. However, even if businesses producing such crystals are selling their practicality and versatility, in the remaining years, the scientific community is thinking approximately their real utility. As Chalker-Scott from Washington State University talked about in her publications at the topic, considering that the typically used watering crystals are created from non-renewable materials, whose monomers can be toxic (e.g., Acrylamide), the potential dangers in their usage are manner higher than the blessings of water storage and controlled release that can, in addition, be obtained in many different methods with lower environmental impact [32].

6.2 Perfume Delivery

During the nineties patents describing risky species delivery technologies started out to grow in number. In particular, the most vastest patented inventions in the field seem to be issued through Procter & Gamble, processing the fragrances into cyclodextrin complexes [33–36]. The standard aim becomes to develop gadgets capable of slowly dispensing fragrances to the surroundings in the long-time period and update the conventional salt-based (sodium dodecylbenzenesulphonate) capsules with new, more sensible, and, let us say it, fancier house care solutions. The position of hydrogels inside the method revolves around, as soon as again, their swelling properties that can be exploited in materials, "wherein launch of a perfume scent is induced through dynamic swelling force of the polymer while the polymer is wetted" [34]. These devices launch risky particles thanks to osmotic diffusion of the species from the swollen hydrogel to new water in the environment.

Cosmetic enterprise is a market continuously increasing in size and product offer. One of the motives of this behavior is related to the long course to approval vital for medical gadgets, clinical procedures, tablets, or biomolecules, which will make allowed to be sold in the marketplace. So, while on the opposite hand, the beauty marketplace requires less time and money costs for the approval of a product, it is not unusual to test new innovations and devices in cosmetics earlier than and in more risky clinical applications. For a product to be authorized in cosmetics, the most vital parameter to be assessed is Primary Irritation Index (PII). This index is simple to obtain and exist both for skin and eyes, and indeed every stage of PII corresponds a determinate effect [37, 38]. Considering that almost all of hydrogels used on this field are appropriate for cells subculture and for different biomedical applications, it is not always unexpected that their irritation index is a number of the lowest. Thus, with especially a small investment groups are able to launch available on the market new cosmetic products totally based on hydrogels, which includes so-called "beauty masks." Usually made with engineered collagen (Masqueology TM through SEPHORA USA Inc., BioCollagen Cosmeceuticals by way of NOVOSTRATA UK Ltd.), hyaluronic acid (SEPHORA USA Inc.), or polyvinylpyrrolidone (Pecogel®), these mask declare to hydrate the skin, repair its elasticity, and sell anti-getting older actions [39]. Pecogel through Phoenix Chemicals Inc., is a big choice of hydrogels, based totally on polyvinylpyrrolidone, with differences in composition and/or crosslinking method. Pecogels are suitable for cosmetic purposes, which includes sunscreen cream or mascara [40]. Furthermore, in a number of the commercially to be had compounds which include hydrogel face masks by way of Fruit & Passion Boutiques Inc., the moisturizing action of those natural polymeric gels is coupled with more complicated drug-delivery systems evolved to release of biomolecules like vitamin C or B3. The beauty enterprise is on the reducing edge of hydrogels, indeed a pH-sensitive material P(MAA-co-EGMA) has been developed for release of cosmetics drugs like arbutin, adenosine, and niacinamide, well understanding molecules for wrinkle remedy and for skin whitening [41]. This hydrogels alternate is permeability responding to the pH changes: At pH 4.0, it holds the pharmaceuticals within the matrix, when in contact with pores and skin, at pH 6 and above, the permeability upward thrust and the medication will be delivered. The authors said that this conduct is due by means of the ionization/deionization of MAA carboxylic groups.

6.3 Plastic Surgery

From their first development into the clinical research subject, hydrogel were seen as good materials for application in contact with the human body due to their extracellular matrix-like (ECM-like) houses [7, 42]. This is the main reason why attempts have been made to introduce hydrogels like new substances for plastic reconstruction. On this path, for plenty years, hyaluronic acid (HA) became thought just like the panacea for every pain [43] is not surprising, considered that HA has been studied to be applied in tissue-filling applications. One brilliant business enterprise operating in the area is Macrolane TM. Starting in 2008, Macrolane's remedies and products were specially studied to enhance breast length and shape and provide a more biocompatible alternative to traditional and aggressive silicone prosthesis. Anyhow, quickly the medical community began to factor out the controversial conduct of these strategies in latter mammographies: briefly, HA labored as a defensive agent, appearing as denser tissue, and consequently ruining the final results of the exam. Thus, in recent times, Macrolane TM is used for diversely located filling except breasts. The compound is injected inside the frame with a syringe and allow it gels restoring the volume. This procedure started to receive criticisms by way of the medical community; however, yet in the interim, there is still no longer a enough amount of courses investigating the residences and possible risks of the gel, concerns commenced to rise specifically approximately the long-term side effects of this treatment and extra mainly approximately HA's function in cancer control and development [44, 45]. Another promising use of hydrogels is bulking agents for remedy of urinary incontinence: smart injectable gels can be involved in clinical approaches in which these substances may be used to tighten the urethral channel and reduce patient's incontinence. With this type of easy solution, it is miles possible to erase or at least reduce a regular social handicap and assist sufferers to keep an ordinary life [46]. An instance of product used in this area is Bulkamid® through Contura International, a polyacrylamide hydrogel is a commercial product advanced to help ladies with incontinence narrowing the conduct. There are few reason for the incontinence, and an exhaustive category is made by way of Lose et al. where they subdivide each class, Bladder/Urethra in continent, in overactivity/underactivity, of their work they wrote that bulk injectable structures are promising to enhance urethral coaptation. Problematics like infections and frame response had been evaluated. In a sample composed by 130 women, researchers have found ten patients with infections, five with injection Web site online pain, two with incontinence, and simplest one with injection site laceration.

6.4 *Environmental Applications*

Over past the years, countries gradually commenced to care approximately environmental troubles and pollutants. Many governments decided to choose greener and more secure for the surroundings policies. Water pollutants is one of the biggest troubles afflicting especially poor areas of Africa, Asia, and South America. Thanks to their affinity for water, hydrogels is probably used in two unique ways to treat water source. First the matrix may be used as a holder for purifying microorganism. Many thrilling studies, on this unique path, had been developed by encapsulating microorganisms inside various carrier materials [47]. Chlorella and Spirulina are the most used ones. These microorganisms are already used to do away with pollution chemical substances from water resources. The idea is to maintain the bacteria inside the network and consequently protect and manage the bacteria-culture while cleaning the Web site online of depuration. Both synthetic and herbal hydrogels had been used. The great operating hydrogels in literature seem like alginate derived [47] or alternatively carrageenan and agar [48]. A second thrilling manner to solve the problem of pollution is to alter the hydrogels to let them capture and maintain the pollutant in the networks. Many authors have attempted this way to seize steel ions: the organization of Irani has dispatched to submit a paper in which they speak a new composite hydrogel for PbI(II) removal. Briefly, they created a polyethylene-g-poly (acrylic acid)-co-starch/OMMT (LLDPE-g-PAA-co-starch/OMMT) hydrogel composite, which use it like an adsorbent pollutant (Pb(II)) remover. They placed the hydrogel in an answer containing lead acetate and then measured the adsorption with an atomic absorption spectrometer (AAS); after that they did a de-absorption phase and repeat for numerous cycles. The electrostatic attractions, ion alternate, and chelation are feasible reason for the metallic adsorption that took place all through the experiments. They stated that the equilibrium adsorption statistics of the hydrogel was regular with Langmuir isotherm and the 430 mg/g adsorption ability was in line with other common place adsorbents [49, 50].

Another feasible manner to gain a thrilling water filtering is explained in a paper by Yan et al., where the group achieved etherification and consequent functionalization of chitosan beads that allows you to achieve carboxy-methylated chitosan with a stronger adsorption of metallic ions. This has been proven to enhance selective adsorption of particular ions like Cu(II), Pu(II), and Mg(II) [50]. It is thrilling properties that can be exploited in dye elimination application and that has been exhibited to be viable by magnetical doping of hydrogel microspheres with interpenetrated network (IPN) structures [51]. Hydrogels could be both used like a probe to locate heavy metal ions like inside the work by means of Wang et al. [52]. Moreover, other utility of polyacrylamide gels is a flood control device referred to as WATER GEL BAG® produced by means of TaiHei Co., Ltd. One of the biggest environmental troubles currently tough to solve is for the positive lack of oil in seas and in different water sources. In the past years, the quantity of oil substance dispersed into the hydrosphere around the sector is risen dramatically. Food processing, hydrocarbons enterprise, and refining manner have improved the danger of pollution. It was observed that effluents receiving wastewaters from enterprise have 40.000 mg/L oil awareness [53]. Different form of depolluting device have been tried like bentonite organically [54, 55], palygorskite [56], a magnesium aluminum phyllosilicate, and activated carbon. In 2010 authors tried to develop a hydrogel to hold water with oil-pollutant molecules [57]. They claim that a very promising hydrogel is chitosan, thanks to an excessive presence of amino corporations and hydrogen which could react with vinyl monomers. Authors explained use of polyacrylamide grafted at the polysaccharides rises the functionality to draw and maintain stable particles which stay in suspension in water, called flocculants. The better crosslinking degree, the lower the retaining functionality. This is because of a smaller distance between two nodes inside the matrix network. Reducing this distance, the swelling degree will decrease. Furthermore the initial concentration of wasted oil within the environment is another essential parameter because of its belongings of affecting the adsorption kinetic. In truth, a higher concentration corresponds to quicker swelling kinetics.

As already talk for the environmental programs section, hydrogels can hold inner their matrix a large wide variety of microorganism for purification of water, for production of biomolecules, or for easy way of life of microorganism by way of themselves. Indeed, agar is famous because of the golden well-known substrate for bacterial way of life in biotechnological applications [58]. Since it is indigestible through a splendid variety of microorganisms, it provides an excellent surroundings for their lifestyle on a solid substrate [59, 60]. Different kinds of agar have been studied, every capability use for likewise one-of-a-kind varieties of bacterial. Between them, brucella agar, columbi agar, schaedler agar, or trypicasesoy agar are the most common. None of those gels has superior outcomes in comparison with the others; instead, all people are suitable for distinct programs in motive of their exclusive professionals and their cons [61, 62].

6.5 Electrophoresis and Proteomic

Gel electrophoresis presently represents one among the most widespread techniques for protein separation. In addition to the most normally employed polyacrylamide crosslinked hydrogels, acrylamide-agarose copolymers had been proposed as promising systems for separation matrices in two-dimensional (2-D) electrophoresis because of the coolest decision of both excessive and occasional molecular mass proteins made viable by means of cautious control and optimization of the hydrogel pore structure. As a count of fact, a thorough information of the nature of the hydrogel pore structure in addition to of the parameters through which it is prompted is critical for the design of hydrogel structures with most suitable sieving properties. What is the proteomics? The time period proteomics changed into first coined in 1995 [63]. Proteomics is to take a look at the proteome, and is the huge-scale observation of proteins, particularly their structures and functions, in a given type of cell or organism, at a given time, underneath described conditions, to make an analogy with genomics, the examining of the genome. The proteome is the complete set of proteins, produced or modified by means of an organism or system. This varies with time and wonderful requirements, or stresses that a cellular or organism undergo. Proteomics is an interdisciplinary domain formed on the idea of the research and development of the Human Genome Project. While proteomics commonly refers to the huge-scale experimental analysis of proteins, it is miles often especially used for protein purification and mass spectrometry. In the field of proteomics, the ability to come across a massive wide variety of proteins in unmarried evaluation represents a key problem to obtain fast and green operation [64]. In this context, the blended use of 2D gel electrophoresis coupled with mass spectrometry has allowed sizeable advances during last few a long time and has emerge as nowadays one of general tactics for proteins separation and identification [65, 66].

Among the substances used for 2D gel electrophoresis, polyacrylamide crosslinked hydrogels have been appreciably investigated inside the literature [67], due to the fact their tunable mesh length porosity appears to be perfect for keeping apart proteins and DNA samples. Typically, acrylamide concentrations higher than 5% are used to shape the separation matrix, with the acrylamide concentration being selected to maximize decision of the range of proteins of interest. Lower acrylamide concentrations are vital when resolution of big high molecular mass (HMM) proteins (>500 kDa) is sought but at the expense of poor mechanical stability of the gel matrix that regularly yields difficulties in handling those media [54]. Other polymeric structures alternative to polyacrylamide gels have additionally been proposed as separation matrices for 2D electrophoresis which include agarose, changed polyacrylamide gels, and acrylamide-agarose copolymers [68–70]. In particular, the blessings of the acrylamide-agarose machine particularly lay inside the possibility of improving the decision of large HMM proteins without compromising the decision of low molecular mass proteins, partly because of the optimal average pore size of these substances. Indeed, the electrophoresis migration technique via the polymeric gel matrix is driven by way of the interactions between the protein fragments and the porous network of the gel, causing the excellent of protein resolution to be pretty structured on special structural parameters function of the gel matrix [71]. Among these, implying gel pore size, pore length distribution, and stiffness of the gel plays essential role. In order to achieve advanced separation performance in 2D electrophoresis applications, it is miles therefore important to recognize the unique nature of the pore shape of the gel and the parameters through which this pore shape may be managed and manipulated [72, 73]. In particular, it is far of first-rate interest to analyze structure-property relationships of these hydrogels in the attempt to optimize their functional overall performance. A wide big variety of hydrogel chemical compositions became studied and their effect on structural and functional houses of the hydrogel become elucidated. By using dynamic rheological checks and creep-healing exams, a correlation became discovered among the rheological response of these hydrogels and their sieving homes. More specifically, the assessment of the crosslinking density by using dynamic assessments and using viscoelastic fashions for figuring out the resistance of non-everlasting crosslinks to move in the network structures shed mild at the pore structure of the hydrogel matrix and helped to clarify its influence at the electrophoretic separation overall performance. The mechanical stability of the crosslinked hydrogels was additionally investigated by way of tensile tests and correlated with the crosslinking density of the gel matrix.

6.6 Hydrogels for Tissue Engineering

Hydrogels are three-dimensional polymer scaffolds used in several packages of tissue engineering. A particularly crucial group of techniques is the so known as in-vivo tissue regeneration. In this case, a patient's very own cells are combined with the polymer and held in-vitro till geared up to be implanted. The hydrogel acts as a natural extra-mobile matrix that eventually promotes cellular proliferation and tissue reboom. The pseudo-extra-cellular matrix, comprised of growth factors, metabolites, and other materials, brings cells collectively and controls tissue shape with the last goal of replacing the natural tissue that become lost or damaged. When components of the complete of sure tissues or organs fail, there are numerous alternatives for treatment, including repair, replacement with a synthetic or natural replacement, or regeneration. The figure underneath indicates how tissue or organ injury, ailment, or failure has evolved to reach the field of tissue engineering. Tissue repair or replacement with an artificial substitute is limited to those situations where surgical methods and implants have finished success. Although implants have been a fairly successful alternative, tissue engineering holds out exceptional promise for regeneration of the failed tissue. The first choice of the diseased or injured organs is extracorporeal treatment, wherein blood is circulated thru polymeric membrane change devices. These gadgets are normally passive alternate structures; however, more recently experimental structures may incorporate entrapped or encapsulated cells from other human or animal sources. Those latter structures are referred to as bio-artificial or bio-hybrid organs. Total replacement of the diseased or malfunctioning organ or tissue with a natural replacement requires transplantation of an acceptable, wholesome replacement, and there is a restricted supply of such organs and tissues. Thus, tissue engineering holds out fantastic promise for regeneration of organs (determine 4). Hydrogels have end up more and more studied as matrices for tissue engineering [66].

Hydrogels designed for use as tissue engineering scaffolds may also comprise pores massive enough to accommodate living cells, or they may be designed to dissolve or degrade away, releasing boom factors and developing pores into which residing cells can also penetrate and proliferate. One huge advantage of hydrogels as tissue engineering matrices vs. extra hydrophobic options along with PLGA is the convenience with which one may additionally covalently incorporate cell membrane receptor peptide ligands, so as to stimulate adhesion, spreading, and boom of cells inside the hydrogel matrix. However, a large downside of hydrogels is their low mechanical strength, posing considerable problems in handling [74]. Sterilization issues are also very challenging. It is clear that there are both good sized advantages and downsides to the use of hydrogels in tissue engineering, and the latter will need to be overcome before hydrogels which becomes sensible and beneficial in this interesting field. It needs to be cited that a gel used as a tissue engineering matrix may additionally never be dried, but the overall water inside the gel is still made from certain and loose water.

6.6.1 Merits of Hydrogels in Tissue Engineering

Aqueous surroundings can shield cells and fragile drugs (peptides, proteins, oligonucleotides, DNA)

- Good shipping of nutrient to cells and merchandise from cells
- May be without problems modified with cell adhesion ligands
- Can be injected in-vivo as a liquid that gels at body temperature
- Usually biocompatible.

6.6.2 Demerits of Hydrogels in Tissue Engineering

- Can be hard to handle

- Usually robotically weak
- May be difficult to load capsules and cells after which crosslink in-vitro as a prefabricated matrix
- May be difficult to sterilize.

In-vivo, the formation of organs and tissues is based on the coordination in time and area of cell differentiation, polarity, shape, division, and death. This coordination is predicated at the mobile integration of indicators from microenvironment, mainly such as the extracellular matrix (ECM) and intercellular communication [75]. The transduction of a typical mixture of those elements coupled to specific cytoplasmic additives can result in the three modern steps in differentiation. First, stem cells are specified toward a certain fate, then they shift from a specified country to a determined state, wherein the mobile fate cannot be reversed and subsequently reach their differentiated kingdom. The main venture of tissue engineering is to reconstitute in-vitro an environment that induces the differentiation of cells and their organization in an ordered purposeful tissue. The cell substrate is of specific importance, due to the fact in-vivo the extracellular area is occupied by means of the ECM. The chemical composition of the ECM and the ensuing mechanical residences are both essential aspects, because the transduction of both chemical and bodily alerts via cellular adhesion molecules affect mobile form, polarization, migration, and differentiation [76]. Furthermore, the ECM topography orients tissue polarity and the morphogenesis of new organs. Molecular hydrogels hold massive potential for cells subculture and tissue engineering; peptide-based totally molecular hydrogels, specifically those of long peptides, could provide appropriate environments for cellular growth, division, and differentiation. Zhang group and the Stupp organization have confirmed that peptide-based totally hydrogels may want to manual the differentiation of stem cells [77]. The project in this area became a way to split cells from gels post-subculture such as using quick-peptide-based totally molecular hydrogels formed by using biocompatible techniques for 3D cell subculture, stem cellular managed differentiation and cellular delivery [78]. Responsive molecular hydrogels for the recuperation of cells post-way of life were additionally studied [79]. Collagen is the maximum ample protein mammals, making up about 30% of the general frame protein content. In order to imitate collagen nanofibers, a serial of quick peptides bearing collagen repeating tripeptide of Gly-Xaa-4-Hyp (GXO, X was Lys (K), Glu (E), Ser (S), Ala (A), or Pro (P)) turned into synthesized [80]. There are two most important bottlenecks right now in tissue engineering. The first one is correlated with the opportunity of obtaining, in-vitro, advanced vascular tissue [81, 82]. In second instance, once a pseudo-physiological angiogenesis is achieved, the ambitious aim is to create engineered complete organs [83]. First attempts on this path in which made through printing 3D organ-like systems with ECM and cells, or through decellularization and seeding. Under these circumstances, hydrogels are normally studied as transient substitutes of the extracellular matrix (ECM) due to comparable physico-chemical properties, which includes stiffness and hydrophilicity [43]. Hydrogels used as scaffold in TE ought to be biocompatible and elicit the smallest response by means of the body. This is a very complex problem, indeed, for the subculture of eukaryotic cells implies numerous precautions. Cells morphology, metabolism, and universal phenotype are without delay correlated with the indicators they receive, to start with from the physical and chemicals homes of substrate. The micro- and nanostructure, the porosity, and the stiffness of the surfaces are all crucial signals for the cells (topography and mechanical residences) [83]. A very crucial phenomenon, for instance, is the touch guidance, introduced by using Weiss in 1934 [84], for which cells aligns to the shape and the microstructure of a given surface. Especially for some kind of cell culture (e.g., myocytes), this phenomenon is essential for the creation of latest working tissue [85, 86]. Hydrogels for TE must be advanced trying to fit this necessity. For the purpose, different substances have been investigated. Actually, studies are focused specifically on degradable scaffold to allow mobile migration while the matrix degenerates. Researches verified the significance of engineered degradation on non-degradable scaffolds by way of viability test [87, 88]. These hydrogels can be divided through the starting place of the polymer that they are made of: either if it is far herbal or synthetic. In the prevailing article, the eye will be targeted on natural derived hydrogels.

1. **Dextran**: Dextran is herbal polysaccharide received from the digestion of amylopectin. Low molecular weight dextran is been used like a plasma expander, thanks to the distinctly inert and nontoxic conduct of this polysaccharide [89]. The maximum interesting function of dextran are its protein rejection properties, the so called non-fouling [74], coupled with first-rate biocompatibility due to its glycocalyx mimic conduct [90]. This property is beneficial to create an ECM-like hydrogels for tissue engineering, and for instance, Yunxiao and collaborators created a copolymer between methacrylate-aldehyde-bifunctionalized dextran (DEXMA-AD) and gelatin B. This fabric was obtained via ultraviolet (UV)-crosslinking among a methacrylate organizations on Dex-MA-AD and the aldehyde corporations, allowing the inclusion of gelatin inside the matrix that granting enzymatically degradation and cellular adhesive homes. Researchers reveal that this sort of hydrogels ought to promote adhesion of vascular endothelial cells [91, 92].
2. **Gelatin**: Gelatin is the denatured form of collagen, one of the major component of ECM. Collagen but contains immunogenicity problems because of the presence of antigens from the unique tissue. Gelatin is a protein materials with a long α helix with a high content in glycine (≈25%) [93]. Gelatin exist in two specific shape, processed in acid solution (kind A, commonly porcine), or in alkali solution (kind B, generally bovine) [94]. In its natural shape, it is quickly dissolved in water, but it can be crosslinked to achieve a hydrogel with higher mechanical residences and degradation rate. The crosslinking can be achieved in many exceptional ways, but from the physical side, we ought to use UV reticulation or the chain polarity. From the chemical one, it is very commonplace to polymerize, with enzymes too, the chains with a sure on the side agencies of amino acids. Lysine and glutamic acid are the most employed for it. Thanks to his similarity to the herbal ECM, many compounds were been evolved with

gelatin-coating or covered element. Chitosan-gelatin, fibroin-gelatin, alginate-gelatin, and dextran-gelatin are very commonplace. Das and collaborators tried to create a fibroin-gelatin biomaterial for bioprinting cells laden in a 3D tissue constructs. They evolved two kinds of hydrogels: one crosslinked by sonication and the alternative crosslinked using tyrosinase enzyme. The results showed that sonication-gelatin fibroin hydrogel proven better osteogenic differentiation at the same time as tyrosinased-gelatin-fibroin supported better chondrogenic and adipogenic differentiation [95].

3. **Chitosan**: Chitosan is a polysaccharides from chitin of the crustacean skeleton. It is composed with the aid of the repetition of N-glucosamine units. An essential index to assess chitosan's properties is the degree of acetylation, described as the quantity of amine in the cloth. For instance, it is far proved that chitosan ought to decrease the adsorption of protein and the binding of bacteria [96]. Some studies file a primary period in which chitosan repelled cell adhesion, and a second wherein cells starts to bind to it. There is a non-uniformity in consequences about this refractory-period, anyhow it can be exploited to seed unique form of cells. For example, Tao Jang et al. posted a look at wherein they used photopolymerization on chitosan with a view to deposit cells between the chitosan patterns. They endorse that after the refractory-period, while cells are disposing in no chitosan lined regions, it is miles viable to seeds every other sort of cells for the improvement of a more complex system [97]. Take a look at every other chitosan that has been coupled with gelatin to create an in-situ gel for cell seeding and/or drug delivery. In particular, the group evaluated the distinction in crosslinking with the aid of two distinctive enzyme [98].
4. **Hyaluronic acid**: Hyaluronic acid—HA—is a glycosaminoglycan GAG enclosed inside the herbal ECM, and middle of the fabric is a polysaccharide with high affinity for water. Usually, to growth the mechanical houses of this biomaterial, a covalent cross-linking between chains is done. Researchers have mentioned that a too high degree of modification and crosslinking ought to influence the biocompatibility property of the fabric [99]. In-vivo HA can have unique molecular weights. Low and excessive molecular weight HA cause an opposite cells behavior [100]. HA macromolecules shown an anti-inflammatory, immunosuppressive properties, and blocks angiogenesis, even as cleaved small fragments set off the contrary conduct, permitting endothelial cells migration and angiogenesis [101, 102]. Indeed, low molecular weight HA had been correlated with a few cancers, like prostatic one [100, 103]. HA hydrogels were received in an effort to exploit the angiogenic power of the molecule in the course of the degradation of a cloth. Kisiel's organization, for example, crosslinked HA with protease-degradable peptides and brought mobile adhesion ligands to improve the cell spreading on the material [104]. On every other hand, Shu et Al. attempted to mimic the ECM copolymerizing HA and gelatin. They introduced a thiol institution to the Hyaluronan, and in this way, they were able to cross-link modified Hyaluronan with modified gelatin through disulfide bonds [105].
5. **Pectin**: Another polysaccharides utilized in tissue engineering hydrogels is pectin. It is received from cells walls after a low pH, high temperature processing.

Unfortunately, till now, researchers have not reached the intention to standardize this product in an economically sustainable way [106]. Based on its esterification degree, pectin is classed from low methoxyl to excessive methoxyl, tuning this property adjustment the mechanical conduct of the cloth. Reticulation can occur by means of lowering pH to obtain bodily gel, or the use of divalent or trivalent ions to obtain water-insoluble gel [107, 108]. In the tissue engineering field, pectin seems very interesting because of its promotion of nucleation of mineral phase while immersed in a selected organic solution [106, 109].

6. **Alginate**: Derived from brown algae, alginate is a polysaccharide composed of beta-D-mannuronic acid and alfa-L-gluronic acid. Its reticulation can also occurs by divalent cations (Ca^{2+}, Fe^{2+}, Ba^{2+}) [110]. In tissue engineering, alginate can be used as an immunoisolation barrier [92]. Alginate scaffold for the regeneration of annulus fibrosus are also been advanced. Alginate scaffold for the regeneration of annulus fibrosus also are been developed. This hydrogels have formed reminiscence capability, are cytocompatible, and help proliferation and metabolic activity [111]. Moreover, alginate hydrogels were used to reduce liver cells death [112] and with silk fibroin as a substrate for stem cells culture [113].

6.7 Cardiac Applications

In the last decade, advancements were made toward growing injectable hydrogels [114] for the purpose of cardiac repair. Hydrogel injections by myself have been shown to attenuate the decline in cardiac feature and left ventricular remodeling commonly seen after myocardial infarction in both large and small animal models. Furthermore, hydrogels have additionally been shown to improve cell retention while co-injected for mobile cardiomyoplasty and to lengthen launch of therapeutics while used as a transport vehicle.

6.8 Dental Applications

Pulp regeneration therapy is essential to overcome the constraints of conventional remedy to result in reparative dentinogenesis. Presently, dentists have not any choice but to eliminate the complete dental pulp with an endodontic method while a dentin disorder with pulp exposure reaches an important size ensuing in an irreversible pulp condition. To triumph over this limitation, it is far considered critical to develop pulp regeneration therapy as well as clarify the mechanisms of pulp wound healing. Pulp wound recovery and regeneration have not unusual processes, and results of some of research have indicated that pulp wound recovery includes initial inductions of apoptosis of broken pulp cells [110], followed by way of reactionary dentinogenesis by surviving odontoblasts and reparative dentinogenesis through odontoblast-like cells [115, 116]. Reactionary or reparative dentin is formed toward the residual dental pulp, however, not in the area wherein the dentinpulp complicated has been lost. To achieve the regeneration of the dentinpulp complicated, induction of suitable pulp wound healing and formation of recent dentin in dentin defects are essential, and a few research have reported crucial pulp treatments to shape new dentin in defects [117]. Fibroblast growth factor-2 (FGF-2), which is normally stored inside the extracellular matrix and launched through enzymatic degradation of extracellular matrix molecules, performs a function in physiologic conditions such as enamel and dentin formation of the tooth germ [118], in addition to pathologic situations [119]. It was previously proven that a sluggish and persistent launch of biologically energetic FGF-2 was performed through in-vivo biodegradation of gelatin hydrogels that included FGF-2 [120]. Furthermore, a controlled launch of FGF-2 from gelatin hydrogels triggered neovascularization and regeneration of numerous tissues, along with bone [121], periodontal tissues [122], and others [123].

7 Conclusions

The role of green hydrogels is very important in order to increase the output of corps. Hydrogels triggers the neovascularization of tissues along with bone to regenerate the new tissues. FGF-2 helps to reduce the dentin formation of the tooth germ. Chitosan decreases the adsorbtion of protein and the binding of bacteria. It is concluded that the field of green hydrogels is very fast and lot of scope to do research to contribute to the world.

References

1. Coinlogitic. Hydrogel Consumption Market Analysis by Current Industry Status and Growth Opportunities. 2018. Available https://coinlogitic.com/hydrogel-consumption-market-research-report/51472/. Last accessed on 11 Oct 2018
2. Hoare TR, Kohane DS (2008) Hydrogels in drug delivery: progress and challenges. Polymer 49:1993–2007
3. Caló E, Khutoryanskiy VV (2015) Biomedical applications of hydrogels: a review of patents and commercial products. Eur Polym J 65:252–267
4. Van Noorden R (2014) Global scientific output doubles every nine years. Nature news blog
5. Chirani NA, Dembahri Z, Tokarski C, Rolando C (2011) Newely designed polyacrylamide gels for electrophoresis protein separation: synthesis and characterization. Polym Int 60:1024–1029
6. Campccia D, Doherty P, Radice M et al (1998) Semisunthetic resorbable materials from hyaluronan esterfication. Biomaterials 19:2101–2127
7. Seow WY, Hauser CAE (2014) Short to ultra short dehydrogels for biomedical uses. Mater Today 17:381–388
8. Lee SC, Kown IK, Park K (2013) Hydrogels for delivery of bioactive agents: a historical persepective. Adv Drug Delivery 65:17–20
9. Kopecek J (2007) Hydrogel biomaterials: a smart future? Biomaterials 28:5185–5192
10. Buwlada SJ, Boere KW et al (2014), Hydrogels in a historical perspective: From simple networks to smart materials. J Control Release 190:254–273

11. Yom-Tov O, Neufeld L, Slektar D (2014) A novel design of injectable porous hydrogels with in situ pore formation. Acta Biomater 10:4236–4246
12. Abebe DG, Fuziwara T (2012) Controlled thermo responsive hydrogels by stereo complexed PLA-PEG-PLA prepared via hybrid micelles of premixed co polymers with different PEG lengths. Bio Macro Mol 13:1828–1836
13. Chung HJ, Lee Y, Park TG (2008) Thermo-sensitive and biodegradable hydrogels primarily based on stereocomplexed Pluronic multi-block copolymers for managed protein transport. J Control Release 127:22–30
14. Kirakci K, Šícha V, Holub J, Kubát P, LangK (2014) Luminescent hydrogel particles organized by way of self-meeting of β-cyclodextrin polymer and octahedral molybdenum cluster complexes. Inorg Chem 53:13012–13018
15. Kono H, Teshirogi T (2015) Cyclodextrin-grafted chitosan hydrogels for managed drug delivery. Int J Biol Macromol 72:299–308
16. Peppas NA (1987) Hydrogels in medicine and pharmacy, vol 21. CRC Press, p 184
17. Park K, Shalaby WSW, Park H (1993). Biodegradable hydrogels for drug delivery. Technomic
18. Harland RS, Prud'homme RK (1992) Polyelectrolyte gels: properties, preparation and application. American Chemical Society
19. Lim F, Sun AM (1980) Microencapsulated islets as bioartificial endocrine pancreas. Science 210:908–910
20. Sefton MV, May MH, Lahooti S, Babensee JE (2000) Making microencapsulation work: conformal coating, immobilization gels and in vivo overall performance. J Control Release 65:173–186. Griffith LG (2000) Polymeric biomaterials. Acta Mater 48:263–277
21. Okay O (2010) General properties of hydrogels. In: Gerlach G, Arndt KF (eds) Hydrogel sensors and actuators, engineering and technology. Springer Series on Chemical Sensors and Biosensors. Springer, Berlin
22. Byju AG, Kulkarni A, Gundiah N (2013) Mechanics of Gelatin and Elastin based hydrogels as Tissue Engineered Constructs. In: International conference on fracture
23. Mondal M, Trivedy K, Nirmal Kumar S (2007) The silk proteins, serin and fibroin in silkworm Bombyxmori. Caspian J Env Sci 5:63–76
24. Menard KP (2008) Dynamic mechanical analysis: a practical introduction, 2nd edn. CRC Press
25. Morrison FA, Understanding rheology. Oxford University Press
26. ASTM International (2004) F2540 Standard practice standard guide for assessing microstructure of polymeric scaffolds for use in tissue engineered medical products
27. Assarsson PG, King PA, Yen SN (1975) US Patent 3901236 A. Disposable absorbent articles containing hydrogel composites having improved fluid absorption efficiencies and processes for preparation
28. Gross JR, Mcfadden RT (1975) US Patent 3926891 A Method for making a crosslinkable aqueous solution which is useful to form soft, water-swellablepolyacrylate articles
29. Statista, The Statistics Portal. https://www.statista.com. Accessed Feb 2015
30. Umachitra G, Bhaarathidhurai (2012) Disposable baby diaper—a threat to the health and environment. J Environ Sci Eng 54:447–452
31. Chalker-Scott L (2015) The myth of polyacrylamide hydrogels
32. Trinh T, Gardlik JM (1990) European Patent EP 0392608 A2. Solid consumer product compositions containing small particle cyclodextrin complexes
33. Subkowski T, Bollschweiler C, Wittenberg J, Siegel W, Pelzer R (2013) International Patent WO 1999004830 A1. Low molecular weight modulators of the cold-menthol receptor trpm8 and use thereof
34. Peppas NA, Peppas LB (1997) Controlled release of fragrances from polymers. J Appl Polym Sci 66:509–513
35. NFPA Guidelines & Definitions. Oklahoma State University (2008). Accessed Feb 2015
36. Mondal S, Das S, Nandi AK (2020) A review on recent advances in polymer and peptide hydrogels. Soft Matter 16:1404–1454

37. An SM, Ham H, Choi EJ, Shin MK et al (2014) Primary irritation index and safety zone of cosmetics: retrospective analysis of skin patch tests in 7440 Korean women during 12 years. Int J Cosmet Sci 36:62–67
38. https://phoenix-chem.com/ESW/Files/Pecogel_Brochure.pdf
39. Lorenz DH (1994) US patent 5306504 A Skin adhesive hydrogel, its preparation and uses
40. Lee E, Kim B (2011) Smart delivery system for cosmetic ingredients using pH-sensitive polymer hydrogel particles. Korean J Chem Eng 28:1347–1350
41. Slaughter BV, Khurshid SS, Fisher OZ, Khademhosseini A, Peppas NA (2009) Hydrogels in regenerative medicine. Adv Mater 21:3307–3329
42. Rhee SM, You HJ, Han SK (2014) Injectable tissue-engineered soft tissue for tissue augmentation. J Korean Med Sci 29(Suppl 3):S170-175
43. Pienaar WE, McWilliams S, Wilding LJ, Perera IT (2011) The imaging features of MACROLANE™ in breast augmentation. Clin Radiol 66:977–983
44. Yamaguchi S, Nagumo Y, Niwa K (2013) Efficacy and safety of Macrolane™ for breast enhancement: a 12-month follow-up study in Asian women. J Plast Surg Hand Surg 47:191–195
45. Joint United Nations Programme on HIV/AIDS. UNAIDS report on the global AIDS epidemic; https://www.unaids.org/globalreport,2010
46. Wen Y, Collier JH (2015) Supramolecular peptide vaccines: tuning adaptive immunity. Curr Opin Immunol 35:73–79
47. Joint I, Mühling M, Querellou J (2010) Culturing marine bacteria—an essential prerequisite for biodiscovery. Microb Biotechnol 3:564–575
48. Irani M, Ismail H, Ahmad Z, Fan M (2015) Synthesis of linear low-density polyethylene-g-poly (acrylic acid)-co-starch/organo-montmorillonite hydrogel composite as an adsorbent for removal of Pb(II) from aqueous solutions. J Environ Sci 27:9–20
49. Yan H, Dai J, Yang Z, Yang H, Cheng R (2011) Enhanced and selective adsorption of copper(II) ions on surface carboxymethylated chitosan hydrogel beads. ChemEng J 174:586–594
50. Ahmad H, Nurunnabi M, Rahman MM, Kumar K et al (2014) Magnetically doped multi stimuli-responsive hydrogel microspheres with IPN structure and application in dye removal. Colloids Surf Physicochem Eng Asp. 459:39–47
51. Wang X, Ye G, Wang X (2014) Hydrogel diffraction gratings functionalized with crown ether for heavy metal ion detection. Sens Actuators B Chem 193:413–419
52. Arcadio PS, Gregoria AS (2003) Physical–chemical treatment of water and wastewater. IWA Publishing, CRC Press
53. Panpanit S, Visvanathan C (2001) The role of bentonite in UF flux enhancement mechanisms for oil/water emulsion. J Membr Sci 184:59–68
54. Hadi M, Viraraghavan T (1999) Removal of oil from water by bentonite organo-clay. In: Hazardous & industrial wastes—proceedings of the mid-Atlantic industrial & hazardous Waste, pp 187–196
55. Qiu Z, Zhang Y, Fang Y (1995) Removal of oil from concentrated wastewater by attapulgite and coagulant. Water Qual Res J 30:89–99
56. Sokker HH, El-Sawy NM, Hassan MA, El-Anadouli BE (2011) Adsorption of crude oil from aqueous solution by hydrogel of chitosan based polyacrylamide prepared by radiation induced graft polymerization. J Hazard Mater 190:359–365
57. Heginbothom M, Fitzgerald TC, Wade WG (1990) Comparison of solid media for cultivation of anaerobes. J Clin Pathol 43:253–256
58. Smith A (2012) History of the agar plate. Laboratory News
59. Cover TL (2012) Perspectives on methodology for in vitro culture of *Helicobacter pylori*. Methods Mol Biol 921:11–15
60. Morales A, Jalel L, Patricia G (2020) Assessment of green approaches for the synthesis of physically crosslinked lignin hydrogels. J Ind Eng Chem 81:475–487
61. Murray PR (1978) Growth of clinical isolates of anaerobic bacteria on agar media: effects of media composition, storage conditions, and reduction under anaerobic conditions. J Clin Microbiol 8:708–714

62. Rabilloud T, Vaezzadeh AR, Potier N, Lelong C, Leize-Wagner E et al (2009) Power and limitations of electrophoretic separations in proteomics strategies. Mass Spectrom Rev 28:816–843
63. Rabilloud T, Chevallet M, Luche S, Lelong C (2010) Two-dimensional gel electrophoresis in proteomics: past, present and future. J Proteomics 73:2064–2077
64. Rogowska-Wrzesinska A, Le Bihan MC, Thaysen-Andersen M, Roepstorff P (2013) 2D gels still have a niche in proteomics. J Proteomics 88:4–13
65. Chirani NA, Dembahri Z, Tokarski C, Rolando C, Benmouna M (2011) Newly designed polyacrylamide/dextran gels for electrophoresis protein separation: synthesis and characterization. Polym Int 60:1024–1029
66. Suh KS, Mutoh M, Gerdes M, Yuspa SH (2005) CLIC4, an intracellular chloride channel protein, is a novel molecular target for cancer therapy. J Investig Dermatol Symp Proc 10:105–109
67. Greaser ML, Warren CM (2012) Protein electrophoresis in agarose gels for separating high molecular weight proteins. Methods Mol Biol 869:111–118
68. Greppi GF, Roncada P (2005) La componente proteica del latte caprino. In: Pulina G (ed) L'alimentazione della capra da latte. Avenue Media Publisher, pp 71–99
69. Wang J, Ugaz VM (2006) Using in situ rheology to characterize the microstructure in photopolymerized polyacrylamide gels for DNA electrophoresis. Electrophoresis 27:3349–3358
70. Righetti PG, Gelfi C (1997) Electrophoresis gel media: the state of the art. J Chromatogr B Biomed Sci Appl 699:63–75
71. Anseth KS, Bowman CN, Brannon-Peppas L (1996) Mechanical properties of hydrogels and their experimental determination. Biomaterials 17:1647–1657
72. Myung D, Waters D, Wiseman M, Duhamel PE, Noolandi J et al (2008) Progress in the development of interpenetrating polymer network hydrogels. Polym Adv Technol 19:647–657
73. Kim SJ, Yoon SG, Kim SI (2004) Synthesis and characterization of interpenetration polymer network hydrogels composed of alginate and poly(diallyldimethylammonium chloride). J Appl Polym Sci 91:3705–3709
74. Zhou M, Smith AM, Das AK, Hodson NW, Collins RF et al (2009) Self-assembled peptide-based hydrogels as scaffolds for anchorage-dependent cells. Biomaterials 30:2523–2530
75. Yang C, Li D, Liu Z, Hong G, Zhang J et al (2012) Responsive small molecular hydrogels based on adamantane-peptides for cell culture. J Phys Chem B 116:633–638
76. Wang H, Yang Z (2012) Short-peptide-based molecular hydrogels: novel gelation strategies and applications for tissue engineering and drug delivery. Nanoscale 4:5259–5267
77. Kaivosoja E, Barreto G, Levón K, Virtanen S, Ainola M et al (2012) Chemical and physical properties of regenerative medicine materials controlling stem cell fate. Ann Med 44:635–650
78. Khademhosseini A, Langer R (2007) Microengineered hydrogels for tissue engineering. Biomaterials 28:5087–5092
79. Mironov V, Boland T, Trusk T, Forgacs G, Markwald RR (2003) Organ printing: computer-aided jet-based 3D tissue engineering. Trends Biotechnol 21:157–161
80. Tanzi MC, Bianchi A, Farè S, Mantero S, Raimondi MT et al (2013) Gruppo Nazionale di Bioingegneria n°32: Approccio Integrato per la Medicina Rigenerativa. PàtronEditore
81. Weiss P (1934) In vitro experiments on the factors determining the course of the outgrowing nerve fiber. J Exp Zool 69:393–448
82. Clark P, Connolly P, Curtis AS, Dow JA, Wilkinson CD (1991) Cell guidance by ultrafine topography in vitro. J Cell Sci 99:73–77
83. Saito AC, Matsui TS, Ohishi T, Sato M, Deguchi S (2014) Contact guidance of smooth muscle cells is associated with tension- mediated adhesion maturation. Exp Cell Res 327:1–11
84. Elisseeff J, Anseth K, Sims D, McIntosh W, Randolph M et al (1999) Transdermal photopolymerization for minimally invasive implantation. Proc Natl Acad Sci U S A 96:3104–3107
85. Elisseeff J, McIntosh W, Anseth K, Riley S, Ragan P et al (2000) Photoencapsulation of chondrocytes in poly(ethylene oxide)-based semi-interpenetrating networks. J Biomed Mater Res 51:164–171

86. Massia SP, Stark J, Letbetter DS (2000) Surface-immobilized dextran limits cell adhesion and spreading. Biomaterials 21:2253–2261
87. Holland NB, Qiu Y, Ruegsegger M, Marchant RE (1998) Biomimetic engineering of non-adhesive glycocalyx-like surfaces using oligosaccharide surfactant polymers. Nature 392:799–801
88. Liu Y, Chan-Park MB (2010) A biomimetic hydrogel based on methacrylated dextran-graft-lysine and gelatin for 3D smooth muscle cell culture. Biomaterials 31:1158–1170
89. Ward AG, Courts A (1997) The science and technology of gelatin, vol 66. Academic Press, pp 373–374
90. Liu HJ, Hu YH, Wang HM, Wang JY, Kong DL et al (2011) Soft Matter 7:5430–5436
91. Type A & B Process Definition (2009) Vyse Gelatin Company. https://www.vyse.com/. Accessed Feb 2015
92. Das S, Pati F, Choi YJ, Rijal G, Shim JH et al (2015) Bioprintable, cell-laden silk fibroin-gelatin hydrogel supporting multilineage differentiation of stem cells for fabrication of three-dimensional tissue constructs. Acta Biomater 11:233–246
93. Ignatova M, Manolova N, Rashkov I (2013) Electrospun antibacterial chitosan-based fibers. Macromol Biosci 13:860–872
94. Jiang T, Deng M, James R, Nair LS, Laurencin CT (2014) Micro- and nanofabrication of chitosan structures for regenerative engineering. Acta Biomater 10:1632–1645
95. Chen T, Embree HD, Brown EM, Taylor MM, Payne GF (2003) Enzyme-catalyzed gel formation of gelatin and chitosan: potential for in situ applications. Biomaterials 24:2831–2841
96. Tezel A, Fredrickson GH (2008) The science of hyaluronic acid dermal fillers. J Cosmet Laser Ther 10:35–42
97. Lam J, Truong NF, Segura T (2014) Design of cell-matrix interactions in hyaluronic acid hydrogel scaffolds. Acta Biomater 10:1571–1580
98. West DC, Kumar S (1989) Hyaluronan and angiogenesis. Ciba Found Symp 143:187–201
99. Tempel C, Gilead A, Neeman M (2000) Hyaluronic acid as an anti-angiogenic shield in the preovulatory rat follicle. Biol Reprod 63:134–140
100. Slevin M, West D, Kumar P, Rooney P, Kumar S (2004) Hyaluronan
101. Angiogenesis and malignant disease. Int J Cancer 109:793–794
102. Kisiel M, Martino MM, Ventura M, Hubbell JA, Hilborn J et al (2013) Improving the osteogenic potential of BMP-2 with hyaluronic acid hydrogel modified with integrin-specific fibronectin fragment. Biomaterials 34:704–712
103. Shu XZ, Liu Y, Palumbo F, Prestwich GD (2003) Disulfide-crosslinked hyaluronan-gelatin hydrogel films: a covalent mimic of the extracellular matrix for in vitro cell growth. Biomaterials 24:3825–3834
104. Munarin F, Tanzi MC, Petrini P (2012) Advances in biomedical applications of pectin gels. Int J Biol Macromol 51:681–689
105. McKenna BA, Nicholson TM, Wehr JB, Menzies NW (2010) Effects of Ca, Cu, Al and La on pectin gel strength: implications for plant cell walls. Carbohydr Res 345:1174–1179
106. Munarin F, Guerreiro SG, Grellier MA, Tanzi MC, Barbosa MA et al (2011) Pectin-based injectable biomaterials for bone tissue engineering. Biomacromol 12:568–577
107. Brandt KA, Goldman SA, Inglin TA (1988) US Patent RE32649 E. Hydrogel-forming polymer compositions for use in absorbent structures
108. Jang J, Seol YJ, Kim HJ, Kundu J, Kim SW et al (2014) Effects of alginate hydrogel cross-linking density on mechanical and biological behaviors for tissue engineering. J Mech Behav Biomed Mater 37:69–77
109. Guillaume O, Daly A, Lennon K, Gansau J, Buckley SF et al (2014) Shape-memory porous alginate scaffolds for regeneration of the annulus fibrosus: effect of TGF-β^3 supplementation and oxygen culture conditions. Acta Biomater 10:1985–1995
110. Shteyer E, Ben Ya'acov A, Zolotaryova L, Sinai A, Lichtenstein Y et al (2014) Reduced liver cell death using an alginate scaffold bandage: a novel approach for liver reconstruction after extended partial hepatectomy. Acta Biomater. 10:3209–3216

111. Ziv K, Nuhn H, Ben-Haim Y, Sasportas LS, Kempen PJ et al (2014) A tunable silk-alginate hydrogel scaffold for stem cell culture and transplantation. Biomaterials 35:3736–3743
112. Burdick JA, Murphy WL (2012) Moving from static to dynamic complexity in hydrogel design. Nat Commun 3:1269
113. Douezan S, Dumond J, Brochard-Wyart F (2012) Wetting trasitions of cellular aggregates induced by substrate rigidity. Soft Matter 8:4578–4583
114. Tse JR, Engler AJ (2010) Preparation of hydrogel substrates with tunable mechanical properties. Curr Protoc Cell Biol Chapter 10: Unit 10
115. Kawagishi E, Nakakura-Ohshima K, Nomura S, Ohshima H (2006) Pulpal responses to cavity preparation in aged rat molars. Cell Tissue Res 326:111–122
116. Nakashima M, Iohara K, Zheng L (2006) Gene therapy for dentin regeneration with bone morphogenetic proteins. Curr Gene Ther 6:551–560
117. Madan AK, Kramer B (2005) Immunolocalization of fibroblast growth factor-2 (FGF-2) in the developing root and supporting structures of the murine tooth. J Mol Histol 36:171–178
118. Annabi B, Naud E, Lee YT, Eliopoulos N, Galipeau J (2004) Vascular progenitors derived from murine bone marrow stromal cells are regulated by fibroblast growth factor and are avidly recruited by vascularizing tumors. J Cell Biochem 91:1146–1158
119. Yamamoto M, Ikada Y, Tabata Y (2001) Controlled release of growth factors based on biodegradation of gelatin hydrogel. J Biomater Sci Polym Ed 12:77–88
120. Nakayama J, Fujioka H, Nagura I, Kokubu T, Makino T et al (2009) The effect of fibroblast growth factor-2 on autologous osteochondral transplantation. Int Orthop 33:275–280
121. Nakahara T, Nakamura T, Kobayashi E, Inoue M, Shigeno K (2003) Novel approach to regeneration of periodontal tissues based on in-situ tissue engineering: effects of controlled release of basic fibroblast growth factor from a sandwich membrane. Tissue Eng 9:153–162
122. Laçin NT (2014) Development of biodegradable antibacterial cellulose based hydrogel membranes for wound healing. Int J Biol Macromol 67:22–27
123. Lee YH, Chang JJ, Yanga MC, Chienc CT, Lai WF (2012) Acceleration of wound healing in diabetic rats by layered hydrogel dressing. Carbohyd Polym 88:809–819
124. Jiang LY, Luo Y (2013) Guided assembly of endothelial cells on hydrogel matrices patterned with microgrooves: a basic model for micro vessel engineering. Soft Matter 9:1113–1121
125. Zhu J (2010) Bioactive modification of poly(ethylene glycol) hydrogels for tissue engineering. Biomaterials 31:4639–4656

Chapter 9
Green Composites from Renewable Sources

Aravind Kumar, T. Krithiga, D. Venkatesan, D. Joshua Amarnath, and S. Sathish

1 Introduction

Green composites can be considered as the sustainable materials envisaged for the next generation gaining rapid interest among various industries and also academic institutions. This detailed chapter is just an overview of the snapshot to provide a complete insight into the recent progress in fabrication of green composites, testing, characterization and at last few applications. Two or more materials are gathered through physical/chemical means with an interface available betwixt such materials to develop a new substance known as composite. The distinct materials can be represented as matrix form and dispersed form of phase substance which may be referred to reinforcement [1].

Such substance may act as load bearing component, whereas the adjacent matrix form completely bonds the attached reinforcement together and divides the load evenly among them. In spite of different properties possessed by matrix as well reinforcement, but on combining together yields a superior material with excellent properties when compared to their parent materials. Composites are graded as various types. (i) Metal matrix type, (ii) ceramics type and (iii) polymer-based type of matrix. Out of these, polymer matrix is of growing concern nowadays owing to its maximum stiffness, less weight, fabrication easiness and excessive strength. In case of polymer composites, the matrix occluded in composite is a polymeric material and usually it is based on aliphatic/aromatic hydrocarbons. The material of reinforcement can be synthetic type fibres generally carbon or else glass fibres. Thus, the polymers

A. Kumar (✉) · D. Venkatesan · D. Joshua Amarnath · S. Sathish
Department of Chemical Engineering, Sathyabama Institute of Science and Technology, Chennai, India
e-mail: aravind.chem@sathyabama.ac.in

T. Krithiga
Department of Chemistry, Sathyabama Institute of Science and Technology, Chennai, India

S. Thomas and P. Balakrishnan (eds.), *Green Composites*,
Materials Horizons: From Nature to Nanomaterials,
https://doi.org/10.1007/978-981-15-9643-8_9

are substituting the metal matrix materials in various engineering fields such as automobile, sports, aerospace equipment and so on [2].

Synthetic composites are primarily based on petroleum products. Owing to growing trend of petroleum resources as well as its minimum reserves and also to deduce the carbon footmark, research scientists are carefully identifying the alternatives to substitute the petroleum-derived composites [3]. Few major issues accounted with synthetic polymeric units are

1. Surplus quantity of plastic waste generation and disposal problems owing to its biodegradable nature.
2. Underdeveloped recycling techniques for degrading polymer composites.
3. More heat/pressure requirement involved during the synthesis of composites along with synthetic fibres.
4. Deficit for petroleum resources.
5. Rising demand for petroleum-based products [4].

Due to represented factors above, nowadays, research people aim at polymers derived from sustainable and eco-friendly base materials being more surplus in availability. This keeps a concern on natural fibres and fibre matrices allocated for synthesis of polymeric composite units. Many research studies forecasted that the involved energy in natural fibre development technique is definitely greater than fifty per cent of consumed energy that could be attributed to generation of synthetic fibres. In spite of higher energy consumption, natural fibres possess certain merits which include lighter weight, non-irritating, non-toxic, non-abrasive and extremely biodegradable [5].

The definite selection of natural fibres for the purpose of reinforcement material can be probably recalled back to nearly 12,000 years. Plant fibres were involved as reinforcement material that can be viewed in brick construction materials during the Egyptian era. Henry Ford designed an automobile body completely out of natural fibre–hemp. Such natural fibres possess good thermal insulation and better acoustic properties. Perhaps, these materials were used for structural panel construction units and also as sandwich vertical beams recommended in housing project units. The verdict of such naturally available fibre materials in construction technology is quite large owing to its maximum performance that can be expressed in terms of reliability, durability, stability, low cost maintenance as well as cost-effectiveness [6].

2 An Insight to Green Composites

The polymeric composite units were composed of both natural fibre units and biopolymers said to be known as green composites. Few natural fibres were coir pith, sisal, hemp, flax units, etc. The biopolymer material includes lactic acid, poly-lactic acid (PLA), starch, furan, etc.

2.1 Classification of Fibres

The coir, cotton kapok unit, jute, hemp, flax and manila type sisal belongs to the category of cellulosic vegetable fibres (Fig. 1). Wild silk material, crossbred unit and horse, rabbit, camel and alpaca can be obtained using aforementioned origin. Fibres generated from such cellulosic background can be comprised based on its origin as seed, leaf, bast, stalk and green grass. The resulted fibres from leaf, stem and bast are developed in to bundles and hence said to be known as fibre bundles, whereas fibre resulted from the seed were known as fibres. The three essential components occluded within the natural fibre units are lignin, cellulose and hemicellulose [7].

Out of this, cellulose is recommended as the primary component responsible for providing inherent strength and excellent stability to such fibre, whereas hemicellulose is associated with fibre structure. In general, natural fibre units are exclusively meant for certain non-structural application purpose which encompasses bag, rope, broom and even furniture in the rural areas. Moreover, apart from having coarse texture and available from white to deep brown colour, these can also be utilized effectively in roofing and insulation purpose.

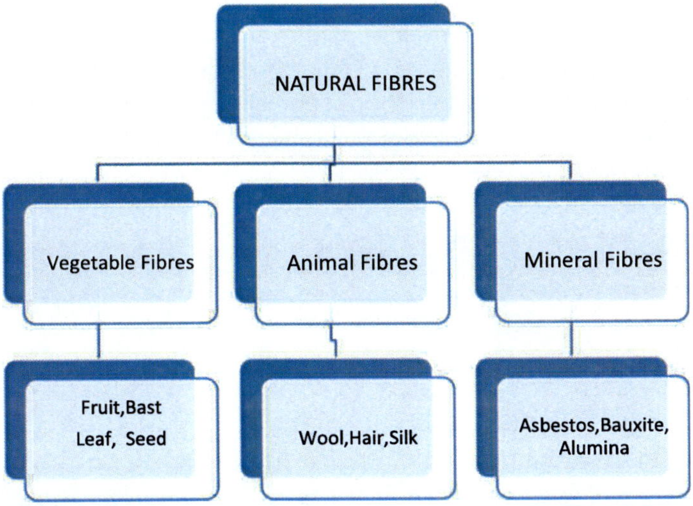

Fig. 1 Classification of natural fibre units *Source* Author

2.2 Extensive Role of Fibre Units

1. Coir

Coir fibre is considered to be extracted from rice husk, coconut shell exterior cover, and hence as a consequence, it may be utilized as a seed-driven fibre unit (Fig. 2-1). Coir is extremely water resistant and even resistant towards salt water damages. During harvest of fibre unit, the presence of coconut at premature stage generates a soft and fine pliable fibre (white in colour), whereas a strong brown fibre is obtained during fully mature stages which is even less flexible than the former one [8].

2. Bamboo

The fast growing plant belonging to grass variety is available in monsoon climate regions. Possess maximum inherent strength and stability to maintain temperature level, especially during the coupling of additive agents. Such bamboo fibres (Fig. 2-2) were involved in furniture making, surf board development and interiors to automobile components [9].

3. Hemp

It is obtained as a plant extract from cannabis family of species. These hemp units are explicitly utilized in textile manufacturing processes and also in paper industry (Fig. 2-3). During earlier centuries, they have found applications in manufacturing the sails of canvas ships, nets, rigging weirs and stable caulk owing to its excellent strength inbuilt with salt water resistance [10].

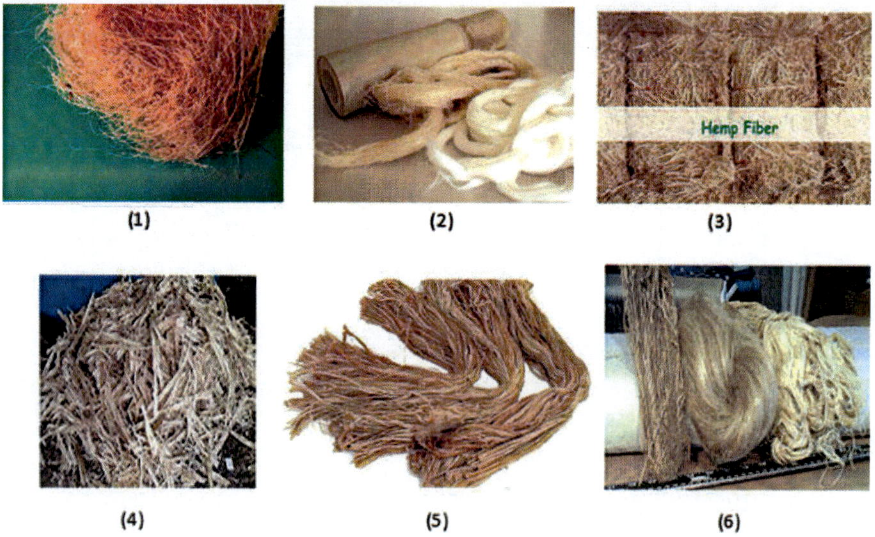

Fig. 2 Natural fibres. *Source* Google

4. Bagasse

Such fibre is developed from the squeezing remains of sugarcane and it has also been used as material for undergoing combustion operations in sugarcane industry itself also in pulp production industries to manufacture paper cloth (Fig. 2-4).

5. Jute

One of the cost-effective fibres generated in surplus quantities in India as well as China and Bangladesh is the jute fibre (Fig. 2-5). The associated specific modulus parameter of such jute fibre nearly approaches to that of glass fibre material and can even remain stable up to a temperature level of 20 °C without incurring any kind of damage to the inbuilt properties [11].

6. Flax

It is one of the very ancient fibres in the universe, such fibres can be definitely woven into various types of fabric units easily which are further utilized to manufacture excellent composites containing different properties (Fig. 2-6). Moreover, such flax material is eco-friendly in nature, economical to involve in process operations and also can be rivalled to modern synthetic fibres as per previous research studies.

Apart from above wood derivatives, various other natural fillers of organic basis have also been prompted to identify application in certain sectors. Few examples include cellulose, sisal, cotton, jute, kenaf, hemp and starch. Perhaps, environment friendliness can be established using post-consuming recyclable plastic materials instead of modern polymer matrices [12, 13].

Such wood flour matrix and fibre materials are much interesting due to its salient features of low cost, elastic modulus factor, dimensional stability but with less improvement in their tensile properties (Table 1). The main drawbacks are its lower toughness, poor adhesive nature associated betwixt filler materials and that of polymer matrix and thermal dissociation developed at operating temperatures beyond 200 °C [14].

Anyhow, hemp, sisal, kenaf and flax are geometrically similar to each other and essentially fibres of maximum length derived from bast of various plant species. These can be absolutely used as filler materials that can be made by means of proper segmenting into short or lengthier fibres. Starch is an organic polysaccharide that may exist in various plant species which aids as a source of energy. It is composed of monomers of glucose derivatives bonded with alpha linkages [15].

The proper addition of granular starch material to a polymer matrix results in a reduced property of large elongation at breakpoint and simultaneously enhanced elastic modulus factor due to high proportion of starch content [16]. A notable shortcoming of such filler material is the ability to absorb large amount of water due to its maximum surface distribution and hygroscopic nature. Few other natural fillers of organic form utilized in less extent are nutshells, rice husk, its ash, fibres of empty palm fruit, corn plants, coconut shells, etc. [16].

Table 1 Physical characteristics of few plant-derived fibres

S. No.	Fibre type	Fibre	Elongation %	Tensile strength (MPa)	Density (kg/m^3)	Young's modulus (GPa)
1	Stem	Jute	1.25–1.52	400–800	1400	15–55
		Hemp	1.5	575–900	1480	25–80
		Flax	1.4–3.2	350–2000	1500	20–80
		Bamboo	1.35	200–600	600–910	20–45
		Kenaf	1.58	160–940	1450	22–62
2	Fruit	Oil palm	5–19	50–410	700–1500	1–9
3	Wood	Softwood	–	–	1100	25
		Hardwood	–	–	1200	39
4	Leaf	Sisal	3.2–7.1	470–710	1500	9.5–22.2
		Banana	2–3.5	162–790	720–880	7.6–9.5
5	Synthetic	Glass	2.9	4500	2500	80
		Aramid	3.5	3125	1500	65

2.3 *Polymer Matrices*

The primary polymer matrix materials involved in the branch of green composites could be narrated in the following section.

2.3.1 Polypropylene

Several literatures reported the combination of polypropylene with natural filler materials extracted from flax, wood, sisal, kenaf, hemp and starch. The attention towards alternative cellulosic sources trace a fact of justification that these compounds are much more renewable compared to wood. Certain adhesive promoters are also well applied [16]. Few mentioned examples in previous literatures involve polypropylene grafted with maleic anhydride (PPgMA), styrene-styrene rubber embedded with ethylene/butadiene supported with maleic anhydride (SSEBgMA), silane-based components, copolymer formed out of propylene, ethylene and diene grafted with matrix of maleic anhydride (PEDgMA) which definitely allowed a significant enhancement in its morphological and mechanical characteristics.

Quite interesting phenomenon have been traced out of wood fibres impact on morphological behaviour and crystallization trends of polypropylene based organic composites. It is proven that wood fibres do not account to kinetics of crystal growth effectively, whereas it can render an elevated phenomenon of nucleation, still it is not evident whether the experimental improvement in mechanical characteristics could be primarily attributed to enhanced polymer-wood adhesion or even better wood supported dispersion or otherwise both. It is henceforth marked that the existence of amorphous portions within the polymer geometry can play a prominent factor to reduce the mechanical behaviour [17].

2.3.2 Polyethylene

The published literatures highlight polyethylene-derived composites involving various fillers such as corn starch [18], rice starch [19], kenaf/sisal fibres [13, 20], wood chips [19] and other fibres. Moreover, papers were being subjected to existence regardless post-consumed recyclable polyethylene, HDPE available from sealed milk bottles [21], wood fibres through food packing containers, polyethylene arising out of greenhouse filmy materials, sago starch and olive stones [22].

Wood fibres are generally blended with 50–70% by molar weight of polymer supported matrix. Thereby, a marked enhancement in stiffness, flexural strength and ductility reduction was also observed [23]. This drawback can be overcome by adapting polar adhesive promoters that is listed as maleic anhydride supported polypropylene/polyethylene (MASPP/MASPE) or even a copolymer constituting ethylene and acrylic acid [24].

Others:

Many classes of polymers when combined with wood such as sisal fibres/polystyrene, starch or wood flour, phenolic resins, polycaprolactone, natural fibres, palm flour, isora fibres, rubber and polyester resins were listed along with natural rubber [25].

2.4 Fibre Extraction Procedure

Retting technique and decortication procedure are generally applicable to disseminate the fibre bundles from agglomerated bast and leaf part of plants. Retting can be a process of subjecting the crop or deseeded part of straw to either chemical/biological treatment in order to assure the fibre bundles to be separated easily from the woody stem so as to ascertain the complete dissemination of fibre bundles. Dew entails of retting leaves the stem part in the agricultural field to undergo rot. Hence, such a process is to be carefully monitored to confirm the separation of bast fibres from the entire core of attached plants without loss in its inbuilt quality. Such dew technique is much famous in European countries in spite of its explicit dependency on geographical distribution and this develops coarser fibres of low quality than being obtained using water retting procedures. Water retting issues entails the discrete soaking of plant stem parts in any water source [26].

A decorticator is an equipment used to firmly strip aggregated bundles of fibre from leaf or stem. Leaves are certainly crushed and hardly beaten by a reciprocating wheel embedded through blunt knives such that the fibres alone remain. The left out leaf parts were washed using water. Such decorticated parts are slowly washed before subjecting them to drying via solar energy or hot air. The resulted dry fibres are then properly combed and arranged into various grade forms.

2.5 Pre-treatment of Fibres

Pre-treatment procedures conducted on fibres using both physical/chemical steps drastically improve the procured adhesion that occurs at the interface of fibre-matrix and thereby deduces moisture adsorption of organic fibres. Further, the chemical step involved in fibres also tend to aid in enhancing mechanical characteristics under the impact of controlled environmental ageing factors.

2.5.1 Physical Modification

Such modification has been always done with the aid of certain instruments to modify the surface and structural fibre properties with the possibility of enhancing the mechanical strength of involved fibres. The hydrophobic nature of fibres, thereby strongly impacts the mechanical bonding embedded onto matrix material. The treatments followed during decades are thermal method, stretching and calendaring process [27].

The thermal treatment is a traditional procedure to modify the organic fibres. When such fibres are exposed to heating temperature above transition range prevailing for lignin content, it is emphasized that softening along with migration of the same occurs on the fibre surface. The transition point of kraft lignin is 145 °C and starts to degrade in its content at 215 °C. Therefore, the prolonged heating beyond 200 °C would make it to soften. Surface disruptions through discharge treatment techniques such as plasma heating at low temperatures, sputtering step and corona decomposition are of much interest with a relative aim to improve the functional properties involved in organic fibres to a successful manner. The sputtering etch provides surface roughness and thereby enhances adhesion. This etch may also be possible by lengthier treatment operations which may also leads to better dye ability and surface wettability. Plasma heating may reduce the fibre strength which may also be affirmed due to elevated operating times involved in such treatment [28].

2.5.2 Chemical Modification

Chemical method involves the addition of chemical agents to disrupt the fibre surface or the entire fibre throughout. Modification comprises oxidation, mercerization, grafting, crosslink and coupling treatment as five main methods.

The above-mentioned research is definitely emphasized on filler surface modification in order to acquire enhancement of interstitial adhesion between the polymeric macromolecular matrix (hydrophobic) and filler materials (hydrophilic) and their essential dispersion in the polymer matrix. This is considered as a serious issue such that the gradual addition of organic filler materials may prone to undesirable mechanical behaviours which can be ultimately true when low strength/diameter ratio of fillers are used. Trendy changes in mechanical properties using mild addition of 25% by weight ratio of natural fillers to recyclable polyethylene are listed below [28].

Oxidation can take place under the influence of mild chemical condition with the introduction of aldehyde, carboxyl and ketone group along the cellulose chains by appropriate oxidation of both primary/secondary hydroxyl radicals. Mercerization is an alkaline addition for fibres using sodium hydroxide. Such treatment renders the fibre to swell which leads to nearly 30% breakage of hydrogen bonds during post-treatment procedures. Such bonds will reunite and produce various effects (i) decrease in spiral angle with increase of molecular detection, (ii) axial split of elementary fibres which thereby elevates fibre density and further structural collapse of cellular matrices and (iii) modification of primary cellulose to secondary cellulose.

As a whole, the overall advantage of green composites lies upon its extra-ordinary stiffness and excellent thermos-mechanical resistance in spite of its deduced ductility parameter and tensile strength. Hence, it is oblique that the modification of chemicals or adhesive promoter's usage can be effective means to excel the occluded mechanical properties. In general, modification process completely relies on physical/chemical steps primarily focused on successful grafting of chemical units that are viable to enhance the interstitial interactions. Few steps can be recalled as follows [29–38].

1. Acetylation

The selected fibres are completely dipped in a bath of glacial acetic acid for about an hour along with addition of acetic anhydride with careful blending of conc. sulphuric acid to few minutes, further filtrated, then washed and dried in an oven. This step is an esterification reaction which abruptly stabilizes the cell walls which could be pronounced in facts of humidity acquisition as well as simultaneous geometric variation.

2. Mercerization

Also known as alkali treatment, it is tested on short fibres by slow heating at nearly 60 °C in 12% NaOH solution for 4–5 h. Further, it is washed and dried via ventilated oven used to generate better fibre quality with an improvement in fibre wetting.

3. Benzylation

After alkali addition, the opted fibres are treated with benzoyl chloride for an hour and followed by filtration, slow washing and drying. Further, it is blended with ethanol for 2 h followed by gradual rinsing and oven drying such a process reduces the hydrophilic character of fibres.

4. Stearic acid addition

Mixture of ethanol and stearic acid was added dropwise to fibres based on its total weight and further dried through hot oven.

5. Peroxide addition

Fibres are saturated with benzoyl peroxide in acetone solution for thirty minutes, further decanted and then dried. Such addition has shown elevated improvement in mechanical properties as proven by recent research studies.

6. TDI treatment

Immersion of fibres in chloroform bath followed by little addition of a catalyst based on di-butylin di-laurate derivative followed by 2 h continuous stirring along with addition of toluene di-isocyanate, and at last dipped in acetone and then only oven dried.

7. Permanganate treatment

Fibres are strictly immersed in acetone-permanganate mixture (conc may vary from 0.005 to 0.25%) not less than a minute followed by decantation and drying. Hydrophilic behaviour is decreased by such treatment.

8. Anhydride treatment

Maleic anhydride mixed with toluene solution is utilized in such treatment, in which the fibres are entirely dipped for impregnation combined with hydroxyl reactions on the fibre surface. This step consequently reduces water absorption capacity.

9. Isocyanate treatment

Such isocyanate compound reacts with hydroxyl units onto fibre surface, thereby impacting better association of interstitial adhesion towards the polymer matrix molecule. Mostly, it performed at moderate temperatures to about 50 °C for an hour.

10. Silane treatment

Fibres are cohesively mixed with 2:3 water alcohol mixture comprising a silane type adhesive promoter for 2 h at a pH of 4 and then subjected to drying. Such silanes undergo reaction with hydroxyl units thereby results in surface quality improvement.

11. Plasma treatment

It enhances significant modification onto fibre surface. Anyhow, chemical as well as morphological modification may be quite heterogeneous which in turn depends on process conditions. Hence, such a step is difficult to materialize resulting in critical attention towards process control and also the final surface dissemination step.

2.6 Biopolymers

Polymers resulted from non-conventional resources can be typically segregated to three broad groups (i) natural organic polymers—starch/cellulose, (ii) artificial polymers resulting from natural organic monomers—poly-lactic acid (PLA), (iii) polymers produced out of microbial fermentation—poly-hydroxyl butyrate (PHB). Poly-lactic acid, one among the distinguished biodegradable polymers can be synthesized from natural source of feedstock including corn starch and also produced from rice, sugar beet, potatoes and other wastes obtained from agricultural activities. PLA production comprises feedstock raw material conversion into dextrose, therein undergoes further conversion into lactide/lactic acid through fermentation reaction in presence of catalyst. The formed lactide is ready for further processing so that the monomer can be purified and aftermath polymerization, PLA can be obtained through polymerization reaction which takes place in presence of a suitable catalyst.

2.6.1 Poly-L-Lactic Acid

PLLA has gained much attention and available on a wider basis which possesses high ordered melting point, wherein the process parameters are relatively similar to polypropylene. It possesses high grade of mechanical properties and also exhibit strong anisotropic effect such that the fibre will orient itself in four principal directions (0–135 °C).

2.6.2 Polyesters

This broad group includes poly-hydroxyl alkanoates and poly-dicarboxylates of alkylene form. They are synthetically leaded by condensation reactions leaded owing to chemical reaction betwixt dicarboxylic acids along with diols. PLA is a structural material due to its tendency to polymerize with high molecular weight. Also, hydrophobic and vapour permeable such permeability renders sufficient lifetime to achieve mechanical competence without drastic hydrolysis. This provided the composting capacity such that all other techniques are also employed [35].

2.6.3 Starch

It is a complex substance comprising amylo-pectin and amylose polysaccharides. The properties do not vary with amylo-pectin/amylose ratio and also according to plant source. A major propaganda of starch is corn but also extracted using rice, potato and wheat. Polymer behaves crystalline nature due to amylo-pectin presence. Poor mechanical properties and water solubility were two major demerits such organic polymer requires short form of durability where rapid degradation is conceivably advantageous. It is processed to a substance as foam which may be an alternate to polystyrene and involved in loose packing filter, moulded parts and food trays.

2.6.4 Cellulose Acetate

Cellulose acetate is an organic polysaccharide which is obtained through a chemical reaction betwixt anhydrous acid and cellulosic products extracted from wood pulp, cotton linters, sugarcane or recycled paper. The techniques involved in cellulose acetate manufacture were first recognized during the last periods of nineteenth century and this was used in films, filaments and organic lacquers, which also expose maximum toughness and moreover high transparency [37].

2.7 *Bio-based Thermosetting Materials*

Most of these were obtained through vegetable oils due to graft reaction of acrylate, hydroxyl group and that of maleate compounds or else the combined form of these substances onto triglycerides of fatty acids. Few were derived from cashew nut liquid from its shell obtained from cashew nut shells as an aggregated by-product from nut processing industry. This liquid contains large content of anarcardic acid, and during heat extraction operation, it may be converted to cardanol. This is being polymerized using free radical type of polymerization.

2.7.1 Liquid of Cashew Nut Shell

This liquid was a waste product of cashew extraction process. A viscous and dark liquid resulted from honey comb structure available in cashew shell which has been a topmost option for obtaining cost-effective and environmental-friendly resins. It comprises of anacardic acid with a few amount of cardol and cardanol in raw form, but when heated as well distilled, it possibly remains in technical grade [39].

2.7.2 Poly Butylene Succinate (PBS)

It obtained from succinic acid, 1,4-butane diol and PBS, which is a biodegradable polymeric material that is having greater attention in the area of bio-composites. Owing to its wider temperature aspect for polymeric processing, it is applicable for injection/extrusion moulding, film blowing and thermos forming.

2.7.3 Cellulose

The structural component exists in plant structures and it may be available in basic or esterified using cotton, trees, sugarcane or else recycled paper material. Examples of such esters are cellulose acetate (CA) and CA butyrate. It is mostly plasticized using citrate material to render it as a polymer matrix such that it can be effectively used as a polymer resin.

2.7.4 Merits of Biopolymer Techniques

1. Relative decrease of fossil fuel importing dependency.
2. Reducing petrol consumption.
3. Possible reduction in emission of greenhouse gases.
4. Simultaneous reduction of solid waste contamination.
5. Reducing the polymer waste.
6. Tremendous enhancement in agricultural sectors employment [40].

2.8 Parameters Impacting Composite Strength

The associated properties rely on characteristics of constituent components such as resins and fibres. The mechanical strength along with stiffness carries the entire load and its volume proportion. The resin component facilitates its relative position within the biocomposite and also used to transfer the adsorbed load from bottom of fibres to its intact portion. The three essential factors of resin/fibre properties along with its interfacial behaviours are more critical to design the strengthened composites. Certainly, modification also improves moisture resistance and also composite characteristics [8]. In addition to that, processing techniques also have specific impact on the mechanical behaviours of such composites. These associated properties may vary with respect to harvest quality, body and age of the plant from which it is derived, extraction operations and the ambient conditions of environmental site [10].

2.9 Rheology and Processing Techniques

With a focus to processing of polymer–organic filler composites, when correlated with neat polymer, it may be proposed as a common rule that viscosity enhancement is observed as a consequence of reduction in processing techniques which may also occur due to higher filler content. Anyhow such processability does not compromise owing to moderate increase in viscosity and torque along with higher shear expressions [41].

There are only few backups which deals with rheological studies of polymer ingredients that are filled with organic materials. Early discoveries prompted viscosity increase in both polyethylene and polypropylene-derived composites embedded with wood waste. In few cases, viscosity undergoes a gradual enhancement by adding more filler content [42]. It is also demonstrated the varying nature of HDPE composites by changing the filler addition employing a capillary tube rheometer. Moreover, wall slipping phenomena is present and extremely dependent on weight % and nature of fillers. Flow analysis by elongation depicted that viscosity relies on filler content rather than nature of filler. A selective analysis of Trouton's factor ratio confirmed it to be effectively reluctant upon type of fibre along with the observation of consistent interaction that occurs through the fibres [43].

It is being noted that rheological characteristics of PP embedded onto wood flour bio-composites with/without adding PP grafted maleic anhydride adhesion promoter. Such promoter aids in mild processing and anyway by keeping an increase in filler quantum, rheology tends to modify significantly owing to the development of bonds between filler-polymers, assisted by the promoter material [44].

2.10 *Merits*

Composites possess excellent mechanical properties per unit volume and its durability allows successful manufacture of large and complex shapes. They may be formed into complicated shapes and can be cut to desired length easily. These comprise jute, wood fibres, sisal, bamboo, banana leaves and coconut shells. These fibres can be blended either alone or in combined form as in hybrid varieties and sometimes partial addition to industrial fibres. These are non-conventional, non-abrasive to process instruments and may be subjected to possible incineration during the end of its lifecycle to recover energy owing to its maximum calorific value. Plant fibres possess less density, high stiffness, low production cost and highly renewable which adds to its extreme advantage when compared to fossil source derived fibres (47).

Plant fibres are carbon neutral and thus eco-friendly. Low specific density of cellulose-derived fibres prone to weight compensation in the manufacture of composite inclusive of direct merits on transportation. Higher amount of fibre volume proportions of various plants as compared to fossil-based reinforcements will lead to discrete cost savings of material due to less amount of investment rendered on plant fibres than binding matrices.

2.11 *Demerits*

The application of natural fibre polymer composites is being strictly prohibited owing to the natural fibre inbuilt properties inclusive of moisture absorbing ability, poor adhesion, reduced wettability along with synthetic counterparts and lesser thermal stability encountered during productive operation. Mechanical properties occluded with these natural composites slowly deteriorate on ageing as the fibre interface matrix portion is extensively influenced by moisture. The extent of deterioration and reversible nature of fibre characteristics is extraneously dependent on the degree of moisture absorption. Less amount of thermal influence, in other words, the chance of moderate temperature degradation from 240 to 250 °C [45].

2.12 Applications

In spite of lack in adequate strength accustomed by synthetic composites in most of the cases, various green composites are still then employed in bearing materials without load condition as in case of sports equipment, furniture, internal/external panels for vehicles and proper housing built for electronic goods. Flax is recorded widely to be strongest in its mechanical behaviours that can be legibly manipulated and showcases the properties as compared to glass fibres when rightly woven to conditions at an optimum level. Even though hard, its lightweight tendency makes the flax-based bio-composites to be useful for wider ranges of different products [46].

Of the above-mentioned bio-composites, cashew shell-based ones are mostly utilized in varying conditions. Sisal fibre derived composites with reinforcement have been analysed with a mechanical strength of around 25 MPa and its Young's modulus to be 9 GPa. The bending tests proved their roofing applications owing to its adequate mechanical strength. Plant fibre manufacturing technology formed from bio-composites centre has developed a variety of construction products based on natural feedstock materials. One of those is the Isonat@R fibre of insulation, obtained from hemp developed on UK farming companies and also available from waste fibres of cotton. This contains nearly 15% polyester so as to provide stability and loftiness. It can also be decomposed using incineration or composting without creating harmful effects. It is able to absorb and release the humidity effectively and such a characteristic feature is utilized in moisture control of buildings with no loss in its thermal performance, and also with no impact on insulation durability [47].

With a focus to industry-oriented applications, various modes have been proposed and it can be postulated that wood can be mostly used organic filler, in specific a cheap filler material for poly-olefins. Wood flour can be gathered from sawmill aftermath a simple sieve analysis. Wood fibres were derived from wastes accumulated in a sawmill by means of thermomechanical process under wet condition. The industrial utilizations comprise door/window frames, railroad sleepers, furniture, automotive panels, gardening tools, packaging materials, shelves and in other applications in which it does not involve high resistance in case of mechanical equipment, but inculcate low maintenance and purchasing expenditures [48]. Further, it is much convenient to employ recycled polymeric composites than the virgin ones, thereby enabling enhanced cost efficiency as well as eco-sustainability. Some more include indoor panels, platforms, footboards, upholstery, insulation panels for noise control were absolutely imported by American, Japanese, Italian, British and German firms [40].

In specific, automotive industry plays a prominent role in market. The earlier carmaker using polymer fillers with fibres was Benz in 1990s by adopting door panels constituting jute fibres [49]. This has been a popular set-task for other carmakers who engaged such polymeric composites containing natural fillers as ingredients for parcel shelves, headrests, roof upholstery, door panels, etc., such that they can confirm proven advantages such as less environmental impact, improved weight, high elastic modulus parameter, and less costs. Sometimes, it was necessary to enhance the mechanical characteristics through certain pre-treatment techniques rendered onto fibres (physical/chemical method) based on application specific, for example, such treated fibres may be used in mats, non-woven structural parts, etc. [30].

Few literatures assured that such pre-treatments subjected onto natural fibres could pave the way of developing better quality composites with enhanced mechanical features as compared to glass fibre intruded organic composites [50]. A consequence which would be no way possible to establish otherwise, however, the hydrophilic mechanism of organic filler encourages agglomeration, moisture absorption and also improper adhesive action with that of polymer matrix, and in fact, large measures were adopted to avoid the issue of interfacial adhesion [49]. Among these, experimental observations on silane mediated adhesion promoters or alkaline/resin addition to fibres prone to the chance of semi-conductor applications if the fibres are having longer lengths.

2.13 Moving Towards Environmental Sustainability

Some recent studies envisaged that, unfortunately, such green composite materials are not completely eco-compatible owing to certain limitations concerned with its recyclability (if processing temperature for recycling goes beyond 200 °C, the main characteristics will worsen due to degradation process) and even their biodegradability depends on filler and not based on conventional polymer matrices [51].

For such reasons, research studies mainly emphasized on the development of 100% eco-friendly green composites by alternating non biodegradable ingredients with biodegradable materials during the last decade. Many biodegradable polymers derived from nature exist as proteins (albumin, casein, elastin, silk, etc.), polysaccharides (cellulose, starch, collagen, chitin, etc.), poly-lactic acid, polyesters, lipids, lignin, natural rubber, few polyamides, poly vinyl acetates/alcohols, and poly lactones. Most of the cases involve degradation of above polymers through enzyme controlled chemical reactions under suitable humid environments [52].

In general, biodegradable polymers are classified on the basis of origin as microbial-assisted polymer (PHA), agro-polymers (starch), agro-monomers synthesized via chemical method (PLA) or traditional monomers (man-made polyesters). Many examples were reported in literature. For instance, Japanese scientists have demonstrated green composites developed from bamboo and starch fibres. In certain cases, they undergo alkaline pre-treatments and also chemical grafting technique. The interesting fact on Monsanto Biopol (a poly-hydroxy butyrate copolymer) synthesized from bananas and jute fibres [53].

Soya protein-based works were also made to produce polymer matrices. For example, Netravalli with his group mates involved soya proteins along with natural fibres yielding superior quality green composites which tend to show universal characteristics proximately better than large wood types. Soya and corn oil procured as feedstock materials for polymer production was an interesting note to few automotive applications inbuilt with excellent resistance, lightness, extreme flexibility and evergreen durability [52]. Few examples for these are seats, panels, furnishing, packaging, etc. Few concerns were working on artificial silk manufacture through the principle of genetic engineering that could also be applied to biodegradable composite materials as well [54].

Takagi have experimented polymeric composites with regards to dispersion class biodegradable cellulose and resin nanofibres, employing a stirring mechanism revolving at low speed along extended times and also upon varying the resin mould pressure. This process significantly enhanced flexural modulus and mechanical strength due to elevated mould pressure that can be compared to control composites, wherein there was no stirring process [47].

2.14 Considerations on Environmental Impact

It has been proposed that usage of organic fillers derived from nature executes certain advantages when compared to mineral-based inorganic constituents. Ultimately, such an environmental impact can be absolutely reduced by moving to renewable sources rather than fossil derived sources. Moreover, biodegradable polymers usage fully establishes biodegradable systems, thereby reducing the issues of solid material production every-day and handling of surplus quantity of polymeric waste [55].

The benevolent aspects were further analysed by invoking the concluded data by which total life cycle analysis (LCA) may provide a complete evidence. For instance, Joshi et al. collected and discussed LCA studies on composited supported with natural fillers as well as glass fibres. It is well known that LCA analysis permits the determination of total environmental impacts associated with products/processes via "cradle to grave" mechanism, thereby evaluating overall mass/energy flows obtained from product manufacturing stage to the effective use in its life cycle [56]. All such steps are associated with direct/indirect environmental implications which have to be evaluated quantitatively. The complete LCA process is governed by ISO 14043 and ISO 14040 regulatory standards [57].

The life cycle process of glass fibre reinforcement comprises both monomer and glass production and then to polymerization whilst glass is to be treated with a view to yield fibres. The feedstock materials are to be processed to get the final product. Then, the product will be used for some time period till its cycle gets over which can follow waste disposal technique, incineration or else recycling. Hence, it is understood that every step comprises of both mass flow and energy calculations that are not easy to quantify, with prevalent ecological impacts and obviously represented using suitable environmental indicators occluded with overall energy pattern for CO_2, SO_2, CH_4, CO, NO_x emissions, phosphates, nitrates, sulphates, BOD/COD emissions in water, etc.

The aforementioned information clearly examines the various outcomes of salient LCA studies demonstrated in the previous literature. The earlier one agrees the selection of polypropylene blend filled with glass/hemp fibres for automotive insulations, predicting that organic composites pave the way to definite deduction of conventional energy demand and also air pollutant emissions [58]. The next one compared the design of car panels using epoxy/ABS resins with hemp organic fibres. Notable decrease in energy utilization along with carbon emissions was also evaluated. Anyhow, NO_x, phosphates were not fully reduced owing to inclusion of man-made fertilizers employed for hemp growth [59].

At last, another one compared the pellets out of polypropylene along with bamboo/glass fibres. The outcomes envisaged a gradual enhancement of the primary environmental indicators: air pollutant emission, BOD/COD emissions and other carcinogenic/toxic impact [60]. The most worsened indicators of nitrate/phosphate emissions owing to artificial fertilizers usage in agricultural fields. Another author revealed four type of interesting indicators amidst NRF and GRF composites mentioning, (1) natural fibre wins due to energy consumption strategy along with lower emission rates excluding nitrate, (2) substitution of higher % of polymers leading to excellent gain in mechanical properties, thereby higher fractions of NF volumes used resulting in minimal usage of oil derived polymeric units, (3) minimum specific weights of NF exposes to reduced emissions along with decreased energy consumption directly than GFR composites, (4) combined credits of carbon as well energy owing to profitable incineration process of NF at the end of its life offers a double merit of maximum energy recovery since theoretical output of CO_2 emission just equals to the amount that is utilized by plants during photosynthesis [57].

Recently, few authors have examined LCA studies on green composites too, without considering certain aspects. Such gap factors may compromise certain stages in life cycle studies itself such as filler production, use or manufacture or disposal, energy input from fertilizers or other environmental parameters [60].

Specifically, the front part of bonnett was taken as the utilitarian unit and the related effects were: cancer-causing agents, respiratory organics, respiratory inorganics, environmental change, radiation, ozone layer, eco-poisonousness, fermentation/eutrophication, land use, minerals and petroleum products. Reusing, burning and landfill situations were considered [61]. They found that the jute fibre composites permit improving the natural execution. Specifically, fuel utilization ends up being lower because of the weight decrease of the vehicle, albeit some obscure effects in the creation and removal stages were discovered and identified with the co-ordinations of jute strands transportation and reusing situation. Besides, the authors featured that stages, for example, development, gathering, mercerizing, drying and fibre refining were credited without any effects, revealing that jute creation for a few little rancher networks along the Amazon River was thought of, with the stream giving additionally the humus and the entirety of the supplements required [62].

Three situations were considered, in particular, (1) mulling over a reuse for the half of materials in any case burned, (2) one portion of the composite material is viewed as arranged municipally, (3) breaking down the commitment of a theoretical powder filled composite having a similar explicit load as sugarcane bagasse composite. Effect classifications were abiotic consumption, fermentation potential, eutrophication potential, an earth-wide temperature boost potential (100 years' premise), ozone layer exhaustion potential and photochemical ozone creation potential. The examination was precise, since it considered helper procedures, for example, diesel, phosphate, nitrogen, potassium, phosphorus and lime creation/sources; nonetheless, still no land use impacts were unequivocally included [63]. The general examination recommended that the green composites are better that the powder filled composites in car applications, particularly, when weight decrease is especially significant. It is additionally proposed that sugarcane bagasse fibre creation prompts lower natural effects contrasted with powder creation, the composites are lighter, sugarcane has a positive commitment as far as carbon credits and the reuse of the finish of-life material is the ideal method to limit the ecological effects.

Similar authors assessed the specialized presentation and natural effects of similar composites in contrast with flawless polypropylene. Diverse finish of life alternatives included cremation, reusing and landfill. The composites indicated better natural execution during the whole life cycle, particularly, in the development stage [64]. Additionally, for this situation, reusing with monetary benefit of the composites was the best choice to limit the natural effects. For additional inside and out examination, an intriguing survey by Jorge on ecological effect and LCA of language cellulosic-determined items can be considered.

3 Conclusion

It was notable that various deformations such as compression, bending, tension, shear, torsion, abrasion, flexing and wear were accounted in common on fibres. Moreover, the usage of polymeric-based composites blended with organic fillers in place of mineral-based inorganic fillers is of greater attention in view of simultaneous reduction in the usage of petroleum-based conventional resources and in more common a diverse consumption of both financial and environmental resources. Instead such green composite materials find several applications in industrial sectors mentioning process ability, ductility, as well dimensional stability. Along universe, researchers aim at bringing suitable solutions which involve chemical alteration of filler along with usage of adhesive promoters/additives. Anyhow, a complete biodegradability with less environmental impact can only be suitable by using biodegradable natural ones for traditional polymeric materials. In such cases, several limitations arise, and right now, the research investigation has been analysing the effective selection of most prominent biodegradable matrix along with simultaneous optimization involved in the synthesis and processing of suitable material. Still, the market is an open phase, and effort can be put-forth in exploring novel applications, improvement in salient properties, the physical appearance, durability and marketing ability of these natural composite materials. All these pertinent issues require, also continues to avail prompt and benevolent research efforts in a view to explore new formulations (virgin type of recycled varieties, still-more biodegradable substances, type, attracting appearance, quality and composition of fillers), structurally characterizes them, utilize them for most specific applications, and in normal, to refine operating techniques for process. If the prevailing competition for such green composites arises, the governing market demand will also enhance leading to simultaneous reduction in cost and also greater improvement in quality can also be achieved.

References

1. Krishan K (2013) Composite materials, science and engineering, 3rd edn.
2. Julia C, Nadezda S, Alena S, Jozef J (2013) Some aspects of lightweight composites durability. Chem Eng Trans 32:212–224
3. Giuseppe C, Alberta L, Giuseppe R, Gianluca C (2014) Composites based on natural fibre fabrics, Woven Fabric Engineering
4. Westman MP, Laddha SG, Fifield LS, Kafentzis TA, Simmons KL (2010) Natural fiber composites: a review. The National Technical Information Service, U.S. Department of Commerce
5. Mwaikambo LY (2010) Review of the history, properties and application of plant fibres. Afr J Sci Technol (AJST) Sci Eng Ser 7:120–133
6. Ticoalu A, Aravinthan T, Cardona F (2010) A review of current development in natural fiber composites for structural and infrastructure applications. In: Southern region engineering conference, 11–12 Nov, Toowoomba, Australia
7. Keller A, Zerlik H, Wintermantel E (1999) Biodegradable polyester matrix reinforced with hemp fibres—effect of fibre refinement by steam explosion. Swiss Federal Research Station for Agricultural Economics and Engineering, CH-8356, Taenikon, Switzerland
8. Tara S, Jagannatha Reddy HN (2011) Various industrial applications of hemp, kinaf, flax and ramie natural fibres. Int J Innov Manage Technol 2:142–155
9. Samuel OD, Agbo S, Adekanye TA (2012) Assessing mechanical properties of natural fibre reinforced composites for engineering applications. J Miner Mater Charact Eng 11:780–784
10. Joshi SV, Drzal LV, Mohanty AK, Arora S (2004) Are natural fiber composites environmentally superior to glass fiber reinforced composites? Compos: Part A: Appl Sci Manuf 4:371–376
11. Petinakis E, Yu L, Simon G, Katherine D (2013) Natural fibre bio-composites incorporating poly (lactic acid). Fiber Reinf Polym Technol Appl Concr Repair 11:21–35
12. Nair KCM, Kumar RP, Thomas S, Schit SC, Ramamurthy K (2003) Rheological behavior of short sisal fiber-reinforced polystyrene composites. Compos Part A 31:1231–1240
13. Chen HL, Porter RS (1994) Composite of polyethylene and kenaf, a natural cellulose fiber. J Appl Polym Sci 54:1781–1793
14. Carroll DR, Stone RB, Siringano AM, Saindon RM, Gose SC, Friedman MA (2001) Structural properties of recycled plastic/sawdust lumber decking planks. Res Cons Recycl 31:241–251
15. Bagheri R (1999) Effect of processing on the melt degradation of starch-filled poly-propylene. Polym Int 48:1257–1263
16. Rozman HD, Lai CY, Ismail H, Mohd Ishak ZA (2000) The effect of coupling agents on the mechanical and physical properties of oil palm empty fruit bunch poly-propylene composites. Polym Int 49:1273–1278

17. Ahmad Fuad MY, Zaini MJ, Jamaludin M, Mohd Ishak ZA, Mohd Omar AK (1994) Determination of filler content in rice husk ash and wood-based composites by thermogravimetric analysis. J Appl Polym Sci 51:1875–1882
18. Wool RP, Raghavan D, Wagner GC, Billieux S (2000) Biodegradation dynamics of polymer–starch composites. J Appl Polym Sci 77:1643–1657
19. Danjaji ID, Nawang R, Ishiaku US, Ismail H, Mohd IZA (2001) Sago starch-filled linear low-density polyethylene (LLDPE) films: their mechanical properties and water absorption. J Appl Polym Sci 79:29–37
20. Joseph K, Thomas S, Pavithran C (1996) Effect of chemical treatment on the tensile properties of short sisal fibre-reinforced polyethylene composites. Polymer 37:5139–5149
21. Yam KL, Gogoi BK, Lai CC, Christopher C, Selke SE (1990) Composites from compounding wood fibers with recycled high density polyethylene. Polym Eng Sci 30:693–699
22. Selke SE, Childress J (1990) Wood fiber/high density polyethylene composites: ability of additives to enhance mechanical properties. Wood fiber–polymer composites: fundamental concepts, processes and material options. Forest Product Society, Madison, Wisconsin, pp 109–111
23. Liao B, Huang YH, Cong GM (1997) Influence of modified wood fibers on the mechanical properties of wood fiber-reinforced polyethylene. J Appl Polym Sci 66:1561–1568
24. Myers GE, Chahyadi IS, Gonzales C, Coberly CA (1993) Wood flour and polypropylene or high-density polyethylene composites: influence of maleated polypropylene concentration and extrusion. In: Wolcott MP (ed) Wood fiber–polymer composites: fundamental concepts, processes and material options. Forest Product Society, Madison, Wisconsin, pp 49–56
25. Mantia LFP, Morreale M (2011) Green composites: a brief review. Compos: Part A 42:579–588
26. Debnath K, Singh I, Dvivedi A, Kumar P (2013) Natural fibre-reinforced polymer composites for wind turbine blades: challenges and opportunities. Recent Adv Compos Mater Wind Turbine Blades 43:211–215
27. Shekhar HSS, Ramachandra M (2018) Mater Today: Proc 5:2518–2526
28. Technical Progress Report (2008) Cellu-wood, technologies and products of natural fibre composites, CIP-EIP-Eco-Innovation-Pilot and market replication projects
29. Lovely M, Joseph KU, Rani J (2004) Isora fibres and their composites with natural rubber. Prog Rubber Plastics Recycl Technol 20:337–349
30. Kalia S, Kaith BS, Kaur I (2009) Pretreatments of natural fibers and their application as reinforcing material in polymer composites—a review. Polym Eng Sci 49:1253–1272
31. Khalil HPSA, Ismail H, Rozman HD, Ahmad MN (2001) The effect of acetylation on interfacial shear strength between plant fibres and various matrices. Eur Polym J 37:1037–1045
32. Dipa R, Sarkar BK, Rana AK, Bose NR (2001) Effect of alkali treated jute fibres on composite properties. Bull Mater Sci 24:129–135
33. Dominkovics Z, Dányádi L, Pukánszky B (2007) Surface modification of wood flour and its effect on the properties of PP/wood composites. Compos Part A 38:1893–1901
34. Holbery J, Houston D (2006) Natural-fiber-reinforced polymer composites in automotive applications. J Miner Met Mater Soc 58:80–86
35. Zafeiropoulos NE (2008) Engineering the fibre-matrix interface in natural-fibre composites, In: Properties and performance of natural-fibre composites. Woodhead Publishing in Materials, pp 43–55
36. Rowell RW (2008) Natural fibers: types and properties. In: Properties and performance of natural-fibre composites. Wood-head Publishing in Materials
37. Thomason JL (2009) Why are natural fibres failing to deliver on composite performance? In: Proceedings of the 17th international conference on composite materials, Edinborough, UK D9, pp 45–65
38. Chard JM, Creech G, Jesson DA, Smith PA (2010) 18th international conference on composite materials—Proceedings, vol 2, pp 12–15
39. Ugoamadi CC (2013) Comparison of cashew nut shell liquid (CNSL) resin with polyester resin in composite development. Niger J Technol Dev 10(2):34–55

40. Mitra BC (2014) Environment friendly composite materials: bio-composites and green composites. Def Sci J 64(3):244–261
41. Khalil HPSA, Rozman HD, Ahmad MN, Ismail H (2000) Acetylated plant-fiber reinforced polyester composites: a study of mechanical, hygrothermal, and aging characteristics. Polym Plast Tech Eng 39:757–781
42. Li TQ, Wolcott M (2004) Rheology of HDPE–wood composites. I. Steady state shear and extensional flow. Compos Part A 35:303–311
43. Natov M, Wassilewa S, Kowatschewa-Welewa S (1995) Rheologische eigenschaften von mit holzmehl gefülltem polypropylen. Angew Makromol Chem 225:73–81
44. Rana AK, Mandal A, Bandyopadhyay S (2003) Short jute fiber reinforced polypropylene composites: effect of compatibiliser, impact modifier and fiber loading. Compos Sci Technol 63:801–806
45. La Mantia FP, Morreale M (2006) Mechanical properties of recycled polyethylene ecocomposites filled with natural organic fillers. Polym Eng Sci 46:1131–1139
46. Dauda M, Yoshiba M, Miura K, Takahashi S (2007) Processing and mechanical property evaluation of maize fiber reinforced green composites. Adv Comp Mater 16:335–347
47. Takagi H, Asano A (2008) Effects of processing conditions on flexural properties of cellulose nanofiber reinforced "'green"' composites. Compos Part A 39:685–689
48. Ferreira RL, Furtado CRG, Visconte LY, Leblanc JL (2006) Optimized preparation techniques for PVC-green coconut fiber composites. Int J Polym Mater 55:1055–1064
49. Marsh G (2003) Next steps for automotive materials. Mater Today 6:36–43
50. Prachayawarakorn J, Khunsumled S, Thongpin C, Kositchaiyong A, Sombatsompop N (2008) Effects of silane and MAPE coupling agents on the properties and interfacial adhesion of wood-filled PVC/LDPE blend. J Appl Polym Sci 108:3523–3530
51. Netravali AN, Chabba S (2003) Composites get greener. Mater Today 6:22–26
52. Lodha P, Netravali AN (2002) Characterization of interfacial and mechanical properties of "green" composites with soy protein isolate and ramie fiber. J Mater Sci 37:3657–3665
53. Rodriguez-Gonzalez FJ, Ramsay BA, Favis BD (2003) High performance LDPE/thermoplastic starch blends: a sustainable alternative to pure polyethylene. Polymer 44:1517–1526
54. Gould P (2002) Exploiting spiders' silk. Mater Today 5:42–47
55. Puglia D, Tomassucci A, Kenny JM (2003) Processing, properties and stability of biodegradable composites based on Mater-Bi and cellulose fibres. Polym Adv Technol 14:749–756
56. Lim ST, Chang EH, Chung HJ (2001) Thermal transition characteristics of heat–moisture treated corn and potato starches. Carbohyd Polym 46:107–115
57. Joshi SV, Drzal LT, Mohanty AK, Arora S (2004) Are natural fiber composites environmentally superior to glass fiber reinforced composites? Compos Part A 35:371–376
58. Corbiere-Nicollier T, Gfeller Laban B, Lundquist L, Leterrier Y, Manson JAE, Jolliet O (2001) Life cycle assessment of biofibres replacing glass fibres as reinforcement in plastics. Resour Conserv Recycl 33:267–287
59. Wotzel K, Wirth R, Flake R (1999) Life cycle studies on hemp fibre reinforced components and ABS for automotive parts. Angew Makromol Chem 272:121–127
60. Vidal R, Martínez P, Garraín D (2009) Life cycle assessment of composite materials made of recycled thermoplastics combined with rice husks and cotton linters. Int J Life Cycle Assess 14:73–82
61. Alves C, Ferrao PMC, Silva AJ, Reis LG, Freitas M, Rodrigues LB et al (2010) Eco-design of automotive components making use of natural jute fiber composites. J Clean Prod 18:313–327
62. Luz SM, Caldeira-Pires A, Ferrão PMC (2010) Environmental benefits of substituting talc by sugarcane bagasse fibers as reinforcement in polypropylene composites: eco-design and LCA as strategy for automotive components. Resour Conserv Recycl 54:1135–1144
63. Caldeira Jorge F (2010) Reducing negative environmental impacts from the manufacturing and utilization of lignocellulosics-derived materials: an overview on research in 2007–2009. Molec Cryst Liq Cryst 522:328–335
64. Luz SM, Ferrão PMC, Alves C, Freitas M, Caldeira-Pires A (2010) Eco-design applied to components based on sugarcane fibers composites. Mater Sci Forum 636–637:226–232

Chapter 10
Recent Trends in the Surface Modification of Natural Fibers for the Preparation of Green Biocomposite

G. L. Devnani

1 Introduction

Recent economic growth and technological developments are inspiring scientific community to look for novel materials which can compete with the cutting-edge technology and at the same time should be environment friendly and sustainable too. Natural fibers reinforced in different polymer matrix offer excellent mechanical and thermal properties. That is why they are getting significant attention in this decade as a replacement of synthetic fibers as a reinforcing material in bio and green composites. Biocomposites are the composite materials in which either one or both the constituents of the composite material are biodegradable while green composites contain both the biodegradable components. Figure 1 shows various categories of biocomposites in which either the matrix is biopolymer and fiber is synthetic, fiber is natural and matrix is synthetic or both the constituents are biodegradable [42].

There are few issues with natural fibers which need to addressed, like poor adhesion and compatibility with hydrophobic polymer matrix, thermal stability, tendency to aggregates and poor resistance to moisture. Substantial amount of work on these lignocellulosic fibers and their reinforcement in different polymer matrix has been done, and a seminal number of reviews have been written. Faruk et al. reviewed all the studies done in the field of natural fiber-reinforced composites from 2000 to 2010 [18]. They discussed about various natural fibers like flax, hemp, jute, kenaf, sisal, abaca, pineapple, ramie, coir, bamboo, rice husk, oil palm, bagasse, their chemical composition, structure and properties. They also compiled various treatment methods, physical as well as chemical to improve the tensile strength, thermal properties and morphology of these fibers. They also collected the papers on different polymer matrix, thermoset as well as thermoplastic used with different natural fibers.

G. L. Devnani (✉)
Department of Chemical Engineering, Harcourt Butler Technical University, Kanpur, U.P., India
e-mail: gldevnani@hbtu.ac.in

S. Thomas and P. Balakrishnan (eds.), *Green Composites*,
Materials Horizons: From Nature to Nanomaterials,
https://doi.org/10.1007/978-981-15-9643-8_10

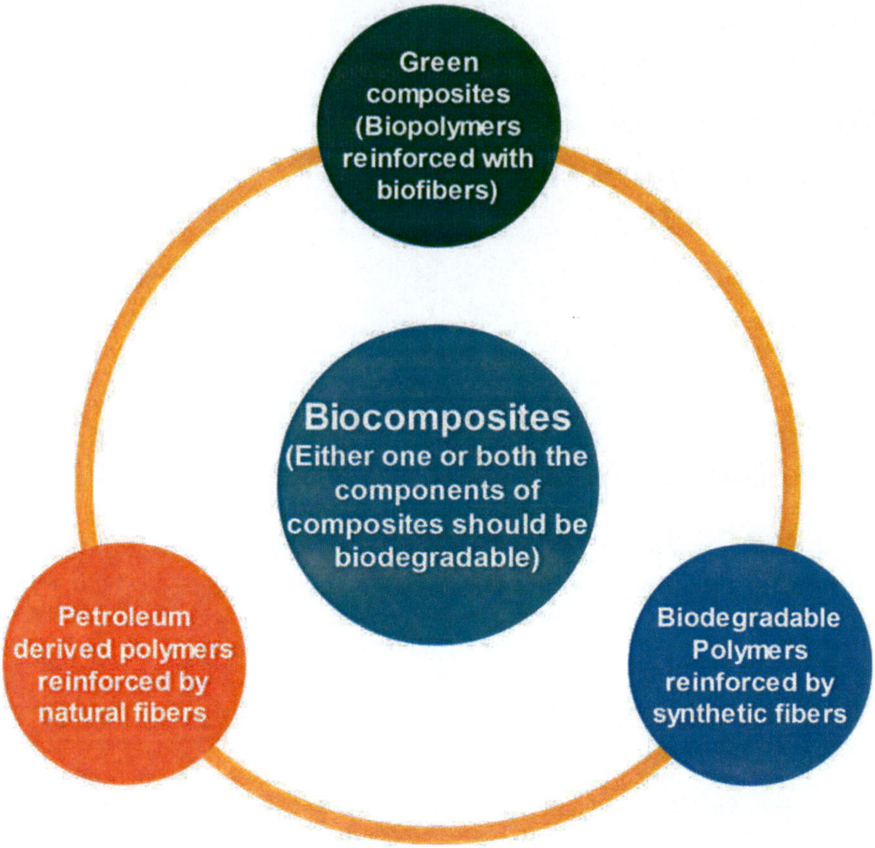

Fig. 1 Various categories of biocomposites. *Source* Author

They also discussed the factors influencing the processing of these composites, various fabrication procedures and their advantages and disadvantages. The characterization techniques like tensile testing, flexural and impact properties were also elaborated in this study. Pickering et al. discussed about recent developments and specifically mechanical performance of these green composites. They analyzed the effect of various factors like fiber and matrix selection, interface strength, geometry and orientations of fiber, fiber dispersion, porosity on the mechanical characteristics of natural fiber-reinforced polymer composites [50]. They also examined the effect of hybridization on the performance of these composites. Influence of moisture and weathering along with applications were also discussed in this review. Sanjay et al. did a comprehensive review on properties and characterization of natural fiber-reinforced polymer composites. They emphasized on tensile, flexural, impact, interlaminar, hardness and water absorption properties. Along that they also compiled the studies done on thermal and tribological properties of these novel materials.

FTIR, XRD and SEM characterizations were also discussed [55]. Ku et al. considered majorly tensile properties of these composites in their review paper [31]. They told about density, tensile strength, elongation, elastic modulus, of these natural fibers. They also collected information on density, water absorption, heat deflection temperature, coefficient of thermal expansion, tensile strength and elastic modulus of different polymer matrix like PP, LDPE, HDPE and PS. Mathematical modeling, rule of mixtures, transverse rule of mixtures, Halpin–Tsai equation, shear log theory and their application in composites were also discussed. Koronis et al. provided the application aspects of these composite materials in the field of automobile industry [30]. They discussed about green interior and green exterior composites in automobile industry. Different reinforcing elements like abaca, kenaf, hemp and flax, ramie jute, their mechanical performance and major issues and challenges regarding the application of natural fibers as reinforcement were talked in this study. The importance of matrix materials, mechanical characteristics of natural resins and concerns related to use of bio-based matrices were also analyzed. Sydow and Bienczak brought diversity of application of these natural fiber-based composites in food packaging in their review [61]. They discussed the barrier properties of these composites which is a very important aspect when we use these composite materials in food packaging. A comprehensive summary of the literature based on these diverse review papers has been presented in Table 1.

2 Issues with Natural Fibers

The main disadvantage of natural fibers is the poor compatibility and adhesion between fiber and matrix and the poor resistance to moisture absorption. Good adhesion between fiber and matrix is the key factor which ultimately decides the final properties of composites. Being lignocellulosic in nature, natural fibers are hydrophilic while the polymer matrix in which they are reinforced is basically hydrophobic in nature so compatibility between two opposite natured material is a key challenge for academicians and researchers. Lot of physical and chemical treatment methodologies have been suggested to improve the compatibility and adhesion between the fiber and matrix which is classified in Fig. 2. Mohanty et al. [43] gave an overview on different surface modification methods and their effect on the quality of biocomposites. They discussed about production, chemical composition and various properties along with advantages and disadvantages of natural fibers. Mechanism and chemistry of surface treatment methods like alkali treatment, graft copolymerization, etherification, acetylation, treatment with isocynate and maleated polypropylene were explained [43]. Li et al. [35] also compiled information on various chemical treatment methodologies of natural fibers for the application in natural fiber-reinforced composites. They reported the outcomes and findings of papers based on alkaline treatment, silane treatment, acetylation of fibers, benzoylation treatment, acrylation and acrylonitrile grafting maleated coupling, permanganate treatment, peroxide treatment, isocynate treatment and other chemical treatments [35]. John and Anandjiwala [24]

Table 1 Seminal review available on diverse aspects of natural fiber-based biocomposites

S. No.	Title	Key aspects covered	References
1	Green composites: A brief review	Summary of both academic and industrial research findings on this topical subject. Processability and rheology were also explained	[41]
2	Biocomposites reinforced with natural fibers: 2000–2010	Compilation of compositional, treatment, characterization studies	[18]
3	A review of recent developments in natural fiber composites and their mechanical performance	Mechanical performance of green composites	[50]
4	Characterization and properties of natural fiber polymer composites: A comprehensive review	Detailed review on mechanical properties and characterization studies done for these novel materials	[55]
5	A review on the tensile properties of natural fiber-reinforced polymer composites	Apart from tensile properties, modeling studies were also compiled	[31]
6	Green composites: A review of adequate materials for automotive applications. *Composites*	Technical performance of different biopolymers and their green reinforcements was compiled for automobile applications	[30]
7	The overview on the use of natural fiber-reinforced composites for food packaging	Biodegradable polymers like starch, cellulose, polyhyroxyalkanoate, chitosan, polylactic acid (PLA) and their composites reinforced by natural fibers were discussed with emphasis on food packaging applications	[61]
8	A review on synthesis and characterization of commercially available natural fibers: Part II	Less common and new natural fibers were discussed	[38]
Based on individual fibers			

(continued)

Table 1 (continued)

S. No.	Title	Key aspects covered	References
9	A review on natural areca fiber-reinforced polymer composite materials	Literature survey related to polymer composites reinforced with areca fibers, different characteristics of areca fibers, different surface treatment methodologies composite development with different matrices, characterization along with acoustics properties	[28]
10	Flax (*Linum usitatissimum* L.) fiber-reinforced polymer composite materials: A review on preparation, properties and prospects	Various factors like plant growth, harvesting, surface treatment methodologies along with structural property relationship of flax fibers were reviewed	[52]
11	Bagasse composites: A review of material preparation, attributes and affecting factors	Factors involved in modification and preparation of bagasse fibers and its reinforced biocomposites were analyzed	[64]
12	Mechanical properties evaluation of sisal fiber-reinforced polymer composites	Fiber percentage, fiber morphology, surface modification and hybridization aspects of sisal fiber in various polymer matrix were discussed	[57]
13	Recent developments in bamboo fiber-based composites: a review	Recent advancements in processing methodology and various applications related to bamboo fibers	[45]
14	Banana and plantain fiber-reinforced polymer composites	From fiber extraction to reinforcement and comparison with other fibers, most of the aspects	[3]
15	A review of coir fiber-reinforced polymer composites	The synthesis and development of various matrices reinforced by coir natural fiber and the mechanical and thermal characteristics have been studied	[4]

(continued)

Table 1 (continued)

S. No.	Title	Key aspects covered	References
16	Review on mechanical properties evaluation of pineapple leaf fiber (PALF)-reinforced polymer composites	Compilation of mechanical properties evaluation of PALF and several parameters affecting performance like type of fiber, its length, matrix type, fiber geometry	[63]

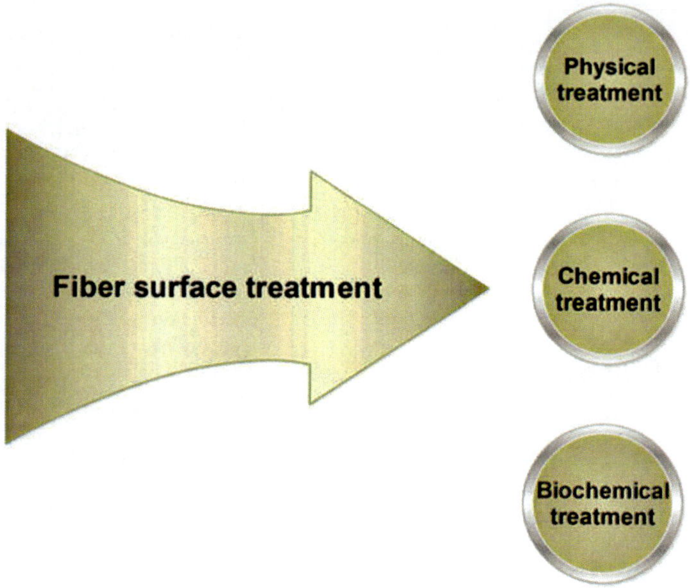

Fig. 2 Different treatment methodologies. *Source* Author

described about the latest developments of chemical modifications and characterization techniques in the field of natural fiber-based composites. They discussed the effect of chemical modification on the performance of aspen fiber composites, abaca fiber composites, bagasse fiber composites, bamboo fiber composites, banana fiber composites, coir fiber composites, date palm fiber composites, flax fiber composites, hemp fiber composites, henequen fiber composites, isora fiber composites, jute fiber composites, kapok fiber composites, kenaf fiber composites, luffa fiber composites, oil palm fiber composites, pineapple fiber composites, ramie fiber composites and sisal fiber composites [24]. Many other papers and compilation work are available in literature which is summarized in Table 2.

Table 2 Compilation of available work on surface treatment methodologies of natural fibers

S. No.	Title	Key aspects covered/discussed	References
1	Surface modifications of natural fibers and performance of the resulting biocomposites: An overview	Various treatment methodologies for natural fiber reinforcement in synthetic as well as biodegradable polymers were discussed	[43]
2	The surface modification of cellulose fibers for use as reinforcing elements in composite materials	Different physical treatment methodologies like plasma, laser, corona, radiation for natural fibers along with chemical methods such as grafting and alkali treatment were discussed extensively	[9]
3	Chemical treatments of natural fiber for use in natural fiber-reinforced composites: A review	Chemical treatments including silane alkali, benzoylation, acetylation, maleated coupling, acrylation, treatment with isocyanates, permanganate and others were discussed	[35]
4	Recent developments in chemical modification and characterization of natural fiber-reinforced composites	Various treatment methodologies to make natural fiber more compatible and less hydrophilic	[24]
5	Pretreatments of natural fibers and their application as reinforcing material in polymer composites—a review	Plasma treatment and graft copolymerization were emphasized along with discussion of other treatment methodologies of natural fibers	[26]
6	Physical modification of natural fibers and thermoplastic films for composites—A review	Steam explosion, thermomechanical methods, plasma, corona treatment were analyzed in detail along with other physical methods	[46]
7	Chemical treatments on plant-based natural fiber-reinforced polymer composites: An overview	Along with traditional treatment methodologies, information on some new chemical treatments like stearic acid, triazine sodium chlorite was presented	[25]
8	Surface modification of plant fibers using environment-friendly methods for their application in polymer composites, textile industry and antimicrobial activities	Recommended environment-friendly methods such as enzyme treatment, plasma treatment, fungi and bacteria	[27]
9	Surface treatments of natural fibers—a review: Part 1	Special discussion on electric discharge, mercerization and graft copolymerization	[2]

(continued)

Table 2 (continued)

S. No.	Title	Key aspects covered/discussed	References
10	Interface and bonding mechanisms of plant fiber composites: an overview	In-depth analysis of structure property relationship of fiber and matrix and according to that selection of suitable treatment methodology	[65]
11	Hydrophobic treatment of natural fibers and their composites—A review	Different treatment methods were suggested to decrease the moisture absorption and degradation of fiber	[5]
12	Radiation-induced modifications in natural fibers and their biocomposites: opportunities for controlled physicochemical modification pathways	Effect of electron beam and gamma (γ) radiation on natural fibers were compiled	[32]
13	Effects of fiber treatment on mechanical properties of kenaf fiber-reinforced composites: a review	Effect of treatment methodologies on a specific fiber kenaf and its reinforced polymer composites	[21]

3 Different Surface Treatment Methodologies for Natural Fiber

From Table 2, it is clear that lot of experimentation and research work had been done to improve the performance and compatibility of natural fibers with polymer matrix but with the advent of newer characterization techniques like high precision microscopy, surface characterization and innovation in knowledge like nanotechnology and biotechnology there is always a scope of improvement in the existing performance of natural fiber-based polymer composites. All the treatment methodologies and their recent developments have been discussed below in a systematic manner.

3.1 Physical Treatment

Physical treatment is applied on natural fibers to basically serve two purposes (a) to separate the bundles of fibers into filaments (b) modification of fiber surface to make it more compatible with polymer matrix [9, 46]. Following are the different treatment schemes used for this purpose (Fig. 3).

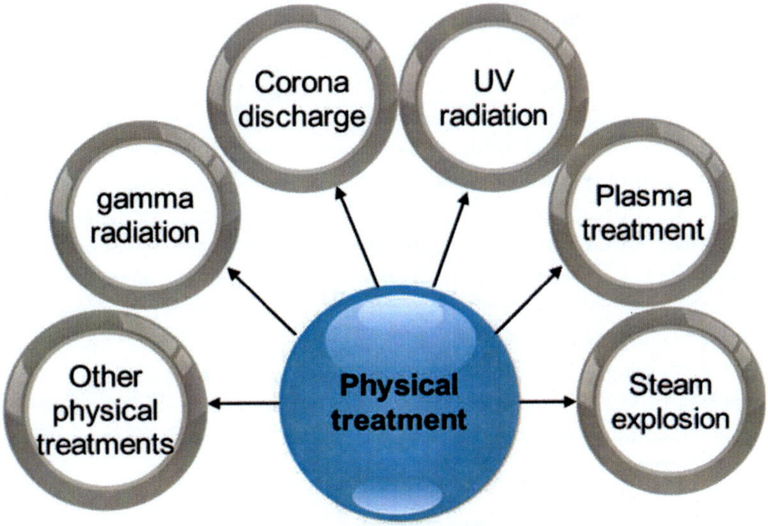

Fig. 3 Different physical treatment for natural fibers. *Source* Author

3.1.1 Steam Explosion

Steam explosion method was developed by Mason as a biomass pre-treatment technique. By using saturated water steam, raw materials are heated 160–260 °C for seconds to minutes time duration at some specific pressure and then suddenly, pressure reduced to atmospheric pressure. At high-pressure and high-temperature condition, raw material expanded because of overheated liquid and gap which would be filled by steam, when pressure is reduced then vaporization of overheated intracellular leads to cellulose explosion due to quick expanding volume, several holes formed so that molecular substances can release from cell. This method got popularity among researchers because of its tremendous advantages, such as no pollution, economical and energy efficient. Steam explosion technique is a fast pre-treatment for affecting plant cell with a sharp change in pressure at high temperature. By this treatment, lignocellulosic materials spilt into its main constitutes such as cellulose, hemicellulose and amorphous lignin. Cellulose percentage increased during steam explosion and increment in crystallinity of stem exploded banana fiber (54.1), pineapple leaf (63.7) and jute fiber (52.9) was obtained [1]. Steam explosion on commercial cellulose was used at temperature (150–200 °C) and reaction time 5–30 min, in initial characterization 91.9% cellulose content was obtained, no degradation was seen in thermal property, and steam explosion method is recommended better substitute over other mechanical treatments [36]. Review of properties of PALF fiber using different methods also suggested steam explosion method for removal of non-cellulosic content from fiber by exposing fiber in steam with its additive at high temperature and high pressure [63]. Lei and fang also improved the performance of sisal fibers by continuous steam explosion [33].

3.1.2 Plasma Treatment

Sir William Crookes described plasma as the fourth state of mater in 1879 and in 1929, first time "plasma" term was used by Langmuir. Plasma consists of electron, neutral atoms, positively and negatively charged molecules. Plasma contain characteristics of visible glow discharge which has colors, and range of blue white to dark purple depends on gas type. Plasma treatment is considered as an eco-friendly treatment for modification of fiber's surface. This treatment has advantages such as that it requires very less chemical because of less consumption during treatment [60]. After this process, no more drying is needed as it is a dry process, does not destroy bulk properties, by this method material can easily be modified which are hard to modify, it does not produce any waste, and this process takes very less time. Surface modifications of cellulose and poly vinyl alcohol were done by using non-thermal argon plasma, and this treatment did not hamper the arrangement of various polymer chains in the composites [13].

3.1.3 Treatment by UV Radiation

The effect of UV radiation on the performance of natural fiber-based composites depends on several parameters like intensity of radiation, exposure duration and wavelength of exposure. Under optimized conditions, it improves the performance of natural fiber-based composites. Mahajan et al. compiled the research work related to UV radiation effect on natural and synthetic fiber-based composites [39].

3.1.4 Corona Discharge Treatment

This technique uses low-temperature corona discharge to induce changes in the fiber surfaces to make it more compatible with polymer matrix. Corona discharges are comparatively lower power electrical discharges that occur in the range of 1 atmospheric pressure range. The corona is produced by heavy electric fields linked with small cross-sectional wires, needles on an electrode [10] used corona treatment on flax natural fiber along with other treatments. They applied 1 kW of discharge power at the rate of 4 m min^{-1}, 5 times of each side and found that contaminants were removed and surface roughness had increased [10].

3.1.5 Treatment by γ Radiation

Gama radiations consisting very high energy are capable of modifying the properties of polymer surfaces. The optimum exposure of gamma radiation improves the tensile properties of composite materials in some extent for the use of materials in different practical purposes. The cause of this enhancement in tensile properties was the high-energy gamma radiation, capable of making cross-link among the molecules. A significant amount of improvement due to gamma radiations in various composites like jute polyester and jute pp was experimented by Kabir et al. [59].

3.1.6 Ultrasonic Treatment

Ultrasonic techniques are not so popular or used in comparisons with other methods. But this is also an effective method for removal of various substances and pollutant, without using surfactants from fiber surface. It is an electromagnetic radiation which has range from 10 to 400 nm. From past some years, ultrasonic method has been used in several industries [12]. Renouard et al. showed that ultrasonication can be a useful method to modify lignocellulosic material composition. They used coir, hemp and short flax fibers for their work. After 24 h, they observed the optimal degradation and found that ultrasonic application on these fibers only degraded hemicelluloses present in the fibers [54].

3.1.7 Other Physical Treatments

Apart from abovementioned treatments, there are few other physical treatments like simple mechanical treatments which involve stretching, rolling, etc., which can improve the tensile strength of the fiber but at the same time can increase elongation [66]. Solvent extraction is also used some times to get high content of cellulose in plant fibers however a precaution is required that this can not degrade the fiber surface [51]. Thermal treatment, fiber beating and laser treatment are few other physical treatment methods that can be used depending on situation and properties of fibers [40].

3.2 Chemical Treatment Methodologies

These lignocellulosic fibers contain basically cellulose, hemicelluloses and lignin. Cellulose is the main component which is responsible for providing strength to the fiber. Different chemicals are applied for treatment of natural fibers. The main purpose is to increase the cellulose content of the fiber and to remove the unwanted impurities like wax and oil. Chemical treatment also improves the adhesion between fiber and matrix by creating some new bonds and disturbing old structure. Different chemical treatment methodologies are shown in Fig. 4. The detailed description of various chemical treatment methods used is given below, and recent developments and findings have been covered in Table 3.

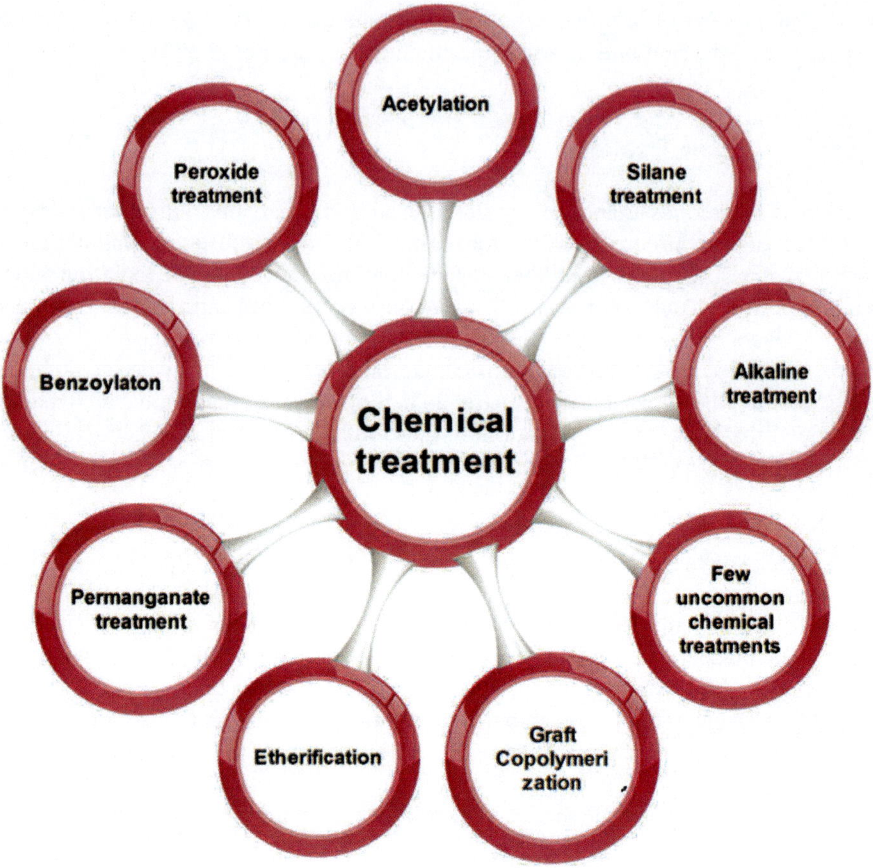

Fig. 4 Different chemical treatment for natural fibers. *Source* Author

3.2.1 Alkaline Treatment

Treatment of natural fibers with optimum concentration of alkali is very economical and promising treatment methodology. The chemical changes occur according to the reaction.

$$\text{Fiber} - \text{OH} + \text{NaOH} = \text{Fiber} - \text{O} - \text{Na} + \text{H}_2\text{O} + \text{surface impurities}$$

The alkali treatment removes a certain amount of lignin, hemicellulose, oils and waxes. The removal of these cementing materials causes the crystallinity index and cellulose content of the fibers to increase. The tensile strength also increases due to increased cellulose content. The improvement in performance of composites because of alkali treatment of natural fibers which have been studied by several researchers. The increasing roughness and fibrillation are assumed to be responsible for better

Table 3 Compilation of some recent chemical treatment of natural fibers

S. No.	Fiber	Matrix	Treatment method	Observation/key findings	References
1	Milki weed	PVA	5% NaOH at 50–60 °C for different time	PVA composites were prepared by hand layup method. Tensile properties of alkali treated composites were significantly high as compared to untreated samples	[56]
2	Kans grass filler	HDPE and LDPE	Maleic anhydride grafted compatibilizer	Performance of kans grass filler-based polyethylene and polypropylene composites has been improved at 10% filler loading	[7, 8]
3	Hemp fiber	PLA and epoxy	Silane and acetic anhydride	PLA composites based on hemp fibers were developed, and 10–13% more activation energy was observed in case of acetic anhydride treated composites while epoxy composites observed improved performance in case of silane treated composites	[49, 58]
4	Kenaf	PLA	Hydrogen peroxide in alkaline medium	Treatment improved the adhesion between Kenaf fiber and PLA matrix	[53]
5	Polyethylene fibers	Natural rubber	Potassium permanganate	Several microfibrillation between the microfiber and the natural rubber matrix was observed	[34]

(continued)

Table 3 (continued)

S. No.	Fiber	Matrix	Treatment method	Observation/key findings	References
6	Alpha fiber	Polypropylene	Etherification	Significant improvement in the properties of composites as compared to untreated one	[6]
7	Bagasse	Polypropylene	Ionic liquid	Ionic liquid treated composites upgraded the properties of composites	[47]
8	Agave Americana	–	Stearic acid	Treatment improved the properties of fiber	[37]

adhesion between natural fiber and polymer matrix. Devnani and Sinha [16] applied different concentrations of alkali that is 5 and 10% on African Teff straw fiber and found that there is significant improvement in the properties of fiber in terms of strength, morphology and crystallinity [16]. They also worked on kans grass fiber and got excellent results with 5% treatment of NaOH [17].

3.2.2 Graft Copolymerization

Modification of natural fibers through graft copolymerization is another very popular method to improve the physical and chemical properties of natural fiber-based composites. Through grafting, a suitable polymer is attached to the fiber surface whose solubility parameter is same with the polymer matrix. Grafting does not affect the biodegradability of cellulosic fiber. Grafting of methyl methacrylate and acrylonitrile is the well-established practices for natural fibers [20]. Maleic anhydride grafted polyethylene and polypropylene were used as a compatibilizer to improve the adhesion between kans grass filler and HDPE as well as PP composites [7, 8].

3.2.3 Silane Treatment

The chemical formula of silane is SiH_4, and these coupling agents can reduce the hydrophilic tendency of natural fibers by removing the hydroxyl groups. The hydrolyzable alkoxy groups form silanol in the presence of moisture. The hydroxyl group present in fiber then reacts with forming stable covalent bonds.

$$CH_2CHSi(OC_2H_5)_3 + H_2O = CH_2CHSi(OH)_3 + 3C_2H_5OH$$

$$CH_2CHSi(OH)_3 + Fiber - OH = CH_2CHSi(OH)_2O - Fiber + H_2O$$

Silane treatment in the case of hemp fibers was reported more effective as compare to alkali treatment [58].

3.2.4 Acetylation

Acetylation is another important surface treatment method for natural fibers to make it more hydrophobic. It also reduces the swelling tendency of natural fibers in the presence of water. The acetyl group reacts with hydroxyl group of water and reduces the hydrophilic behavior of natural fibers. This method also improves the dimensional stability.

$$Fiber - OH + CH_3 - C(=O) - O - C(=O) - CH_3 = Fiber - OCOCH_3 + CH_3COOH$$

Flax fibers were acetylated by putting in solution consisting of 250 ml toluene, 125 ml acetic anhydride and a small amount of catalyst perchloric acid (60%). The temperature was kept at 60 °C. More than 25% enhancement in strength was experienced in case of flax-based polypropylene composites [11].

3.2.5 Peroxide Treatment

Being a convenient method, peroxide treatment is also attracting researchers to improve the mechanical properties of natural fibers. **Benzoyl peroxide** and **dicumyl peroxide** are the common organic peroxides which have been used extensively for the natural fibers to improve their performance. During the process, decomposition of organic peroxide takes place in free radicals and this free radical reacts with cellulose of the fiber. Treatment of kenaf fiber was done in alkaline medium containing hydrogen peroxide. Substantial improvement in the morphology and strength was observed in its PLA-based composites [53].

3.2.6 Benzoylation

Importance of benzoylation is to decrease hydrophilic nature of natural fiber and to increase adhesion between fiber and matrix. Benzoyl chloride is the common reagent used for this purpose, and it reacts with hydroxyl group of the cellulose and improves the performance of the natural fiber. Remarkable improvement in mechanical properties of sugar palm fiber was observed in case of alkali pretreated and benzoylated fibers [44].

3.2.7 Permanganate Treatment

Potassium permanganate is the chemical which is used for this method. The Mno_4^- group of the potassium permanganate reacts with cellulosic group of the natural fiber and forms a complex ion. Highly reactive Mn ion induces the graft polymerization. $KMnO_4$ treatment enhanced the roughness and fibrillation in polythene fibers which significantly improved the properties of its rubber composites [34].

3.2.8 Etherification

Etherification is a chemical treatment which facilitates fiber to react easily with polymer chain of matrix by grafting bifunctional monomers [48]. A significant improvement was noted in thermal stability of alfa fiber-reinforced polypropylene composites in case of etherified fibers [6].

3.2.9 Few Uncommon Chemical Treatments

Apart from these treatments, there are few other uncommon treatments which researchers have experimented and got encouraging results. **Ionic liquid treatment** by ChOAc (choline acetate) in case of bagasse powder improved the performance of its polypropylene composites [47]. Similarly, 1% **stearic acid** solution in ethanol significantly removes the amorphous portion of the Agave Americana fiber [37]. An eco-friendly **sodium bicarbonate** treatment was also applied on sisal fibers, and improvement of performance was observed in its epoxy composites as compare to untreated one [19].

3.3 Biochemical Treatment (Enzymatic and Fungi)

The abovementioned chemical treatment used for surface modification has many advantages but in spite of that they are toxic in nature which can create problems for the environment. A new biochemical treatment method is **enzyme treatment** which is getting importance day by day because of its environment-friendly nature. Enzymes attack basically on non-cellulosic components of fiber moreover enzymes can be recycled and cost effective too. Ramie fiber was treated by bacterial cellulase. Fibers modified enzymatically were incorporated as reinforcing material for the fabrication of poly(butylene succinate) biocomposites. Due to enzymatic treatment, polysaccharide layers and gum from the surface of ramie fibers were removed and hence contributed toward the enhanced compatibility between fibers and poly butylene succinate [62]. Use of enzymes as alternate of dew retting was also analyzed. Enzyme retting with polygalacturonas enzyme was experimented for flax fibers and improvements were observed in its biocomposites [14, 15]. Pectane lyase enzyme pre-treatment was used for the development of bagasse/PLA composites, and effect was observed [22]. Another eco-friendly biochemical method is fungi treatment. White rot fungi treatment caused the pits on hemp fiber surface which not only provided the roughness on fiber surface but also increased the adhesion between fiber and matrix [23, 29].

4 Conclusion

Bio-based green composites are the promising option as a replacement of synthetic fiber and matrix-based composites whose sustainability is dwindling because of dependence on non-renewable petroleum feed stock. Natural fiber-based green composites are not only environment friendly but also offering versatile properties, at the same time they are creating potential of employment opportunities in the rural sector where these fibers originate. To overcome the existing issues like poor adhesion, compatibility and thermal degradation, various treatment methodologies have been suggested by researchers including physical, chemical and biochemical route. Compilation of existing literature including alkali treatment, graft copolymerization along with few uncommon treatment methodologies like ionic liquid treatment have been discussed. Suitable treatment methodology along with optimum concentration and exposure can definitely improve the properties of natural fiber-based green composites and overcome the existing shortcomings with these novel materials. With suitable treatment method, natural fiber-based green composites have huge potential to replace the existing synthetic composites.

References

1. Abraham E, Deepa B, Pothan LA, Jacob M, Thomas S, Cvelbar U, Anandjiwala R (2011) Extraction of nanocellulose fibrils from lignocellulosic fibres: a novel approach. Carbohyd Polym 86(4):1468–1475. https://doi.org/10.1016/j.carbpol.2011.06.034
2. Adekunle KF (2015) Surface treatments of natural fibres—a review: part 1. Open J Polym Chem 05(03):41–46. https://doi.org/10.4236/ojpchem.2015.53005
3. Adeniyi AG, Ighalo JO, Onifade DV (2019) Banana and plantain fiber-reinforced polymer composites. J Polym Eng 39(7):597–611. https://doi.org/10.1515/polyeng-2019-0085
4. Adeniyi AG, Onifade DV, Ighalo JO, Adeoye AS (2019) A review of coir fiber reinforced polymer composites. Compos B Eng 176(August):107305. https://doi.org/10.1016/j.compositesb.2019.107305
5. Ali A, Shaker K, Nawab Y, Jabbar M, Hussain T, Militky J, Baheti V (2018) Hydrophobic treatment of natural fibers and their composites—a review. J Ind Text 47(8):2153–2183. https://doi.org/10.1177/1528083716654468
6. Arrakhiz FZ, Elachaby M, Bouhfid R, Vaudreuil S, Essassi M, Qaiss A (2012) Mechanical and thermal properties of polypropylene reinforced with Alfa fiber under different chemical treatment. Mater Des 35:318–322. https://doi.org/10.1016/j.matdes.2011.09.023

7. Barman A, Shrivastava NK, Khatua BB, Ray BC (2015) Effect of filler content on the properties of polypropylene/*Saccharum spontaneum* green composite. Polym-Plast Technol Eng 54:1231–1240. https://doi.org/10.1080/03602559.2014.1003233
8. Barman A, Shrivastava NK, Khatua BB, Ray BC (2015) Green composites based on high-density polyethylene and *Saccharum spontaneum*: effect of filler content on morphology. Therm Mech Prop. https://doi.org/10.1002/pc.23126
9. Belgacem MN, Gandini A (2005) The surface modification of cellulose fibres for use as reinforcing elements in composite materials. Compos Interfaces 12(1–2):41–75. https://doi.org/10.1163/1568554053542188
10. Bessa J, Matos J, Mota C, Cunha F, Araújo I, Silva L, Pinho E, Fangueiro R (2017) Influence of surface treatments on the mechanical properties of fibre reinforced thermoplastic composites. Procedia Eng200:465–471. https://doi.org/10.1016/j.proeng.2017.07.065
11. Bledzki AK, Mamun AA, Lucka-Gabor M, Gutowski VS (2008) The effects of acetylation on properties of flax fibre and its polypropylene composites. Exp Polym Lett 2(6):413–422. https://doi.org/10.3144/expresspolymlett.2008.50
12. Czaplicki Z, Ruszkowski K (2014) Optimization of scouring alpaca wool by ultrasonic technique. J Nat Fibers 11(2):169–183. https://doi.org/10.1080/15440478.2013.864577
13. Dalei G, Das S, Jena SR, Nayak J, Samanta L, Das SP (2019) Surface modification of cellulose/polyvinyl alcohol biocomposites by non-thermal argon plasma: applications towards biological relevance. Cellulose 26(4):2437–2451. https://doi.org/10.1007/s10570-019-02243-0
14. De Prez J, Van Vuure AW, Ivens J, Aerts G, Van de Voorde I (2019) Effect of enzymatic treatment of flax on fineness of fibers and mechanical performance of composites. Compos A Appl Sci Manuf 123(January):190–199. https://doi.org/10.1016/j.compositesa.2019.05.007
15. De Prez J, Van Vuure AW, Ivens J, Aerts G, Van de Voorde I (2020) Flax treatment with strategic enzyme combinations: effect on fiber fineness and mechanical properties of composites. J Reinf Plast Compos 39(5–6):231–245. https://doi.org/10.1177/0731684419884645
16. Devnani GL, Sinha S (2018) African Teff Straw as a potential reinforcement in polymer composites for light-weight applications: mechanical, thermal, physical, and chemical characterization before and after alkali treatment. J Nat Fibers 1–15. https://doi.org/10.1080/15440478.2018.1546640
17. Devnani GL, Sinha S (2019) Extraction, characterization and thermal degradation kinetics with activation energy of untreated and alkali treated *Saccharum spontaneum* (Kans grass) fiber. Compos B 166(February):436–445. https://doi.org/10.1016/j.compositesb.2019.02.042
18. Faruk O, Bledzki AK, Fink HP, Sain M (2012) Biocomposites reinforced with natural fibers: 2000–2010. Prog Polym Sci 37(11):1552–1596. https://doi.org/10.1016/j.progpolymsci.2012.04.003
19. Fiore V, Scalici T, Nicoletti F, Vitale G, Prestipino M, Valenza A (2016) A new eco-friendly chemical treatment of natural fibres: effect of sodium bicarbonate on properties of sisal fibre and its epoxy composites. Compos B 85:150–160. https://doi.org/10.1016/j.compositesb.2015.09.028
20. Fuqua MA, Huo S, Ulven CA (2012) Natural fiber reinforced composites. Polym Rev 52(3–4):259–320. https://doi.org/10.1080/15583724.2012.705409
21. Hamidon MH, Sultan MTH, Ariffin AH, Shah AUM (2019) Effects of fibre treatment on mechanical properties of kenaf fibre reinforced composites: a review. J Mater Res Technol 8(3):3327–3337. https://doi.org/10.1016/j.jmrt.2019.04.012
22. Huerta-Cardoso O, Durazo-Cardenas I, Longhurst P, Simms NJ, Encinas-Oropesa A (2020) Fabrication of agave tequilana bagasse/PLA composite and preliminary mechanical properties assessment. Ind Crops Prod 152(June 2019):112523. https://doi.org/10.1016/j.indcrop.2020.112523
23. Jafari M (2007) The effect of *Pleurotus* spp. fungi on chemical composition and in vitro digestibility of rice straw. Pak J Biol Sci 10(15):2460–2464
24. John MJ, Anandjiwala RD (2008) Recent developments in chemical modification and characterization of natural fiber-reinforced composites. Polym Compos. https://doi.org/10.1002/pc.20461

25. Kabir MM, Wang H, Lau KT, Cardona F (2012) Chemical treatments on plant-based natural fibre reinforced polymer composites: an overview. Compos B Eng 43(7):2883–2892. https://doi.org/10.1016/j.compositesb.2012.04.053
26. Kalia S, Kaith BS, Kaur I (2009) Pretreatments of natural fibers and their application as reinforcing material in polymer composites—a review. Polym Eng Sci 1253–1272. https://doi.org/10.1002/pen.21328
27. Kalia S, Thakur K, Celli A, Kiechel MA, Schauer CL (2013) Surface modification of plant fibers using environment friendly methods for their application in polymer composites, textile industry and antimicrobial activities: a review. J Environ Chem Eng 1(3):97–112. https://doi.org/10.1016/j.jece.2013.04.009
28. Kamath SS, Sampathkumar D, Bennehalli B (2017) A review on natural areca fibre reinforced polymer composite materials. Ciencia E Tecnol Dos Mater 29(3):106–128. https://doi.org/10.1016/j.ctmat.2017.10.001
29. Khoshnava SM, Rostami R, Ismai M, Valipour A (2014) The using fungi treatment as green and environmentally process for surface modification of natural fibres. Appl Mech Mater 554(October 2015):116–122. https://doi.org/10.4028/www.scientific.net/AMM.554.116
30. Koronis G, Silva A, Fontul M (2013) Composites: part B green composites: a review of adequate materials for automotive applications. Compos B 44(1):120–127. https://doi.org/10.1016/j.compositesb.2012.07.004
31. Ku H, Wang H, Pattarachaiyakoop N, Trada M (2011) A review on the tensile properties of natural fiber reinforced polymer composites. Compos: Part B 42:856–873. https://doi.org/10.1016/j.compositesb.2011.01.010
32. Le Moigne N, Sonnier R, El Hage R, Rouif S (2017) Radiation-induced modifications in natural fibres and their biocomposites: opportunities for controlled physico-chemical modification pathways? Ind Crops Prod 109(August):199–213. https://doi.org/10.1016/j.indcrop.2017.08.027
33. Lei B, Feng Y (2020) Sustainable thermoplastic bio-based materials from sisal fibers. J Clean Prod 265:121631. https://doi.org/10.1016/j.jclepro.2020.121631
34. Li W, Meng L, Ma R (2016) Effect of surface treatment with potassium permanganate on ultra-high molecular weight polyethylene fiber reinforced natural rubber composites. Polym Testing 55:10–16. https://doi.org/10.1016/j.polymertesting.2016.08.006
35. Li X, Tabil LG, Panigrahi S (2007) Chemical treatments of natural fiber for use in natural fiber-reinforced composites: a review. J Polym Environ 15(1):25–33. https://doi.org/10.1007/s10924-006-0042-3
36. Lorenzo-Hernando A, Martín-Juárez J, Bolado-Rodríguez S (2018) Study of steam explosion pretreatment and preservation methods of commercial cellulose. Carbohyd Polym 191(February):234–241. https://doi.org/10.1016/j.carbpol.2018.03.021
37. Madhu P, Sanjay MR, Jawaid M, Siengchin S, Khan A, Pruncu CI (2020) A new study on effect of various chemical treatments on Agave Americana fiber for composite reinforcement: physico-chemical, thermal, mechanical and morphological properties. Polym Testing 85:106437. https://doi.org/10.1016/j.polymertesting.2020.106437
38. Madhu P, Sanjay MR, Senthamaraikannan P, Pradeep S, Saravanakumar SS, Yogesha B (2019) A review on synthesis and characterization of commercially available natural fibers: Part II. J Nat Fibers 16(1):25–36. https://doi.org/10.1080/15440478.2017.1379045
39. Mahajan S (2016) Preface: international conference on recent trends in physics (ICRTP 2016). J Phys: Conf Ser 755(1). https://doi.org/10.1088/1742-6596/755/1/011001
40. Manna S, Saha P, Chowdhury S, Thomas S (2017) Alkali treatment to improve physical, mechanical and chemical properties of lignocellulosic natural fibers for use in various applications. Lignocellul Prod Ind Appl 47–63. https://doi.org/10.1002/9781119323686.ch3
41. La Mantia FP, Morreale M (2011) Composites: part A green composites: a brief review. Compos A 42(6):579–588. https://doi.org/10.1016/j.compositesa.2011.01.017
42. Mitra BC (2014) Environment friendly composite materials: biocomposites and green composites. Def Sci J 64(3):244–261. https://doi.org/10.14429/dsj.64.7323

43. Mohanty AK, Misra M, Drzal LT (2001) Surface modifications of natural fibers and performance of the resulting biocomposites: an overview. Compos Interfaces 8(5):313–343. https://doi.org/10.1163/156855401753255422
44. Mohd Izwan S, Sapuan SM, Zuhri MYM, Mohamed AR (2020) Effects of benzoyl treatment on NaOH treated sugar palm fiber: tensile, thermal, and morphological properties. J Mater Res Technol 9(3):5805–5814. https://doi.org/10.1016/j.jmrt.2020.03.105
45. Muhammad A, Rahman MR, Hamdan S, Sanaullah K (2019) Recent developments in bamboo fiber-based composites: a review. Polym Bull 76(5):2655–2682. https://doi.org/10.1007/s00289-018-2493-9
46. Mukhopadhyay S, Fangueiro R (2009) Physical modification of natural fibers and thermoplastic films for composites—a review. J Thermoplast Compos Mater 22(2):135–162. https://doi.org/10.1177/0892705708091860
47. Ninomiya K, Abe M, Tsukegi T, Kuroda K, Omichi M, Takada K, Noguchi M, Tsuge Y, Ogino C, Taki K, Taima T (2017) Ionic liquid pretreatment of bagasse improves mechanical property of bagasse/polypropylene composites. Ind Crops Prod 109(May):158–162. https://doi.org/10.1016/j.indcrop.2017.08.019
48. Ouarhim W, Zari N, Bouhfid R, Qaiss AEK (2018) Mechanical performance of natural fibers-based thermosetting composites. Mech Phys Testing Biocompos Fibre-Reinf Compos Hybrid Compos 43–60. https://doi.org/10.1016/B978-0-08-102292-4.00003-5
49. Oza S, Ning H, Ferguson I, Lu N (2014) Effect of surface treatment on thermal stability of the hemp-PLA composites: correlation of activation energy with thermal degradation. Compos B Eng 67:227–232. https://doi.org/10.1016/j.compositesb.2014.06.033
50. Pickering KL, Efendy MGA, Le TM (2016) A review of recent developments in natural fibre composites and their mechanical performance. Compos A Appl Sci Manuf 83:98–112. https://doi.org/10.1016/j.compositesa.2015.08.038
51. Płotka-Wasylka J, Rutkowska M, Owczarek K, Tobiszewski M, Namieśnik J (2017) Extraction with environmentally friendly solvents. TrAC-Trends Anal Chem 91:12–25. https://doi.org/10.1016/j.trac.2017.03.006
52. Ramesh M (2019) Flax (*Linum usitatissimum* L.) fibre reinforced polymer composite materials: a review on preparation, properties and prospects. Prog Mater Sci 102:109–166. https://doi.org/10.1016/j.pmatsci.2018.12.004
53. Razak NIA, Ibrahim NA, Zainuddin N, Rayung M, Saad WZ (2014) The influence of chemical surface modification of kenaf fiber using hydrogen peroxide on the mechanical properties of biodegradable kenaf fiber/poly(Lactic Acid) composites. Molecules 19(3):2957–2968. https://doi.org/10.3390/molecules19032957
54. Renouard S, Hano C, Doussot J, Blondeau JP, Lainé E (2014) Characterization of ultrasonic impact on coir, flax and hemp fibers. Mater Lett 129:137–141. https://doi.org/10.1016/j.matlet.2014.05.018
55. Sanjay MR, Madhu P, Jawaid M, Senthamaraikannan P, Senthil S, Pradeep S (2018) Characterization and properties of natural fiber polymer composites: a comprehensive review. J Cleaner Prod 172:566–581. https://doi.org/10.1016/j.jclepro.2017.10.101
56. Sayanjali Jasbi M, Hasani H, Zadhoush A, Safi S (2018) Effect of alkali treatment on mechanical properties of the green composites reinforced with milkweed fibers. J Text Inst 109(1):24–31. https://doi.org/10.1080/00405000.2017.1320816
57. Senthilkumar K, Saba N, Rajini N, Chandrasekar M, Jawaid M, Siengchin S, Alotman OY (2018) Mechanical properties evaluation of sisal fibre reinforced polymer composites: a review. Constr Build Mater 174:713–729. https://doi.org/10.1016/j.conbuildmat.2018.04.143
58. Sepe R, Bollino F, Boccarusso L, Caputo F (2018) Influence of chemical treatments on mechanical properties of hemp fiber reinforced composites. Compos B Eng 133:210–217. https://doi.org/10.1016/j.compositesb.2017.09.030
59. Shahriar Kabir M, Hossain MS, Mia M, Islam MN, Rahman MM, Hoque MB, Chowdhury AMS (2018) Mechanical properties of gamma-irradiated natural fiber reinforced composites. Nano Hybrids Compos 23(January 2019):24–38. https://doi.org/10.4028/www.scientific.net/nhc.23.24

60. Sun D (2016) Biodegradable green composites. Biodegrad Green Compos 19–38. https://doi.org/10.1002/9781118911068.ch1
61. Sydow Z, Bieńczak K (2019) The overview on the use of natural fibers reinforced composites for food packaging. J Nat Fibers 16(8):1189–1200. https://doi.org/10.1080/15440478.2018.1455621
62. Thakur K, Kalia S (2017) Enzymatic modification of ramie fibers and its influence on the performance of ramie-poly(butylene succinate) biocomposites. Int J Plast Technol 21(1):209–226. https://doi.org/10.1007/s12588-017-9178-3
63. Todkar SS, Patil SA (2019) Review on mechanical properties evaluation of pineapple leaf fibre (PALF) reinforced polymer composites. Compos B Eng 174(May):106927. https://doi.org/10.1016/j.compositesb.2019.106927
64. Xiong W (2018) Bagasse composites: a review of material preparation, attributes, and affecting factors. J Thermoplast Compos Mater 31(8):1112–1146. https://doi.org/10.1177/0892705717734596
65. Zhou Y, Fan M, Chen L (2016) Interface and bonding mechanisms of plant fibre composites: an overview. Compos B Eng 101:31–45. https://doi.org/10.1016/j.compositesb.2016.06.055
66. Zimniewska M et al (2016) In: Rana R, Sohel F (eds) Fibrous and textile materials for composite applications. Springer, Berlin

Chapter 11
Lignin Nanoparticles and Their Biodegradable Composites

Rizwan Nasir, Tazien Rashid, Khuram Maqsood, Danial Qadir, Dzeti Farhah Mohshim, Abulhassan Ali, Humbul Suleman, Hafiz Abdul Mannan, Hilmi Mukhtar, and Aymn Abdulrahman

R. Nasir (✉) · K. Maqsood · A. Ali · A. Abdulrahman
Department of Chemical Engineering, University of Jeddah, Asfan Road, Jeddah, Saudi Arabia
e-mail: rnasir@uj.edu.sa

K. Maqsood
e-mail: kmaqsood@uj.edu.sa

A. Ali
e-mail: aquddusi@uj.edu.sa

A. Abdulrahman
e-mail: aabdulrahman@uj.edu.sa

T. Rashid
Department of Chemical Engineering, NFC Institute of Engineering and Fertilizer Research, Jaranwala Road, Faisalabad, Pakistan
e-mail: trtazien8@hotmail.com

D. Qadir
School of Chemical Engineering, The University of Faisalabad, West Canal Road, Faisalabad, Pakistan
e-mail: danial2715@gmail.com

D. F. Mohshim
Department of Petroleum Engineering, Universiti Teknologi PETRONAS, Seri Iskandar 32610, Malaysia
e-mail: dzetifarhah.mohshim@utp.edu.my

H. Suleman
School of Computing, Engineering and Digital Technologies, Teesside University, Middlesbrough TS1 3BX, United Kingdom
e-mail: h.suleman@tees.ac.uk

H. A. Mannan
Department of Polymer Engineering & Technology, University of the Punjab, Lahore, Pakistan
e-mail: hmannan.pet.ceet@pu.edu.pk

S. Thomas and P. Balakrishnan (eds.), *Green Composites*,
Materials Horizons: From Nature to Nanomaterials,
https://doi.org/10.1007/978-981-15-9643-8_11

1 Overview of Natural Bioresources

Natural bioresources are renewable and biodegradable material that can satisfy the basic and essential needs of humans. These needs consist mainly of critical socio-economic parameters like food, shelter, fuel, etc. The living species and their population primarily represent the bioresources. The survival and further development of the human race require enhanced resources of food, material, and energy [1]. The bioresources close the material cycle at the end of their useful life and break into the water, minerals, and consumable carbon dioxide [2]. The importance of bioresources is not defined by how much it is needed for humans due to inexhaustible nature. Every plant and animal is also an integral part of the natural environment. Each bioresource is not only crucial for the ecological cycle but also for the aesthetic point. The bioresource safety of a country mainly depends upon their use and conservation. Thus, it is necessary to know clearly about the resources and socio-economic conditions.

Plants and animals provide material requirements for human survival. Twenty different species of plants and animals offer 80% of the human food requirements [3]. With the increase in the human population, genetically modified plants and vegetables are necessary to meet the needs. The production can be further enhanced by improving the forestry and by application of research and technology advancements in vegetable, animal farming, and fisheries.

A single bioresource can produce one major and multiple co-products. The process is normally represented as multi-branched or cascade utilization of bioresource. The production of one primary and several co-products is called a multi-branched process. Cascade utilization of bioresource is the use of the material separated from the main product line to make valuable products. The economical combination of multi-branched and cascade usage makes a process optimized and sustainable [4].

A significant share of chemical feedstock comes from bioresources, and it is now mandatory to increase this share by using better technology and finding more suitable feedstocks [5]. The use of bioresources for producing energy and heat is primitive, but with the growth of population, and depleting fossil fuel reserves has forced the researchers to give priority to biofuels. The energy demand of the world is increasing with the improvement in GDP and living lifestyle.

The main contributors to the energy share are fossils, mainly crude oil, natural gas, and coal. Shale oil and coal tar are now also contributing to the world's energy in the present scenario. The global energy consumption was 370 EJ in 2017. The oil shares 40% of the energy consumption, whereas 20% of coal and gas each. Fossil fuels were the main contributor to the energy with an 80% share in total global energy production. Among renewable energy sources, bioenergy is the most significant source. Bioenergy accounted for 70% of renewable energy consumption,

H. Mukhtar
Department of Chemical Engineering, Universiti Teknologi PETRONAS, Seri Iskandar 32610, Malaysia
e-mail: hilmi_mukhtar@utp.edu.my

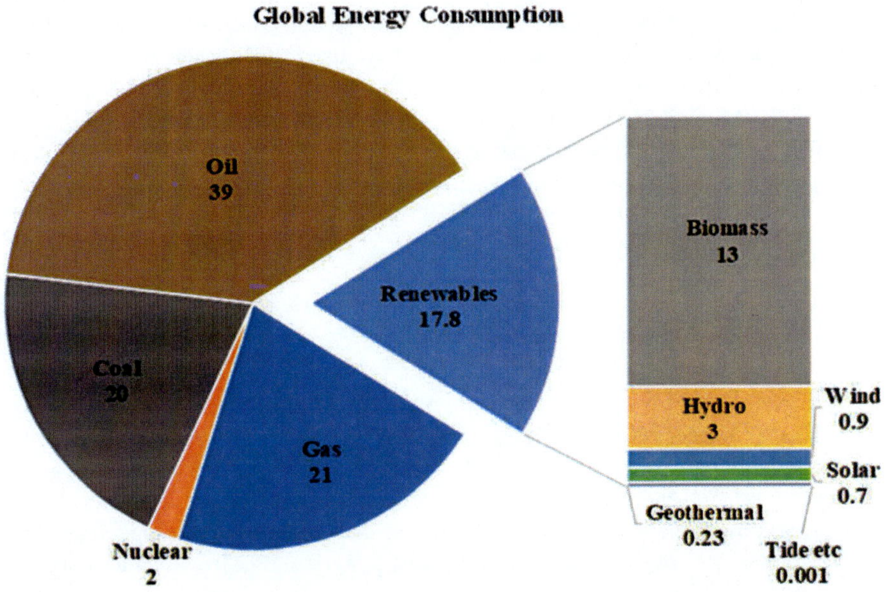

Fig. 1 Energy consumption in 2017 [6] "author"

and half of the contribution of this energy comes from Africa. Figure 1 shows the energy consumption pattern of the world in 2017 [6].

Renewable energy technologies have made considerable progress in decarbonizing the electricity sector. Hydro-systems produced most of the renewable electricity, followed by wind and biomass. As per the International Energy Agency (IEA), 154 billion liters of biofuel was produced in 2018 worldwide [7].

The primary use of biomass is as a solid biofuel, which includes wood chips, wood pellets, fuelwood for cooking and heating, etc. This type of biomass is usually used in space heating for residential, commercial, and industrial processes and establishments. Secondly, biomass can be used as liquid biofuels, mainly in the form of bioethanol and biodiesel. The contribution of biomass-based biofuels (bioethanol, biodiesel, etc.) is increasing in the transport sector as an alternative option for replacing the traditional fuels. The research has been carried out to find the most suitable feedstocks with a technology that can make the process more economical to compete with fossil-based fuels. The biomass required to produce these fuels comes from the three sources. The residue of agricultural waste is one primary source of the biomass, which approximately supplies biomass for 3% of the overall bioenergy production and a significant contributor to global power if adequately utilized, as it is quickly perishable. The major crops which can contribute to this type of biomass are mainly wheat, rice, and maize. The forestry sector is the most promising sector for the bioenergy mix globally. It is the major contributor of the bioenergy worldwide. The final biofuel of the forestry origin is mostly in the form of wood, pellets, charcoal, and wood chips. Unfortunately, most of the time, it is used for inefficient, traditional burning/heating purposes. Africa and Asia are the primary users of forestry biomass. One of the leading products from forests that are used for bioenergy production is wood fuel. Municipal and industrial waste is the third-largest resource of the bioenergy supply. Municipal waste mainly consists of the waste generated in the urban areas in the form of solid or sewage. Whereas, industrial waste is mostly related to the paper and pulp and food industry. Finally, aquatic bioresources form the edible food source, like fish, prawns, etc., and value-added bio-products, like algae-derived biodiesel [8].

1.1 Classification of Bioresources

Natural bioresources refer to all kinds of naturally occurring resources of biological origin. Humans can use bioresources for numerous purposes: to produce value-added products, food, and energy carriers. They can be categorized into primary, secondary, tertiary, and quaternary bioresources based on their quantity and usage. The main sourcing sectors of the bioresources are forestry, agriculture, livestock and poultry, and waste. Figure 2 shows the types of bioresources. Primary bioresources are mainly directly related to food and animals. The amount and availability of the primary bioresource depend upon geographical location and other natural conditions. Wheat, corn, fruits, vegetables, and animals are the main participants. Figure 2 shows the

Bioresources categories		Forestry	Agriculture	Live stock, Poultry	Aqua Culture	Waste
Primary	*Unprocessed*	harvesting residues	corn, wheat, fruits	cattle, camels etc.	fish, prawns	-
	Commercial / Non industrial	stem wood	corn grains, rice, jam	meat, eggs	meat and oil	-
	Industrial	wood for paper/cardboard, wood processing, furniture	corn oil, flour	animal fat, fillet,	alga, oil, fish fillet	-
Secondary	*Maintenance sectors*	-	leaf, straw	-	-	-
	Domestic/non industrial	-	used lawn cuttings	-	-	-
	Industrial	paper/cardboard offcuts	rice husk, animal skin and fat, press cake	-	-	-
Tertiary	*Maintenance*	weeds, roots	weeds	refuse	refuse	-
	Domestic	vegetation or plant residue	used cooking oil, food residual	spoiled food	aquariums, seafood residual	-
	Industrial residual	waste water, waste pulp, wood chips	fruits/ vegetables residual	spoiled products	spoiled/ damaged products	-
Quaternary	*Short term*		farm excrements	animal dungs	farm excrements	feces, urine sewage sludges
	Mid term	packaging material	-	-	-	packaging material
	Long Term	wood, furniture	-	-	-	consumer goods with biobased parts

Fig. 2 Types of bioresources Reprinted and reuse with permission from Körner et al. [9] Elsevier Copyright (2015)

different types of primary bioresources. The main sectors can further be divided into raw or unprocessed resources, commercial, and industrial products. Non-processed and commercial products are produced after limited processing, whereas the industrial products need detailed processing to convert them into final products. Primary bioresources are used primarily to meet food requirements rather than energy. These bioresources are produced as by-products of the processing stage of the primary bioresources. Different physical or chemical processes can be used to generate secondary resources. The quantities are significant, and the amount of impurities is lower. For example, rice husk is obtained once the rice is separated and sent to the next section. The residue collected as leaves, branches, or lawn cutting from the parks and residential areas can be regarded as secondary bioresources. The industrial section provides us with paper or cardboard, which can easily be recycled. Tertiary bioresources are a kind of residue, which is separated from the main bioresources. The primary sources are the post-harvesting waste and household waste, which is separated from the fruits or vegetables includes the peels, waste oil, and other food residues. This type of bioresource is generally considered as the throw-away part of a bioresource and can be very useful to produce biofuels like biogas and biodiesel. Waste pulp, waste cooking oil, and spoiled foods can form the cheapest feed sources of such biofuels. Quarterly bioresources can be divided into short-, medium- and long-term categories based on their availability after the first use. Short-term bioresources include human feces, urine, and animal dung. This kind of bioresources can be available shortly after the use of primary resources. Mid-term bioresource can be available after weeks or months; packaging boxes are the example of mid-term bioresources. The long-term bioresources are generally available after years of their first use, for instance, wood furniture or homes constructed by woods [4, 9]. The primary bioresources have always remained in high demand irrespective of geological boundaries or historical timeframe. However, with an ever-mounting increase in human consumption and improved lifestyle, secondary, tertiary, and quaternary bioresources must be optimally utilized. Technologies and processes should be developed to convert them into useful products like biogas, biodiesel, chemicals, and bio-fertilizer.

1.2 Agricultural by-Products

Biomass originating from agriculture forms a base material for many secondary energy carriers, e.g., ethanol, bio-methane, and hydrogen. Alternative energy can be complemented by using microbial action on biomass to produce these chemicals. Abundant biomass resources can be derived from lignocellulosic substrates that mainly contain 60–70% holocellulose which includes cellulose and hemicellulose and 10–25% lignin [10]. So, the next sections will focus on lignin, lignin nanoparticles (NPs) and their biodegradable composites.

2 Overview of Lignin and Lignin Nanoparticles (NPs)

2.1 Lignin

The three primary amorphous tridimensional polymer units are containing syringyl (S), guaiacyl (G), and p-hydroxyphenyl (H), as shown in Fig. 3 present in lignin and combined by C − C and ether linkages. All units comprise of a phenyl group and a propyl side chain; therefore, the representative aromatic unit in lignin is generally called a phenylpropane unit [11].

The number of methoxy groups in syringyl, guaiacyl, and p-hydroxyphenyl unit are the elementary difference in these units. The contents of these units are related to the structure of the plant. The soft wood lignin consists of guaiacyl units, hardwood has both guaiacyl and syringyl units, and the grass lignin consists of all units. Mostly, the ether and C − C relation are present in monomeric lignin units. The aliphatic side chain carbon atoms are presented as α, β, and γ and number 1 and 6 is assigned to carbon of aromatic area. From the β-O-4 connection type, a bond of β carbon of the aliphatic side chain exists with the oxygen atom of the C_4 of the aromatic region as shown in Fig. 4. The amount of β − 1, β − β, etc., linkages varies from source to source, depending upon the environmental factors [11, 12].

A concise lignin model to demonstrate the usual connections in lignin is shown in Fig. 5. These relations define the lignin reactivity to chemical digestion. The β − O−4 relation is the most important in lignin. The monomer units present consist of C–C and C–O connection, as can be seen in Fig. 5. These monomer units may vary in different lignin sources (softwood or hardwood), and the structure and rigidity of lignin depend on the degree of crosslinking and substitution between the monomer units [13]. Lignin is rich in phenolic and aliphatic hydroxyl groups. These hydrophobic and hydrophilic functional groups can be used for further modification and polymerization purposes, which make lignin an appropriate and attractive polymer for other value-added applications [14, 15].

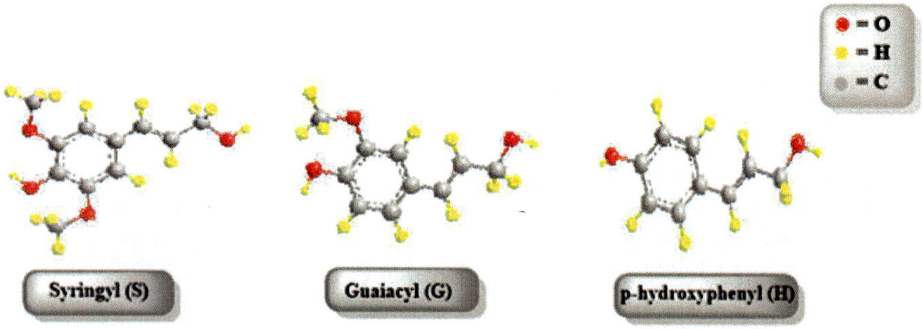

Fig. 3 Primary units present in lignin reprinted (adapted) with permission from Li et al. [11]. Copyright (2015) American Chemical Society

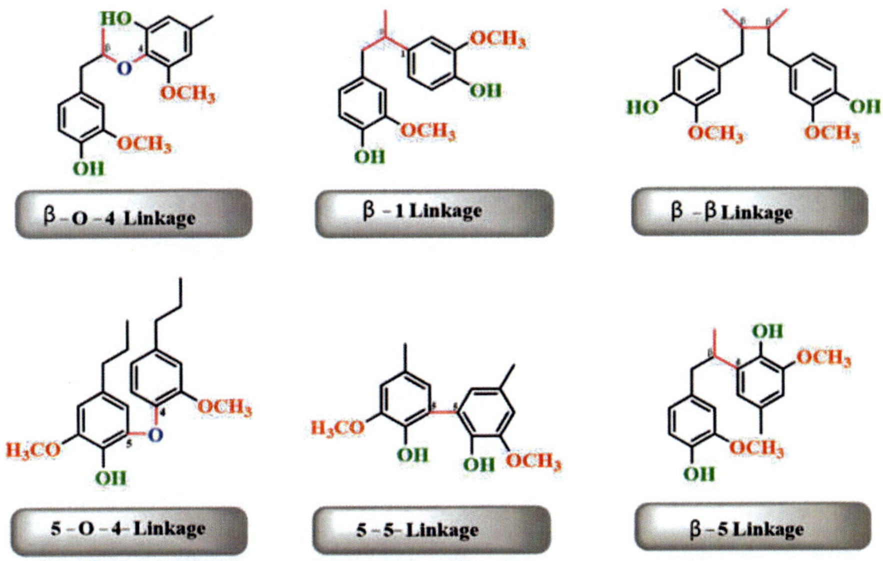

Fig. 4 Typical linkages between the primary units of lignin reprinted (adapted) with permission from Li et al. [11]. Copyright (2015) American Chemical Society

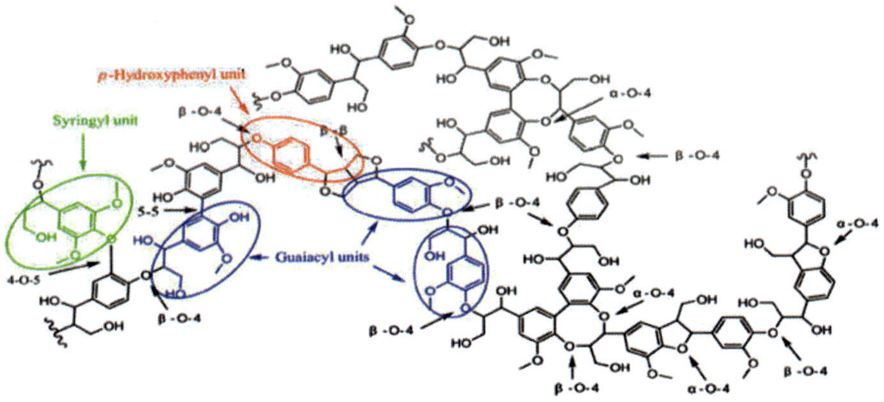

Fig. 5 Lignin descriptive structure model reprinted (adapted) with permission from Li et al. [11]. Copyright (2015) American Chemical Society

Lignin's relatively small solubility and complex structure with a very large molecular weight and microstructure distribution are possible reasons to find lignin as a biomass waste product. Hence, only 3% is employed as binders (pellets for feed, bricks, pottery, etc.) in agricultural and other sectors [16, 17]. The increased demand for conventional applications (paper processing, biomass fuels, composting, feed for animals, etc.), new applications, and potential lignin markets has been identified.

The majority of the products are graded as low-value products that are based on the source and the techniques adopted [18, 19]. Some isolation and purification techniques for lignin separation are described in the subsequent sections.

2.1.1 Separation and Extraction of Lignin

Lignin is separated and extracted from various biomass sources like sweetgrass, corn stover and wheat straw, etc. Lignin is mostly a by-product of the pulp and paper industry, where the cellulose forms the primary product of choice, while lignin is burnt as boiler fuel or wasted. However, with the growing trends of lignin utilization and valorization in industry, numerous methods have been proposed for the selective extraction of lignin with an emphasis on preserving the actual structure of lignin. Still, these methods have their respective advantages and disadvantages [20]. Also, partial cleavage of lignin bonds takes place in the extraction process, which may vary in different extraction methods hence affecting the functional groups, properties, and molecular weight of the isolated lignins. Biomass pretreatment can be divided into main groups, namely mechanical fractionation (ball milling), chemical fractionation (acidic, alkaline), solvent fractionation (organosolv, ionic liquid pretreatment, etc.), and biological fractionation (enzymatic) [21]. All of these techniques have certain benefits and limitations associated with them; for example, ionic liquids have replaced the conventional solvents and have addressed several drawbacks of conventional methods regarding equipment corrosion, high operating conditions (extraction temperature, extraction time). Still, their industrial implementation has been limited because of expensive raw materials required for their synthesis and involvement of more complicated procedures and steps [22]. A better understanding of these pretreatment methods is essential to consider lignin as a significant starting point for biofuel and renewable sources.

2.1.2 Mechanical Pretreatment

The mechanical pulping is the oldest process employed for the extraction of cellulose fiber from biomass. In mechanical pulping, intense mechanical strength is applied to separate cellulose fibers from biomass. Large abrasive grinding wheels are used for grinding the biomass, and the resulting mixture is used to make pulp by dissolving it in water [23]. Although this process has a high yield of cellulose fiber, a significant amount of lignin is also included in the pulp which results in low strength and altered properties of cellulose fiber. The cellulose obtained by this method is majorly used in the production of newspapers, books, and magazines [24].

2.1.3 Chemical Pretreatment

Chemical pretreatment (acid and alkaline) is the most common method to achieve commercial-scale lignin [25, 26]. These pretreatments lead to high sulfur contents present in lignin with extensive structural changes depending on the pulping additives and process parameters [27, 28]. Every commercial isolation method has certain reagents involved that affect the molecular weight and purity of the end product, along with the scope of processing for the new chemicals [28]. Cellulose-rich pulp/paper processing is currently the leading manufacturer for commercial-grade lignin, but in these techniques, the emphasis is given on only cellulose production [29].

Sulfite pulping: Sulfite pulping (lignosulfonate process) is the oldest acid pulping technique, which was initially implemented for cellulose extraction during the paper production cycle. The process proceeds in a regulated oxygen atmosphere, and the sulfur combustion leads to the formation of sulfur dioxide (SO_2), which reacts with water to form sulfuric acid. The reaction takes place at 125–150 °C and 3–7 h [30]. The sulfuric acid (H_2SO_4) is used to digest wood; the pH of the cycle is maintained between 1.5 and 5 affected by the bases used in the process. The isolated polymer contains 5% sulfur having carbohydrate residues as impurities. The obtained lignin is water-soluble, large distribution of molecular weight and a relatively small volume of ash. Sulfite pulping requires a complex isolation procedure due to its water-soluble characteristics. A complex separation strategy is adopted to deal with water-soluble lignosulphonates, in which water-insoluble product is formed by long-chain alkylamine that needs to be extracted by using some organic solvents. Eventually, lignosulfonate regeneration takes place by subsequent addition of a base [30, 31].

Kraft pulping: Kraft pulping is the form of chemical pulping used to degrade and dissolve lignin by using chemical reactions to remove pure cellulose from the wood [32]. In this approach, sodium hydroxide (NaOH) and sodium sulfide (NaS_2) both act as a nucleophile to cleave β-aryl bonds present between lignin-carbohydrate linkages leading to the production of chemically resistant cellulose [33]. This process is carried out at a high temperature of 175 °C for about 2 h. The final product, which is known as black liquor contains a mixture of pulping chemicals, with small quantities of sodium/calcium carbonate, sodium thiosulfate, hemicellulose [33, 34]. Unlike the sulfite pulping, kraft lignin is rich in phenolic hydroxyl groups due to extensive cleavage of β-aryl bonds during pulping process. Kraft lignin contains less sulfur, less carbohydrate residues and inorganic impurity, and a high degree of purity relative to sulfite lignin. Nevertheless, kraft lignin is considered as the by-product in paper manufacturing and major application is limited to power generation at the pulp mill [14, 29, 35].

Soda/Alkaline pulping: The alkaline pulping process is used to produce cellulose and lignin worldwide due to high energy efficiency. Alkaline pulping is the oldest techniques for lignin production. In general, the process is extended to non-woody biomass like bagasse and straw. This technique is used to treat the lignocellulose at a temperature of around 150–200 °C with dilute NaOH (≈ 14 wt%). Major delignification steps occur by the saponification of the ester bonds present among hemicellulose and lignin, whereas lignin is deconstructed partially by the cleavage of the α- and β-ether bonds. Lignins derived from soda processes can, therefore, be more valuable and hence can be more beneficial for the value-added production of composite and bioplastic products [36].

Organosolv lignin: The organosolv cycle employs the removal of lignin and hemicellulose from lignocellulosic biomass using organic solvents. It involves the pretreatment of biomass with an organic solvent (carboxylic acids, alcohols, ketones, cyclic compounds, acetone, ketones, and amines) and occasionally binary solvents are used to achieve high-quality sulfur-free lignin. Organosolv lignin is substantially pure as compared to other methods, as the functional groups and properties are near to the native lignin [37, 38]. The lignin obtained is of lower molecular weight, free of sulfur, alkali/alkaline metals, etc. The recovery of expensive solvents at higher temperatures involved in the process is the major constrain [39]. The lignin resulting from the organosolv process is a good source for polymers and resins; however, it is too costly. One major drawback of this process is that lignin is recovered by extensive washing using water at the end of the process. Moreover, an acid catalyst is added into the pulping process, which results in a breakdown of ether linkages and promotes intramolecular condensation reactions, resulting in a more complex structure of the organosolv lignin [40, 41]. Furthermore, the acid utilization in the process raises material corrosion and environmental concerns [14, 42].

2.2 Lignin Nanoparticles (NPs)

Nanomaterials and nanotechnology lead to the implementation of new approaches for the valorization of biomass [43, 44]. Lignin nanoparticles (lignin NPs) exhibit novel properties, i.e., increased surface area and enhanced properties relative to the original materials. Moreover, surface modification of lignin NPs can be performed easily as lignin is enriched with many functional groups such as; aliphatic, hydroxyl, and phenolic groups. This modification of lignin to lignin NPs could result in a chemically changed structure and improved properties to enhance their uses [45]. The following section addresses some of the methods adopted for the synthesis and preparation of lignin NPs.

3 Conventional Synthesis Methods of Lignin NPs

3.1 *Acid/Alkaline Precipitation Method*

The acid precipitation method is used to synthesize the non-toxic and biodegradable lignin NPs. This method addressed two different techniques to prepare lignin NPs, which have variances instability when pH is modified. The first method involves acidic conditions (hydrochloric acid), and ethylene glycol is used for the precipitation of low-sulfonated lignin. This allows highly stable lignin NPs to be recovered in various pH ranges as the synthesized lignin NPs possess densely packed lignin domains. In the second process, lignin is acid-precipitated using nitric acid from high pH aqueous solution (sodium hydroxide) leading to the formation of lignin NPs, which only have stability at low pH ranges and the synthesized lignin NPs are porous [46].

Gutiérrez-Hernández et al. developed a method for alkaline precipitation, [47] using Agave tequila lignin obtained by organosolv pulping for the synthesis of lignin NPs. Briefly lignin (17 wt%) is mixed with a predetermined amount of sodium hydroxide (NaOH) for 60 min. After 120 min, NH_4OH is added to the mixture and is stirred at high speed of 25,000 rpm for about 5 min. Once the mixing is complete, active formaldehyde is added into the mixture, which raises the temperature of the mixture to 85 °C. The mixture is placed for 120 min under magnetic stirring to allow crosslinking and lignin NPs formation. Using this method, six different types of lignin NPs were synthesized depending upon the wt% of active formaldehyde used and the lignin source. Zinc oxide (ZnO) has been extensively used in sunscreens to block UV radiation. The synthesized lignin NPs can be blended with ZnO as it was revealed that the blend of both materials possessed excellent qualities of photoprotection.

3.2 *Self-assembly*

The mechanism producing a pattern or organized structure due to the specific interactions (incoherent configuration) of the pre-existing components usually takes place in the absence of some external direction. Dai et al. [48] developed a first-of-a-kind framework for green NPs using lignin. In preparing the spherical nanoparticles with good dispersion characteristics, alkali lignin (AL) was used as a lignin source along with resveratrol (RSV, a bioactive molecule) and Fe_3O_4 NPs. A basic self-assembly method was followed in this preparation step, and AL/RSV/Fe_3O_4 NPs prepared resulted in a stable nano-drug carrier, and the schematic presentation of the process is shown in Fig. 6. The AL/RSV/Fe_3O_4 NPs demonstrated important anticancer effects and improved the release and stabilization of in vitro resveratrol, and the required tumor reduction in cytological and animal studies. In another study, lignin NPs were prepared by Liu et al. [49] using ethanol as well as several stages of catalysts for selective lignin dissolution from biomass, using sequential organosolv

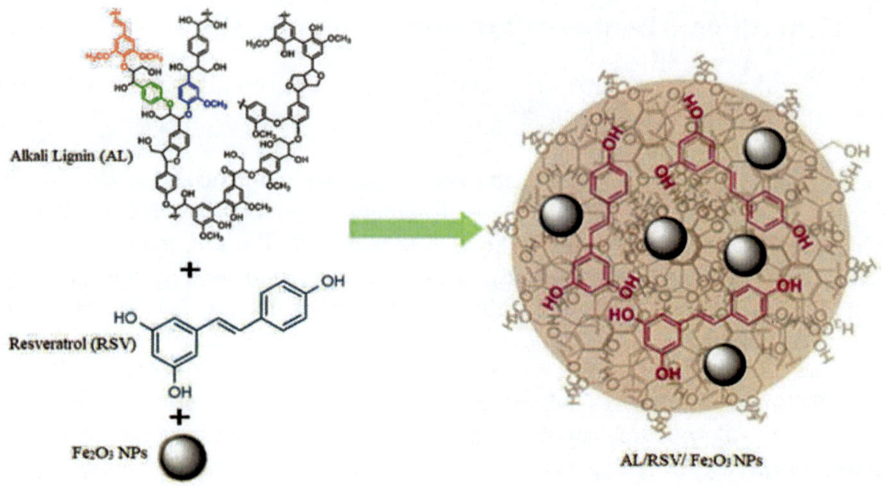

Fig. 6 Basic self-assembly method followed for the preparation of AL/RSV/Fe_3O_4 NPs reprinted (adapted) with permission from Qian et al. [50]. Copyright (2017) American Chemical Society

fragmentation approach (SOFA) as the basis of the process. Several uniform lignin streams were prepared, and physicochemical characteristics, and the reactivity of lignin, were explored to fabricate lignin NPs of expected features and characteristics via self-assembly. In this way, high-quality spherical lignin NPs, small effective diameters, and good stability have been developed using SOFA to boost the self-assembly process and thus contribute to the productivity of biorefineries. This represents a sustainable means to upgrade the low-value lignin.

3.3 *Polymerization Followed by Surface Modification*

The series of ionic responsive poly(N-isopropylacrylamide) (PNIPAM) surface-modified lignin nanofiber mats were synthesized by employing Atom Transfer Radical Polymerization (ATRP) technique. Lignin NPs grafted PNIPAM brushes with different molecular sizes, thicknesses, and grafting densities were immobilized in the electrospun lignin nanofiber mats by modifying the initial monomer concentration and the surface initiator densities. The findings have been verified by the various characterization techniques such as ATR-FTIR, scanning electronic microscopes (SEM), thermogravimetric analysis (TGA), XPS, and water-contact angle. This study showed that lignin NPs possessed a lower critical solution temperature (CST), which was further like that of PNIPAM and demonstrated ionic responsive characteristics. The water-contact angle of the lignin NPs improved correspondingly with the increasing ionic concentrations. Figure 3.6 depicts the SEM images of lignin NPs, and lignin NPs grafted PNIPAM brushes after modification using ATRP. It is evident from Fig. 3.6 that after modification through ATRP, there is a strong increase in the diameter of lignin nanofiber mat. Therefore, it was suggested that these nanofibrous materials could serve as a platform to build thermo-responsive separating and purifying devices.

In another study, Qian et al. [50] modified water-insoluble lignin by grafting 2-(diethylamino)ethyl methacrylate (DEAEMA) through ATRP. These lignin NPs grafted DEAEMA possessed a size range of 237–404 nm. They demonstrated that lignin NPs grafted DEAEMA can be used as surfactants in pickering emulsion.

3.4 Compressed CO_2 Anti-solvent Strategy

Compressed/supercritical fluid-based strategy has been evolved as an emerging technology used in polymer NPs growth during the last two decades. Its features, such as size distribution and morphology, make it essential in the production of pharmaceutical and drug delivery applications. This is a multistage patented process which involves several dissolutions and precipitation stages. The commercial-grade lignin from black liquor is dissolved in dioxane. The mixture is centrifuged at high speed to achieve undissolved lignin. Using a supercritical carbon dioxide system and following an anti-solvent strategy, lignin NPs can be obtained from the lignin-dioxane solution. Due to its physicochemical properties, CO_2 has gained considerable interest in the recent past year. It has excellent features, i.e., ($TC = 304.2$ K and $PC = 7.4$ MPa), low-cost, non-toxic, non-flammable, and available in plentiful quantities. However, in macromolecules, namely polymers, it is a weak solvent since it is sometimes referred to as the best anti-solvent in precipitation processes, in which precipitates can be controlled by temperature and pressure controls [51]. The key characteristics of the process should be considered in this method, such as that the raw material is a waste-pollutant that can address the problems of contamination of the paper industry; no leakage exists as the supercritical anti-solvent strategy is employed. Those are all the approaches used to provide a wider viewpoint for broader lignin applications.

In a study by Myint et al. [52], lignin NPs were produced from kraft lignin by adopting a compressed CO_2 anti-solvent strategy. Kraft lignin was dissolved in N, N-Dimethylformamide (DMF) first, after which lignin NPs were precipitated using compressed liquid CO_2 as an anti-solvent. This resulted in an identical and quasi-sphere lignin NPs ($\approx$ 38 nm) formation that demonstrated improved water solubility and water stability compared to raw kraft lignin. It was observed that the pressure and solution flow rate has been decreased with an increase in temperature and the lignin concentration.

3.5 *Chemo-Mechanical Methods*

The chemo-mechanical methods offer an advantage by allowing the development of uniform particle sizes between 10 and 70 nm with an average particle size of 30 nm. Moreover, it is a simple processing tool, most importantly, without contamination to the atmosphere [53]. An additional patent has been granted for lignin NPs preparations with a controllable particle size, which includes: (a) preparation of alkali lignin, (b) grinding the alkali lignin to prepare liquid suspension, (c) conversion of alkali lignin into lignin solution, and (d) drying to obtain lignin NPs [54].

4 Sustainable Synthesis of Lignin NPs

4.1 *Microwave-Assisted Strategy*

Although many strategies have been adopted for lignin NPs synthesis and modification, only a few green and simple methods are being explored to generate high yield as well as regularly formed lignin NPs so far. A study was first attempted by Wang et al. [55] and co-workers, lignin was modified in a microwave-assisted strategy by using acetic anhydride. Acetylation of lignin was processed without catalyst, and the solvent used acted both as reaction reagent and the dispersion medium. A solvent treatment subsequently gave the standard lignin NPs in conjunction with the ultrasound method (US). Meanwhile, the solvent used can be recycled and reused, the process can be streamlined, and costs and industrial growth sizes are minimized. The lignin NPs were quickly formed without dialysis and can be easily centrifuged. The maximum yield of lignin NPs achieved was as high as $\approx 83\%$. Meanwhile, ultrasound therapy improved lignin NPs uniformity and dispersal characteristics. The modified lignin NPs were found to be good UV-absorbents, and it was suggested that several green chemistry principles can be achieved.

4.2 *Ultrasonication*

As discussed above, the structure of lignin is very complicated, which strongly depends upon the separation process and plant species used. In a recent study Gilcal et al. [56] introduced a physical method to generate lignin NPs using ultrasonication of wheat straw and Sarkanda grass lignin. A dilute aqueous suspension of lignin (0.7%) was prepared and was ultrasonicated for 1 h. A uniform stable nano dispersion was finally achieved. Lignin was dried in mild conditions for further study in the samples that induced sonication. The synthesized lignin NPs were 10–50 nm, the dimensions and morphology of the prepared lignin NPs were characterized to assess any unusual changes in the structure and composition. For potential applications, ultrasonication is even more desirable because no toxic solvents are employed in this method. Later in a research, Gonzalez et al. [57] developed ultrasonic treatment of lignin NPs for different periods (i.e., 2, 4, and 6 h) of treatment by utilizing kraft lignin as the lignin source. The obtained lignin NPs were around 10–50 nm in size. Ultrasonication is considered as an easily implementable and straightforward method for lignin NPs synthesis.

4.3 Biological Methods

Enzymatic pretreatment for the saccharification and fermentation of lignocellulosic biomass has been potentially used in recent years. The enzymatic hydrolysis not only proved to be efficient in lignin to lignin NPs production, but it also improved the properties of the lignin NPs being synthesized. Recently, Rangan et al. [58] synthesized lignin NPs through the breakup of the lignin-cellulose complex with different enzymes. The enzymes degradation helped to achieve lignin-rich nanoparticles from Luffa cylindrica a lignocellulosic fiber. Later in a study, Juikar and co-workers [59] in 2018 developed lignin NPs by controlled microbial hydrolysis using Aspergillus sp a lignin-degrading fungal isolate. The mixture was shaken at 10 rpm for 15 days at ambient temperature. The method has been tracked, and samples were analyzed on alternative days. The samples were centrifugated for 15 min at about 1000 rpm to remove biomass and unhydrolyzed lignin. The supernatant has been filtered, and the filtrate consisting of lignin NPs has been used for further analysis using characterization techniques such as field emission gun-canning electron microscope (FEG-SEM), transmission electron microscopy (TEM), and atomic force microscopy (AFM). The lignin NPs size ranged from 2 to 150 μm with a mean value of 55 μm. The yield of lignin NPs was compared with the homogenization (high shear homogenizer used for nano-lignin NPs synthesis) and the ultrasonication process. It was found that the biological method yield was lower, i.e., $\approx$ 59% compared with 82 and 65% in the latter, respectively. Figure 7 depicts lignin NPs micrographs using FEG-SEM, TEM, and AFM produced by the three techniques, i.e., homogenization, ultrasonication, and biological methods, respectively. It can be concluded by the micrographs that the lignin NPs produced by the biological method is well dispersed as compared to the lignin NPs by homogenization and ultrasonication process which were found to be aggregated and scattered randomly. They reported that the biological method for lignin NPs synthesis is energy and environmentally friendly as compared to the conventional methods.

4.4 Flash Precipitation

Flash nanoprecipitation (FNP) is a process in which an insoluble, low molecular weight compound is rapidly mixed and stabilized in a polymer-stabilized delivery

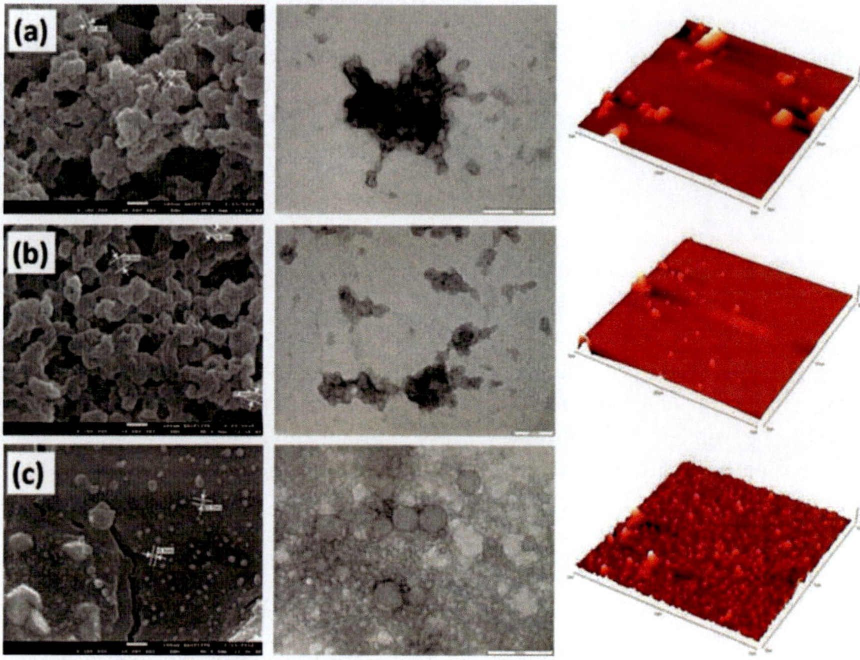

Fig. 7 Micrographs of FEG-SEM (first column), TEM (second column) and AFM (last column) of lignin NPs synthesized by **a** homogenization, **b** ultrasonication and **c** biological methods, reprinted and reuse with permission from Juikar et al. [59] Elsevier © (2017)

vehicle. In a recent study, Richter et al. [60] employed the green synthesis process based on dissolved lignin polymer flash precipitation, which enabled the manufacture of lignin NPs ranging from 45-250 nm in size. In this method, a predetermined amount (0.25 gm) of commercially available organosolv lignin is dissolved in 50 ml of acetone in a batch synthesis assembly. The solution is whisked for half an hour and syringe filtered using a 0.45 μm filter. The addition of 9.2 ml of water leads to lignin supersaturation in the solution and phase separation by NPs of lignin. The obtained samples are subsequently diluted with 0.05 wt% water, and dynamic light scattering (DLS) technique is adopted for the particle growth monitoring regularly after a fixed interval of time.

Additionally, by coating it with cationic polyelectrolyte, poly (diallyldimethylammonium chloride), the surface characterizations and properties of the lignin NPs were refined and controlled. This method helped to synthesize lignin NPs to be stable in extreme pH conditions. Ionic liquids (aprotic ionic liquids and protic ionic liquids) and deep eutectic solvents (DES) have been recently used as an emerging solvent for lignocellulosic biomass pretreatment. Lou et al. [61] isolated lignin NPs from wheat straw by selectively extracting lignin using choline chloride-lactic acid-based DES. The lignin NPs yield was about 85%, and the particles ranging from 70-90 nm in size were synthesized. It was stated that DES could not break the ether bonds in lignin but also can break the linkages between lignin and the hemicellulose at the same time. Hence, green solvents can produce purified lignin NPs by dissolution and following a simplified flash precipitation method.

5 Biodegradable Composites

Mankind is very familiar with biodegradable composites since 121 BC. The composites were made stronger by mixing red willow reeds and gravel in 209 BC and later further mixed with stone, clay, sand, reeds, and willow branches in 206–221 BC. But with the awareness hike, the researchers are focused on developing such materials that are eco-friendly and biodegradable. This approach highlights the research progress to convert nonrenewable to renewable degradable materials. This progress leads to the development of new materials with desired properties and structures [62]. These new materials (bio-composites) reveal the improvement in thermal stability, increased durability, tearing and tensile strength, and reduced water absorption tendency. The classification of biodegradable composites depends upon manufacturing methods, applications, densities, and some other systems. Due to these factors, the composites can be divided into two major groups: matrix (biodegradable polymers) and natural bioresources (natural fiber and agricultural waste) [63, 64].

Due to the versatility demands of bio-composites utilization, no general procedure exists for their synthesis. Therefore, researchers have reported several types of bio-composites like wood/plastic, cement composites, glass/fiber, filament/wound composites, etc. It is pertinent to mention that a variety of production methods are practiced to produce bio-composites, which include extrusion, injection, molding, thermoforming, etc., [63, 65]. Currently, the research is focused on the improvement and utilization of such biodegradable composites suited for engineering applications.

5.1 Lignin Biodegradable Composites

The discussion mentioned above highlighted the utilization of renewable resources to develop materials, fuels, and chemicals. Natural bioresources are abundantly available biomass, and from all fractions of biomass, the lignin is not being used enough due to complex structure [66]. Many functional groups are present in the lignin and provide the sites for the modification to meet the product requirements. By these modifications in structures and properties, the lignin composites can be developed for various practical applications [67, 68]. Various approaches have been deployed to convert lignin to value-added materials by combining them with other materials. Particularly, to develop the eco-friendly composite materials, the lignin is added into various biodegradable polymers [69]. Lignin composites were also recently used in the vanadium redox flow battery for high ion selectivity. Ye et al. synthesized the sulfonated poly (ether ether ketone) (SPEEK)—lignin composite membranes.

Table 1 Brief list of lignin biodegradable composites "author"

Name of lignin biodegradable composite	Applications	Reference
Chitosan-alkali lignin composites	Adsorption of dye and metal ion from wastewater	[71]
Lignin@TiO_2 composites	Photocatalysis	[72]
Lignin-lactide copolymer bio-based composites	To enhance UV absorption and reduce brittleness without a sacrifice in the modulus of elasticity	[73]
Liginin-poly(butylene succinate) (PBS)	Due to their antibacterial and antioxidant properties it is used in biomedical applications	[74]
Fe_3O_4 and lignosulfonate composite	To remove dyes from water efficiently	[75]
Biorefinery lignin and polyhydroxybutyrate composites	3D printing	[76]
Gellan gum, 2-hydroxyethyl cellulose, and lignin	Food packaging and biomedical application	[77]
Poly(lactic acid)/lignin-based biodegradable composites		[78]

They found the outstanding ion selectivity and the superb protons conductivity of the membrane by the addition of lignin and shows ultrahigh coulombic efficiency (over 99.5%) and energy efficiency (over 83.5%) [70]. In Table 1, a brief list of some biodegradable lignin composites is presented.

5.2 *Lignin NPs Biodegradable Composites*

The use of nano-lignin in macro-scale compounds, i.e., natural and synthetic, is an important technique to develop suitable composites. In recent literature, the nano-lignin as filler was used to develop polymer composites [79]. Some studies highlighted the utilization of lignin NPs as antioxidants, reinforcement, antimicrobial, antimutagenic agents, and UV adsorbent for the synthesis of nanocomposites. These nanocomposites can be used in different applications [51, 69, 80]. It was noticed in many studies that the mechanical properties of nanocomposites are far better than micro- and macro-composites [81, 82]. Lignin NPs have been used in different biodegradable polymer matrix, i.e., wheat gluten [83], PVA [84], polylactic acid (PLA) [85], natural rubber [86] and bio-PPT [87], etc., to make mechanically strong nanocomposites.

The polylactic (acid) (PLA)/lignin NPs composites were developed by solvent casting and melt extrusion. The hydrochloric acidolysis was used to synthesize the lignin NPs. They found that the uniform dispersion of lignin NPs in the PLA matrix by using the melt extrusion method. Meanwhile, the solvent casting method exhibited a weak interaction of NPs and PLA, along with aggregated dispersion of NPs.

Moreover, the higher loadings of NPs gave rougher surface structures and higher degradation rates [88].

In another study, the biodegradable poly(butylene adipate-co-terephthalate) (PBAT) has been used with lignin NPs to develop biodegradable composite. The lignin NPs and composites have been developed by using acidic precipitation of lignin and injection molding, respectively. Modified NPs are well dispersed in the PBAT matrix at a low loading of NPs, exhibiting improved mechanical properties and enhanced processing ability due to a decrease in the viscosity by the addition of NPs in the PBAT matrix. It was also found that low compatibility decrease in modulus when in a molten state [89].

The biodegradable polymer waterborne polyurethane (PU) is used to developed biodegradable composite with lignin NPs. The lignin NPs were developed by ultrasonication treatment. The prepared lignin NPs at different loadings has been incorporated in the PU matrix to prepare the composite. The morphological analysis of biodegradable composite confirmed the excellent dispersion of NPs in the matrix along with improved thermal stability (@ > 400 °C). Moreover, due to the ultrasonication process for the development of NPs, the noncovalent interactions are developed between NPs and PU matrix, and due to this interaction, mechanical properties of developed composites have been improved [57].

The PLA is also used with NPs to develop the composite film. The PLA/NPs composite film and lignin NPs were prepared by hydrochloric acidolysis and the pickering emulsions method, respectively. The results confirmed the improved dispersion of NPs in the PLA matrix. The mobility of PLA has been enhanced by the addition of NPs, which lead to greater crystallinity. The PLA/NPs composite films exhibited an increase in Young's modulus and a decrease in elongation at break and tensile strength as compared to pure PLA film. These improved properties by adopting pickering emulsions method confirm that this composite film is advantageous for industrial applications [90].

The agro-polymer wheat gluten (WG) is used with lignin NPs to develop the bio-nano composites films [91]. The lignin NPs is synthesized by acidolysis. The developed NPs have uniform size distribution and better thermal stability. The synthesized bio-nano composites films exhibit improved mechanical strength and thermal stability, homogenous distribution of NPs, and reduced water sensitivity. Moreover, the loading of NPs in WG films caused a decrease in the transparency of the film. This decrease in film transparency and excellent UV resistance help to use these bio-nanocomposites films in food packaging and agricultural bag manufacturing [83].

Recently, one study focused on the functionalization of NPs to enhance the properties of the composites. Chollet et al. prepared NPs via the dissolution-precipitation process from Kraft lignin microparticles (LMP) and used diethyl chlorophosphite and diethyl (2-(triethoxysilyl)ethyl) phosphonate-functionalized NPs as the additive flame-retardant additive for polylactide (PLA). We observed that the functionalization of NPs reduced the degradation of PLA during melt processing and improved the thermal stability of the nanocomposites. Hence, the functionalized lignin NPs can be used as a flame retardant additive even at low concentration [92].

The biodegradable lignin NPs/poly(vinyl alcohol) (PVA) nanocomposite films were prepared by a simple solution-casting method. The T-shaped microchannel reactor was used to prepare the lignin NPs (average size 13 nm). The NPs showed uniform dispersion in the PVA matrix. Also, these nanocomposite films improve UV efficacy by 13.3%. These films can be used for medicine and packaging [93].

6 Value-Added Applications of Lignin NPs

Lignin which is known to be the second most abundant biopolymer is a by-product of cellulose with low value-added applications. However, in the last few decades, due to its abundant sources, researchers have started to investigate its application. Lignin usage was started with its application in the polymer chemistry perspective [94]. They also mentioned that lignin is the main renewable source of aromatic, which makes it suitable to be used as polymer and other aromatic components like phenol and monomers. The usage of lignin could reduce the dependability on petroleum feedstock. However, the use of lignin as polymer-based materials is a great challenge due to its complex macromolecular structure. Not that long time after that, scientists have moved the synthesis of lignin as nanoparticles due to its non-toxic which could lead toward global sustainability [46] and high-value of technical lignin [43]. Synthesis of general nanomaterials is usually harming the environment and costly. Thus, a cost-effective, fast yet easy, and green path of nanomaterials synthesis is now preferred. Nanoparticles from renewable resources like lignin are known to be biodegradable, biocompatible, antioxidant and can be used in various applications [60]. This section will discuss the immense and valuable lignin applications, including biomedical, environmental, engineering, and technology fields.

A lot of successful production of lignin nanoparticles via different methods is found targeted on the biotechnology and nanotechnology applications [95–97]. Nanoparticle lignin based is known to be an eco-friendly option and thus stimulates its utilization in the nanocomposite, medicals, and biomedical applications depending on their source, chemical modifications, and physicochemical properties [98, 99].

6.1 *Medicine and Pharmaceuticals*

Drug-related industries are among the highest number of areas that are using lignin nanoparticles. Lignin NP is embedded in medicine and pharmaceuticals, mainly due to its antimicrobial and inherent antioxidant properties. Besides, lignin also possesses biological activities, i.e., reducing cholesterol, antidiabetic, viral infections, and cancers [100]. One of the important aspects of a drug is its release [101]. This is normally achieved by its inherent properties like highly biodegradable, stable, non-toxic, and inexpensive: properties that support potent drug delivery. Kraft lignin that is also a fat adsorbent could control and prevent obesity. Lignin known as lignophenols has the ability of reducing cholesterol as it is a stable and act as an antioxidant at the same time [102]. Lignin–carbohydrate–protein complexes (LC) were tested against some viruses like herpes and measles and found to be a positive outcome [103]. Another lignin lignosulfonic has the anti-HIV that applicable in low dosage and found to be approved in clinical trials [104]. Lignin was also used as an antimicrobial application or as a cancer treatment as water-dispersed lignin could carry silver ions [46].

6.2 Metallic Nanoparticles Synthesis

With its abundant sources, lignin is also used in making greener and high-cost metallic nanoparticles, which is opposite to its inherent economic image. Akali lignin that is known to be water insoluble was also used in the ball milling method for crushing, and the results show an increase in surface area by 10-fold. Based on similar studies, alkali lignin and hemicellulose were a good candidate in producing precious metals like Pt, Au, and Pd nanoparticles [105]. In addition, lignin was again used as a reducing and capping agent in producing CuO nanoparticles with a particle size of 100–200 nm [106].

6.3 Crosslinking Agent

A self-healing hydrogel used lignin nanoparticles in its preparation along with polyvinyl alcohol (PVA) and cellulose nanofibrils (CNF), where the lignin acts as a spacer and found to have improved in the Gs values due to its anti-aggregation and elasticity properties [107]. These hydrogels are widely used in several field applications, including pressure sensors, soft machines, and tissue scaffolds. Lignin nanoparticles were also found used as crosslinking junctions to synthesize high mechanical hydrogels like polyacrylamide (PAM)/lignin nanoparticle nanocomposite hydrogels. The resulted hydrogels were found to have high tensile strengths and highly compressive at megapascal measurement [108] due to their strong hydrogen bonding [109].

6.4 Antimicrobial

The hybrid of lignin together with nano-silver or nano-chitosan was used as an antibacterial agent [110]. Another study used switchgrass lignin showed good promise for bacteria flocculation in wastewater treatment [111]. Traditional adhesive hydrogels are known to have poor mechanical strength, less adhesiveness effective, and most importantly low antibacterial activity. Another plant-based adhesive hydrogel was then delivered and prepared from silver-lignin nanoparticles [112]. It showed significant cell affinity and, most importantly, high antibacterial activity due to its catechol groups and antibacterial activity of silver-lignin nanoparticles.

6.5 *Others*

An interesting invention in making nano-colorant from nanoparticles lignin was found to produce more fluorescent nano-colorants as compared to pristine lignin [97], since lignin nanoparticles have the antioxidant ability, the solubility of nanoscale lignin improved tremendously as compared to the conventional lignin. It also exhibited excellent UV protection, finding applications in cosmetic products [47]. Nanoparticles were also used in emulsion microencapsulation. Lignin nanoparticles also are suitable for this application where it was embedded with sodium dodecyl sulfate (SDS) [113]. Lignin extracted from wheatgrass was employed with an ionic liquid (IL) producing very fine lignin nanoparticles, ranging from 200 nm to 1.5 μm. In another study, lignin nanoparticles were grafted with diethylaminoethyl methacrylate as applied as surfactant for CO_2/N_2 pickering emulsion [114]. Lignin nanoparticles was also applied for the immobilization of oxidative enzymes due to their electrochemical responsiveness [115]. Hence, these studies show that lignin nanoparticles have found various applications over the years. This is potentially promising as this material is inexpensive and non-toxic.

7 Industrial Applications

Lignin and its derivatives have been used for a variety of purposes in industrial and research purposes. Lignin and its derivatives have been used mainly for biofuels, chemical reagents and polymer purposes. However, numerous other applications such as adsorbents, drug delivery, fertilizer release, antifoulant additive, adhesives, coatings, nano-catalyst for oxidation reaction, and energy storage composites have also been found and investigated by the researchers. Table 2 highlights the main uses of lignin and its derivatives for different applications on an industrial scale [94, 116–119].

M. S. Alqahtani et al. synthesized lignin nanoparticles using the citric acid crosslinking technique to enhance the oral drug delivery system of a naturally occurring compound, curcumin, which is a low soluble with limited bioavailability, naturally occurring compound. Their studies proved that lignin nanoparticles (LNPs) improved the LNP-loaded curcumin had high stability, higher cellular uptake, slow-release, increased permeability (fivefold), and increased bioavailability (up to

Table 2 Potential applications of lignin and its composites "author"

Application type	Uses
Biofuels	Diesel, syngas, hydrogen, char
Chemicals	Dispersant, binder, polymer, adsorbents, adhesives, nano-catalyst, antifouling agent, binders, sizing agents, antioxidant, anticoagulant, corrosion inhibitors, fire retardant, agglomeration agent
Agriculture	Controlled release of fertilizers, Sequestering agents of heavy metals
Biomedical	Tissue engineering/regeneration, antiproliferation agent, drug delivery, pickering for microencapsulation, antiviral
Battery	Thermal energy storage (B-70)
Polymer	Thermal stability, mechanical stability, adsorbing agent, a reactive component in resin

tenfold) [120]. P. Figueiredo et al. derived peptin functionalized LNPs from dentin phosphophoryn for increased cellular uptake into different types of cancer cells. They concluded that functionalized LNPs showed a great commitment in cellular uptake with improved stability and antiproliferative effect for three types of cancer cell lines [121]. T. Zou et al. also studied the potential of LNPs for controlled drug delivery of ciprofloxacin and reported that antioxidative lignin particles hold great potential for the rapid release of ciprofloxacin when encapsulated with chitosan [122]. Y. Li et al. in their study discussed that alkali lignin-based colloidal spheres resulted in the controlled release of avermectin, and it helped in protecting the avermectin from UV irradiation even after 96 h [123]. Apart from examples quoted above, numerous other investigations have reported improved efficiency in terms of drug release, encapsulation, cellular uptake, and bioavailability for biomedical applications of lignin-based nanoparticles [48, 124, 125]. P. Figueiredo et al. have recently published a comprehensive review of such an application, and it can be consulted for further reading [119].

Lignin nanoparticles, their composites, and derivatives have also found their wider applications in greener fuel production. Many studies have recently been conducted for the production of hydrogen, syngas, biogas, biodiesel [126], and biochar from lignin and its derivatives [127–131]. Pei et al. reported the production of hydrogen gas through the gasification of lignin in supercritical water. Their analysis revealed that by controlling the tube wall temperature lignin-based gasification could produce a higher production of hydrogen [132]. Similar results were also reported by Kang et al., when they used nano-catalyst of lignin for improved production of hydrogen [133]. S. Kang et al. analyzed the production of hydrochar using lignin, cellulose, D-Xylose, and wood meal. They concluded that lignin could be effectively employed for the production of solid fuel and adsorbent, albeit controlling parameters to produce such fuels would be of great importance during the process [134].

Another important application of lignin material and its derivatives is in energy storage devices. Recent advances in thermal energy storage (TES) devices have been shifted to carbon-based materials, particularly materials with greener synthesis techniques because of stringent environmental laws [135–137]. M. H. Sipponen et al. employed fatty acid-lignin hybrid nanocapsules for TES. Their results proved the vitality of lignin-based nanocapsules since these capsules prevented fragmentation of high heating–cooling cycles up to 290. They also concluded that such high efficacy of these lignin-based nanocapsules could go beyond TES systems [138]. H. Li et al. reported the development of supercapacitors using a lignin-based carbon monolith. They inferred that these additive-free electrodes depicted the highest volumetric capacitance of 3 F cm^{-2}/97.1 F cm^{-3} and energy density of 0.16 mWh cm^{-2} at 1.75 mW cm^{-2} with appreciable cycling performance after 10,000 cycles [139]. W. Zhang also reported similar results for high capacitance applications using lignin-based hierarchical porous carbon with porous morphology. They stated that specific capacitance of such carbon was 286.7 F g^{-1} at a charge–discharge current density of 0.2 A g^{-1} [140].

Numerous application of lignin and lignin-based materials have also been investigated which have shown enhanced properties ranging from anticorrosive behavior [141], antimicrobial character [142], high mechanical and thermal strength [88, 143], hydrophilic character [144]. Frollini et al. synthesized a thermal isolating structural foam, which showed that the use of lignin could minimize the thermal conductivity of such foams effectively. They also claimed that lignin-based cellulosic fibers could act as a reinforcing agent by enhancing the impact of phenolic matrices up to 35 folds [145]. Zheng et al. used lignin coatings to improve the fire retardancy of substrate materials. Their study showed that sulfonated lignin had better performance than kraft lignin, hence providing the usefulness of lignin materials as a fire retardant [146]. P. K. Mishra and R. Wimmer synthesized lignin-based colloidal particles for their applications in cosmetics and drug delivery applications. They suggested considerable stability in terms of coating layers for synthesized colloidal lignin particles [147]. Lee et al., used lignin spherical particles in preparation of sunscreen for UV protection. They emphasized that the inclusion of spherical lignin nanoparticles produced sunscreen with higher sun protection factors and UVA exceptional protection factors. They also stated that inclusion of light color lignin nanoparticles could improve the aesthetic aspect of sunscreen too [148]. Similar results for the improved performance of sunscreen were reported by other researchers also [149–151]. Rehman et al. used the lignin nanoparticles as an anticorrosive material for the protection of carbon steel. They reported enhanced protection against the corrosion of carbon steel when lignin nanoparticles were employed as a protective coating on it [141]. Dastpak et al., also claimed enhanced (up to 3 order magnitude) anticorrosive property of stainless steel when coated with lignin nanoparticles [152].

8 Issues and Challenges

Despite wider applications and potential uses of lignin nanoparticles and lignin derivatives, their commercial usage is hindered by very peculiar challenges such as color, sustainability, stability, degradation, greener synthesis methods, and high polydispersity. Currently, many improved techniques and resources have been used to derive the lignin from natural resources with minimum hazards; however, such techniques have not been reported yet for any scale-up studies [60]. Application of lignin nanoparticles, specifically in cosmetics products, can also be improved if the color of lignin could be improved using some greener techniques which do not affect the structure and properties of lignin nanoparticles [117]. Another challenge that researchers are facing persistently is the solubility of lignin nanoparticles or lignin-based materials into solvents specifically for analytical purposes such as NMR. Such difficulty makes it harder for a complete investigation of these materials in developing products, and it could pave the way for the researchers to either find newer greener solvents or modification techniques to make lignin materials soluble into these solvents [153]. Furthermore, isolation of lignin from its source is still an issue to be taken seriously and vigorously, and this area particularly holds the key to its global acceptance as a greener filler nanomaterial. Finally, it can be concluded that the wider application of lignin nanoparticles is highly dependent on our quest to improve its developing methods, which are inexpensive, greener, and quicker.

9 Future Recommendations

The physical characteristics of lignin, which is used as a precursor for lignin NPs depend majorly on biomass origins, heterogeneity, and methods employed for pretreatment. It is important to adopt safer, greener, and eco-friendly methods for synthesizing a wide range of lignin NPs. Moreover, major challenges such as complexity, subsequent molecular properties, and lignin structures can be fulfilled by replacing the conventional harsh solvents (acid/alkaline) with room temperature ionic liquids (RTILs) and deep eutectic solvents (DES). In addition, various biological techniques like enzyme treatment, gel-electrophoresis, and bio-membranes can be utilized as a medium for the separation and production of lignin-based materials.

Forecasting energy and environmental issues cause the effective usage of natural resources to be re-evaluated. Researchers are also convincing to identify potential applications of sustainable technology for natural and renewable resources. Therefore, lignin has tremendous potential and can satisfy the modern world's existing energy demand. It is important to find alternative sources for liquid fuels in a world where there is excessive reliance on petroleum. Alternatively, progress in biotechnology may help to discover and characterize new enzymes and to generate them in homologous or heterologous systems. Low-cost conversion of lignin to biofuels and biochemicals would be the goal. The key focus of future work, which could lead to the previously stated activities, will include the introduction of nanotechnology, green biotechnology, and new technological developments in lignin processing and techniques.

The biodegradable composites are the potential replacements for plastics in the food-based industry, e.g., food packaging and single-use cutlery. However, these cannot be suitably developed until core properties like color, sustainability, stability, and degradation are fine-tuned to the market demand. This sector can find extensive applications worldwide. The cost of the lignin-based materials is a constraint in replacing many industrial and domestic processes. If low-temperature photo-catalytic and immiscible solvent-based extraction processes can be developed, the price of separation will be substantially reduced, economizing the overall costs. Moreover, the catalytic nature of the process will improve product recovery, increase process kinetics, and enhance the quality of the product.

Researchers need to consider how the processing conditions alter the chemistry and material properties of lignin. Here are discussed common commercial lignin-producing methods including sulfite, Kraft, and soda pulping as the key commercial extraction processes. The current research in the sector has been limited to low technology readiness levels. At the current state of knowledge, many processes can be developed further into commercialized technologies with minimal investment and funding support.

Acknowledgements All authors are thankful to their affiliated institutions for providing of the academic and support facilities for the completion of this work.

References

1. Ioelovich M (2015) Recent findings and the energetic potential of plant biomass as a renewable source of biofuels—a review, vol 10, p 36, 2015-01-08 2015
2. Krozer Y, Narodoslawsky M (2019) Economics of bioresources: concepts, tools, experiences. Springer International Publishing Imprint, Cham
3. Jin J (1987) Protecting biological resources to sustain human progress. Ambio 16:262–266
4. Abdullah MFF, Ali MTBM, Yusof FZM (2018) Bioresources technology in sustainable agriculture: biological and biochemical research. Apple Academic Press, Waretown, NJ
5. Höfer R, Selig M (2012) 10.02 - Green chemistry and green polymer chemistry. In: Matyjaszewski K, Möller M (eds) Polymer science: a comprehensive reference. Elsevier, Amsterdam, pp 5–14
6. W. B. Association (2019) Global bioenergy statistics 2019
7. I. I. E. Agency. Renewables and waste. Available: https://www.iea.org/data-and-statistics/data-tables/?country=WORLD&year=2017&energy=Renewables%20%26%20waste
8. Abdel-Shafy HI, Mansour MSM (2018) Solid waste issue: sources, composition, disposal, recycling, and valorization. Egyptian J Petroleum 27:1275–1290, 2018/12/01/
9. Körner I (2015) Chapter 7—Civilization biorefineries: efficient utilization of residue-based bioresources. In: Pandey A, Höfer R, Taherzadeh M, Nampoothiri KM, Larroche C (eds) Industrial biorefineries & white biotechnology. Elsevier, Amsterdam, pp 295–340
10. Gaurav N, Sivasankari S, Kiran G, Ninawe A, Selvin J (2017) Utilization of bioresources for sustainable biofuels: a review. Renew Sustain Energy Rev 73:205–214
11. Li C, Zhao X, Wang A, Huber GW, Zhang T (2015) Catalytic transformation of lignin for the production of chemicals and fuels. Chem Rev 115:11559–11624
12. Pandey MP, Kim CS (2011) Lignin depolymerization and conversion: a review of thermochemical methods. Chem Eng Technol 34:29–41
13. Doherty WOS, Mousavioun P, Fellows CM (2011) Value-adding to cellulosic ethanol: Lignin polymers. In: Industrial crops and products, vol 33, pp 259–276, 2011/03/01/
14. Rashid T, Kait CF, Regupathi I, Murugesan T (2016) Dissolution of kraft lignin using Protic Ionic Liquids and characterization. In: Industrial crops and products, vol 84, pp 284–293

15. Zakzeski J, Bruijnincx PC, Jongerius AL, Weckhuysen BM (2010) The catalytic valorization of lignin for the production of renewable chemicals. Chem Rev 110:3552–3599
16. Calvo-Flores FG, Dobado JA (2010) Lignin as renewable raw material. Chemsuschem 3:1227–1235
17. Lora JH, Glasser WG (2002) Recent industrial applications of lignin: a sustainable alternative to nonrenewable materials. J Polym Environ 10:39–48
18. Gargulak J, Lebo S. Commercial use of lignin-based materials. ACS Publications
19. Stewart D (2008) Lignin as a base material for materials applications: chemistry, application and economics. Industrial Crops Prod 27:202–207
20. Boerjan W, Ralph J, Baucher M (2003) Lignin biosynthesis. Ann Rev Plant Biol 54:519–546
21. Perez-Cantu L, Schreiber A, Schütt F, Saake B, Kirsch C, Smirnova I (2013) Comparison of pretreatment methods for rye straw in the second generation biorefinery: effect on cellulose, hemicellulose and lignin recovery. Bioresource Technol 142:428–435
22. Liu Y, Zheng J, Xiao J, He X, Zhang K, Yuan S et al (2019) Enhanced enzymatic hydrolysis and lignin extraction of wheat straw by triethylbenzyl ammonium chloride/lactic acid-based deep eutectic solvent pretreatment. ACS Omega 4:19829–19839
23. Agbesola Y (2013) Sustainability of municipal solid waste management in Nigeria: a case study of Lagos
24. Salah M, El-Haggar PE (2007) Chapter 5—Sustainability of municipal solid waste management. In: Sustainable industrial design and waste management, pp 149–196
25. da Costa Sousa L, Chundawat SP, Balan V, Dale BE (2009) 'Cradle-to-grave' assessment of existing lignocellulose pretreatment technologies. Curr Opinion Biotechnol 20:339–347
26. Kumar P, Barrett DM, Delwiche MJ, Stroeve P (2009) Methods for pretreatment of lignocellulosic biomass for efficient hydrolysis and biofuel production. Ind Eng Chem Res 48:3713–3729
27. Tutt M, Kikas T, Olt J (2012) Influence of different pretreatment methods on bioethanol production from wheat straw. Agronomy Res 10:269–276
28. Bensah EC, Mensah M (2013) Chemical pretreatment methods for the production of cellulosic ethanol: technologies and innovations. Int J Chem Eng
29. Rashid T, Gnanasundaram N, Appusamy A, Kait CF, Thanabalan M (2018) Enhanced lignin extraction from different species of oil palm biomass: kinetics and optimization of extraction conditions. In: Industrial crops and products, vol 116, pp 122–136
30. Doherty WO, Mousavioun P, Fellows CM (2011) Value-adding to cellulosic ethanol: Lignin polymers. Industrial Crops Prod 33:259–276
31. Azadi P, Inderwildi OR, Farnood R, King DA (2013) Liquid fuels, hydrogen and chemicals from lignin: a critical review. Renew Sustain Energy Rev 21:506–523
32. Sixta H, Potthast, Antje, Krotschek, Andreas W (2008) Chemical pulping processes. Handbook of Pulp. Wiley VCH Verlag GmbH, vol. Sections -4.2.5 pp 109–229
33. Lourençon TV, Hansel FA, da Silva TA, Ramos LP, de Muniz GIB, Magalhães WLE (2015) Hardwood and softwood kraft lignins fractionation by simple sequential acid precipitation. In: Separation and purification technology, vol 154, pp 82–88
34. Lu F, Ralph J (2010) Chapter 6—Lignin. In: Sun R-C (ed) Cereal straw as a resource for sustainable biomaterials and biofuels. Elsevier, Amsterdam, pp 169–207
35. Sjostrom E (2013) Wood chemistry: fundamentals and applications. Elsevier
36. Ragauskas AJ, Beckham GT, Biddy MJ, Chandra R, Chen F, Davis MF et al (2014) Lignin valorization: improving lignin processing in the biorefinery. Science 344:1246843
37. Erdocia X, Prado R, Corcuera MA, Labidi J (2014) Effect of different organosolv treatments on the structure and properties of olive tree pruning lignin. J Industrial Eng Chem 20:1103–1108
38. Behling R, Valange S, Chatel G (2016) Heterogeneous catalytic oxidation for lignin valorization into valuable chemicals: what results? What limitations? What trends? Green Chem 18:1839–1854
39. Gomes FJB, Santos FA, Colodette JL, Demuner IF, Batalha LAR (2014) Literature review on biorefinery processes integrated to the pulp industry. Natural Resources 05(09):14

40. SjÖStrÖM E (1993) Chapter 4—LIGNIN. In: SjÖStrÖM Ed (ed) Wood Chemistry, 2nd edn. Academic Press, San Diego, pp 71–89
41. Shimada K, Hosoya S, Ikeda T (1997) Condensation reactions of softwood and hardwood lignin model compounds under organic acid cooking conditions. J Wood Chem Technol 17:57–72
42. Zhao X, Cheng K, Liu D (2009) Organosolv pretreatment of lignocellulosic biomass for enzymatic hydrolysis. Appl Microbiol Biotechnol 82:815–827
43. Mishra PK, Ekielski A (2019) A simple method to synthesize lignin nanoparticles. Colloids Interfaces 3:52
44. Gao W, Fatehi P (2019) Lignin for polymer and nanoparticle production: current status and challenges. Canadian J Chem Eng 97:2827–2842
45. Chauhan PS (2020) Lignin nanoparticles: eco-friendly and versatile tool for new era. Bioresource Technol Rep 9:100374
46. Frangville C, Rutkevičius M, Richter AP, Velev OD, Stoyanov SD, Paunov VN (2012) Fabrication of environmentally biodegradable lignin nanoparticles. ChemPhysChem 13:4235–4243
47. Gutiérrez-Hernández JM, Escalante A, Murillo-Vázquez RN, Delgado E, González FJ, Toríz G (2016) Use of Agave tequilana-lignin and zinc oxide nanoparticles for skin photoprotection. J Photochem Photobiol B: Biol 163:156–161
48. Dai L, Liu R, Hu L-Q, Zou Z-F, Si C-L (2017) Lignin nanoparticle as a novel green carrier for the efficient delivery of resveratrol. ACS Sustain Chem Eng 5:8241–8249
49. Liu Z-H, Hao N, Shinde S, Pu Y, Kang X, Ragauskas AJ et al (2019) Defining lignin nanoparticle properties through tailored lignin reactivity by sequential organosolv fragmentation approach (SOFA). Green Chem 21:245–260
50. Qian Y, Zhang Q, Qiu X, Zhu S (2014) CO_2-responsive diethylaminoethyl-modified lignin nanoparticles and their application as surfactants for CO_2/N_2-switchable pickering emulsions. Green Chem 16:4963–4968
51. Beisl S, Friedl A, Miltner A (2017) Lignin from micro-to nanosize: Applications. Int J Mol Sci 18:2367
52. Myint AA, Lee HW, Seo B, Son W-S, Yoon J, Yoon TJ et al (2016) One pot synthesis of environmentally friendly lignin nanoparticles with compressed liquid carbon dioxide as an antisolvent. Green Chem 18:2129–2146
53. Zhiming L, Chao L, Haiying W (2013) Preparation method of nanolignin with controllable particle size. Chinese Patent CN. 103145999A., Dated June 12, 2013
54. Zhiming L, Guochao L, Haiying W (2013) Preparation method of nanolignin with controllable particle size. Chinese patent CN A, vol. 103145999
55. Wang B, Sun D, Wang H-M, Yuan T-Q, Sun R-C (2019) Green and facile preparation of regular lignin nanoparticles with high yield and their natural broad-spectrum sunscreens. In: ACS sustainable chemistry & engineering, vol 7, pp 2658–2666
56. Gilca IA, Popa VI, Crestini C (2015) Obtaining lignin nanoparticles by sonication. Ultrasonics Sonochemistry 23:369–375
57. Garcia Gonzalez MN, Levi M, Turri S, Griffini G (2017) Lignin nanoparticles by ultrasonication and their incorporation in waterborne polymer nanocomposites. J Appl Polym Sci 134:45318
58. Rangan A, Manchiganti MV, Thilaividankan RM, Kestur SG, Menon R (2017) Novel method for the preparation of lignin-rich nanoparticles from lignocellulosic fibers. In: Industrial crops and products, vol 103, pp 152–160
59. Juikar SJ, Vigneshwaran N (2017) Extraction of nanolignin from coconut fibers by controlled microbial hydrolysis. In: Industrial crops and products, vol 109, pp 420–425
60. Richter AP, Bharti B, Armstrong HB, Brown JS, Plemmons D, Paunov VN et al (2016) Synthesis and characterization of biodegradable lignin nanoparticles with tunable surface properties. Langmuir 32:6468–6477
61. Lou R, Ma R, Lin K-T, Ahamed A, Zhang X ()2019 Facile extraction of wheat straw by Deep Eutectic Solvent (DES) to produce lignin nanoparticles. In: ACS sustainable chemistry & engineering, vol 7, pp 10248–10256

62. Satyanarayana KG, Arizaga GG, Wypych F (2009) Biodegradable composites based on lignocellulosic fibers—an overview. Prog Polym Sci 34:982–1021
63. Dada OR, Abdulrahman KO, Akinlabi ET (2019) Production of biodegradable composites from agricultural waste. Biodegradable Compos Mater Manuf Eng 10:39
64. Rowell RM (1995) Composite materials from agricultural resources. In: Research in industrial application on non-food crops, I. plant fibres, pp 27–41
65. Ren H, Zhang Y, Zhai H, Chen J (2015) Production and evaluation of biodegradable composites based on polyhydroxybutyrate and polylactic acid reinforced with short and long pulp fibers. Cellul Chem Technol 49:641–652
66. Khan A, Colmenares JC, Gläser R (2020) Lignin based composite materials for photocatalysis and photovoltaics. In: Lignin Chemistry. Springer, Berlin, pp 1–31
67. Kazzaz AE, Feizi ZH, Fatehi P (2019) Grafting strategies for hydroxy groups of lignin for producing materials. Green Chem 21:5714–5752
68. Xiong S-J, Pang B, Zhou S-J, Li M-K, Yang S, Wang Y-Y et al (2020) Economically-competitive biodegradable PBAT/lignin composites: effect of lignin methylation and compatibilizer. In: ACS Sustainable Chemistry & Engineering
69. Kai D, Tan MJ, Chee PL, Chua YK, Yap YL, Loh XJ (2016) Towards lignin-based functional materials in a sustainable world. Green Chem 18:1175–1200
70. Ye J, Cheng Y, Sun L, Ding M, Wu C, Yuan D et al (2019) A green SPEEK/lignin composite membrane with high ion selectivity for vanadium redox flow battery. J Membrane Sci 572:110–118
71. Nair V, Panigrahy A, Vinu R (2014) Development of novel chitosan–lignin composites for adsorption of dyes and metal ions from wastewater. Chem Eng J 254:491–502
72. Morsella M, d'Alessandro N, Lanterna AE, Scaiano JC (2016) Improving the sunscreen properties of TiO_2 through an understanding of its catalytic properties. ACS Omega 1:464–469
73. Chung Y-L, Olsson JV, Li RJ, Frank CW, Waymouth RM, Billington SL et al (2013) A renewable lignin–lactide copolymer and application in biobased composites. ACS Sustain Chem Eng 1:1231–1238
74. Domínguez-Robles J, Larrañeta E, Fong ML, Martin NK, Irwin NJ, Mutjé P et al (2020) Lignin/poly(butylene succinate) composites with antioxidant and antibacterial properties for potential biomedical applications. Int J Biol Macromolecules 145:92–99
75. Hu L, Guang C, Liu Y, Su Z, Gong S, Yao Y et al (2020) Adsorption behavior of dyes from an aqueous solution onto composite magnetic lignin adsorbent. Chemosphere 246:125757, 2020/05/01/
76. Vaidya AA, Collet C, Gaugler M, Lloyd-Jones G (2019) Integrating softwood biorefinery lignin into polyhydroxybutyrate composites and application in 3D printing. Mater Today Commun 19:286–296
77. Rukmanikrishnan B, Ramalingam S, Rajasekharan SK, Lee J, Lee J (2020) Binary and ternary sustainable composites of gellan gum, hydroxyethyl cellulose and lignin for food packaging applications: biocompatibility, antioxidant activity, UV and water barrier properties. Int J Biol Macromolecules 153:55–62
78. Kumar Singla R, Maiti SN, Ghosh AK (2016) Crystallization, morphological, and mechanical response of poly (lactic acid)/lignin-based biodegradable composites. Polymer-Plastics Technol Eng 55:475–485
79. Feldman D (2016) Lignin nanocomposites. J Macromolecular Sci Part A 53:382–387
80. Yang W, Rallini M, Wang D-Y, Gao D, Dominici F, Torre L et al (2018) Role of lignin nanoparticles in UV resistance, thermal and mechanical performance of PMMA nanocomposites prepared by a combined free-radical graft polymerization/masterbatch procedure. In: Composites Part A: applied science and manufacturing, vol 107, pp 61–69
81. Alexandre M, Dubois P (2000) Polymer-layered silicate nanocomposites: preparation, properties and uses of a new class of materials. Mater Sci Eng R Rep 28:1–63
82. Cai X, Riedl B, Zhang S, Wan H (2008) The impact of the nature of nanofillers on the performance of wood polymer nanocomposites. Compos Part A Appl Sci Manuf 39:727–737

83. Yang W, Kenny JM, Puglia D (2015) Structure and properties of biodegradable wheat gluten bionanocomposites containing lignin nanoparticles. Industrial Crops Prod 74:348–356
84. Yang W, Owczarek J, Fortunati E, Kozanecki M, Mazzaglia A, Balestra G et al (2016) Antioxidant and antibacterial lignin nanoparticles in polyvinyl alcohol/chitosan films for active packaging. Industrial Crops Prod 94:800–811
85. Yang W, Fortunati E, Dominici F, Giovanale G, Mazzaglia A, Balestra G et al (2016) Synergic effect of cellulose and lignin nanostructures in PLA based systems for food antibacterial packaging. Eur Polym J 79:1–12
86. Jiang C, He H, Jiang H, Ma L, Jia D (2013) Nano-lignin filled natural rubber composites: preparation and characterization. Express Polym Lett 7
87. Gupta AK, Mohanty S, Nayak S (2015) Influence of addition of vapor grown carbon fibers on mechanical, thermal and biodegradation properties of lignin nanoparticle filled bio-poly (trimethylene terephthalate) hybrid nanocomposites. RSC Adv 5:56028–56036
88. Yang W, Fortunati E, Dominici F, Kenny J, Puglia D (2015) Effect of processing conditions and lignin content on thermal, mechanical and degradative behavior of lignin nanoparticles/polylactic (acid) bionanocomposites prepared by melt extrusion and solvent casting. Eur Polym J 71:126–139
89. Yang X, Zhong S (2020) Properties of maleic anhydride-modified lignin nanoparticles/polybutylene adipate-co-terephthalate composites. J Appl Polym Sci 49025
90. Li X, Hegyesi N, Zhang Y, Mao Z, Feng X, Wang B et al (2019) Poly (lactic acid)/lignin blends prepared with the Pickering emulsion template method. Eur Polym J 110:378–384
91. Kayserilioğlu BŞ, Bakir U, Yilmaz L, Akkaş N (2003) Drying temperature and relative humidity effects on wheat gluten film properties. J Agric Food Chem 51:964–968
92. Chollet B, Lopez-Cuesta J-M, Laoutid F, Ferry L (2019) Lignin nanoparticles as a promising way for enhancing lignin flame retardant effect in polylactide. Materials 12:2132
93. Ju T, Zhang Z, Li Y, Miao X, Ji J (2019) Continuous production of lignin nanoparticles using a microchannel reactor and its application in UV-shielding films. RSC Adv 9:24915–24921
94. Duval A, Lawoko M (2014) A review on lignin-based polymeric, micro- and nano-structured materials. Reactive Functional Polym 85:78–96
95. Duval A, Lawoko M (2014) A review on lignin-based polymeric, micro-and nano-structured materials. Reactive Functional Polym 85:78–96
96. Si M, Zhang J, He Y, Yang Z, Yan X, Liu M et al (2018) Synchronous and rapid preparation of lignin nanoparticles and carbon quantum dots from natural lignocellulose. Green Chem 20:3414–3419
97. Pillai MM, Karpagam K, Begam R, Selvakumar R, Bhattacharyya A (2018) Green synthesis of lignin based fluorescent nanocolorants for live cell imaging. Mater Lett 212:78–81
98. Gillet S, Aguedo M, Petitjean L, Morais A, da Costa Lopes A, Łukasik R et al (2017) Lignin transformations for high value applications: towards targeted modifications using green chemistry. Green Chem 19:4200–4233
99. Beisl S, Miltner A, Friedl A (2017) Lignin from micro-to nanosize: production methods. Int J Mol Sci 18:1244
100. Vinardell MP, Mitjans M (2017) Lignins and their derivatives with beneficial effects on human health. Int J Mol Sci 18:1219
101. Fernández-Pérez M, Villafranca-Sánchez M, Flores-Céspedes F (2007) Controlled-release formulations of cyromazine-lignin matrix coated with ethylcellulose. J Environ Sci Health 42:863–868
102. Shankar S, Rhim J-W (2017) Preparation and characterization of agar/lignin/silver nanoparticles composite films with ultraviolet light barrier and antibacterial properties. Food Hydrocolloids 71:76–84
103. Lee J-B, Yamagishi C, Hayashi K, Hayashi T (2011) Antiviral and immunostimulating effects of lignin-carbohydrate-protein complexes from Pimpinella anisum. Biosci Biotechnol Biochem 1101242356-1101242356
104. Karim QA, Karim SSA, Frohlich JA, Grobler AC, Baxter C, Mansoor LE et al (2010) Effectiveness and safety of tenofovir gel, an antiretroviral microbicide, for the prevention of HIV infection in women. Science 329:1168–1174

105. Lin X, Zhao J, Wu M, Kuga Sh HY (2016) Green synthesis of gold, platinum and palladium nanoparticles by lignin and hemicellulose. J Microbiol Biotechnol 5:14–18
106. Li P, Lv W, Ai S (2016) Green and gentle synthesis of Cu_2O nanoparticles using lignin as reducing and capping reagent with antibacterial properties. J Exp Nanosci 11:18–27
107. Bian H, Jiao L, Wang R, Wang X, Zhu W, Dai H (2018) Lignin nanoparticles as nano-spacers for tuning the viscoelasticity of cellulose nanofibril reinforced polyvinyl alcohol-borax hydrogel. Eur Polym J 107:267–274
108. Iravani S, Varma RS (2019) Plants and plant-based polymers as scaffolds for tissue engineering. Green Chem 21:4839–4867
109. Chen Y, Zheng K, Niu L, Zhang Y, Liu Y, Wang C et al (2019) Highly mechanical properties nanocomposite hydrogels with biorenewable lignin nanoparticles. Int J Biol Macromolecules 128:414–420
110. Klapiszewski Ł, Rzemieniecki T, Krawczyk M, Malina D, Norman M, Zdarta J et al (2015) Kraft lignin/silica–AgNPs as a functional material with antibacterial activity. Colloids Surf B Biointerfaces 134:220–228
111. Yiamsawas D, Beckers SJ, Lu H, Landfester K, Wurm FR (2017) Morphology-controlled synthesis of lignin nanocarriers for drug delivery and carbon materials. ACS Biomater Sci Eng 3:2375–2383
112. Gan D, Xing W, Jiang L, Fang J, Zhao C, Ren F et al (2019) Plant-inspired adhesive and tough hydrogel based on Ag-Lignin nanoparticles-triggered dynamic redox catechol chemistry. Nature Commun 10:1–10
113. Pang Y, Wang S, Qiu X, Luo Y, Lou H, Huang J (2017) Preparation of lignin/sodium dodecyl sulfate composite nanoparticles and their application in pickering emulsion template-based microencapsulation. J Agricultural Food Chem 65:11011–11019
114. ur Rahman O, Shi S, Ding J, Wang D, Ahmad S, Yu H (2018) Lignin nanoparticles: synthesis, characterization and corrosion protection performance. New J Chem 42:3415–3425
115. Milczarek G, Nowicki M (2013) Carbon nanotubes/kraft lignin composite: characterization and charge storage properties. Mater Res Bull 48:4032–4038
116. Wang X, Han G, Shen Z, Sun R (2015) Fabrication, property, and application of Lignin-Based nanocomposites. In: Eco-friendly polymer nanocomposites. Springer, Berlin, pp 73–99
117. Iravani S, Varma RS (2020) Greener synthesis of lignin nanoparticles and their applications. Green Chem 22:612–636
118. Norgren M, Edlund H (2014) Lignin: recent advances and emerging applications. Curr Opin Colloid Interface Sci 19:409–416
119. Figueiredo P, Lintinen K, Hirvonen JT, Kostiainen MA, Santos HA (2018) Properties and chemical modifications of lignin: towards lignin-based nanomaterials for biomedical applications. Progr Mater Sci 93:233–269
120. Alqahtani MS, Alqahtani A, Al-Thabit A, Roni M, Syed R (2019) Novel lignin nanoparticles for oral drug delivery. J Mater Chem B 7:4461–4473
121. Figueiredo P, Sipponen MH, Lintinen K, Correia A, Kiriazis A, Yli-Kauhaluoma J et al (2019) Preparation and characterization of dentin phosphophoryn-derived peptide-functionalized lignin nanoparticles for enhanced cellular uptake. Small 15:1901427
122. Zou T, Sipponen MH, Österberg M (2019) Natural shape-retaining microcapsules with shells made of chitosan-coated colloidal lignin particles. Front Chem 7:370
123. Li Y, Yang D, Lu S, Lao S, Qiu X (2018) Modified lignin with anionic surfactant and its application in controlled release of avermectin. J Agricultural Food Chem 66:3457–3464
124. Ciolacu D, Oprea AM, Anghel N, Cazacu G, Cazacu M (2012) New cellulose–lignin hydrogels and their application in controlled release of polyphenols. Mater Sci Eng, C 32:452–463
125. Figueiredo P, Ferro C, Kemell M, Liu Z, Kiriazis A, Lintinen K et al (2017) Functionalization of carboxylated lignin nanoparticles for targeted and pH-responsive delivery of anticancer drugs. Nanomedicine 12:2581–2596
126. Bhat AH, Dasan YK, Khan I (2015) Extraction of lignin from biomass for biodiesel production. In: Hakeem KR, Jawaid M, Alothman OY (eds) Agricultural biomass based potential materials. Springer International Publishing, Cham, pp 155–179

127. Sato T, Furusawa T, Ishiyama Y, Sugito H, Miura Y, Sato M et al (2006) Effect of water density on the gasification of lignin with magnesium oxide supported nickel catalysts in supercritical water. Industrial Eng Chem Res 45:615–622
128. Osada M, Sato O, Watanabe M, Arai K, Shirai M (2006) Water density effect on lignin gasification over supported noble metal catalysts in supercritical water. Energy Fuels 20:930–935
129. Van de Velden M, Baeyens J, Brems A, Janssens B, Dewil R (2010) Fundamentals, kinetics and endothermicity of the biomass pyrolysis reaction. Renew Energy 35:232–242
130. Yaman S (2004) Pyrolysis of biomass to produce fuels and chemical feedstocks. Energy Conversion Manage 45:651–671
131. Bai X, Kim KH (2016) Biofuels and chemicals from lignin based on pyrolysis. In: Fang Z, Smith JRL (eds) Production of biofuels and chemicals from lignin. Springer, Singapore, pp 263–287
132. Pei JCGLL, Xin AG (2007) Experimental investigation on hydrogen production by gasification of lignin in supercritical water. Acta Energiae Solaris Sinica 9
133. Kang K, Azargohar R, Dalai AK, Wang H (2015) Noncatalytic gasification of lignin in supercritical water using a batch reactor for hydrogen production: an experimental and modeling study. Energy Fuels 29:1776–1784
134. Kang S, Li X, Fan J, Chang J (2012) Characterization of hydrochars produced by hydrothermal carbonization of lignin, cellulose, D-xylose, and wood meal. Industrial Eng Chem Res 51:9023–9031
135. Budnyak TM, Slabon A, Sipponen MH (2020) Lignin–inorganic interfaces: chemistry and applications from adsorbents to catalysts and energy storage materials. ChemSusChem
136. Sipponen MH, Lange H, Ago M, Crestini C (2018) Understanding lignin aggregation processes. A case study: budesonide entrapment and stimuli controlled release from lignin nanoparticles. In: ACS sustainable chemistry & engineering, vol 6, pp 9342–9351
137. Li J, He Y, Inoue Y (2001) Study on thermal and mechanical properties of biodegradable blends of poly(ε-caprolactone) and lignin. Polym J 33:336–343
138. Sipponen MH, Henn A, Penttilä P, Österberg M (2020) Lignin-fatty acid hybrid nanocapsules for scalable thermal energy storage in phase-change materials. Chem Eng J 124711
139. Li H, Yuan D, Tang C, Wang S, Sun J, Li Z et al (2016) Lignin-derived interconnected hierarchical porous carbon monolith with large areal/volumetric capacitances for supercapacitor. Carbon 100:151–157
140. Zhang W, Zhao M, Liu R, Wang X, Lin H (2015) Hierarchical porous carbon derived from lignin for high performance supercapacitor. Colloids Surf A: Physicochem Eng Aspects 484:518–527
141. Rahman OU, Shi S, Ding J, Wang D, Ahmad S, Yu H (2018) Lignin nanoparticles: synthesis, characterization and corrosion protection performance. New J Chem 42:3415–3425
142. Setälä H, Alakomi H-L, Paananen A, Szilvay GR, Kellock M, Lievonen M et al (2020) Lignin nanoparticles modified with tall oil fatty acid for cellulose functionalization. Cellulose 27:273–284
143. Anwer MA, Naguib HE, Celzard A, Fierro V (2015) Comparison of the thermal, dynamic mechanical and morphological properties of PLA-Lignin & PLA-Tannin particulate green composites. Compos Part B Eng 82:92–99
144. Park Y, Doherty WOS, Halley PJ (2008) Developing lignin-based resin coatings and composites. In: Industrial crops and products, vol 27, pp 163–167
145. Frollini E, Paiva JMF, Trindade WG, Tanaka Razera IA, Tita SP (2004) Plastics and composites from lignophenols. In: Wallenberger FT, Weston NE (eds) Natural fibers, plastics and composites. Springer US, Boston, MA, pp 193–225
146. Zheng C, Li D, Ek M (2019) Improving fire retardancy of cellulosic thermal insulating materials by coating with bio-based fire retardants. Nordic Pulp Paper Res J 34:96–106
147. Mishra PK, Wimmer R (2017) Aerosol assisted self-assembly as a route to synthesize solid and hollow spherical lignin colloids and its utilization in layer by layer deposition. Ultrasonics Sonochemistry 35:45–50

148. Lee SC, Yoo E, Lee SH, Won K (2020) Preparation and application of light-colored lignin nanoparticles for broad-spectrum sunscreens. Polymers 12:699
149. Qian Y, Qiu X, Zhu S (2016) Sunscreen performance of lignin from different technical resources and their general synergistic effect with synthetic sunscreens. ACS Sustain Chem Eng 4,:4029–4035, 2016/07/05
150. Morsella M, Giammatteo M, Arrizza L, Tonucci L, Bressan M, d'Alessandro N (2015) Lignin coating to quench photocatalytic activity of titanium dioxide nanoparticles for potential skin care applications. RSC Adv 5:57453–57461
151. Qian Y, Qiu X, Zhong X, Zhang D, Deng Y, Yang D et al (2015) Lignin reverse micelles for UV-absorbing and high mechanical performance thermoplastics. Industrial Eng Chem Res 54:12025–12030
152. Dastpak A, Yliniemi K, De Oliveira Monteiro MC, Höhn S, Virtanen S, Lundström M et al (2018) From waste to valuable resource: lignin as a sustainable anti-corrosion coating. Coatings 8:454
153. Henn A, Mattinen M-L (2019) Chemo-enzymatically prepared lignin nanoparticles for value-added applications. World J Microbiol Biotechnol 35:125

Chapter 12
Recent Trends in Surface Modification of Natural Fibres for Their Use in Green Composites

Mariana D. Banea, Jorge S. S. Neto, and Daniel K. K. Cavalcanti

1 Introduction

In recent years, many industries have invested in sustainable processes aiming to reduce their dependency of petroleum-based productivity chains, creating new biodegradable materials that are less harmful to the environment [1–4]. Currently, numerous types of natural fibres are used in various industrial sectors due to their numerous advantages such as low cost, low density, renewability, biodegradability, easy availability and worldwide abundance [5]. However, they have some important drawbacks, such as poor interfacial adhesion between fibres (hydrophilic) and matrix (hydrophobic), moisture absorption, poor fire resistance, low impact strength and low durability. In order to surpass these drawbacks, surface modification of natural fibres is commonly used. It was shown in the literature that the fibre surface modification can improve the fibre–matrix interfacial bonding, roughness, wettability, the hydrophilic nature of fibres and also decrease the moisture absorption [6–16]. The surface modification methods can be divided into two main groups: chemical and physical approaches (see Fig. 1). It is seen from the literature that chemical treatments are widely used to improve the interfacial adhesion between fibre and matrix and improve the durability, while the physical approaches are used to enhance the mechanical properties of the composites.

Figure 1 presents a classification of the main methods of surface modification of natural fibres. These methods will be briefly presented in the next sections.

M. D. Banea (✉) · J. S. S. Neto · D. K. K. Cavalcanti
Federal Center of Technological Education of Rio de Janeiro (CEFET/RJ), Av. Maracanã, 229, Rio de Janeiro 20271-110, Brazil
e-mail: mdbanea@gmail.com

S. Thomas and P. Balakrishnan (eds.), *Green Composites*,
Materials Horizons: From Nature to Nanomaterials,
https://doi.org/10.1007/978-981-15-9643-8_12

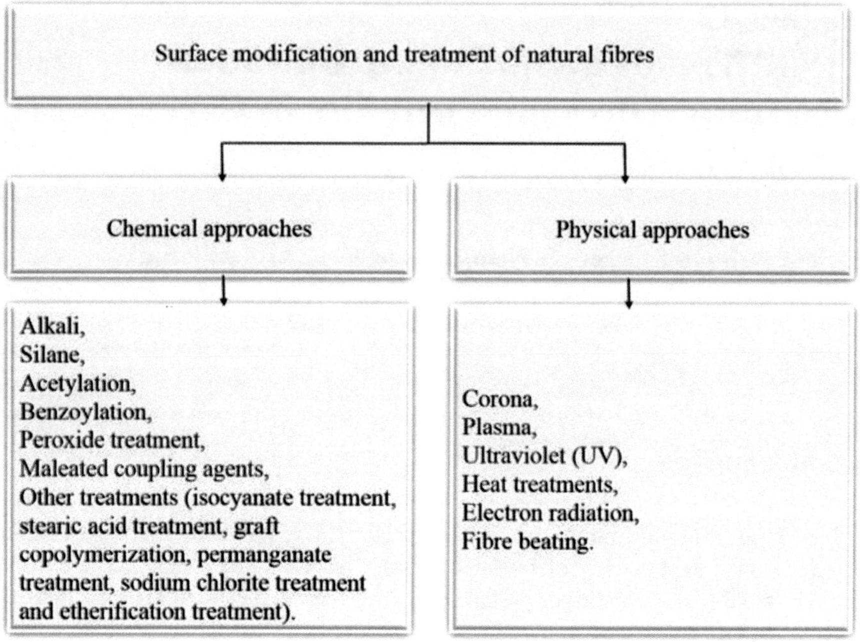

Fig. 1 Scheme of natural fibres surface modification classification. *Source* Author

2 Chemical Approaches for Surface Modification of Natural Fibres

Chemical treatments can increase the interface adhesion between the fibre and matrix by chemical bonding or mechanical interlocking at the interface and decrease the water absorption of fibres [17, 18]. Several compounds were used to promote adhesion by chemically coupling the fibres to the matrix, such as sodium hydroxide, silane, acetic acid, acrylic acid, maleated coupling agents, isocyanates, potassium permanganate and peroxide. The most common chemical treatments used to treat/modify the surface of natural fibres will be briefly discussed in this section.

2.1 *Alkaline Treatment*

Alkali treatment, also called mercerization, is one of the most popular chemical treatments of natural fibres mainly because of its simplicity and low cost [3]. Sodium hydroxide (NaOH) is commonly used in this method. It was shown in the literature that the alkalization treatment removes the impurities and organic components such as lignin, pectin and hemicellulose from the surface of the natural fibres, increasing in this way the roughness and wettability of the fibres [17]. This method not only removes impurities but also exposes the fibres fibrils and functionalizes its surface with –OH groups, thus increasing fibre–matrix interactions. However, the efficiency of alkali treatment depends on the type and concentration of the alkaline solution, time of treatment and the temperature used. It was shown that if the alkali solution concentration is higher than the optimum condition, an excess of the delignification of the fibre can occur, which results in weakening or damaging the fibre. Various studies from the literature [17, 19–26] reported improvements in interfacial shear strength (IFSS), improvement of fibre strength and improved interfacial bonding for alkali modified natural fibre. Table 1 summarizes a number of studies on the effect of alkaline treatment on the main characteristics of natural fibres and interfacial adhesion.

2.2 Silane Treatment

Silane (SiH_4) is used as a coupling agent to modify the surface of natural fibres in green composites [28]. Silane treatment consists of hydrolysis of alkoxy groups on silane with water to form silanol (Si–OH) groups which can further react with hydroxyl groups on the fibre surface. Figure 2 shows the reaction mechanism of silane with the functional group of natural fibres. It was shown that silane treatment increases the hydrophobicity of natural fibres and strength of composites with higher increases occurring when stable covalent bonding occurs between silane and the matrix [6].

It was also shown in the literature that the silane treatment can benefit from a previous alkalinization process [3, 29, 30]. For example, Sepe et al. [30] analysed the effect of alkalinization (1 and 5% NaOH) and silane (1.5 and 20 wt% of silane coupling agent) chemical treatments in epoxy composites reinforced with hemp fibres, and the test results showed that silane treatment of hemp fibres improved both tensile and flexural properties of the composites. Figure 3 shows the effect of the silane treatment on hemp fibre-reinforced composites.

2.3 Acetylation Treatment

Acetylation is a chemical reaction in which a hydrogen atom is substituted for an acetyl group ($CH_3C{=}O$ group) in a compound. During acetylation of natural fibres, esterification occurs by reaction of acetyl groups (CH_3CO–) with hydroxyl groups (–OH) (see Fig. 4). The substitution of the polar OH with the less polar CH_3CO will render the fibre a more hydrophobic nature, thus promoting better compatibility with the non-polar matrix [16]. Commonly, the acetylation treatment of the natural fibres is preceded by an alkaline treatment [5]. It was showed that it improves interfacial bonding, tensile and flexural strength, stiffness, as well as thermal stability of

Table 1 Effect of alkali treatment on the natural fibre characteristics and interfacial adhesion

Fibre type	Matrix	Alkali treatment wt. NaOH	Effect of treatment	References
Hemp	Polypropylene	5% and 10% for 1 h	The treatment resulted in the removal of surface impurities and a rougher surface of fibres	[19]
Curauá	Polypropylene	5% for 1 h	The chemical treatment improved the interaction between the fibre and matrix, increasing the mechanical and thermal properties	[20]
Abaca	Epoxy	5, 10 and 15% for 2 h	5% of NaO produced an increase in cellulose crystallinity, improved roughness of fibres and increase in interfacial shear strength (IFSS)	[22]
Jute	Epoxy	Ethanol and toluene de-waxing followed by 1%	Significant increase in interfacial adhesion	[21]
Flax	PLA	10%, 20% and 30% for 20 min	Increase in surface roughness of fibres	[25]
Sisal	Epoxy	10% for 24, 120 and 240 h	120 h was the optimum time for treating sisal fibre to achieve the highest interfacial adhesion and mechanical properties	[24]
Kenaf	Epoxy	6% for 48 h and 144 h	Increase in fibre–matrix interfacial adhesion. The 144 h treatments were detrimental to tensile properties	[23]
Bamboo	Epoxy	1, 4, and 7%	Interfacial shear strength increased after treatment at 4% NaOH concentration, but decreased after 7% alkali treatment	[26]
Palm	Epoxy	3%, 6% and 9% for 24 h	6% NaOH was the optimum concentration which provides acceptable fibre strength and surface characteristics	[27]

Source Author

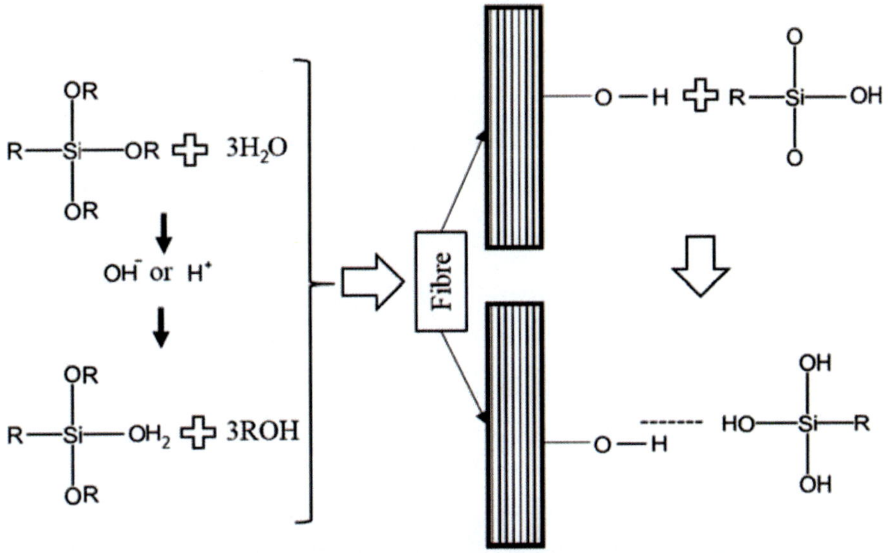

Fig. 2 Schematic of the chemical reaction between the silane groups and the functional groups of the natural fibres. *Source* Author

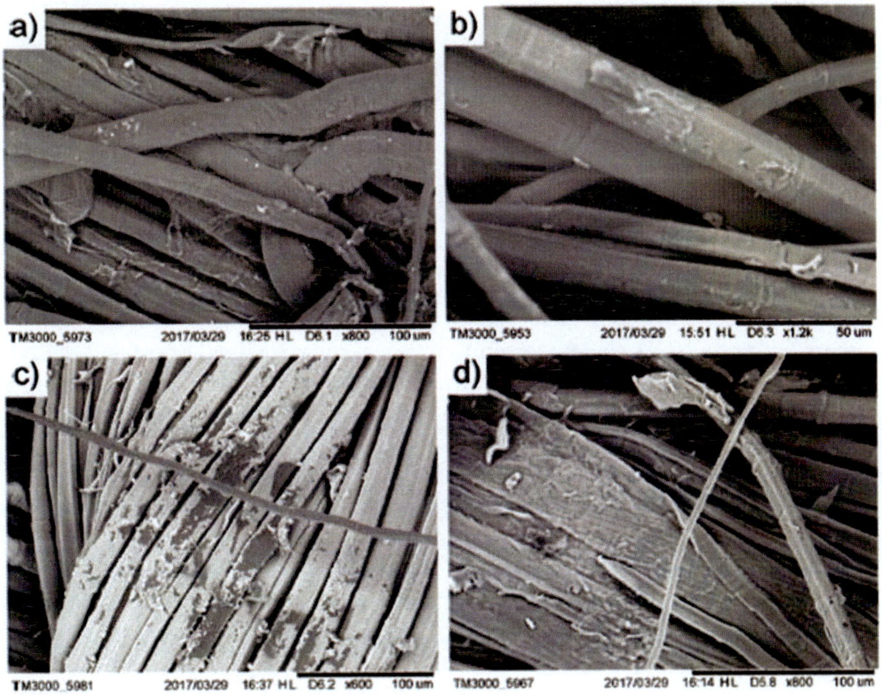

Fig. 3 SEM images of the surfaces of **a** untreated, **b** 1% silane-treated, **c** 5% silane-treated, and **d** 20% silane-treated hemp fibre-reinforced composites (Reproduced from Sepe et al. [30])

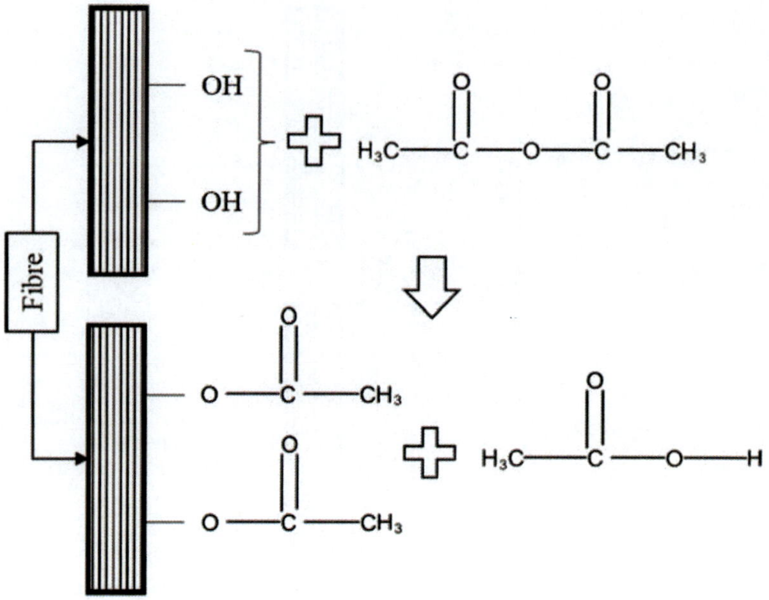

Fig. 4 Schematic of the reaction of acetic anhydride with natural fibre. *Source* Author

composites [31]. Several researchers report that depending on the concentration of acetylation, it can prevent up to 65% of moisture absorption [16, 32]. However, overtreatment can be deleterious to mechanical properties, due to degradation of cellulose and cracking of fibres which can occur with the catalysts used in this process [5].

2.4 Benzoylation Treatment

Benzoyl chloride ($C_6H_5C{=}O$) is used to improve interfacial adhesion and thermal stability of natural fibres by reacting to hydroxyl groups (OH) from the fibre surface [9, 16]. This results in addition of benzoyl groups on the cell wall of fibre. However, the condition that this reaction takes place is that several constituents (i.e. waxes, oils, hemicellulose and lignin) are removed using pre-treatments in order to expose the reactive OH groups on the fibres surface [33, 34]. After pre-treatment, the OH groups are replaced by benzoyl groups, as can be seen in Fig. 5.

2.5 Peroxide Treatment

The peroxide chemical treatment of natural fibre to be used in green composites has attracted the interest of several researchers [6, 35, 36]. The chemical compound

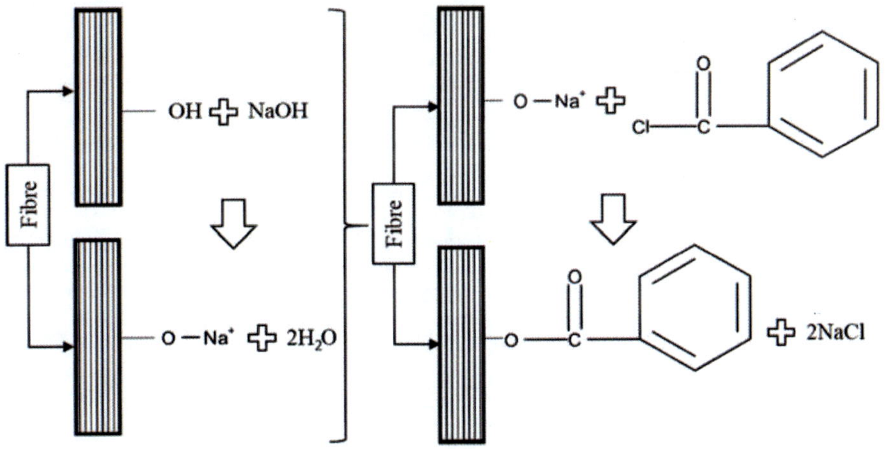

Fig. 5 Schematic of the reactions involved in the benzoylation process using a NaOH solution. *Source* Author

uses the functional group (ROOR) which contains divalent ion bond O–O which is conducive to dissociate free radicals (RO), in addition to having the tendency to react with the hydrogen group of cellulose fibres and matrix [6, 9, 18]. During peroxide treatment, the natural fibres are treated with 6% dicumyl peroxide or benzoyl peroxide (see Eqs. 1–3) in acetone solution after they were previously pre-treated by alkalinization [11].

$$\text{RO - OR} \rightarrow \text{2RO,RO + Polymer} \rightarrow \text{ROH + Polymer} \quad (1)$$

$$\text{RO} + (C_6H_{10}O_5)_n \rightarrow \text{ROH} + (C_6H_{10}O_5)_n \quad (2)$$

$$\text{Polymer} + (C_6H_{10}O_5)_n \rightarrow \text{Polymer - }(C_6H_{10}O_5)_n \quad (3)$$

2.6 *Maleated Coupling Agents*

Maleated coupling is another technique used for enhancing the properties of green composites reinforced with natural fibres. The main difference compared to the other methods is that this method modifies not only the fibre surface, but also the matrix as the maleic anhydride (MA) is commonly grafted to the same polymer used as the matrix to ensure compatibility between the matrix and the coupling agent [16, 28]. MA reacts with the hydroxyl groups of the natural fibres and also forms C–C bonds with the polymeric chains of the polymer matrix, which increases the compatibility

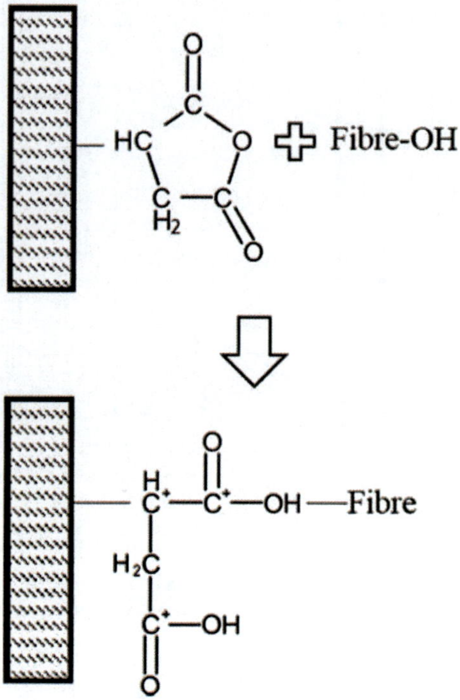

Fig. 6 Schematic of the chemical process of the maleated coupling. *Source* Author

between the fibre and matrix and in this way the interfacial adhesion [12, 16, 28]. Figure 6 illustrates a schematic of the chemical process of the maleated coupling.

One of the most common polymers used in the literature is the polypropylene (PP) matrix, which in the presence of MA results in grafted maleic anhydride polypropylene (MAPP) [28, 37]. Baykus et al. [37] investigated the alkaline treatment and the mixed treatment (alkaline + maleic anhydride) on jute fabric-based composites with polypropylene (PP) and polyethylene (PE) matrices. They found that the mixed treatment produced an increase in tensile strength of 90% in PP matrix composites and 40% in PE matrix composites, when compared to composites without treatment. González et al. [38] studied the incorporation of 10 to 40% by weight of agave fibre in polylactic acid (PLA) and maleated PLA (MAPLA) and showed that MAPLA presented an increase in tensile strength and Young´s modulus, in addition to decreasing the absorption coefficient and diffusion of water.

2.7 *Other Treatments*

Several other chemical treatments (i.e. isocyanate treatment, stearic acid treatment, graft copolymerization, permanganate treatment, sodium chlorite treatment and etherification treatment) were used in the literature to reduce the hydroxyl groups of natural fibres, in addition to improving the interfacial adhesion of the fibre/matrix [9, 28, 39, 40]. For instance, Tayfun et al. [39] evaluated the influence of isocyanate treatment on polyurethane composites reinforced with flax fibres, and the results indicated that the composites treated with isocyanate exhibited the best results in terms of tensile strength and Young's modulus when compared to untreated fibre-reinforced composites. Zafeiropoulos et al. [41] used stearic acid ($CH_3(CH_2)_{16}COOH$) to treat flax fibres. They reported that this treatment removed non-crystalline constituents of the fibres, altering the fibre surface topography. They also found that treated flax fibres were more crystalline than the untreated ones and stearation decreased the fibre surface-free energy.

3 Physical Treatment Modifications

Physical treatments (i.e. plasma, corona, ultraviolet (UV), heat treatments electron radiation and fibre beating) have been also used to modify the surface of natural fibres for their use in green composites [16, 28, 42]. These treatments improve the adhesion between the fibre and matrix by changing the surface properties of the fibres without changing their structural composition [9, 43]. However, physical treatments are more expensive compared to chemical ones, mainly because of the equipment involved in the surface modification processes [43].

3.1 Plasma Treatment

The plasma treatment is a technique that modifies the surface of natural fibres through a gas ionization process [44, 45]. Several gases can be used, such as oxidizing gases (H2O, O2 and N2O), inert gases (Ar or He), active gases (NH_3) which generate amino groups and polymerizing gases (monomer gases for direct polymerization, Ar or He pre-treated) [18]. Plasma treatment can be carried out under atmospheric pressure and vacuum pressure and at low and high temperatures [16, 18, 44]. However, for natural fibres low temperatures are preferred because of the flammable constituents of the natural fibres [16, 28]. It was showed that plasma treatment induces surface etching which improve the surface roughness of the fibre and result in a better interface with the matrix through mechanical interlocking. In addition, it can introduce various functional groups on the natural fibre surface, which form strong covalent bonds with the matrix, thus resulting in strong fibre–matrix interface [28, 33]. Figure 7 illustrates the effect of the plasma treatment on the surface of the fibre by surface etching and functionalizing.

An important factor to be considered in plasma treatment is the exposure time, since the longer the exposure period, the higher the possibility of the fibre surface to suffer degradation and damage [44]. As an example, Fig. 8 shows the effect that plasma treatment has on coir fibre surface over time.

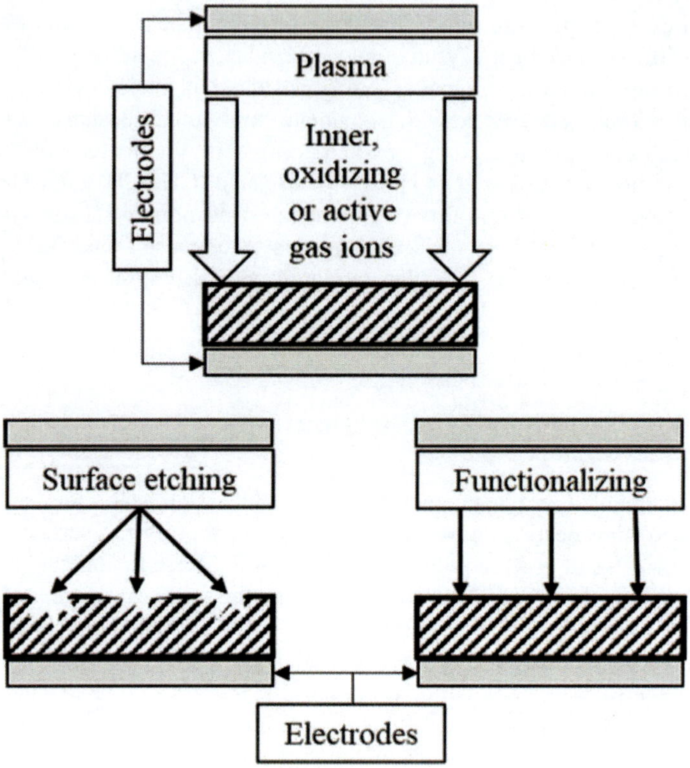

Fig. 7 Schematic of the effect of plasma treatment on the surface of fibres by surface etching and functionalizing. *Source* Author

3.2 *Corona Treatment*

Corona treatment is a surface modification technique that uses plasma generated by the application of a high voltage to sharp electrode tips separated by quartz and causes the formation of electromagnetic fields that oxidize the surfaces of the fibres [13]. The corona discharge produces chemical and physical changes of fibres as well as increased surface polarity (due to increased carboxyl and hydroxyl groups) and increased fibre roughness [12, 13, 16]. However, this treatment has some disadvantages, such as the low depth of penetration of the electrical discharge and the difficulty to be applied to three-dimensional surfaces including natural fibres [46]. On the other hand, the great advantage in relation to the other physical treatments, mainly plasma treatment, is the fact that the process has low cost and energy consumption, as do not need vacuum chambers for low temperatures. Gassan and Gutowski [46] used corona plasma to treat jute fibres to be used in epoxy composites. The treatments increased the polarity of fibres but decreased fibre strength and tenacity leading to reduced composite strength with corona treatment. Oudrhiri et al. [47] treated

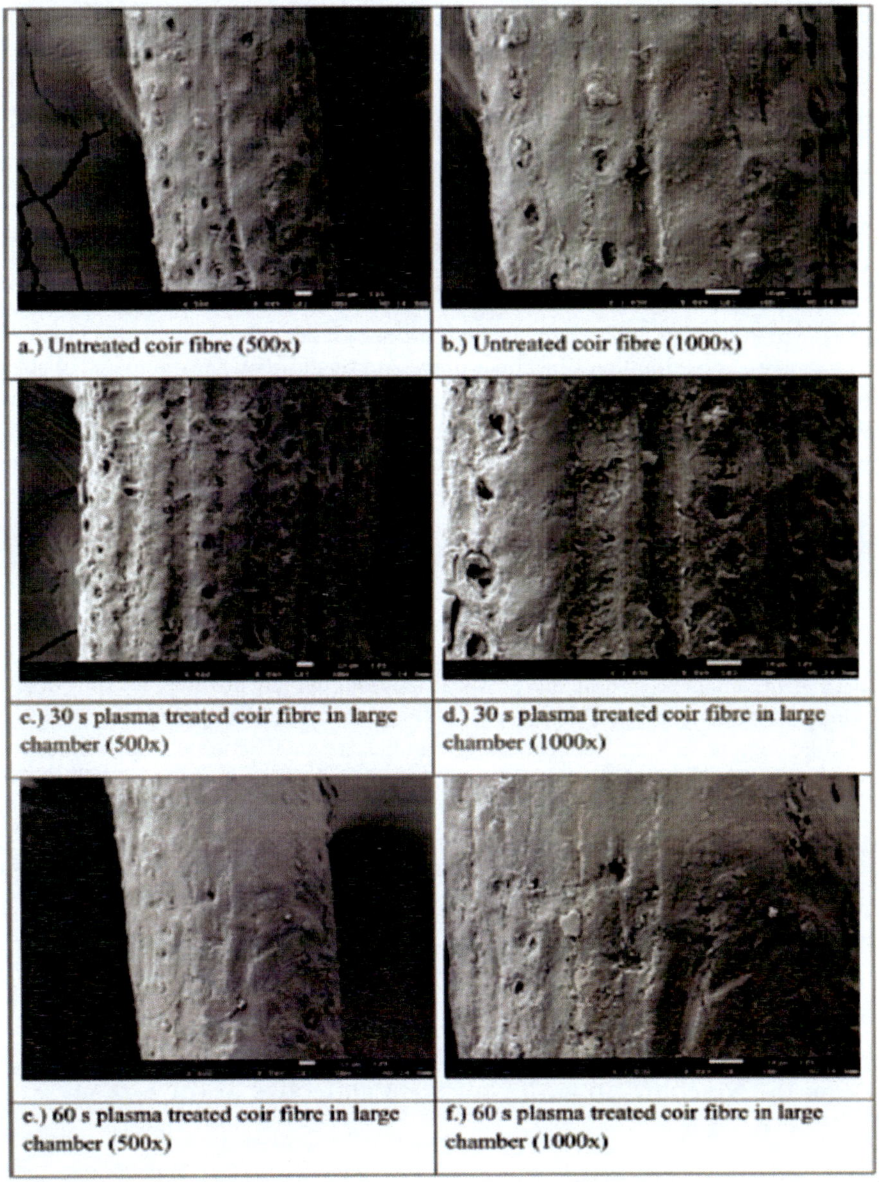

Fig. 8 SEM analysis of untreated coir fibres (**a**, **b**), 30 s plasma treated (**c**, **d**) and 60 s plasma treated (**e**, **f**) coir fibres (Reproduced from [44])

aloe vera fibres with corona discharge during 5, 10 and 15 min. and observed the degradation of fibres due to etching mechanism caused by corona.

3.3 Ultraviolet Treatments

The ultraviolet treatment is a technique that uses electromagnetic radiation ranging from 10 to 400 nm [18]. The process causes chemical changes (C–C, C–O, C–F, C–Si) in the surface of the natural fibres and promotes the cleaning of fatty acids, waxes and other constituents exposed on the fibre surface [7, 28, 48]. The main advantage of this technique is its simplicity, low investment and versatility when compared to other physical treatments. Benedetto et al. [48] investigated the effects of UV irradiation on banana composites for 7 and 15 days. They found better mechanical and thermal properties of the composites. Gassan and Gutowski [46] used UV to treat jute fibres and found 30% increase in the flexural strength of composites.

3.4 Thermal Treatments

Thermal treatment is one of the physical methods which involves heating of the natural fibre between the temperatures 100 and 200 °C. This process reduces the moisture content, and the lignocelluloses bundles are detached into single filaments. By heating the fibres at moderate temperatures, the crystallinity of the fibre also increases. It was shown in the literature that the thermal treatment can increase the physical adhesion between the fibres and the matrix and also increase the mechanical properties of the composites [49]. However, longer thermal exposure can alter the chemical and physical compositions which may be prejudicial for the composites properties [18].

4 The Effect of Surface Modification of Natural Fibres on the Mechanical and Thermal Properties of the Green Composites

4.1 Mechanical Properties

Numerous researchers have investigated the effect of fibre modification on the mechanical properties of green composites [22–24, 37–39, 50]. All these studies have highlighted that the adhesion between the fibre and matrix plays a significant role in the final mechanical properties of the composites since the stress transfer between matrix and fibre determines the reinforcement efficiency. The type of fibre and matrix and fibre loading are also key factors responsible for the mechanical properties of green composites.

Table 2 presents a summary of the several recent studies on the effect of fibre treatment on the mechanical properties of natural fibre composites. It is evident that the fibre modification methods have different efficacy in improvement of adhesion between the fibre and matrix and consequently the mechanical properties of the green composites. In general, chemical modification of natural fibres enhances the mechanical properties of the composites. However, the improvement depends on the concentration of the chemical used and exposure time, as it was showed that the mechanical properties deteriorate beyond a critical concentration and exposure time [23]. Furthermore, combination of treatments with two different chemicals proved better mechanical properties than the individual treatments in some cases [37].

In some studies, the impact strength of the polymer composites decreased after fibre treatment, as the composite tends to absorb less impact energy after fibre treatment due to the enhanced fibre/matrix adhesion [36].

Table 2 shows that physical treatments contribute significantly to improving of the mechanical properties of the composites. However, it was shown that prolonged physical treatment will damage the fibres and consequently the mechanical properties of the composites [48].

4.2 Thermal Properties

Thermal stability of natural fibres is an important aspect to be considered since the processing temperature plays a key role in the green composites manufacturing. Thermogravimetric analysis (TGA), differential scanning calorimetry (DSC) and differential mechanical thermal analysis (DMA) are widely used in the literature to investigate the thermal stability of composites. The percentage of weight loss, the degradation temperature, T_g and viscoelastic properties (storage modulus, loss modulus, and the damping factor) are the most common thermal properties determined by these techniques. The effect of fibre modification on the thermal properties of green composites was intensively investigated in the literature [26, 29, 40, 52–58].

Table 3 summarizes several recent studies on the effect of chemical treatments on the thermal properties of natural fibre-reinforced composites. It can be seen that in general, chemical modification of natural fibres improves the thermal stability of the composites. However, sometimes, the gain in thermal stability affects negatively the mechanical properties of composites, for example, tensile and flexural strength. The balance in improvements in thermal properties depends on fibre treatment type and time. It was also showed that combination of treatments (mixed treatments) promotes increased thermal stability and composite conductivity [52, 59].

Table 2 Effect of fibre modification on the mechanical properties of natural fibre-reinforced composites

Composite	Treatment	Effects	Ref
Abaca-epoxy	Alkaline treatment (5, 10, 15% NaOH for 2 h)	8% increase in tensile strength and 36% increase in tensile modulus	[22]
Hemp-epoxy	Alkaline (1% and 5% NaOH for 0.5 h) and Silane treatment (1, 5, 20% for 1 h)	Increase in tensile modulus of approx. 15% compared to the untreated composite and 10% compared to the alkaline treated composite	[30]
Bamboo-epoxy, polyester and vinyl ester	Alkaline treatment (10% NaOH for 48 h)	Increase of approx. 119.4 in tensile strength for the epoxy composites Increase of approx. 304.6% in flexural strength, when compared to the untreated composites	[51]
Kenaf-epoxy	Alkaline (6% NaOH for 48 and 144 h)	Treated unidirectional fibres presented an increase of 11% and 3.5% in tensile strength and modulus, compared to their untreated counterparts. The 144 h treatments were detrimental to tensile properties	[23]
Sisal-epoxy	Alkaline (10% for 24, 120 and 240 h)	120 h was the optimum time for treating sisal fibre to achieve highest mechanical properties (increase of approx. 197.9% and 115.0% in tensile strength and modulus for the 120 h treated composite)	[24]
Jute-polypropylene (PP)/ polyethylene (PE)	Alkaline and mixed treatment (alkaline + maleic anhydride)	The mixed treatment produced an increase in tensile strength of 90% in PP matrix and 40% in PE matrix composites, when compared to untreated composites	[37]
Hemp-epoxy	Alkaline (1 and 5% NaOH) and silane (1.5 and 20%)	Silane treatment of hemp fibres improved both tensile and flexural properties of the composites	[30]

(continued)

Table 2 (continued)

Composite	Treatment	Effects	Ref
Banana-polypropylene (PP)	Alkaline and mixed treatments (Alkaline + acetylation)	Composites with acetylation after the alkali pre-treatment showed better interfacial shear strength and mechanical properties compared to the alkali-treated and untreated fibre composites. The tensile strength and modulus of PP/acetylated BBF composites increased up to 40% of fibre loading The alkali treatment enhanced the impact strength by 72.75% at 40 wt % fibre loading as compared to untreated fibres The acetylated BBF into PP increased the impact strength by 65.64% compared to alkali-treated fibres	[31]
Flax-polypropylene	Acetylation treatment for 1–3 h	Tensile and flexural strength presented an increase by approx. 20% to 35% as a function of acetylation degree	[32]
Vetiveria zizanioides-polyester	Benzoylation treatment (10% 10% for 30 min)	The treatment increased tensile strength by 113%, compressive strength by 56.78% and impact by 95% when compared to the untreated composite	[35]
Vetiveria zizanioides-polyester	Peroxide treatment (10% Hydrogen peroxide for 3 h)	Tensile strength presented an increase of 57.26% when compared to the sodium hydroxide treated composite. Flexural strength increased by 56.13%, when compared to the untreated composite	[35]
Jute-PLA	Peroxide treatment (6% benzoyl peroxide acetone solution for 30 min after alkali pre-treatment)	The tensile and flexural properties were improved Izod impact strength was reduced	[36]
Jute-polypropylene (PP)/ Polyethylene (PE)	Maleated coupling agents	The treatment produced an increase in tensile strength of 90% in PP matrix and 40% in PE matrix composites, when compared to untreated composites	[37]

(continued)

Table 2 (continued)

Composite	Treatment	Effects	Ref
Agave fibre in polylactic acid (PLA) and maleated PLA (MAPLA)	Maleated coupling agents	The tensile strength of treated fibre-reinforced composites increased by approx. 68% (from 25 to 41 MPa) and the tensile modulus increased by 32% (from 1.30 to 1.74 GPa) compared to untreated fibre composites The flexural properties also increased for the treated fibres (flexural strength increase from 50 to 55 MPa and flexural modulus from 2.4 to 3.0 GPa)	[38]
Flax-polyethylene	Plasma treatment (30 mtorr and 1 min at 30 W)	Plasma treated fibres presented a considerably improved tensile strength and Young's modulus (tensile strength and Young's modulus increased by 18.6% and 32%, respectively)	[45]
Jute-epoxy	Corona discharge (40-100 W, 10–88 cm/min) UV (53 mm focal point; 80 W)	Flexural strength was negatively impacted by the corona treatment. However, the UV treatment presented an increase of up to 30%	[46]
Aloe vera	Corona discharge (10 kV, 10 kHz, 1 μs pulse) 5, 10 and 20 min treatment time	Mechanical properties were negatively impacted by the treatment	[47]
Jute-epoxy	Ultraviolet treatment	30% increase in the flexural strength of composites	[46]
Banana-epoxy	UV (7 and 15 days)	The 7 day case presented significant increases in tensile strength and modulus. However, the 10 day case presented negative effects	[48]

Source Author

Table 3 Effect of treatments on the thermal properties of natural fibre-reinforced composites

Composite	Method	Treatment	Effect on the thermal properties	Ref
Flax-epoxy	DMA	5% and 10% of sodium bicarbonate	By increasing the concentration of sodium bicarbonate, negligible changes in the T_g and reductions in the tan δ peak heights were found	[50]
Jute-polyester	DMA	5% of alkali treatment, poly (lactic acid)-coated and mixed treatment	The mixed and alkaline treatment improved the T_g	[59]
Kenaf-PLA	TG and DTG	Acetylation (0.5, 1, 2 and 3 h)	The thermal stability was improved as treatment time was increased	[60]
Banana-PP	TG and DSC	Untreated, alkaline and acetylated treatment	The acetylation treatment improved the thermal stability and raised up the crystallization temperature	[31]
Jute-PLA	TG and DMA	6% benzoyl peroxide acetone solution for about 30 min after alkali pre-treatment	Higher storage modulus and lower tan δ compared to untreated composite	[36]
Kenaf-PU	DMA	Acetylation, blocked isocyanate, maleic anhydride and permanganate treatment	The acetylated treatment provided better values for storage modulus (E') when compared to all other treatments	[53]
Banana + sisal-PLA	TG and DSC	Untreated and benzoyl peroxide treatment	The peroxide treatment improved the thermal stability by delaying the thermal degradation No significant changes in the (T_g) and melting temperature (T_m)	[61]
Jute + curauá-epoxy	TG and DMA	Untreated, alkali, silane and mixed treatment	The chemical treatment improved both T_g and thermal stability of the composites. tan δ value decreased	[29]

(continued)

Table 3 (continued)

Composite	Method	Treatment	Effect on the thermal properties	Ref
Bamboo-epoxy, polyester and vinyl ester	TG	Alkali (0 h-0%, 24 h-10%, 48 h-5, 10, 15% and 72 h-10%)	The bamboo-epoxy composite treated with the 10% of NaOH for 48 h presented the highest thermal stability	[51]
Cotton-LDPE	DSC	Alkaline, silane, mixed treatment, maleic anhydride, and alkali–maleic anhydride coupling agents	The mixed treated composite presented higher T_g	[52]
Jute-epoxy	DMA	Enzyme treatment, ozone treatment and laser treatment	The laser treated composites presented higher T_g values in relation to the other treatments	[62]
Ramie fibre-reinforced poly(lactic acid) (PLA)	TG and DSC	Maleic anhydride (MA)	The thermal stability increased but the T_g decreased	[55]

Source Author

5 Conclusions

Poor adhesion between natural fibres and matrix, thermal stability and moisture absorption of natural fibres are the principal challenges that need to be addressed in order to increase the use of green composites in industry. The modifications of fibres and/or matrix in order to overcome these limitations are critical steps in future development of green composites. This chapter presents the main chemical and physical methods used to modify natural fibres to be used in green composites. It was evident from the literature that natural fibres that undergo chemical and physical surface modifications reveal improved fibre–matrix interfacial adhesion, improved fibre roughness and wettability. In addition, various treatments help to minimize the hydrophilic nature and decrease the moisture absorption of natural fibres.

The fibre modification approaches discussed in this chapter have different efficacy. In general, chemical approaches are more studied in the literature than physical approaches and provide better improvements in properties. The improvement in mechanical and thermal properties depends on concentration of the chemical used and exposure time. Moreover, combination of treatments with two different chemicals showed better mechanical and thermal properties than the individual treatments in some cases.

It can be concluded that the selection of the appropriate fibre modification technique will depend on the particular fibre/matrix used and the composite end-use application. It is fundamental to understand the interfacial properties and bonding mechanisms of fibre–matrix, and this requires significant research efforts in order to maximize green composites applications.

References

1. Pereira AL, Banea MD, Neto JS, Cavalcanti DK (2020) Mechanical and thermal characterization of natural intralaminar hybrid composites based on sisal. Polymers 12(4):866
2. Lau K-t, Hung P-y, Zhu M-H, Hui D (2018) Properties of natural fibre composites for structural engineering applications. Compos B Eng 136:222–233
3. Cavalcanti D, Banea M, Neto J, Lima R, da Silva L, Carbas R (2019) Mechanical characterization of intralaminar natural fibre-reinforced hybrid composites. Compos B Eng 175:107149
4. de Queiroz H, Banea M, Cavalcanti D (2020) Experimental analysis of adhesively bonded joints in synthetic-and natural fibre-reinforced polymer composites. J Compos Mater 54(9):1245–1255
5. Pickering KL, Efendy MA, Le TM (2016) A review of recent developments in natural fibre composites and their mechanical performance. Compos Part A Appl Sci Manuf 83:98–112
6. Ali A, Shaker K, Nawab Y, Jabbar M, Hussain T, Militky J, Baheti V (2018) Hydrophobic treatment of natural fibers and their composites—a review. J Ind Text 47(8):2153–2183
7. Al-Maharma AY, Al-Huniti N (2019) Critical review of the parameters affecting the effectiveness of moisture absorption treatments used for natural composites. J Compos Sci 3:1–27
8. Azwa Z, Yousif B, Manalo A, Karunasena W (2013) A review on the degradability of polymeric composites based on natural fibres. Mater Des 47:424–442

9. Gholampour A, Ozbakkaloglu T (2020) A review of natural fiber composites: properties, modification and processing techniques, characterization, applications. J Mater Sci 1–64
10. John MJ, Anandjiwala RD (2008) Recent developments in chemical modification and characterization of natural fiber-reinforced composites. Polym Compos 29(2):187–207
11. Kalia S, Kaith B, Kaur I (2009) Pretreatments of natural fibers and their application as reinforcing material in polymer composites—a review. Polym Eng Sci 49(7):1253–1272
12. Koohestani B, Darban A, Mokhtari P, Yilmaz E, Darezereshki E (2019) Comparison of different natural fiber treatments: a literature review. Int J Environ Sci Technol 16(1):629–642
13. Liu M, Thygesen A, Summerscales J, Meyer AS (2017) Targeted pre-treatment of hemp bast fibres for optimal performance in biocomposite materials: a review. Ind Crops Prod 108:660–683
14. Mahzan S, Fitri M, Zaleha M (2017) UV radiation effect towards mechanical properties of natural fibre reinforced composite material: a review. In: IOP conference series: materials science and engineering, vol 1. IOP Publishing, p 012021
15. Sood M, Dwivedi G (2018) Effect of fiber treatment on flexural properties of natural fiber reinforced composites: a review. Egyptian J Petroleum 27(4):775–783
16. Zhou Y, Fan M, Chen L (2016) Interface and bonding mechanisms of plant fibre composites: an overview. Compos B Eng 101:31–45
17. de Araujo Alves Lima R, Kawasaki Cavalcanti D, de Souza e Silva Neto J, Meneses da Costa H, Banea MD (2020) Effect of surface treatments on interfacial properties of natural intralaminar hybrid composites. Polym Compos 41(1):314–325
18. Verma A, Parashar A, Jain N, Singh V, Rangappa SM, Siengchin S (2020) Surface modification techniques for the preparation of different novel biofibers for composites. In: Biofibers and biopolymers for biocomposites. Springer, pp 1–34
19. Sullins T, Pillay S, Komus A, Ning H (2017) Hemp fiber reinforced polypropylene composites: the effects of material treatments. Compos B Eng 114:15–22
20. Marques MDV, Melo RP, Araujo RD, Lunz JD, Aguiar VD (2015) Improvement of mechanical properties of natural fiber-polypropylene composites using successive alkaline treatments. J Appl Polym Sci 132(12):41710
21. Brodowsky H, Mäder E (2012) Jute fibre/epoxy composites: Surface properties and interfacial adhesion. Compos Sci Technol 72(10):1160–1166
22. Cai M, Takagi H, Nakagaito AN, Li Y, Waterhouse GI (2016) Effect of alkali treatment on interfacial bonding in abaca fiber-reinforced composites. Compos Part A Appl Sci Manuf 90:589–597
23. Fiore V, Di Bella G, Valenza A (2015) The effect of alkaline treatment on mechanical properties of kenaf fibers and their epoxy composites. Compos B Eng 68:14–21
24. Fiore V, Scalici T, Nicoletti F, Vitale G, Prestipino M, Valenza A (2016) A new eco-friendly chemical treatment of natural fibres: effect of sodium bicarbonate on properties of sisal fibre and its epoxy composites. Compos Part B Eng 85:150–160
25. Aydın M, Tozlu H, Kemaloglu S, Aytac A, Ozkoc G (2011) Effects of alkali treatment on the properties of short flax fiber–poly (lactic acid) eco-composites. J Polym Environ 19(1):11–17
26. Wang F, Lu M, Zhou S, Lu Z (2019) Effect of fiber surface modification on the interfacial adhesion and thermo-mechanical performance of unidirectional epoxy-based composites reinforced with bamboo fibers. Molecules (Basel, Switzerland) 24(15)
27. Alsaeed T, Yousif BF, Ku H (2013) The potential of using date palm fibres as reinforcement for polymeric composites. Mater Des 43:177–184
28. Ferreira DP, Cruz J, Fangueiro R (2019) Surface modification of natural fibers in polymer composites. In: Green composites for automotive applications. Elsevier, pp 3–41
29. Neto J, Lima R, Cavalcanti D, Souza J, Aguiar R, Banea M (2019) Effect of chemical treatment on the thermal properties of hybrid natural fiber-reinforced composites. J Appl Polym Sci 136(10):47154
30. Sepe R, Bollino F, Boccarusso L, Caputo F (2018) Influence of chemical treatments on mechanical properties of hemp fiber reinforced composites. Compos B Eng 133:210–217

31. Zaman HU, Khan RA (2019) Acetylation used for natural fiber/polymer composites. J Thermoplastic Compos Mater. https://doi.org/10.1177/0892705719838000
32. Bledzki A, Mamun A, Lucka-Gabor M, Gutowski V (2008) The effects of acetylation on properties of flax fibre and its polypropylene composites. Express Polym Lett 2(6):413–422
33. Amiandamhen S, Meincken M, Tyhoda L (2020) Natural fibre modification and its influence on fibre-matrix interfacial properties in biocomposite materials. Fibers Polym 21:677–689
34. Kalia S, Kaushik VK, Sharma RK (2011) Effect of benzoylation and graft copolymerization on morphology, thermal stability, and crystallinity of sisal fibers. J Nat Fibers 8(1):27–38
35. Vinayagamoorthy R (2019) Influence of fiber surface modifications on the mechanical behavior of Vetiveria zizanioides reinforced polymer composites. J Nat Fibers 16(2):163–174
36. Goriparthi BK, Suman KNS, Mohan Rao N (2012) Effect of fiber surface treatments on mechanical and abrasive wear performance of polylactide/jute composites. Compos A Appl Sci Manuf 43(10):1800–1808
37. Baykus O, Mutlu A, Doğan M (2016) The effect of pre-impregnation with maleated coupling agents on mechanical and water absorption properties of jute fabric reinforced polypropylene and polyethylene biocomposites. J Compos Mater 50(2):257–267
38. González-López ME, Pérez-Fonseca AA, Cisneros-López EO, Manríquez-González R, Ramírez-Arreola DE, Rodrigue D, Robledo-Ortíz JR (2019) Effect of maleated PLA on the properties of rotomolded PLA-agave fiber biocomposites. J Polym Environ 27(1):61–73
39. Tayfun U, Dogan M, Bayramli E (2017) Investigations of the flax fiber/thermoplastic polyurethane eco-composites: influence of isocyanate modification of flax fiber surface. Polym Compos 38(12):2874–2880
40. Premalatha N, Saravanakumar SS, Sanjay MR, Siengchin S, Khan A (2019) Structural and thermal properties of chemically modified luffa cylindrica fibers. J Nat Fibers 1–7. https://doi.org/10.1080/15440478.2019.1678546
41. Zafeiropoulos NE, Williams DR, Baillie CA, Matthews FL (2002) Engineering and characterisation of the interface in flax fibre/polypropylene composite materials. Part I. Development and investigation of surface treatments. Compos Part A Appl Sci Manuf 33(8):1083–1093
42. Beg MDH, Pickering KL (2008) Mechanical performance of Kraft fibre reinforced polypropylene composites: influence of fibre length, fibre beating and hygrothermal ageing. Compos A Appl Sci Manuf 39(11):1748–1755
43. Latif R, Wakeel S, Zaman Khan N, Noor Siddiquee A, Lal Verma S, Akhtar Khan Z (2019) Surface treatments of plant fibers and their effects on mechanical properties of fiber-reinforced composites: a review. J Reinf Plast Compos 38(1):15–30
44. Praveen K, Thomas S, Grohens Y, Mozetič M, Junkar I, Primc G, Gorjanc M (2016) Investigations of plasma induced effects on the surface properties of lignocellulosic natural coir fibres. Appl Surf Sci 368:146–156
45. Enciso B, Abenojar J, Martínez M (2017) Influence of plasma treatment on the adhesion between a polymeric matrix and natural fibres. Cellulose 24(4):1791–1801
46. Gassan J, Gutowski VS (2000) Effects of corona discharge and UV treatment on the properties of jute-fibre expoxy composites. Compos Sci Technol 60(15):2857–2863
47. Oudrhiri Hassani F, Merbahi N, Oushabi A, Elfadili MH, Kammouni A, Oueldna N (2020) Effects of corona discharge treatment on surface and mechanical properties of Aloe Vera fibers. Mat Today: Proc 24(1):46–51
48. Benedetto RMD, Gelfuso MV, Thomazini D (2015) Influence of UV radiation on the physical-chemical and mechanical properties of banana fiber. Mater Res 18:265–272
49. Rong MZ, Zhang MQ, Liu Y, Yang GC, Zeng HM (2001) The effect of fiber treatment on the mechanical properties of unidirectional sisal-reinforced epoxy composites. Compos Sci Technol 61(10):1437–1447
50. Fiore V, Scalici T, Valenza A (2018) Effect of sodium bicarbonate treatment on mechanical properties of flax-reinforced epoxy composite materials. J Compos Mater 52(8):1061–1072
51. Chin SC, Tee KF, Tong FS, Ong HR, Gimbun J (2020) Thermal and mechanical properties of bamboo fiber reinforced composites. Mater Today Commun 23:100876

52. Bodur MS, Bakkal M, Sonmez HE (2016) The effects of different chemical treatment methods on the mechanical and thermal properties of textile fiber reinforced polymer composites. J Compos Mater 50(27):3817–3830
53. Datta J, Kopczyńska P (2015) Effect of kenaf fibre modification on morphology and mechanical properties of thermoplastic polyurethane materials. Ind Crops Prod 74:566–576
54. Gupta M, Singh R (2019) PLA-coated sisal fibre-reinforced polyester composite: water absorption, static and dynamic mechanical properties. J Compos Mater 53(1):65–72
55. Yu T, Jiang N, Li Y (2014) Study on short ramie fiber/poly(lactic acid) composites compatibilized by maleic anhydride. Compos A Appl Sci Manuf 64:139–146
56. Zin MH, Abdan K, Norizan MN (2019) The effect of different fiber loading on flexural and thermal properties of banana/pineapple leaf (PALF)/glass hybrid composite. In: Structural health monitoring of biocomposites, fibre-reinforced composites and hybrid composites. Elsevier, pp 1–17
57. Chen H, Zhang W, Wang X, Wang H, Wu Y, Zhong T, Fei B (2018) Effect of alkali treatment on wettability and thermal stability of individual bamboo fibers. J Wood Sci 64(4):398–405
58. Jena PK, Mohanty JR, Nayak S (2020) Effect of surface modification of vetiver fibers on their physical and thermal properties. J Nat Fibers 1–12. https://doi.org/10.1080/15440478.2020.1726249
59. Gupta M (2020) Investigations on jute fibre-reinforced polyester composites: Effect of alkali treatment and poly (lactic acid) coating. J Ind Text 49(7):923–942
60. Chung T-J, Park J-W, Lee H-J, Kwon H-J, Kim H-J, Lee Y-K, Tai Yin Tze W (2018) The improvement of mechanical properties, thermal stability, and water absorption resistance of an eco-friendly PLA/kenaf biocomposite using acetylation. Appl Sci 8(3):376
61. Asaithambi B, Ganesan GS, Ananda Kumar S (2017) Banana/sisal fibers reinforced poly (lactic acid) hybrid biocomposites; influence of chemical modification of BSF towards thermal properties. Polym Compos 38(6):1053–1062
62. Jabbar A, Militký J, Wiener J, Karahan M (2016) Static and dynamic mechanical properties of novel treated jute/green epoxy composites. Text Res J 86(9):960–974

Chapter 13
Biodegradable Polymeric Materials for Medicinal Applications

Sobhi Daniel

1 Introduction

The design and development of biocompatible and biodegradable materials is a promising and challenging task in the fabrication of materials targeting medicinal applications. The 1982 Consensus Development Conference Statement of the National Institute of Health (NIH) defines a biomaterial as any substance (other than drug) or combination of substances, synthetic or natural in origin, which can be used for any period, as a whole or as a part of a system which treats, augments or replaces any tissue, organ or function in the body [1].

Biodegradable materials can be defined as those materials which modify their chemical and physical structure when these materials are in contact with the biological environment. If a material is biodegradable, then, by extending the right conditions and in the presence of microorganisms, fungi, or bacteria, it will ultimately break down to its basic elements and reverse back to earth. Preferably, these substances degrade without leaving any toxins behind. Biodegradation is a term used to designate the process of breakdown of a material by nature; however, in the case of medical purpose, biodegradation emphasized in the biological processes that cause a steady dissection of the material inside the living body. Biomaterials degradation is a very significant feature to contemplate when they are used for the medical purpose since their capability to function for a convinced application depends on the length of time that it is essential to keep them in the body. Biodegradation in a biotic environment may be defined as a gradual collapse of a material intervened by a specific biological activity. When materials are exposed to the body fluids, they may experience alterations in their physicochemical properties as a result of chemical, physical, mechanical, and biological correlations between the material and the neighboring

S. Daniel (✉)
Maharaja's Technological Institute, Government of Kerala, Thrissur 680020, India
e-mail: sobhi.daniel@gmail.com

S. Thomas and P. Balakrishnan (eds.), *Green Composites*,
Materials Horizons: From Nature to Nanomaterials,
https://doi.org/10.1007/978-981-15-9643-8_13

environment [2]. A very crucial component in biodegradation is the interaction with the immune system and the specialized cells. An ideal biodegradable biomaterial should have degradation products that are non-toxic and easily metabolized and cleared from the body.

Frequently used biodegradable materials used in medicinal applications include ceramics, metal composites, and polymeric materials. Biodegradable polymers are designated as the "materials of today" [3]. In recent times, the biodegradable polymers have become highly significant in the field of biomaterials and tissue engineering, due to the unnecessary supplementary surgery to eliminate the implants or scaffolds. In medicine, biodegradable polymers offer the great capability for controlled drug delivery and wound management (e.g., clips, staples, adhesives, sutures, and surgical meshes), for orthopedic devices (ligaments, tacks, screws, pins, and rods), nonwoven materials and scaffolds for tissue engineering. The other important applications of biodegradable polymers embrace implantable large devices, in cardiovascular applications, intestinal applications, in dental applications, drug delivery systems, personal protection equipment, surgical instruments, biological liquids transfer and storage systems, artificial organs, carriers of cells or enzymes, microfluidic devices, biosensors, in vitro diagnosis tools, etc.

2 Classification of Biodegradable Polymeric Materials

The general classification of biodegradable polymers is schematically represented in Fig. 1. They are broadly classified in to natural, semi-synthetic, and synthetic based on their occurrence and synthetic strategies.

1. Natural polymers: These are the polymers which are derived directly from the natural sources
   (a) Polysaccharides (starch, cellulose, chitin and chitosan, hyaluronic acid, alginic acid, etc.).

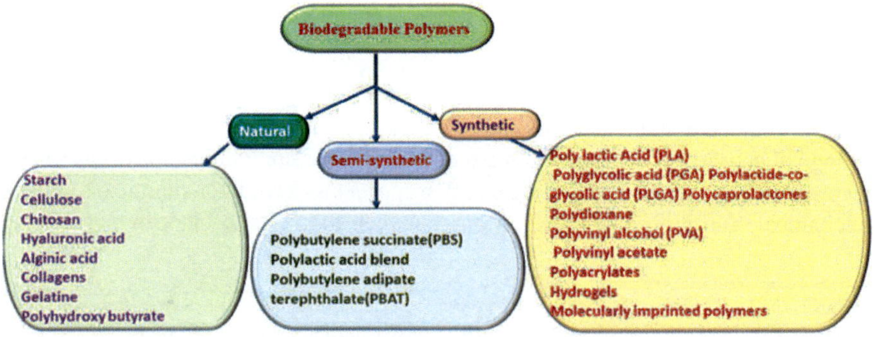

Fig. 1 Classification of biodegradable polymers. *Source* author

(b) Polypeptides such as collagens, gelatine.
(c) Bacterial polyesters, e.g., polyhydroxy butyrate.

2. Semi-synthetic polymers: These are the polymers in which raw material is obtained from nature, but polymerization takes place after some chemical modification.
3. Synthetic polymers: These are purely synthetic polymers derived from chemical synthesis.
   (a) Polymers having hydrolyzable backbone, e.g., polylactic acid copolymer (PLA), polyglycolic acid (PGA), polylactide-co-glycolic acid (PLGA), polycaprolactones, polydioxane, etc.
   (b) Polymers having carbon back bonds such as polyvinyl alcohol (PVA), polyvinyl acetate, polyacrylates, etc.

3 Commonly Used Natural and Synthetic Polymers

Natural biodegradable polymers are generally generated in nature by all living organisms. Even though the process of degradation of naturally occurring polymers is slow, they truly represent renewable resources since they are biodegradable. The extensively used natural polymers include polysaccharides, such as cellulose and starch. Other important classes of natural polymers are chitosan, gelatin, hyaluronic acid, collagen silk, etc. Their biodegradability, flexibility, and bioactivity make polysaccharides very promising natural biomaterials. The chemical structures of the commonly used natural, semi-synthetic and synthetic biodegradable polymers are shown in Fig. 2.

4 Mechanisms of Biodegradation in Polymers

The degradation of polymeric materials inside the body occurs mostly by three general mechanisms, oxidation, hydrolytic, and enzymatic mechanism. The foremost significant organisms in biodegradation are fungi, bacteria, and algae. Natural polymers (i.e., proteins, polysaccharides, nucleic acids) are decayed in biological systems by oxidation and hydrolysis and biodegradable materials decompose into biomass, greenhouse gas, and methane. In the case of synthetic polymers, microbial consumption of its carbon backbone as a carbon source is required [4]. Polymer degradation within the living body could also be interpreted as an interaction between the organism tissue components and a polymeric material acting as a distant body at the implantation. The key compounds causing polymer degradation in our body are water, salt, trace elements, and the enzymes present in our body fluids [5]. Biodegradation of polymeric materials within the biological environment commences with the "biodeterioration" process. During this step, the polymeric materials are first disintegrated into smaller components by the action of microbial organisms and

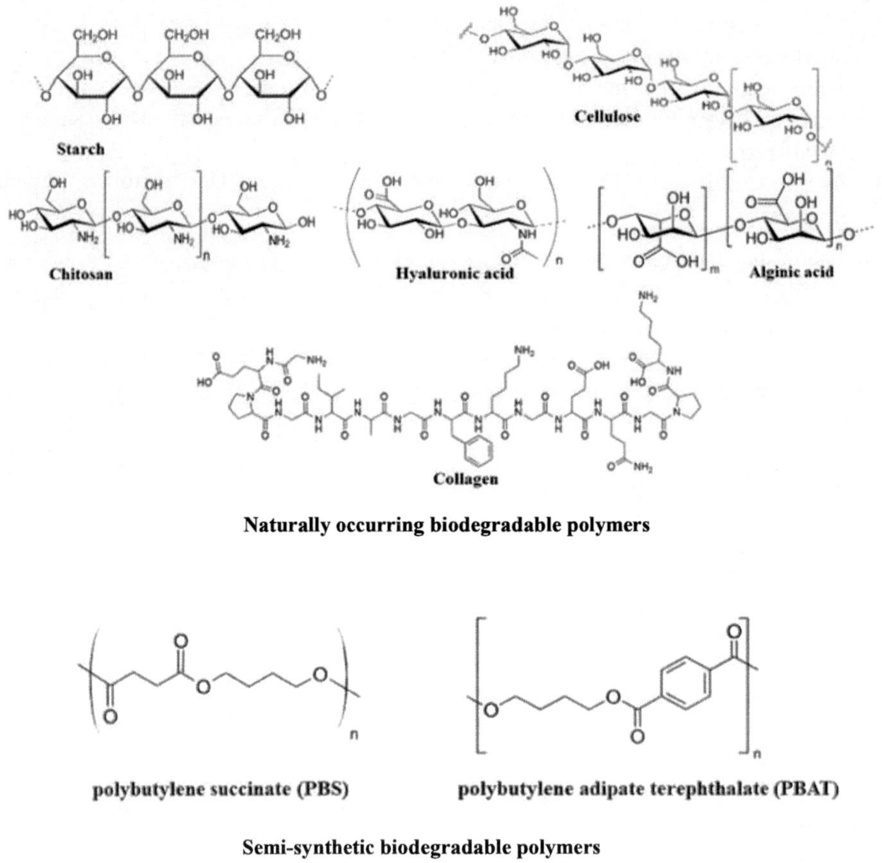

Fig. 2 Commonly used natural, semi-synthetic, and synthetic biodegradable polymers. *Source* author

abiotic agents present within the specific surroundings. Consequently, the polymeric backbones are cleaved by different hydrolytic enzymes (as well as other catalytic agents like free radicals) produced by various biodegrading microorganisms. This ends up in the progressive reduction of the relative molecular mass of the polymer. Several degradation products will be assimilated by the microorganisms resulting in mineralization of organic compounds and results in the generation of biomass [6]. Biocompatible polymers that change their properties in several physiological environments are extensively accustomed to design and develop smart drug delivery systems.

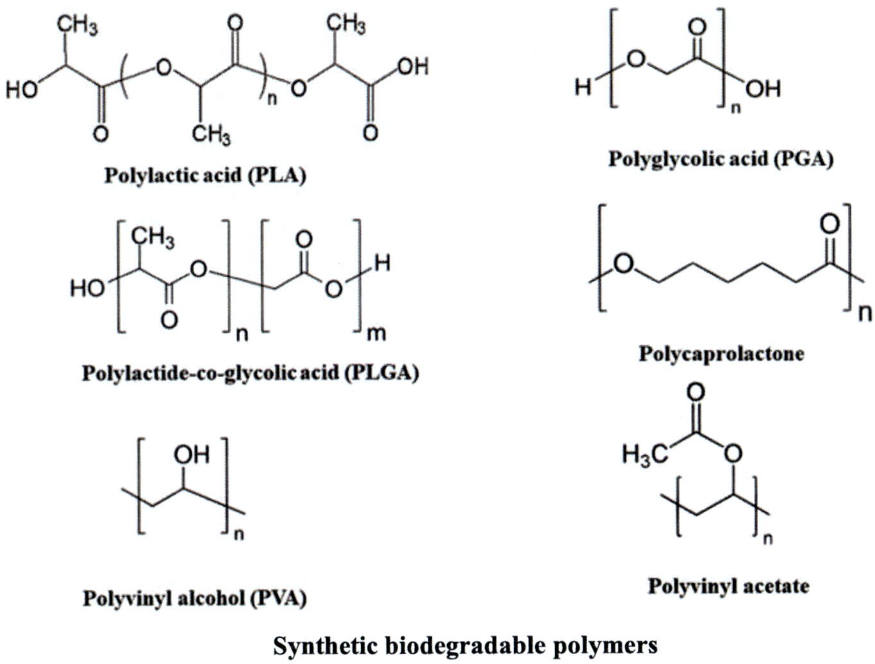

Fig. 2 (continued)

4.1 Oxidative Mechanism

One of the key degradation mechanisms of polymeric biomaterials that happens in our biological system is through the oxidative degradation process. This mechanism generally operates through the formation of highly reactive oxygen species (ROS) like superoxide (O^{2-}), per oxide (H_2O_2), gases (e.g., NO), and acid (HOCl) when the polymeric materials are exposed to the body fluids and tissues. The oxidative effect of those ROS may cause the fragmentation of the polymeric backbone and eventually ends up in their degradation. Also, in the presence of oxygen in air or water the formation of ROS is feasible and resulting in the assembly of hydroxyl free radicals ($OH^{\bullet}$) and singlet oxygen, which might deteriorate several critical biomolecules. For instance, superoxide could accelerate the degradation of aliphatic polyesters by the cleavage of ester bonds via the nucleophilic attack of O_2. It is also reported that polyurethanes are attacked initially by neutrophils which secrete reactive oxygen species (ROS) and HOCl, one among the foremost oxidative compounds [7].

4.2 Hydrolysis

Hydrolytic degradation of polymers will be defined as the fission of chemical bonds in the polymer backbone by the attack of water to make oligomers and at last to corresponding monomers. This process occurs in polymers which are having water-sensitive active functional groups. Also, usually, these varieties of polymers possess heteroatoms within the main or side chains. Biodegradable polymers undergo hydrolytic bond cleavage to make water-soluble degradation products which will dissolve in an aqueous environment, leading to polymer erosion [8]. For instance, the degradation of the polymers through hydrolysis of ester bonds leads to the alteration of their chemical structure.

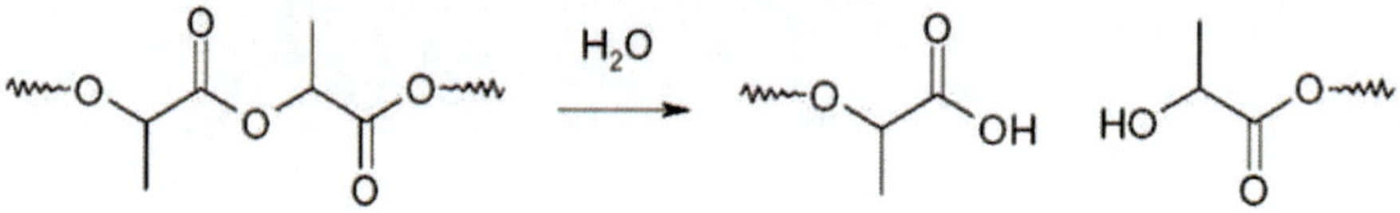

Usually, biodegradable polymers encompass hydrolyzable bonds such as glycosides, esters, orthoesters, anhydrides, carbonates, amides, urethanes, ureas, etc. The hydrolytic cleavage can be catalyzed by the presence of acids, bases, or salts. The acid-labile chemical bonds are stable at neutral pH but were degraded or hydrolyzed in acidic media. This unique property is having immense application in the development of drug delivery devices. The acid-labile functional groups include acetal, orthoester, hydrazine, imine, cis-aconyl bonds, etc. [9]. Important labile chemical bonds present in the main chain of the biodegradable polymers is shown in Fig. 3. During the process of hydrolytic degradation, initially, the hydrolytic cleavage of the main polymeric chains occurs and which in turn leads to a decrease in the molecular weight of the material. While degradation proceeds, the molecular weight of degradation products is reduced by further hydrolysis which allows them to diffuse from the bulk material to the surface and then to the solution, causing significant weight loss [10].

Examples of such synthetic polymers that are used in biomedical devices and drug delivery platforms include polylactic acid (PLA), polycaprolactone (PCL), poly (glycolic acid) (PGA), and polydioxanone (PDO).

4.3 Enzymatic Degradation

Polymeric biomaterials cast-off in the biomedical applications will be get tarnished in the body fluids and tissues by several enzymes either by the process of oxidation or hydrolysis. Enzymes are known for their biological catalytic activity, and they will accelerate the degradation reaction rates in living organisms without undergoing any permanent change. Generally, the degradation of polymers in the living

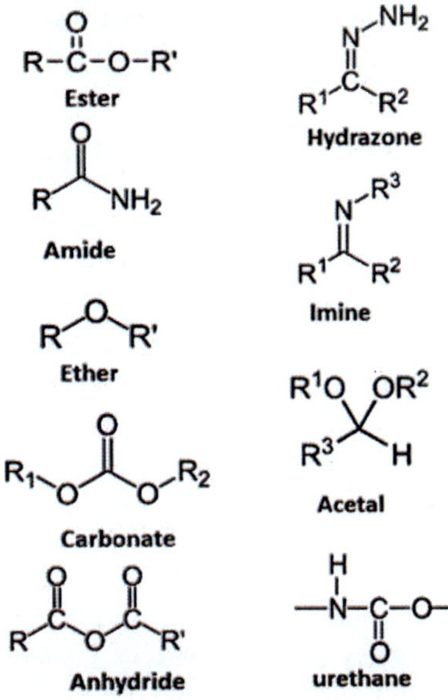

Fig. 3 Important labile chemical bonds present in the main chain of the biodegradable polymers. *Source* author

body may take place under the action of hydrolases and oxidases enzymes [11]. Hydrolases enzymes include proteases, esterases, glycosidases, and phosphatases, among others. For example, the degree of biodegradation of polyurethanes, in the presence of cholesterol esterase enzyme, is about 10 times higher than in the presence of buffer alone. Initially, the microorganisms will be attracted to the surface of the polymer and, the enzymes secreted by the microorganism converted the polymer macromolecule into their tiny components and eventually converted them into CO_2 and H_2O.

The mode of degradation mechanisms in the case of enzymatic process primarily occur through the diffusion of the enzyme from the bulk solution to the solid surface and are tracked by the adsorption of the enzyme on the substrate, resulting in the establishment of the enzyme–substrate complex. After the diffusion process, the enzyme-catalyzed hydrolysis occurs, and ultimately by the diffusion of the soluble degradation products from the solid substrate to the solution will occur. Enzymatic degradation of natural origin polymers proceeds through the action of specific enzymes.

Most materials employed for the fabrication of medical devices comprise synthetic polymers such as polyurethane, polyethylene, polypropylene, polystyrene, polyester, polycarbonate, polyvinyl chloride, polyacrylate, elastomers, fluoropolymers, silicone, or polyethylene terephthalate. For the construction of drug delivery devices, generally, polyacrylates, polyesters, isopropylacrylamides, poly(2-oxazoline)s, polyethyleneimine as well as naturally-based polymers (collagen and chitosan) and poly(ethylene glycol)s for hydrogels formulation are used [12].

5 Design Principles of Biodegradable Polymeric Materials for Medicinal Applications

The ultimate properties and function of the polymeric biomaterial are governed by its chemical, structural, and mechanical features of the polymer and how it will interact with the biological environment inside the human body [13]. The main principle behind the production of biodegradable polymer comprehends the introduction of weak linkages inside the polymeric backbone, in such a way that these linkages can easily be cleaved inside our body by the action of either temperature, pH, the concentration of the salt contents, the action of microbials or in the presence of highly reactive free radicals [14]. Thus, several synthetic strategies are adopted to formulate biodegradable polymers of varying degradation rates. The incorporation of functional groups such as carbonyl, ester, imine, acetal, hydrazone, etc., as a linking group in the main chain, which can be easily cleaved by chemical hydrolysis, aids the biodegradation process in a quite faster pathway.

The universal guidelines that have to be tracked while designing a biodegradable polymeric material targeting medicinal applications include an increased ratio of hydrophilic/hydrophobic ratio, selection of heteroatom-containing polymeric chain instead of the fully integrated carbon chain, branched-chain polymers, condensation polymers instead of addition polymers, replacement of macro-size polymers with lower molecular weight oligomers, lower surface area, water solubility, and purity. Another important design principle behind the biodegradable polymer includes the surface modification of polymeric chains, which can create meticulous densities of hydroxyl groups on the material surface, and then these groups can provide sites for the covalent attachment of specific biomaterials such as proteins or peptides [15]. In this case, the design of the materials is based on principles of surface segregation of a component with the lower surface energy. Thus, the surface science techniques can have a major role in the fabrication of biodegradable polymers, as evidenced by the application of these degradable polymers.

The prime factors which have to be taken into account during the design of biodegradable polymeric materials include

(1) The polymeric materials should not induce a sustained inflammatory response.
(2) possess a degradation time overlapping with their function.
(3) have suitable mechanical properties for their expected use.
(4) produce non-toxic degradation products that can be instantly resorbed or expelled and.
(5) include appropriate absorptivity and processability for proposed application [16].

Consequently, the biodegradable polymers should have the possibility to become part of new medical devices with precise and distinctive physical, chemical, and mechanical properties, such as electrical conductivity, optical properties, chemical reactivity, and mechanical strength [17]. In conclusion, the variables like the structure of the polymer, its chemical composition, distribution of the monomeric units, presence of functional groups, nature of main chain and side chains, structural configurations, molecular weights and polydispersity of the polymer, morphology, annealing effects, storage history, etc., might be considered while selecting and designing a biodegradable polymer. The synthesis of biodegradable polymers can be done by any one of the following synthetic reactions such as ring-opening, polycondensation, bulk synthesis, dehydrative coupling, transesterification, and polymerization.

6 Biomedical Applications of Biodegradable Polymers

An ideal biopolymer for a medicinal application should satisfy the following criteria.

- The mechanical properties of the material should match with the application and will remain sufficiently strong until the surrounding tissue is healed.
- It should not appeal an inflammatory or toxic response within the body
- The material should be metabolized in the body after satisfying its purpose, leaving no trace.
- The material should be easily processable into the final product form.
- It should have an acceptable shelf life and must be easily sterilized.

The important applications of biodegradable polymeric materials in medicine are schematically represented in Fig. 4. Beholden to their unique properties, biodegradable polymers now replace metals, alloys, and ceramics for use as biomaterials.

Biodegradable polymeric materials have been extensively used in various medical applications which include the applications focusing both in vivo and invitro studies over the past decade. The major important applications include the administration of pharmaceuticals and biomedical devices, controlled drug delivery systems, different forms of implants and devices for fracture repairs, ligament reconstruction, surgical dressings, dental repairs, artificial heart valves, contact lenses, cardiac pacemakers, vascular grafts, tracheal replacements and organ regeneration, implants in the blood vessels and as absorbable clinical sutures, etc. [18–20].

7 Drug Delivery Devices

The design and development of novel drug delivery systems lead to the development of novel biodegradable polymers with good biocompatibility and target specific capabilities. The oral drug delivery route is known as the golden standard for the consumption of drug administration. An ideal drug delivery device can be termed as a highly biocompatible material and will be releasing bioactive agents at the precise rate at the exact site and simultaneously maintaining the optimum level of drug to

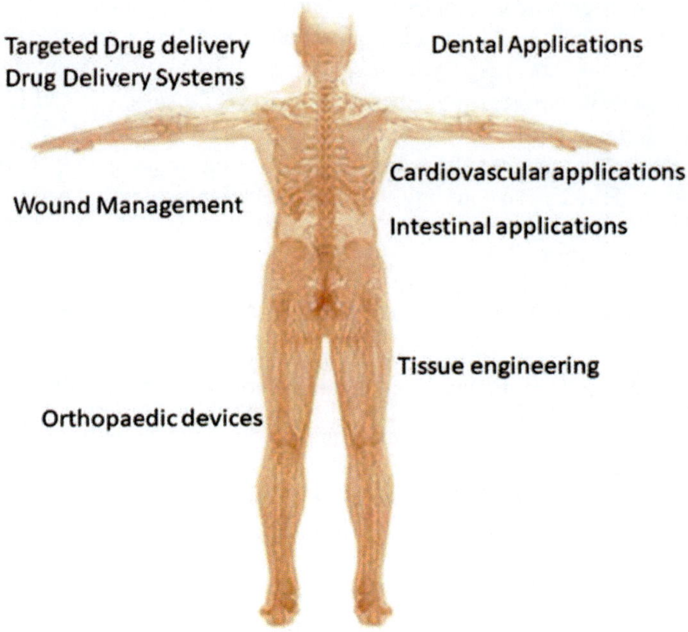

Fig. 4 Applications of biodegradable polymeric materials in medicine. *Source* author

avail maximum efficacy and minimize associated side effects [21]. Advanced materials used in drug delivery applications are either in the form of micro-/nanogels, microparticle/nano-particles, hydrogels or as imprinted polymers. Stimulus responsive nanomaterials which can be termed as intelligent or smart materials were found to be highly potential compared to the traditional drug delivery devices [22]. The most widely used biodegradable hydrogels include hydroxypropyl methyl methacrylate (HPMA), polyethylene glycol (PEG), polyglutamic acid (PGA), poly(L-lysine), polyethyleneimine (PEI), dextran, dextrin, chitosan, etc. [23]. Chitosan, a cationic polymer has been examined as an excipient in controlled delivery formulations and mucoadhesive dosage forms because of its gelling and adhesive properties. Chitosan can hypothetically be used as a drug carrier, a tablet excipient, delivery platform for drug formulations, disintegrate, and tablet coating [24].

One resolution is to design pharmaceutical polymers that endure physiochemical changes in retort to environmental incentives such as temperature, pH, electric or magnetic field, enzymes, solvent polarity, etc. The stimuli-responsive biodegradable polymers, which can release the drugs at the desired site with respect to the change in pH or temperature are of potential applications in cancer drug delivery devices since the cells around the cancer cells normally have a lower pH value between 5 and 6 and a higher temperature ~42 °C because of the increased metabolic activity near the cancer cells.

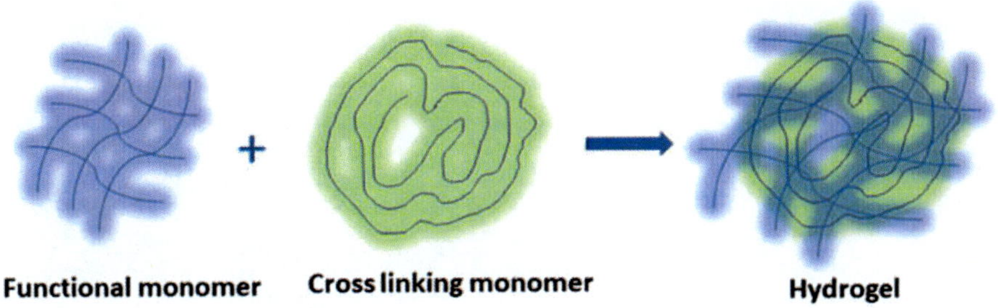

Fig. 5 Schematic representation of formation of hydrogels. *Source* author

An important category of stimuli-responsive cross-linked polymers are hydrogels. These polymers are centered on hydrophilic, water-soluble oligomeric or polymeric chains that have been bonded to each other to establish a covalent network. The synthesis of a hydrogel-based drug involves the crosslinking of linear polymer or simultaneous polymerization of monofunctional monomers and crosslinking with polyfunctional monomers. Usually, polymers containing hydroxyl, amine, amide, ether, carboxylate and sulfonate as functional groups in their side chain are explored for the synthesis of hydrogels [25]. The synthesis of a hydrogel is schematically shown in Fig. 5.

Controlled drug release systems based on molecular imprinting strategy is an advanced drug delivery system and have gained much recognition in recent years. Molecularly imprinted polymers are artificial polymers with recognition binding sites which are able to recognize a target analyte or its structural analogues from a complex. Molecularly imprinted polymers are prepared by the co-polymerization of functional and crosslinking monomers in the presence of the imprint molecule. The functional monomers initially forms a complex with the target molecule and following polymerization; the functional groups are apprehended in location by the highly cross-linked polymeric structure. Subsequent ejection of the imprint molecule exposes the binding sites that are correlative in size and shape of the analyte. Thus, a specific molecular memory is launched in to the polymeric matrix proficient of rebinding the targeted analyte with high specificity and selectivity [26–28]. The synthetic strategy adopted for a molecularly imprinted polymer is shown in Fig. 6.

8 Surgical and Orthopedic Devices

Biodegradable polymers are extensively used in surgical devices such as implants, suture materials, and staples, which are the major domain of polymers in general surgery. These materials for surgical purposes are mainly concerned in diverse areas such as in orthopedics and traumatology, maxillofacial surgery, vascular surgery,

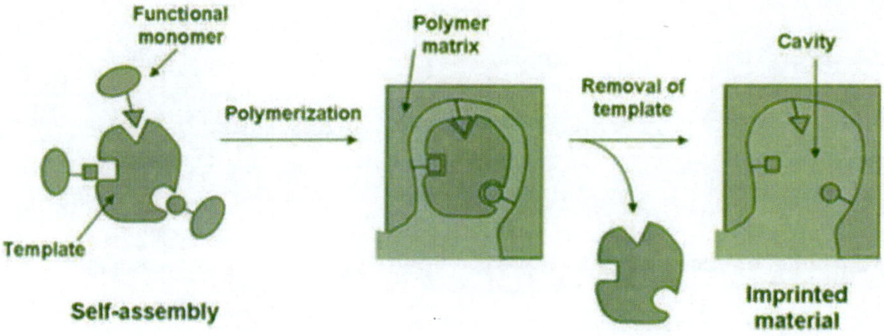

Fig. 6 Synthetic strategy for a molecularly imprinted polymer. *Source* author

microsurgery, neurosurgery, etc. [29, 30]. The suture materials will provide artificial fiber support until the natural fiber (collagen) is synthesized in our biological system. The important parameters that must be considered in surgical suture materials depend on tensile strength, friction/trauma to tissue, degradability, and the stability of knots. Degradable biological suture materials are fabricated with collagen-based materials, catgut, silk or cellulose (cotton). Synthetic materials widely explored are PGA, a copolymer of a small amount of lactide with glycoside (PGA/PLLA), etc. Recently, several clinical studies have been published using PGA, PGA/PLLA and PDS sutures or bands in the repair of tendons, ligaments and dislocations of joints. Reinforced composites of PGA and PGA/PLLA-polylactide/polyurethane (PLLA/PU) composites are widely employed for artificial skin and vascular prosthesis [31–33]. Many biodegradable materials, such as poly(P-hydroxybutyrate), poly(valerolactone), poly(p-dioxanone) and poly(amino acid), have been considered as potential candidates for the fixation process, and the most important ones employed in surgery are poly(L-lactide) (PLLA), polyglycolide (PGA) and their copolymers [34].

The emerging applications of biodegradable polymers in orthopedic applications comprised of in the fabrication of fixation devices, which in turn include pins, rods, screws, tacks, and ligaments. In orthopedic treatments, the effective use of a bone morphogenic protein will catalyze the healing process after a fracture or may help to avert osteomyelitis following surgery. Most of the commercially available biodegradable devices are polyesters composed of homopolymers or copolymers of glycolide and lactide. There are also products made from copolymers of trimethylene carbonate, e-caprolactone and polydioxanone. Synthetic biodegradable polymers explored in orthopedic devices have advantages over metal implants and reduce the need for a consequent operation for implant removal. These polymeric materials have found immense applications where lower-strength materials are adequate, such as in the fixation screws of the ankle, knee, hand areas as interference, tacks, and pins for ligament attachment and meniscal repair, suture anchors, and rods and pins for fracture fixation.

Middleton et al. [35] has compiled a review of the synthetically biodegradable polymers offered in orthopedic devices. The other important categories of biodegradable polymeric materials include polyaryletherketones (PAEKs), which are progressively utilized as biomaterials for orthopedic, trauma, and spinal implants. Polyether ether ketones (PEEK) are used in trendy settings for spinal implants, femoral stems, bearing materials for hip and knee replacement, and hip resurfacing. Polyurethane (PU) biomaterials have been investigated for compatible orthopedic-bearing materials. Silicones, synthetic polymers, are broadly used in health care and orthopedic applications. The most common orthopedic applications of silicone are hand and foot joint implants [36].

Bioabsorbable polymers like polyglycolic acid (PGA) is used in the fabrication of pins and screws. Polylactic acid (PLA) is also used in a variety of implants including pins, rods, tacks, screws, and plates. Extensively explored implants in orthopedic devices are made from the copolymers of PLA and PGA. Other bioabsorbable materials used are poly(orthoesters), poly(glycolide-co-trimethylene carbonate), poly (p-dioxanone), poly(ε-caprolactone) (PCL), poly (hydroxybutyrate) (PHB), and PHB hydroxy valeric acid, etc.[37]. PDS has also been effectively used in tissue regeneration and fracture repair applications [38]. Composites produced using calcium phosphates and collagens are believed as the most biomimetic system for bone substitute and re-forming applications. Chitosan and collagen incorporated chitosan composite micro-granules can be effectively utilized as bone substitutes. Chitosan membranes engineered with silica xerogels also have been appraised for applications in the bone regeneration process [39–41].

9 Wound Dressings

A perfect wound dressing should accomplish the healing of the wound with infinitesimal time and cost-effective. It should possess qualities such as simple and painless to apply, antimicrobial, maintains a moist wound environment, and requires minimal dressing modifications and protecting the wound. Biodegradable and antimicrobial wound dressing will decrease the indisposition by diminishing interaction during wound care, eradicating the need for repeated dressing shifts and dispensing drugs that can reduce the risks of infection. Due to the advancement in the field of synthetic biodegradable polymers, the wound management system has been modernized in several corridors in recent times [42–46]. Numerous investigations have detailed the use of polymers in dressings and in medical devices for prospective advancement in regulating the wound healing process.

Natural polymer films such as alginate, fucoidan, silk sericin, keratin, chitosan, and hyaluronic acid are widely exploited in wound dressing procedures. Chitosan and chitosan coupled with hyaluronic acid were found to be excellent wound healing systems that have been used for the development of dressing for skin ulcers. Polymeric forms like films, foams, hydrogel, or hydrocolloids are the major configurations in the wound management process. Polyurethane (PU) is utilized in numerous semi-permeable dressings owing to its ability to furnish great fences and penetrability to oxygen [47]. Hydrocolloids are polymers having inherent hydrophilic characteristics due to the occurrence of many hydroxyl groups. The hydrocolloids can be developed either synthetically or naturally. The important category of hydrocolloids used in wound healing applications is generally polysaccharides. Agar, alginate, pectin, gelatin, etc., are examples of natural hydrocolloids. The current flea market for hydrocolloid dressings explored its uses in healing diabetic foot ulcers [48].

Hydrogels synthesized using crosslinked polymers (hydrophilic), for example polyvinylpyrrolidone, polyacrylamide, and polyethylene oxide, are also having widespread use in wound dressing applications. Hydrogels are employed as wound dressings in the form of elastic films or amorphous gels. Also, poly(lactide-co-glycolide) (PLGA), polyethylene glycol (PEG), polyurethane (PU), polycaprolactone (PCL), etc. are the other important category of synthetic biodegradable polymers employed [49–51] in wound dressing fabrication. Polymer blend solution of poly(lactic-co-glycolic acid) and poly(ethylene glycol) (PLGA/PEG) sprayed with a portable airbrush were found to be inherently adhesive polymer fibers and can accumulate on and heal the wound. Recently, Mir et al. [52] reported the effective usage of silver salts incorporated (PLGA/PEG/Ag) as a sprayable and antimicrobial wound dressing agent.

10 Dental Applications

Current advances in biodegradable polymers have led to the development of novel materials for dental applications and have extended their use in protective, rejuvenating, and regenerative treatments. The utilization of biomaterials in dentistry is more broadened similar to other medical fields in terms of both quantity and quality. Polymers derived from natural sources have been applied in oral and maxillofacial surgery as extra-cellular matrix (ECM) substitute in skeletal muscle, bone, and periodontal regeneration [53]. The natural biodegradable polymers explored in dental applications include chitosan, collagen, hyaluronic acid, and alginate. The most frequent synthetic biodegradable polymeric materials employed for periodontal regeneration are poly(lactic acid) (PLA), polyglycolic acid, (PGA, polylactide-co-glycolide (PLGA), polycaprolactone (PCL), etc. Polylactic acids are utilized for dental pulp and dentin restoration, and bioactive polymers are explored in advanced drug delivery systems. PLGA is used in a variety of dental applications ranging from creating screws for bone fixation, handling periodontal pathogens, and generating buccal mucosa or indirect pulp-capping procedures. PLGA can also be utilized for periodontal treatment, antibiotic delivery, and in the forms of PLGA implants, disks, and dental films [54]. Also, gel composite fabrics of PLGA can be used in bone regeneration, The other important category of biodegradable polymeric materials normally used in dentistry is polyethylene (PE), polymethyl methacrylate (PMMA), polycarbonate (PC), polyethylene glycol (PEG), polydimethylsiloxane, polyurethane (PU), poly(ε-caprolactone) (PCL), polypyrrole(PPy), hexamethyl disilazane (HMDC), N-isopropyl acrylamide, N-test-butyl acrylamide and hydrogels of different natural and synthetic polymers. Recently, Agata Szczesio-Wlodarczyk et al. [55] reviewed the dental composites based on methacrylate resins. Hydrogels based on micro-cellulose (MC), chitosan, and its composites, the natural well-known polymer gelatine, and its combination with hydroxyapatite (HA) were also explored in dentistry.

Polymeric materials (PMs) and polymeric films (PMFs) have extended its horizons in medicine and dentistry. PMs and PMFs are tapped in dentistry owing to their advanced antimicrobial, drug delivery properties, and corrosion resistance [56]. Novel techniques include the use of antibacterial polymer coatings for inhibiting bacterial growth on artificial tooth surfaces in other dental materials and dental composite kits increasing the restoration's longevity. Examples of such antibacterial coatings include copolymers of acrylic acid, alkyl methacrylate and polydimethylsiloxane copolymers, pectin-coated liposomes, and carbopol. PMFs like acrylic acid copolymers are employed as a dental adhesive. Innumerable polymeric films were found to be protecting the teeth against erosion by preventing the direct contact of the acidic environment in the oral cavity with the teeth. Beyer et al. [57] analyzed the proficiency of a polymer-modified citric acid solution of propylene glycol alginate to reduce tooth erosion. Chitosan is also found to be an excellent candidate for the protection against erosion and enamel demineralization process and can be attributed to its ability to form a protective multilayer on the tooth surface in the presence of mucin from saliva [58]. For controlled drug release applications, different pectin-coated liposomes and polymers such as polycarbonate micelles have also been employed. Also, polycarbonate, blended with polyethylene glycol (PEG) and antimicrobial agents were explored in controlled drug release applications. Travan et al. [59] reported an antimicrobial nanocomposite containing lactose-modified chitosan incorporated with Ag-NPs and polymerized PMMA in the dental industry.

Biomimetic titanium surfaces smeared with nanohydroxyapatite (n-HA) and poly(lactic-co-glycolic acid) (PLGA)/collagen nanofibers have been analyzed for dental and bone-implant surfaces to improve osseointegration [60]. Biocompatible modified polymeric films have been coated on NiTi alloy wires to increase corrosion resistance and improve mechanical properties [61].

11 Cardiovascular Applications

Biodegradable polymeric materials are extensively explored in numerous areas in cardiovascular applications. These applications mainly focused on drug delivery, stent and graft preparation, fabrication of artificial valve, and tissue regeneration [62]. Treatments for cardiac ailment include tactics ranging from medicines to surgical intrusions. Popular surgical treatments encompass skirting the injured tissue like bypass grafts, or substituting them, as in heart transplants. Frequent biodegradable polymers that include both natural and synthetic ones have been used to control cells that are related to cardiovascular tissue engineering applications. The natural polymers embrace cellulose, hyaluronic acid (HA), chitosan, collagen, gelatin, etc. Synthetic biomaterials explored in cardiovascular applications principally involve polymers, metals, or a combination of both the synthetic biodegradable polymers and mainly embrace poly(glycolic acid) (PGA), poly(lactic acid) (PLA), poly(ε-caprolactone) (PCL), poly(ethylene glycol) (PEG), polyhydroxyalkanoate (PHA), and their copolymers [63]. Typical synthetic biopolymers that have been used for myocardial tissue engineering include polyglycolic acid (PGA), polylactic-l-acid (PLA), polylactic glycolic acid (PLGA), and polyurethane. Polyurethanes (PUs) are typically utilized in cardiac pacing leads as an insulator. PU is also being examined as a substrate in cardiac stem cell therapy and durable devices to biodegradable scaffolds [64]. The cytocompatibility of synthetic polymers can be improved by different chemical modification processes.

Polytetrafluoroethylene (ePTFE) is one of the popular polymeric material employed in routine cardiovascular applications due to its superior material performance. These characteristic features of ePTFE have made it an excellent selection for producing shunts, renovation, and valve repair and have even been used for casing implantable devices to minimize inflammation [65]. Polyethylene terephthalate (PET), PET, is recommended for fabricating vascular grafts. Polymers, possessing the inherent properties such as biodegradation, drugs releasing capacity, or biomimicking, are of great concern in the development of cardiovascular implants.

Biocompatible and bioabsorbable polymeric stents have enticed much consideration as substitutes to metal stents. Endovascular stents have become the most reliable medical devices for treating coronary artery diseases. The stents overcome the limitations and drawbacks of bypass surgery and balloon angioplasty by enabling scaffolding, widening, and supporting the blocked vessels. Also, these types of stents displayed many advantages like short duration of post-stenting, avoid chronic inflammatory processes as well as the ability of the vessel remodeling. Moreover, bioresorbable stents are recommended for tracheomalacia treatment in infants because removal surgery is not necessary. Thus, the advanced progress in the field of polymeric materials makes the stent-based polymers as an attractive tool in cardiovascular medicine. Cardiovascular tissue engineering techniques embrace injectable biomaterials, cell therapies, and artificial organ fabrication. Injectable materials mostly employed in cardiovascular applications are hydrogels composed of alginate, fibrin, chitosan, collagen or matrigel and self-assembling peptides generally in the form of nanofibers. Injectable polymeric materials formerly engendered interest due to their biocompatibility, the capability to offer advantageous chemical surroundings, and their potential non-invasive delivery routes [66].

Biodegradable polymers are not only intended for stents but also effectively utilized as a drug carrier in drug-eluting stent (DES). Generally, polymers employed as stent platforms and coating matrices for drug-eluting stents (DES), are vascular stents that allow the delivery of antiproliferative drugs to inhibit vascular smooth muscle cell (SMC) growth. Biomimetic polymers, such as phosphorylcholine (PC), poly(vinylidene fluoride), and hexafluoropropylene (PVDF-HFP), do not interfere with the stent and are presently explored in the second- and third-generation DES [67]. Integrating a biocompatible polymeric coating with poly(L-lactic-acid) (PLA) is one of the most predominant approaches to reduce the bio-corrosion of biodegradable metals and maintaining their biocompatibility. Polymeric coating can normalize the metal corrosion by isolating it from the corrosive environment or by conquering the dissolution of metal or equivalent cathodic reaction. Polymeric coating can control the metal corrosion by isolating it from the corrosive environment or by suppressing the dissolution of metal or corresponding cathodic reaction. Incorporating a polymeric coating not only can act as a corrosion barrier, but also it can be encumbered with drugs that can be released in controlled amounts to prevent post-surgery inflammations or restenosis [68].

In parallel to cardiovascular applications, a large number of biodegradable polymers are effectively employed in intestinal applications such as colon-specific drug delivery, its formulation aspects, and in the fabrication of stents.

12 Tissue Engineering Applications

Tissue engineering is an effective pathway for the fabrication of biological alternatives that repair, retain, or enhance the proper functioning of tissues. It can be described as a tool by which we can assess the structure–function relationship of individual tissues. In tissue engineering, biomaterials have to be designed to substitute or regenerate completely (or partly) the injured tissue. The biomimetic materials will act as a three-dimensional matrix as a scaffold in the regeneration process. An essential role for biomimetic materials is to furnish a three-dimensional matrix as a scaffold and guaranteeing the preservation of cells and signals for redeveloping the tissue or organ [69]. Creating physiologically functional artificial tissues and organs is an essential requirement of tissue engineering, and technological advancements in tissue engineering repeatedly strengthen its progress. Several synthetic and natural biodegradable polymers and their composite materials have been cast-off to engineer scaffolds for bone tissue engineering, nerve rejuvenation, controlled drug release, dental structure regeneration, guided tissue regeneration (GTR), strengthening of dental composite, bone and cartilage rejuvenation [70]. Nano-configuring of biomaterial scaffolds from nanoparticles, nanocomposites, and organic–inorganic hybrid polymer materials has also exhibited progress in organ regeneration and tissue engineering applications.

The most frequently explored synthetic polymers for tissue engineering include aliphatic polymers, poly(lactic acid) (PLA), poly(glycolic acid) (PGA), poly(lactic-co-glycolide) (PLGA), poly(ε-caprolactone) (PCL), poly(p-dioxanone), copolymer soft trimethylene carbonate and glycolide. These materials are smart biomaterials to construct scaffolds and have been widely applied with favorable results in regenerative medicine.

Blends of D-PLA and L-PLA (PDLLA), PLA, PGA, and PLGA have been utilized clinically to cure patients facing damaged or lost organs or tissues. Natural biopolymers encompass polysaccharides (e.g., starch, alginate, chitin/chitosan, hyaluronic acid derivatives) or proteins (e.g., soy, collagen, fibrin gels, silk) [71–74]. These materials operate as fundamental prototypes for cell connection and expansion owing to their intrinsic biocompatibility. Collagen and fibrin have been widely probed in cardiac tissue engineering due to its ability of natural interaction with cells inside the human body. Biodegradable synthetic polymers that have been effectively employed for myocardial tissue engineering include polyglycolic acid (PGA), poly(lactic acid) (PLA), polylactic glycolic acid (PLGA) and polyurethane. polyurethanes have also marked its potential applications in dermal regeneration. Poly(ε-caprolactone) (PCL) is an appealing material when long-term implants are preferred. Another group of thermoplastic polymers that have been developed recently for tissue engineering includes multi-block copolymers comprising poly(ethylene oxide) (PEO) and poly(butylene terephthalate) (PBT) (PEOT/PBT)). Fascinating class of other degradable materials explored in tissue engineering is polyphosphate-esters (PPEs), polyphosphates), polyanhydrides (PAs) and polymorpho-esters (POEs).

The augmentation of precursors like polyols and macromonomers based on polyesters emerged as novel candidates in injectable and situ curable polymer formulations. Poly(propylene fumarate) is an example of an injectable polymer system used in tissue engineering applications Polyurethanes also offer many advantages in the design of injectable and biodegradable polymer compositions.

Recently hydrogels have revolutionized the field of tissue engineering where they are engineered as scaffolds to monitor the growth of new tissues. The design and application of biodegradable hydrogels have significantly enhanced the prospective power of hydrogel materials in the biomedical field and facilitated the advancement of exciting materials focused on tissue engineering applications. Examples of hydrogel-forming polymers of natural origin are collagen, gelatin, fibrin, HA, alginate, and chitosan, and the synthetic polymers are PLA, PPF-derived copolymers PEG derivatives, and PVA [75]. Nanofibers have also emerged as scaffolds for musculoskeletal tissue engineering, which include bone, cartilage, ligament, and skeletal muscle, skin, vascular, neural tissue engineering. Natural polymers and synthetic polymers explored as a fibrous scaffold in biomedical applications. For the fabrication of nanofibers includes collagen, gelatin, chitosan, HA, silk fibroin, PLA, PU, PCL, PLGA, PEVA, and PLLA-CL.

13 Conclusions

The design and development of biodegradable polymers is an emerging area in the field of medicinal chemistry. The fabrication of newer and newer smart materials targeting human cells is highly demanded in the current scenario. The mechanistic aspects of degradation, its sustainability in the human physiological conditions, and the longevity of the materials are the major concerns and challenges to the researchers in this area. Certainly, the present area must be explored in different dimensions of human life that persist on the planet earth. The filed tissue engineering, fabrication of implants, artificial valves and stents, and artificial organs, bioabsorbable wound healing devices are among the few in the category of biodegradable materials. By adopting the natural mechanisms of degradation, the naturally biodegradable polymeric materials can be derivatized and by the incorporation of nanomaterials, the mechanical strength and the corrosion resistance of the composite materials can be enhanced. The synthetic materials widely explored can be fine-tuned and can be configured to newer formats to meet the advanced applications such as in the development of novel drug delivery devices, stimuli-responsive targeted drug delivery applications, antibacterial and anti-viral agents, biodegradable safety equipment, etc. We hope by the advancement of newer technologies and the emergence of novel materials, applicatisons of biodegradable polymers will surely open broad horizons in the medicinal field.

References

1. Chammy R (ed) (2013) Biodegradation, life of science. Intech open. https://doi.org/10.5772/52777
2. Elisa T, Ariadna R (2013) Biodegradation of medical purpose polymeric materials and their impact on biocompatibility. Intech Open https://doi.org/10.5772/56220
3. Kunduru KR, Basu A, Domb AJ (2016) Biodegradable polymers: medical applications. Encyclopedia Polym Sci Technol 1–22. https://doi.org/10.1002/0471440264
4. Katarzyna L, Grażyna L (2010) Polymers biodegradation and biodegradable polymers—a review. Polish J Environ 19:255–266
5. Moiseev YV, Daurova TT et al (2007) The specificity of polymer degradation in the living body. J Polym Sci Polym Symp 66:269–276
6. Banerjee A, Chatterjee K et al (2014) Enzymatic degradation of polymers: a brief review. Mater Sci Technol 30:567–573
7. Mythili P, Janis L et al (2017) Biodegradable materials and metallic implants—a review. J Funct Biomater 8:44–58
8. Chhaya E, Jigisha P et al (2011) Review on hydrolytic degradation behavior of biodegradable polymers from controlled drug delivery system. Trends Biomater Artif Organs 25:79–85
9. Roongnapa S (2013) Novel strategic innovations for designing drug delivery system using molecularly imprinted micro/nanobeads 44:235–268
10. Juan L, Yuran H et al (2014) pH sensitive nano systems for drug delivery in cancer therapy 32:693–710
11. Yanhua L, Wenping W (2013) pH sensitive polymeric micelles triggered drug release for extracellular and intracellular drug targeting delivery 8:159–167
12. Yu V, Moiseev T et al (1979) The specificity of polymer degradation in the living body. J Polym Sci Polym Symp 66:269–276
13. Bernard M, Jubeli E et al (2018) Biocompatibility of polymer-based biomaterials and medical devices—regulations. In: Vitro screening risk-management. https://doi.org/10.1039/C8BM00518D
14. Ozdil D, Wimpenny HM et al (2016) Biocompatibility of biodegradable medical polymers. In: Science and principles of biodegradable and bioresorbable medical polymers. https://doi.org/10.1016/B978-0-08-100372-5.00013-1
15. Pillai CKS (2013) Recent advances in biodegradable polymeric materials. Mater Sci Technol 30:558–566
16. Ha CS, Gardella JA (2005) Surface chemistry of biodegradable polymers for drug delivery systems. Chem Rev 105:4205–4232
17. Bret D, Ulery et al (2011) Biomedical applications of biodegradable polymers. J Polym Sci Part B: Polym Phys 49:832–864

18. Edgar M, Maria I et al (2013) Critical evaluation of biodegradable polymers used in nanodrugs. Int J Nanomed 8:3071–3091
19. Neeraj K, Aviva ES et al (2002) Biodegradable polymers, medical applications. Encyclopaedia Polym Sci Technol 5:263–285 (John Wiley & Sons Inc)
20. Santosh K, Sumit S et al (2017) Structural features and biomedical applications of biodegradable polymers. Int J Eng Dev Res 5:761–765
21. Manfred FM (2015) Applications of synthetic polymers in clinical medicine. Biosurface Biotribol 1:161–176
22. Tejas V, Shah et al (2019) A glimpse of biodegradable polymers and their biomedical applications. e-Polymers 19:385–410
23. Liu Y, Wang W et al (2013) pH-sensitive polymeric micelles triggered drug release for extracellular and intracellular drug targeting delivery. Asian J Pharm Sci 8:159–167
24. Liu J, Huang Y et al (2014) Sensitive nano-systems for drug delivery in cancer therapy. Biotechnol Adv 32:693–710
25. Gupta P, Vermani K et al (2002) Hydrogels: from controlled release to pH-responsive drug delivery. Drug Discovery Today 7:569–579
26. Piyush G, Kavitha V et al (2002) Hydrogels:from controlled release to pH responsive drug delivery 7:569–579
27. Curk T, Dobnikar J et al (2016) Rational design of molecularly imprinted polymers. Soft Matter 12:35–44
28. Haupt K (2002) Imprinted polymers—tailor-made mimics of antibodies and receptors. Chem Commun 2:171–178
29. Chen L, Wang X et al (2016) Molecular imprinting: perspectives and applications. Chem Soc Rev 45:2137–2211
30. Javed F, Al-Askar M et al (2012) Tissue reactions to various suture materials used in oral surgical interventions. ISRN Dent. 2012:762095. https://doi.org/10.5402/2012/762095
31. Nakamura T, Shimizu Y et al (1992) A novel bioabsorbable monofilament surgical suture made from (Poly-caprolactone, L-lactide)copolymer.In: Dauner M, Renardy M, Planck H (eds) Degradation phenomena of polymeric biomaterials. Springer, Stuttgart
32. Jain R, Shah NH et al (1998) Controlled drug delivery by biodegradable poly (ester) devices: different preparative approaches. Drug Dev Indian Pharm 24:703–727
33. Jain RA (2000) The manufacturing techniques of various drug loaded biodegradable poly(lactide-co-glycolide) (PLGA) devices. Biomaterials 21:2475–2490
34. Kovanya M, Viness P et al (2012) Oral drug delivery systems comprising altered geometric configurations for controlled drug delivery. Int J Mol Sci 13:18–43
35. John CM, Arthur JT (2000) Synthetic biodegradable polymers as orthopedic devices. Biomaterials 21 2335–2346
36. Kevin L, Ong BM et al (2015) New biomaterials for orthopedic implants. Orthopedic Res Rev 7:107–130
37. Zeeshan S, Shariq N et al (2015) Biodegradable materials for bone repair and tissue engineering applications. Materials 8:5744–5794
38. Rhee SH, Lee JD et al (2000) Nucleation of hydroxyapatite crystal through chemical interaction with collagen. Journal of Am Ceram Soc 83:2890–2892
39. Hsu FY, Chueh SC et al (1999) Microspheres of hydroxyapatite/reconstituted collagen as supports for osteoblast cell growth. Biomaterials 20:1931–1936
40. Brodie J, Goldie E et al (2005) Osteoblast interactions with calcium phosphate ceramics modified by coating with typeI collagen. J Biomed Mater Res A 73:409–421
41. Rodrigues C, Serricella P et al (2003) Characterization of a bovine collagen-hydroxyapatite composite scaffold for bone tissue engineering. Biomaterials 24:4987–4997
42. Vainionpää S, Rokkanen P et al (1989) Surgical applications of biodegradable polymers in human tissues. Prog Polym Sci 14:679–716
43. Zhang X, Mattheus G et al (1993) Biodegradable polymers for orthopedic applications. J Macromolecular Sci Part C Polym Rev 33:81–102

44. Daristotle JL, Lau L et al (2019) Sprayable and biodegradable, intrinsically adhesive wound dressing with antimicrobial properties. Bioeng Translational Med. https://doi.org/10.1002/btm2.10149
45. Sun G, Zhang X et al (2011) Dextran hydrogel scaffolds enhance angiogenic responses and promote complete skin regeneration during burn wound healing. Proc Natl Acad Sci 108:20976–20981
46. Zhao X, Lang Q et al (2016) Photocrosslinkable gelatin hydrogel for epidermal tissue engineering. Adv Healthc Mater 5:108–118
47. Konieczynska MD, Villa-Camacho JC et al (2016) On-demand dissolution of a dendritic hydrogel-based dressing for second-degree burn wounds through thiol-thioester exchange reaction. Angew Chem Int Ed 55:9984–9987
48. Zahedi P, Rezaeian I et al (2009) A review on wound dressings with an emphasis on electrospun nanofibrous polymeric bandages. Polym Adv Technol 21:77–95
49. Katti DS, Robinson KW et al (2004) Bioresorbable nanofiber-based systems for wound healing and drug delivery: optimization of fabrication parameters. J Biomed Mater Res 70B:286–296
50. Masutani K, Kimura Y (2015) Chapter 1 PLA synthesis. From the monomer to the polymer. In: Poly(lactic acid) science and technology: processing, properties, additives and applications. The Royal Society of Chemistry, London, UK, pp 1–36
51. Zhong XD, Dehghani F (2010) Solvent free synthesis of organometallic catalysts for the copolymerisation of carbon dioxide and propylene oxide. Appl Catal B Environ 98:101–111
52. Mariam M, Murtaza NA et al (2018) Synthetic polymeric biomaterials for wound healing: a review. Progr Biomater 7:1–21
53. Elisa B, Elena V et al (2011) Degradable polymers may improve dental practice. J Appl Biomater Biomech 9:223–231
54. Maria JRV, Daniela M et al (2015) Current uses of Poly(lactic-co-glycolic acid) in the dental field: a comprehensive review. J Chem Volume, Article ID 525832, 12 pages
55. Agata SW, Jerzy S et al (2020) Ageing of dental composites based on methacrylate resins—a critical review of the causes and method of assessment. Polymers 12:882–893
56. Dinesh R, Viritpon S et al (2018) Polymeric materials and films in dentistry: an overview. J Adv Res https://doi.org/10.1016/j.jare.2018.05.001
57. Beyer M, Reichert JHE et al (2010) Pectin, alginate and gum arabic polymers reduce citric acid erosion effects on human enamel. Dent Mater 26:831–839
58. Andra D, Maria L et al (2005) Mucin-chitosan complexes at the solid-liquid interface: multilayer formation and stability in surfactant solutions. Langmuir 21:9502–9509
59. Andrea T, Eleonora M et al (2011) Silver–polysaccharide nanocomposite antimicrobial coatings for methacrylic thermosets. Acta Biomater 7:337–346
60. Mona AO, Yuncang L et al (2019) A comprehensive review of biodegradable synthetic polymer-ceramic composites and their manufacture for biomedical applications. Bioactive Mater 4:22–36
61. Rajeswari R, Clarisse CH et al (2012) Biomimetic surface modification of titanium surfaces for early cell capture by advanced electrospinning. Biomed Mater 7 015001 (16pp)
62. Xifeng L, Shanfeng W et al (2015) Tissue engineering, cardiovascular: biodegradable polymers. In: Mishra M (ed) Encyclopedia of biomedical polymers and polymeric biomaterials. Taylor & Francis, New York, USA. https://doi.org/10.1081/E-EBPP-120051253
63. Mai TL, Joseph CW (2012) Biomaterial applications in cardiovascular tissue repair and regeneration. Expert Rev Cardiovasc Ther 10:1039–1049
64. Valeria C, Pamela M et al (2014) Polyurethane-based scaffolds for myocardial tissue engineering. Interface Focus 4:20130045
65. Ian CA, Norsyahidah MH et al (2014) Enhancing expanded poly(tetrafluoroethylene) (ePTFE) for biomaterials applications. J Appl Polym Sci. https://doi.org/10.1002/APP.40533
66. Helen MN, Elazer RE (2003) Tissue engineering therapy for cardiovascular disease. Circ Res 92:1068–1078
67. Anne S, Raila B (2015) Polymers for cardiovascular stent coatings. Int J Polym Sci Article ID 782653, 11 pages

68. Ratchapol J, Phruedsaporn T et al (2015) Recent trend in applications of polymer materials to stents. Gastrointest Interv 4:83–88
69. Peter X Ma (2008) Biomimetic materials for tissue engineering. Adv Drug Deliv Rev 60:184–198
70. Isabella CPR, Andreas K et al (2018) Cardiac tissue engineering: current state-of-the-art materials, cells and tissue formation. Einstein (São Paulo) 16: RB4538
71. Helen M, Nugent et al (2003) Tissue engineering therapy for cardiovascular disease. Circ Res 92:1068–1078
72. Naseer I, Abdul SK et al (2018) Recent concepts in biodegradable polymers for tissue engineering paradigms: a critical review. Int Mater Rev. https://doi.org/10.1080/09506608.2018.1460943
73. Pathiraja A, Gunatillake et al (2003) Biodegradable synthetic polymers for tissue engineering. P Eur Cells Mater 5:1–16
74. Van Dijkhuizen R, Moroni L et al (2008) Degradable polymers for tissue engineering. Tissue Eng 193–221 (Chapter 7). https://doi.org/10.1016/b978-0-12-370869-4.00007-0
75. Brahatheeswaran D, Yasuhiko Y et al (2011) Polymeric scaffolds in tissue engineering application: a review, Int J Polym Sci Volume, Article ID 290602, 19 pages. https://doi.org/10.1155/2011/2906

Chapter 14
Applications of Biodegradable Green Composites

Ayfer Yildirim and Hilal Acay

1 Introduction

Increased uncontrolled environmental pollution in recent years has increased concerns all over the world. The fact that the oil reserves are limited and the recycling of petroleum-based polymers is difficult has led to the need to turn to the production of new polymers that can replace these materials. Green polymers and green composites have been used in many areas to produce solutions to rapidly increasing environmental problems. Materials called biodegradable green composites consisting of matrices and reinforcers made entirely from natural resources, besides being lightweight, can be used for mechanical, thermal, etc. They can also fulfill the features [1].

Biocomposites can supplement and replace petroleum-based composite materials in many applications. This can offer powerful advantages. Since biocomposites are derived from renewable resources, material costs can be reduced significantly with their large-scale use. Also, the green composites being biocompatible and biodegradable have widespread their use in biomedical applications such as tissue engineering. Enzymatic or hydrolytically degradable implants used in the regeneration and repair of damaged tissues are becoming more and more preferable in the medical sector today [2]. Besides being used frequently in the medical sector, green polymers and composites are also preferred in short-lived disposable product applications such as

A. Yildirim
Mardin Artuklu University, Vocational School of Health Services, 47200 Mardin, Turkey
e-mail: ayferyildirim@artuklu.edu.tr

H. Acay (✉)
Department of Nutrition and Dietetics, Faculty of Health Science, Mardin Artuklu University, 47200 Mardin, Turkey
e-mail: hilalacay@gmail.com

S. Thomas and P. Balakrishnan (eds.), *Green Composites*,
Materials Horizons: From Nature to Nanomaterials,
https://doi.org/10.1007/978-981-15-9643-8_14

packaging, bottles, beverage cans, and textile products [1]. The fact that the automotive sector, other than the medical and packaging sector, has regulations regarding sustainability, encourages this sector to use renewable materials as well. The introduction of green composites in the automotive industry took place in 1940 when Henry Ford produced the first green automotive part with soybean extract. However, cheaper production of petroleum-based polymers at that time reduced interest in biopolymer-based materials, but later the need to reduce vehicle weight and use sustainable materials began to increase interest in green materials [3]. The samples of green composite applications in the literature were bio-based roof [4], hemp reinforced sunglasses, PLA-based golf ball holder, PLA/kenaf composite prototype car roof, door panel (Kestrel Hemp Car). Biomobile vehicle is given. Besides, some natural polymers and their composites, especially cellulose, chitosan, starch, alginate, etc., are biodegradable and sustainable and are easily used in adsorption [5, [6]], electronic devices [7], and construction [8] area because of their attractive properties such as environmentally friendly, disposal, or composting.

As the application of green composites in different industries increases, it is known that the interest in using these multifunctional materials as an alternative to conventional materials will increase. Therefore, the purpose of this book chapter is to provide guidelines for the selection and application of some important composite materials. To provide a better overview of the progress in this area, it is to classify composites according to their applications. This specific chapter aims in providing detailed information about various applications of green biodegradable composites in biomedical applications, food packaging, adsorption, electronics, construction, and some other areas.

2 Applications of Green Composites

2.1 Biomedical Applications

It is known that biopolymers are already widely used in medical applications [9]; however, the inclusion of new green composite materials, which offer unmatched performance and functionality, has recently taken place [10]. The new developments to be achieved are exciting in terms of offering life-changing medical treatments, for example, soy-derived polymers as bone fillers; bacterial nanocellulose has also been reported to be beneficial for artificial blood vessels. Besides, nanofiller can be applied to clinical medicine as biodegradable composite materials. Biodegradable nanocomposites are very useful in tissue engineering for regeneration of primary tissue structures [11–13] Various applications of different polymer composite biomaterials [14] are given in Fig. 1. Since biomedical applications are broadly given in the other part of the book, they will not be mentioned further.

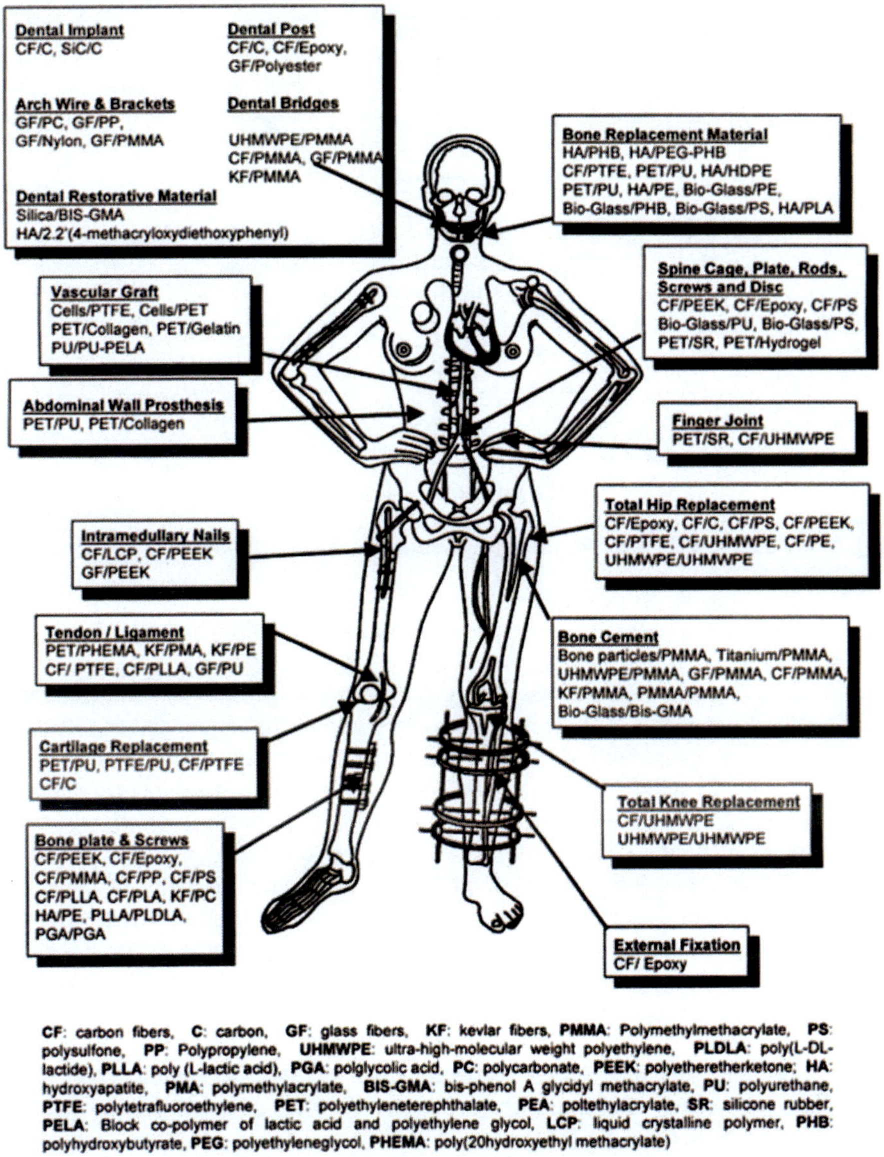

Fig. 1 Various applications of different polymer composite biomaterials [14]

2.2 *Food Packaging*

Basically, a food package should delay moisture gain or loss due to its mechanical, optical, and thermal properties, prevent microbial contamination, and provide barrier properties against the penetration of compounds such as water vapor, oxygen, carbon

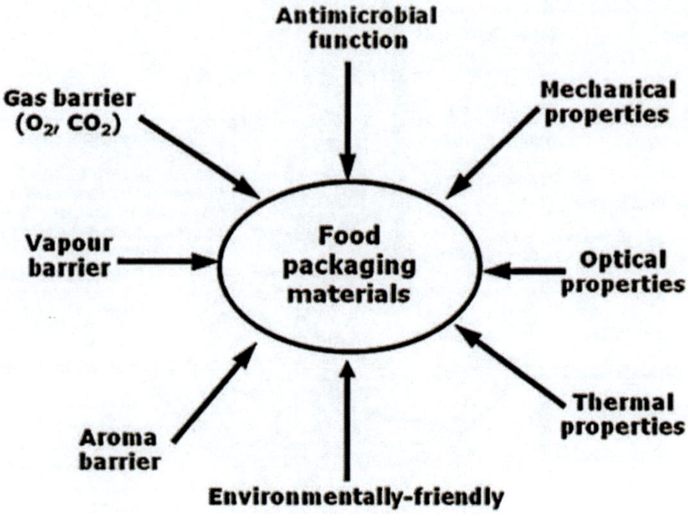

Fig. 2 Required properties of food packaging materials (Rhim et al. 2013)

dioxide, aroma, and paint [15]. The properties of food packaging materials are given in Fig. 2 (Rhim et al. 2013).

Today, consumer demands are directed toward less processed, food-free, and healthy food products. Packaging used for food is an important factor affecting food quality and preserving its shelf life. There has been an increasing consumer demand for better quality, fresh-like, and convenient products; therefore, food packaging becomes important more than ever to provide safe products and minimize food losses. Most of the food packaging materials are based on nondegradable synthetic polymers, thus representing a serious global environmental problem. Besides, the dependency on fossil resources brings the sustainability problem for raw materials of food packaging production. Biopolymer-based packaging materials represent an alternative to plastic films, and they are originated from naturally renewable resources as polysaccharides, proteins, and lipids; from chemical synthesis of bio-derived monomers, such as polylactate; and from polymers naturally produced by microorganisms, such as polyhydroxybutyrate and polyhydroxyvalerate [16]. Table 1. summarizes the advantages of natural biopolymer films over traditional synthetic films (Source: Rhim and Ng [17]).

The well-known application of biodegradable polymers in food packaging is as edible films that are used for individual coating of small food products or placed within the food. Biopolymer films can also improve the quality of food products and act as an efficient carrier agent for incorporating various additives including antimicrobials, antioxidants, coloring agents, and other nutrients [16]. The economic production of bio-based food packaging materials requires using raw material abundant in nature.

Table 1 Advantages and application areas of natural biopolymer-based packaging materials

• They are biodegradable
• They can be used as edible coatings
• They can increase the nutritional value of foods enhance its characteristics like appearance, odor, and flavor
• They can be used as active packaging with the incorporation of antimicrobial agents and antioxidants
• They can control the transfer of moisture, gases, lipids, and solutes
• They can be used for microencapsulation and controlled release of antimicrobial agents, antioxidants, and active ingredients
• They may be component of a multilayer food packaging materials with non-edible films
• They have low cost
• They are abundant and annually renewable resources
• They are suitable for individual packaging of particulate food such as nuts
• Using them lead to reduced packaging volume, weight, and waste
• They can extend shelf life and improve the quality of usually non-packaged items

Source: Rhim and Ng [17]

Being a natural polysaccharide, chitosan can be used in many fields such as pharmaceutical, agriculture, food, cosmetics, textile and water treatment, directly or indirectly, as well as its biodegradability and biocompatibility properties [18–20]. Also, good film-forming and mechanical properties make chitosan an important edible film component and can create transparent films that can meet a variety of packaging needs [21].

Jridi et al. (2014) investigated the physicochemical and mechanical properties of chitosan (obtained from shrimps), cuttlefish gelatin, and composite films in their work. The results showed that the chitosan film had higher tensile strength and lower tensile elongation compared to other films. In their study, Boran et al concluded that the films they obtain by producing laboratory-scale chitosan and pectin-based films from industrial wastes provide practical applications in food products as coating or packaging materials and may reveal many desired features in the future. [22] have prepared chitosan and gelatin-based biodegradable films that can be used as packaging materials. In a study conducted by Sun et al. [23], it was aimed to characterize physical, mechanical, and bioactive chitosan films combined with thinned immature apple phenols. In this study, they concluded that chitosan films combined with immature apple phenols can be an alternative to synthetic materials and can contribute to the extending shelf life of foodstuffs [23].

In addition to composite materials consisting of biopolymers such as chitosan, nanotechnology is seen from studies that have been able to provide a new lightweight material with stronger packaging barriers that protect food quality during transportation, prolong the freshness of fruits and vegetables during storage, and protect meats or poultry from pathogens. Today, it is known that these nanomaterials have been used in a variety of food contact packaging and containers as a new alternative additive to improve the polymeric properties of packaging materials [24].

Recently, nanoparticles have been used as additives to improve polymer performance. Nanoclays (layered silicates) [25], cellulose nanowhiskers [26], ultra-fine layered titanate [27], and carbon nanotubes can be counted as various nanoreenforms currently being developed [28]. However, among them, it is stated that only layered inorganic silicates such as clay have attracted great interest by the packaging industry [29]. Nanotechnological advances can enable more environmentally friendly economic degradable materials in this regard.

2.3 Adsorption Applications

Uncontrolled discharge of industrial waste and sewage waters leads to increased amounts of undesirable inorganic and organic pollutants in water resources. Due to their very high toxicity, these contaminants that cause many symptoms such as nephritis, acute diarrhea, gastrointestinal ulceration, skin irritation, severe headaches, and cancer in the digestive system become a serious problem. Different efforts have been promoted to develop various physical and chemical methods to eliminate this problem. Because of its low-cost, high efficiency, and activity for removing a large number of organic pollutants, dyes and heavy metals from wastewater, adsorption is one of the most favorable appreciable methods compared to various conventional methods used such as ion exchange, precipitation, reverse osmosis, solvent extraction, coagulation, and filtration [18, 30, 31].

For an ideal adsorbent to be preferred for removing water contaminants, it should have the advantages of high adsorption capacity, efficiency, low cost, environmental safety, and easy regeneration [6]. Therefore, due to their biocompatibility and biodegradability, a great deal of research is supported for natural biomaterials and their composites. In particular, those that are nontoxic, biodegradable biocompatible, and natural biopolymers that are abundant in natüre are preferred as biosorbents for the treatment of various wastewater systems including industrial wastewater.

Among the possible adsorbents to be used, biodegradable green materials such as cellulose, chitosan, alginate, starch, and agricultural wastes are suitable adsorbents media shown in Fig. 3. When their economic and environmental importance is made remarkable, biodegradable green materials have been extensively preferred as green efficient adsorbents for wastewater treatment fields regarding the economic feasibility and environmental importance of them. It is of great importance in their preference that they have superior properties such as easy modification, high biodegradability, low cost, non-toxicity, biocompatibility, and environmental friendliness. However,

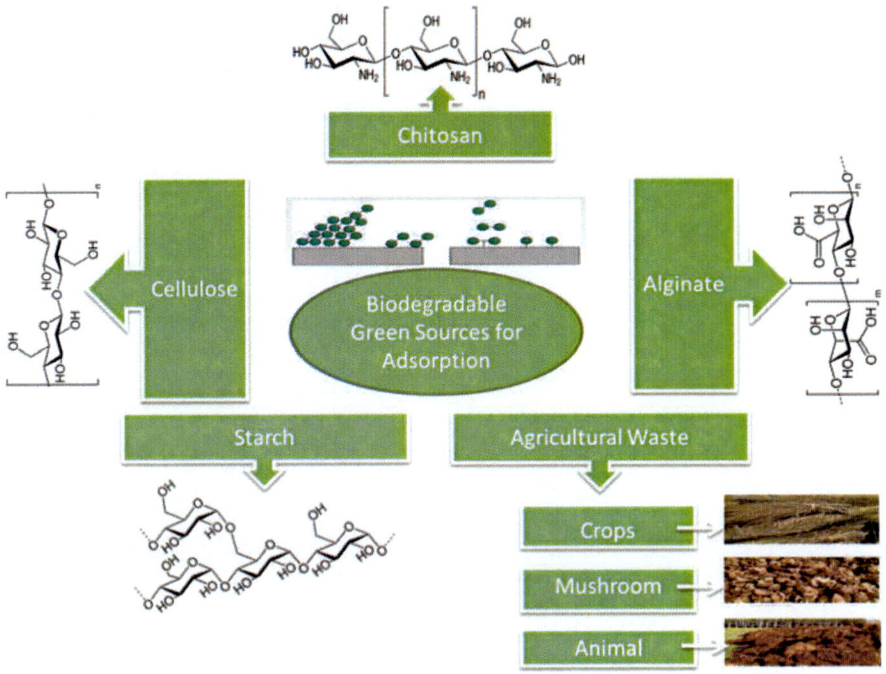

Fig. 3 Biodegradable green sources for adsorption mechanism

their application is limited due to their low surface area, dissolution, and low mechanical strength. Due to overcome of the above-mentioned disadvantages, their transformations into composites have become much more attractive by providing physical and chemical modification processes to increase the adsorption capacities of these structures and provide better conditions for their weak mechanical properties [32, 33].

Many partially/completely green biodegradable composites have been used to enhance the adsorption capacity of adsorbents to remove pollutants from wastewater. Composite materials can generally enhance adsorption capacities compared to the capacity of elements of composite alone. Table 2 shows the summary of possible green and biodegradable composites and their adsorption capacities for the treatment of wastewater pollutants.

Cellulose is the most common organic compound and biopolymer in the world. Approximately, 33% of plants, 90% of cotton, and 50% of wood are composed of cellulose. Biodegradable cellulose has a wide range of uses due to its natural availability on Earth. Cellulose can be obtained from cotton, trees, straw plant sources, and also some bacteria sources [36, 55]. Natural cellulose has a high surface area and tensile strength, transparency, biocompatibility, biodegradability, and impressive mechanical properties. These properties make natural cellulose a very important choice in composite formation as backing for green composites. Cellulose ingredients

Table 2 Biodegradable green composites for removal of wastewater pollutants

Composite	Type of removed material	Biosorption capacity (mg/g)	References
Cellulose extracted (rhizomes of Alpinia nigra)/organophillic montmorillonite composite	Eosin Y dye	199.9	[34]
Cellulose/gelatin composite hydrogel	Cu(II)	28.4	[35]
Bacterial cellulose/gram-negative species of bacteria composite	Pb(II)	52.00	[36]
Cellulose/montmorillonite hydrogels	MB	277 mg/g	[37]
Magnetic nanocomposites of cellulose	Pb (II)	21.5	[38]
Chitosan-based composite hydrogels	RB MB	21.74 9.66	[39]
Chitosan-blended/polyvinyl alcohol composite	Pb(II)	76.60	[40]
Chitosan/polyacrylic acid/bentonite composites	Malachite green	454.55	[18]
Xanthate-modified chitosan/poly(N-isopropylacrylamide) composite hydrogel	Cu(II) Pb(II) Ni(II)	115.1 172.0 66.9	[41]
Poly(1-vinylimidazole)-modified-chitosan composite	Cr (VI)	196.1	[42]
Chitosan–montmorillonite nanocomposite	Fe(III)	78.13	[43]
Polyurethane/chitosan bio-based composite	Food Red 17	267.24	
Magnetic glutamic acid/chitosan and silica-coated composite	MB CV CLY 7GL	180.01 375.4 217.3	
Nanochitin-contained chitosan nanocomposite hydrogels	Cu(II)	64.9	
Chitosan/polyaniline composite	Tartrazine	584.0	[44]
Magnetic alginate/rice husk beads biocomposite	MB	274.9	[45]
Calcium alginate/clay hybrid composites	Organic acid anions	3.6	[46]
Organic montmorillonite-sodium alginate composites	PAH	1.2	[47]
N-doped carbon dots/alginate composite	Gd (III)	201.21	[48]
Alginate/natural bentonite composite	MB CR	1171 95.55	[49]
A silica sand/anionized-starch composite	MB CV Cu(II)	653.31 1246.40 383.08	[50]

(continued)

Table 2 (continued)

Composite	Type of removed material	Biosorption capacity (mg/g)	References
Corncob biochar-based montmorillonite composite	Pb(II) At	139.78 86.86	[51]
White-rot edible fungi Pleurotus ostreatus-based-chitosan nanocomposite	RO16	65.5	[52]
Argania spinosa tree nutshells bio sourced composite	Diclofenac carbamazepine sulfamethoxazole	153.8 105.3 125.0	[53]
Encapsulated cellulose-based modified citrus peels/calcium alginate composite	MB CV	881.36 923.07	[54]

can be widely used in packaging industries, cigarette filters, textile fibers, coatings, membrane filtration, lamination, nano-macro composites, various medical and pharmaceutical products [56–58]. Besides, its large surface area and multiple functional groups make cellulose attractive in adsorption applications. The functional groups on the cellulose structure have structural auxiliaries such as reactive surfaces and the ability to change surface chemistry, which empowers it as a promising additive for various composites for biosorbents. Nowadays, biodegradable natural cellulose-based composites are getting increasing attention as nontoxic sorbents toward the removal of the wastewater contaminants.

Goswami and Das investigated the biosorption performance of cellulose extracted from rhizomes of *Alpinia nigra*/organophilic montmorillonite composite onto the Eosin Y toxic dye [34]. Biodegradable cellulose/gelatin composite hydrogel has been synthesized as biosorbent for copper (II) (Cu(II)) ions biosorption with 79.5% biodegradability rate [35]. Also, cellulose/montmorillonite hydrogels have been prepared for adsorption of methylene blue (MB) dye [37]. In another study, biodegradable bi-functional cellulose derivatives and cellulose/clay composites have been used to remove calcium (Ca(II)), magnesium (Mg(II)), iron (Fe(II)), lead (Pb(II)), and Cu(II) metal ions from widespread underground water. The removal result order was as Pb(II) > Mg(II) > Fe(II) > Cu(II) > Ca(II) [59]. Also, bacterial cellulose/gram-negative species of bacteria composite has been prepared and investigated for biosorption of Pb(II) metal ion [36].

Another substance used as a biosorbent is chitosan.. It is the second-most abundant biological material in nature, which is biologically and chemically compatible with low cost. This is because the macroporous chitosan has extreme biological properties such as biocompatibility, biodegradability, anti-antigenic, nontoxic, excellent film-forming ability, bio-adsorptive, as well as super absorbency. All these important features make chitosan a very interesting ingredient that can be used in environmental and biological fields [60, 61]. Chitosan is an ecologically interesting and promising natural polysaccharide for removing many pollutants from the aqueous wastewater system through the presence of a large number of hydroxyl and primary amine groups. There are many studies in the literature where chitosan-based composites are prepared and used as a biosorbent, which shows a very good adsorption capacity in the purification of pollutants that cause water pollution. For example, rhodamine B (RB) and MB dye adsorption were examined by chitosan (CS) / by graphene oxide, carbon nanotubes, and layered double hydrotalcite composites [39]. In another study, chitosan-blended polyvinyl alcohol composite was prepared for adsorption of Pb(II) metal ions, and adsorption capacity was found as 76.60 mg/g [40]. Xanthate-modified chitosan/poly(N-isopropylacrylamide) composite hydrogel was used for the adsorption of Cu(II), Pb(II), and Ni(II) metal ions [41]. Poly(1-vinylimidazole)-modified-chitosan composite was performed for adsorption of chromium (VI) (Cr (IV)) ions as adsorption capacity 196.1 mg/g [42]. Chitosan–montmorillonite nanocomposite for ferric (III) (Fe(III)) ions adsorption was investigated [43]. A full biodegradable magnetic adsorbent based on glutamic acid-modified chitosan and silica-coated Fe_3O_4 composite was prepared and used for adsorption of MB, crystal violet (CV), and cationic light yellow 7GL (CLY 7GL).

Another conspicuous substantial nature biosorbent is alginate that could be extracted from brown seaweed as a natural and renewable polysaccharide polymer material, owning affluently hydroxyl and carboxyl functional groups. Due to its superior properties such as affinity, hydrophilicity, easy separation, non-toxicity, biocompatibility, and strong alginate, it has a wide range of applications such as tissue repair, drug release, wastewater treatment, and adsorption [59, 62]. By combining all these extraordinary properties with some other substances, green biosorbents with extraordinary selective detection and adsorption have been obtained in the application of adsorption.

Many studies have been done with this biosorbent in the past. To illustrate, magnetic alginate/rice husk beads, an eco-friendly and low-cost biocomposite, are used for MB removal with adsorption capacity as 274.9 mg/g [45]. In another study, organic montmorillonite-sodium alginate composites were prepared and the removal performance of polycyclic aromatic hydrocarbons (PAH) from aqueous medium was investigated [47]. Alginate/natural bentonite composite beads were prepared for adsorption of MB and congo red (CR) dyes [49]. N-doped carbon dots/alginate composite were greenly generated for adsorption of gadolinium (III) (Gd (III) [48].

Starch, the major dietary source of carbohydrates and can be extracted from corn, rice, potato, wheat, cassava, tapioca, and other crops, is abundantly available, highly biodegradable, renewable, and cheap. At high temperatures and pressures, starch loses its crystallinity and becomes thermoplastic, but the films cast from this material are brittle, which limits their applications [63, 64]. Therefore, it has been becoming important to improve its thermal and mechanic properties to make its use areas more extensive by transforming to composite structure. In a conducted research, a silica sand/anionized-starch composite was synthesized and used for adsorption of MB, crystal violet (CV), and Cu(II) metal ions from water. Adsorption capacities were found 653.31, 1246.40, and 383.08 mg/g, for MB, CV, and Cu(II), respectively [50].

Biodegredable agricultural waste-based composites have attracted excellent attention for adsorption process. It provides a convenient way to collect agricultural wastes such as tree nutshell, mushroom, citrus peels, and corncob that are extraordinarily advantageous and biodegradable, and to use them for a greener environment, thus increasing the ability of agricultural wastes to adsorb pollutants in wastewater. Biological waste-based composites are usually impregnated with other biologic materials that can improve the adsorption capacity by increasing their specific surface area, porosity, mechanical and thermal cavity to provide appropriate modifications. In a related study, corncob biochar-based montmorillonite composite was synthesized for the adsorption of Pb(II) and a pharmaceutical emerging organic contaminant Atenolol (AT) [51]. White-rot edible fungi *Pleurotus ostreatus*-based-chitosan nanocomposite was synthesized for investigation of reactive orange 16 (RO16) dye [52]. In another approach presented by Mauchtari et al. [53], high-surface-area-activated carbon was prepared with *Argania spinosa* tree nutshells bio-sourced composite as activated carbon/TiO2 was used for removing pharmaceuticals, diclofenac (DCF), carbamazepine (CBZ), and sulfamethoxazole (SMX) from aqueous solution. Encapsulated cellulose-based modified citrus peels/calcium alginate composite was prepared and used as effective adsobent for MB and CV dyes. The good adsorption capacities were found as 881.36 and 923.07 mg/g for CV and MB, respectively [54].

2.4 Electronics

Electronics that have become an indispensable part of daily life will grow even more in the future and are expected to cover almost every aspect of our life in the future. In this section, we present a versatile and general understanding of electronic applications of different biodegradable green composite materials. Electronics have many applications such as communications, optics, imaging, sensing, energy storage, energy collection, artificial muscles, neuroscience, and bioengineering [65] (Fig. 4) (Yıldırım and Acay). With the rapid development in electronic biomaterials and related production techniques, the use of biodegradable green composite materials in electronic materials is becoming very important. In recent years, great efforts have been made to investigate the synthesis and properties of environmentally friendly biodegradable composites. By combining with some other biological materials, creating composites, the mechanical properties of these green biomaterials can be improved on a large scale, which is promising in the development of electronic devices [66].

Applications of electronic devices such as computers, mobile phones, and medical devices are becoming more and more widespread in our lives with the rapid development of science and technology and provide us with great convenience. Electronics, which are widely used in many fields such as telecommunications, entertainment, and health services, leave a deep impact on people. In particular, the choice of green biodegradable electronics is important to eliminate environmental problems such as electronic waste storage [7]. Preliminary results of synthesized bio-based

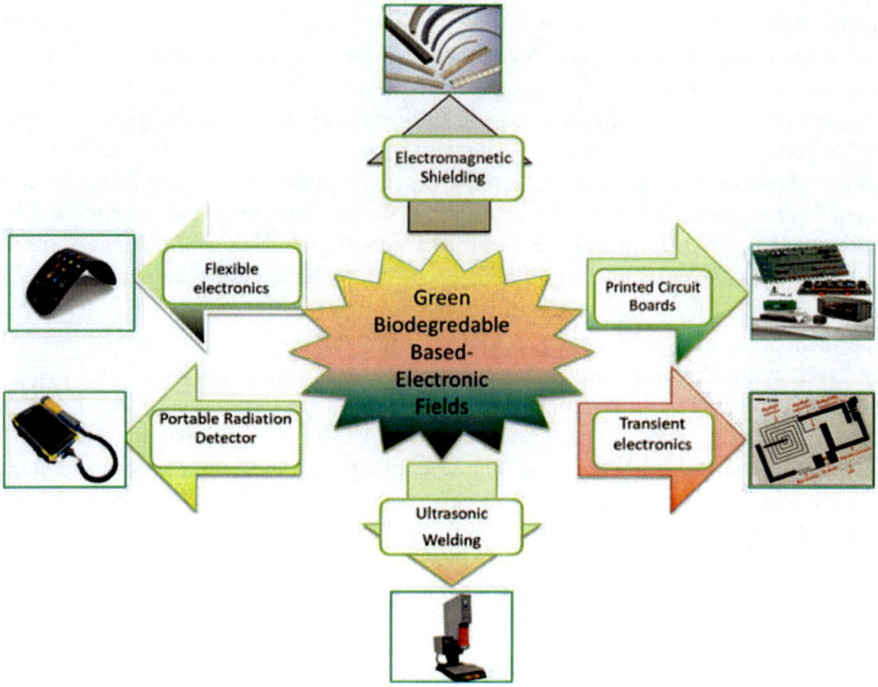

Fig. 4 Green biodegradable-based electronic fields (Yıldırım and Acay)

biodegradable composites have been studied for new and different electronic application fields such as ultrasonic welding, electromagnetic interference shielding, flexibility, transient bioelectronics, and printed circuit boards, as a promising technology for environmental improvement.

Recently, ultrasonic welding used as a suitable fusion bonding technique to combine various engineering components in the electrical and electronics industry has attracted attention. In this technique, which is mostly used for thermoplastic bonding, high-frequency ultrasonic vibrations are provided and shredding is produced by generating heat at the joint interface. Moreover, the advantages of this method are that they do not require additional parts and materials, they are very cheap and environmentally friendly in terms of cost, and they also save time. In a previous study on this field, [47, 67] in their study investigated the ultrasonic welding behavior of the poly(lactic) acid (PLA)/bamboo fully degradable green composite by examining the effect of the welded sample on the tensile failure load with welding parameters (holding time, welding time, and holding pressure) by application of high-frequency (20 kHz) ultrasonic vibration.

With the developing technology, portable and wearable electronics are very popular due to the rapid growth of the internet and digitalization and the development of innovative functional materials in parallel. Besides, the increase in telecommunication systems causes electromagnetic interference difficulties and consequently, data pollution, device reliability, and health negatively affect. In order to prevent such problems, using environmentally friendly solutions that can only benefit humanity by using sustainable and biodegradable all-green composites, also versatile, affordable, and high-performance shielding solutions of electromagnetic interference (EMI) have become inevitable. Emitting electromagnetic fields at various frequencies by electrical devices cause low performance/serious damage to this equipment. Electromagnetic shielding developed for the reflection and/or absorption of electromagnetic radiation by a material is needed against the protection of electrical devices from EMI. In addition, the composites that are especially important for wearable devices offer reasonably low mass densities, often with good stretchability and flexibility. Tolvanen et al. [68], have examined EMI shielding capacity with green composites using biochar obtained from pine nuts, graphite as filler, and poly (lactic acid) as host and highlight that these green composites show rather high efficiency for shielding (shielding effectiveness >32 dB). As a result, they emphasize that the materials developed are suitable for use in wearable and portable radiation-sensitive electronic device applications. In another similar study, biodegradable poly(lactic acid) cheap conductive carbonaceous fillers were developed [69]. In addition, the study has been concluded that the prepared composite shows good mechanical strength and low-density functional properties and thus can be sufficiently used in packaging applications and against electromagnetic interference.

Research has been conducted into the use of biodegradable natural polysaccharide sodium alginate immersed into the $CaCl_2$ crosslinker for producing the alginate fiber [70]. The impedance experiments were performed for conductivity change. Since these and similar materials prepared are biodegradable and do not require conductive cables, they can create versatile applications for environmentally friendly and low-cost flexible electronics.

Liu et al. researched the synthesis of poly(vinyl alcohol)/chitosan composites by evaluation of dynamic mechanical analysis, tensile testing, and thermogravimetric analysis and concluded that obtained composite can be an alternative application in sustainable and transient bioelectronics [71].

Production of printed circuit boards (PCB) is increasing in parallel with the increasing demand for electronic devices. In this application, the approach of using environmentally friendly biodegradable materials is in great demand. Zhan and Wool prepared the bio-based composites from soybean oil resins, chicken feathers, and E-glass fibers for using the printed circuit boards in PCB. They investigated the electrical, thermal, and mechanical properties, also flammability, and peel strength of biocomposites and found that these bio-based materials showed comparable values with traditional materials [72]. In addition, 35, used bio-based composites (natural protein and natural cellulose fiber) from agricultural wastes, natural cellulose fibers extracted from banana stems and wheat gluten, in order to provide the materials for completely biodegradable printed circuit boards and other electronic applications. In the study, it was emphasized that biocomposites did not experience any loss in their performance even after exposure to 90% humidity for 48 h and 100 °C for 8 h [73].

2.5 Construction Applications

When designing construction materials, it is very important to consider issues such as safety, health, and the environment while using the products. Because of this, the development and use of biodegradable green polymers and their composites are considered as one of the substantial ways for reducing the environmental challenge from the use of non-biodegradable petroleum-based polymers in the construction industry.

Recently, various biopolymers as starch, cellulose, lignosulphonate, some water-soluble polysaccharides, and agricultural residues are used in a wide range of applications of construction materials including grouts, concrete, mortars, plasters, and plasterboards, paints, and oil well fluids of drilling. Also, natural fiber-reinforced composites have been gaining a lot of popularity in non-structural construction applications using door and window frames for wall insulation and floor lamination [74]. In addition, biodegradable fiber like bamboo and coconut are studied for both indoor and outdoor applications. Several studies have been carried out over recent years to obtain eco-friendly, biodegradable, low-cost, and lightweight natural polymer composites in this sector.

The use of agricultural residues and by-products for construction directly promotes sustainability and green buildings. Green buildings that use bio-based materials and technologies to reduce the harmful effects of the production, use, and disposal of cement and concrete on the environment are becoming important all over the world. Sugarcane bagasse, abundant and easily available agricultural residues, has been valorized by developing completely biodegradable composites combined with wheat gluten for construction applications [75]. In their study [75], pointed out the conclusion that bagasse-gluten composites have the potential to be an ideal alternative for gypsum-based ceiling coverings. In addition, fiber cement composites containing natural cellulosic fibers are important materials that appear in the building industry. Everaert et al. [76] have investigated the biodegradability of the cellulosic fiber cement composites; thus, the effect of the composite's particle size and material aging has been evaluated. The study carried out by Živkovic et al. [77], highlighted the two types of natural fibers, flax and basalt, and their hybridization composite (flax fiber reinforcement in the central zone, basalt fiber reinforcement in the outer layers) has been used for the construction of boats and yachts. Another study fiber-based composite was examined by Chandekar et al. [78]. They have performed chemical treatment by improving jute fibers, fiber–matrix interface adhesion, and reviewed composites obtained with different polymeric matrices that provide a basic direction for future construction material research of these jute-reinforced composites.

2.6 Other Applications

It is expected that technical innovations for green composites will continue, permanent political and environmental pressures will increase with the definition of new applications and investment in new methods for collecting fibers and processing biological fibers [79].

Toys are one of the potential application areas for green composites. Examples of biodegradable green composite toys (except wooden toys) are limited. However, studies on toys made from recycled plastic with cellulose reinforcement are exciting. This results in both consumer and commercial interest in the use of materials with environmental credentials in toy applications.

It is seen that Boat Hulls and Canoes offer great opportunities for biocomposite applications in the marine sector. The excellent sound behavior of plant fibers makes them a promising material for musical applications. The use of natural composites in snowboards/skis and surfboards is already a reality, and several examples can be found. Biocomposites also offer immense opportunities for an increasing role as alternate material, especially as wood substitutes in the furniture market [2].

The agricultural sector is another area where composite material is evaluated. In addition to PLA, which is the most used polymer in this field, it is used as a mixture of polybutylene adipate terephthalate (PBAT) and PBS polymer or as a composite component [80].

The main biopolymers used in the cosmetics industry are proteins and polysaccharides. While proteins and peptides are used in applications especially on hair and skincare, peptides are also used in topical moisturizer applications [81]. Increasing demands from different markets will tend to increase the application areas of biocomposite materials.

3 Conclusion

The use of biodegradable polymers as matrices and the use of natural fibers as reinforcements in composite materials appear to contribute to the development of green composites in terms of performance and sustainability. Green composites provided important commercial markets especially for value-added products in the packaging industry. Applications in the automotive, construction, and electronics industry, usage for furniture, luggage, grinding disks, and safety helmets have also been proposed by scientists and manufacturers. Also, using the adsorption technique for the removal of contaminants from the aquatic environment by using bio-based composites is a great research area. However, to cover other areas, scientists and researchers still need to speed up work on scaling up products. Therefore, launching laboratory-scale ideas is an expected effort from the scientific community. In the future, depending on further development and research of these green composites, an increase is expected in different areas of use, such as structural applications. However, it seems that several problems need to be resolved before green composites become fully competitive with synthetic fiber composites. In recent years, great progress has been the establishment of nanocomposites (i.e., the use of nanocellulose as fibers made from crystals or natural fibers). These nanomaterials are thought to compete with components made from conventional materials. Nanotechnology can offer many opportunities to improve the properties of green composite products. For example, the use of cellulose nanocrystal and cellulose nanofibers has been researched for a variety of uses as it is stronger than steel and harder than aluminum. As more bio-based composites are developed and designed, used, and disposed of in the future, green composites are expected to enter daily life with enormous environmental benefits, and this will last for years.

References

1. La Mantia FP, Morreale M (2011) Green composites: a brief review. Compos Part A: Appl Sci Manufact 42:579–588
2. Georgios K, Silva A, Furtado S (2016) Applications of Green composite materials. Biodegrad Green Compos 16:312–337
3. Drzal LT, Mohanty A, Misra M (2001) Bio-composite materials as alternatives to petroleum-based composites for automotive applications. Magnesium 40:1–3
4. Dweib M, Hu B, Shenton H, Wool R (2006) Bio-based composite roof structure: manufacturing and processing issues. Compos Struct 74:379–388
5. Baran MF, Duz MZ (2019) Biosorption of Pb2+ from aqueous solutions by *Bacillus licheniformis* isolated from Tigris river with a comparative study. Int. J. Int J Latest Eng Manage Res 4(5):108–121
6. Jiang L, Zhang J (2011) Biodegradable and biobased polymers. Appl Plast Eng Handbook, pp 145–158
7. Li R, Wang L, Kong D, Yin L (2018) Recent progress on biodegradable materials and transient electronics. Bioact Mater 3:322–333
8. Zhang J, Chevali VS, Wang H, Wang CH (2020) Current status of carbon fiber and carbon fiber composites recycling. Compos B Eng 193:108053
9. Nair L, Laurencin C (2006) Polymers as biomaterials for tissue engineering and controlled drug delivery. Tissue Eng I 102:47–90
10. Dicker MPM, Duckworth PF, Baker AB, Francois G, Hazzard MK, Weaver PM (2014) Green Composites: A review of material attributes andcomplementary applications. Composites: Part A: Appl Sci Manufact 56:280–289
11. Giavaresi G, Fini M, Salvage J, Nicoli Aldini N, Giardino R, Ambrosio L, Nicolais L, Santin M (2009) Bone regeneration potential of a soybean-based filler: experimental study in a rabbit cancellous bone defects. J Mater Sci Mater Med 21:615–626
12. McCullen SD, Ramaswamy S, Clarke LI, Gorga RE (2009) Nanofibrous composites for tissue engineering applications. Wiley Interdiscip Rev Nanomed Nanobiotechnol 1:369–390
13. Tang J, Li X, Bao L, Chen L, Hong FF (2017) Comparison of two types of bioreactors for synthesis of bacterial nanocellulose tubes as potential medical prostheses including artificial blood vessels. J Chem Technol Biotech 92:1218–1228
14. Ramakrishna S, Mayer J, Wintermantel E, Leong KW (2001) Biomedical applications of polymer-composite materials: a review. Compos Sci Technol 61(9):1189–1224
15. Rhim JW, Park HM, Ha CS (2013) Bio-nanocomposites for food packaging applications. Agricu Agric Sci Procedia 2:296–303

16. Wihodo M, Moraru CI (2013) Physical and chemical methods used to enhance the structure and mechanical properties of protein films: a review. J Food Eng 114:292–302
17. Rhim JW, Ng PK (2007) Natural biopolymer-based nanocomposite films for packaging applications. Crit Rev Food Sci Nutr 47:411–432
18. Acay H, Yildirim A, Güzel EE, Kaya N, Baran MF (2020) Evaluation and characterization of *Pleurotus eryngii* extract-loaded chitosan nanoparticles as antimicrobial agents against some human pathogens. Prep Biochem Biotechn (Online Publication). https://doi.org/10.1080/10826068.2020.1765376
19. Kaya M, Khadem S, Çakma YS, Mujtaba M, Ilk S, Akyuz L, Salaberria AM, Labidi J, Abdulqadira AH, Deligoz E (2018) Antioxidative and antimicrobial edible chitosan films blended with stem, leaf and seed extracts of pistacia terebinthus for active food packaging. RSC Adv 8:3941–3950
20. Niamsa N, Baimark Y (2009) Preparation and characterization of highly flexible chitosan films for use as food packaging. Am J Food 4(4):162–169
21. Singh TP, Chatli MK, Sahoo J (2015) Development of chitosan based edible films: process optimization using response surface methodology. J Food Sci Technol 52(5):2530–2543
22. Ahmed S, Ikram S (2016) Chitosan and Gelatin based biodegradable packaging films with UV-light protection. J Photochem Photobiol Biol 163:115–124
23. Sun L, Sun J, Chen L, Niu P, Yang X, Guo Y (2017) Preparation and characterization of chitosan film incorporated with thinned young apple polyphenols as an active packaging material. Carbohyd Polym 163:81–91
24. Bumbudsanpharoke N, Ko S (2015) Nano-food packaging: an overview of the market, migration research, and safety regulations. J Food Sci 80(5):910–923
25. Pinnavaia TJ, Beall GW (2000) Polymer clay nanocomposites. Wiley, New York, p 370
26. Anglès MN, Dufresne A (2001) Plasticized starch/tunicin whiskers nanocomposite materials. Mech Behav Macromolecules 34, 9, 2001, 2921–2931.
27. Hiroi R, Ray SS, Okamoto M, Shiroi T (2004) Organically modified layered titanate: a new nanofiller to improve the performance of biodegradable polylactide. Macromol Rapid Commun 25(15):1349–1349
28. Sreekumar TV, Liu T, Min BG, Guo H, Kumar S, Hauge RH, Smalley RE (2004) Polyacrylonitrile single-walled carbon nanotube composite fibers. Adv Mater 16(1):58–61
29. Tang XZ, Kumar P, Alavi S, Sandeep KP (2012) Recent advances in biopoly mers and biopolymer-based nanocomposites for food packaging materials. Crit Rev Food Sci Nutr 52:426–442
30. Acay H, Baran MF (2019) Biosynthesis and characterization of silver nanoparticles using king oyster (Pleurotus eryngii) extract: effect on some microorganisms. Appl Ecol Environ Res 17(4):9205–9214
31. Baran MF, Yildirim A, Acay H, Keskin C, Aygun H (2020) Adsorption performance of *Bacillus licheniformis* sp. bacteria isolated from the soil of the Tigris River on mercury in aqueous solutions. Int J Environ Anal Chem xx:1–16
32. Mobarak M, Mohamed EA, Selim AQ, Eissa MF, Seliem MK (2019) Experimental results and theoretical statistical modeling of malachite green adsorption onto MCM–41 silica/rice husk composite modified by beta radiation. J Mol Liq 273:68–82
33. González-López ME, Pérez-Fonseca AA, Arellano M, Gómez C, Robledo-Ortíz JR (2020) Fixed-bed adsorption of Cr(VI) onto chitosan supported on highly porous composites. Environ Technol Innov 19:100824
34. Goswami M, Das AM (2019) Synthesis and characterization of a biodegradable cellulose acetate-montmorillonite composite for effective adsorption of Eosin Y. Carbohyd Polym 206:863–872
35. Teow YH, Kam LM, Mohammad AW (2018) Synthesis of cellulose hydrogel for copper (II) ions adsorption. J Environ Chem Eng 6(4):4588–4597
36. Wan Y, Wang J, Gama M, Guo R, Zhang Q, Zhang P, Yao F, Luo H (2019) Biofabrication of a novel bacteria/bacterial cellulose composite for improved adsorption capacity. Compos A 125:105560

37. Wang Q, Wang Y, Chen L (2019) A green composite hydrogel based on cellulose and clay as efficient absorbent of colored organic effluent. Carbohyd Polym 210:314–321
38. Xiong R, Wang Y, Zhang X, Lu C (2014) Facile synthesis of magnetic nanocomposites of cellulose@ultrasmall iron oxide nanoparticles for water treatment. RSC Advances 4(43):22632
39. Wang R, Zhang X, Zhu J, Bai J, Gao L, Liu S, Ji T (2020) Facile preparation of self-assembled chitosan-based composite hydrogels with enhanced adsorption performances. Colloids and Surfaces a: physicochemical and Engineering Aspects xxx:124860.
40. Lv X, Liu Y, Zhang J, Zhao M, Zhui K (2019) Study on the adsorption behavior of glutaric acid-modified Pb(II) imprinted chitosan-based composite membrane to Pb(II) in aqueous solution. Mater Lett 251:172–175
41. Wu S, Wang F, Yuan H, Zhang L, Mao S, Liu X, Alharbi NS, Rohani S, Lu J (2018) Fabrication of xanthate-modified chitosan/poly(N-isopropylacrylamide) composite hydrogel for the selective adsorption of Cu(II), Pb(II) and Ni(II) metal ions. Chem Eng Res Des 139:197–210
42. Islam MN, Khan MN, Mallik AK, Rahman MM (2019) Preparation of bio-inspired trimethylsilyl group terminated poly(1-vinylimidazole)-modified-chitosan composite for adsorption of chromium (VI) ions. J Hazard Mater 379:120792
43. Kenawy IMM, Eldefrawy MM, Eltabey RM, Zaki EG (2019) Melamine grafted chitosan-montmorillonite nanocomposite for ferric ions adsorption: Central composite design optimization study. J Cleaner Prod 241:118189
44. Sahnoun S, Boutahala M (2018) Adsorption removal of tartrazine by chitosan/polyaniline composite: kinetics and equilibrium studies. Int J Biol Macromol 114:1345–1353
45. Alver E, Metin AÜ, Brouer F (2020) Methylene blue adsorption on magnetic alginate/rice husk bio-composite. Int J Biol Macromol 154:104–113
46. Edathil AA, Pal P, Kannan P, Banat F (2020) Total organic acid adsorption using alginate/clay hybrid composite for industrial lean amine reclamation using fixed-bed: parametric study coupled with foaming. Int J Greenhouse Gas Control 94:102907
47. Dai WJ, Wu P, Liu D, Hu J, Cao Y, Liu TZ, Okoli CP, Wang B, Li L (2020) Adsorption of Polycyclic Aromatic Hydrocarbons from aqueous solution by Organic Montmorillonite Sodium Alginate Nanocomposites. Chemosphere 251:126074
48. Guo Z, Li Q, Li Z, Liu C, Liu X, Liu Y, Dong G, Lan T, Wei Y (2020) Fabrication of efficient alginate composite beads embedded with N-doped carbon dots and their application for enhanced rare earth elements adsorption from aqueous solutions. J Colloid Interface Sci 562:224–234
49. Oussalah A, Boukerroui A, Aichour A, Djellouli B (2019) Cationic and anionic dyes removal by low-cost hybrid alginate/natural bentonite composite beads: adsorption and reusability studies. Int J Biol Macromol 124:854–862
50. Li P, Gao B, Li A, Yang H (2020) Evaluation of the selective adsorption of silica-sand/anionized-starch composite for removal of dyes and Cupper(II) from their aqueous mixtures. Int J Biol Macromol 149:1285–1293
51. Fu C, Zhang H, Xia M, Lei W, Wang F (2020) The single/co-adsorption characteristics and microscopic adsorption mechanism of biochar-montmorillonite composite adsorbent for pharmaceutical emerging organic contaminant atenolol and lead ions. Ecotoxicol Environ Saf 187:109763
52. Yildirim A, Acay H, Baran F (2020) Synthesis and characterisation of mushroom-based nanocomposite and its efficiency on dye biosorption via antimicrobial activity. International Journal of Environmental Analytical Chemistry xx:1–18
53. El Mouchtari EM, Daou C, Rafqah S, Najjar F, Anane H, Piram A, Hamade A, Briche S, Wah-Chung PW (2020) TiO_2 and activated carbon of Argania Spinosa tree nutshells composites for the adsorption photocatalysis removal of pharmaceuticals from aqueous solution. J Photochem Photobio A: Chem 388:112183
54. Aichour A, Zaghouane-Boudiaf H (2019) Single and competitive adsorption studies of two cationic dyes from aqueous mediums onto cellulose-based modified citrus peels/calcium alginate composite. Int J Bio Macromolecules 154:1227–1236
55. Srasri K, Thongroj M, Chaijiraaree P, Thiangtham S, Manuspiya H, Pisitsak P, Ummartyotin S (2018) Recovery potential of cellulose fiber from newspaper waste: an approach on magnetic cellulose aerogel for dye adsorption material. Int J Biol Macromol 119:662–668

56. Zhang Y, Jiang Y, Han L, Wang B, Xu H, Zhong Y, Zhang L, Mao Z, Sui X (2018) Biodegradable regenerated cellulose-dispersed composites with improved properties via a pickering emulsion process. Carbohyd Polym 179:86–92
57. Barbosa RFS, Souza AG, Maltez HF, Rosa DS (2020) Chromium removal from contaminated wastewaters using biodegradable membranes containing cellulose nanostructures. Chem Eng J 395:125055
58. Kalka S, Huber T, Steinberg J, Baronian K, Müssig J, Staiger MP (2014) Biodegradability of all-cellulose composite laminates. Compos: Part A 59:37–44
59. Abd El-Aziz ME, Kamal KH, Ali KA, Abdel-Aziz MS, Kamel S (2018) Biodegradable grafting cellulose/clay composites for metal ions removal. Int J Biol Macromol 118:2256–2264
60. Alves DCS, Coseglio BB, Pinto LAA, Cadaval TRS (2020) Development of Spirulina/chitosan foam adsorbent for phenol adsorption. J Mol Liq 309:113256
61. Islam MN, Khan MN, Mallik AK, Rahman MM (2019) Preparation of bio-inspired trimethylsilyl group terminated poly(1-vinylimidazole)-modified-chitosancomposite for adsorption of chromium (VI) ions. J Hazard Materi 379:120792
62. Wan Y, Wang J, Gam M, Guo R, Zhang Q, Zhang P, Yao F, Luo H (2019) Biofabrication of a novel bacteria/bacterial cellulose composite for improved adsorption capacity. Compos a Appl Sci Manuf 125:105560
63. Aygün A, Uslu MK, Polat S (2017) Effects of starch sources and supplementary materials on starch based foam trays. J Polym Environ 25(4):1163–1174
64. Ji Y, Lin X, Yu J (2020) Preparation and characterization of oxidized starch-chitosan complexes for adsorption of procyanidins. LWT 117:108610
65. Koh LD, Yeo J, Lee YY, Ong Q, Han M, Tee BCK (2018) Advancing the frontiers of silk fibroin protein-based materials for futuristic electronics and clinical wound-healing (Invited review). Mater Sci Eng C 86:151–172
66. Wang X, Liu Z, Zhang T (2017) Flexible sensing electronics for wearable/attachable health monitoring. Small 13:1602790–16027108
67. Choudhury MR, Debnath KN (2020) Analysis of tensile failure load of single-lap green composite specimen welded by high-frequency ultrasonic vibration. Mater today: Proc 28:739–744
68. Tolvanen J, Hannu J, Hietala M, Kordas K, Jantunen H (2019) Biodegradable multiphase poly(lactic acid)/biochar/graphite composites for electromagnetic interference shielding. Compos Sci Technol 181:107704
69. Frackowiak S, Ludwiczak J, Leluk K, Orzechowski K, Kozlowski M (2015) Foamed poly(lactic acid) composites with carbonaceous fillers for electromagnetic shielding. Mater Des 65:749–756
70. Liakos IL, Mondini A, Filippeschi C, Mattoli V, Tramacere F, Mazzolai B (2017) Towards ultra-responsive biodegradable polysaccharide humidity Sensors. Mater Today Chem 6:1–12
71. Liu L, Liang H, Zhang J, Zhang P, Xu Q, Lu Q, Zhang C (2018) Poly(vinyl alcohol)/Chitosan composites: physically transient materials for sustainable and transient bioelectronics. J Cleaner Prod 195:786–795
72. Zhan M, Wool RP (2013) Design and evaluation of bio-based composites for printed circuit board applications. Compos: Part A 47:22–30
73. Guna VK, Murugesan G, Basavarajaiah BH, Ilangovan M, Olivera S, Krishna V, Reddy N (2016) Plant-based completely biodegradable printed circuit boards. IEEE Trans Electron Devices 63(12):4893–4898
74. Corradi S, Isidori T, Corradi M, Soleri F (2009) Composite boat hulls with bamboo natural fibres. Int J Mater Prod Technol 36(1):73–89
75. Guna V, Ilangovan M, Hu C, Venkatesh K, Reddy N (2019) In the construction industry, many additives are used to stabilize the soil, and the compressive strength of the soil can be improved using the correct stabilization method. Ind Crops Prod 131:25–31
76. Everaert M, Broos K, Nielsen P, Dun WV, Boone M, Quaghebeur M (2019) Carbonation is affecting biodegradability testing of fiber-cement composites. Sci Total Environ 686:888–892

77. Živković I, Fragassa C, Pavlović A, Brugo T (2017) Influence of moisture absorption on the impact properties of flax, basalt and hybrid flax/basalt fiber reinforced green composites. Compos B Eng 111:148–164
78. Chandekar H, Chaudhari V, Waigaonkar S (2020) A review of jute fiber reinforced polymer composites. Mater Today: Proc 26:2079–2082
79. Satyanarayana KG, Gregorio GCA, Fernando W (2009) Biodegradable composites based on lignocellulosic fibers-an overview. Prog Polym Sci 34:982–1021
80. Yoruç ABH, Uğraşkan V (2017) Green polymers and applications. Afyon Kocatepe Univ J Sci Eng 17(017102):318–337
81. Augustine R, Rajendran R, Cvelbar U, Mozetič M, George A (2013) Biopolymers for health, food, and cosmetic applications. Handbook of biopolymer-based materials: from blends and composites to gels and complex networks, pp 801–849
82. Yan YZ, An QD, Xiao ZY, Zheng W, Zhai SR (2017) Flexible core-shell/bead-like alginate@PEI with exceptional adsorption capacity, recycling performance toward batch and column sorption of Cr(VI). Chem Eng J 31:475–486
83. Mohammadi R, Azadmehr A, Maghsoudi A (2019) Fabrication of the alginate-combusted coal gangue composite for simultaneous and effective adsorption of Zn(II) and Mn(II). J Environ Chem Eng 7(6):103494

Chapter 15
Mechanical Properties of Flax-Cotton Fiber Reinforced Polymer Composites

Ashwin Sailesh and K. Palanikumar

1 Introduction

A natural fiber composite material consists of natural fibers and polymeric resin which are glued together under optimum operating conditions. A proper knowledge on the properties of reinforcing fiber, polymeric matrix, the process of fabricating the composite material, and proper bonding at the interface is a crucial aspect which contributes a lion's share in determining the properties of the material [1]. In general, a variety of natural fibers are available for the use of reinforcement. Among them, flax fiber proves to be the best in terms of mechanical properties, acoustical, and vibration properties. The most notable aspect of a flax fiber reinforced composite is that the composite material weight is less than that of glass fiber reinforced composite material [2].

Omkar Nath et al. [3] tested the behavior of polymer composites reinforced with jute-cotton fiber embedded in a polyester resin. The results of the various testing carried out revealed that input parameters such as fiber loading and fiber orientation are the critical parameters which have to be concentrated upon. Adding to this, the water absorption rate of the composite gradually elevates in accordance with fiber loading.

Sabinesh et al. [4] conducted experimental investigation on the behavior of cotton fiber reinforced isophthalic polyester composites under tension and flexural conditions. A range of fiber fractional volume of the reinforcing fiber was considered. The author reported that the tensile behavior of the composite material increases with

A. Sailesh · K. Palanikumar (✉)
Department of Mechanical Engineering, Sri Sai Ram Institute of Technology, Chennai 600044, India
e-mail: palanikumar_k@yahoo.com

A. Sailesh
e-mail: ashwinmech9191@gmail.com

S. Thomas and P. Balakrishnan (eds.), *Green Composites*,
Materials Horizons: From Nature to Nanomaterials,
https://doi.org/10.1007/978-981-15-9643-8_15

an increase in the fiber fractional volume. Menderes Koyuncu et al. [5] explored on the potentiality of the cotton fabric as a possible reinforcement in composite material and the effect of alkali treatment on composite material properties. The results revealed that the modification of the fabric by means of alkali treatment enhance the mechanical property significantly when compared with the untreated fabric. The studies also suggested that the cotton fabric can be utilized as potential reinforcement in composite material.

Pickering et al. [6] reviewed the recent developments in natural fiber composites and their mechanical performance. It was postulated that mechanical behavior of the natural fiber composites are in par with glass fiber. Ajith Gopinath et al. [7] experimented on the mechanical properties of natural fiber reinforced composite material with jute fiber as the reinforcement and two different polymeric matrices, namely epoxy resin and polyester resin. The reinforcing fiber was chemically modified by using proper quantity of sodium hydroxide solution. The behavior of the prepared composite material was explored by subjecting the material to destructive testing. The author concluded that the strength of jute-epoxy composite material was superior to that of jute-polyester composites indicating a poor bonding in the interfacial region of the latter.

Cai et al. [8] explored the process of fiber modification and its influence on the interfacial bonding and the resulting composite material properties. Abaca fibers and epoxy resin were considered as the reinforcement and matrix, respectively. The abaca fibers were subjected to NaOH treatment. A varied concentration of NaOH was utilized for this purpose. It was concluded that the interfacial bonding was superior at a low concentration of chemical treatment when compared to a higher concentration. Also, the chances of fibrillation to occur are more at elevated levels of NaOH concentration.

Yahaya et al. [9] experimented on the layering pattern of the fiber on the behavior of the composite material. Kevlar and kenaf fibers were reinforced in the epoxy matrix. Two different sequences of layering of the reinforcing fibers were followed. In the first sequence, kenaf fiber was considered as the skin layer, and in the other, kevlar was the skin layer. The fabricated composite samples were tested for their mechanical properties, and the test results showed some significant trends. The composite material with kevlar as the skin layer possessed superior tensile and flexural strength, whereas the impact strength was not satisfactory. On the other hand, the behavior in the kenaf reinforced sample was in contrast. It was concluded that the behavior exhibited by the material depends on the layering sequence of the reinforcement.

Sood et al. [10] reviewed the effect of fiber treatment on the flexural properties of natural fiber reinforced composite materials and postulated that alkali treatment of natural fibers along with the inclusion of coupling agents improves the flexural behavior. The review also suggested that the treatment of the matrix material with the aid of coupling agents can enhance the overall characteristics of the composite material. Fragassa et al. [11] explored on the mechanical behavior of flax and basalt fiber embedded in a vinyl ester. Composite samples were destructively tested so as to evaluate the behavior of the material during the application of mechanical loading conditions. The test results concluded that in a hybrid natural fiber reinforced composite material, the bonding between the different fibers and the matrix material is a crucial aspect that has to be concentrated upon. The good bonding between the different fibers and the matrix material can ameliorate the mechanical properties of the composite material.

Ku et al. [12] studied the tensile behavior of a range of natural fiber composites. It was concluded that the tensile properties of a natural fiber reinforced composite material depends largely on the interfacial adhesion between the reinforcement and the matrix material. In addition to this interesting aspect, the author evidenced that as the fiber content increases the tensile strength of the composite follows the same trend up to an optimum level and then follows a decreasing trend. Ramesh et al. [13] experimented on the characteristics and fabricating techniques of flax fiber reinforced polymer matrix composites. An intensive exploration of the features of the flax fiber composites was made, and it was revealed that the modulus of elasticity of the flax fiber followed a decreasing trend with an elevation in the flax fiber diameter. This shows that an optimum fiber diameter is essential to achieve a satisfactory modulus value. It was also suggested that modification of the fiber surface has a negative influence on the thermal conductivity and diffusivity to a great extent.

Aruchamy et al. [14] fabricated and experimented on the mechanical properties of woven natural fiber reinforced polymeric composites materials. Cotton and bamboo fibers were weaved as a single fabric mat and embedded in the epoxy resin. Hot press compression molding technique was employed for the fabrication. Analysis of the mechanical behavior and interfacial relationship was carried out as per American standard specifications. The author highlighted that tensile behavior of the prepared sample follows an increasing trend for a certain fiber weight percentage, after which there was a gradual dip in the tensile strength, which is an important observation. The same trend was observed in the case of flexural properties as well. The result also revealed that the hybridization of fibers weaved in a single fabric would yield better results compared to the non-hybridization of fibers.

Boopathi et al. [15] explored the various mechanical and physical characteristics of untreated and chemically modified Borassus fruit fiber. Alkali treatment of the reinforcing fiber using sodium hydroxide solution was performed. The influence of alkali treatment on the fiber was studied. The test results revealed that the fiber-matrix interface was enhanced on alkali treatment of the fiber. Also, the alkali treatment of the fiber led to the formation of a strong hydrogen bond, which ameliorates the fiber's properties and could be used as reinforcement in a composite material. On a similar note, Tran Huu Nam et al. [16] evaluated the influence of alkali treatment on the behavior of coir fiber reinforced poly (butylene succinate) resin. It was evident that the chemical modification of coir fiber in a suitable alkali medium leads to the formation of strong mechanical interlocking between the reinforcement and the matrix material. This contributes to superior strength under tension.

Ramesh et al. [17] experimented on the characteristics of natural fiber composite material reinforced with sisal-jute-glass fibers in a polymer matrix. Composite samples were fabricated by the conventional hand lay-up technique. The fabricated composite material was tested according to the ASTM standards. The test results concluded that the combination of sisal, jute fiber, and glass fiber improved the overall mechanical property.

Ghosh et al. [18] prepared a natural fiber composite material using banana fiber as the reinforcing material and vinyl ester was distributed over the reinforcing material. The influence of the volume fraction of the fiber on the characteristics of the material was evaluated. The fabricated composite samples were according to the ASTM standards and were subjected to mechanical testing following the same guidelines. Based on the evaluation made, it was highlighted that an increase in the volume fraction of the reinforcing fiber, there was an unexpected dip in the tensile strength, which subsided and followed an increasing trend.

The amount of moisture absorbed by a composite material pose a vital role in determining the mechanical behavior and the application area of the composite material. Yamini et al. [19] experimented on the characteristics of the composite material with coir and aloe vera fibers as the reinforcing fiber embedded with epoxy resin. The composite material was prepared as per the American standards. The samples were tested for the quantity of moisture being absorbed and reported that the composite material absorbs more moisture due to the presence of more number of microvoids. The same phenomenon was also evidenced by Venkateshwaran et al. [20], wherein the formation of microvoids during the fabrication of the composite sample has a more considerable influence on the overall mechanical behavior of the composites. Hence, researchers prefer compression molding than the conventional technique of fabrication. The notable aspect in compression molding is the possibility of formation of microvoids is very minimal because of the distribution of the matrix material being uniform throughout. Hence, the mechanical behavior of the fabricated sample is precise. Also, the interfacial adhesion is one step ahead compared to those in the conventional hand lay-up technique of fabrication.

In a natural fiber composite material, the property of resistance to vibration and energy-absorbing capability is greatly influenced by the order in which the fibers are laid. This is evidenced by the experiment conducted by Senthil Kumar et al. [21]. This feature also contributes to the overall characteristics of the natural fiber composite, which is evident from the investigations made by Sathiskumar et al. [22]. Dhaka et al. [23] experimented on the influence of fiber volume fraction on the moisture absorption and the overall mechanical properties of the fabricated composites. The investigations revealed that the interfacial adhesion decreases with an elevation in the fiber volume fraction. As a result, there is a dip in the flexural properties of the composite. As the volume fraction of the reinforcing material increases, the cellulose content of the fabricated composite material increases. As a result, moisture-absorbing capacity of the composite material is increased.

Kushwaha et al. [24] experimented on natural fiber composites with bamboo fiber being reinforced with a polymeric resin. The fibers were subjected to chemical modification by pre-impregnation. Mechanical and water absorption properties were determined by subjecting the material to the required testing. The test revealed that pre-impregnation had a positive influence on the overall mechanical characteristics and the water-absorbing properties of the composite material. Hristov et al. [25] studied and made a comparison on the influence of aging on the strength retention capability of flax fiber reinforced composite material with glass fiber reinforced composite material. Adequate samples were fabricated. The fabricated composite samples were subjecting to varied environmental conditions for different durations. Then, the samples were subjected to mechanical loading to have a quantitative measure of the strength. The results posed that flax fiber reinforced composite material possesses a good strength retention property than the glass fiber reinforced composite material. Adding to this, flax fiber reinforced composite material proved to be environmentally friendly than that of glass fiber reinforced composite material.

Since the combination of cotton and flax as a single fabric mat has not been studied, this experimental work deals with the fabrication and evaluation of the behavior of the flax-cotton fiber reinforced polymer composite under different conditions of loading. The behavior of the material thus obtained will be correlated with the interfacial images.

2 Materials and Methods

2.1 Materials

Based on the literature studies carried out, flax and cotton fibers are used as the reinforcement in the current experimental work. Flax and cotton fibers possess superior mechanical properties. Hybridization of the reinforcement is performed by weaving both the reinforcing fibers into a single fabric mat. A fabric mat consists of two sets of thread or fiber which are woven together. They are technically termed as the weft and warp. The weft is a set of fibers placed horizontally, and the warp is a set of fibers placed vertically at the time of weaving using a power loom or handloom. Proper selection of the loom is significant; this decides the properties of the fabric mat. A power loom is preferred than the handloom due to a faster rate of production. Here, flax fiber is placed on the weft direction and the cotton fibers in the warp direction.

Moving on to the matrix phase, finely blended epoxy resin and hardener mixture are used as the polymeric matrix. Faster and more comfortable process of curing makes epoxy to top the list. Thorough mixing of the resin and the hardener under optimum condition is essential. The hardener acts as a catalyst and enhances the curing process. Curing is the most critical step, which is to be concentrated upon else the interfacial adhesion will be in vain.

2.2 Fabrication Method of the Composite Samples

Compression molding method was employed for the fabrication of the sample. Though the cost of fabricating by the conventional hand lay-up technique is less, the formation of microvoids and uneven distribution of the resin makes the hand lay-up

Table 1 ASTM standards and test specimen dimensions [26–29]

S. No.	Name of the testing	ASTM standards	Dimensions of the test specimen (in 'mm')
1	Tensile testing	ASTM D3039	250 × 25 × 3
2	Flexural testing	ASTM D790	127 × 13 × 3
3	Impact testing	ASTM D256	66 × 13 × 3

technique less preferable when compared to compression molding. A notable aspect of compression molding is that the rate at which curing is done is tremendously high when compared to the hand lay-up process, which makes the compression molding technique even more preferable than its competitor.

The main requirement in compression molding is that the reinforcement should be of the same size and shape of the mold. The size of the mold used in the present experimental work is 300 mm × 300 mm. Hence considering the permissible allowances, the woven fabric mat was cut as a perfect square shape of 270 mm × 270 mm. The number of layers of the fabric to be used depends on the sample's thickness as per the ASTM standards. Accounting to this fact, 14 layers of the woven fabric was used with epoxy matrix coated in between the fabric.

On completing the fabrication process, a proper check on the curing process was done by visual inspection. The test specimens were prepared in accordance with the American standards. The dimensions of the test specimen are tabulated below (Table 1). The dimensional sketch of the test specimen is shown from Figs. 1, 2 and 3.

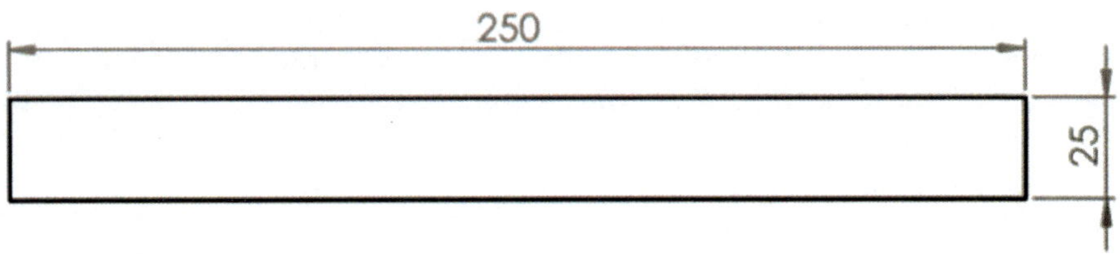

Fig. 1 Dimensional sketch of the tensile test specimen as per ASTM D3039

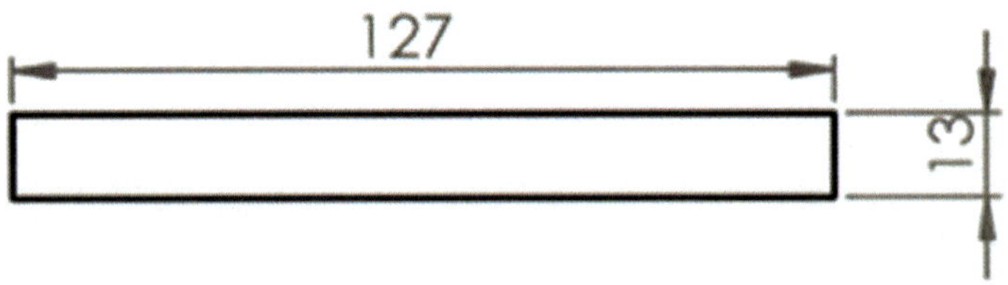

Fig. 2 Dimensional sketch of the flexural test specimen as per ASTM D790

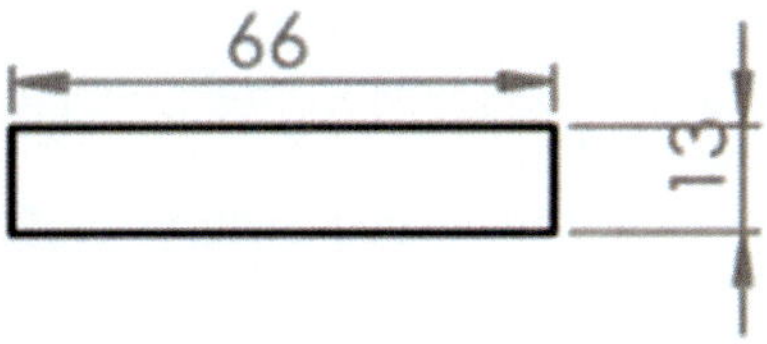

Fig. 3 Dimensional sketch of the impact test specimen as per ASTM D256

3 Mechanical Testing of the Composite Samples

On the successful preparation of the specimens, the mechanical tests were carried out. Tensile and the bending tests were performed using different fixture setups in a computer-controlled universal testing machine (UTM). Prepared test specimen was placed along the longitudinal direction in between the fixtures of the UTM. Load was applied as a pulling load, which created induced internal stress within the specimen and under the peak load or the maximum load, and the maximum strength of the specimen was recorded. This can be interpreted as the maximum load where the composite specimen can withstand under pulling load conditions (Fig. 4).

Flexural strength of the composite sample prepared was measured using the UTM with a standard three-point bending test setup attached to it. This arrangement looks

Fig. 4 Tensile test setup with the composite specimen loaded on to the UTM

similar to that of a simply supported beam with a point at the center. As the specimen is rested horizontally on the supports of the apparatus, the load was applied at the center of the test specimen. As the ends of the sample were arrested, the applied load caused maximum deflection in the center portion of the composite sample. The load corresponding to the maximum deflection or the displacement was recorded. This can be interpreted as the maximum load where the composite specimen can withstand under bending conditions (Fig. 5).

Impact testing of the composite material is conducted to determine the maximum energy absorbed by the material due to sudden shock loads. A composite material should possess sufficient impact strength so as to use it in the automobile industry. The prepared composite test specimen was placed on the work holder of the Charpy impact tester. The impact testing sample was an un-notched sample and was placed vertically on the Charpy impact tester (Fig. 6).

During the process of testing the fabricated composite samples, the chances of error occurrence are more. The sources of the errors are always uncertain. To account for the errors that could arise during the testing, it is always preferred to have five test specimens for each of the above-mentioned testing procedures. The average value of the results shall be considered for further processing.

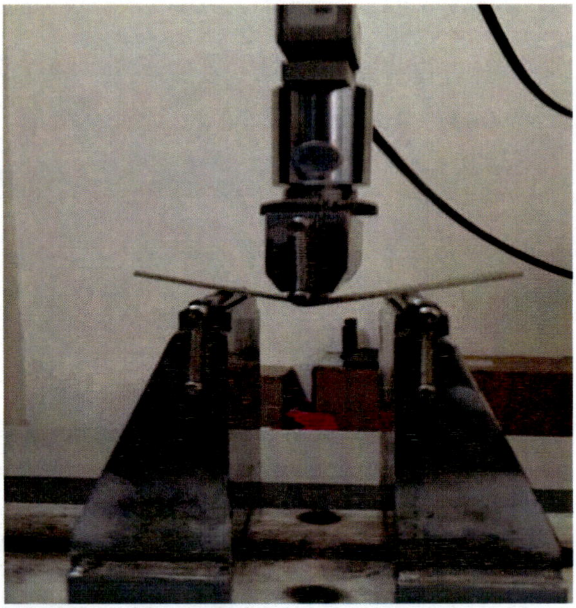

Fig. 5 Flexural test setup attached in the UTM with the composite specimen loaded

Fig. 6 Impact testing setup with the composite specimen loaded

4 Results and Discussion

The results of the mechanical testing carried out are tabulated below (Table 2). The detailed analysis is done subsequently (Fig. 7).

Table 2 Results of the mechanical testing

Sample No.	Ultimate tensile strength (MPa)	Flexural strength (MPa)	Impact strength (kJ/mm^2)
1	129.914	102.181	29
2	128.746	96.571	32
3	128.884	89.171	38
4	129.524	90.280	40
5	129.775	91.355	36
Average	129.368	93.911	35

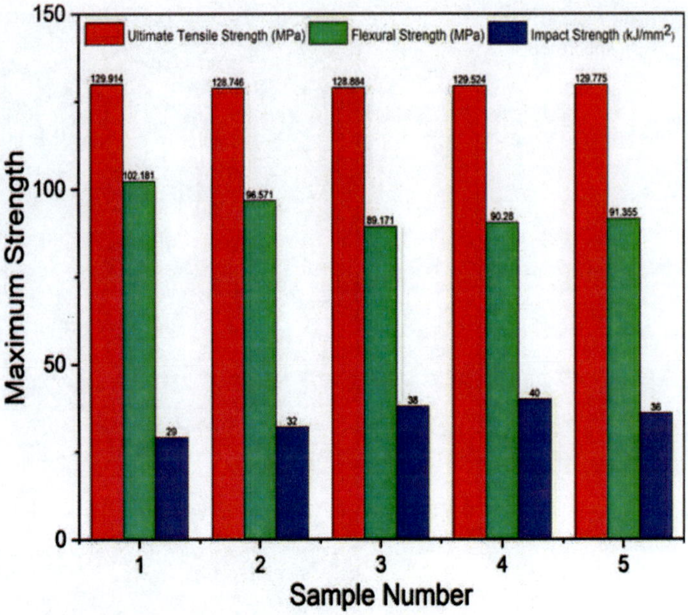

Fig. 7 Comparison of the results of the mechanical testing

5 Tensile Behavior

Factors such as the orientation of the fiber, fiber length, number of layers of the fabric material, the interfacial bonding, and the weaving pattern largely contribute to the tensile strength of the fabricated composite sample. In general, the maximum resistance offered by a material to external load is the ultimate tensile strength of that material, and this holds good for a natural fiber composite material as well. The average tensile strength and the average breaking load are recorded in Table 3 and Fig. 8, respectively. These values indicate that if the composite material is loaded beyond this threshold value will eventually lead to rupture, which is a sign of failure.

Table 3 Breaking load of the tested composite samples

Sample No.	Breaking load (N)
1	9743.55
2	9655.95
3	9666.30
4	9714.30
5	9733.12
Average	9702.64

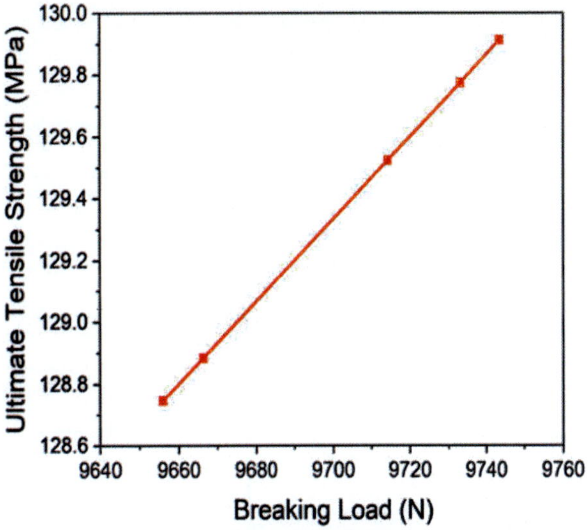

Fig. 8 Breaking load versus ultimate tensile strength

Knowledge about the mechanism of rupture is essential when applying the fabricated material to an industrial application. The reason for the rupture of the fabricated composite could be attributed to the weakening of the bonding between the reinforcement and the polymeric matrix. In the present experimental studies, as the fabricated material is loaded against the peak value of the internal resistance of the material, first, the matrix material, i.e., the epoxy resin, starts deteriorating. Hence, the interfacial adhesiveness gets gradually reduced. As a result of the loss of interfacial adhesiveness, formation of cracks in the form of microvoids takes place on the internal surface of the fabricated composite material and as the applied load increases the formed microvoids gets enlarged which is a sign of crack propagation, and at one particular point which is technically the UTS point, a peculiar phenomenon known as the fiber pull-out takes place which is a sign of complete failure of the fabricated material (Fig. 9).

As the test material is loaded in the longitudinal direction up to a particular amount of load, the fibers resist, but after the safe loading limit, the individual fibers in the longitudinal direction, i.e., the warp side, start expanding gradually. This can also be attributed to the weakening of the adhesive force. The weakening of the adhesive force gradually increases and the interfacial contact becomes loose. As a result, the warp set of fibers expands at a drastic rate, and this is termed as the % elongation of the composite material. Since the warp and the weft set of fibers were weaved in an interlocking manner, the weft set of fiber also loses its strength. At one particular point, a drastic pullover of the fiber takes place, and the corresponding load is known as the breaking load. Beyond this load, the material will surely fail under any operating circumstances (Fig. 10).

It can be interpreted from the plot (Fig. 11) that the maximum stress that the composite material can withstand is 129.914 MPa. On reaching this maximum stress

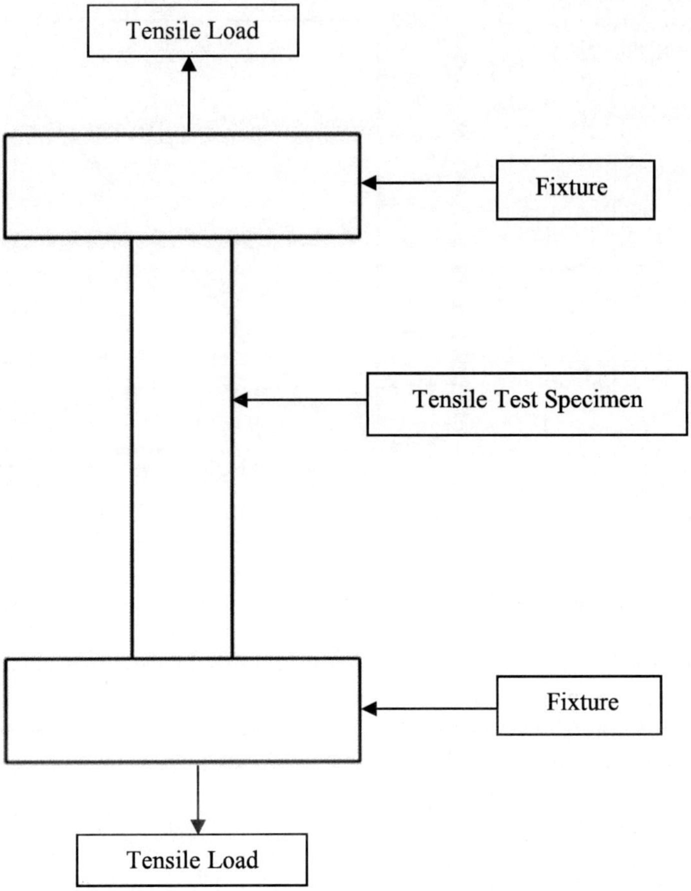

Fig. 9 Schematic representation of the test specimen for tensile testing

value, the fiber and the matrix get separated. Beyond this particular value, the material undergoes severe elongation, and the stress value dips drastically, which indicates that the material fails.

The debonding between the reinforcing material and the matrix is clearly indicated in the SEM image. Debonding is a phenomenon in which the adhesive loses its adhesiveness, and thus the interfacial contact between the reinforcement and the matrix becomes weak, paving its way to ultimate failure (Fig. 12).

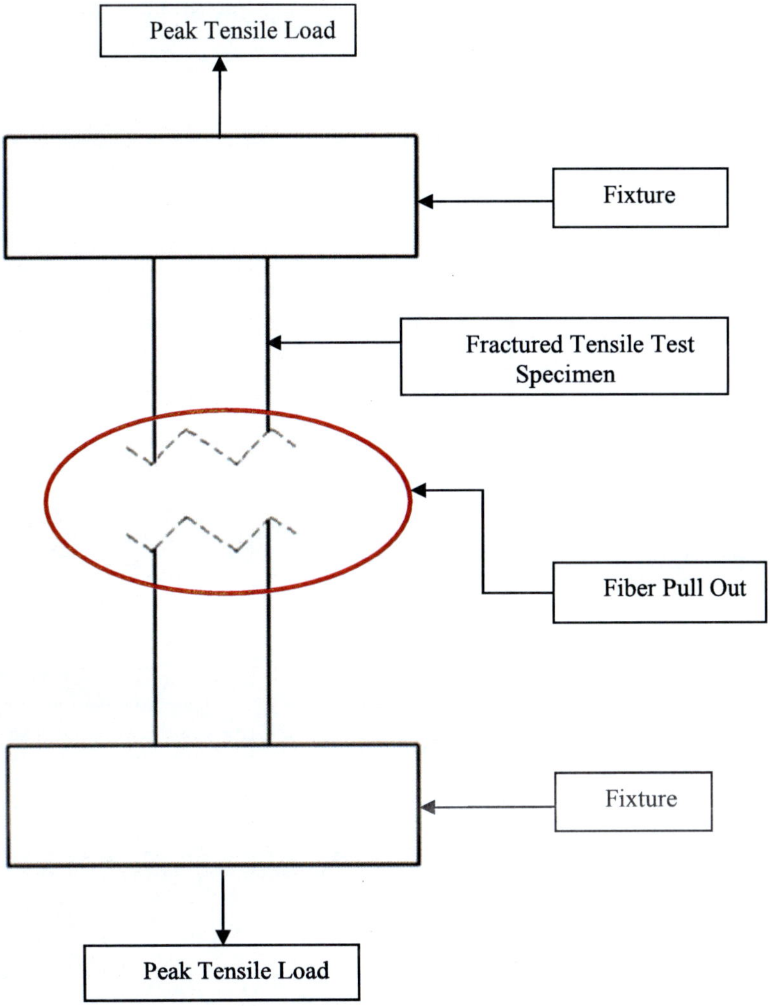

Fig. 10 Schematic representation of fiber pull-out

6 Flexural Behavior

Flexural strength of the composite material is a measure of the material to withstand bending loads. The flexural property of the material is equally important as that of the tensile property. A three-point flexure test attachment is attached to the UTM, and the test sample is placed on the supports. The load is given perpendicular to the fibers, i.e., in the transverse direction. As the material is loaded transversely, the failure mechanism in the flexural loading is a clear distinct featured when compared to the tensile testing.

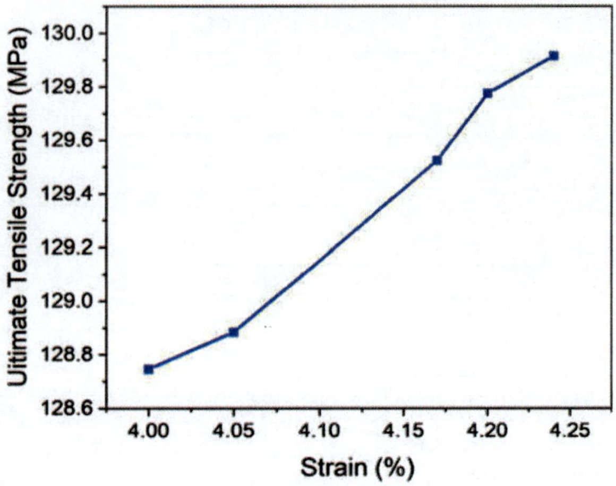

Fig. 11 Stress versus strain plot of the tensile tested sample

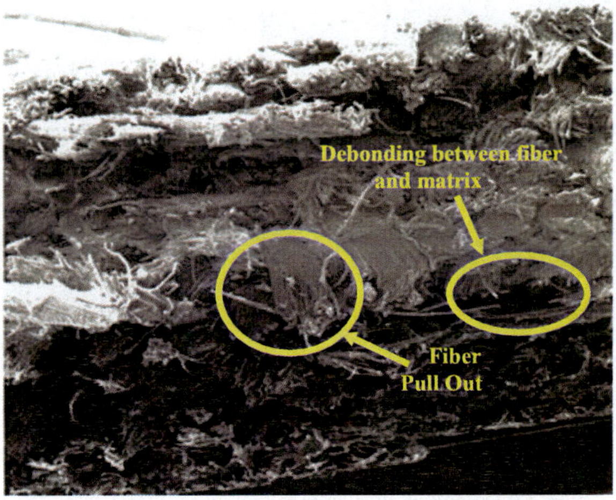

Fig. 12 Scanning electron microscopy image of the fractured tensile test specimen

In the current experimental work, the test specimens were tested for its behavior against a bending load, and the corresponding strength values are tabulated in Table 2. The average value of the flexural strength of 93.911 MPa indicates the maximum resistance that the fabricated composite material can offer to bend load beyond which the composite material will fail.

It is observed that the flexural strength of the fabricated composite material largely depends on the interfacial relationship. In order to establish a good interfacial relation, it is required that the resin should be uniformly distributed over the reinforcing fabric mat. In order to ensure proper and uniform distribution of the polymeric matrix, a

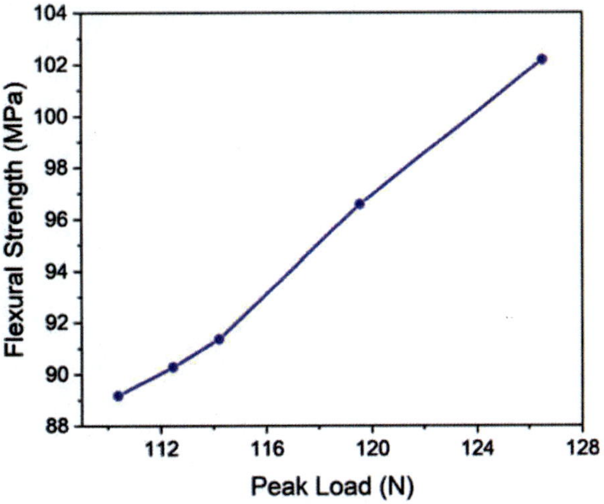

Fig. 13 Plot illustrating the maximum value of the load and flexural strength of the samples

pressure of 1500 psi was used in the compression molding setup during the fabrication of the composite (Fig. 13).

The failure mechanism in the flexural loaded sample varies that of the tensile loaded sample. In a flexural loaded sample, as the load increases gradually the formation of voids initiates. The size of the void initially being in the microscopic range propagates and thus leads to failure. An important aspect to be noted in a flexural test sample is that the microvoid propagates along the direction perpendicular to the applied bending load, which is the horizontal direction. As the load is applied transversely, the maximum bending displacement is observed in the bottom-most layer of the composite material. The SEM image in Fig. 14 clearly shows the presence of void in the composite sample. The propagation in the size of the void indicates the weakening of the interfacial strength; hence, the process of debonding between the fiber and the matrix takes place.

The flexural strength of the composite material also depends on the dimensional accuracy of the sample. In a three-point flexure test, a certain portion of the material will be overhanging away from the supports of the flexure attachment. So, the effective test area is the area within the supports. As this effective test area is vital, the dimensional accuracies at the time of sample preparation should be followed properly, failing, which will lead to erratic results, and the material selection process for an application will not be foolproof.

Adding to this, the type of fiber yarn shares an equal contribution to the final flexural strength of the composite. Based on the literature review made, it was evident that yarn of 30 s count proves to be beneficial, and hence the same count of cotton fiber and the flax fiber was used during the fabrication of the composite samples. The variation in the flexural strength values as recorded in Table 3 can be attributed to the uncertain minor errors that would have been aroused due to dimensional instability during the preparation of the test specimen. Hence, it is always safe to have a larger

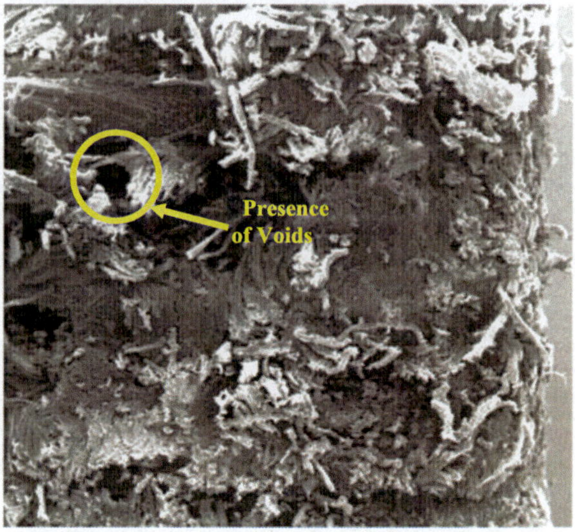

Fig. 14 Scanning electron microscopy image of the fractured flexural test specimen

number of experimental runs so that the overall value of a particular property could be asserted from its average value.

7 Impact Behavior

Determining the material behavior toward sudden shock load is important as most of the materials are subjected to impact loads in the practical scenario. The impact testing of the fabricated composite samples was tested using a Charpy impact tester. The impact strength of the fabricated composite sample is recorded in Table 2.

During an impact load condition, a material, in general, should absorb maximum energy possible before failure. This measure is significant in developing real-time products that are vulnerable to shock loads such as a spare part for an automobile. With reference to the present work, the values of the impact strength of the fabricated composites indicate that only a minimum amount of energy is absorbed before failure, which is a severe concern, and evaluating the cause for the same is important. The decrease in the impact strength can be attributed to the length of the fiber being pulled out during the destructive impact testing. Fiber pull-out is the common mechanism of failure in the case of a fiber reinforced polymeric composites (Fig. 15).

In the case of epoxy-based fiber composites, the length of the fiber pull-out is less owing to the fact that epoxy resin and hardener mixture when properly distributed over the reinforcement establishes a superior interfacial bonding. Hence, as the interfacing bonding is more, the length of the fiber pull-out is less and hence attributed to the decline in the impact property. This can be elevated by a perfect bonding between the

Fig. 15 Scanning electron microscopy image of the fractured impact test specimen

reinforcement and the matrix. Dimensional stability throughout the sample is very important in computing the impact properties of the composite sample.

8 Conclusion

In the present experimental work, flax-cotton fiber reinforced polymer composite material was fabricated by compression molding technique. Hybridization of the fibers was followed where the flax and the cotton fibers were weaved as a single fabric mat with the cotton fiber placed in the warp direction and the flax in the weft direction. Keeping the quality of the fabric mat as the top most priority, the fibers were weaved using a power loom.

Epoxy resin blended with a suitable hardener (HY951) mixed in an optimum ratio was utilized as the continuous phase (matrix). Mechanical testing of the prepared samples was carried out to evaluate the behavior of the material subjected to tension, flexural and impact load conditions. Taking the various errors that may arise during testing into consideration, the average of five test specimens was taken as the best fit for all the mechanical testing. The average tensile strength of the composite material was recorded at 129.914 MPa, the flexural strength at 93.911 MPa, and the impact strength at 35 kJ/mm^2.

The interfacial studies of the fractured samples were carried out. Based on the analysis of the mechanical testing and the SEM images, the following interpretations are noted, as shown below.

1. Interfacial bonding contributes a major share in influencing the mechanical characteristics of the fabricated material. The relationship between the reinforcement and the matrix material should be established properly. Many factors contribute

to a good interfacial bonding. Uniform distribution of the matrix tops the list of factors contributing to interfacial bonding. A wise selection of techniques of fabrication ensures a uniform distribution of the matrix. Hence, the compression molding technique was used for fabricating the composite samples. As a thumb rule, uniform distribution contributes to superior interfacial bonding and hence enhanced wettability property of the matrix. Wettability is an important property of a matrix that is used to establish and maintain proper contact with the reinforcement.
2. The next important parameter to be considered is the type of fiber yarn. In this context, the tenacity of the yarn and the count of the fiber yarn are important. This contributes an equal share toward the overall behavior of the fabricated material. It is always wise to use bleached yarn for the weaving of the fibers. Bleaching is a process in which the impurities present in the normal yarn are removed, and hence proper weaving of the fibers is possible. Chemical treatment of the fiber after weaving is also important as this modifies the surface roughness of the fiber and ensures superior contact with the matrix. Chemical modification using NaOH solution and silane treatment tops the chart.
3. Modification of the matrix material by the addition of polymer enhancing material has a wide scope in improving the properties of the composite material.
4. Dimensional stability of the composite material is an important factor to be taken care of. Dimensional instability may arise during the sample preparation for mechanical testing. Hence, the proper care should be given to ensure a constant property.
5. Proper selection of warp and weft side fibers is essential during hybridization. As previously stated, warp refers to the vertical side of the fabric, and the weft refers to the horizontal side of the fabric. Hence, it is wise to place a strong fiber in the longitudinal direction (warp) and the other in the transverse direction (weft). Proper interlocking between the warp and weft has to be ensured during the weaving of the fabric.

References

1. Sailesh A, Arunkumar R, Saravanan S (2018) Mechanical properties and wear properties of kenaf-aloe vera-jute fiber reinforced natural fiber composites. Mater Today Proc 5:7184–7190
2. Yan L, Chouw N, Jayaraman K (2014) Flax fiber and its composites—a review. Compos: Part B, 56:296–317
3. OmkarNath, Ziaulhaq M (2017) Investigating the mechanical properties of polyester natural fiber composite. Int Res J Eng Technol 4(7):3502–3506
4. Sabinesh S, Thomas Renald CJ, Sathish S (2014) Investigation on tensile and flexural properties of cotton fiber reinforced isophthallic polyester composites. Int J Current Eng Technol 2:213–219
5. Koyuncu M, Karahan M, Karahan N, Shaker K, Nawab Y (2016) Static and dynamic mechanical properties of cotton/epoxy green composites. Fibres Text Eastern Eur 24, 4(118):105–111
6. Pickering KL, Aruan Efendy MG, Le TM (2016) A review of recent developments in natural fibre composites and their mechanical performance. Compos A Appl Sci Manuf 83:98–112

7. Gopinath A, Senthil Kumar M, Elayaperumal A (2014) Experimental investigations on mechanical properties of jute fiber reinforced composites with polyester and epoxy resin matrices. Procedia Eng 97:2052–2063
8. Cai M, Takagi H, Nakagaito AN, Li Y, Waterhouse GIN (2016) Effect of alkali treatment on interfacial bonding in abaca fiber reinforced composites. Compos: Part A, 90:589–597
9. Yahaya R, Sapuan SM, Jawaid M, Leman Z, Zainudin ES (2015) Effect of layering sequence and chemical treatment on the mechanical properties of woven kenaf-aramid hybrid laminated composites. Mater Des 67:173–179
10. Sood M, Dwivedi G (2018) Effect of fiber treatment on flexural properties of natural fiber reinforced composites: a review. Egypt J Pet 27:775–783
11. Fragassa C, Pavlovic A, Santulli C (2018) Mechanical and impact characterisation of flax and basalt fibre vinylester composites and their hybrids. Compos Part B, 137:247–259
12. Ku H, Wang H, Pattarachaiyakoop N, Trada M (2011) A review on the tensile properties of natural fiber reinforced polymer composites. Compos: Part B, 42:856–873
13. Ramesh M (2019) Flax (*Linum usitatissimum* L.) fibre reinforced polymer composite materials: a review on preparation. Propert Prospects Prog Mater Sci 102:109–166
14. Aruchamy K, Subramani SP, Palaniappan SK, Sethuraman B, Kaliyannan GV (2020) Study on mechanical characteristics of woven cotton/bamboo hybrid reinforced composite laminates. J Mater Res Technol 9(1):718–726
15. Boopathi L, Sampath PS, Mylsamy K (2012) Investigation of physical, chemical and mechanical properties of raw and alkali treated borassus fruit fiber. Compos B 43:3044–3052
16. Huu Nam T, Ogihara S, Tung NH, Kobayashi S (2011) Effect of alkali treatment on interfacial and mechanical properties of coir fiber reinforced poly (Butylene Succinate) biodegradable composites. Compos: Part B, 42:1648–1656
17. Ramesh M, Palanikumar K, Hemachandra Reddy K (2013) Mechanical property evaluation of sisal-jute-glass fiber reinforced polyester composites. Compos: Part B, 48:1–9
18. Ghosh R, Reena G, Rama Krishna A, Lakshmipathi RB (2011) Effect of fiber volume fraction on the tensile strength of Banana fiber reinforced Vinyl Ester resin composites. Int J Adv Eng Sci Technol 4(1):89–91
19. Yamini S, Shanugasundaram K (2015) Study on the mechanical behaviour of natural based composite using coir and aloe vera. IJETCSE 13(4):278–280
20. Venkateshwaran N, Elayaperumal A, Sathiya GK (2012) Prediction of tensile properties of hybrid-natural fiber composites. Compos: Part B, 43:793–796
21. Senthil Kumar K, Siva I, Rajini N, Winowlin Jappes JT, Amico SC (2016) Layering pattern effects on vibrational behaviour of coconut sheath/banana fiber hybrid composites. Mater Des 90:795–803
22. Sathishkumar TP, Navaneethakrishnan P, Shankar S, Kumar J (2012) Mechanical properties of randomly oriented snake grass fiber with banana and coir fiber-reinforced hybrid composites. J Compos Mater 47(18):2181–2191
23. Dhakal HN, Zhang ZY, Richardson MOW (2007) Effect of water absorption on mechanical properties of hemp fiber reinforced unsaturated polymer composites. J Compos Technol 67:1674–1683
24. Pradeep K, Kushwaha RK (2011) Influence of chemical treatments on the mechanical and water absorption properties of bamboo fiber composites. J Reinf Plast Compos 30(1):73–85
25. Hristozov D, Wroblewski L, Sadeghian P (2016) Long-term tensile properties of natural fibre reinforced polymer composites: comparison of flax and glass fibers. Compos: Part B, 95:82–95
26. Balachandar M, Vijayaramnath B, Barath R, Bharath Sankar S (2019) Mechanical characterization of natural fiber polymer composites. Mater Today Proc 6(2):1006–1012
27. ASTM D3039 standard for tensile testing of composite—ASTM international
28. ASTM D790 standard for flexural testing of composite–ASTM international
29. ASTM D256 standard for impact testing of composite—ASTM international

Chapter 16
Green Composite Film Synthesized from Agricultural Waste for Packaging Applications

Shobhit Dixit and Vijay Laxmi Yadav

1 Introduction

The extensive uses of non-biodegradable plastics have crossed the permissible limits of plastics wastes and created environment imbalance condition all over the world. The incinerated or land filled plenty available agro-wastes released greenhouse gases, smoke and particulate matters directly in to the environment and breached environment laws. So, researchers should find the appropriate way in order to reduce the plastics wastes generation by using alternate biodegradable material such as agricultural wastes, i.e., wheat straw, rice straw, hemp, kenaf, jute, sisal, etc. [1–4]. The cellulosic property of agro-waste has motivated the researchers to synthesize biodegradable composites due to their renewable properties, provided remarkable mechanical and thermal stabilities as compared to non-biodegradable plastics. Generally, polymers such as polyethylene, polylactic acid, polypropylene, starch and agro-wastes have remarkable stability for green packaging applications. The present chapter explores the reliable use of abundantly available biodegradable agricultural wastes for packaging applications such as food, sprout and active packaging's. The perfect use of agricultural wastes as a reinforcing agent in polymer matrix is the eye opener for current researchers. Many researchers promoted the use of agro-waste to synthesize biodegradable packaging film due to their cheap, lightweight, high tensile strength and benchmark thermal stability. Enhanced cellulose percentage area in agro-waste encourages suitability of this fiber to minimize the contribution of polymers with green agro-waste. Thus, the suitable use of agro-waste to synthesize green composites for packaging application is need of the hour [5, 6].

S. Dixit · V. L. Yadav (✉)
Department of Chemical Engineering and Technology, IIT BHU, Varanasi, Uttar Pradesh 221005, India
e-mail: vlyadav.che@itbhu.ac.in

S. Thomas and P. Balakrishnan (eds.), *Green Composites*,
Materials Horizons: From Nature to Nanomaterials,
https://doi.org/10.1007/978-981-15-9643-8_16

Native-agro-waste incorporated polymeric films have low mechanical and impact strengths as compared to petroleum-derived materials due to their hydrophilic nature. Inherent composition (lignin and hemicellulose) does not permit agro-waste to reinforce the polymeric film successfully. This behavior of agro-waste showed incompatibility found between the materials resulting in low desirable properties present in synthesized film. Moreover, this signified the less mechanical stability of the agro-waste reinforced polymeric film. However, many chemical modification techniques have enhanced surface property of agro-waste for appropriate blending in polymer matrix. This strategy helped to remove undesirable contents from agro-waste and created some void openings at the surface for improving blending characteristics. Chemical, physical and biological treatments are some of the frequently used popular techniques for reducing recalcitrance nature of agro-waste. The findings of many scientists elucidated that the mechanical and thermal characteristics of modified agro-waste incorporated polymeric film had considerably higher as compared to native-agro-waste reinforced polymeric film [7]. So the surface advancement in agro-waste is an emerging field in the polymeric composite research (Fig. 1). The prepared composite film is reliable as compared to synthetic packaging film in terms of contact angle, water vapor migration rate, flexural strength, water vapor permeability, mechanical stability and impact strength for green composite packaging application.

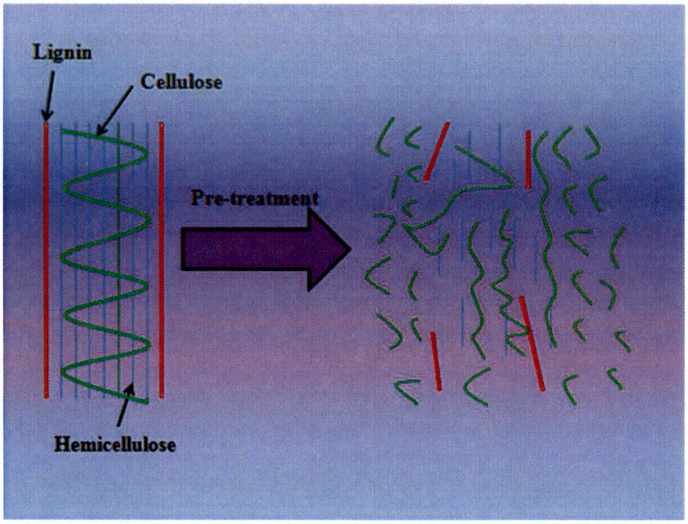

Fig. 1 Pretreatment of agricultural waste (*Source* Author)

1.1 Agro-waste

Agricultural wastes are generally produced from various from agricultural activities by the farmers. Agro-waste is waste which has no economic value. In India, the annual production of agro-waste is around 910 metric tons. Moreover, agro-waste has lightweight, readily available, comparatively low cost product and simple to process material. So, the safe disposal and better utilization of agro-waste are the key prospects for green life. Many scientists claimed that agro-waste had a property to use as filler in polymer matrix to increase composites tensile stability for various applications such as automotive, packaging, household and building industries [8]. Thus, it will build a new green environment for human life.

1.2 Classification of Agricultural Waste

In present juncture, the environment aspects encouraged researchers to increase the use of agro-waste as reinforcing agent in polymer matrix and maintain the environment balance condition. Agro-waste can be categorized in two ways, i.e., field residues and process residues (Fig. 2). Field residue is classified such as stems, stalks, leaves and seed pods and process residues are classified such as husks, seeds, roots and bagasse. [9]. Agro-wastes embedded products are required low maintenance with long life durable ability. However, these virtues are attracted industrialist to surging use of agro-waste for green composite preparation.

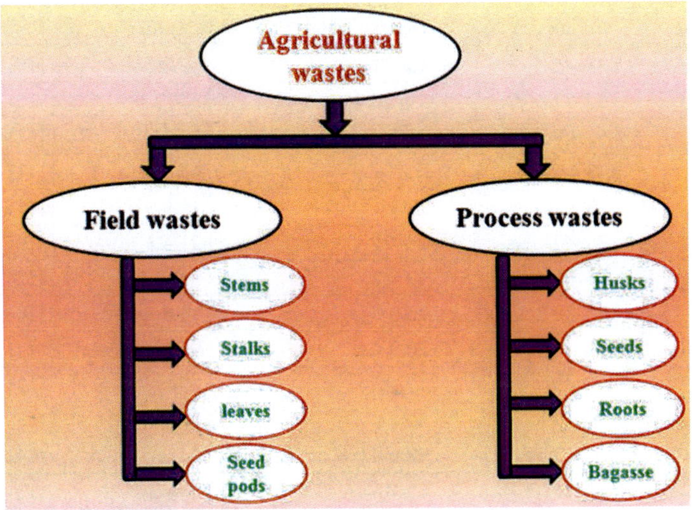

Fig. 2 Classification of agricultural wastes (*Source* Author)

1.3 Chemical Composition of Agricultural Waste

Agro-waste is composed of cellulose tightly embedded with hemicellulose protected by a strong wall of lignin. The percentage of cellulose, lignin and hemicellulose in agro-waste is varying with the type of agro-waste [9]. In order to attain the desirable properties, the higher cellulosic percentage in agro-waste has provoked researchers to use them for synthesizing green packaging film. The chemical composition of different types of agro-waste is depicted in Table 1.

2 Green Packaging Film Synthesized from Agricultural Waste

The required combination of agro-waste and polymers, i.e., green composites sustained a promising alternative to replace the non-biodegradable petroleum-derived materials for packaging applications. The main key factor is reduced the consumption of synthetic plastics, while researchers are moving toward renewable sources that are agricultural wastes to synthesize green plastics [4]. Recently, polymers are used for several applications like packaging, automotive and furniture goods. Thus, the use of agricultural wastes opens the way for biodegradable composite film and prepares a more sustainable waste-based packaging film that also helps to boost our economic sector. The various polymers such as polyethylene, polypropylene, polylactic acid

Table 1 Chemical composition of several types of agricultural waste

Agro-waste	Cellulose (%)	Hemicellulose (%)	Lignin (%)	References
Wheat straw	38–45	15–31	12–20	[10]
Hemp	57–77	14–22.4	3.7–13	[10]
Banana	53.45	28.56	15.46	[11]
Coir	43.44	0.25	45.84	[11]
Jute	59–61	22.1	15.9	[11]
Seed flax	43–47	24–26	21–23	[10]
Sisal	65–68	10–22	9.9–14	[12]
Bagasse	40–55.2	25.3	16.8	[10]
Kenaf	72	20.3	9	[13]
Rice husk	35–45	19–25	20	[13]
Rice straw	41–57	33	8–19	[13]
Flax	62–72	18.6–20.6	2–5	[14]
Ramie	68–76	13–16	0.6–0.7	[13]
Bamboo	26–65	30	5–31	[14]

Source Author

and polyvinyl alcohol have been frequently used for the packaging sector. Moreover, many agricultural wastes, i.e., wheat straw, hemp fiber, rice straw, kenaf, banana, jute etc. have been used as a filler in polymer matrix due to their considerable mechanical strength [4]. Although, the use of biocomposites for packaging applications made from lignocellulosic fiber is the most adaptable technique for green market (Table 2).

3 Some Improving Interfacial Interaction Techniques for Polymer/Agro-waste Composites

Pretreatment is the imperative technique involved in the effective reinforcement of agricultural waste in the polymer matrix. Several strategies are available such as physical, chemical, biological and physiochemical treatments for enhancing the interfacial interaction between fiber and polymer [30]. In the present scenario, these strategies have been reported in many literatures to attain the higher mechanical and thermal characteristics of the agro-waste incorporated green composite film (Fig. 3). Moreover, literature confirmed that treated-agro-waste was a promising substitute of polymer to synthesize green composites for packaging applications and serve as a potential alternative of non-biodegradable petroleum-derived materials [31]. In other words, these interfacial interaction improving techniques are the possible solution of poor mechanical stability of native agricultural waste-based polymer composite film.

3.1 Physical Pretreatment

In order to change the specific surface area or degree of polymerization of agro-waste, many physical treatment processes such as grinding, milling and chipping machines are used. The main focus of using this treatment is to avoid the wastage of chemicals use for requiring changes in lignocellulosic biomass. This treatment has reduced the particle size that increases the bulk density of treated-agro-waste and makes it appropriate for blending in the polymer matrix. The main disadvantage of this treatment is required higher energy consumption to operate the process [32].

3.2 Chemical Pretreatment

This pretreatment involved the reaction changes in agricultural waste. This treatment is easily removed undesirable materials (hemicellulose and lignin) from agro-waste and increases its cellulose digestibility for polymer adhesion. Many chemicals such

Table 2 Green packaging films synthesized from agricultural waste

Raw materials (reference)	Parameters	Pre-treatment	Method
Polystyrene (PS) [15]	Wheat straw (WS) (0–20%)	Chemical pre-treatment	Solvent casting method
Polyethylene (PE), polypropylene (PP) [4]	Wheat straw (0–20%)	Alkali treatment	Solvent casting method
Starch, polyvinyl alcohol (PVA) [16]	Banana fiber (BF) (0–30%)	NaOH treatment	Solvent casting method
Polylactic acid (PLA) [17]	Kenaf (KF)	Silane, alkaline treatment	Melt blending method
Cinnamon essential oil (CE), Bisphenol-A-based benzoxazine (BOZ) [5]	Hemp fiber (HF) (6–24%)	NaOH treatment	Hydraulic hot-press
PLA [18]	Glycidyl methacrylate-grafted (GMA) (1%), tert-butyl perbenzoate (TBPB) (0.5%), hemp hurd (HH) (10–30%)		Extrusion and injection
High density polyethylene(HDPE) [19]	HF (10–30%), BF (10–40%), maleic anhydride modified high density polyethylene (MAPE) (0.8–1.2%)		Injection
PP [20]	HF (5–15%)	NaOH treatment	Hand lay-up

(continued)

Table 2 (continued)

Raw materials (reference)	Parameters	Pre-treatment	Method
Polyhydroxy butyrate (PHB) [21]	Almond shell, rice husk, seagrass (10–20%)	NaOH treatment	Melt blending
Polylactic acid (PLA) [22]	WS (0–30%)	Solid state shear milling process (SSSM)	Extrusion
Green epoxy [23]	Sisal	Mechanical pressing	Unidirectional single layer laminates
Polyvinyl alcohol [24]	Rice straw (RS) (1–3%)	NaOH treatment	Solvent casting method
Hydroxy propyl methyl cellulose [25]	Jute fiber (0–3%)		Solvent casting method
Starch [26]	Rice husk fiber (2–10%)	Sulphuric acid treatment	Solvent casting method
Polylactic acid (PLA), poly (butylene succinate) PBS [27]	Cellulose crystals (CNC) (extracted from Hemp fiber	Sulphuric acid treatment	Solvent casting method
Low density polyethylene (LDPE) [28]	Doum fiber (DF) (5–30%)	NaOH treatment	Extrusion
PP [29]	Alfa fiber (AF)	Chemical treatment	Hot-press

(continued)

Table 2 (continued)

Results				Applications
PS Tensile strength (TS) ≈ 3 MPa Elongation at break (EAB) ≈ 210%	**PS/20% WS** TS ≈ 2.3 MPa EAB ≈ 110%	**PS/20% Alkali treated-WS** TS ≈ 3.56 MPa EAB ≈ 140 %	**PS/20% Acid treated-WS** TS ≈ 3.14 MPa EAB ≈ 130%	Packaging
PE/PP TS ≈ 46.5 MPa EAB ≈ 122% Contact angle ≈ 121° WVTR ≈ 47.50 g m^{-2} day^{-1}	**PE/PP/30% Native-WS** TS ≈ 34 MPa EAB ≈ 116.20% Contact angle ≈ 92° WVTR ≈ 132 g m^{-2} day^{-1}		**PE/PP/30% Alkali treated-WS** TS ≈ 45.010 MPa EAB ≈ 119.98% Contact angle ≈ 117° WVTR ≈ 51.89 g m^{-2} day^{-1}	Packaging
PVA/Starch/BF TS ≈ 30.8 MPa	**PVA/Starch/20% Alkali treated-BF** TS ≈ 34.4 MPa		**PVA/Starch/20% Acid treated-BF** TS ≈ 19.2 MPa	Packaging
PLA TS ≈ 55 MPa Impact strength (IS) ≈ 38 J/m		**PLA/Alkali treated-KF** TS ≈ 8 MPa IS ≈ 70 J/m		Packaging
CE/BOZ TS ≈ 28.05 MPa Tensile modulus (TM) ≈ 1.60 GPa Strain ≈ 2.10% Water absorption (WAB) ≈ 0.75%		**CE/BOZ/Treated-HF (20%)** TS ≈ 55.74 MPa TM ≈ 3.47 GPa Strain ≈ 1.38% WAB ≈ 5%		Packaging

(continued)

Table 2 (continued)

Results				Applications
PLA TS ≈ 65 MPa EAB ≈ 104.5% Flexible strength (FS) ≈ 100 MPa	**PLA/30% HH** TS ≈ 45MPa EAB ≈ 102% FS ≈ 60 MPa		**GMA-g-PLA/30% HH** TS ≈ 50 MPa EAB ≈ 102.1% FS ≈ 80 MPa	Packaging and disposable goods
HDPE TS ≈ 20 MPa Strain ≈ 30%	**65% HDPE/30% BF/5% MAPE** TS ≈ 50 MPa Strain ≈ 7.25%	**65% HDPE/15% HF/15% BF/5% MAPE** TS ≈ 47.5 MPa Strain ≈ 7.5%	**65% HDPE/30% HF/5% MAPE** TS ≈ 37.5 MPa Strain ≈ 7.5%	Packaging
PP TS at 10 mm thickness ≈ 40 MPa Fracture toughness (FT) at 10 mm Thickness ≈ 4.2 MPa $m^{0.5}$	**95% PP/5% Alkali treated-HF** TS at 10 mm thickness ≈ 60 MPa FT at 10 mm thickness ≈ 4.3 MPa $m^{0.5}$		**85% PP/15% Alkali treated-HF** TS at 10 mm thickness ≈ 44 MPa FT at 10 mm thickness ≈ 3.95 MPa $m^{0.5}$	Automobiles, military, construction, packaging, aerospace, railway coach
PHB TS ≈ 40 MPa EAB ≈ 101.7% WVP ≈ (2.1) × 10^{-15} kg s^{-1} Pa^{-1} m^{-1}	**PHB/20 % Almond shell** TS ≈ 35 MPa EAB ≈ 101.25% WVP ≈ 5 × 10^{-15} kg s^{-1} Pa^{-1} m^{-1}	**PHB/20% Rice husk** TS ≈ 29 MPa EAB ≈ 101.12% WVP ≈ 8 × 10^{-15} kg s^{-1} Pa^{-1} m^{-1}	**PHB/20% Seagrass** TS ≈ 40 MPa EAB ≈ 100.75% WVP ≈ 1.75 × 10^{-15} kg s^{-1} Pa^{-1} m^{-1}	Packaging
PLA TS ≈ 70 MPa EAB ≈ 107%	**PLA/30% WS** TS ≈ 60 MPa EAB ≈ 103.85%		**PLA/30% WS (SSSM)** TS ≈ 68 MPa EAB ≈ 103.1%	Packaging

(continued)

Table 2 (continued)

Results				Applications
Epoxy/73% native-SF TS ≈ 320 MPa		**Epoxy/77% treated-SF** TS ≈ 410 MPa		Packaging
PVA TS ≈ 41 MPa Transparency ≈ (68–95)%		**PVA/NaOH treated-RS** TS ≈ 98 MPa Transparency ≈ (52–90)%		Packaging
Hydroxypropylmethyl cellulose film TS ≈ 41.56 MPa EAB ≈ 141% Transparency at 600 nm ≈ 89% WVTR ≈ 0.053 g/cm^2/24 h		**3% Jute fiber based hydroxypropylmethyl cellulose film** TS ≈ 57.86 MPa EAB ≈ 145% Transparency at 600 nm ≈ 78% WVTR ≈ 0.039 g/cm^2/24 h		Packaging
Starch TS ≈ 2.5 MPa Water uptake ≈ 250%		**Starch/10% cellulose nanocrystals** Tensile strength ≈ 3.4 MPa Water uptake ≈ 100%		Packaging
PLA WVP ≈ 0.071 × 10^{-11} g mm kPa^{-1} h^{-1} m^{-2} Transmittance at 450 nm ≈ 88.2%	**PLA/20% PBS** WVP ≈ 0.042 × 10^{-11} g mm kPa^{-1} h^{-1} m^{-2} Transmittance at 450 nm ≈ 86%		**PLA/20% PBS/3% CNC** WVP ≈ 0.042 × 10^{-11} g mm kPa^{-1} h^{-1} m^{-2} Transmittance at 450 nm ≈ 81%	Packaging
LDPE TS ≈ 12.75 MPa EAB ≈ 140%		**70% LDPE/30% TDF** TS ≈ 10 MPa EAB ≈ 40%		Packaging
PP Young modulus (YM) ≈ 700 MPa TS ≈ 35 MPa	**Raw-AF/PP** YM ≈ 1000 MPa TS ≈ 31 MPa	**Alkali treated-AF/PP** YM ≈ 1200 MPa TS ≈ 31 MPa	**Esterified-AF/PP** YM ≈ 1400 MPa TS ≈ 32 MPa	**Etherified-AF/PP** YM ≈ 900 MPa TS ≈ 32 MPa

Source Author

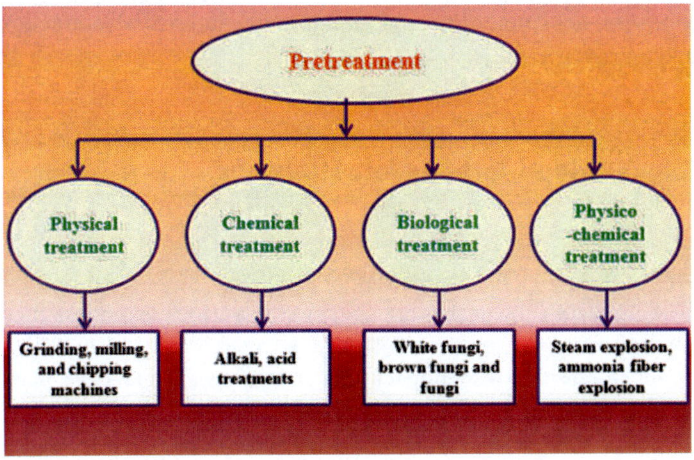

Fig. 3 Types of pretreatment (*Source* Author)

as alkalis (KOH, NaOH, $Ca(OH)_2$ etc.), acids (HCl, H_2SO_4, H_3PO_4, etc.) and ionic liquids have been commonly used for hydrolysis of hemicellulose with delignification process. This pretreatment can be considered in one of the most effective treatments for making more cellulose accessible [32].

3.3 Biological Pretreatment

Biological pretreatment is considered as a minimum energy required process in which agro-waste is kept under a control atmosphere with preferred microorganism. That microorganism has created the desirable surface changes of the agro-waste that makes biomass more acceptable for polymer adhesion. In this treatment, many organisms, i.e., white, brown and fungi have been used for enhancing the suitability of agro-waste in the polymer matrix. But this process has required larger residence time with an effective control atmosphere represent this treatment less attractive as compared to other pretreatment processes [33].

3.4 Physicochemical Pretreatment

This category of treatment is classified as a combination of physical and chemical changes. In this treatment, agro-waste is treated at higher temperatures and pressure which breaks the recalcitrance structure of agro-waste and creates some desirable surface changes in the agro-waste. Many physicochemical treatment processes are steam explosion, ammonia fiber explosion, liquid hot water hydrothermal treatment, wet oxidation, etc., are commonly used for improving interfacial interaction between polymer and agro-waste [32].

4 Various Testing Methods for Analyzing the Characteristics of Green Packaging Film

Several types of testing methods are used for analyzing the properties of agro-waste-based green composites for packaging application. Various testing methods such as water vapor transmission rate (WVTR), water vapor permeability (WVP), tensile test, optical characteristic test, impact test and contact angle have been used by researchers for analyzing the characteristics of packaging film [15].

4.1 Water Vapor Transmission Rate and Water Vapor Permeability

WVTR is an essential property for analyzing the quality of packaging film. This test explores the water vapor migration rate through the green composite film. A low value of this rate signifies synthesized film is applicable to keep warm products warm. In this test, a glass cup of 100 ml was taken and filled with distilled water. After that, cup is covered by synthesized film tightly and observed the weight of that wet cup chamber. This chamber is placed in an incubation chamber at a known temperature with fixed relative humidity. The changes in a total weight of wet cup chamber are analyzed at a regular interval of a time period. Water vapor transmission rate can be determined using the following equation.

$$\mathrm{WVTR} = \frac{\mathrm{WC}_1 - \mathrm{WC}_2}{\mathrm{WC}_1 * A * \mathrm{day}} \tag{1}$$

where WC_1 and WC_2 represent the initial and final weights of wet cup chamber; A represents the exposing area of wet cup chamber [34].

Water vapor permeability is also determined using the following equation.

$$\mathrm{WVP} = \frac{\mathrm{WVTR} * \text{thickness of green film}}{P(\mathrm{RH}_1 - \mathrm{RH}_2)} \tag{2}$$

where P is saturation vapor pressure at constant temperature, and RH_1 and RH_2 are relative humidity of inside and outside of the wet cup chamber.

4.2 Tensile Test

This test is demonstrated the mechanical stability of green composite film using universal testing machine. This property depends on materials thickness, testing speed and method of preparation. In this test, film is cut in strip form at fixed gauge length and gauge width according to ASTM. Generally, many researchers have been used ASTMD 0882 for materials thickness less than 1 mm. A sample is fixed between the grips of testing machine and allowed to elongate a material at a known testing speed. Further, the machine calculated the ultimate tensile strength and maximum flexibility of the green film that plays a vital role in observing the mechanical stability of the material [4].

4.3 Dart Impact Test

Dart impact strength is an important property for industries of packaging sectors to access the durability of the packaging film. A sample is cut in required dimensions according to ASTM. Commonly, ASTMD1709 has been used frequently for analyzing the impact strength of the green packaging film. In this impact test, a dart of known weight is free fell on the surface of the film at a fixed height and observed that film is punctured or not. If it is not punctured, again increase the weight of dart and repeated the process. The phenomena helped to understand the impact strength of the green packaging film. The energy required to fracture the surface of film is equivalent to drop a known dart impact failure weight at fixed height on the sample.

4.4 Contact Angle Test

In order to elucidate the hydrophobic property of green film, water contact angle is measured at vapor–liquid interface meets at solid surface of sample using drop shape analyzer machine. Sessile drop method is the simplest way to observe the water repelling quality of the prepared film. This test is measured by the wettability of the surface of the sample. In this test, a drop of distilled water is dropped on the surface of the composite film and examined the surface hydrophobicity of the sample. Higher contact angle (greater than 90°) signified the hydrophobic property of the green composite film [4].

4.5 Optical Characteristics Test

This test visualized the transparency of the green composite film for packaging applications using UV–Vis spectrophotometer machine. Green film is cut in strip form and placed in a cuvette using blank cuvette as a reference in the machine and scanned over the wavelength of 400–800 nm visible range. This test is provided the transparency of the sample over visible range and demonstrated the optical characteristics of the green film [35].

4.6 Thermal Stability Test

Thermal stability test is provided information about ability of the packaging film to resist the action of heat using PerkinElmer thermogravimetric analyzer. This test represents the maximum temperature at which our materials can sustain their mechanical property. In this test, a known weight of green film is placed in a machine and allowed to heat from 30 to 800 °C at required heating rate. This process is done in a nitrogen-controlled atmosphere for avoiding the undesirable thermal cracking of the green film. Thermal analyzer provided the graph between mass (%) with temperature. This graph helped to understand the main thermal degradation temperature range of the green composite film [21].

5 Conclusion

In summary, the effect of agricultural wastes on the packaging properties of the polymeric film was evaluated. This chapter has enlightened the perfect use of agricultural waste as a reinforcing agent in the polymer matrix for packaging applications. Various improving interfacial interaction techniques, i.e., physical, chemical, biological and physiochemical treatments have been discussed in order to enhance the suitability of agricultural waste in the polymer matrix in terms of water vapor migration rate, transparency, mechanical and thermal stabilities. Published literature confirmed the importance of various strategies for agricultural waste for polymer blending. This chapter is also explored the testing methods, i.e., water vapor transmission rate, water vapor permeability, contact angle, tensile test, impact test, optical characteristics test, thermal stability test and some ASTM standards for tensile and impact tests. This chapter described a brief introduction for the utilization of abundantly available agricultural wastes as filler in the polymer matrix for green packaging application.

References

1. Dixit S, Yadav VL (2019) Synthesis of green thermally resistant composite: a review. Indian J Chem Technol (IJCT) 26:494–503
2. Dixit S, Yadav VL (2020) Biodegradable polymer composite films for green packaging applications. In: Kharissova OV, Martínez LMT, Kharisov BI (eds) Handbook of nanomaterials and nanocomposites for energy and environmental applications. Springer International Publishing, Cham, pp 1–17
3. Dixit S, Joshi B, Kumar P, Yadav VL (2020) Novel hybrid structural biocomposites from alkali treated-date palm and coir fibers: morphology, thermal and mechanical properties. J Polym Environ
4. Dixit S, Yadav VL (2019) Optimization of polyethylene/polypropylene/alkali modified wheat straw composites for packaging application using RSM. J Cleaner Prod 240:118228
5. Zegaoui A, Derradji M, Ma R-K, Cai W-A, Medjahed A, Liu W-B, Dayo AQ, Wang J, Wang G-X (2018) Influence of fiber volume fractions on the performances of alkali modified hemp fibers reinforced cyanate ester/benzoxazine blend composites. Mater Chem Phys 213:146–156
6. Dayo AQ, Gao B-C, Wang J, Liu W-B, Derradji M, Shah AH, Babar AA (2017) Natural hemp fiber reinforced polybenzoxazine composites: curing behavior, mechanical and thermal properties. Compos Sci Technol 144:114–124
7. Dixit S, Yadav VL (2019) Comparative study of polystyrene/chemically modified wheat straw composite for green packaging application. Polym Bull 1–20
8. Chawla K, Bastos A (1979) The mechanical properties of jute fibers and polyester/jute composites. Mech Behav Mater 3:191–196
9. Sadh PK, Duhan S, Duhan JS (2018) Agro-industrial wastes and their utilization using solid state fermentation: a review. Bioresour Bioprocess 5:1
10. Chaitanya S, Singh AP, Singh I (2017) Natural fiber-reinforced biodegradable and bioresorbable polymer composites, p 163
11. Shahinur S, Hasan M (2019) Jute/Coir/Banana fiber reinforced bio-composites: critical review of design, fabrication, properties and applications. In: Reference module in materials science and materials engineering. Elsevier, Amsterdam
12. Senthilkumar K, Saba N, Rajini N, Chandrasekar M, Jawaid M, Siengchin S, Alotman OY (2018) Mechanical properties evaluation of sisal fibre reinforced polymer composites: a review. Constr Build Mater 174:713–729
13. Faruk O, Bledzki AK, Fink H-P, Sain M (2012) Biocomposites reinforced with natural fibers: 2000–2010. Prog Polym Sci 37:1552–1596
14. Santhosh Kumar S, Hiremath S (2019) Natural fiber reinforced composites in the context of biodegradability: a review. In: Reference module in materials science and materials engineering. Elsevier, Amsterdam
15. Dixit S, Yadav VL (2020) Comparative study of polystyrene/chemically modified wheat straw composite for green packaging application. Polym Bull 77:1307–1326
16. Srivastava KR, Singh MK, Mishra PK, Srivastava P (2019) Pretreatment of banana pseudostem fibre for green composite packaging film preparation with polyvinyl alcohol. J Polym Res 26:95
17. Islam MS, Rahman MM, Hasan M (2019) Kenaf fiber based bio-composites: processing, characterization and potential applications
18. Khan BA, Na H, Chevali V, Warner P, Zhu J, Wang H (2018) Glycidyl methacrylate-compatibilized poly (lactic acid)/hemp hurd biocomposites: processing, crystallization, and thermo-mechanical response. J Mater Sci Technol 34:387–397
19. Sarasini F, Tirillò J, Sergi C, Seghini MC, Cozzarini L, Graupner N (2018) Effect of basalt fibre hybridisation and sizing removal on mechanical and thermal properties of hemp fibre reinforced HDPE composites. Compos Struct 188:394–406
20. Madhusudhana H, Desai B, Venkatesha C (2018) Experimental investigation on parameter effects on fracture toughness of hemp fiber reinforced polymer composites. Mater Today Proc 5:20002–20012

21. Sánchez-Safont EL, Aldureid A, Lagarón JM, Gámez-Pérez J, Cabedo L (2018) Biocomposites of different lignocellulosic wastes for sustainable food packaging applications. Compos B Eng 145:215–225
22. Yang S, Bai S, Wang Q (2018) Sustainable packaging biocomposites from polylactic acid and wheat straw: enhanced physical performance by solid state shear milling process. Compos Sci Technol 158:34–42
23. Zuccarello B, Marannano G, Mancino A (2018) Optimal manufacturing and mechanical characterization of high performance biocomposites reinforced by sisal fibers. Compos Struct 194:575–583
24. Perumal AB, Sellamuthu PS, Nambiar RB, Sadiku ER (2018) Development of polyvinyl alcohol/chitosan bio-nanocomposite films reinforced with cellulose nanocrystals isolated from rice straw. Appl Surf Sci 449:591–602
25. Orasugh JT, Saha NR, Rana D, Sarkar G, Mollick MMR, Chattoapadhyay A, Mitra BC, Mondal D, Ghosh SK, Chattopadhyay D (2018) Jute cellulose nanofibrils/hydroxypropylmethylcellulose nanocomposite: a novel material with potential for application in packaging and transdermal drug delivery system. Ind Crops Prod 112:633–643
26. Kargarzadeh H, Johar N, Ahmad I (2017) Starch biocomposite film reinforced by multiscale rice husk fiber. Compos Sci Technol 151:147–155
27. Luzi F, Fortunati E, Jiménez A, Puglia D, Pezzolla D, Gigliotti G, Kenny JM, Chiralt A, Torre L (2016) Production and characterization of PLA_PBS biodegradable blends reinforced with cellulose nanocrystals extracted from hemp fibres. Ind Crops Prod 93:276–289
28. Arrakhiz FZ, El Achaby M, Malha M, Bensalah MO, Fassi-Fehri O, Bouhfid R, Benmoussa K, Qaiss A (2013) Mechanical and thermal properties of natural fibers reinforced polymer composites: doum/low density polyethylene. Mater Des 43:200–205
29. Arrakhiz F, Elachaby M, Bouhfid R, Vaudreuil S, Essassi M, Qaiss A (2012) Mechanical and thermal properties of polypropylene reinforced with Alfa fiber under different chemical treatment. Mater Des 35:318–322
30. Agbor VB, Cicek N, Sparling R, Berlin A, Levin DB (2011) Biomass pretreatment: fundamentals toward application. Biotechnol Adv 29:675–685
31. Dhakal HN, Sarasini F, Santulli C, Tirillò J, Zhang Z, Arumugam V (2015) Effect of basalt fibre hybridisation on post-impact mechanical behaviour of hemp fibre reinforced composites. Compos A Appl Sci Manuf 75:54–67
32. Jędrzejczyk M, Soszka E, Czapnik M, Ruppert AM, Grams J (2019) Physical and chemical pretreatment of lignocellulosic biomass. In: Second and third generation of feedstocks. Elsevier, Amsterdam, pp 143–196
33. Fengel D, Wegener G (1984) Wood: chemistry, ultrastructure. Reactions 613:1960–1982
34. Sirviö JA, Kolehmainen A, Liimatainen H, Niinimäki J, Hormi OE (2014) Biocomposite cellulose-alginate films: promising packaging materials. Food Chem 151:343–351
35. Naskar A, Khan H, Sarkar R, Kumar S, Halder D, Jana S (2018) Anti-biofilm activity and food packaging application of room temperature solution process based polyethylene glycol capped Ag-ZnO-graphene nanocomposite. Mater Sci Eng C 91:743–753

Chapter 17
Green Composites for Application in Antistatic Packaging

Leonardo de Souza Vieira, Isabela Cesar Oyama, Larissa Stieven Montagna, Mirabel Cerqueira Rezende, and Fabio Roberto Passador

1 Introduction

1.1 Antistatic Packaging

The generation and accumulation of electrostatic charges are the main problems encountered in the production and operation processes in the electronics industry. Electronic components, boards, and integrated circuits are very sensitive to electrostatic voltages generated by friction with their packaging and can be completely damaged when transported, stored, or handled incorrectly [22, 57]. The generation of electrostatic charges by friction (or triboelectrification), when not properly controlled, can even lead to the risk of fires or small explosions if there are gases, vapors, or flammable powders in the factory environment [64, 69]. Estimates show that, in the USA, 5% of the market related to the sale of electronic devices is affected by losses caused by electrostatic discharge (ESD) damages [57].

In this way, the development of new materials and production processes which mitigate the problem of generation and accumulation of electrostatic charges in the electronics industry represent a reduction in operating risks and production costs. Moreover, it is obtained an improvement in product quality and reliability, guaranteeing plain customer satisfaction and resulting in fewer expenses in repairing goods, which increases the profits obtained by the company.

The packaging used to protect electronic devices against ESD is called *antistatic packaging* and is used in large volumes in the electronics industry [57]. Antistatic packaging must present a sufficiently low electrical resistivity in order to dissipate electrical charges through their structure [54]. Antistatic materials present electrical

L. de S. Vieira · I. C. Oyama · L. S. Montagna · M. C. Rezende · F. R. Passador (✉)
Polymer and Biopolymer Technology Laboratory (TecPBio), Federal University of São Paulo (UNIFESP), 330 Talim St., São José dos Campos Zip code 12231-280, Brazil
e-mail: fabio.passador@unifesp.br

S. Thomas and P. Balakrishnan (eds.), *Green Composites*,
Materials Horizons: From Nature to Nanomaterials,
https://doi.org/10.1007/978-981-15-9643-8_17

resistivity in the range 10^3–10^{10} Ω m [18]. The norms regarding the establishment of standards for the control and protection of equipment susceptible to damage by electrostatic discharge are developed and disseminated by different organizations, including the ESD Association. This organization is accredited by the American National Standards Institute (ANSI) and represents the USA in the International Electrotechnical Commission (IEC), an institution organized by the international community for the development of a series of documents regarding the establishment of standards for the ESD control. These documents receive the general designation IEC 61340 and are technically equivalent to the ANSI/ESD S20.20 standard, developed by the ESD Association [18].

Antistatic packages are generally produced with non-biodegradable polymers of fossil resource, such as polyethylene (PE) [58] and poly(ethylene terephthalate) (PET) [64]. Thus, the production of packaging obtained from renewable resources and biodegradable materials constitutes a promising research area, attracting several companies and generating large investments in the development of more sustainable alternatives [6, 49].

For example, a promising alternative to the use of conventional low-density polyethylene (LDPE) obtained from fossil resources is the use of green low-density polyethylene (green LDPE), in which the main raw material is sugar cane, that is, a material from a renewable resource. The growth process of sugar cane consumes CO_2 present in the atmosphere and, in this way, contributes to reduce the amount of this greenhouse gas in the environment. Green LDPE presents similar properties and performance presented by petrochemical-based PE, which favors industrial interests in this material [46].

Among the biodegradable polymers available to produce biodegradable packagings are polycaprolactone (PCL) [27], poly(lactic acid) PLA [65], and poly(hydroxybutyrate-*co*-hydroxyvalerate) (PHBV) [12]. Besides green LDPE, PHBV is also a material obtained from renewable resources that is widely studied due to its good biodegradability and biocompatibility. Furthermore, PHBV has physical and mechanical properties similar to PE and polypropylene (PP) [2, 32, 35, 71].

Antistatic packages produced from insulating polymeric matrices, such as green LDPE and PHBV, need modification in their formulation in order to decrease the electrical resistivity of the material. So, it can dissipate the electrical charges formed by the friction between the electronic device and the packaging during transport and storage of that electronic device. There are many ways to achieve this goal, and the most common technique is the addition of electrically conductive particles that leads to the formation of electron-conducting paths in the polymeric matrix [58, 64]. These electrically conductive particles may be generically called *antistatic agents* [57]. The increase in electrical conductivity with the addition of increasing levels of the conductive material has the characteristic of being nonlinear. There is a certain amount of antistatic agent that, when added to the polymeric matrix, ensures that its particles establish physical contact with each other and form a network of continuous electron conduction throughout the composite structure, with an increase of several decades of magnitude in its electrical conductivity [62]. This critical amount is called the *electrical percolation threshold* [53]. A low electrical percolation threshold is desirable, avoiding major changes in the mechanical and rheological properties of the polymeric matrix [58, 62].

Carbon black (CB) is one of the materials most employed as antistatic agent mainly due to its relatively low cost [62, 68, 73]. Silva et al. [58] developed PLA/CB composites reinforced with different contents of CB (5, 10 and 15 wt%) using a thermokinetic mixer. Mechanical tests showed that the Izod impact strength of the composites presented a slight decrease when compared to neat PLA. Thermogravimetric analysis (TGA) showed that the addition of 10 wt% of CB increased the degradation temperature (T_{onset}) by up to 32 °C when compared to neat PLA, which was attributed to the formation of char that thermally insulated polymer chains not yet thermally degraded. The electrical conductivity test by impedance spectroscopy showed that the addition of 10 wt% of CB increased the electrical conductivity in 9 decades of magnitude which qualifies the material as a promising alternative for the production of ESD protective packaging.

Nonetheless, CB has the drawback of presenting a high electrical percolation threshold in many polymer matrices. Due to its low electrical conductivity and its low tendency to disperse, it is necessary to add a relatively high amount of this carbon material in the polymer matrix to establish an effective electron-conducting path in the polymer matrix [64, 68, 73]. Wang et al. [68] explained that, in order to obtain satisfactory antistatic properties in poly(vinyl chloride) (PVC)/CB composites, it is necessary to add up to 25 wt% of CB to the PVC matrix, which may negatively interfere with the mechanical properties of the material and impair its rheological and processing characteristics.

Another disadvantage of the use of CB is the unhealthy aspect related to its production and handling, which involves the generation of many nanoparticles suspended in the air and respiratory problems in those who inhale them [44].

In this way, there are various alternatives of reducing electrical resistivity of insulating polymeric materials available in the literature, which includes the use of surfactant molecules [16], metallic particles [76], oxide whiskers [75], polymer electrolytes [14], the addition of intrinsically conducting polymers such as polypyrrole and poly (3, 4-ethylenedioxythiophene) [33], use of conductive ionomers [39], and the addition of carbon materials such as carbon nanotubes [9, 19], graphene [74], and graphite nanoparticles [47].

1.2 Methods to Obtain Polymeric Materials with Antistatic Properties

An interesting way of increasing the electrical conductivity of electrically insulating polymers constitutes the addition of organic antistatic agents such as ionic and non-ionic surfactants [5]. The molecules of surfactants with low molar weight migrate to the outermost layers of the polymer and absorb water molecules present in the atmosphere due to the hydrophilicity of its chemical groups, conferring conductive properties to the surface of the material [67]. An example of an antistatic agent used in the production of materials with dissipative properties is ionic liquids, which basically consist of molten salts at room temperature. Ionic liquids have interesting properties because they are not volatile and flammable and present high thermal stability and high ionic density [16]. Ding et al. [16] developed a material with promising antistatic characteristics from the addition of 1-n-tetradecyl-3-methylimidazolium bromide ([C14mim] Br) in PP polymer matrix with a weight ratio of 100/3. A good dispersion of [C14mim] Br in the polymer matrix was reported, indicating good chemical compatibility between the two phases. The addition of [C14mim] Br molecules increased the impact resistance of PP, which was attributed to its effect of a nucleating agent in the polymeric matrix, refining the structure of its spherulites. Electrical conductivity tests showed that the addition of the antistatic agent led a decrease of 7 decades of magnitude in the electrical resistance of PP (from $2.67 \times 10^{14}\ \Omega$ to $2.60 \times 10^{7}\ \Omega$), obtaining a material with promising antistatic properties. However, this technique presents a reduction in the dissipative properties over time as the molecules of the antistatic agent migrate and deteriorate on the surface [67]. Another negative issue is the effectiveness dependence of the packaging in relation to the relative humidity of the air [69].

Zhou et al. [75] prepared PVC/zinc oxide whiskers (ZnOw) composites with different ZnOw contents (1, 2, 3, 4, 5, and 6 wt%) to obtain materials with antistatic properties. Zinc oxide whiskers (ZnOw) are ceramic materials with a semiconductive character and are shaped like fine needles that grow in four different directions and form a three-dimensional structure. Within the polymer, adjacent needles that overlap are able to guarantee a flow of electrical charges through the composite and increase the electrical conductivity of the polymer matrix. Results of electrical tests indicated that the addition of 6 wt% of ZnOw can lead to a reduction in the electrical resistivity of 6 decades of magnitude and can be a promising material for ESD protection. The addition of ZnOw in another polymeric matrix (PP, in this case) allows obtaining composites with equally attractive properties. The addition of 3 wt% of ZnOw decreases the electrical resistivity of neat PP from $10^{16}\ \Omega$ cm to $10^{9}\ \Omega$ cm, which represents a considerable reduction of 7 decades of magnitude in this electrical property [60].

The formation of gel polymeric electrolytes constitutes another alternative of obtaining materials with electrically dissipative characteristics. Gel polymer electrolytes are dimensionally stable and are developed basically by the introduction of an ion-conducting mechanism inside electrically insulating polymers [14]. Che et al. [14] prepared composites with PVC as polymer matrix and the addition of bis [2-(2-methoxyethoxy ethyl] phthalate (BMEP doped with sodium thiocyanate salt (NaSCN. Composites were processed using a torque rheometer, and different contents of BMEP/NaSCN (or AP) were investigated. The Na^{+} cation dissociates from the salt and establishes a coordination effect with the oxygen atoms of the ether groups present in the BMEP structure. With the appearance of a potential difference, mobile cations migrate through percolative paths established by the network of BMEP molecules mixed inside the PVC matrix, which can collaborate to increase the electrical conductivity of the material. Electrical conductivity tests showed a low electrical percolation threshold (~20 AP phr), and a decrease of 7 decades of magnitude in the electrical resistivity of composites was obtained (10^{15} Ω sq^{-1} (0 phr AP) to 10^{8} Ω sq^{-1} (60 phr AP)). The effect of the addition of 40 phr of dibutyl phthalate (DBP) plasticizer to the PVC/AP composite was also verified. The addition of DBP promoted a further reduction in the electrical resistivity of PVC/AP composite. This phenomenon was attributed to the plasticizing effect that DBP imparts to the PVC polymeric chains, reducing its rigidity and facilitating the transfer of charge carriers (Na^{+}). Electrical tests were also carried out by varying the relative humidity (RA) from 50 to 0%, and it was verified that the electrical resistivity of the composites reduced only one decade of magnitude, indicating that an advantage of this method consists in obtaining materials that do not depend on this environmental condition to acquire dissipative properties.

Among the methods previously mentioned to produce antistatic packages, Maki et al. [39] evaluated the production of antistatic packages using a different way, which is without mixing or coating the polymer with an antistatic agent. Low resistivity was reached using an ionomer of a copolymer of ethylene with potassium as the metal ion type with other host resins. Generally, the ionomer has sodium or zinc ions in its chain that do not promote antistatic properties. The change with the addition of potassium helps to increase electrical conductivity. The authors developed a random ethylene copolymer (derived from polymerization of ethylene-comethacrylic and acid isobutyl acrylate) in which carboxylic groups (about 80%) were neutralized with potassium resulting in the production of the ionomer. A multi-layer cast film machine was used to produce films with single and multi-layer of potassium ionomer, metallocene linear low-density polyethylene (mLLDPE) with an antistatic agent and potassium ionomer/mLLDPE so as to characterize antistatic properties. The static decay time, surface resistivity and volumetric resistivity and space charge distribution (pulsed electroacoustic method) tests were evaluated. Static decay time test proved that potassium ionomer has rather good static decay properties compared to mLLDPE with an antistatic agent. It means that antistatic properties can be found on the surface of the film if potassium ionomer is added in any desired layer. The same behavior occurs with mLLDPE containing the antistatic agent. The stable performance of the ionomer was gain in the multi-layer structure instead of the single layer. The study of a two-layered film (one layer is the potassium ionomer, and the other is mLLDPE) was essential to understand the reason why multi-layered structures present good antistatic properties. The authors concluded that the use of potassium ionomer in antistatic packaging is satisfactory because it showed antistatic properties equal to polymeric composites with antistatic agents.

Jonas and Heywang [33] developed a method to coat plastic films with polypyrrole. Polypyrrole is an infusible and insoluble conductive polymer due to branching in its chain, so it was of great meaning finding a methodology to produce antistatic packaging from plastic film coated with polypyrrole. Antistatic packaging must have low electrical surface resistivity, and its antistatic characteristic cannot change by atmospheric humidity and need to be stable for at least four weeks at 70 °C. Besides that, the authors produced films of polypyrrole-coated polycarbonate following three steps. The materials used were a film of polycarbonate, polyvinyl acetate with ferric salt, and pyrrole monomer. The production of polypyrrole-coated polycarbonate follows the following procedure. The pyrrole monomer was polymerized at the polyvinyl acetate surface, and a conductive polymer was created and coated the polycarbonate film; it is important to frame that the polymer resulted from coating process is thermoformable. The coating procedure enabled an electrical surface resistivity of 10^3–10^4 $\Omega\ m^{-1}$; thus, this low surface resistivity allows the use of the film in antistatic packaging.

In general, the mechanism of surfactant antistatic agents in packages is based on the migration of them to the film surface when humidity is absorbed on the polymeric surface. Therefore, humidity is essential to trigger the antistatic role, even at low percentages. Besides electrical properties, antistatic packaging demands flexibility in order to facilitate processing and minimizing defects of the final product. Up to recent years, plasticizer derived from phthalate had been used mainly, but it has the problem associated with its toxicity, which can lead to utterly hazardous diseases.

Linking the necessity of a non-harmful plasticizer and antistatic charge, Wang et al. [70] engineered an eco-friendly biodegradable antistatic agent and plasticizer to produce antistatic composites with the polymeric matrix. PVC/citrate electrolyte-based antistatic plasticizer (CEAP)/tributyl citrate (TBC) composite was produced. TBC is a biodegradable and non-toxic plasticizer. Foremost, the synthesis of CEAP was made reacting citric acid, 2-butoxyethano, and sodium thiocyanate through an esterification reaction. The composites were prepared in a torque rheometer with a temperature of 160 °C, rotor speed of 30 rpm, and blending time set of 6 min. In general, a decrease of two decades of magnitude in superficial resistivity at room temperature was observed. Besides that, a reduction of one decade of magnitude in electrical superficial resistivity was reached increasing the temperature (40–80 °C). No significant modification at surface resistivity occurred changing relative humidity. The composite without TBC showed insufficient antistatic properties; thus, TBC is essential as a plasticizer and contributes to ion's mobility in polymer chain decreasing electrical surface resistivity.

Another way to produce antistatic packaging is the addition of metal fillers in the polymeric matrix due to their permanent electrical conductivity and low cost [76].

The addition of carbon materials to insulating polymers is the most used technique for obtaining antistatic packaging. Carbon materials are suitable for the most varied applications since carbon atoms are capable of forming materials with distinct crystalline structures and versatile and attractive properties [62]. Among the carbon materials studied and employed in dissipative materials to replace carbon black are the carbon nanotubes [9, 19], graphene [74], graphite [47], and, more recently, glassy carbon [46, 54, 63].

Carbon nanotubes and graphene are relatively recent and promising alternatives to replace carbon black since they confer excellent mechanical, thermal, and electrical properties to polymers [68]. Huang et al. [26] developed cellulose nanocomposites reinforced with different contents of carbon nanotubes (CNT) (1, 3, 5, 8 and 10 wt%) using a sequential solution dispersion, gelation, and hot-press drying process. The use of cellulose in the production of composites for applications in antistatic packaging is very promising, aiming at the substitution of polymers from non-renewable resources. The electrical percolation threshold occurs with the addition of 0.71 wt%, and the addition of 5 wt% of CNT to the cellulose increased the electrical conductivity from 10^{-13} S m^{-1} to only 5.6 S m^{-1}, obtaining an electrically dissipative material. Therefore, it is possible to obtain antistatic properties with a much lower amount of filler compared to carbon black.

Notwithstanding its good properties, carbon nanotubes and graphene present a high specific surface area (due to its nanoscale dimension) and a strong van der Waals interaction established between its nanofillers, which increases the tendency of agglomeration and hinders the distribution and dispersion processes of these materials in the polymer matrix, impairing the properties of composites [24]. Another disadvantage presented by carbon nanotubes is their high stable surface and their hard surface, which impairs the interaction between the polymer matrix and the nanofiller and the transfer of mechanical stress among these phases. For this reason, it is common the use of chemical/physical treatments in CNT to modify its surface properties by the introduction of polar functional groups to the nanofiller. The increase in the polarity of CNT may facilitate its interaction with polymeric chains in polymeric composites, increasing the mechanical properties of the material [36]. However, it may lead to a loss in the electrical and mechanical properties of CNT.

In this way, the need to search for an alternative material to the carbon black fosters the interest of novel materials that can be employed in the production of polymeric materials with antistatic properties. Although there are few studies in the literature on the use of glassy carbon (GC) as a filler in polymeric matrices, it is verified that this material gives good mechanical and electrical properties to polymers, making it a promising material in the production of antistatic packaging [46, 54].

Szeluga et al. [63] prepared composites with 5, 10, and 20 wt% of GC in epoxy resin (EP) and hybrid composites with the addition of GC and multiwall carbon nanotubes (MWCNT). It was verified that GC presents a good adhesion to the polymeric matrix, which explains its good dispersion and distribution. It was verified that the binary composites of EP/GC with 10 wt% GC presented an electrical resistivity of 3.02×10^7 Ω cm, 7 decades of magnitude lower than the neat EP (4.82×10^{14} Ω cm). The addition of 0.25 wt% MWCNT led to a further decrease in the electrical resistivity of the material (1.68×10^3 Ω cm), indicating a synergistic effect between the carbon fillers.

The use of GC in thermoplastic polymer matrix composites for the production of antistatic packaging has also shown to be a very promising alternative. Santos et al. [54] produced composites of low-density polyethylene (LDPE) with different contents of GC (0.5, 1, 5, 10, 15, and 20 wt%) using a thermokinectic mixer. It was observed that GC particles are well-dispersed and distributed in the polymer matrix, making it possible to obtain a relatively low electrical percolation threshold with the addition of 0.5 wt% GC. A decrease of 2 decades of magnitude in the electrical resistivity was also observed. The composite with 0.5 wt% GC presented the best optimization of mechanical and thermal properties, comparable to neat LDPE, which demonstrates that GC is a promising alternative of an antistatic agent in replacement to CB.

Oyama et al. [46] verified that low electrical resistivity was reached using contents of GC below 0.5 wt% in a green polymeric matrix, green LDPE. GC contents of 0.1, 0.3, and 0.5 wt% were used in the green LDPE matrix. The materials were mixed in a thermokinetic homogenizer. A reduction of 6 decades of magnitude in electrical resistivity (from 10^{14} to 10^{8} Ω m) was reached with the addition of only 0.1 wt% of GC in green LDPE/GC composite. A reduction of 7 decades of magnitude in the electrical resistivity was verified with the addition of 0.3 and 0.5 wt% GC. It indicates that small contents of GC are able to produce conductive paths in the green polymeric matrix.

As it is perceived in the examples previously mentioned, there are many works present in the literature involving the obtainment of materials with antistatic characteristics produced with polymers originated from non-renewable resources, such as PVC, PE, and PP. However, it is also noticed the use of polymers obtained by renewable resources and/or biodegradable polymers.

In this chapter, we will show the effectiveness of using green composites for the production of antistatic packaging. The antistatic agent chosen is the glassy carbon because it is from a renewable resource. The polymeric matrices used for the production of the green composites were a biodegradable polymer and a polymer from a renewable resource.

1.3 Glassy Carbon: A Green Antistatic Agent

Glassy carbon (GC), also named as vitreous carbon, is a carbon and isotropic material that generally presents both high hardness and elastic modulus, elevated chemical inertness, and temperature resistance. It is commonly obtained from the carbonization of certain thermoset resins, such as phenolic and polyfurfuryl alcohol (PFA) resins, at temperatures normally above 1000 °C in inert atmosphere [11]. Polyfurfuryl alcohol resin, also named as furfuryl resin, is originated through the polymerization of furfuryl alcohol, which can be obtained from the catalytic hydrogenation process of furfural. Furfural consists of a heterocyclic and aromatic aldehyde produced by dehydrating pentoses present in lignocellulosic biomass of plants and residual materials, such as corncobs, rice husks, and sugar cane bagasse [41]. In this way, GC produced from furfuryl resin constitutes a green material, since it is originated from renewable resources. When treated at high temperatures, the precursor resins of GC lose considerably the content of oxygen, hydrogen, and nitrogen present in their polymer chains, leaving high carbon content in their chemical composition (at least 90% by mass). The resulting carbon material has the term vitreous in its designation for having a shiny appearance when polished and for having a fracture surface similar to that presented by the glass [21].

GC belongs to the carbon materials class that is arranged by stacking of hexagonal planes of C with sp^2 hybridization, which is called graphitic materials [28]. Graphite presents the most well-organized basic structure among graphitic materials. The distance between its graphitic planes is 0.3354 nm, and the distance between the C

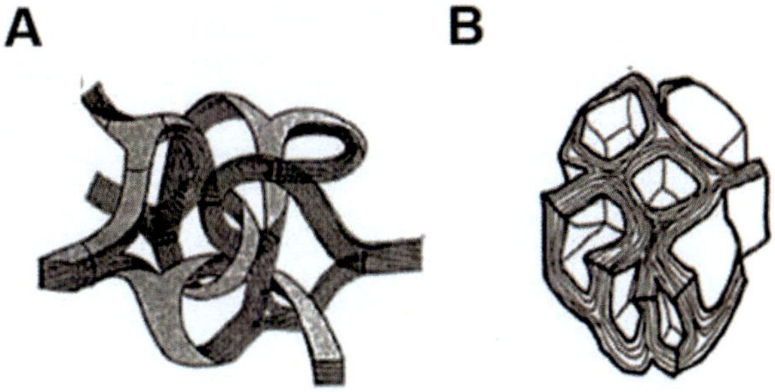

Fig. 1 Schematic representation of the GC structure according to **a** the Ribbon Model and **b** the Shell Model [28]

atoms in the plane is 0.1412 nm [28]. Carbon materials obtained under heat treatment at temperatures below 1300 °C have reduced levels of regularity in the stacking of the graphitic planes. Thus, the length of these plans tends to be small, as well as the number of hexagonal plans stacked. The arrangement of atoms formed in these materials, with lower order than that present in graphite, is called *turbostratic structure* [28]. The submission of graphitizable carbon materials to heat treatments at temperatures above 1700 °C can lead to an improvement in the organization of the turbostratic structure and an increase in the size and number of stacked planes. This process is called *graphitization.* However, in carbon materials classified as *non-graphitizable* (e.g., GC), the stacking of the graphite planes is highly disorganized and linked by disorganized carbons, and thus, it is not possible to obtain a graphite structure even when subjected to heat treatments at high temperatures [28].

Among the structural models proposed for the GC are the *Ribbon Model*, proposed by Jenkins and Kawamura (Fig. 1a), and the Shell Model, proposed by Shiraishi (Fig. 1b) [28, 30]. The Ribbon Model proposes that this carbon material consists of a network of folded and twisted carbon layers [30]. This model, however, is ineffective in explaining some GC properties, such as its high porosity (closed pores) and its low gas permeability. The Shell Model, however, elucidates this issue by presenting the microstructure of the GC as layers of hexagonal carbon arrangements that form closed structures (pores) [28].

GC has good chemical and mechanical resistance, high hardness, high modulus of elasticity (20–40 GPa), high electrical (~10^4 S m^{-1}) and thermal conductivities, and good gas impermeability [29, 45, 56]. These characteristics make this material suitable for a wide variety of applications, including heart valves, electrodes, refractory applications, catalyst supports, and others [8, 29, 34].

GC can be obtained, basically, in the form of reticulated glassy carbon (RGC) or monolithic glassy carbon (MGC) (Fig. 2a, b, respectively). RGC consists of a macroporous material with many accessible pores (as shown in the zoom of Fig. 2a) and a high surface area in relation to its volume, presenting high chemical stability, lightweight, and low cost [45]. The most common applications for RGC include water purifiers, gas separators, electrodes, natural gas stores, and biological growth supports [25, 51].

Another common form of GC is MGC, whose structure is characterized by the presence of a predominant amount of micro- and mesopores [21]. The diffusion

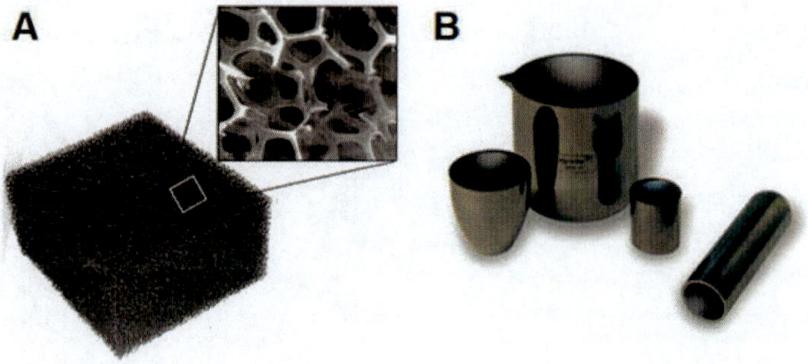

Fig. 2 Reticulated glassy carbon (RGC) (**a**) and monolithic glassy carbon (MGC) (**b**) [20], SPI [61]

of volatiles and the shrinkage of the material under heat treatment can lead to the formation of tensions and discontinuities in its structure. In this way, the heating rate must be strictly controlled, limiting the thickness of the piece obtained [8]. Its high purity and its physicochemical properties make MGC an interesting choice for the production of special crucibles (Fig. 2b) and tubes, catalyst supports, gaskets, electrodes, heart valves, among others [31].

1.4 Green Polymer for the Production of Antistatic Packaging

The production of polymeric materials from renewable resources gained great notoriety in the 1970s during the oil crisis. At this time, the price of oil barrels had a great increase, as well as the price of the products obtained from petroleum [36]. Since then, the production of polymers from renewable resources has received massive investments in science and in the industrial sector [6, 49].

Technological development in the area of biodegradable polymers has also obtained remarkable prominence in recent years due to the potential risk that the disposal of plastics can cause to the environment and to the most diverse ecosystems around the world. The application of plastic is widespread in practically all areas of the productive sector, and it is especially important in the packaging industry since it consumes around 44.8% of all plastic produced annually [23]. The antistatic packaging generally constitutes a single-use product. PP and PE are the most used polymers in the production of packaging and present high chemical stability and durability. This high chemical stability is a common characteristic of conventional polymers of fossil resource, making the biodegradation rate of these materials relatively low. It is estimated that, since 1950, about 79% of the plastic produced in the world is still present in landfills or in the environment [52].

Table 1 General properties presented by Green LDPE (SBF0323HC grade) [10]

General properties	Value	Unit
Melt flow rate (190 °C/2.16 kg)	0.32	g/10 min
Density	0.923	g/cm^3
Tensile strength at break	20	MPa
Elongation at break	390	%

1.4.1 Green Low-Density Polyethylene (Green LDPE)

Polyethylene is one of the most used polymers in the world due to its excellent mechanical properties and low density. The main applications include the production of many types of packaging, car fuel tanks, wire, and pipes.

A Brazilian petrochemical company developed a different route to produce PE changing the oil-base origin to a green one [43]. Green LDPE is produced by renewable feedstock, sugar cane, and has the same properties, processing and recyclability possibilities as petrochemical LDPE. The green LDPE presents low electrical conductivity, excellent thermal, mechanical properties, and great water vapor impermeability. Table 1 shows green LDPE properties. Summarizing, the process consists in producing ethylene monomer from ethanol extracted from sugar cane. This is an eco-friendly environmental process due to the capture of an enormous quantity of CO_2 from the atmosphere by sugar cane [43]. Besides this, sugar cane has been also using to produce fuels for automobiles [15]. In this way, this material participates in a green cycle.

Ethanol can be obtained from a variety of sugary feedstock like sugar cane, cornstarch, cassava, potato, and sugar beet. Brazil is one of the leaders of sugar cane production. This fact facilitates the continuous production of green polymers due to the abundance of feedstock material, which reduces the cost linked to production, and ease of logistics. Ethanol's obtaining route consists of fermentation of sugar cane stalk, followed by a catalytic dehydration reaction to produce ethylene, which has exactly the same chemical structure as ethylene from fossil origin. Then ethylene is polymerized to obtain green LDPE. It is worth mentioning that the same polymerization reactor used for petrochemical LDPE can be used to polymerize green LDPE [43].

1.4.2 Biodegradable Polymer for the Production of Antistatic Packaging

Many biodegradables polymers have been proposed to replace petroleum-based polymers in industry, such as cellulose [26], PCL [27], PLA [65], and polyhydroxyalkonates (PHAs) [12, 35, 66].

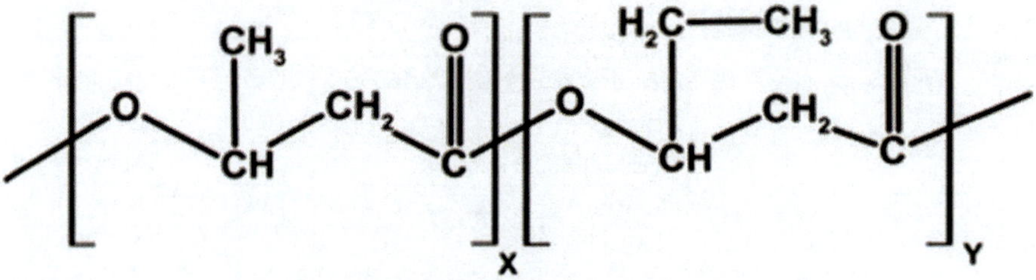

Fig. 3 Schematic representation of the PHBV chemical structure. *Source* Author

Table 2 General properties presented by PHBV (PHI 002)

General properties	Value	Unit
Melt flow rate (190 °C/2.16 kg)	5 to 10	g/10 min
Density	1.23	g/cm^3
Tensile strength at yield	18.7	MPa
Tensile elongation at yield	1	%
Tensile maximal strength	39.6	MPa
Tensile strength at break	39.6	MPa
Elongation at break	3.2	%

PHAs are basically polyesters of high molar weight that possess good biodegradability and biocompatibility [1]. Among PHAs is the poly(hydroxybutyrate-*co*-hydroxyvalerate) (PHBV), a polymer that presents properties similar to those observed for conventional polymers, such as PE e PP [2, 32, 35, 71].

Poly (Hydroxybutyrate-co-Hydroxyvalerate) (PHBV)

PHBV constitutes a statistical copolymer of mere hydroxybutyrate (HB) with random segments of mere hydroxyvalerate (HV) (Fig. 3) [35]. PHBV is produced during the bacterial fermentation of sugars for the production and accumulation of energy reserves in intracellular granules [50]. In industry, PHBV can be synthesized in large bioreactors through bacterial fermentation of propionic acid and glucose. The originated polymer can be captured and purified through the use of a specific solvent, obtaining a solid and dry final product [48, 50]. The general properties presented by PHBV are shown in Table 2.

PHBV presents a high rate of biodegradation since its ester bonds are easily hydrolyzed by the action of bacteria and fungi, generating compounds of smaller molar mass [13]. Biodegradation is an irreversible process of deterioration of a material, causing structural modification and alteration in the chemical and mechanical properties [38]. This process occurs through the enzymatic action of microorganisms (such as bacteria and fungi) that use the product of degradation as resources of energy, electrons, and atoms (e.g., carbon, nitrogen, oxygen, phosphorus, sulfur, etc.), which are necessary for the maintenance of their cellular activities [37]. Among the bacteria involved in the biodegradation process of PHAs, it is possible to mention the species *Alcaligenes faecalis T1, Pseudomonas lemoignei, Pseudomonas fluorescens GK13, Streptomyces exfoliates K10, Comamonas testosterone, Comamonas acidovorants*, etc. As an example of fungi capable of biodegrading PHAs are *Aspergillus penicilloides, Aspergillus fumigatos, Penicillium funiculosum, Penicillium daleae, Paecilomyces marquandii, Candida guilliermondii*, etc. [36].

The biodegradation process can occur inside the cells of the microorganisms that naturally produce PHAs or outside the cells of microorganisms that do not produce PHAs [36]. The high molar weight of the polymeric chains and their low hydrophilicity make it difficult for whole macromolecules to enter the cells of these microorganisms; therefore, it is necessary that the degradation occurs initially in an external environment. The process begins with the formation of a layer of water and excreted enzymes (*depolymerases*) called biofilm, where the polymeric chain is broken down [36]. In PHBV biodegradation, the ester bonds of the polymer chains are repeatedly hydrolyzed forming monomers HV and HB, which are soluble in water and have a molar weight small enough to enter the cell wall through passive diffusion. In general, monomers are metabolized inside cells during the tricarboxylic acid cycle and through β-oxidation, generating energy for cellular functioning [55].

In the presence of O_2, the complete process of biodegradation (mineralization) leads to the metabolization of polymer macromolecules into carbon dioxide (CO_2), water, and new cell biomass. In the absence of O_2 (anaerobic biodegradation), the production of CO_2 is replaced by the generation of methane (CH_4) [49].

2 Experimental Section: Production of Green Composites for Antistatic Packaging and Properties Analysis

2.1 *Obtainment Process of Glassy Carbon, Milling and Structural Characterization*

GC was obtained by the cure of PFA resin using as catalyst a 3% wt/wt aqueous solution of p-toluenesulfonic acid (APTS) (60% wt/vol.). The PFA cure was realized in four steps: at 60 °C for 24 h, at 80 °C for 2 h, at 110 °C for 2 h, and at 180 °C for 6 h. The cured PFA resin was subsequently subjected to a thermal carbonization treatment from room temperature up to 1000 °C at 10 °C min^{-1} under an inert N_2 atmosphere (1.0 L h^{-1}). The sample was kept at final temperature for 30 min and then cooled to room temperature.

GC plates (Fig. 4a) were ground in an IKA mini mill (model A11), for approximately 10 min to produce GC particulate (Fig. 4b). GC with particle size of around 75 μm was obtained using 200 mesh metal sieves.

Particulate GC was structurally analyzed on a Rigaku Ultima IV X-ray diffractometer (PANalytical, X'pert Powder model), operating at 40 kV and 30 mA with

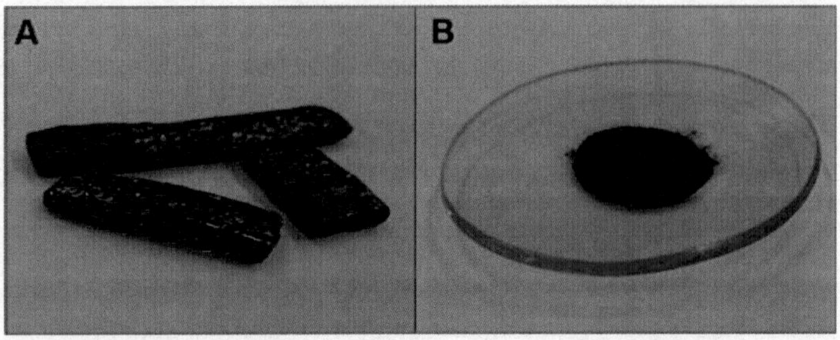

Fig. 4 Glassy carbon, in plate form (**a**), and the particulate material obtained after the milling process (**b**). *Source* Author

CuKα radiation ($\lambda = 1.54056$ Å) at a scan rate of 5° min^{-1} varying 2θ from 5° to 70°.

2.2 *Processing of Green LDPE/GC Composites*

Neat green LDPE (SBF0323HC grade, Braskem, Brazil) and LDPE/GC composites with 0.1, 0.3, and 0.5 wt% of GC were prepared in a thermokinetic homogenizer (MH50-H, MH Equipamentos Ltda—Brazil) with rotation of 3000 rpm and 185 °C during 30 s. After this processing, thin films (with 150 μm thickness) were produced by hydropneumatic press (PR8HP, MH Equipamentos—Brazil) at 180 °C with a pressure of 2 bar for 3 min. Thin films (0.6 mm) were used for electrical and thermal tests. Also, standardized specimens (3.2 mm) to tensile test were produced at the same equipment and temperature, with a pressure of 5 bar for 5 min.

2.3 *Processing of PHBV/GC Composites*

Neat PHBV (PHI 002, Nature Plast, France) and PHBV/GC composites with 1.0, 2.5, and 5,0 wt% of GC were prepared in a co-rotational extruder (AX Plastics, model AX16:40DR—Brazil), with $L/D = 40$ and $D = 16$ mm (screw thread). The thread rotation was 100 rpm, and the feeder rotation was 30 rpm. The temperature profile set in the extruder for each heating zone was 165 °C, 170 °C, 170 °C, 170 °C, and 165 °C. The material obtained was subsequently pelletized. After this processing, thin films (with 150 μm thickness) were produced by hydropneumatic press (PR8HP, MH Equipamentos—Brazil) at 190 °C with a pressure of 1 bar for 3 min.

2.4 Mechanical, Thermal, Electrical, and Biodegradability Tests

The mechanical behavior of the composites was evaluated by tensile tests according to ASTM D638-14 [3] for green LDPE composites and ASTM D 882-18 [4] for PHBV composites.

Specimens were performed in an MTS Criterion testing machine (model 45, Brazil), with a load cell of 50 kN and strain rate of 40 mm/min. The films were performed in an MTS Criterion testing machine (model 42, Brazil), with a load cell of 250 N and an initial strain rate of 0.1 mm/mm/min. At least seven samples of each composition were tested.

The thermal behavior composites were evaluated by differential scanning calorimetry (DSC) in a DSC Q2000 (TA Instruments, Brazil) equipment using small amounts (<10 mg) of dried samples that were placed into aluminum pans. The heating cycle was performed from room temperature to 200 °C with a heating rate of 10 °C min^{-1} under N_2 atmosphere. The degree of crystallinity (X_c) was calculated according to Eq. 1:

$$X_c(\%) = \frac{\Delta H_m}{\Delta H_m^\circ} \times \frac{1}{(1-\varphi)} \times 100 \tag{1}$$

where X_c is the degree of crystallinity, ΔH_m is the fusion enthalpy obtained by DSC, ΔH_m° is the fusion enthalpy of the 100% crystalline polymer (for PHBV $\Delta H_m^\circ =$ 146 J g^{-1} [40] and for green LDPE $\Delta H_m^\circ = 293$ J g^{-1} [72]) and φ is the content of GC added to the composite.

The electrical characterization was performed on a Solartron Impedance/gain–phase impedance analyzer (model SI 1260, Brazil) to measure alternating current electrical resistivity by impedance spectroscopy. It produced a metal-nanocomposite-metal structure by depositing a thin layer of gold/palladium alloy on both sides of the samples using a metallizer (MED020 Bal-tec, Brazil) in order to form the electrical contact. The electrical resistivity of the materials was calculated using the thickness (l), contact area (A) values of the samples, and the impedance value (Z) obtained by the analysis, as indicated by Eq. 2:

$$\rho = \frac{Z \times A}{l} \tag{2}$$

The biodegradation test in aqueous medium was performed only for neat PHBV and PHBV/GC composites based on the methodology developed by Silva et al. [59]. The microorganisms involved in the biodegradation test were extracted from vegetal soil. In summary, samples of neat PHBV and PHBV/GC composites were immersed in 20 mL of a mineral solution containing mineral salts necessary to microorganisms cell development. Subsequently, an amount of a microbial solution (containing the microorganisms extracted from the soil) necessary to obtain a final solution with microorganism concentration (cell mass/L) of 0.025 g L^{-1} was added to the mineral solution. The test was carried out in Falcon tubes (50 mL) under stirring (120 rpm) at 30 °C in a shaker New Brunswick™ (model Innova® 44/44R) with semi-open Falcon tubes for gas exchange. Samples were taken every 15 days during a total period of 45 test days of biodegradation in aqueous solution. It used four samples for each composition. After removal, the samples were washed with distilled water to remove material adhered to the surface and were vacuum dried.

3 Results and Discussion

Figure 5 shows the X-ray pattern obtained for GC. It is possible to observe the presence of two diffraction peaks characteristic of a turbostratic structure. The first peak is located at $2\theta = 23.6°$ and is attributed to the plane (002), which is related to the distance between the graphitic planes. The second peak is located at $2\theta = 43.4°$, attributed to the plane (10), which is associated with the in-plane structure [34, 45].

Figure 6a, b show the graphs of elastic modulus and degree of crystallinity as a function of the GC content for neat LDPE and green LDPE/GC composites and neat PHBV and PHBV/GC composites, respectively.

It is possible to verify, in relation to green LDPE/GC composites, that the increase in the GC content leads to an increase in the elastic modulus and a slight decrease in the degree of crystallinity of the green LDPE. Probably, the adhesion between the LDPE and the GC was strong enough to allow the transfer of mechanical stress from the matrix to the reinforcing phase, increasing the stiffness of the material.

There is also a decrease in the degree of crystallinity of green LDPE (up to 2.6%) with the increase in the GC content. Probably, the GC particles hindered the ordering

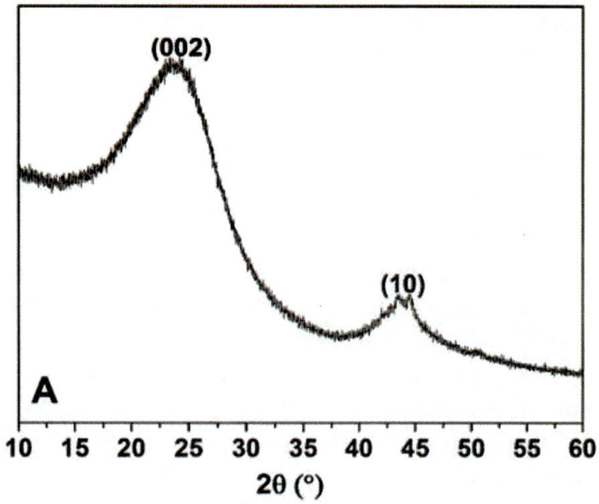

Fig. 5 X-ray pattern of GC. *Source* Author

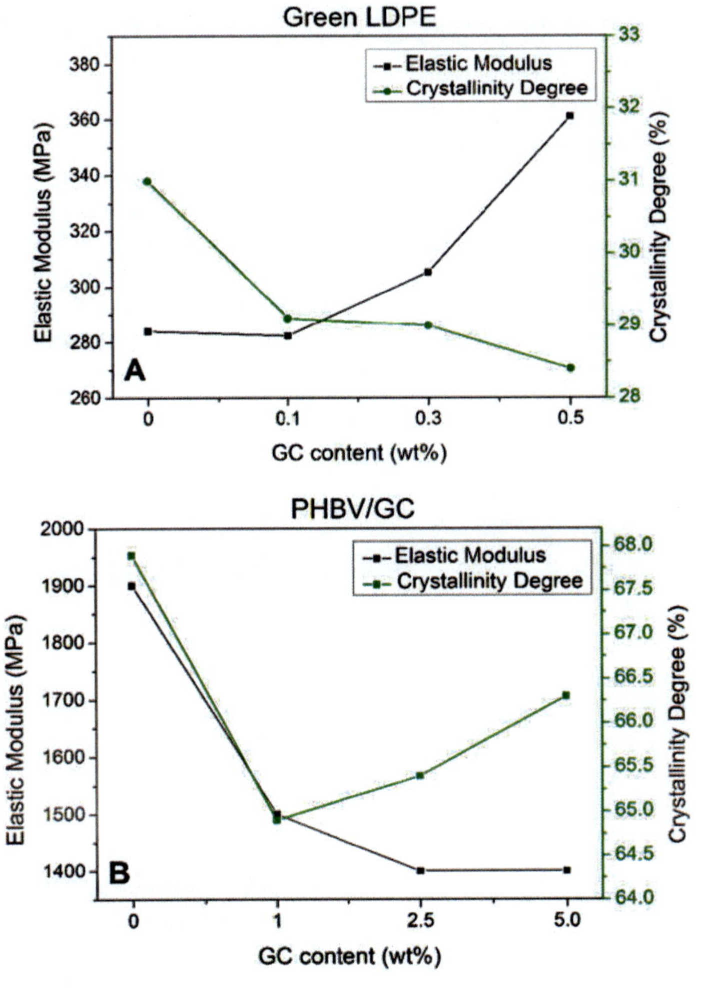

Fig. 6 Graphs of elastic modulus and crystallinity degree as a function of GC content for **a** LDPE/GC and **b** PHBV/GC composites. *Source* Author

of polymeric LDPE chains in crystallites during the crystallization process, restricting relaxation. It is suggested that a similar phenomenon occurred in PHBV present in PHBV/GC composites since in these composites the degree of crystallinity decreased up to 3% (Fig. 6b). The reduction in the crystallinity degree is the possible cause for the decrease in the elastic modulus in the material, which dropped from 1900 to 1400 MPa (PHBV/2.5 GC). The decrease in the stiffness of PHBV constitutes a positive factor for its use in the production of antistatic packaging since it is necessary a certain flexibility in the material to facilitate the handling of the package. Thus, regarding the flexibility requirements, green LDPE/GC composites have a certain advantage over PHBV/GC composites, since they have stiffness about four times lower.

Figure 7a, b show the graphs of electrical resistivity for neat green LDPE and green LDPE/GC and for neat PHBV and PHBV/GC composites as a function of the GC content present in the composites.

The addition of 0.1 wt% GC decreased the electrical resistivity of green LDPE by 6 decades of magnitude, ranging from 1.4×10^{14} Ω m to 1.7×10^{8} Ω m. Green LDPE/GC composites with 0.3 and 0.5 wt% GC showed an electrical resistivity of 4.8×10^{7} Ω m and 2.1×10^{7} Ω m, respectively, that is, 7 decades of magnitude lower than neat green LDPE. In these composites, the GC content was sufficient for the particles of the antistatic agent to come into contact with each other, forming an

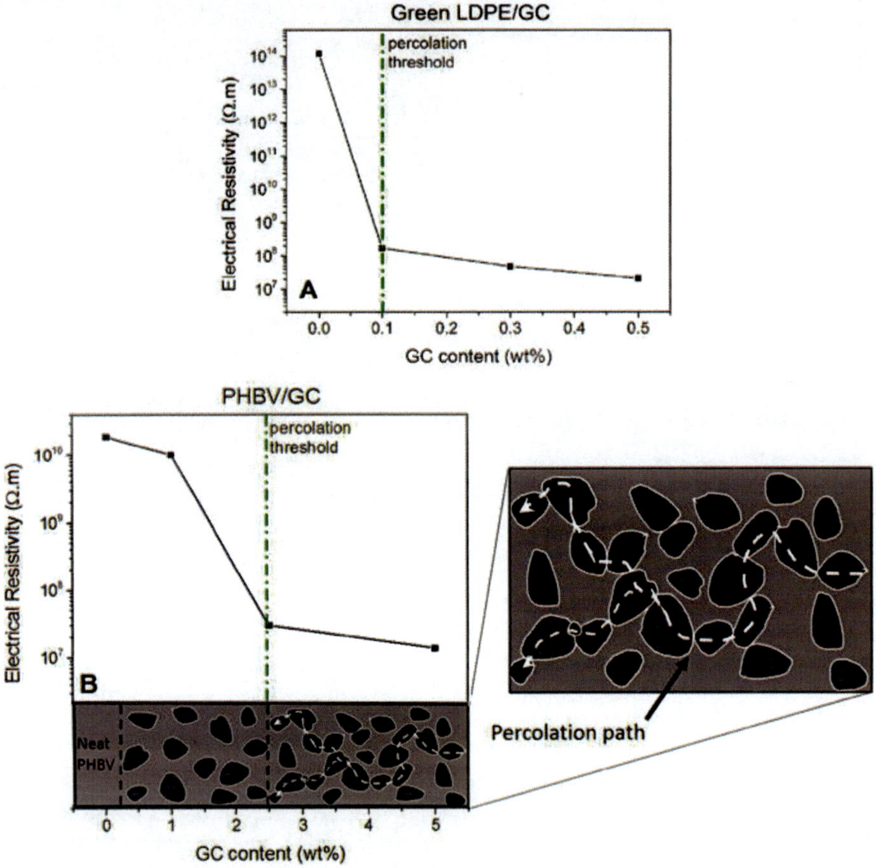

Fig. 7 Graphs of electrical resistivity as a function of GC content for **a** green LDPE/GC and **b** PHBV/GC composites with the scheme of formation of the electric percolation path. *Source* Author

electron percolation path in the material, resulting in a drastic reduction in electrical resistivity.

It can be seen in Fig. 7b that the percolation threshold of GC in the PHBV matrix is greater than that observed in the green LDPE matrix. According to Montanheiro et al. [42], the electrical resistivity of neat PHBV is around 2.0×10^{10} Ω m; that is, the electrical resistivity of the material is located in the upper boundary of the electrical resistivity range of antistatic materials (10^3–10^{10} Ω m) [17]. The addition of 1.0 wt% of GC to the PHBV matrix did not lead to any significant decrease in the electrical resistivity of PHBV. However, the addition of 2.5 wt% and 5.0 wt% GC content was responsible for the formation of an electron conduction path in the material, reducing the electrical resistivity of PHBV to 3.0×10^7 and 1.4×10^7 Ω m, respectively. The formation of electrical conduction path in PHBV composites is shown in the zoom of Fig. 7b. In this way, it is verified that the composites of green LDPE/GC with 0.1, 0.3, and 0.5 wt% GC and the composites of PHBV/GC with 2.5 and 5.0 wt% GC are materials that satisfy the electrical requirements to be classified as antistatic materials and may be used in the production of antistatic packaging. It is also verified that the 2.5 wt% GC content is the electrical percolation threshold of GC in PHBV matrix, which is slightly higher than the electrical percolation threshold of GC in green LDPE (0.1 wt%); that is, it is necessary to use smaller amounts of GC in green LDPE matrix to obtain materials with antistatic properties.

Santos et al. [54] produced composites of conventional LDPE with different GC contents (0.5, 1, 5, 10, 15, and 20 wt%) and also observed a relatively low electrical percolation threshold for GC (0.5 wt% GC) in the polymeric matrix, corroborating with the interesting results verified for the green LDPE/GC composites.

The biodegradation test in aqueous medium was performed only for neat PHBV and PHBV/2.5 GC composite, since it contained the lowest GC content necessary to impart antistatic properties to PHBV. Although green LDPE/GC composites present greater flexibility and a lower electrical percolation threshold than PHBV/GC composites, green LDPE/GC composites are recyclable, but not biodegradable materials. The PHBV/2.5 GC composites, on the other hand, present a good biodegradability, as shown in Fig. 8. Control samples (Fig. 8a, c) present homogeneous surface coloration and full physical integrity, while samples exposed to biodegradation in aqueous medium (Fig. 8b, d) present a lack of brightness and a loss in its surface homogeneity with the presence of cavities caused by the deterioration of the polymer matrix promoted by the action of the microorganisms.

Figure 9 shows the graph of the residual weight of neat PHBV and PHBV/2.5 GC composite as a function of biodegradation time. The samples of neat PHBV and PHBV/2.5 GC submitted to the biodegradation process in aqueous medium have a gradually decreasing weight over time. It is also observed that both neat PHBV and PHBV/2.5 GC composite show approximately the same decrease in weight (~52%) after 45 days of biodegradation in liquid medium, suggesting that the hydrophobic character of GC particles may not interfere in the microorganism's adhesion and proliferation on the PHBV matrix [7]. It is possible to suggest that in this biodegradation process, the microorganisms converted high molar mass polymer chains into

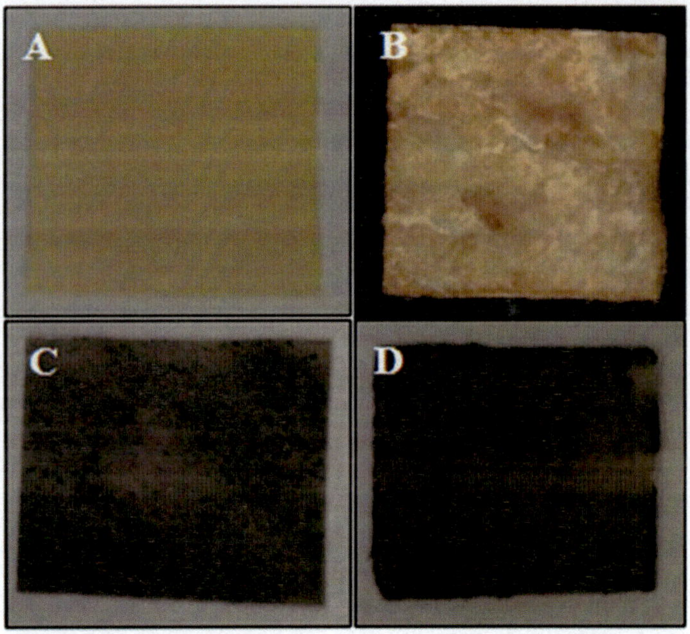

Fig. 8 Visual macroscopic analysis for neat PHBV: control neat PHBV sample (**a**) and neat PHBV exposed to biodegradation in aqueous medium during 45 days; control PHBV/2.5 GC (**c**) and PHBV/2.5 GC (**d**) exposed to biodegradation in aqueous medium during 45 days. *Source* Author

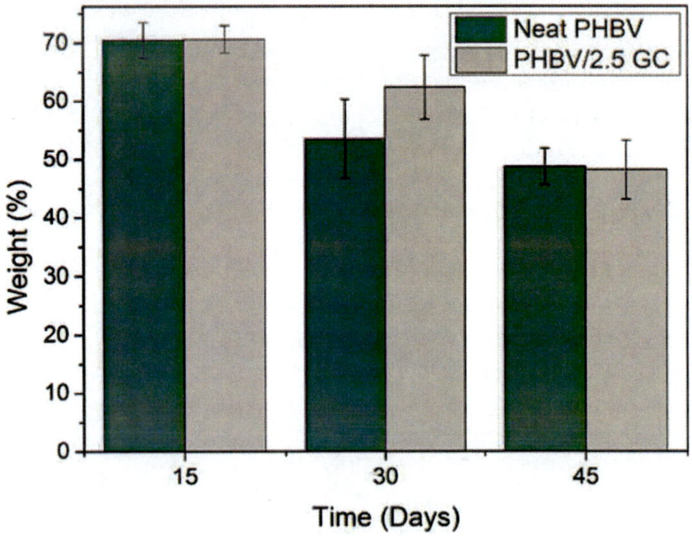

Fig. 9 Graph of residual weight of neat PHBV and PHBV/2.5 GC composite as a function of biodegradation time. *Source* Author

products of lower molar mass, such as water molecules and carbon dioxide, justifying the observed loss of weight [49].

4 Conclusions

The use of green composites is an interesting alternative for the production of antistatic packaging. These packaging must have low electrical resistivity to dissipate electrical charges through their structure and avoid damages in the electronic component.

Many techniques are studied to produce antistatic packaging like the addition of organic antistatic agents, such as ionic and non-ionic surfactants, metallic particles, metal oxide whiskers, the addition of intrinsically conducting polymers, use of conductive ionomers, the formation of polymeric electrolytes, and the addition of carbon materials like carbon black, carbon nanotubes, and glassy carbon (GC) in polymeric matrices. Among all these techniques, the use of an antistatic agent is an alternative that can be sustainable. GC can be used as an antistatic agent and is characterized as a carbon material originated from a renewable resource since it is obtained from the biomass of plants.

A sustainable way to produce antistatic packaging is using biodegradable or green polymers from renewable resources. Green LDPE and PHBV are promising alternatives to produce antistatic packaging with GC. The use of different polymeric matrices is an important step toward the diversification of the plastic market.

Mechanical, thermal, and electrical results confirm the good properties of these composites, and the low electrical resistivity can be reached with low contents of GC in green LDPE matrix (0.1–0.5 wt%) and PHBV matrix (2.5–5.0 wt%), especially when compared to carbon black (10–15 wt%). A reduction in electrical resistivity, on average of 99%, was achieved in both cases.

The use of biodegradable and green polymeric matrices is an advantage for the sector, contributing to the environment through the production of eco-friendly packaging.

Acknowledgements The authors are rather grateful to FAPESP (process 2018/09531-2), CNPq (Conselho Nacional de Desenvolvimento Científico e Tecnológico, process 310196/2018-3, 305123/2018-1), and the Coordenação de Aperfeiçoamento de Pessoal de Nível Superior—Brasil (CAPES)—Finance Code 001. The authors also thank Braskem (Brazil) for the donation of the Green LDPE.

References

1. Akaraonye E, Keshavarz T, Roy I (2010) Production of polyhydroxyalkanoates: the future green materials of choice. J Chem Technol Biotechn 85:732–743. https://doi.org/10.1002/jctb.

2392
2. Ambrosio-Martín J, Gorrasi G, Lopez-Rubio A et al (2015) On the use of ball milling to develop poly(3-hydroxybutyrate-co-3-hydroxyvalerate)-graphene nanocomposites (II)-Mechanical, barrier, and electrical properties. Appl Polym Sci 132: 42217. https://doi.org/10.1002/app.42217
3. ASTM (2014) D638-14: Standard Test Method for Tensile Properties of Plastics
4. ASTM (2018) D882-18: Standard Test Method for Tensile Properties of Thin Plastic Sheeting
5. Bao L, Lei J, Wang J (2013) Preparation and characterization of a novel antistatic poly(vinyl chloride)/quaternary ammonium based ion-conductive acrylate copolymer composites. J Electrostat 71:987–993. https://doi.org/10.1016/j.elstat.2013.09.001
6. Bélard L, Poncin-Epaillard F, Dole P et al (2013) Plasma-polymer coatings onto different biodegradable polyesters surfaces. Eur Polym 49:882–892. https://doi.org/10.1016/j.eurpolymj.2012.11.022
7. Belu AM, Guerrouani N, Baldo A et al (2007) Fluorinated-plasma coating on polyhydroxyalcanoate PHBV: effect on the biodegradation. J Fluor Chem 128:925–930. https://doi.org/10.1016/j.jfluchem.2007.03.018
8. Botelho EC, Scherbakoff N, Rezende MC (2001) Porosity control in glassy carbon by rheological study of the furfuryl resin. Carbon 39:45–52. https://doi.org/10.1016/S0008-6223(00)00080-4
9. Braga NF, LaChance AM, Liu B et al (2019) Influence of compatibilizer and carbon nanotubes on mechanical, electrical, and barrier properties of PTT/ABS blends. Adv Ind Eng Polym Res 2:121–125. https://doi.org/10.1016/j.aiepr.2019.07.002
10. Braskem (2020) Low density polyethylene SBF0323HC. https://www.braskem.com.br/busca-de-produtos?p=232. Accessed 10 Apr 2020
11. Callstrom MR, Neenan TX, McCreery RL et al (1990) Doped glassy carbon materials (DGC): low-temperature synthesis, structure, and catalytic behavior. J Am Chem Soc 112:4954–4956
12. Chan CM, Vandi LJ, Pratt S et al (2018) Mechanical properties of poly(3-hydroxybutyrate-co-3-hydroxyvalerate)/wood flour composites: effect of interface modifiers. J Appl Polym Sci 135:46828. https://doi.org/10.1002/app.46828
13. Chandra R, Rustgi R (1998) Biodegradable polymers. Prog Polym Sci 23:1273–1335
14. Che R, Yang W, Wang J et al (2010) Electrolyte-based antistatic plasticizer for soft poly(vinyl chloride) composites. J Appl Polym Sci 116:1718–1724. https://doi.org/10.1002/app.31600
15. Coutinho PLA, Morita AT, Cassinelli LF et al (2013) Braskem's ethanol to polyethylene process development. In: Imhof P, Waal JC (eds) Catalytic process development for renewable materials, 1st edn. Wiley-VCH, Weinheim, pp 149–166
16. Ding Y, Tang H, Zhang X et al (2008) Antistatic ability of 1-n-tetradecyl-3-methylimidazolium bromide and its effects on the structure and properties of polypropylene. Eur Polym J 44:1247–1251. https://doi.org/10.1016/j.eurpolymj.2008.01.030
17. ESD Association (2018) Part 3: basic ESD control procedures and materials. https://www.esda.org/esd-overview/esd-fundamentals/part-3-basic-esd-control-procedures-and-materials/. Accessed 10 Apr 2020
18. ESD Association (2020) Part 6: ESD Standards. https://www.esda.org/esd-overview/esd-fundamentals/part-6-esd-standards/. Accessed 10 Apr 2020
19. Feng X, Wang J, Zhang C (2018) Fabrication and characterization of antistatic epoxy composite with multi-walled carbon nanotube-functionalized melamine foam. RSC Adv 8:14740–14746. https://doi.org/10.1039/c8ra01044g
20. Ferrari PE, Rezende MC (1998) Carbono polimérico: processamento e aplicação. Polímeros 8:22–30. https://doi.org/10.1590/S0104-14281998000400005
21. Gaefke CB, Botelho EC, Ferreira NG et al (2007) Effect of furfuryl alcohol addition on the cure of furfuryl alcohol resin used in the glassy carbon manufacture. J Appl Polym Sci 106:2274–2281. https://doi.org/10.1002/app.26938
22. Ge C, Devar G (2017) Formation of polyvinyl alcohol film with graphene nanoplatelets and carbon black for electrostatic discharge protective packaging. J Electrostat 89:52–57. https://doi.org/10.1016/j.elstat.2017.07.004

23. Geyer R, Jambeck JR, Law KL (2017) Production, use, and fate of all plastics ever made. Sci Adv 3:25–29. https://doi.org/10.1126/sciadv.1700782
24. Gojny FH, Kopke U, Wichmanna MHG et al (2004) Carbon nanotube-reinforced epoxy-composites: enhanced stiffness and fracture toughness at low nanotube content. Compos Sci Technol 64:2363–2371. https://doi.org/10.1016/j.compscitech.2004.04.002
25. Harikrishnan G, Patro TU, Khakhar DV (2007) Reticulated vitreous carbon from polyurethane foam-clay composites. Carbon 45:531–535. https://doi.org/10.1016/j.carbon.2006.10.019
26. Huang HD, Liu CY, Zhang LQ et al (2015) Simultaneous reinforcement and toughening of carbon nanotube/cellulose conductive nanocomposite films by interfacial hydrogen bonding. ACS Sustain Chem Eng 3:317–324. https://doi.org/10.1021/sc500681v
27. Huang Y, Liu H, He P et al (2010) Nonisothermal crystallization kinetics of modified bamboo fiber/PCL composites. J Appl Polym Sci 116:2119–2125. https://doi.org/10.1002/app
28. Inagaki M, Kang F (2014) Materials science and engineering of carbon: fundamentals. Butterworth-Heinemann, Oxford
29. Jacobsen AJ, Mahoney S, Carter WB et al (2010) Vitreous carbon micro-lattice structures x. Carbon 49:1025–1032. https://doi.org/10.1016/j.carbon.2010.10.059
30. Jenkins GM, Kawamura K (1971) Structure of glassy carbon. Nature 231:175–176. https://doi.org/10.1038/231175a0
31. Jenkins GM, Kawamura K (1976) Polymeric carbons: carbon fibre, glass and char. Cambridge University Press, Cambridge
32. Jiang N, Abe H (2015) Crystallization and mechanical behavior of covalent functionalized carbon nanotube/poly (3-hydroxybutyrate-co-3-hydroxyvalerate) nanocomposites. J Appl Polym Sci 132:42136. https://doi.org/10.1002/app.42136
33. Jonas F, Heywang G (1994) Technical applications for conductive polymers. Electrochim Acta 39:1345–1347
34. Kalijadis A, Jovanović Z, Laušević M et al (2011) The effect of boron incorporation on the structure and properties of glassy carbon. Carbon 49:2671–2678. https://doi.org/10.1016/j.carbon.2011.02.054
35. Lemes AP, Soto-Oviedo MA, Waldman WR et al (2010) Effect of lignosulfonate on the thermal and morphological behavior of poly(3-hydroxybutyrate-co-3-hydroxyvalerate). J Polym Environ 18:250–259. https://doi.org/10.1007/s10924-010-0170-7
36. Lemes AP, Montanheiro TLA, Passador FR et al (2014) Nanocomposites of polyhydroxyalkanoates reinforced with carbon nanotubes: chemical and biological properties. In: Thakur VK, Thakur MK (eds) Eco-friendly polymer nanocomposites: processing and properties. Springer, New Delhi, pp 79–108
37. Lucas N, Bienaime C, Belloy C et al (2008) Polymer biodegradation: mechanisms and estimation techniques. Chemosphere 73:429–442. https://doi.org/10.1016/j.chemosphere.2008.06.064
38. Luckachan GE, Pillai CKS (2011) Biodegradable polymers—a review on recent trends and emerging perspectives. J Polym Environ 19:637–676. https://doi.org/10.1007/s10924-011-0317-1
39. Maki N, Nokano S, Sasaki H (2004) Development of a packaging material using non-bleed-type antistatic ionomer. Packag Technol Sci 17:249–256. https://doi.org/10.1002/pts.653
40. Modi S, Koelling K, Vodovotz Y (2011) Assessment of PHB with varying hydroxyvalerate content for potential packaging applications. Eur Polym J 47:179–186. https://doi.org/10.1016/j.eurpolymj.2010.11.010
41. Montané D, Salvadó J, Torras Č et al (2002) High-temperature dilute-acid hydrolysis of olive stones for furfural production. Biomass Bioenerg 22:295–304. https://doi.org/10.1016/S0961-9534(02)00007-7
42. Montanheiro TLA, Cristóvan FH, Machado JPB et al (2014) Effect of MWCNT functionalization on thermal and electrical properties of PHBV/MWCNT nanocomposites. J Mater Res 30:55–65. https://doi.org/10.1557/jmr.2014.303
43. Morschbacker A, Campos CES, Cassiano LC et al (2014) Bio-polyethylene. In: Oksman K, Mathew AP, Bismarck A et al (eds) Handbook of green materials, vol 4. World Scientific Publishing Company, California, pp 89–104

44. Nel A, Xia T, Mädler L et al (2006) Toxic potential of materials at the nanolevel. Science 311:622–627. https://doi.org/10.1126/science.1114397
45. Oishi SS, Botelho EC, Rezende MC et al (2017) Structural and surface functionality changes in reticulated vitreous carbon produced from poly(furfuryl alcohol) with sodium hydroxide additions. Appl Surf Sci 394:87–97. https://doi.org/10.1016/j.apsusc.2016.10.112
46. Oyama IC, de Souza GPM, Rezende MC et al (2020) A new eco-friendly green composite for antistatic packaging: green low-density polyethylene/glassy carbon. Polym Compos. https://doi.org/10.1002/pc.25572
47. Pandit JA, Sudarshan K, Athawale AA (2016) Electrically conductive epoxy-polyester-graphite nanocomposites modified with aromatic amines. Polymer 104:49–60. https://doi.org/10.1016/j.polymer.2016.09.084
48. Raza ZA, Abid S, Banat IM (2018) Polyhydroxyalkanoates: characteristics, production, recent developments and applications. Int Biodeterior Biodegrad 126:45–56. https://doi.org/10.1016/j.ibiod.2017.10.001
49. Rizzarelli P, Carroccio S (2014) Modern mass spectrometry in the characterization and degradation of biodegradable polymers. Anal Chim Acta 808:18–43. https://doi.org/10.1016/j.aca.2013.11.001
50. Rupp B, Ebner C, Rossegger E et al (2010) UV-induced crosslinking of the biopolyester poly(3-hydroxybutyrate)-co-(3-hydroxyvalerate). Green Chem 12:1796–1802. https://doi.org/10.1039/c0gc00066c
51. Sakintuna B, Yürüm Y (2005) Templated porous carbons: a review article. Ind Eng Chem Res 44:2893–2902. https://doi.org/10.1021/ie049080w
52. Salomez M, George M, Fabre P et al (2019) A comparative study of degradation mechanisms of PHBV and PBSA under laboratory-scale composting conditions. Polym Degrad Stab 167:102–113. https://doi.org/10.1016/j.polymdegradstab.2019.06.025
53. Sandler JKW, Kirk JE, Kinloch IA et al (2003) Ultra-low electrical percolation threshold in carbon-nanotube-epoxy composites. Polymer 44:5893–5899. https://doi.org/10.1016/S0032-3861(03)00539-1
54. Santos MS, Montagna LS, Rezende MC et al (2018) A new use for glassy carbon: development of LDPE/glassy carbon composites for antistatic packaging applications. J Appl Polym Sci 136:47204. https://doi.org/10.1002/app.47204
55. Shah AA, Hasan F, Hameed A et al (2008) Biological degradation of plastics: a comprehensive review. Biotechnol Adv 26:246265. https://doi.org/10.1016/j.biotechadv.2007.12.005
56. Sharma S (2018) Glassy Carbon: a promising material for micro and nanomanufacturing. Materials 11:1857. https://doi.org/10.3390/ma11101857
57. Silva LN, dos Anjos EGR, Morgado GFM et al (2019) Development of antistatic packaging of polyamide 6/linear low-density polyethylene blends-based carbon black composites. Polym Bull. https://doi.org/10.1007/s00289-019-02928-3
58. Silva TF, Menezes F, Montagna LS et al (2018) Preparation and characterization of antistatic packaging for electronic components based on poly(lactic acid)/carbon black composites. J Appl Polym Sci 136:47273. https://doi.org/10.1002/app.47273
59. Silva AP, Montanheiro TLA, Montagna LS et al (2019) Effect of carbon nanotubes on the biodegradability of poly (3-hydroxybutyrate-co-3-hydroxyvalerate) nanocomposites. J Appl Polym Sci 136:48020. https://doi.org/10.1002/app.48020
60. Sun Q, Li W (2016) Inorganic-Whisker-reinforced polymer composites: synthesis, properties and applications. CRC Press, London
61. SPI Supplies (2020) Glassy carbon. https://www.2spi.com/category/labware-crucibles-glassy-carbon/labware/. Accessed 15 Apr 2020
62. Szeluga U, Kumanek B, Trzebicka B (2015) Synergy in hybrid polymer/nanocarbon composites. A review. Compos Part A 73:204–231. https://doi.org/10.1016/j.compositesa.2015.02.021
63. Szeluga U, Pusz S, Kumanek B et al (2016) Influence of unique structure of glassy carbon on morphology and properties of its epoxy-based binary composites and hybrid composites with carbon nanotubes. Compos Sci Technol 134:72–80. https://doi.org/10.1016/j.compscitech.2016.08.004

64. Tian Y, Zhang X, Geng HZ et al (2017) Carbon nanotube/polyurethane films with high transparency, low sheet resistance and strong adhesion for antistatic application. RSC Adv 7:53018–53024. https://doi.org/10.1039/c7ra10092b
65. Turan D, Sirin H, Ozkoc G (2011) Effects of POSS particles on the mechanical, thermal, and morphological properties of PLA and plasticised PLA. J Appl Polym Sci 121:1067–1075. https://doi.org/10.1002/app
66. Vidhate S, Innocentini-Mei L, D'Souza NA (2012) Mechanical and electrical multifunctional poly(3-hydroxybutyrate-co-3-hydroxyvalerate)-multiwall carbon nanotube nanocomposites. Polym Eng Sci 56:1367–1374. https://doi.org/10.1002/pen.23084
67. Wang Y, Zhang C, Du Z et al (2013) Synthesis of silver nanoparticles decorated MWCNTs and their application in antistatic polyetherimide matrix nanocomposite. Synth Met 182:49–55. https://doi.org/10.1016/j.synthmet.2013.09.006
68. Wang H, Xie G, Fang M et al (2015) Electrical and mechanical properties of antistatic PVC films containing multi-layer graphene. Compos Part B Eng 79:444–450. https://doi.org/10.1016/j.compositesb.2015.05.011
69. Wang J, Bao L, Zhao H et al (2012) Preparation and characterization of permanently anti-static packaging composites composed of high impact polystyrene and ion-conductive polyamide elastomer. Compos Sci Technol 72:976–981. https://doi.org/10.1016/j.compscitech.2012.03.006
70. Wang J, Che R, Yang W et al (2011) Biodegradable antistatic plasticizer based on citrate electrolyte doped with alkali metal salt and its poly(vinyl chloride) composites. Polym Int 60:344–352. https://doi.org/10.1002/pi.2952
71. Weng YX, Wang Y, Wang XL et al (2010) Biodegradation behavior of PHBV films in a pilot-scale composting condition. Polym Test 29:579–587. https://doi.org/10.1016/j.polymertesting.2010.04.002
72. Wypych G (2012) Handbook of polymers. ChemTec Publishing, Toronto
73. Zhang X, Liu J, Wang Y et al (2017) Effect of polyamide 6 on the morphology and electrical conductivity of carbon black-filled polypropylene composites. R Soc Open Sci 4:170769. https://doi.org/10.1098/rsos.170769
74. Zheng W, Chen WG, Ren SX et al (2019) Interfacial structures and mechanisms for strengthening and enhanced conductivity of graphene/epoxy nanocomposites. Polymer 163:171–177. https://doi.org/10.1016/j.polymer.2018.12.055
75. Zhou Z, Chu L, Tang W et al (2003) Studies on the antistatic mechanism of tetrapod-shaped zinc oxide whisker. J Electrostat 57:347–354. https://doi.org/10.1016/S0304-3886(02)00171-7
76. Zhu Y, Zhao Y, Zhang X et al (2016) Metal filaments/nano-filler filled hybrid polymer fibers with improved conductive performance. Mater Lett 173:26–30. https://doi.org/10.1016/j.matlet.2016.03.007

Chapter 18
Green Preparation and Environmental Applications of Some Electrospun Fibers

Juanjuan Yin, Qingrui Zhang, Lexin Zhang, Jingxin Zhou, and Tifeng Jiao

1 Introduction

Electrospinning technology is a technology for preparing polymer micro–nano-fibers [1–3]. This technology is simple and easy to operate and has high production efficiency. A thin-film composite of continuous nanoscale fibers can be prepared from a polymer solution. The electrostatic spinning device is mainly composed of three parts: high voltage power supply, spinning head, and receiving screen. Under the action of electric field force, the charges on the surface of the droplet at the tip of the spinning needle are concentrated and repel each other, and the droplet is elongated to form a Taylor cone [4, 5]. After spraying out, it solidifies in the air and finally forms disorderly arranged nanofibers on the receiving device. Under the impetus of the stepping pump, the spinning continued, and the fine stream ejected from the needle was continuously deposited on the receiving screen and finally formed a film product woven by nanofibers [6, 7]. In recent decades, the research on electrospinning has mainly focused on the development of electrospinning nanofiber raw materials, multi-component polymer electrospinning, and electrospinning jet instability models.

Electrospinning technology has many outstanding advantages, such as no complicated and expensive equipment and instruments, and lower experimental costs. In addition, the operation of electrospinning is simple and easy, and the applicable raw materials are widely free from harsh requirements. The most important thing is that the prepared fibers are at the nanometer level, and the average fiber diameter is generally between tens and hundreds of nanometers. Therefore, the resulting film has the characteristics of controllable fiber diameter [8], high tensile strength [9,

J. Yin · Q. Zhang · L. Zhang · J. Zhou · T. Jiao (✉)
Hebei Key Laboratory of Applied Chemistry, School of Environmental and Chemical Engineering, Yanshan University, Qinhuangdao 066004, China
e-mail: tfjiao@ysu.edu.cn

S. Thomas and P. Balakrishnan (eds.), *Green Composites*,
Materials Horizons: From Nature to Nanomaterials,
https://doi.org/10.1007/978-981-15-9643-8_18

10], and very large surface area [11–17]. Therefore, these advantages make the electrospun nanofiber film have a wide range of potential applications in many fields, including tissue engineering [18], drug sustained-release [19], nanosensors [20], energy applications [21], biochips and catalyst loading.

At present, the construction of nanostructures of various materials (such as polymers, inorganics, and multi-component composites) has been achieved through electrospinning technology [22–24], which has shown great abilities in the fields of catalysis, drug carriers, and filtration potential [25, 26]. Liu et al. have successfully prepared a new composite film based on silver nano-particles modified polyvinyl alcohol/polyacrylic acid/carboxyl functionalized graphene oxide (PVA/PAA/GO-COOH@AgNPs) as Efficient Dye Photocatalyst Materials for Wastewater Treatment [27]. Panaitescu et al. processed modified poly(3-hydroxybutyrate) by using nanocellulose and plasma technology as a candidate material for food packaging applications [28]. Rauwel et al. introduced several nanomaterials in the book "Application and Behavior of Nanomaterials in Water Treatment", such as graphene/CNT and nanostructured Prussian blue hybrid nanocomposites, controlled growth of LDH films and rod-shaped MnO nanocomposites, and so on. It is applied to sewage treatment, and a good purification effect is obtained [29]. In addition, the fiber material shows good adsorption capacity due to its microporous structure. This chapter provides a comprehensive review of the preparation process of various electrospun fiber materials and applies these composite nanofiber materials to the fields of adsorption, catalysis, and air filtration. It also provides challenges and prospects to inspire more exciting developments in the future.

2 Preparation of Electrospinning Nanocomposite Membrane and Its Adsorption and Degradation of Organic Dyes in Wastewater

The preparation of electrospun nanocomposite membranes has been a hot topic in many fields, such as chemistry, materials science, and nanotechnology. In this section, we will focus on the application of different types of polymer nanofilms in the adsorption and degradation of organic dyes: (i) bioinspired polydopamine sheathed nanofibers containing carboxylate graphene oxide nanosheet for high-efficient dyes scavenger, (ii) preparation of TiO_2 nanoparticles modified electrospun nanocomposite membranes toward efficient dye degradation for wastewater treatment, (iii) polydopamine-coated electrospun poly(vinyl alcohol)/poly(acrylic acid) membranes as efficient dye adsorbent with good recyclability, (iv) fabrication and highly efficient dye removal characterization of beta-cyclodextrin-based composite polymer fibers by electrospinning, and (v) self-assembled AgNP-containing nanocomposites constructed by electrospinning as efficient dye photocatalyst materials for wastewater treatment.

2.1 Bioinspired Polydopamine Sheathed Nanofibers Containing Carboxylate Graphene Oxide Nanosheet for High-Efficient Dyes Scavenger

Firstly, Xing et al. successfully prepared a new type of hierarchical bioinspired nanocomposite materials of poly(vinyl alcohol)/poly(acrylic acid)/carboxylate graphene oxide nanosheet@polydopamine (PVA/PAA/GO-COOH@PDA) by electrospinning technique, thermal treatment, and polydopamine modification [30]. The obtained composite membranes are composed of polymeric nanofibers with carboxylate graphene oxide nanosheets, which are anchored on the fibers by the heat-induced crosslinking reaction. The preparation process demonstrates an eco-friendly and controllable manner. These as-formed nanocomposites were characterized by various morphological methods and spectral techniques. Due to the unique polydopamine and graphene oxide containing structures in composites, the as-obtained composite demonstrates well-efficient adsorption capacity toward dye removal, which is primarily due to the specific surface area of electrospun membranes and the active polydopamine/graphene oxide components [31–33]. In addition, it should be noted that the adsorption capacities of the as-obtained PVA/PAA/GO-COOH@PDA nanocomposites on MB show better performance than two other used dyes. The main reason for the difference can be speculated to the matched strong π–π stacking and electrostatic interactions between nanocomposites and MB molecules. At the same time, the abundant amino and hydroxyl groups in the PDA surface give more adsorbent activity points for dye molecules, which demonstrate the tailored strategy to improve the absorption performance (Figs. 1 and 2).

Therefore, the present work is expected to open a new avenue for the design and preparation of eco-friendly electrospun composites loaded with functional GO and nanoparticles, which could enhance the practical application in wastewater treatment by using functionalized composite nanofibers materials.

2.2 Preparation of TiO_2 Nanoparticles Modified Electrospun Nanocomposite Membranes Toward Efficient Dye Degradation for Wastewater Treatment

Firstly, Hou et al. prepared nano-polyvinyl alcohol/polyacrylic acid/carboxyl functionalized graphene oxide nanocomposite films by using electrospinning technology [34]. The composite membrane prepared by electrospinning technology is composed of polymer nanofibers with carboxyl functionalized graphene oxide sheets, which are anchored to the fibers through a thermally induced crosslinking reaction. TiO_2 nanoparticles are deposited uniformly and uniformly on the surface of the resulting composite film. The preparation process of the nano-composite membrane is simple, green, and environmentally friendly, and it is easy to operate and control. The prepared composite membranes exhibited an effective photocatalytic ability for dye

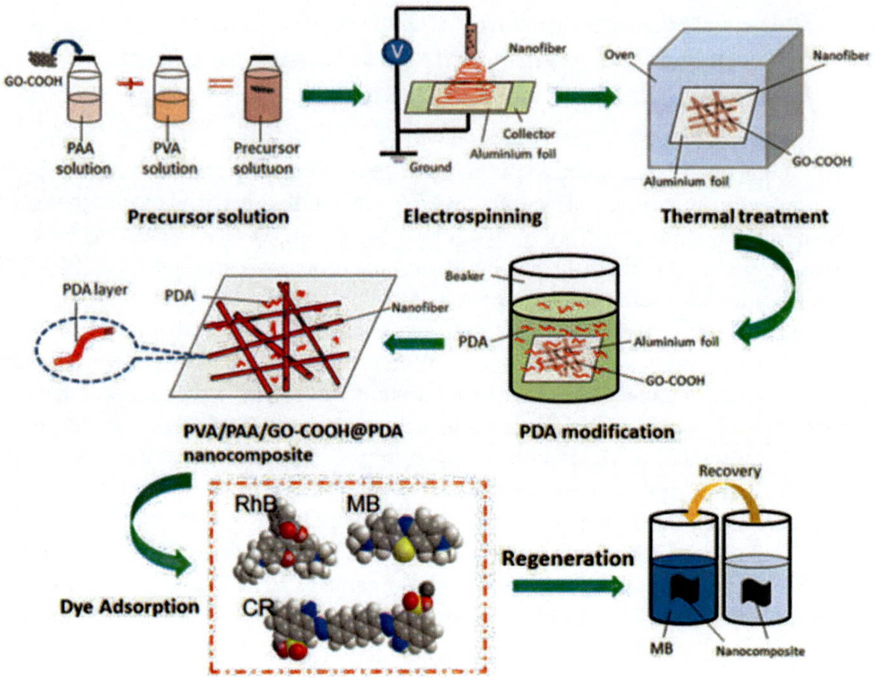

Fig. 1 Schematic illustration of the fabrication and dye adsorption of PVA/PAA/GO-COOH@PDA nanocomposites by electrospinning and thermal treatment. Reproduced with permission from Ref. [30]. Copyright 2017, American Chemical Society

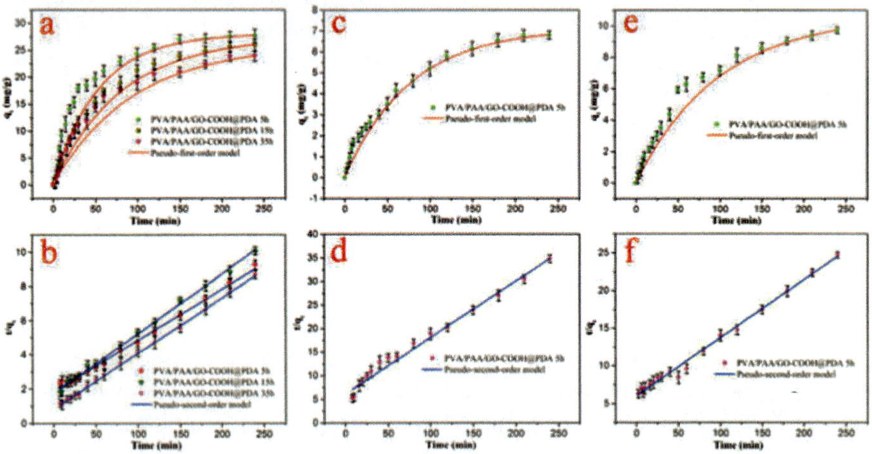

Fig. 2 Adsorption kinetics curves of as-prepared PVA/PAA/GO-COOH@PDA nanocomposites on MB (**a**, **b**), RhB (**c**, **d**), and CR (**e**, **f**) at 298 K. Reproduced with permission from Ref. [30]. Copyright 2017, American Chemical Society

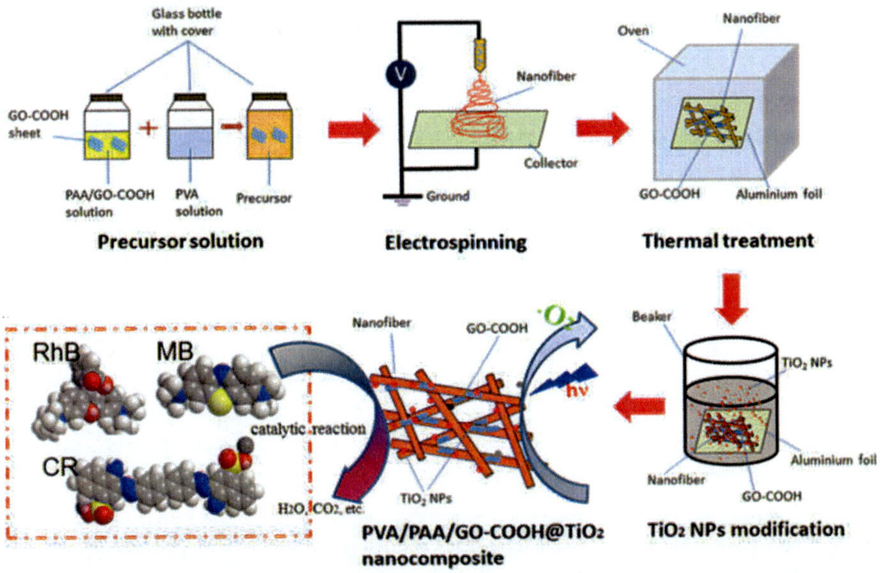

Fig. 3 Schematic illustration of the fabrication of PVA/PAA/GO-COOH@TiO_2 nanocomposite by electrospinning and thermal treatment. Reprinted with permission from Ref. [34]. Copyright 2017, Elsevier Ltd

degradation, which was mainly attributed to the specific surface area of electrospun membranes and the photoactivity of TiO_2 nanoparticles. In addition, the composite membrane reported here is easy to regenerate, which indicates potential large-scale applications in wastewater treatment and dye removal (Figs. 3, 4 and Table 1).

At the same time, the author also studied the photocatalytic performance of the prepared PVA/PAA/GO-COOH@TiO_2 nanocomposite membrane to three model dye solutions (CR, RhB, and MB). The degradation program currently studied was characterized by placing the obtained PVA/PAA/GO-COOH@TiO_2 nanocomposites in different aqueous dye solutions [35–40]. In addition, this catalytic experiment was measured and repeated three times. The degradation kinetics experiments of the prepared PVA/PAA/GO-COOH@TiO_2 nanocomposites were carried out. The results are shown in Fig. 4. This work is expected to open up a new way for the design and preparation of environmentally friendly electrospun composite materials containing functional GO and nanoparticles. The use of functional composite nanofiber materials can enhance the practical application of wastewater treatment.

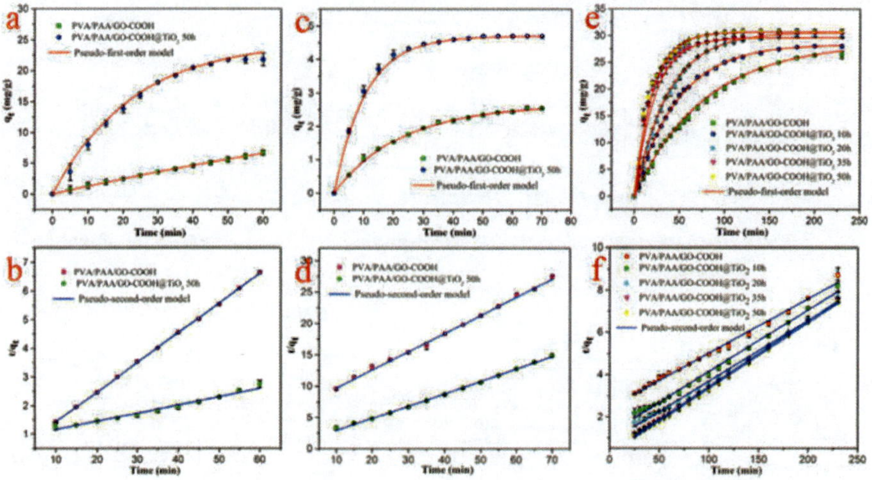

Fig. 4 Photocatalytic kinetics curves of as-prepared PVA/PAA/GO-COOH@TiO_2 nanocomposite on CR (**a** and **b**), RhB (**c** and **d**), and MB (**e** and **f**) at 298 K. Reprinted with permission from Ref. [34]. Copyright 2017, Elsevier Ltd

2.3 *Polydopamine-Coated Electrospun Poly(Vinyl Alcohol)/poly(Acrylic Acid) Membranes as Efficient Dye Adsorbent with Good Recyclability*

Inspired by the characteristics of adhesion proteins in marine mussels, dopamine, which is a catecholamine, can self-polymerize under alkaline reaction conditions and form a polydopamine (PDA) film on almost all types of substrates [41, 42]. The PDA coating can form a highly stable polymer layer on the target surface and shows special adhesion in the presence of residual catechol groups on the PDA layer, which helps to further react with appropriate molecules; it also provides the possibility for customized PDA coatings for various applications [43]. For example, PDA membranes are often used as modifiers to improve the hydrophilicity and reactivity of target substrates (such as clay [44], polystyrene nanofibers [45], and even polytetrafluoroethylene [46]). In particular, Gao et al. prepared a PDA-functionalized graphene hydrogel in a one-step process and found that the PDA-coated graphene hydrogel has good adsorption capacity for heavy metals and organic dyes in wastewater [47]. Recently, nanoparticles decorated with PDA (such as Fe_3O_4 [48] and natural zeolite [49]) have also been synthesized and used to remove various pollutants. However, considering the above-environmental issues, we should focus on the use of polydopamine as an adsorbent to remove dye contaminants in water for more research.

Table 1 Kinetic parameters of PVA/PAA/GO-COOH nanocomposite and PVA/PAA/GO-COOH@TiO_2 nanocomposite for CR, RhB, and MB (e and f) degradations and removal at 298 K (experimental data from Fig. 4)

CR	Pseudo-first-order model			Pseudo-second-order model		
	q_e (mg/g)	R^2	k_1 (min^{-1})	q_e (mg/g)	R^2	k_2 (g/mg min)
PVA/PAA/GO-GOOH	8.661	0.964	0.1204	9.662	0.9998	0.0272
PVA/PAA/G0-C00H@$Ti0_2$ 50 h	24.52	0.9955	0.0447	33.57	0.9779	0.0011
RhB	Pseudo-first-order model			Pseudo-second-order model		
	q_e (mg/g)	R^2	k_1 (min^{-1})	q_e (mg/g)	R^2	k_2 (g/mg min)
PVA/PAA/GO-GOOH	2.642	0.9981	0.0442	3.431	0.9976	0.0125
PVA/PAA/G0-C00H@$Ti0_2$ 50 h	4.691	0.9997	0.096	5.112	0.9983	0.0387
MB	Pseudo-first-order model			Pseudo-second-order model		
	q_e (mg/g)	R^2	k_1 (min^{-1})	q_e (mg/g)	R^2	k_2 (g/mg min)
PVA/PAA/GO-GOOH	28.56	0.9989	0.0129	38.79	0.9939	2.73×10^{-4}
PVA/PAA/G0-C00H@$Ti0_2$ 10 h	28.42	0.998	0.02096	4.29	0.9949	6.86×10^{-4}
PVA/PAA/G0-C00H@$Ti0_2$ 20 h	30.56	0.9971	0.02671	34.77	0.994	1.03×10^{-3}
PVA/PAA/G0-C00H@$Ti0_2$ 35 h	29.66	0.9879	0.04191	32.16	0.9995	2.89×10^{-3}
PVA/PAA/G0-C00H@$Ti0_2$ 50 h	30.45	0.9903	0.05228	32.22	0.9989	3.10×10^{-3}

Reprinted with permission from Ref. [34]. Copyright 2017, Elsevier Ltd

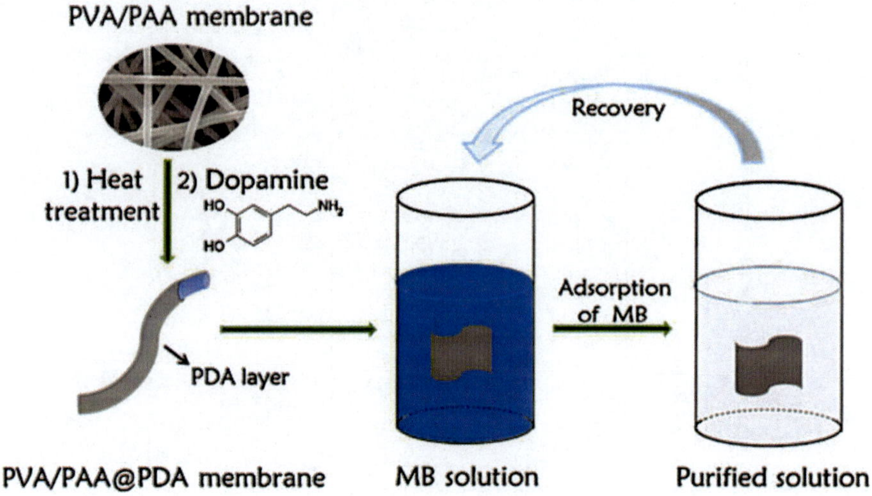

Fig. 5 Schematic illustration of the preparation of PVA/PAA@PDA membranes and their applications for dye adsorption. Reprinted with permission from Ref. [50]. Copyright 2015, Elsevier Ltd

Yan et al. prepared a free-standing polyvinyl alcohol/polyacrylic acid (PVA/PAA) film with polydopamine (PDA) coating based on electrospinning and self-polymerization of dopamine [50]. The preparation process is simple, green, controllable, and low energy consumption, and there are no strict restrictions on the reaction conditions. Thanks to the high specific surface area of electrospun membranes and the rich "adhesive" functional groups of polydopamine, the membranes produced exhibited effective adsorption properties for methyl blue, with an adsorption capacity of up to 1147.6 mg g^{-1}. Moreover, compared with other nanoparticle adsorbents, the prepared free-standing membrane has high flexibility and is easy to handle and recycle, and the most important thing is easy to elute and regenerate so that it has potential applications in wastewater treatment prospect (Figs. 5 and 6).

2.4 Fabrication and Highly Efficient Dye Removal Characterization of Beta-Cyclodextrin-Based Composite Polymer Fibers by Electrospinning

As a class of cyclic oligosaccharides, cyclodextrins have hollow cones, which have external hydrophilicity and internal hydrophobicity. This special structure has many special physical and chemical properties. It can selectively bind small organic molecules in aqueous solution, and the formed inclusion complex has different degrees of stability [51–53]. Therefore, cyclodextrins and its derivatives are widely

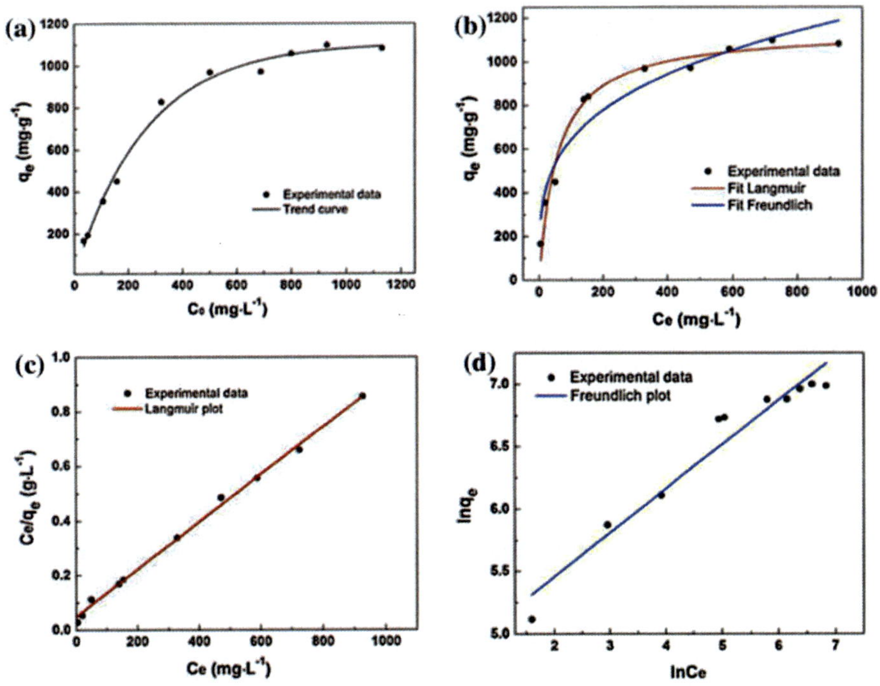

Fig. 6 Adsorption isotherms (**a** and **b**) and the corresponding Langmuir plot (**c**), Freundlich plot (**d**) for MB adsorption on PVA/PAA@PDA-15 membrane. Reprinted with permission from Ref. [50]. Copyright 2015, Elsevier Ltd

used in medicine, food, chemical engineering materials, especially wastewater treatment [54–58]. For example, Li et al. prepared a material based on cyclodextrins to remove malachite green [59]. The adsorption results are in accordance with the Langmuir model, and the maximum adsorption capacity reaches 91.9 mg/g. Yilmaz et al. synthesized two polymers based on β-cyclodextrin with the aid of 4, 4′-methylene-bis-phenyldiisocyanate (MDI) or hexamethylene diisocyanate (HMDI) [60]. These materials can remove azo dyes and aromatic amines, and the main adsorption mechanism is the host–guest interaction. At present, some cyclodextrin-based fiber systems by electrospinning have been reported. For example, Cui et al. described the use of plasma-treated polyethylene oxide-β-cyclodextrin nanofibers to enhance antibacterial activity [61]. Celebioglu et al. demonstrated electrospinning of polymer-free nanofiber structures formed from inclusion compounds between hydroxypropyl-β-cyclodextrin and vitamin E [62]. The prepared vitamin E-containing fiber web provides enhanced photostability of sensitive vitamin E by inclusion complexes even after exposure to UV light.

Guo et al. prepared a new type of composite fiber adsorption material composed of ε-polycaprolactone (PCL) and β-cyclodextrin-based polymer (PCD) by electrospinning, used to solve the increasing pollution of azo dye problem [63]. More importantly, PCD was chosen because of its infinite long chain and cavity structure. Due to a large number of free cyclodextrin pores on the fiber surface, the resulting membrane helps to form more host–guest interactions. Therefore, according to previous reports [64], its excellent selective adsorption capacity can be foreseen. During the electrospinning process, long-chain polymer molecules may occupy some cyclodextrin cavities, which is not conducive to the interaction between the host and the guest and even leads to the reduction of poor adsorption. However, our PCL/(n%) PCD composite fiber has countless holes, which can guarantee selective adsorption capacity. Therefore, it is obvious that the PCL/(n%) PCD composite fibers obtained can exhibit significant adsorption capacity for azo dyes and have host–guest interactions. Moreover, the introduction of a β-cyclodextrin polymer can effectively improve the mechanical strength and stability of the membrane. This shows that the obtained composite material has great potential, which can help the problem of azo dye pollution in wastewater treatment (Figs. 7, 8 and 9).

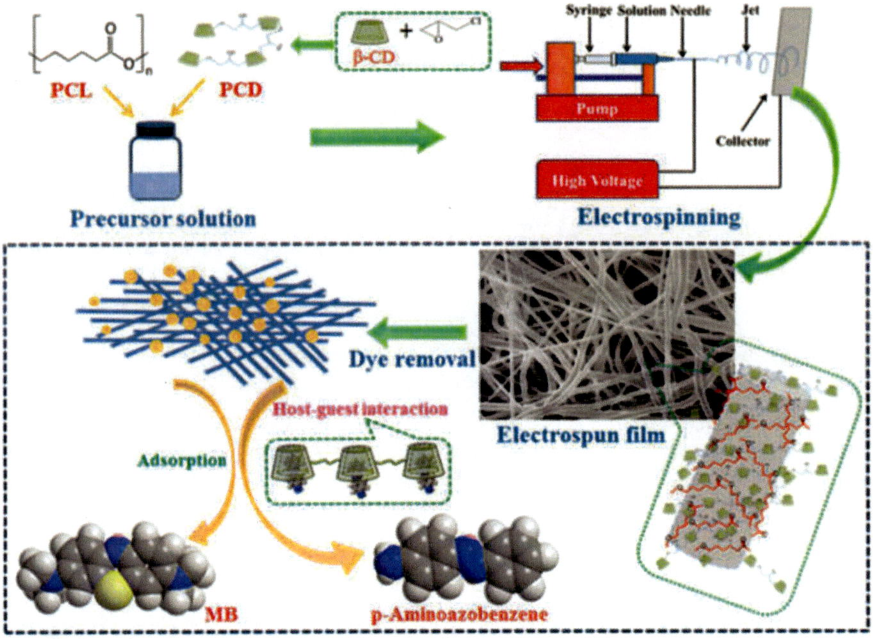

Fig. 7 Schematic illustration of preparation and application in dye removal of PCL/(n%)PCD composite fibers by electrospinning. Reprinted with permission from Ref. [63]. Copyright 2019, MDPI Ltd

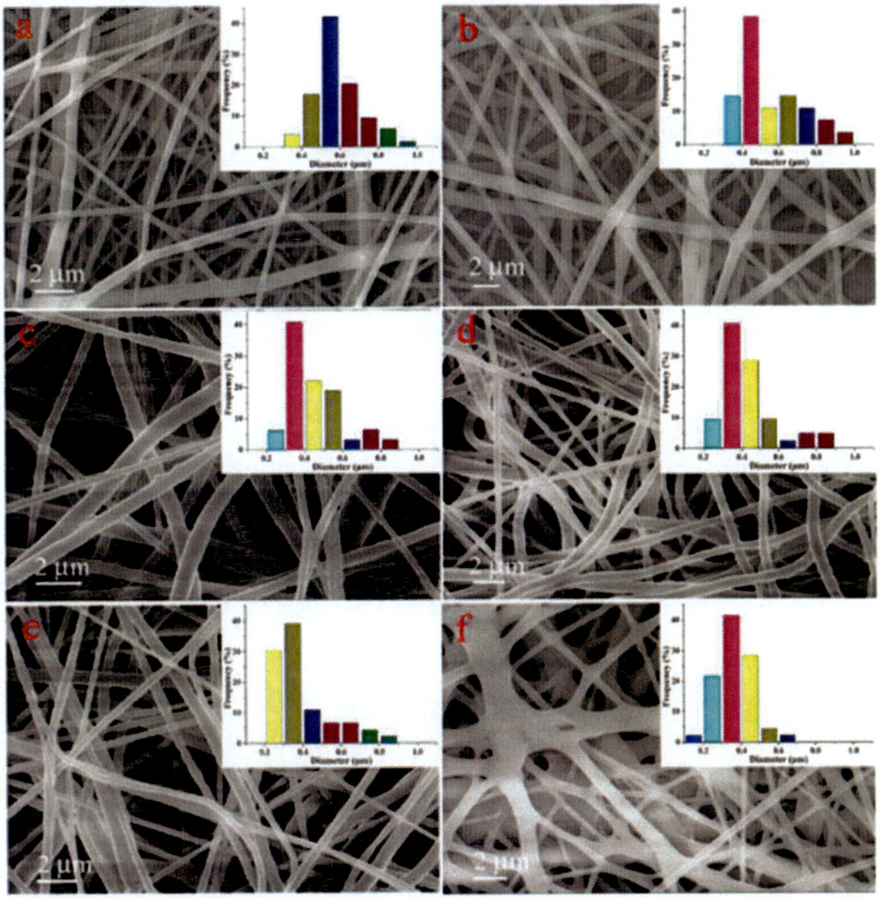

Fig. 8 SEM images and diameter distribution histograms of neat PCL fibers (**a**), PCL/(10%)PCD(**b**), PCL/(20%)PCD(**c**), PCL/(30%)PCD (**d**), PCL/(40%)PCD (**e**), and PCL/(50%)PCD (**f**). Reprinted with permission from Ref. [63]. Copyright 2019, MDPI Ltd

2.5 Self-assembled AgNP-Containing Nanocomposites Constructed by Electrospinning as Efficient Dye Photocatalyst Materials for Wastewater Treatment

Graphene oxide (GO)-based nanocomposite fibers have attracted great interest due to their adjustable dispersibility, large oxygen-containing functional groups, and special chemical modification and preferred reaction positions/sites [65]. Liu et al. have successfully prepared a new composite film based on silver nanoparticles-modified polyvinyl alcohol/polyacrylic acid/carboxyl functionalized graphene oxide (PVA/PAA/GO-COOH@AgNPs) [27]. The new composite membrane utilizes the advantages of graphene oxide (GO) nanocomposites and many novel advantages of

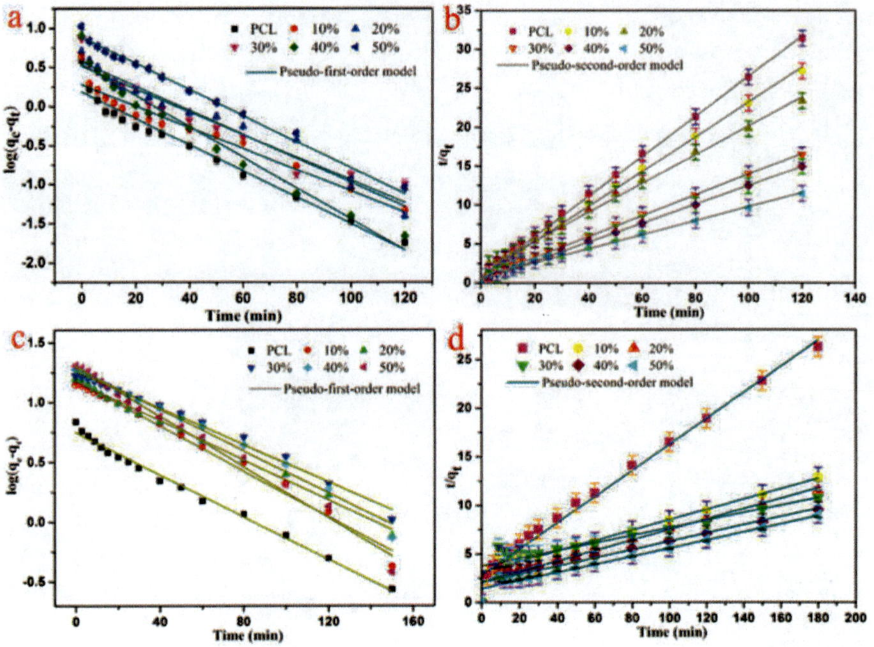

Fig. 9 Kinetic adsorption of MB (**a**,**b**) and 4-aminoazobenzene (**c**, **d**). Reprinted with permission from Ref. [63]. Copyright 2019, MDPI Ltd

electrospun fibers. At the same time, the Ag nanoparticles produced by the reduction of ascorbic acid solution to $AgNO_3$ are firmly fixed in the nanometer using hydrogen bonding and electrostatic interaction with the surface of the fiber. First, due to its chemical and mechanical properties and large specific surface area, the prepared PVA/PAA/GO-COOH system was selected as the matrix material for electrospinning [66]. Secondly, compared with conventional organic solvents, deionized water is used as a solvent for preparing electrospinning solutions, which has the characteristics of low cost and environmental friendliness. Most importantly, Ag nanoparticles show that they can achieve good stability between the flexible surface of the fiber and the effective GO nanosheets and function under visible light. The powerful π–π force in the GO sheet can make various dyes in water have strong adsorption force. In addition, the fixed carboxyl groups in GO nanosheets can generate strong electrostatic interactions through their highly negatively charged characteristics, which can promote the diffusion and enrichment of target dyes. For the degradation of MB, the prepared PVA/PAA/GO-COOH@AgNPs nanocomposite membranes still have significant catalytic activity even after eight catalytic degradations at room temperature. Therefore, the author provides a green and novel method to prepare highly efficient dye photocatalytic wastewater treatment materials (Fig. 10).

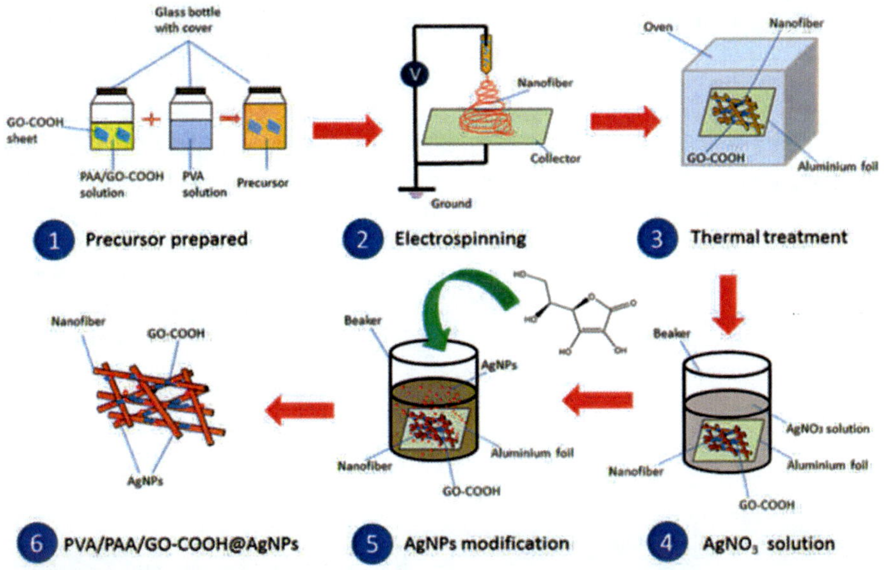

Fig. 10 Schematic illustration of the fabrication of PVA/PAA/GO-COOH@AgNPs nanocomposites by electrospinning and thermal treatment. Reprinted with permission from Ref. [27]. Copyright 2018, MDPI Ltd

3 Preparation of Electrospun Nanofiber and Characterization of Catalytic Performance

3.1 Scalable Fabrication of Nanoporous Carbon Fiber Films as Bifunctional Catalytic Electrodes for Flexible Zn-Air Batteries

With the rapid development of flexible and wearable optoelectronic devices, there is an urgent need to use flexible high-density energy storage devices as the power source [67]. Recently, people have made tremendous efforts to develop flexible lithium-ion batteries and supercapacitors. However, due to the low energy density of the batteries and the limited cycle life, it has become a research difficulty [68]. Because metal–air batteries are said to have high energy capacity [69], they are the next generation of promising wearable optoelectronic products for energy storage devices. Of particular concern is that zinc–air batteries have received extensive research and development attention due to their high energy density, low cost, and high safety [70]. However, most cathodes currently used in zinc–air batteries are bulky and rare. Meet the specific requirements of flexible Zn-air batteries. In addition, the development of highly efficient dual-functional electrocatalysts for the oxygen reduction reaction (ORR) and oxygen release reaction (OER) of flexible rechargeable zinc–air batteries remains a huge challenge [71], although important progress has been made recently used in traditional zinc–air batteries [72] (Figs. 11, 12 and 13).

Liu et al. found that newly developed nanoporous carbon nanofiber films (NCNFs) with the large specific surface areas are flexible and show that they are excellent when used as air cathodes in liquid Zn-air batteries used in ambient air [73]. The performance has a high open-circuit voltage (1.48 V), maximum power density (185 mW cm^{-2}), and energy density (776 Wh kg-1) and has a large specific surface area (1249 m^2 g^{-1}), high electrical conductivity (147 S m^{-1}), moderate tensile strength (1.89 MPa), and tensile modulus (0.31 GPa). They show excellent performance for ORR (initial potential = 0.97 V vs RHE; limiting current density = 4.7 mA cm^{-2}) and OER (initial potential = 1.43 V vs RHE, potential = 1.84 V @ 10 mA cm^{-2}) dual-function electrocatalytic activity.

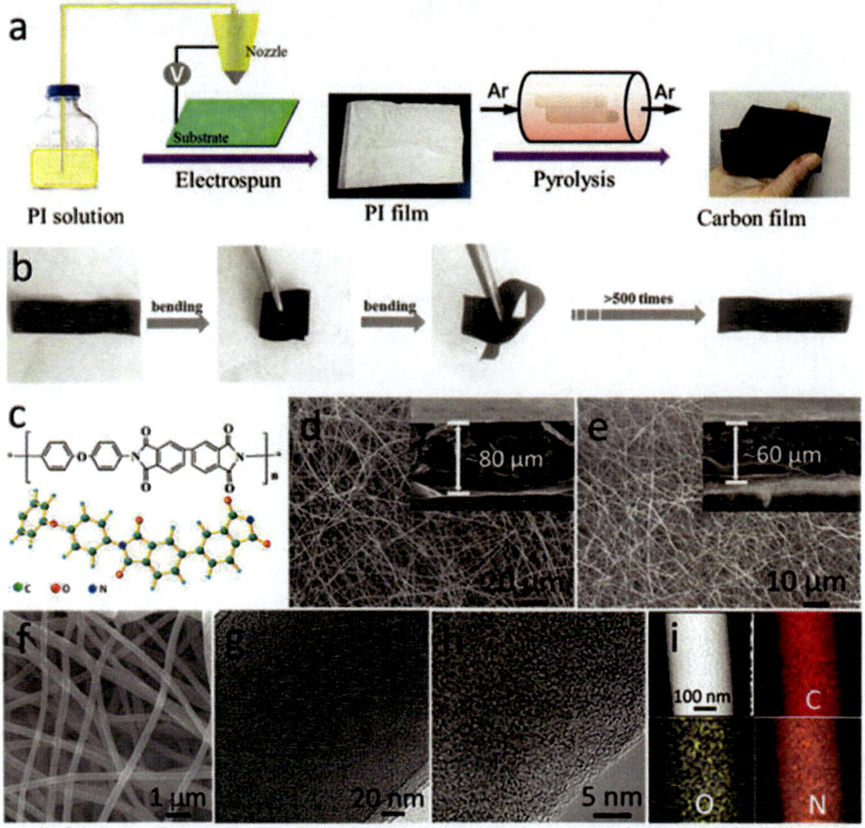

Fig. 11 **a** Schematic representation of the fabrication procedure toward the NCNF. **b** Photographs of the resultant flexible NCNF. **c** Chemical structure of PI polymer. SEM images of **d** the pristine PI film and **e**, **f** NCNF-1000. **g**, **h** HRTEM, **i** STEM images and corresponding elemental mapping images of C, O, N of NCNF-1000. Reprinted with permission from Ref. [73]. Copyright 2016, Wiley-Blackwell

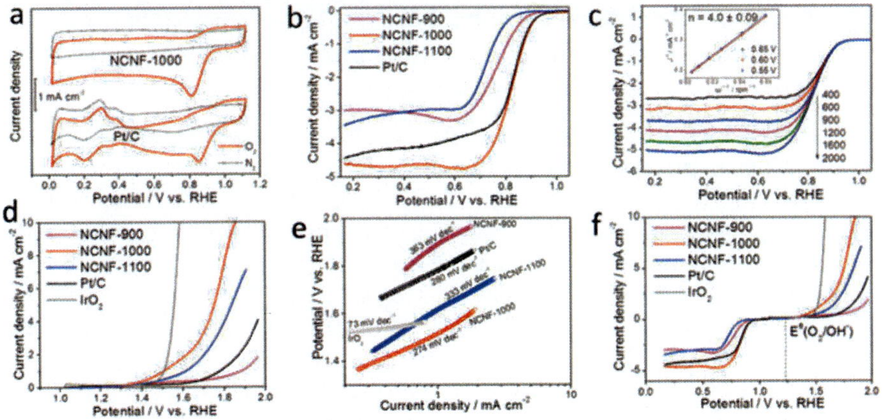

Fig. 12 **a** CV curves of NCNF-1000 and Pt/C, in O_2-saturated (solid line) and N_2-saturated (dotted line) 0.1 M KOH. **b** LSV curves of different catalysts for ORR in O_2-saturated at 1600 rpm. **c** LSV curves of NCNF-1000 for ORR at different rotating speeds and the inset is K-L plots of NCNF-1000 at different potentials including the calculated number of electron transfer (*n*) per O_2. **d** LSV curves of different catalysts for OER at 1600 rpm in 0.1 M KOH. **e** Tafel slopes derived from (**d**). **f** LSV curves of different catalysts for both ORR and OER in 0.1 M KOH at 1600 rpm (scan rate 5 mV s^{-1}). The catalyst loading was 0.1 mg cm^{-2} for all catalysts. Reprinted with permission from Ref. [73]. Copyright 2016, Wiley-Blackwell

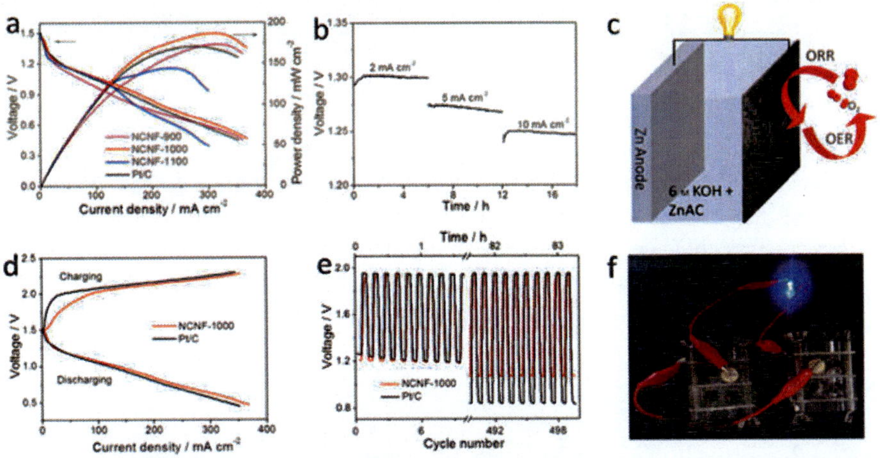

Fig. 13 **a** Polarization and power density curves of the primary Zn-air batteries with different catalysts. **b** Galvanostatic discharge curves of the primary Zn-air battery with NCNF-1000 as a catalyst at different current densities, which was normalized to the area of air–cathode. **c** Schematic representation of the rechargeable Zn-air battery. **d** Charge and discharge polarization curves. **e** Galvanostatic discharge–charge cycling curves at 10 mA cm^{-2} of rechargeable Zn-air batteries with the NCNF-1000 and Pt/C as catalyst, respectively. **f** Photograph of a blue LED (≈3.0 V) powered by two liquid Zn-air batteries with the NCNF-1000 air–cathode connected in series. Reprinted with permission from Ref. [73]. Copyright 2016, Wiley-Blackwell

3.2 *Carbon Nanofiber-Supported PdNi Alloy Nanoparticles as Highly Efficient Bifunctional Catalysts for Hydrogen and Oxygen Evolution Reactions*

Currently, much attention is focused on developing efficient and clean renewable energy technologies to increase energy demand and alleviate environmental problems [74–76]. As we all know, the production of hydrogen through water electrolysis is a key strategy to overcome these energy challenges [77–79]. However, the water decomposition half-reactions, namely the hydrogen evolution reaction (HER) and the oxygen evolution reaction (OER), have a high overpotential and are high energy [80, 81]. An effective electrocatalyst is essential to enhance the electrochemical water-splitting performance. In particular, precious metal catalysts (such as Pt and RuO_2) have been used as the most advanced electrocatalysts for HER and OER, respectively [82–84]. However, the high cost and low global reserves of these metals limit their wide practical application. In addition, a catalyst capable of driving both HER and OER is very needed, which is a basic requirement as a high-efficiency energy conversion device in water decomposition. Unfortunately, it is relatively difficult to develop effective, durable, and cost-effective bifunctional catalysts for HER and OER. Note that highly efficient bifunctional catalysts with high HER and OER activity are still highly desired. Chem et al. prepared carbon nanofiber-loaded PdNi alloy nanoparticles, a highly efficient dual-function catalyst for hydrogen and oxygen release reactions [85].

A new class of the PdNi alloy structure was prepared on CNFs by electrospinning and subsequent carbonization. The as-synthesized PdNi/CNFs, which have a low Pd loading, exhibit superior catalytic activity, and good stability in both the HER and OER and are thus a promising alternative bifunctional electrocatalyst. The highly efficient catalytic performance of the PdNi/CNFs is attributed to the synergistic effects of the PdNi alloy and the properties of the CNF substrate. The combination of these factors enhances the catalytic activity and is expected to enable the realization of cost-effective hydrogen and oxygen generation. Therefore, the catalyst developed in this work could be suitable for use in various water-splitting applications (Figs. 14, 15 and 16).

3.3 *Preparation of Palladium Nanoparticles Decorated Polyethyleneimine/Polycaprolactone Composite Fibers Constructed by Electrospinning with Highly Efficient and Recyclable Catalytic Performances*

In recent years, in the context of rapid technological and economic development, available water sources are increasingly scarce, and people have paid extensive attention to the protection and purification of water. Hazardous organic waste is the main

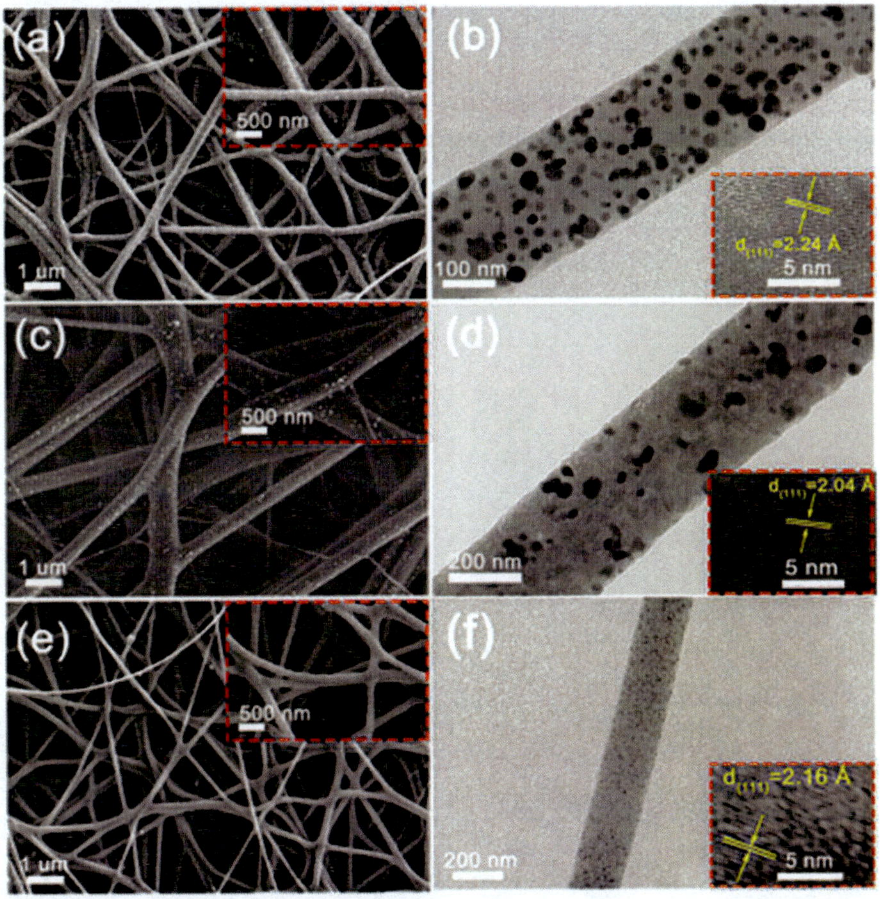

Fig. 14 **a**, **c**, **e** FE-SEM and **b**, **d**, **f** TEM images of the Pd/CNFs (**a**, **b**), Ni/CNFs (**c**, **d**) and PdNi/CNFs-1:2 (**e**, **f**). The insets are the corresponding HRTEM images. Reprinted with permission from Ref. [85]. Copyright 2017, Elsevier Ltd

source of water pollution [86–92]. The conversion of harmful organic chemicals into harmless or low-toxic compounds under mild conditions has become an extremely important research field [93–95]. As we all know, 4-nitrophenol (4-NP) is widely used in the synthesis of dyes and drugs and is one of the chemical products commonly used in the chemical industry [96].

Nano-sized palladium nanoparticles show high catalytic activity due to the tendency of aggregation and have serious limitations in the field of catalysis. A solid substrate with a large specific surface area is an ideal carrier for palladium nanoparticles. Wang et al. successfully designed and prepared polyethyleneimine/polycaprolactone/Pd nanoparticles (PEI/PCL@PdNPs) composite catalyst by electrospinning and reduction method using PEI/PCL electrospun fiber as the carrier [97]. A large number of pit structures increase the specific

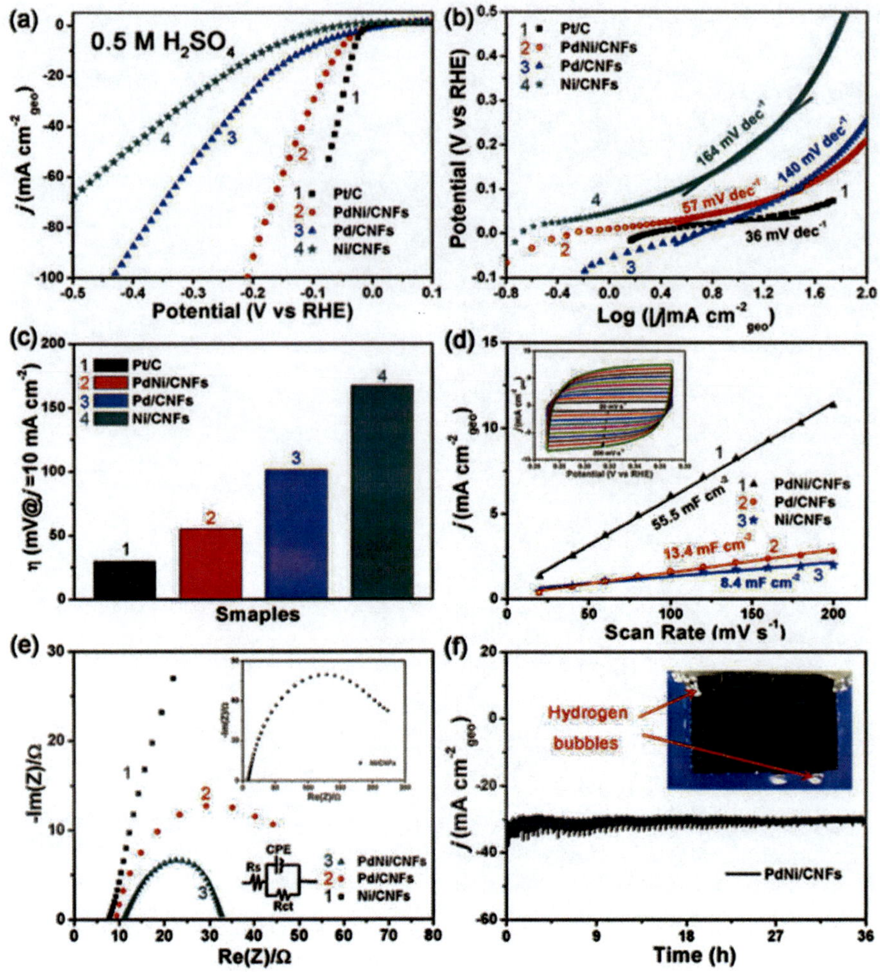

Fig. 15 HER electrocatalysis in 0.5 M H_2SO_4. **a** Polarization curves and **b** corresponding Tafel plots of the Pd/CNFs, Ni/CNFs, PdNi/CNFs-1:2 and commercial Pt/C catalyst; **c** Histograms of overpotentials at $j = 10$ mA cm^{-2} for the catalysts of Pd/CNFs, Ni/CNFs, PdNi/CNFs-1:2 and commercial Pt/C; **d** The linear fit of the capacitive currents of the catalysts versus the scan rates, inset in (**d**) is the electrochemical cyclic voltammograms of Pd/CNFs at potential scanning rates from 20 mV s^{-1} to 200 mV s^{-1}; **e** EIS of the Pd/CNFs, Ni/CNFs and PdNi/CNFs-1:2 at the potential of 0.1 V versus RHE (the insets show the equivalent circuit of the fitted curves and the full EIS spectrum of the Ni/CNFs); **f** chronoamperometric response (j-t) curve of PdNi/CNFs-1:2 at a constant voltage of 50 mV versus RHE (the insets show a digital image of the H2 bubbles formed on the PdNi/CNFs-1:2 membrane during the electrocatalytic process). Reprinted with permission from Ref. [85]. Copyright 2017, Elsevier Ltd

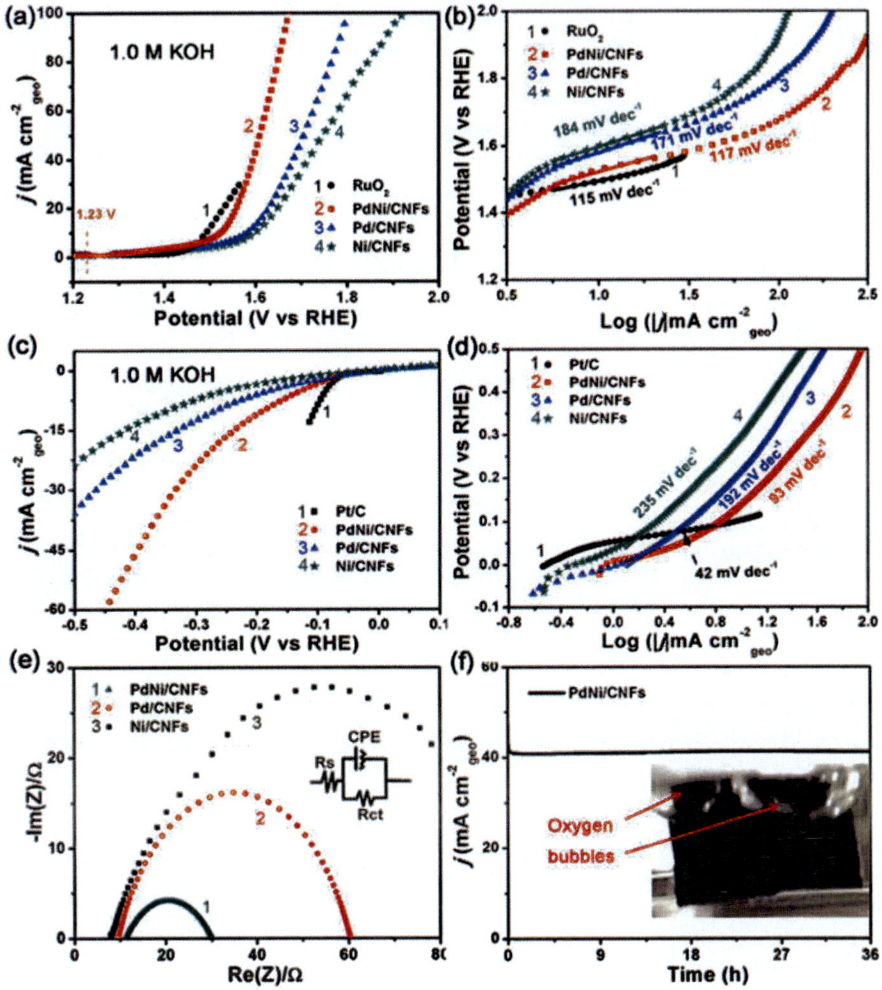

Fig. 16 OER and HER electrocatalysis in 1 M KOH. **a** Polarization curves and **b** corresponding Tafel plots of the Pd/CNFs, Ni/CNFs, PdNi/CNFs-1:2 and commercial RuO_2 catalyst for OER; **c** Polarization curves and **d** corresponding Tafel plots of the Pd/CNFs, Ni/CNFs, PdNi/CNFs-1:2 and commercial Pt/C catalyst for HER in 1 M KOH electrolyte; **e** EIS of the Pd/CNFs, Ni/CNFs and PdNi/CNFs-1:2 at the potential of 1.60 V versus RHE (the inset shows the equivalent circuit of the fitted curves); **f** chronoamperometric response (j-t) curve of PdNi/CNFs-1:2 at a constant potential of 1.60 V versus RHE (the inset shows a digital image of the O_2 bubbles formed on the PdNi/CNFs-1:2 membrane during the electrocatalytic process). Reprinted with permission from Ref. [85]. Copyright 2017, Elsevier Ltd

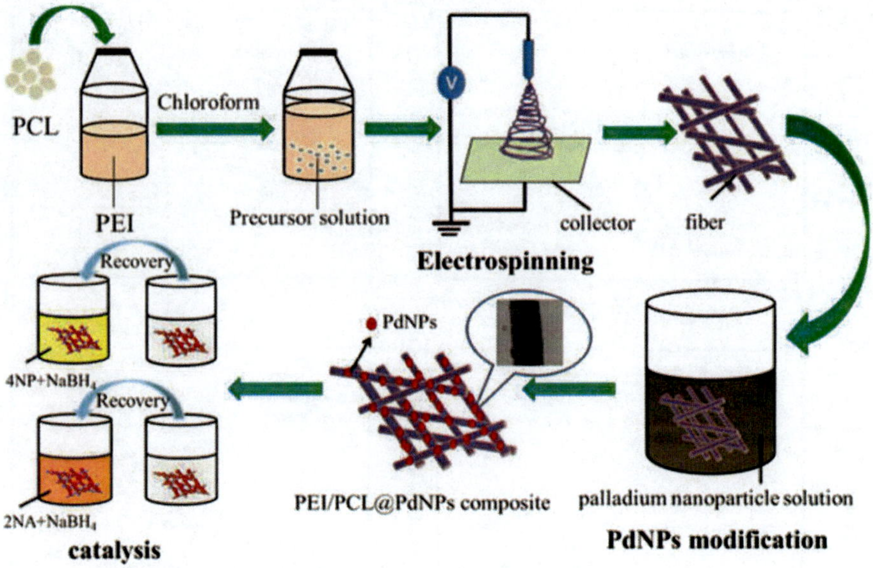

Fig. 17 Preparation process of polyethyleneimine/polycaprolactone/Pd nanoparticles (PEI/PCL@PdNPs) composite and its catalytic performances. Reprinted with permission from Ref. [97]. Copyright 2019, MDPI Ltd

surface area of electrospun fibers and provide an active site for the loading of palladium particles, so the added PEI component effectively adjusts the microscopic morphology of PEI/PCL fibers. The obtained PEI/PCL@PdNPs catalysts for the reduction of 4-nitrophenol (4-NP) and 2-nitroaniline (2-NA) showed very effective, stable, and reusable catalytic performance. It is worth mentioning that the reaction rate constant of 4-NP catalytic reduction is 0.16597 s^{-1}. Therefore, we have developed a highly efficient catalyst with potential application prospects in the field of catalysis and water treatment (Fig. 17 and 18).

The stability and recyclability of the catalyst are another aspect of evaluating the catalyst. Therefore, it is very important and necessary to explore the repetitive catalytic capabilities of PEI/PCL@PdNPs catalysts for 4-NP and 2-NA, as shown in Fig. 19. The catalytic efficiency of PEI/PCL@PdNPs catalysts after repeated catalytic reduction of fresh 4-NP and 2-NA systems for 8 cycles was 95% and 92%, respectively. The above results indicate that the PEI/PCL@PdNPs catalyst has high catalytic activity, stability, and recyclability. Referring to the previous literature, the catalytic efficiency has hardly decreased, which is attributed to the adhesion of organic matter on the catalyst surface and the loss of palladium particles on the catalyst surface during the washing of PEI/PCL@PdNPs with ethanol and ultrapure water [98]. This work provides new research clues for the preparation of composite materials loaded with PdNPs and ideal metal particle-based carriers.

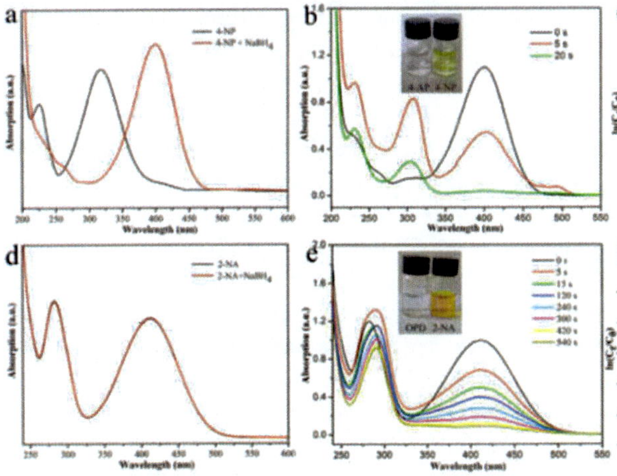

Fig. 18 UV absorption curves of 4-NP and 2-NA before and after adding NaBH4 aqueous solution (**a**, **d**); catalytic reduction of 4-NP and 2-NA with PEI/PCL@PdNPs composite (PEI:PCL, w/w, 35:65) and photographs of 4-NP and 2-NA solution after reduction (**b**, **e**); the linear relationship of the reduction process (**c**, **f**). Reprinted with permission from Ref. [97]. Copyright 2019, MDPI Ltd

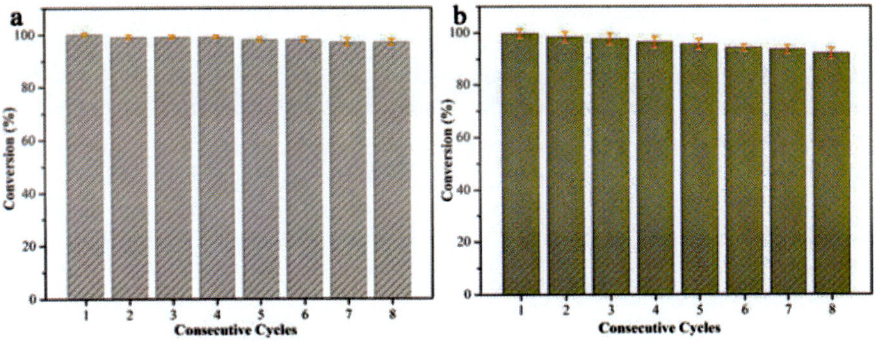

Fig. 19 Recyclability test of PEI/PCL@PdNPs catalyst for the reduction of 4-NP (**a**) and 2-NA (**b**). Reprinted with permission from Ref. [97]. Copyright 2019, MDPI Ltd

4 Hierarchical Electrospun Nanofibers Treated by Solvent Vapor Annealing as Air Filtration Mat for High-Efficiency PM2.5 Capture

In the past few years, environmental issues have received extensive attention due to their danger to human health and the environment [99–101]. Particulate matter (PM) is a very complex mixture of pollutants, with very fine particles and small droplets [102]. PM2.5 is particularly harmful because these small particles can penetrate the lungs and bronchi [30, 103–107]. In an indoor environment, these particles can be filtered through ventilation or central air-conditioning system, and the personal protection of outdoor personnel is not good, because most commercial masks have low PM2.5 removal efficiency because of their small diameter particles. Therefore, how to effectively and conveniently remove these PM pollutants from the air has become an urgent and challenging problem for researchers. For example, designing and developing high-efficiency filter materials that can efficiently remove PM2.5 particles are the central focus: evidence-based, high porosity, high accumulation of filter fibers. When the fiber diameter is small and the fiber diameter is large, the removal efficiency of PM2.5 is higher. As we all know, electrospinning is a relatively simple method for producing continuous fibers with nanometer and submicron diameters. The fibers prepared by electrospinning have high porosity, fine pore size, staggered pore structure, small pore size, and controllable diverse structure and thickness, which make them ideal for air filtration materials.

Huang et al. introduced a new type of high-efficiency air filter mat that can be used for outdoor protection [108]. The nanocomposite of the new high-efficiency air filter was successfully made of poly (ε-caprolactone)/polyethylene oxide (PCL/PEO) using electrospinning technology and solvent vapor annealing (SVA). The SVA treatment gives the fiber surface a wrinkle effect and enhances the PM2.5 capture ability of the protective mask. This nano-wrinkled air filter mat can effectively filter PM2.5 under severe pollution conditions (PM2.5 particle concentration is higher than 225 mg m^{-3}), and the removal efficiency is 80.01%. Field tests have shown that the air filter mat has a high PM2.5 removal efficiency under dense fog. Compared with commercial masks, the manufactured SVA-treated PCL/PEO air filter pads have a simpler, greener, and more environmentally friendly preparation process and have excellent degradation characteristics, with a wide range of potential applications and high filtration efficiency (Figs. 20, 21 and 22).

5 Conclusions and Remarks

In summary, as a simple and widely used technology, electrospinning can satisfy people's desire to quickly prepare nanofiber materials from a wide range of materials. By studying the composition, structure, porosity, surface, and fiber orientation of nanofibers, fiber properties can be selectively tailored for various applications. Although nanofibers have broad application prospects in heterogeneous catalysis and biomedical research, they still face a series of challenges, such as low dispersibility and uncontrollable degradability. With the precise control of the nanofiber microstructure, the preparation of surface-active nanofiber sponges will provide the possibility of developing flexible and recyclable catalytic systems. As a scaffold polymer nanofiber for tissue regeneration, its design, composition, and structure still need to be further optimized in clinical applications. Currently, the preparation of nanofibers is still in the laboratory research stage, the fiber output is low, mass production cannot be achieved, and industrialization is difficult to achieve. Although the electrospinning method is simple and can prepare a variety of nanofibers

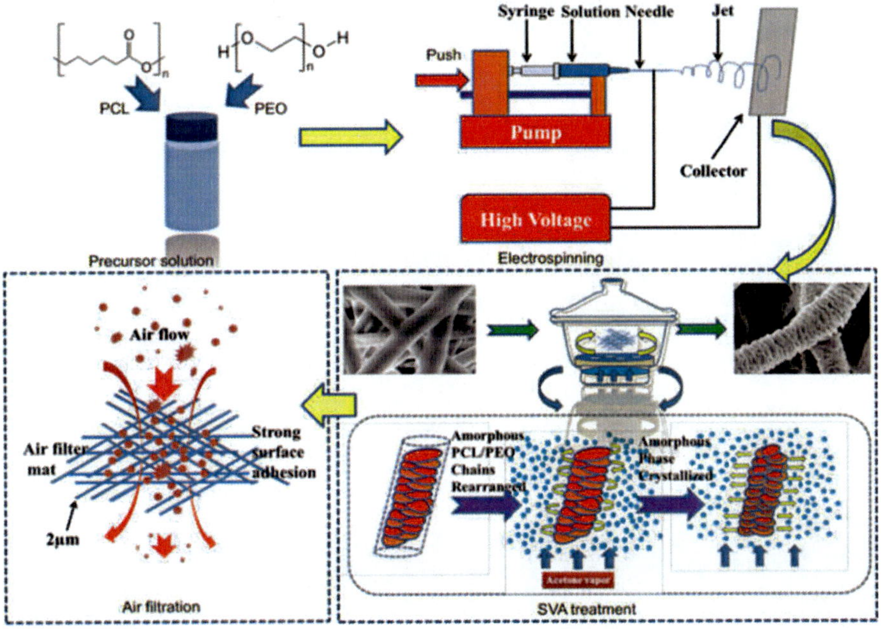

Fig. 20 A schematic illustration of the preparation and filtration of the obtained electrospun nanowrinkled air filtration mat. Reprinted with permission from Ref. [108]. Copyright 2019, Springer Ltd

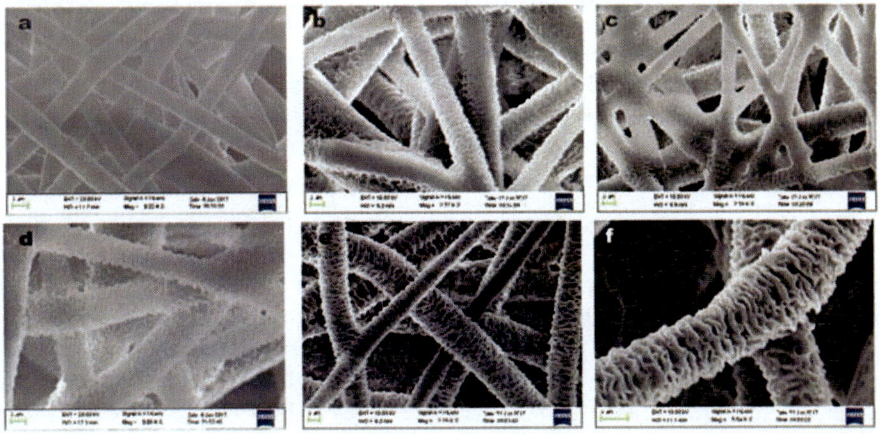

Fig. 21 SEM images of the prepared electrospun PCL/PEO nanofibers (**a**, Sample 1) and SVA treatment at different time intervals: **b** one day; **c** 2 days; **d** 3 days; **e** 4 days; **f** 5 days. Reprinted with permission from Ref. [108]. Copyright 2019, Springer Ltd

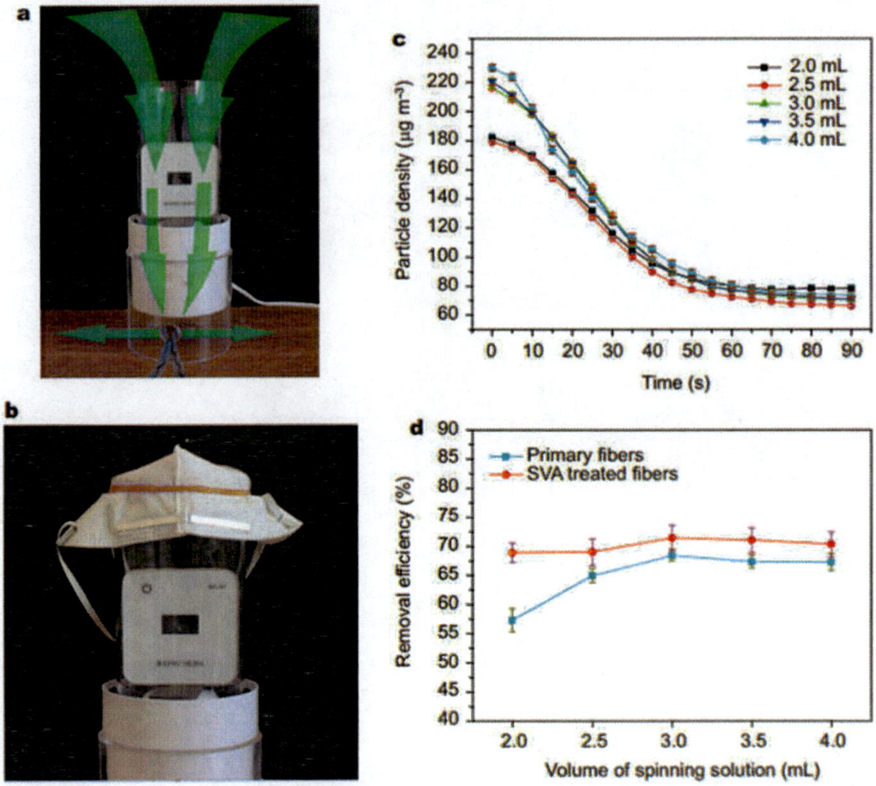

Fig. 22 Schematics of the air filtration mat that captured PM2.5 by airflow (**a**) and working state with mask filtration (**b**); **c** PM2.5 filtration curves for the air filtration mat (Sample 1) with different volumes of spinning precursor solution; **d** PM2.5 removal efficiency plots for primary fibers and SVA-treated fibers. Reprinted with permission from Ref. [108]. Copyright 2019, Springer Ltd

with different compositions, different structures, and different arrangements, there are still many challenges. For example, the nanofibers prepared by electrospinning cannot obtain filaments separated from each other, and the yield is low, and the strength Low, which limits the application. At present, people have made some breakthroughs in these problems. By simulating the complex spatial distribution of natural tissue, the three-dimensional scaffold has functional grading in terms of composition, arrangement, porosity, and pore size, which will make it better for tissue regeneration applications. It is believed that in the near future, electrospinning technology will be fully developed, and eventually from laboratory to industry, from basic research to clinical application.

Acknowledgements We greatly appreciate the financial supports of National Natural Science Foundation of China (No. 21872119), the Talent Engineering Training Funding Project of Hebei Province (No. A201905004), and the Research Program of the College Science and Technology of Hebei Province (No. ZD2018091).

References

1. Bhardwaj N, Kundu S (2010) Electrospinning: a fascinating fiber fabrication technique. Biotechnol Adv 28:325–347
2. Simons H (1966) Process and apparatus for producing patteined non woven fabrics .US3280229
3. Doshi J, Reneker D (1995) Electrospinning process and applications of electrospun fibers. J Electrostat 35:151–160
4. Ju J, Kang W, Li L (2016) Preparation of poly(tetrafluoroethylene) nanofiber film by electro-blown spinning method. Mater Lett 171:236–239
5. Yang D, Li EZ, Guo W (2011) Research and industrial development of nanofibers prepared by electrospinning. Mater Rev 25:64–68
6. Li M, Zhang J, Liu X, Wang Y (2016) Synthesis of high performance SAPO-34 zeolite membrane by a noveltwo-step hydrothermal synthesis dry gel conversion method. Micropor Mesopor Mat 225:261–271
7. Anton F (1975) Process and apparatus for preparing artificial threads: U.S. Patent 504:1934-10-2
8. Yan X, Gevelber M (2017) Electrospinning of nanofibers: characterization of jet dynamics and humidity effects. Part Sci Technol 35:139–149
9. Yao J, Bastiaansen C, Peijs T (2014) High strength and high modulus electrospun nanofibers. Fibers 2:158–186
10. Zhang X, Dai H, Cao Y (2013) Fabrication of high-strength aligned multi-walled carbon nanotubes/polyvinyl alcohol composite nanofibers by electrospinning. Adv Mater Res 796:311–316
11. Wang J, Zhang Z, Zhang Q, Liu J, Ma J (2018) Preparation and adsorption application of carbon nanofibers with large specific surface area. J Mater Sci 53:16466–16475
12. Wang W, Wang H, Wang H, Jin X, Li J, Zhu Z (2018) Electrospinning preparation of a large surface area, hierarchically porous, and interconnected carbon nanofibrous network using polysulfone as a sacrificial polymer for high performance supercapacitors. RSC Adv 8:28480–28486
13. Miao Y, Wang R, Chen D, Liu Z, Liu T (2012) Electrospun self-standing membrane of hierarchical SiO_2@γ-AlOOH (Boehmite) core/sheath fibers for water remediation. ACS Appl Mater Inter 4:5353–5359
14. Greiner A, Wendorff J (2007) Electrospinning: a fascinating method for the preparation of ultrathin fibers. Angew Chem Int Ed 46:5670–5703
15. Wang X, Yu J, Sun G, Ding B (2016) Electrospun nanofibrous materials: a versatile medium for effective oil/water separation. Mater Today 19:403–414
16. Wang Q, Liu S, Fu L, Cao Z, Ye W, Li H, Guo P, Zhao XS (2018) Electrospun γ-Fe_2O_3 nanofibers as bioelectrochemical sensors for simultaneous determination of small biomolecules. Anal Chim Acta 1026:125–132
17. Xue Z, Huang L, Wang C (2016) Sonocrystallization of ZIF-8 on electrostatic spinning TiO_2 nanofibers surface with enhanced Photocatalysis property through synergistic effect. ACS Appl Mater Inter 8:20274
18. Fan L, Xue M, Kang Z, Li H, Qiu S (2012) Electrospinning technology applied in zeolitic imidazolate framework membrane synthesis. J Mater Chem 22:25272–25276
19. Lu W, Sun J, Jiang X (2014) Recent advances in electrospinning technology and biomedical applications of electrospun fibers. J Mater Chem B 2:2369–2380
20. Choi S, Persano L, Camposeo A, Jang J, Koo W, Kim S, Cho H, Kim I, Pisignano D (2017) Electrospun nanostructures for high performance chemiresistive and optical sensors. Macromol Mater Eng 302:1600569
21. Bai Y, Liu Y, Li Y (2017) Mille-feuille shaped hard carbons derived from polyvinylpyrrolidone via environmentally friendly electrostatic spinning for sodium ion battery anodes. Rsc Adv 7:5519–5527

22. Hu X, Liu S, Zhou G, Huang Y, Xie Z, Jing X (2014) Electrospinning of polymeric nanofibers for drug delivery applications. J Control Release 185:12–21
23. Liu J, Bauer A, Li B (2014) Solvent vapor annealing: An efficient approach for inscribing secondary nanostructures onto electrospun fibers. Macromol Rapid Commun 35:1503–1508
24. Lu Q, Yu Y, Ma Q, Chen B, Zhang H (2016) 2D Transition-metal-dichalcogenide-nanosheet-based composites for photocatalytic and electrocatalytic hydrogen evolution reactions. Adv Mater 28:1917–1933
25. Lin J, Ding B, Yang J, Yu J, Sun G (2012) Subtle regulation of the micro-and nanostructures of electrospun polystyrene fibers and their application in oil absorption. Nanoscale 4:176–182
26. Si Y, Wang X, Li Y, Chen K, Wang J, Yu J, Wang H, Ding B (2014) Optimized colorimetric sensor strip for mercury (II) assay using hierarchical nanostructured conjugated polymers. J Mater Chem A 2:645–652
27. Liu Y, Hou C, Jiao T, Song J, Zhang X, Xing R, Zhou J, Zhang L, Peng Q (2018) Self-assembled AgNP-containing nanocomposites constructed by electrospinning as efficient Dye photocatalyst materials for wastewater treatment. Nanomaterials 8:35. https://doi.org/10.3390/nano8010035
28. Panaitescu D, Ionita E, Nicolae C, Gabor A, Dinescu G (2018) Poly(3-hydroxybutyrate) modified by nanocellulose and plasma treatment for packaging applications. Polymers 10:1249
29. Rauwel P, Rauwel E, Uhl W (2019) Application and behavior of nanomaterials in water treatment. Nanomaterials
30. Xing R, Wang W, Jiao T, Ma K, Zhang Q, Hong W, Qiu H, Zhou J, Zhang L, Peng Q (2017) Bioinspired polydopamine sheathed nanofibers containing carboxylate graphene oxide nanosheet for high-efficient dyes scavenger. ACS Sustain Chem Eng 5:4948–4956
31. Qiu H, Liang C, Yu J, Zhang Q, Song M, Chen F (2017) Preferable phosphate sequestration by nano-La(III) (hydr)oxides modified wheat straw with excellent properties in regeneration. Chem Eng J 315:345–354
32. Zhang Q, Teng J, Zou G, Peng Q, Du Q, Jiao T, Xiang J (2016) Efficient phosphate sequestration for water purification by unique sandwich-like MXene/magnetic iron oxide nanocomposites. Nanoscale 8:7085–7093
33. Bao Q, Zhang H, Yang J, Wang S, Tang D, Jose R, Ramakrishna S, Lim C, Loh K (2010) Graphene-polymer nanofiber membrane for ultrafast photonics. Adv Funct Mater 20:782–791
34. Hou C, Jiao T, Xing R, Chen Y, Zhou J, Zhang L (2017) Preparation of TiO_2 nanoparticles modified electrospun nanocomposite membranes toward efficient dye degradation for wastewater treatment. J Taiwan Inst Chem E 78:118–126
35. Luan V, Tien H, Hur S (2015) Fabrication of 3D structured ZnO nanorod/reduced graphene oxide hydrogels and their use for photo-enhanced organic dye removal. J Colloid Inter Sci 437:181–186
36. Zhou W, Ding C, Jia X, Tian Y, Guan Q, Wen G (2015) Self-assembly of Fe_2O_3/reduced graphene oxide hydrogel for high Li-storage. Mater Res Bull 62:19–23
37. Sun Y, Cheng Y, He K, Zhou A, Duan H (2015) One-step synthesis of three-dimensional porous ionic liquid-carbon nanotube-graphene gel and MnO_2-graphene gel as freestanding electrodes for asymmetric supercapacitors. RSC Adv 5:10178–10186
38. Jiao T, Zhao H, Zhou J, Zhang Q, Luo X, Hu J, Peng Q, Yan X (2015) Self-assembly reduced graphene oxide nanosheet hydrogel fabrication by anchorage of chitosan/silver and its potential efficient application toward dyes degradation for wastewater treatments. ACS Sustain Chem Eng 3:3130–3139
39. Huang H, Ren P, Chen J, Zhang W, Ji X, Li Z (2012) High barrier graphene oxide nanosheets/poly(vinyl alcohol) nanocomposite films. J Membrane Sci 409–410:156–163
40. Balandin A, Ghosg S, Bao W, Calizo I (2008) Superior thermal conductivity of single-layer graphene. Nano Lett 8:902–907
41. Lee H, Dellatore S, Miller W, Messersmith P (2007) Mussel-inspired surface chemistry for multifunctional coatings. Science 318:426–430

42. Waite J, Qin X (2001) Polyphosphoprotein from the adhesive pads of mytilus edulis. Biochemistry 40:2887–2893
43. Ryu J, Lee Y, Kong W, Kim T, Park T, Lee H (2011) Catechol-functionalized chitosan/pluronic hydrogels for tissue adhesives and hemostatic materials. Biomacromol 12:2653–2659
44. Phua S, Yang L, Toh C, Huang S, Tsakadze Z, Lau S, Mai Y, Lu X (2012) Reinforcement of polyether polyurethane with dopamine-modified clay: the role of interfacial hydrogen bonding. ACS Appl Mater Inter 4:4571–4578
45. Yang H, Lan Y, Zhu W, Li W, Xu D, Cui J, Shen D, Li G (2012) Polydopamine-coated nanofibrous mats as a versatile platform for producing porous functional membranes. J Mater Chem 22:994–17001
46. Son H, Ryu J, Lee H, Nam Y (2013) Silver-polydopamine hybrid coatings of electrospun poly(vinyl alcohol) nanofibers. Macromol Mater Eng 298:547–554
47. Gao H, Sun Y, Zhou J, Xu R, Duan H (2013) Mussel-inspired synthesis of polydopamine-functionalized graphene hydrogel as reusable adsorbents for water purification. ACS Appl Mater Interfaces 5:425–432
48. Zhang S, Zhang Y, Bi G, Liu J, Wang Z, Xu Q, Lui H, Li X (2014) Mussel-inspired polydopamine biopolymer decorated with magnetic nanoparticles for multiple pollutants removal. J Hazard Mater 270:27–34
49. Yu Y, Shapter J, Popelka-Filcoff R, Bennett J, Ellis A (2017) Copper removal using bio-inspired polydopamine coated natural zeolites. J Hazard Mater 273:174–182
50. Yan J, Huang Y, Miao Y, Tjiu W, Liu T (2015) Polydopamine-coated electrospun poly(vinyl alcohol)/poly(acrylic acid) membranes as efficient dye adsorbent with good recyclability. J Hazard Mater 283:730–739
51. Li L, Guo X, Fu L, Prudhomme R, Lincoln S (2008) Complexation behavior of α-, β-, and γ-cyclodextrin in modulating and constructing polymer networks. Langmuir 24:8290–8296
52. Guo X, Wang J, Li L, Chen Q, Li Z, Pham D, May B, Prud'homme R, Easton C (2010) Tunable polymeric hydrogels assembled by competitive complexation between cyclodextrin dimers and adamantly substituted poly(acrylate)s. AIChE J 56:3021–3024
53. Wang J, Pham D, Guo X, Li L, Pham D, Luo Z, Ke H, Zheng L, Prud'homme R (2009) Polymeric networks assembled by adamantyl and β-cyclodextrin substituted poly(acrylate)s: host-guest interactions, and the effects of ionic strength and extent of substitution. Ind Eng Chem Res 49:609–612
54. Theron J, Walker J, Cloete T (2008) Nanotechnology and water Treatment: applications and emerging opportunities. Criti Rev Microbiol 34:43–69
55. Leudjo T, Pillay K, Yangkou M (2017) Nanosponge cyclodextrin polyurethanes and their modification with nanomaterials for the removal of pollutants from waste water: a review. Carbohyd Polym 159:94–107
56. Khaoulani S, Chaker H, Cadet C, Bychkov E, Cherifaouali L, Bengueddach A, FourmentinS (2015) Wastewater treatment by cyclodextrin polymers and noble metal/mesoporous TiO_2 photocatalysts. C R Chim18: 23–31
57. Yamasaki H, Makihata Y, Fukunaga K (2010) Efficient phenol removal of wastewater from phenolic resin plants using crosslinked cyclodextrin particles. J Chem Technol Biotechnol 81:1271–1276
58. Chai K, Ji H (2012) Dual functional adsorption of benzoic acid from wastewater by biological-based chitosan grafted β-cyclodextrin. Chem Eng J 203:309–318
59. Li J, Loh X (2008) Cyclodextrin-based supramolecular architectures: syntheses, structures, and applications for drug and gene delivery. Adv Drug Deliv Rev 60:1000–1017
60. Yilmaz E, Memon S, Yilmaz M (2010) Removal of direct azo dyes and aromatic amines from aqueous solutions using two β-cyclodextrin-based polymers. J Hazard Mater 174:592–597
61. Cui H, Bai M, Lin L (2018) Plasma-treated poly(ethylene oxide) nanofibers containing tea tree oil/beta-cyclodextrin inclusion complex for antibacterial packaging. Carbohyd Polym 179:360–369
62. Celebioglu A, Uyar T (2017) Antioxidant Vitamin E/cyclodextrin inclusion complex electrospun nanofibers: enhanced water solubility, prolonged shelf life, and photostability of Vitamin E. J Agric Food Chem 65:5404–5412

63. Guo R, Wang R, Yin J, Jiao T, Huang H, Zhao X, Zhang L, Li Q, Zhou J, Peng Q (2019) Fabrication and highly efficient dye removal characterization of beta-cyclodextrin-based composite polymer fibers by electrospinning. Nanomaterials 9:127
64. Lin L, Dai Y, Cui H (2017) Antibacterial poly(ethylene oxide) electrospun nanofibers containing cinnamon essential oil/beta-cyclodextrin proteoliposomes. Carbohyd Polym 178:131–140
65. Narayan R, Kim J, Kim J, Lee K, Kim S (2016) Graphene oxide liquid crystals: discovery, evolution and applications. Adv Mater 28:3045–3068
66. Zeng J, Hou H, Wendorff J, Greiner A (2004) Electrospun poly(vinyl alcohol)/poly(acrylic acid) fibres with excellent water-stability. E-Polymers 4:899–906
67. Wang X, Lu X, Liu B, Chen D, Tong Y, Shen G (2014) Flexible energy storage devices design consideration and recent progress. Adv Mater 26:4763–4782
68. Yin J, Zhan F, Jiao T, Deng H, Zou G, Bai Z, Zhang Q, Peng Q (2020) Highly efficient catalytic performances of nitro compounds via hierarchical PdNPs-loaded MXene/polymer nanocomposites synthesized through electrospinning strategy for wastewater treatment. Chinese Chem Lett. https://doi.org/10.1016/j.cclet.2019.08.047
69. Li L, Peng S, Wu H, Yu L, Madhavi S, Lou X (2015) A Flexible quasi-solid-state asymmetric Electrochemical capacitor Based on hierarchical porous V_2O_5 nanosheets on carbon nanofibers. Adv Energy Mater 5:1500753
70. Li Y, Dai H (2014) Recent advances in zinc-air batteries. Chem Soc Rev 43:5257
71. Nam G, Park J, Choi M, Oh P, Park S, Kim M, Park N, Cho J, Lee J (2015) Single crystalline pyrochlore nanoparticles with metallic conduction as efficient bi-functional oxygen electrocatalysts for Zn-air batteries. ACS Nano 9:6493
72. Jin Y, Chen F (2018) Bifunctional electrocatalysts for Zn–air batteries. Electrochim Acta 2:39–67
73. Liu Q, Wang Y, Dai L, Yao J (2016) Scalable fabrication of nanoporous carbon fiber films as bifunctional catalytic electrodes for flexible zn-air batteries. Adv. Mater. 28:3000–3006
74. Strmcnik D, Lopes P, Genorio B, Stamenkovic V, Markovic N (2016) Design principles for hydrogen evolution reaction catalyst materials. Nano Energy 29:29–36
75. Tahir M, Mahmood N, Zhang X, Mahmood T, Butt F, Aslam I, Tanveer M, Idrees F, Khalid S, Shakir I, Yan Y, Zou J, Cao C, Hou Y (2015) Bifunctional catalysts of Co_3O_4@GCN tubular nanostructured (TNS) hybrids for oxygen and hydrogen evolution reactions. Nano Res. 8:3725–3736
76. Safizadeh F, Ghali E, Houlachi G (2015) Electrocatalysis developments for hydrogen evolution reaction in alkaline solutions. Int J Hydrogen Energy 40:256–274
77. Popczun E, Mckone J, Read C, Biacchi A, Wiltrout A, Lewis N, Schaak R (2013) Nanostructured nickel phosphide as an electrocatalyst for the hydrogen evolution reaction. J Am Chem Soc 135:9267
78. Kanan M, Nocera D (2008) In situ formation of an oxygen-evolving catalyst in neutral water containing phosphate and Co^{2+}. Science 321:1072–1075
79. Mcintyre N, Cook M (1975) X-ray photoelectron studies on some oxides and hydroxides of cobalt, nickel, and copper. Anal Chem 47:2208–2213
80. Bergmann A, Zaharieva I, Dau H, Strasser P (2013) Electrochemical water splitting by layered and 3D cross-linked manganese oxides: correlating structural motifs and catalytic activity. Energ Environ Sci 6:2745–2755
81. Khan S, Khan S, Asiri A (2016) Electro-catalyst based on cerium doped cobalt oxide for oxygen evolution reaction in electrochemical water splitting. J Mater Sci Mater E 27:1–9
82. Lee Y, Suntivich J, May K, Perry E, Shao-Horn Y (2012) Synthesis and activities of rutile IrO2 and RuO2 nanoparticles for oxygen evolution in acid and alkaline solutions. J Phys Chem Lett 3:399–404
83. Kibsgaard J, Jaramillo T (2015) Molybdenum phosphosulfide: an active, acidStable, earth-abundant catalyst for the hydrogen evolution reaction. Angew Chem Int Ed 53:14433–14437
84. Nong H, Oh H, Reier T, Willinger E, Willinger M, Petkov V, Teschner D, Strasser P (2015) Oxide-supported IrNiO(x) core-shell particles as efficient, costeffective, and stable catalysts for electrochemical water splitting. Angew Chem Int Ed Eng 54:2975–2981

85. Chen J, Chen D, Yu D, Zhang M, Zhu H, Du M (2017) Carbon nanofiber-supported PdNi alloy nanoparticles as highly efficient bifunctional catalysts for hydrogen and oxygen evolution reactions. Electrochim Acta 246:17–26
86. Lukatskaya M, Mashtalir O, Ren C, Dall'Agnese Y, Rozier P, Taberna P, Naguib M, Simon P, Barsoum M, Gogotsi Y (2013) Cation intercalation and high volumetric capacitance of two-dimensional titanium carbide. Science 341:1502–1505
87. Jian M, Wang C, Wang Q, Wang H, Xia K, Yin Z, Zhang M, Liang X, Zhang Y (2017) Advanced carbon materials for flexible and wearable sensors. Sci China Mater 60:1206–1062
88. Li Z, Wang L, Sun D, Zhang Y, Liu B, Hu Q, Zhou A (2015) Synthesis and thermal stability of two-dimensional carbide MXene Ti_3C_2. Mater Sci Eng B 191:33–40
89. Naguib M, Mochalin V, Barsoum M, Gogotsi Y (2013) 25th anniversary article: MXenes: a new family of two-dimensional materials. Adv Mater 26:992–1005
90. Li J, Du Y, Huo C, Wang S, Cui C (2015) Thermal stability of two-dimensional Ti_2C nanosheets. Ceram Int 41:2631–2635
91. Miranda A, Halim J, Barsoum M, Lorke A (2016) Electronic properties of freestanding $Ti_3C_2T_x$ MXene monolayers. Appl Phys Lett 108:033102
92. Boota M, Anasori B, Voigt C, Zhao M, Barsoum M, Gogotsi Y (2015) Pseudocapacitive electrodes produced by oxidant-free polymerization of pyrrole between the layers of 2D titanium carbide (MXene). Adv Mater 28:1517–1522
93. Wang H, Zhang J, Wu Y, Huang H, Li G, Zhang X, Wang Z (2016) Surface modified MXene Ti_3C_2, multilayers by aryl diazonium salts leading to large-scale delamination. Appl Surf Sci 384:287–293
94. Karlsson L, Birch J, Halim J, Barsoum M, Persson P (2015) Atomically resolved structural and chemical investigation of single MXene sheets. Nano Lett 15:4955–4960
95. Huang X, Wang R, Jiao T, Zou G, Zhan F, Yin J, Zhang L, Zhou J, Peng Q (2019) Facile preparation of hierarchical AgNP-loaded MXene/Fe_3O_4/polymer nanocomposites by electrospinning with enhanced catalytic performance for wastewater treatment. ACS Omega 4:1897–1906
96. Amezquita-Garcia H, Razo-Flores E, Cervantes F, Rangel-Mendez J (2013) Activated carbon fibers as redox mediators for the increased reduction of nitroaromatics. Carbon 55:276–284
97. Wang C, Yin J, Han S, Jiao T, Bai Z, Zhou J, Zhang L, Peng Q (2019) Preparation of palladium nanoparticles decorated polyethyleneimine/polycaprolactone composite fibers constructed by electrospinning with highly efficient and recyclable catalytic performances. Catalysts 9:559
98. Guo R, Jiao T, Xing R, Chen Y, Guo W, Zhou J, Zhang L, Peng Q (2017) Hierarchical AuNPs-loaded Fe_3O_4/polymers nanocomposites constructed by electrospinning with enhanced and magnetically recyclable catalytic capacities. Nanomaterials 7:317
99. Zhang R, Jing J, Tao J, Hsu S, Wang G, Cao J, Lee C, Zhu L, Chen Z, Zhao Y (2013) Chemical characterization and source apportionment of PM2.5 in Beijing: seasonal perspective. Atmos Chem Phys 13:7053–7074
100. Li K, Jiao T, Xing R, Zou G, Zhou J, Zhang L, Peng Q (2018) Fabrication of tunable hierarchical MXene@AuNPs nanocomposites constructed by self-reduction reactions with enhanced catalytic performances. Sci China Mater 61:728–736
101. Cheng C (2005) Interfacial behaviors of PMMA-PEO block copolymers at the air/water interface. Sci China Ser B 48:567–573
102. Zhao H, Jiao T, Zhang L, Zhou J, Zhang Q, Peng Q, Yan X (2015) Preparation and adsorption capacity evaluation of graphene oxide-chitosan composite hydrogels. Sci China Mater 58:811–818
103. Sun Z, Liao T, Kou L (2017) Strategies for designing metal oxide nanostructures. Sci China Mater 60:1–24
104. Wang R, Liu Q, Jiao T, Li J, Rao Y, Su J, Bai Z, Peng Q (2019) Facile preparation and enhanced catalytic properties of self-assembled Pd nanoparticle-loaded nanocomposites films synthesized via electrospun approach. ACS Omega 4:8480–8486
105. Zhou J, Liu Y, Jiao T, Xing R, Yang Z, Fan J, Liu J, Li B, Peng Q (2018) Preparation and enhanced structural integrity of electrospun poly(ε-caprolactone)-based fibers by freezing amorphous chains through thiol-ene click reaction. Colloids Surface A 538:7–13

106. Wang C, Yin J, Wang R, Jiao T, Huang H, Zhou J, Zhang L, Peng Q (2019) Facile preparation of self-assembled polydopamine-modified electrospun fibers for highly effective removal of organic dyes. Nanomaterials 9:116
107. Wang C, Sun S, Zhang L, Yin J, Jiao T, Zhang L, Xu Y, Zhou J, Peng Q (2019) Facile preparation and catalytic performance characterization of AuNPs-loaded hierarchical electrospun composite fibers by solvent vapor annealing treatment. Colloids Surface A 561:283–291
108. Huang X, Jiao T, Liu Q, Zhang L, Zhou J, Li B, Peng Q (2019) Hierarchical electrospun nanofibers treated by solvent vapor annealing as air filtration mat for high-efficiency PM2.5 capture. Sci China Mater 62:423–436

Chapter 19
Green Composites Films with Antibacterial Properties

Rafael Selgas and Ángel Serrano-Aroca

1 Introduction

The polyhydroxyalkanoates are a family of lineal green biopolyesters composed of units of hydroxyalkanoate (HA) forming a structure obtained by bacterial fermentation [45]. These biopolymers accumulate into polymers that are packed as granular inclusions in the cytoplasm of several Gram-negative and Gram-positive bacteria under conditions of nutritional deficiency [42]. The most widespread and extensively studied member of this family is poly(3-hydroxybutyrate) (PHB) [1]. It is highly biodegradable, non-toxic, and biocompatible. Among its most prominent characteristics are its high degree of crystallinity, its insolubility in water, and that it is relatively resistant to hydrolytic degradation [86]. Moreover, in contrast to PHAs produced by Gram-negative bacteria that can contain high levels of endotoxins, PHB with a low endotoxin level is suitable for biomedical applications [43]. Nevertheless, its applications are limited because of its high fragility and poor mechanical properties [8, 9, 39]. Therefore, the incorporation of various monomers to form a copolymer is the current strategy to improve the properties of PHB. One of the more promising materials for biomedical applications is the poly(3-hydroxybutyrate-co-3-hydroxyvalerate) biopolymer, or PHBV, due to its null toxicity, high biocompatibility with many diverse types of cells, and that nowadays can be produced at large scale [12, 13, 93].

On the other hand, alginate is a naturally occurring anionic green biopolymer typically obtained from brown seaweed. It has been extensively investigated and used for many biomedical applications, due to its biocompatibility, low toxicity, relatively low

R. Selgas · Á. Serrano-Aroca (✉)
Biomaterials and Bioengineering Lab, Centro de Investigación Traslacional San Alberto Magno, Universidad Católica de Valencia San Vicente Mártir, c/Guillem de Castro 94, 46001 Valencia, Spain
e-mail: angel.serrano@ucv.es

S. Thomas and P. Balakrishnan (eds.), *Green Composites*, Materials Horizons: From Nature to Nanomaterials,
https://doi.org/10.1007/978-981-15-9643-8_19

cost, and mild gelation by addition of divalent cations such as Ca^{2+} [25]. Commercially available alginate is typically extracted from brown algae (*Phaeophyceae*), including *Laminaria hyperborea*, *Laminaria digitata*, *Laminaria japonica*, *Ascophyllum nodosum*, and *Macrocystis pyrifera* [41]. It can also be produced from some bacteria such as *Azotobacter* and *Pseudomonas*. The pathway of biosynthesis of alginate from bacteria is generally divided into four stages: a synthesis of precursor substrate, the polymerization and cytoplasmic membrane transfer, the periplasmic transfer and modification, and export through the outer membrane [64]. Furthermore, there is the possibility of alginate production with customized characteristics and wide applications in biomedicine due to recent progress in the regulation of alginate biosynthesis and the relative ease of bacterial modification. There is a wide range of alginate types with different molecular masses and proportions between D-manuronic and L-guluronic or chemically modified groups that give it a broad range of different properties [31]. Alginate hydrogels can be prepared by various crosslinking methods, and their structural similarity to extracellular matrices of living tissues allows wide applications in biomedicine such as wound healing, delivery of bioactive agents such as drugs and proteins, and cell transplantation [41].

However, although these two green and sustainable polymers have excellent properties, they do not have antibacterial activity, which is highly desirable in many industrial applications. In this regard, much work is currently underway to improve the properties of these materials by adding compounds that, in addition to improving their physical properties, also provide antibacterial activity to the final composite. From all the types of compounds that can be used for this purpose, carbon nanomaterials, such as graphene oxide (GO) and carbon nanofibers (CNFs), are very promising in this sense. Thus, our research group has focused the attention on the physical and biological enhancements that can be achieved with the incorporation of these nanomaterials as reinforcement agents of these green made polymers. The antimicrobial activity of new advanced materials can be analyzed by an easy-to-follow protocol, reported by our research group [51], consisting of two complementary procedures based on two existing methods: the agar disk diffusion test and the antimicrobial activity measured on material surfaces according to the ISO 22196:2011 norm. The importance of having detailed and consistent protocols for use in different laboratories is that many of the antimicrobial tests reported in the literature are highly dependent on the tests, and this makes difficult the result reproducibility. This protocol can be used with many types of materials cut into 10-mm-diameter disk shapes. Brittle hydrophilic materials such as alginate hydrogels can be swollen in a suitable solvent such as autoclaved distilled water for 1 h to facilitate the cutting process. Other solvents, such as ethanol, ketone, and dichloromethane, can be used to swell hydrophobic materials for 1 h before cutting them. However, some materials such as poly (3-hydroxybutyrate-co-3 hydroxyvalerate) do not need to be swollen and can be cut directly. Finally, it is important to dry the disks in a vacuum oven and sterilize each sample with ethanol and UV radiation for 1 h to avoid any risk of contamination before being tested [51].

2 Green Biopolymers

There are a great number of diverse industrial applications that depend on the use of biodegradable materials. Additionally, there is a need to discover sustainable solutions to face the environmental devastation caused by plastic contamination. In this regard, polyhydroxyalkanoates (PHAs) constitute a very promising source of bioplastics due to its unique properties. PHAs can be produced in a sustainable way by bacterial fermentation [45] and are biodegradable [28]. In the biomedical field, one of the most promising biopolymers of the PHAs' family is poly(3-hydroxybutyrate-co-3-hydroxyvalerate), usually abbreviated as PHBV or PHBHV. Its null cytotoxicity, easy complete biodegradability and high biocompatibility with different cell types [12, 13] are the most important characteristics that render it a powerful material in tissue engineering applications such as biodegradable implants, tissue patches, biosensors, and scaffolds for bone regeneration [29, 44, 87]. However, in spite of the mentioned excellent properties, PHBV exhibits some disadvantages when comparing it with conventional polymers [26, 40]. Thus, it has low impact resistance, reduced elongation at break, high fragility and does not possess antibacterial capacity [10]. Furthermore, it has a hydrophobic surface, which makes cell adhesion and proliferation difficult [55]. To face theses drawbacks, there is an increasing number of enhancement strategies focused on improving the properties of this microbial polymer [66]. Thus, the reinforcement of PHBV can be achieve by combining it with other materials such as polymers [35, 52, 63] or natural fibers [61, 82, 83]. For example, when mixing PHBV with polymers, such as PLA, an improvement of the mechanical properties is achieved [38, 52]. Materials obtained from the mixture of polyethylene (PE) with up to 30% PHBV exhibited mechanical properties like tensile strength, Young's modulus, and elongation at break, comparable to those of commercial plastics. In addition, the rate of transmission of oxygen was reduced and the rate of transmission of water vapor increased compared to that of pure PE [56]. Furthermore, it has been reported that mixing 20–35% PHBV with PLA achieves high barrier properties for oxygen and water vapor, while maintaining the biocompatibility of the material [35], although the systems produced are immiscible and with a minimal improvement in flexibility [23]. The incorporation of vegetable fibers in biodegradable plastic compounds of natural origin is of great importance today [61]. These materials, in addition to being obtained from sustainable renewable sources, they produce wastes that can be assimilated and degraded by microorganisms, thus preventing their accumulation in the environment. Achieving the development of these biomaterials would be a step forward in replacing petroleum-based plastics [34, 88]. Biocomposites based on PHBV that contain 10–40% w/w of maple wood fiber have been developed [87], and it was concluded that the tensile and flexural modulus of 40% reinforced materials with fibers was improved by 167%, compared to pure PHBV. Bamboo fibers have also been shown to be a valuable reinforcing material for PHBV with 30–40% w/w [83]. In such materials, the tensile modulus increased by 175% over pure PHBV. However, it is important to achieve sufficient interaction between the bamboo fibers and the polymer matrix because otherwise, the

PHBV tensile strength may decrease [68]. Therefore, the reinforcement of PHBV with natural fibers turns out to be a good option, due to the biocompatibility and biodegradability of the obtained composite materials, which are very promising for advanced applications such as sustainable food packaging [5, 6]. In addition, the incorporation of essential oils into PHBV is a feasible option to provide antimicrobial action to the polymer [15]. Moreover, the surface of this biomaterial can be chemically or biologically modified to promote cell adhesion and proliferation through several techniques including treatment with plasma gas (N_2, O_2, NH_3, Ar), chemical treatment, and ultraviolet radiation [21, 37]. However, in addition to all these presented strategies, there is a very promising approach based on the utilization of carbon nanotechnology to reinforce these biopolymers, enhancing simultaneously their physical and antibacterial properties.

On the other hand, alginate is a biopolymer produced from renewable sources such as brown algae and microorganisms [91]. It is composed of D-mannuronic (M) and L-guluronic (G) acid residues and is approved by the US Food and Drug Administration (FDA) for human use in certain biomedical applications [24]. In the presence of divalent cations, such as Ca^{2+}, alginate chains are crosslinked and the resulting hydrogel have many industrial applications: water treatment [16], biodegradable plastic packaging [95], and biotechnology [62] because it is non-toxic, biodegradable, and biocompatible. Also, it is renewable and has much lower cost than many other biopolymers [91]. However, this polymer exhibits weak mechanical properties, especially in the swollen state, which restrict its industrial applications and do not have intrinsic antibacterial activity [50]. Thus, several techniques have been developed to enhance the mechanical properties of polymers: increase of crosslinking density [2, 53, 76, 78], plasma-polymerization onto a hydrophobic substrate [3, 79, 80] and the incorporation of carbon nanomaterials such as carbon nanofibers [49] and graphene oxide [71, 74]. In addition to produce reinforcement, the incorporation of carbon nanomaterials such as CNFs and GO provided antibacterial activity to the composites and that is the green nanocomposite materials in which we are going to focus in this book chapter.

3 Carbon Nanomaterials

Due to the excessive use of antibiotics, hospital-associated infections are increasingly spreading without any suitable treatment [18]. Therefore, there is an urgent need to find new alternative non-toxic antibacterial strategies able to deal with multidrug-resistant bacteria. Common antimicrobial strategies usually include the utilization of antibiotics [58], metal ions and metal oxides [47], antimicrobial peptides (AMPs) [92], and quaternary ammonium compounds (QACs) [32]. However, the use of nanotechnology and in particular, carbon nanomaterials, against multidrug-resistant pathogens has shown to be one of the most promising approach [7, 60, 73]. The antibacterial action of carbon nanomaterials is usually attributed to membrane disruption, bacteria wrapping, electron transfer, and induction of oxidative stress by reactive oxygen species, which hinder the generation of resistance by microorganisms [89, 100]. Thus, carbon nanomaterials (CNMs) such as graphene oxide [50] and carbon nanofibers [70] have demonstrated to be a non-cytotoxic promising choice for a next-generation of antibacterial agents against clinically relevant Gram-positive pathogens [50]. Many studies have also confirmed the antibacterial activity of GO against Gram-negative bacterial strains [46], and thus, it has been proposed as a new nanoweapon to combat multidrug resistance bacteria [94] as a substitute to antibiotics.

In the field of carbon nanomaterials, graphene is a very stable 2D compound exceptionally strong and stiff [57] with a high thermal [4] and electrical conductivity [54]. It has the capacity to improve cell adhesion of several cell types like osteoblasts and mesenchymal cells [36]. Moreover, its elasticity, flexibility, as well as its great mechanical properties make it an excellent reinforcement option for polymers, such as PHBV and alginate, which have poor mechanical properties [99]. Although graphene has exhibited very significant antibacterial activity against a variety of bacterial species [46], it has shown to display cytotoxicity sometimes that could induce apoptosis and reduced cell adhesion [96]. Therefore, it is of great importance to be cautious with the use of graphene and its derivatives in pure form in biomedicine. Thus, the oxidized form of graphene, graphene oxide, is also a two-dimensional nanomaterial (see Fig. 1) with a large surface area but decorated with hydroxyl, carbonyl, and epoxy groups located at the edges and basal planes, which render it hydrophilic and soluble in water [98].

It has imperfect polar surfaces and secondary reactive sites susceptible to chemical modification with organic and inorganic molecules [22]. GO nanosheets possess outstanding physical properties like graphene [14] but easier processing, larger scale production and lower cost [59, 99]. GO exhibits excellent physical properties,

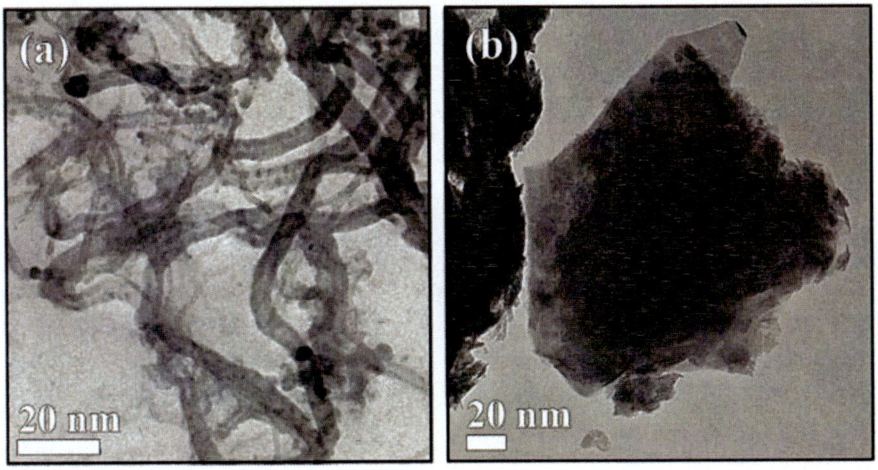

Fig. 1 Electron microscopy images: **a** HR-TEM of one-dimensional carbon nanofibers and **b** HR-TEM of two-dimensional graphene oxide nanosheets [69]

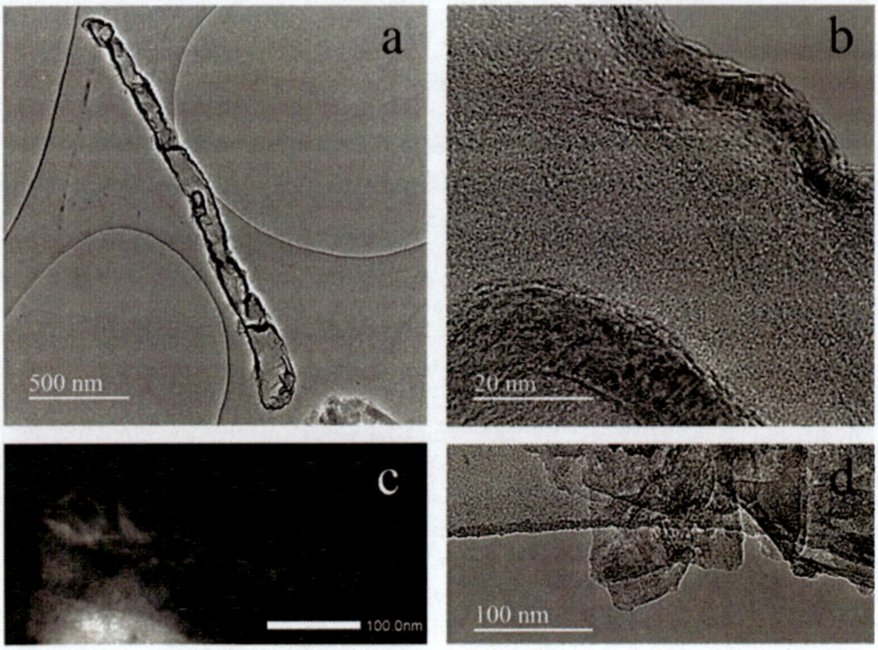

Fig. 2 High-resolution transmission electron microscopy: micrographs at two magnifications (**a** and **b**) with STEM dark field (**c**) of GO crosslinked with Zn^{2+} and uncrosslinked GO nanosheets (**d**) [74]

biocompatibility and nontoxicity [98]. In addition, the antibacterial activity of GO in nanocomposites [33, 81] widens the promising application fields of this nanofiller in the biomedical and bioprocess industries where contamination is difficult to avoid.

GO nanosheets can be crosslinked with divalent cations to form irregular tubes of GO (see Fig. 2) by a green and low-cost synthetic procedure by coordination chemistry [74].

These micrometer length irregular carbon tubes are 3D GO networks formed by folded and linked GO nanosheets. The crosslinking of GO nanosheets is much more effective at room temperature (24 °C) using Zn^{2+} and Sr^{2+} than with Ca^{2+}, which requires the temperature to be raised up to 34 °C to achieve the same crosslinking density. These 3D GO networks can be used to reinforce alginate films [74] and show an example of the potential use of these novel carbon materials to reinforce polymers increasing significantly the storage modulus and producing a more homogeneous structure. Therefore, these 3D networks of GO coordinated with divalent cations (Ca, Zn, Sr, or others) will enable broadening of the field toward the synthesis of new continuous carbon materials and reinforced composites with unique electric, thermal, oxidative, and chemical properties.

Carbon nanofibers (CNFs) are one-dimensional structures (see Fig. 1) in which the carbon atoms are grouped into filiform structures with excellent mechanical, electrical, and thermal properties, suitable for advanced applications, such as regenerative medicine [90]. Besides, CNFs are currently manufactured at large scale, but at a much lower price than GO (about 21 times more economic when purchased from Sigma-Aldrich, Buchs, Switzerland). Thus, nanocomposites with CNFs have shown great promise in certain biomedical applications where electrical conductivity plays an important role such as promotion of cardiomyocyte growth [85] and induction of neural regeneration [90]. In addition, CNFs have been recently shown to possess antibacterial activity in pure form and when they are incorporated into calcium alginate [70]. Moreover, in composite biomaterials, CNFs can significantly enhance the water diffusion and mechanical properties with a minuscule addition of them [49].

In addition, light has shown to be a powerful tool to enhance antibacterial activity [97]. It has been reported that the antibacterial activity of GO against *Escherichia coli* can be improved with simulated sunlight [14]. Following this research line, our research group has reported a nanotechnological strategy consisting of GO or CNFs combined with light-emitting diodes (LED) irradiation, which has shown great potential in a wide range of biomedical applications, as novel technique to kill two clinically relevant Gram-positive multidrug-resistant pathogens: methicillin-resistant *Staphylococcus aureus* (MRSA) and methicillin-resistant *Staphylococcus epidermidis* (MRSE) [17]. These results showed that GO or CNFs exhibited enhanced antibacterial activity and no cytotoxicity in human cells. Therefore, the implementation of these combined technologies opens a wide range of research opportunities in many important fields within the biomedical and bioengineering sector, providing an enhanced strategy for the prevention and treatment of Gram-positive multidrug-resistant infections.

4 Green Alginate-Based Composite Films

The mechanical and antibacterial properties of natural biopolymers such as alginate or PHBV can be improved with the addition of low amounts of CNMs like GO and CNFs [48, 49, 65, 74, 77]. GO is widely used due to its excellent physical properties [98], easy processing, low cost, and large-scale production [59, 99] compared to other graphene family materials. Carbon nanofibers (CNFs) are one-dimensional nanomaterials with excellent mechanical, electrical and thermal properties currently manufactured at large scale with lower price than GO [90]. Therefore, GO and CNFs have shown to be capable of improving the physical properties of alginate and PHBV films and provide them antibacterial capacity [19, 50, 65, 75].

4.1 Green Alginate-Based Composite Films

Green alginate-based composite films have been developed with enhanced physical and biological properties using carbon nanomaterials. Thus, CNMs can improve the mechanical performance, wettability, and thermal properties, among other physical properties. Furthermore, CNMs can improve the biological properties of calcium alginate films in terms of antibacterial activity. In addition, these green composite films are produced keeping the principles of green chemistry and in good agreement with the sustainable non-hazardous synthetic routes with no demands on thermal nor sonic energy consumption.

4.1.1 Physical Properties of Alginate-Based Nanocomposites

Reinforced alginate-based composite films have been produced with calcium chloride as crosslinker and several low graphene oxide (GO) contents ranging from 0 to 1% w/w [50] (see Fig. 3).

Thus, the compression modulus of calcium alginate films increased with increasing GO content almost 9 and 6 times higher (in case of 1% w/w of GO) in the dry and hydrated state, respectively, when compared with neat calcium alginate. The nanostructure of the composites showed GO embedded in the alginate polymer matrix and randomly distributed. In this study, it was observed that calcium alginate/GO composites get darker and darker with increasing GO content. However, the addition of 0.1 or 0.5% w/w of GO produced only a low increase of opacity, which might be very interesting for certain applications requiring transparency and lower-cost materials. Therefore, the incorporation of a low amount of GO (0.5% w/w) can produce low-cost reinforced calcium alginate films with a low reduction of transparency. Moreover, it has been reported that increasing crosslinking density can also increase the diffusion coefficient of alginate hydrogels [77]. However, the maximum capacity of swelling decreases with crosslinking density. Contrary, when the amount of crosslinker is too low, water needs to separate the polymer chains in order to penetrate rendering water diffusion slower but can swell the alginate polymer network much more. The compressive modulus of calcium alginate hydrogels increased with

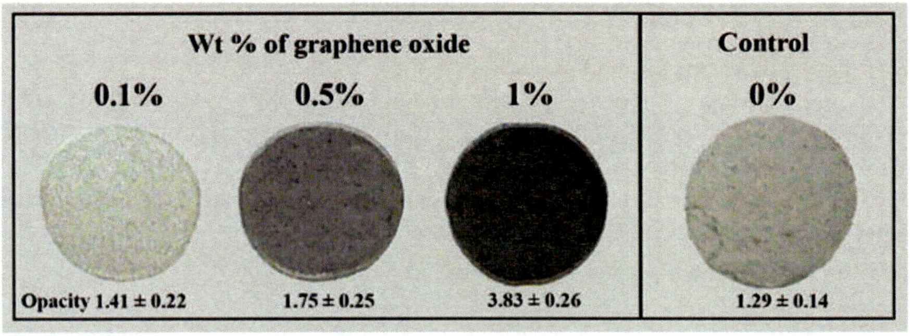

Fig. 3 Calcium alginate composite films with GO contents ranging from 0 to 1% w/w. The opacity (O) values, expressed as mean ± standard deviation, are indicated below each sample film. Reprinted with permission from Elsevier [50]

the addition of a minuscule amount of GO (0.1% w/w) of GO. These results can be explained due to the complexation reaction between the GO nanosheets and the Ca^{2+} cations, which render calcium alginate hydrogels much more resistant because of having a continuous crosslinked GO networks inside the polymer network. However, in those alginate hydrogels that had not enough number of calcium atoms, large GO networks were not able to be formed to improve compression so much. Therefore, the combination of 0.1% w/w GO and high amounts of calcium crosslinker in alginate-based films can produce hydrogels with an increase of water diffusion (almost 6 times faster) and mechanical compressive modulus (more than 4 times higher) [77]. In addition, similar results were obtained with CNFs [49]. In these green alginate-based composites, a significant reduction of water sorption in the calcium alginate hydrogels was observed with increasing crosslinker content and with the small addition of 0.1% w/w of CNFs. These green alginate-based composites produced with the incorporation of only 0.1% w/w of CNFs showed a very significant enhancement of water diffusion compared with the corresponding alginate hydrogels with a similar crosslinker content. The effect of the addition of 0.1% w/w of CNFs into the alginate biopolymer hydrogels became more and more pronounced with increased weight percentage of calcium chloride, as seen also with the incorporation of GO [77]. Furthermore, the addition of CNF resulted in a compressive modulus almost 3 times higher in composites with a high amount of crosslinker [49]. So, the reinforcement of calcium alginate achieved with the use of CNFs in this study is lower than that obtained in that utilizing the same amount of GO and calcium chloride [77]. This is probably because GO possess more oxygen functional groups than CNFs and thus more crosslinking points. In that study, the enhancement of water diffusion achieved with the addition of 0.1% w/w of GO was also higher than that obtained with 0.1% w/w of CNFs due to the presence of more carbon nanochannels in the dry state produced by the higher crosslinking between GO nanosheets. Nonetheless, CNFs have much lower cost than GO (e.g., approximately 21 times more economical than GO when obtained from Sigma-Aldrich) and have attracted great interest in the field of regeneration medicine due to their excellent mechanical, magnetic and electrical properties, which renders these advanced composite materials very promising for certain biomedical and industrial applications in need of low-cost hydrogels with enhanced physical properties.

Since CNFs possess high hydrophobic and tend to agglutinate [11], the morphology of the green alginate-based films fabricated with CNFs are more heterogeneous (see Fig. 4) due to CNFs tend to be randomly distributed in the alginate polymer network forming aggregates [48].

When these alginate-based films are hydrated in distilled water, the closed pores observed in the dry state open in the hydrated state to store water within the alginate polymer matrix (see Fig. 4). Furthermore, swelling at body temperature produces a morphology with larger pores than swelling at 26 °C. Alginate-based composite containing 0.5% w/w of CNFs leads to the highest dynamic mechanical modulus, and just a small amount of CNFs (0.1% w/w) is able to produce a significant increase of compression modulus (approximately 3 times higher) of calcium alginate in the dry state, as already demonstrated in a previous study [49], and a significant increase

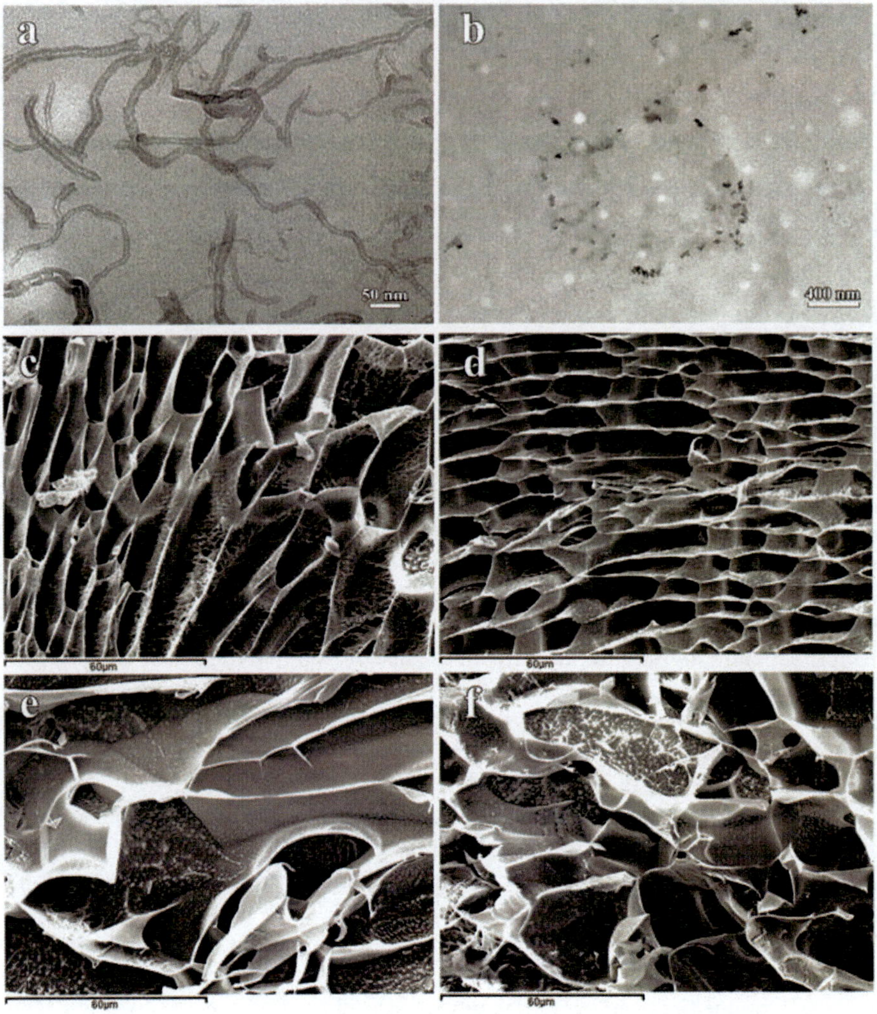

Fig. 4 TEM images of carbon nanofibers (**a**) and ultrathin section of the calcium alginate composite film with 1% w/w of carbon nanofibers (**b**) and cryoSEM of sample neat calcium alginate (**c** and **e**) and CNFs1 (**d** and **f**) hydrated at 26 ± 0.5 °C and 37 ± 0.5 °C, respectively, for 24 h. Reprinted with permission from Elsevier [48]

of tensile modulus. Thus, the tensile modulus and strength in the dry state increased with increasing the percentage of incorporated CNFs. The oxygen-containing groups present on the CNFs enable chemical interactions with the functional groups present in the alginate polymer chains. As a result, OH-bonding can enhance the tensile and compression properties of calcium alginate in the hydrated state as reported for GO in sodium alginate [30] and another carbohydrate polymers such as chitosan [27].

Another factor that affects the physical properties of alginate composite films is the green synthetic route followed in their fabrication [75]. Thus, new synthetic routes have been developed in order to enhance the physical properties of alginate-based composite hydrogel films crosslinked with calcium chloride and 1% w/w of GO. In that study, composites were fabricated crosslinking GO with alginate before and during gelation and results were compared with composites made by the conventional crosslinking technique of immersion in an aqueous solution with calcium cations. In the new fabrication methods, crosslinking occurs between the GO nanosheets, GO with alginate chains and between alginate chains by coordination via divalent calcium atoms. Moreover, when crosslinking with the divalent cations was carried during gelation, the improvement of mechanical properties was higher. Thus, significant increases in liquid water diffusion (>4 times) and tensile properties (up to almost 2.5 times higher) can be achieved by the new synthetic routes in comparison with the conventional one. Furthermore, surface tension, glass transition temperature, and equilibrium water content of these new composite hydrogels also improved significantly [75]. These results show not only the importance of the composition of the material but also the manufacturing method. The improvement of mechanical properties of the sodium alginate/GO films is mainly attributed to compatibility, alignment, and specific interactions of alginate OH groups with GO oxygenated functional groups. However, in the calcium alginate-GO composites synthesized in this work, the enhancement of mechanical properties must also be attributed to the complexation reaction that occurs between GO and the divalent calcium atoms, which render calcium alginate nanocomposites much more reinforced because of the continuous crosslinked GO networks distributed around their calcium alginate polymer structure with additional crosslinking between calcium alginate chains and crosslinked GO networks.

On the other hand, green alginate-based composite films have been developed using irregular tubes of GO, produced by crosslinking of GO nanosheets with divalent cations, as reinforcement materials for biomedical applications [74]. These green composites have exhibited higher dynamic mechanical performance than green alginate-based composite films fabricated with the incorporation of single GO nanosheets and using the same amount of GO and cations.

The synergic effect of GO with other divalent cations such as Zn^{2+} has been hardly explored [20, 67]. The addition of 1% w/w of GO to zinc alginates produced a significant reduction of equilibrium water sorption and increased wettability and water diffusion due to the formation of carbon nanochannels. In addition, the opacity of the zinc alginate films increased dramatically, which should be considered for certain biomedical fields demanding transparent materials such as ophthalmology and odontology. Another study analyzed the effect of the synergic combination of zinc and GO in alginate films in terms of morphology, structural conformation, thermal behavior/degradation, and dielectric properties [67]. In that study, in comparison with sodium alginate samples crosslinked only with Zn^{2+} ions, the composites with GO produced hydrogels with good structural integrity after immersion in water and a high degree of swelling. In addition, the glass transition temperature increased to a higher temperature (above zinc-crosslinked alginate networks) due to reduced mobility induced by additional GO crosslink bonds, while the molecular organization of the compound showed enhanced thermal stability. Charge carrier mobility was hampered by the compact structure produced by crosslinking, which reduces conductivity and leads to increased activation energy to trigger the process [67].

4.1.2 Biological Properties of Alginate-Based Composites

Alginate-based films with carbon nanofibers exhibited negligible toxicity in human keratinocyte HaCaT cells and no cell adhesion properties [48]. Therefore, these composite films displayed similar biological behavior to that of neat alginate but enhanced mechanical properties. Furthermore, the antibacterial activity of the sterilized films of calcium alginate/CNFs in the form of 10 mm disks tested against MRSE showed a high antibacterial activity (Fig. 5) and no toxicity for human keratinocyte HaCaT cells at the same concentration [70].

Regarding the antimicrobial capacity of alginate reinforced with low amounts of GO, only the composite films with at least 0.5% w/w, display a very high antibacterial activity against the life-threatening *Staphylococcus aureus* and MRSE pathogens

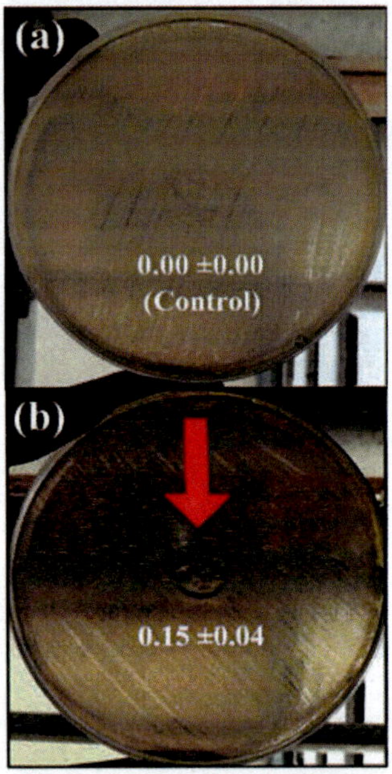

Fig. 5 Antibacterial activity of the calcium alginate films without carbon nanofibers (control sample disk) (**a**) and calcium alginate/carbon nanofibers composite films (**b**) against multidrug-resistant *Staphylococcus epidermidis* by the agar disk diffusion test after 24 h of culture at 37 °C. The red arrow indicates the sample disk showing antibacterial "halo" with its normalized width [70]

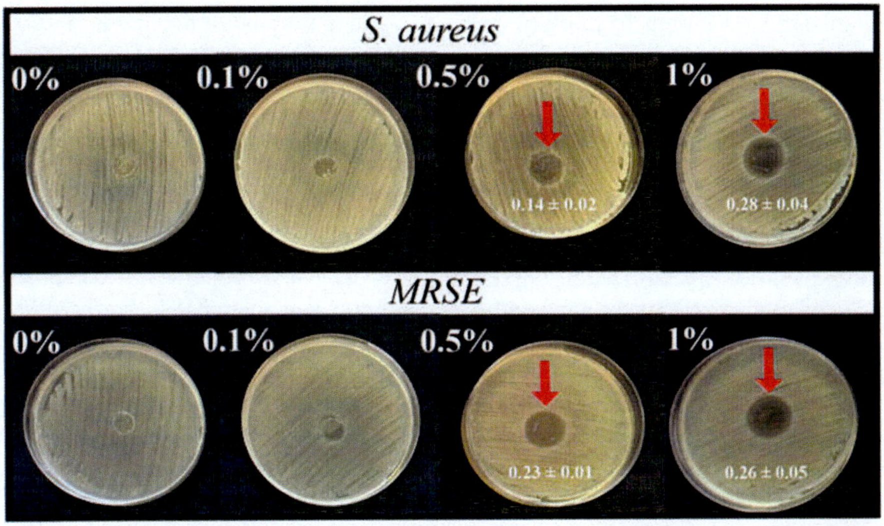

Fig. 6 Representative images of the antibacterial activity of calcium alginate with several GO contents ranging from 0 to 1% w/w against Staphylococcus aureus and methicillin-resistant Staphylococcus epidermidis (MRSE) by the agar disk diffusion test. Red arrows indicate the composite films exhibiting antibacterial "halo" and their normalized width, expressed as mean ± standard deviation, are indicated on the plates. Reprinted with permission from Elsevier [50]

(Fig. 6) without showing any toxicity to human keratinocyte HaCaT cells either [50].

Furthermore, alginate/GO nanocomposite films showed similar biological behavior to neat alginate films in terms of cell adhesion [69].

Nevertheless, the addition of 1% w/w of GO to zinc alginates did not enhanced the antimicrobial activity [20]. In these samples, the release of zinc ions was the mechanism responsible for antibacterial activity and cytotoxicity, being independent of the amount of GO in the composites. However, the GO reinforcement of the zinc alginate films produced a significant increase of water diffusion, wettability, and opacity.

To sum up, the addition of these CNMs into alginate films can produce a significant enhancement of mechanical performance, wettability, water diffusion, and antibacterial activity as mentioned before [49, 50, 70, 75, 77]. In addition, the MTT cytotoxicity assay performed with human keratinocyte HaCaT cells showed no statistically significant differences (~100% viability) between the negative controls and the extracts of the alginate/GO and alginate/CNF composite films. No cell adhesion of HaCat cells on the alginate-based composites was observed independently on de CNF or GO concentration [69]. Both one-dimensional hydrophobic CNFs and two-dimensional hydrophilic GO nanosheets remain embedded in the alginate polymer matrix, and they do not enhance cell adhesion on the composite surface, in good agreement with the FESEM observations. Nonetheless, these composite films show similar promising applications to those of calcium alginate films with enhanced physical and antibacterial properties [49, 50, 70, 75, 77].

4.2 Green PHBV-Based Composite Films

Graphene oxide provides also the possibility of reinforcing PHBV, acting as a nucleating agent for crystallization, improving the mechanical properties of composite materials and increasing the temperature of maximum degradation [84]. Furthermore, its presence does not interfere with PHBV biodegradability, although it could restrict the mobility of PHBV chains in the crystal growth process. Carbon nanofibers can also be employed to enhance the conductivity, thermal, mechanical, and gas barrier properties of thermoplastic biopolyesters, such as PHBV [72]. In this regard, the incorporation of a low percentages of GO and/or CNFs into PHBV through strong intermolecular interactions such as hydrogen bonds can enhance some physical and biological properties [65].

4.2.1 Physical Properties of PHBV-Based Composites

GO and CNFs show different morphology composed of rectangular nanosheets stacked together, in case of GO, and hollow filaments (CNF) with usually a wide range of diameters (21.9 ± 12.5 nm) and lengths (645.8 ± 466.5 nm) respectively [65]. Another difference remains in their distribution in the PHBV-based composite. While the GO nanosheets possess many oxygen-containing functional groups as hydroxyl groups on the basal planes and at the edges, the CNFs are highly hydrophobic and tend to aggregate. Thus, when the CNFs were mixed with the polar PHBV biopolymer using the polar solvent, dichloromethane, the dispersion of these filamentous materials was very poor and a heterogeneous composite structure was obtained in comparison with that achieved with GO [65]. Nevertheless, the addition of GO and CNFs enhanced significantly the wettability properties on the surface of the neat PHBV films, resulting in an increase of surface energy, that is, more hydrophilic. Furthermore, compression strength increased with the addition of both GO nanosheets and CNFs in a similar percentage (about 25%) [65]. Besides, the thermal study of the composite films showed that they exhibited similar double melting peaks as those of neat PHBV. Moreover, the GO and CNFs filling slightly increased the glass transition temperature and heat of fusion of the bulk biopolymer as previously reported in another study of PHBV-based nanocomposites with carbon nanotubes and CNFs [72].

4.2.2 Biological Properties of PHBV-Based Composites

Biocompatibility of the green PHBV-based composite films was tested in canine adipose derived mesenchymal stem cells (cADSCs) at short (24 h) and long (48–72 h) times. The results not only showed that there was no cytotoxicity but also cADSCs seeded on PHBV/GO and PHBV/CNFs composites exhibited a significant increase in proliferation in comparison with plastic condition or neat PHBV at 24 h of incubation [65]. Furthermore, PHBV/GO maintained this increase of proliferation along the time (48–72 h). Confocal evaluation revealed a more homogeneous and improved distribution of the cADSCs onto PHBV/GO with respect to PHBV/CNFs and neat PHBV (see Fig. 7).

The cADSCs spread and invaded the PHBV/GO substrate showing cell-to-cell and cell-to-substrate focal adhesions with a harmonized intracellular distribution of the F-actin fibers labeled with Phalloidin (Fig. 7b). Furthermore, the quantification analysis showed a significant increased number of cADSCs seeded on PHBV/GO biomaterial at all tested times, 24, 48, and 72 h, and in PHBV/CNF at 24 h in comparison with the pure PHBV film (Fig. 7d). Nevertheless, the antimicrobial agar diffusion test revealed that PHBV/CNFs possessed more antibacterial activity than PHBV/GO (see Fig. 8).

Nonetheless, both PHBV-based composite films were able to inhibit the bacterial growth of *S.aureus* around the samples in some extent in comparison with the neat PHBV film.

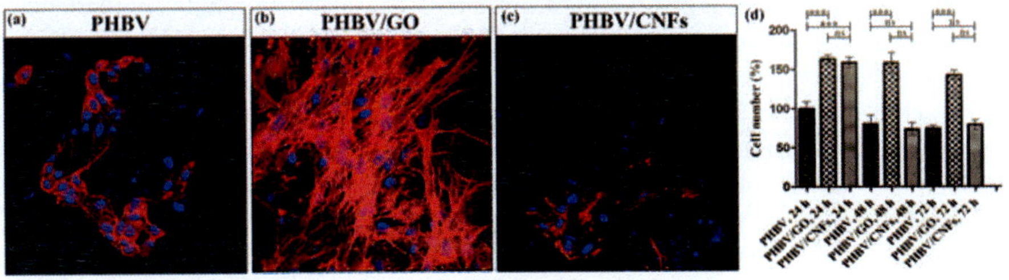

Fig. 7 cADSCs attach and homogenously distribute showing long projections on the PHBV/GO surface (**b**) in contrast to PHBV control sample (**a**) and PHBV/CNFs (**c**). Representative confocal microscope images (20×) shows the distribution of F-actin fibers, stained in red by using Phalloidin conjugated with Texas Red. DAPI was employed for nuclei staining (blue) and cell quantification (**d**). Three independent experiments were performed in quadruplicates. Data was expressed as mean ± SD; *$p > 0.05$; ***$p > 0.01$; ***$p > 0.001$; ns: not significant. Reprinted with permission from Elsevier [65].

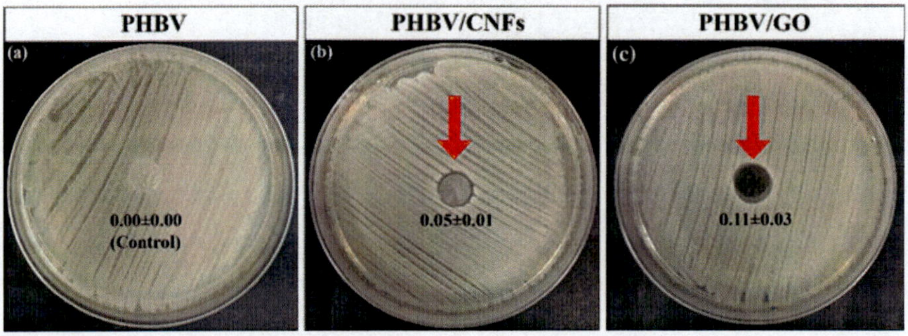

Fig. 8 Antibacterial activity of the films of **a** poly(3-hydroxybutyrate-co-3-hydroxyvalerate) (PHBV, control sample), PHBV with 1% w/w of **b** graphene oxide (PHBV/GO) or **c** carbon nanofibers (PHBV/CNFs) against Staphylococcus aureus by the agar disk diffusion test after 24 h of culture at 37 °C. The red arrow indicates the sample exhibiting antibacterial activity and the normalized widths as indicated in each image. Reprinted with permission from Elsevier [65]

5 Conclusions

Alginate is a naturally occurring anionic green polymer extensively investigated and used for many biomedical applications, due to its biocompatibility, low toxicity, relatively low cost, and mild gelation by addition of divalent cations. Alginate hydrogels can be prepared by various crosslinking methods, and they can be used in a wide range of industrial applications. Other promising natural green material is the poly(3-hydroxybutyrate-co-3-hydroxyvalerate) biopolymer, or PHBV, due to its null toxicity, high biocompatibility, and that can nowadays be produced at large scale. However, these materials in pure state present very poor mechanical properties and do not possess antibacterial activity. Therefore, among many other techniques, the strategy of incorporating carbon nanomaterials, such as graphene oxide (GO) and carbon nanofibers (CNF), is proposed here as a very promising approach that can enhance both mechanical and antibacterial capacity. Thus, the incorporation of low amounts of GO and CNFs into alginate films produced very significant increases of compression modulus when compared with neat alginate. The opacity of calcium alginate can also be affected by the incorporation of CNMs into the polymer matrix, which must be considered for certain industrial applications such as contact lens or dentistry. In addition, the combination of a minuscule amount of GO or CNFs and a high amount of crosslinker can enhance very significantly the water diffusion properties of calcium alginate. However, it was concluded that the reinforcement achieved with CNFs in calcium alginate is lower than that obtained utilizing the same amount of GO in calcium chloride probably because GO possesses more oxygen functional groups than CNFs, and thus more crosslinking points. However, CNFs possess currently the advantage that have much lower cost and much higher electrical conductivity than GO. The green chemical route followed to produce alginate-based composite films has also shown to be a crucial factor in the enhancement of physical properties such as wettability, water diffusion, thermal and tensile properties. The GO and CNFs can improve the biological properties of neat alginate in terms of antibacterial activity against clinically relevant pathogens such as *S. aureus* and MRSE without showing any toxicity in human keratinocyte HaCaT cells. However, the incorporation of GO to zinc alginate does not affect neither the intrinsic antibacterial activity of zinc alginate films nor the cytotoxicity but increases physical properties such as thermal stability. In the same way, GO and CNFs can reinforce and enhance significantly the wettability and thermal properties of neat PHBV films. In addition, GO and CNFs showed not only no toxicity, but also a significant increase in the proliferation of cADSCs seeded in PHBV-based composites. This increase in proliferation is maintained in PHBV/GO composite films over time. Finally, regarding antibacterial activity, both PHBV-based films with incorporated GO and CNFs nanomaterials are capable of inhibiting part of the bacterial growth of *S. aureus*.

Acknowledgements The authors would like to acknowledge the Fundación Universidad Católica de Valencia San Vicente Mártir for the financial support through the 2019-231-003UCV grant (awarded to Á. S-A).

Conflicts of Interest The authors declare no conflict of interest.

References

1. Anderson AJ, Dawes E (1990) Ocurrence, metabolism, metabolic role and industrial uses of bacterial polyhydroxyalkanoates. Microbiol Rev 54:450–472
2. Aroca AS, Pradas MM, Ribelles JLG (2008) Effect of crosslinking on porous poly(methyl methacrylate) produced by phase separation. Colloid Polym Sci 286:209–216. https://doi.org/10.1007/s00396-007-1755-0
3. Aroca AS, Pradas MM, Ribelles JLG, Rault J (2015) Thermal analysis of water in reinforced plasma-polymerised poly(2-hydroxyethyl acrylate) hydrogels. Eur Polym J 72:523–534. https://doi.org/10.1016/j.eurpolymj.2015.05.032
4. Balandin AA (2011) Thermal properties of graphene and nanostructured carbon materials. Nat Mater 10:569–581. https://doi.org/10.1038/nmat3064
5. Berthet MA, Angellier-Coussy H, Chea V, Guillard V, Gastaldi E, Gontard N (2015) Sustainable food packaging: valorising wheat straw fibres for tuning PHBV-based composites properties. Compos Part A Appl Sci Manuf 72:139–147
6. Berthet MA, Angellier-Coussy H, Machado D, Hilliou L, Staebler A, Vicente A, Gontard N (2015) Exploring the potentialities of using lignocellulosic fibres derived from three food by-products as constituents of biocomposites for food packaging. Ind Crops Prod 69:110–122. https://doi.org/10.1016/j.indcrop.2015.01.028
7. Blecher K, Nasir A, Friedman A (2011) The growing role of nanotechnology in combating infectious disease. Virulence. https://doi.org/10.4161/viru.2.5.17035
8. Bucci DZ, Tavares LBB, Sell I (2005) PHB packaging for the storage of food products. Polym Test 24:564–571
9. Bugnicourt E, Cinelli P, Lazzeri A, Alvarez V (2014) Polyhydroxyalkanoate (PHA): review of synthesis, characteristics, processing and potential applications in packaging. Express Polym Lett 8:791–808. https://doi.org/10.3144/expresspolymlett.2014.82

10. Castro-Mayorga JL, Fabra MJ, Lagaron JM (2016) Stabilized nanosilver based antimicrobial poly(3-hydroxybutyrate-co-3-hydroxyvalerate) nanocomposites of interest in active food packaging. Innov Food Sci Emerg Technol 33:524–533
11. Chaba JM, Nomngongo PN (2018) Preparation of V2O5-ZnO coated carbon nanofibers: application for removal of selected antibiotics in environmental matrices. J Water Process Eng 23:50–60. https://doi.org/10.1016/j.jwpe.2018.03.003
12. Chang HM, Wang ZH, Luo HN, Xu M, Ren XY, Zheng GX, Wu BJ, Zhang XH, Lu XY, Chen F, Jing XH, Wang L (2014) Poly(3-hydroxybutyrate-co-3-hydroxyhexanoate)-based scaffolds for tissue engineering. Braz J Med Biol Res 47:533–539. https://doi.org/10.1590/1414-431X20143930
13. Chen GQ, Wu Q (2005) The application of polyhydroxyalkanoates as tissue engineering materials. Biomaterials 26:6565–6578. https://doi.org/10.1016/j.biomaterials.2005.04.036
14. Chong Y, Ge C, Fang G, Wu R, Zhang H, Chai Z, Chen C, Yin JJ (2017) Light-enhanced antibacterial activity of graphene oxide, mainly via accelerated electron transfer. Environ Sci Technol 51:10154–10161. https://doi.org/10.1021/acs.est.7b00663
15. Debiagi F, Kobayashi RKT, Nakazato G, Panagio LA, Mali S (2014) Biodegradable active packaging based on cassava bagasse, polyvinyl alcohol and essential oils. Ind Crops Prod 52:664–670. https://doi.org/10.1016/j.indcrop.2013.11.032
16. Draget K, Skjakbrak G, Smidsrod O (1994) Alginic acid gels: the effect of alginate chemical composition and molecular weight. Carbohydr Polym 25:31–38
17. Elias L, Taengua R, Frígols B, Salesa B, Serrano-Aroca Á (2019) Carbon nanomaterials and LED irradiation as antibacterial strategies against gram-positive multidrug-resistant pathogens. Int J Mol Sci 20:3603. https://doi.org/10.3390/ijms20143603
18. Fernández J, Bert F, Nicolas-Chanoine MH (2016) The challenges of multi-drug-resistance in hepatology. J Hepatol. https://doi.org/10.1016/j.jhep.2016.08.006
19. Frígols B, Martí M, Hernández-Oliver C, Aarstad O, Teialeret Ulset A-S, Inger Sætrom G, Lillelund Aachmann F, Serrano-Aroca Á (n.d.) Graphene oxide in zinc alginate films: antibacterial activity, water sorption, wettability and opacity. Carbohydr Polym (under review)
20. Frígols B, Martí M, Salesa B, Hernández-Oliver C, Aarstad O, Ulset AST, Sætrom GI, Aachmann FL, Serrano-Aroca Á (2019) Graphene oxide in zinc alginate films: antibacterial activity, cytotoxicity, zinc release, water sorption/diffusion, wettability and opacity. PLoS ONE 14:e0212819. https://doi.org/10.1371/journal.pone.0212819
21. García-García JM, Quijada-Garrido I, López L, París R, Núñez-López MT, De La Peña Zarzuelo E, Garrido L (2013) The surface modification of poly(3-hydroxybutyrate-co-3-hydroxyhexanoate) copolymers to improve the attachment of urothelial cells. Mater Sci Eng C 33:362–369. https://doi.org/10.1016/j.msec.2012.08.052
22. Georgakilas V, Otyepka M, Bourlinos AB, Chandra V, Kim N, Kemp KC, Hobza P, Zboril R, Kim KS (2012) Functionalization of graphene: covalent and non-covalent approach. Chem Rev 112:6156–6214. https://doi.org/10.1021/cr3000412
23. Gerard T, Budtova T (2012) Morphology and molten-state rheology of polylactide and polyhydroxyalkanoate blends. Eur Polym J 48:1110–1117. https://doi.org/10.1016/j.eurpolymj.2012.03.015
24. Godbey WT, Atala A (2002) In vitro systems for tissue engineering. Ann NY Acad Sci 961:10–26. https://doi.org/10.1111/j.1749-6632.2002.tb03041.x
25. Gombotz WR, Wee SF (1998) Protein release from alginate matrices. Adv Drug Deliv Rev 31:267–285. https://doi.org/10.1016/S0169-409X(97)00124-5
26. Ha CS, Cho WJ (2002) Miscibility, properties, and biodegradability of microbial polyester containing blends. Prog Polym Sci 27:759–809. https://doi.org/10.1016/S0079-6700(01)00050-8
27. Han D, Yan L, Chen W, Li W (2011) Preparation of chitosan/graphene oxide composite film with enhanced mechanical strength in the wet state. Carbohydr Polym 83:653–658. https://doi.org/10.1016/j.carbpol.2010.08.038

28. Han J, Wu LP, Liu XB, Hou J, Zhao LL, Chen JY, Zhao DH, Xiang H (2017) Biodegradation and biocompatibility of haloarchaea-produced poly(3-hydroxybutyrate-co-3-hydroxyvalerate) copolymers. Biomaterials 139:172–186. https://doi.org/10.1016/j.biomaterials.2017.06.006
29. Hutmacher DW (2000) Scaffolds in tissue engineering bone and cartilage. Biomater Silver Jubilee Compendium 21:175–189. https://doi.org/10.1016/B978-008045154-1.50021-6
30. Ionita M, Pandele MA, Iovu H (2013) Sodium alginate/graphene oxide composite films with enhanced thermal and mechanical properties. Carbohydr Polym 94:339–344
31. Jeon O, Bouhadir KH, Mansour JM, Alsberg E (2009) Photocrosslinked alginate hydrogels with tunable biodegradation rates and mechanical properties. Biomaterials 30:2724–2734. https://doi.org/10.1016/j.biomaterials.2009.01.034
32. Jia Z, Shen D, Xu W (2001) Synthesis and antibacterial activities of quaternary ammonium salt of chitosan. Carbohydr Res 333:1–6. https://doi.org/10.1016/S0008-6215(01)00112-4
33. Jin S, Xu D, Zhou N, Yuan J, Shen J (2013) Antibacterial and anticoagulation properties of polyethylene/geneO-MPC nanocomposites. J Appl Polym Sci 129:884–891
34. John MJ, Thomas S (2008) Biofibres and biocomposites. Carbohydr Polym 71:343–364. https://doi.org/10.1016/j.carbpol.2007.05.040
35. Jost V, Kopitzky R (2015) Blending of polyhydroxybutyrate-co-valerate with polylactic acid for packaging applications—reflections on miscibility and effects on the mechanical and barrier properties. Chem Biochem Eng Q 29:221–246. https://doi.org/10.15255/CABEQ.2014.2257
36. Kalbacova M, Broz A, Kong J, Kalbac M (2010) Graphene substrates promote adherence of human osteoblasts and mesenchymal stromal cells. Carbon 48:4323–4329. https://doi.org/10.1016/j.carbon.2010.07.045
37. Ke Y, Liu C, Zhang X, Xiao M, Wu G (2017) Surface modification of polyhydroxyalkanoates toward enhancing cell compatibility and antibacterial activity. Macromol Mater Eng 302:1700258. https://doi.org/10.1002/mame.201700258
38. Krikštanavičiene K, Stanys S, Jonaitiene V (2014) Comparative investigation of mechanical-physical characteristics of biodegradable and non-degradable yarns. Autex Res J 14:61–72. https://doi.org/10.2478/aut-2014-0001
39. Kushwah BS, Kushwah AVS, Singh V (2016) Towards understanding polyhydroxyalkanoates and their use. J Polym Res 23. https://doi.org/10.1007/s10965-016-0988-3
40. Lagaron JM, Lopez-Rubio A (2011) Nanotechnology for bioplastics: opportunities, challenges and strategies. Trends Food Sci Technol 22:611–617. https://doi.org/10.1016/j.tifs.2011.01.007
41. Lee KY, Mooney DJ (2012) Alginate: properties and biomedical applications. Prog Polym Sci 37:106–126. https://doi.org/10.1016/j.progpolymsci.2011.06.003
42. Lee SY (1996) Plastic bacteria? Progress and prospects for polyhydroxyalkanoate production in bacteria. Trends Biotechnol 14:431–438. https://doi.org/10.1016/0167-7799(96)10061-5
43. Lee SY, Choi JI, Han K, Song JY (1999) Removal of endotoxin during purification of poly(3-hydroxybutyrate) from gram-negative bacteria. Appl Environ Microbiol 65:2762–2764
44. Li W, Ding Y, Rai R, Roether JA, Schubert DW, Boccaccini AR (2014) Preparation and characterization of PHBV microsphere/45S5 bioactive glass composite scaffolds with vancomycin releasing function. Mater Sci Eng C 41:320–328. https://doi.org/10.1016/j.msec.2014.04.052
45. Li Z, Yang J, Loh XJ (2016) Polyhydroxyalkanoates: opening doors for a sustainable future. NPG Asia Mater. https://doi.org/10.1038/am.2016.48
46. Liu S, Zeng TH, Hofmann M, Burcombe E, Wei J, Jiang R, Kong J, Chen Y (2011) Antibacterial activity of graphite, graphite oxide, graphene oxide, and reduced graphene oxide: membrane and oxidative stress. ACS Nano 5:6971–6980. https://doi.org/10.1021/nn202451x
47. Liu Y, Wang X, Yang F, Yang X (2008) Excellent antimicrobial properties of mesoporous anatase TiO_2 and Ag/TiO_2 composite films. Microporous Mesoporous Mater 114:431–439. https://doi.org/10.1016/j.micromeso.2008.01.032
48. Llorens-Gámez M, Salesa B, Serrano-Aroca Á (2020) Physical and biological properties of alginate/carbon nanofibers hydrogel films. Int J Biol Macromol 151:499–507. https://doi.org/10.1016/j.ijbiomac.2020.02.213

49. Llorens-Gámez M, Serrano-Aroca Á (2018) Low-cost advanced hydrogels of calcium alginate/carbon nanofibers with enhanced water diffusion and compression properties. Polymers (Basel) 10:405. https://doi.org/10.3390/polym10040405
50. Martí M, Frígols B, Salesa B, Serrano-Aroca Á (2019) Calcium alginate/graphene oxide films: reinforced composites able to prevent *Staphylococcus aureus* and methicillin-resistant *Staphylococcus epidermidis* infections with no cytotoxicity for human keratinocyte HaCaT cells. Eur Polym J 110:14–21. https://doi.org/10.1016/J.EURPOLYMJ.2018.11.012
51. Martí M, Frígols B, Serrano-Aroca Á (2018) Antimicrobial characterization of advanced materials for bioengineering applications. J Vis Exp e57710. https://doi.org/10.3791/57710
52. Modi S, Koelling K, Vodovotz Y (2013) Assessing the mechanical, phase inversion, and rheological properties of poly-[(R)-3-hydroxybutyrate-co-(R)-3-hydroxyvalerate] (PHBV) blended with poly-(l-lactic acid) (PLA). Eur Polym J 49:3681–3690. https://doi.org/10.1016/j.eurpolymj.2013.07.036
53. Monleón Pradas M, Gómez Ribelles JL, Serrano Aroca A, Gallego Ferrer G, Suay Antón J, Pissis P (2001) Interaction between water and polymer chains in poly(hydroxyethyl acrylate) hydrogels. Colloid Polym Sci 279:323–330. https://doi.org/10.1007/s003960000426
54. Murata H, Nakajima Y, Saitoh N, Yoshizawa N, Suemasu T, Toko K (2019) High-electrical-conductivity multilayer graphene formed by layer exchange with controlled thickness and interlayer. Sci Rep 9:1–5. https://doi.org/10.1038/s41598-019-40547-0
55. Nakagawa M, Teraoka F, Fujimoto S, Hamada Y, Kibayashi H, Takahashi J (2006) Improvement of cell adhesion on poly(L-lactide) by atmospheric plasma treatment. J Biomed Mater Res Part A 77:112–118. https://doi.org/10.1002/jbm.a.30521
56. Norrrahim MNF, Ariffin H, Hassan MA, Ibrahim NA, Nishida H (2013) Performance evaluation and chemical recyclability of a polyethylene/poly (3-hydroxybutyrate-co-3-hydroxyvalerate) blend for sustainable packaging. RSC Adv 3:24378–24388
57. Novoselov KS, Fal'ko VI, Colombo L, Gellert PR, Schwab MG, Kim K (2012) A roadmap for graphene. Nature 490:192–200
58. Pandey H, Parashar V, Parashar R, Prakash R, Ramteke PW, Pandey AC (2011) Controlled drug release characteristics and enhanced antibacterial effect of graphene nanosheets containing gentamicin sulfate. Nanoscale 3:4104. https://doi.org/10.1039/c1nr10661a
59. Papi M, Palmieri V, Bugli F, De Spirito M, Sanguinetti M, Ciancico C, Braidotti MC, Gentilini S, Angelani L, Conti C (2016) Biomimetic antimicrobial cloak by graphene-oxide agar hydrogel. Sci Rep 6:12. https://doi.org/10.1038/s41598-016-0010-7
60. Pelgrift RY, Friedman AJ (2013) Nanotechnology as a therapeutic tool to combat microbial resistance. Adv Drug Deliv Rev. https://doi.org/10.1016/j.addr.2013.07.011
61. Raquez JM, Deléglise M, Lacrampe MF, Krawczak P (2010) Thermosetting (bio)materials derived from renewable resources: a critical review. Prog Polym Sci 35:487–509. https://doi.org/10.1016/j.progpolymsci.2010.01.001
62. Ratner BD, Hoffman AS, Lemons JE (2006) Biomaterials science: an introduction to materials, vol I. In: Basic concepts 3rd annual microscopy workshop in microscopy for materials these topics will be covered: 31, pp 25–27
63. Recco MS, Floriano AC, Tada DB, Lemes AP, Lang R, Cristovan FH (2016) Poly(3-hydroxybutyrate-co-valerate)/poly(3-thiophene ethyl acetate) blends as a electroactive biomaterial substrate for tissue engineering application. RSC Adv 6:25330–25338. https://doi.org/10.1039/c5ra26747a
64. Remminghorst U, Rehm BHA (2006) Bacterial alginates: from biosynthesis to applications. Biotechnol Lett 28:1701–1712. https://doi.org/10.1007/s10529-006-9156-x
65. Rivera-Briso AL, Aachmann FL, Moreno-Manzano V, Serrano-Aroca Á (2019) Graphene oxide nanosheets versus carbon nanofibers: enhancement of physical and biological properties of poly(3-hydroxybutyrate-co-3-hydroxyvalerate) films for biomedical applications. Int J Biol Macromol. https://doi.org/10.1016/j.ijbiomac.2019.10.034
66. Rivera-Briso AL, Serrano-Aroca Á (2018) Poly(3-hydroxybutyrate-co-3-hydroxyvalerate): enhancement strategies for advanced applications. Polymers (Basel). https://doi.org/10.3390/polym10070732

67. Sabater i Serra R, Molina-Mateo J, Torregrosa-Cabanilles C, Andrio-Balado A, Dueñas JMM, Serrano-Aroca Á (2020) Bio-nanocomposite hydrogel based on zinc alginate/graphene oxide: morphology, structural conformation, thermal behavior/degradation, and dielectric properties. Polymers (Basel) 12:702. https://doi.org/10.3390/polym12030702
68. Sahari J, Sapuan SM (2012) Natural fibre reinforced biodegradable polymer composites. Rev Adv Mater Sci 30:166–174
69. Salesa B, Llorens-Gámez M, Serrano-Aroca Á (2020) Study of 1D and 2D carbon nanomaterial in alginate films. Nanomaterials 10:206. https://doi.org/10.3390/nano10020206
70. Salesa B, Martí M, Frígols B, Serrano-Aroca Á (2019) Carbon nanofibers in pure form and in calcium alginate composites films: new cost-effective antibacterial biomaterials against the life-threatening multidrug-resistant staphylococcus epidermidis. Polymers (Basel) 11:453. https://doi.org/10.3390/polym11030453
71. Sánchez-Correa F, Vidaurre-Agut C, Serrano-Aroca A, Campillo-Fernández AJ (2018) Poly(2-hydroxyethyl acrylate) hydrogels reinforced with graphene oxide: remarkable improvement of water diffusion and mechanical properties. J Appl Polym Sci 46158. https://doi.org/10.1002/app.46158
72. Sanchez-Garcia MD, Lagaron JM, Hoa SV (2010) Effect of addition of carbon nanofibers and carbon nanotubes on properties of thermoplastic biopolymers. Compos Sci Technol 70:1095–1105. https://doi.org/10.1016/j.compscitech.2010.02.015
73. Seil JT, Webster TJ (2012) Antimicrobial applications of nanotechnology: methods and literature. Int J Nanomed. https://doi.org/10.2147/IJN.S24805
74. Serrano-Aroca Á, Deb S (2017) Synthesis of irregular graphene oxide tubes using green chemistry and their potential use as reinforcement materials for biomedical applications. PLoS ONE 12:e0185235. https://doi.org/10.1371/journal.pone.0185235
75. Serrano-Aroca Á, Iskandar L, Deb S (2018) Green synthetic routes to alginate-graphene oxide composite hydrogels with enhanced physical properties for bioengineering applications. Eur Polym J 103:198–206. https://doi.org/10.1016/j.eurpolymj.2018.04.015
76. Serrano-Aroca Á, Llorens-Gámez M (2017) Dynamic mechanical analysis and water vapour sorption of highly porous poly(methyl methacrylate). Polymer (Guildf) 125:58–65. https://doi.org/10.1016/j.polymer.2017.07.075
77. Serrano-Aroca Á, Ruiz-Pividal JF, Llorens-Gámez M (2017) Enhancement of water diffusion and compression performance of crosslinked alginate films with a minuscule amount of graphene oxide. Sci Rep 7:1–8. https://doi.org/10.1038/s41598-017-10260-x
78. Serrano Aroca A, Campillo Fernández AJ, Gómez Ribelles JL, Monleón Pradas M, Gallego Ferrer G, Pissis P (2004) Porous poly(2-hydroxyethyl acrylate) hydrogels prepared by radical polymerisation with methanol as diluent. Polymer (Guildf) 45. https://doi.org/10.1016/j.polymer.2004.10.033
79. Serrano Aroca A, Gómez Ribelles JL, Monleón Pradas M, Vidaurre Garayo A, Suay Antón J (2007) Characterisation of macroporous poly(methyl methacrylate) coated with plasma-polymerised poly(2-hydroxyethyl acrylate). Eur Polym J 43:4552–4564. https://doi.org/10.1016/j.eurpolymj.2007.07.026
80. Serrano Aroca A, Monleón Pradas M, Gómez Ribelles JL (2007) Plasma-induced polymerisation of hydrophilic coatings onto macroporous hydrophobic scaffolds. Polymer (Guildf) 48:2071–2078. https://doi.org/10.1016/j.polymer.2007.02.017
81. Shao W, Liu X, Min H, Dong G, Feng Q, Zuo S (2015) Preparation, characterization, and antibacterial activity of silver nanoparticle-decorated graphene oxide nanocomposite. ACS Appl Mater Interfaces 7:6966–6973
82. Singh S, Mohanty AK (2007) Wood fiber reinforced bacterial bioplastic composites: fabrication and performance evaluation. Compos Sci Technol 67:1753–1763. https://doi.org/10.1016/j.compscitech.2006.11.009
83. Singh S, Mohanty AK, Sugie T, Takai Y, Hamada H (2008) Renewable resource based biocomposites from natural fiber and polyhydroxybutyrate-co-valerate (PHBV) bioplastic. Compos Part A Appl Sci Manuf 39:875–886. https://doi.org/10.1016/j.compositesa.2008.01.004

84. Sridhar V, Lee I, Chun HH, Park H (2013) Graphene reinforced biodegradable poly(3-hydroxybutyrate-co-4-hydroxybutyrate) nano-composites. Express Polym Lett 7:320–328. https://doi.org/10.3144/expresspolymlett.2013.29
85. Stout DA, Yoo J, Santiago-Miranda AN, Webster TJ (2012) Mechanisms of greater cardiomyocyte functions on conductive nanoengineered composites for cardiovascular applications. Int J Nanomed 7:5653–5669. https://doi.org/10.2147/IJN.S34574
86. Sudesh K, Abe H, Doi Y (2000) Synthesis, structure and properties of polyhydroxyalkanoates: biological polyesters. Prog Polym Sci 25:1503–1555. https://doi.org/10.1016/S0079-6700(00)00035-6
87. Sultana N, Wang M (2012) PHBV/PLLA-based composite scaffolds fabricated using an emulsion freezing/freeze-drying technique for bone tissue engineering: surface modification and in vitro biological evaluation. Biofabrication 4. https://doi.org/10.1088/1758-5082/4/1/015003
88. Sykacek E, Hrabalova M, Frech H, Mundigler N (2009) Extrusion of five biopolymers reinforced with increasing wood flour concentration on a production machine, injection moulding and mechanical performance. Compos Part A Appl Sci Manuf 40:1272–1282. https://doi.org/10.1016/j.compositesa.2009.05.023
89. Tegou E, Magana M, Katsogridaki AE, Ioannidis A, Raptis V, Jordan S, Chatzipanagiotou S, Chatzandroulis S, Ornelas C, Tegos GP (2016) Terms of endearment: bacteria meet graphene nanosurfaces. Biomaterials. https://doi.org/10.1016/j.biomaterials.2016.02.030
90. Tran PA, Zhang L, Webster TJ (2009) Carbon nanofibers and carbon nanotubes in regenerative medicine. Adv Drug Deliv Rev. https://doi.org/10.1016/j.addr.2009.07.010
91. Vauchel P, Kaas R, Arhaliass A, Baron R, Legrand J (2008) A new process for extracting alginates from laminaria digitata: reactive extrusion. Food Bioprocess Technol 1:297–300
92. Wang L, Chen J, Shi L, Shi Z, Ren L, Wang Y (2014) The promotion of antimicrobial activity on silicon substrates using a "click" immobilized short peptide. Chem Commun (Camb) 50:975–977. https://doi.org/10.1039/c3cc47922f
93. Wang L, Du J, Cao D, Wang Y (2013) Recent advances and the application of poly(3-hydroxybutyrate-co-3-hydroxyvalerate) as tissue engineering materials. J Macromol Sci Part A 50:885–893. https://doi.org/10.1080/10601325.2013.802540
94. Yousefi M, Dadashpour M, Hejazi M, Hasanzadeh M, Behnam B, de la Guardia M, Shadjou N, Mokhtarzadeh A (2017) Anti-bacterial activity of graphene oxide as a new weapon nanomaterial to combat multidrug-resistance bacteria. Mater Sci Eng C https://doi.org/10.1016/j.msec.2016.12.125
95. Zactiti EM, Kieckbusch TG (2009) Release of potassium sorbate from active films of sodium alginate crosslinked with calcium chloride. Packag Technol Sci 22:349–358
96. Zhang Y, Nayak TR, Hong H, Cai W (2012) Graphene: a versatile nanoplatform for biomedical applications. Nanoscale 4:3833–3842. https://doi.org/10.1039/c2nr31040f
97. Zhao E, Chen Y, Wang H, Chen S, Lam JWY, Leung CWT, Hong Y, Tang BZ (2015) Light-enhanced bacterial killing and wash-free imaging based on AIE fluorogen. ACS Appl Mater Interfaces 7:7180–7188. https://doi.org/10.1021/am509142k
98. Zhao J, Liu L, Li F (2015) Fabrication and reduction. https://doi.org/10.1007/978-3-662-44829-8_1
99. Zhu Y, Murali S, Cai W, Li X, Suk JW, Potts JR, Ruoff RS (2010) Graphene and graphene oxide: synthesis, properties, and applications. Adv Mater 22:3906–3924. https://doi.org/10.1002/adma.201001068
100. Zou X, Zhang L, Wang Z, Luo Y (2016) Mechanisms of the antimicrobial activities of graphene materials. J Am Chem Soc 138:2064–2077. https://doi.org/10.1021/jacs.5b11411

Chapter 20
Green Composites from Medicinal Plants

T. Krithiga and Aravind Kumar

1 Introduction

Composites are hybrid materials framed by the incorporation of at least two materials with superior structure and preferable properties over original materials.

1.1 Properties of Composites

Composites have the accompanying properties

1. They have magnificent strength, high stiffness and are lightweight materials
2. They have high specific modulus
3. They are highly resistant to corrosion
4. They can be formed into any shape and size with required mechanical properties by moulding
5. They are recyclable, dependable and devour less energy.

1.2 Applications of Composites

The composites are utilized in various applications due to the excellent above-mentioned properties. A few applications of composites are summarized below

T. Krithiga (✉)
Department of Chemistry, Sathyabama Institute of Science and Technology, Chennai, India
e-mail: krithi_t2001@yahoo.com

A. Kumar
Department of Chemical Engineering, Sathyabama Institute of Science and Technology, Chennai, India

S. Thomas and P. Balakrishnan (eds.), *Green Composites*,
Materials Horizons: From Nature to Nanomaterials,
https://doi.org/10.1007/978-981-15-9643-8_20

1. Assembling propelled aviation structures
2. Lightweight construction materials
3. Fast and fuel-efficient transport vehicles
4. Electronic segments and outdoor supplies.

1.3 Nanocomposites

Nanocomposite is a material in which the nanosized reinforcement phase is dispersed in nanosized matrix phase materials. These materials possess superior properties which are imparted by the reinforcement phase materials that are lacking or lower in the matrix phase. A very low percentage of these nanosized reinforcement phase materials alter or improve the colour, texture, size, thermal stability, tensile strength, flexibility, stiffness, toughness, thermal and electrical conductivity of the nanocomposites. The nanocomposite materials possess a wide range of explicit significant properties makes it suitable for various applications right from catalysis to drug delivery including automotive parts, electronic components, food packing, energy storage, battery components, etc.

1.3.1 Components of Nanocomposites

Nanocomposites comprise two distinct phases, namely the matrix phase (bulk phase) and the reinforcement phase (dispersed phase).

Matrix Phase

The matrix phase is the essential primary phase which provides the lattice for the nanocomposite. It encompasses and holds the individual existing reinforcing elements. Examples: Polymers, Metals, Ceramics.

Functions of Matrix Phase

1. Matrix phase provides stability against surface damage by mechanical abrasion and corrosion by environment.
2. Matrix phase plays a transport mechanism in transferring the load and stress to the reinforcement phase.
3. Matrix phase gives shape and network to the nanocomposites.

Reinforcement Phase

The reinforcement phase is the secondary phase of choice which takes up the load and provides stiffness.

Functions of Reinforcement Phase

1. Reinforcement phase transmits mechanical properties like lightweight, strength, stiffness, toughness and thermal stability to the nanocomposite material.

2. Reinforcement phase plays a significant role as load bearing constituents.

1.3.2 Classification of Nanocomposites

Based on the type of matrix phase used, nanocomposites are classified into three types

1. ***Polymer Matrix Nanocomposites (PMC)***: Polymer matrix nanocomposite is a material that consists of reinforcement phase dispersed in polymer matrix. Polymers such as unsaturated polyesters, epoxides, vinyl esters and phenolics are used as matrix phase. These PMC are lightweight composite materials which provide high tensile strength, high stiffness, high fracture resistance and good corrosion resistance. The main limitations of these materials are low thermal resistance and high coefficient of thermal expansion.
2. ***Metal Matrix Nanocomposites (MMC)***: Metal matrix nanocomposite is a material that consists of reinforcement phase dispersed in metal matrix. Reinforcements materials such as ceramics (oxides (SiO_2, Al_2O_3), carbides (SiC, TiC), borides (TiB_2) or metals (lead, tungsten, Molybdenum, etc.) are dispersed in metal matrix such as aluminium, magnesium, titanium and copper. These MMC are low density materials which possess high strength at elevated temperatures, high stiffness, high thermal stability, good creep resistance and high thermal conductivity.
3. ***Ceramic Matrix Nanocomposites (CMC)***: Ceramic matric nanocomposite is a material that consists of reinforcement phase dispersed in ceramic matrix. The ceramic materials such as ZrO_2, whiskers of SiC and TiB_2 are used as reinforcement materials which are dispersed in ceramic matrix phase such as oxide material (alumina) or non-oxide material (SiC). These CMC are materials which have high strength at elevated temperatures, high stiffness, high thermal stability, high thermal conductivity and high corrosion resistance even at elevated temperatures.

2 Experimental Methods

2.1 Synthetic Methods of Nanocomposites

The nanocomposites have been synthesized by various methods which can be classified as chemical methods [1–4], physical methods [5–8] and biological methods [9–11].

2.1.1 Chemical Methods

The chemical methods widely used are sol–gel method, chemical reduction method, microwave-mediated method, polyol method, microemulsion method, hydrothermal method and pyrolysis method.

Sol–gel method: In this method, a sol of metal precursor is prepared with water which on hydrolysis followed by condensation and polycondensation results in the formation of a gel which is porous network filled with liquid like water and alcohol (by-products of hydrolysis reaction). This gel on super critical drying (drying at critical temperature and critical pressure) results in the formation of aerogel (porous network without water or alcohol) of nanocomposites. If the conditions of super critical drying are not maintained, it results in the formation of xerogel (collapsed porous network of nanocomposite aerogel).

Chemical reduction method: In this method, metal nanoparticles and nanocomposites can be prepared by reduction of metal precursor such as metal halides using reducing agents like $NaBEt_3H$ in the presence of solvent at room temperature. It is an easy and simple method to produce group VI metal nanoparticles and nanocomposites.

Microwave-mediated method: In this method, the energy source required for the synthesis of nanocomposites is microwave radiation. Nowadays, this method is utilized by researchers to synthesize nanocomposites due to its high efficiency and shorter reaction time for the synthesis.

Polyol method: In this method, the nanocomposites are synthesized by adopting polyol as solvent. Here, polyol such as ethylene glycol serves as reducing agent and complex with the precursor material and reduces to the nanocomposites. It further acts as stabilizing agent and stabilizes the system. After the formation of nanocomposites, they are separated from the liquid medium by centrifugation, then washed with ethanol and dried and used for further studies.

Microemulsion method: This method uses microemulsion solution for the synthesis of nanocomposites. These microemulsion solutions are homogeneous mixture of polar, nonpolar and surfactant components which are isotropic in nature and thermodynamically stable solution. On this approach, the microemulsion solution is mixed with metal precursor which results in the formation of precipitate followed by nucleation and coagulation, and nanocomposites are produced.

Hydrothermal method: The method is most widely used technique for the synthesis of nanocomposites. This method involves the liquid phase reaction (mostly water as the solvent) of metal precursor performed at higher temperature and higher pressure resulting in the formation of nanocomposites. This method depends on the reaction conditions and metal precursors used.

Pyrolysis method: This method involves thermal decomposition of precursors resulting in the formation of nanocomposites. Mostly organic metal precursors are used in this method. Generally, high temperature is applied to the reactants in order to cleave the bonds to form the nanocomposites.

2.1.2 Physical Methods

Physical methods do not engage any toxic reagents as in the case of chemical methods. Physical methods require shorter reaction time and tailor-made uniform narrow size distribution of nanocomposites. However, a lot of energy consumption in physical methods becomes the major disadvantage. The common physical methods used are physical vapour deposition, electric-arc discharge method, radio frequency sputtering, direct current sputtering and ball milling methods.

Physical vapour deposition: It is a technique in which solid precursor material is vapourized by either thermal heating, laser or high energy bombardment of ions, and the vapourized material is converted into nanocomposites.

Electric-arc discharge method: In this method, an electric arc is produced by applying a voltage (usually 20–25 V) between the graphite electrodes maintained at a distance 1 mm. This electric arc produces a high temperature that ionizes the anode material which forms a plasma (ions or atoms in vapour state at high temperature) between the electric arcs. These ions will move towards the cathode, deposit and grow as nanocomposites. The conditions such as pressure of inert gas atmosphere inside the furnace where the reaction occurs, voltage, distance between the electrodes should be maintained properly for efficient and high yield of nanocomposites formed.

Sputtering method: Sputtering is one of the physical vapour deposition techniques in which the solid precursor target material is vapourized by high energy bombardment of ions such as argon ions, and the vapourized target material is converted into nanocomposites. Sputtering method is of two types namely radio frequency sputtering (RF sputtering) and direct current sputtering (DC sputtering). In RF sputtering method, radio frequency waves of frequency 13.5 MHz are used to ionize the argon gas atoms (inert gas filled in the reaction chamber) into Ar^+ ions which possess high energy. These energized Ar^+ ions bombard the solid target precursor material and vapourize it which forms the nanocomposites. In DC sputtering method, a DC voltage is applied between the electrodes to ionize the argon gas atoms, and the same procedure is followed similar to the RF sputtering method. Another difference between these two methods is both conducting and non-conducting material targets can be converted into nanocomposites by RF sputtering, but only conductors and metal targets can be used by DC sputtering method.

Ball milling method: This method is associated with mechanical grinding of bulk precursors materials into nanocomposites. This method is a cheap, easy and simple technique which is used to synthesize nanocomposites. The main disadvantage of this method is physical damage which occurs to the nanocomposites.

2.1.3 Biological Methods:

In biological methods, toxic chemicals are not employed, instead nontoxic agents like proteins, enzymes, carbohydrates, etc., and living being like bacteria, fungi, yeast and plants are used. The extensively used biological methods are enzymatic reduction and non-enzymatic reduction. The enzymatic reduction method uses enzymes such as NADPH reductase, and non-enzymatic reduction method utilizes the living being such as bacteria [9], fungi [10], yeast [11] and plants [12]. Although, nanocomposites are synthesized using bacteria, actinomycetes, algae, fungi and yeast, the most widely practised method is utilizing the plant extracts. The attention towards medicinal plant extracts used as reducing and capping agents is increasing due to its medicinal potentials which may be incorporated to the nanocomposite synthesized. Many medicinal plants which are day to day available such as tea, mustard, aloe vera, lemongrass, neem, tulsi and lotus are used in the synthesis of nanocomposites.

2.2 *Green Synthesis of Nanocomposites*

The green synthetic methods are environmentally friendly, simple, efficient, cost-effective and advantageous compared to other chemical and physical methods. Mostly, these methods directly result in the formation of nanocomposites. There are numerous research articles related to the enzymatic and non-enzymatic reduction methods published by the scientific community. Much interest is shown towards the green synthesis of nanocomposites by living being like bacteria, fungi, yeast and plants.

2.2.1 Bacteria

The bacterial cell wall contains certain functional groups due to which they possess an inherent property to reduce heavy metals. This important factor makes bacteria a potential agent to synthesize green nanocomposites. The bacteria such as *Escherichia coli, Bacillus cereus, Klebsiella pneumoniae, Actinobacter* sp., *Lactobacillus* spp., *Corynebacterium* sp. and *Pseudomonas* sp. are used in the green nanocomposite synthesis. ZrO_2 nanoparticles, gold nanoparticles and palladium nanoparticles were prepared by *Actinobacter* sp. [13], *Pseudomonas* [14] and *Desulfovibrio desulfuricans* [15].

2.2.2 Actinomycetes

The cell wall and cell membrane of actinomycetes secrete an enzyme which helps in the reduction of metal precursor to form the nanocomposites. The proteins in the cell wall and cell membrane assist in the capping and stabilizing nanocomposites during synthesis [16].

2.2.3 Algae

Algae possess negative charges on the surface of algal cells which accelerates the nucleation and growth of nanocomposite. This provides shorter reaction time for the synthesis of green composites. Certain algae culture medium secreted metabolites, flavonoids and terpenoids which act as capping and stabilizing agents which plays an important role in the medical field. Silver nanoparticles were biosynthesized using Spirulina platensis, and its antibacterial activity was determined [17].

2.2.4 Fungi

Fungi possess enzymes and proteins which are capable of reducing metal salts into its corresponding metal nanoparticles and act as stabilizing agents [18]. Fungus, *Fusarium oxysporum* were used to synthesize gold nanoparticles extracellularly which was reduced by NADH-dependent reductase present in the reaction mixture released from the fungi [19].

2.2.5 Medicinal Plants

The plants particularly medicinal or therapeutic plants offer additional favourable circumstances, and furthermore, it does not require any unpredictable conventions or procedures. On account of microorganisms, there are multi-stage processes including confinement of expected microorganism, explicit culture preparation, preservation of culture and so forth. Moreover, the procedure by means of plants is relatively simple for scaling up, for enormous creation of nanoparticles. In green synthesis technique, the plant materials such as leaves, root, stem, bark, flower, fruit [20] act as reducing and stabilizing agents. They can be easily blended with metal precursors such as silver nitrate, titanium oxide, zinc oxide, etc to form the nanocomposites. Silver nanoparticles from the medicinal plant, *Tribulus terrestris* L., produce the round moulded NPs having size between 16 and 28 nm. Additionally, the silver nanoparticles from leaf of *T. terrestris* bearing therapeutic properties showed antibacterial movement [21]. The silver nanoparticle prepared from the leaves of *Podophyllum* [22] and Titanium dioxide nanoparticles (TiO_2NPs) were synthesized from *Cissus quadrangular* [23]. Cinnamon was utilized as reductant and stabilizer in green blend of silver nanoparticles [24]. The biosynthesized zinc oxide nanoparticles from leaf of *Justicia wynaadensis* which is a medicinal herb indicated antimitotic and DNA-restricting potential [25]. The concentrate of medicinal plant, *Caesalpinia pulcherrima*, was utilized for silver nanoparticle synthesis [26]. The phytomediated preparation of zinc oxide nanoparticle was prepared by using Berberis aristata leaves extract [27]. Gold and silver nanoparticles were synthesized from the leaf concentrate of Mussaenda glabrata by utilizing green synthetic technique [28]. Cissus quadrangularis was utilized for easy biosynthesis of copper oxide nanoparticles [29]. The plant extract of Nyctanthes arbour-tristis was used in the biosynthesis of zinc oxide nanoparticles [30]. A therapeutic herb Rosmarinus officinalis L. (rosemary) was utilized for green synthesis of iron nanoparticles [31]. The effortless biosynthesis of nanoparticles by utilizing plants particularly therapeutic plants is environmentally safe, shorter reaction time required and monetarily feasible in contrast with other natural living beings like microorganisms and so on [32, 33]. The green orchestrated nanoparticles have expanded antimicrobial removal in contrast with the different nanoparticles. This actuated antimicrobial action might be the outcome synergistic activity of proteins which work as capping and stabilizing the biosynthesized nanoparticles [34].

2.3 Green Synthetic Procedure

The synthesis of nanocomposites in aqueous medium generally contains three main components:

1. Metal precursors—These are the raw materials or the starting materials from which the metal and metal oxide nanoparticles and nanocomposites are synthesized by reduction reaction.
2. Reducing agents—These are agents which reduce the metal precursor into its corresponding metal nanoparticles. The leaf extract possesses certain agents which act as reducing agents, and no external reducing agents need to be added.
3. Stabilizing/capping agents—The stabilizing agent stabilizes the medium and the nanocomposites that are formed during the green synthesis. The capping agents assist in the binding of metal precursors with the active part of the reducing agent in the extract.

2.3.1 Preparation of Leaf Extract

The leaf extract of the medicinal plants is generally prepared by the following procedure. The leaves of medicinal plants are collected, washed with distilled water to remove the dirt, dust and other impurities. The washed leaves were dried at room temperature or water adsorbent paper and cut into small pieces or uniformly grinded into fine powder by a mechanical grinder. About 10–20 g of dried leaves or fine powder of medicinal plant is boiled with distilled water at 60 °C for 10–15 min. The resultant mixture was filtered using a Whatman filter, and leaf extract was collected. For certain medicinal plants, a mixture of distilled water and methanol is used, and the extract was collected. The methanol extract of the medicinal plant can be obtained from extraction using Soxhlet apparatus. The aqueous extract collected can be stored at 4 °C in a refrigerator and used for green synthesis of nanoparticles or nanocomposites. The shelf life of these aqueous or alcohol extract when stored at proper conditions ranges from 1 to 2 weeks. For long period of activity of these extracts, the leaf extract can be concentrated at ambient temperature under reduced pressure with a rotatory evaporator and prepared into a dried extract by using lyophilizer. The solid dried extract can be stored at 4 °C in airtight containers and used for nanoparticle and nanocomposite synthesis.

2.3.2 Synthesis of Nanoparticles

Metal and metal oxide nanoparticles can be prepared by green synthesis using medicinal plant leaf extract. These nanoparticles synthesized by green synthetic methods can be used as such for various applications and also can be used to synthesize green nanocomposites. Researchers have focused on the synthesis of metal and metal oxide nanoparticles such as copper nanoparticle [35], silver nanoparticle [36], TiO_2 nanoparticle [37], gold nanoparticle [38], ZnO nanoparticle [39], CeO_2 nanoparticle [40], Sm_2O_3 nanoparticle [41]), Co_3O_4 nanoparticle [42], Dy_2O_3 nanoparticle [43], CdO nanoparticle [44]), NiO nanoparticle [45]), In_2O_3 nanoparticle [46] and hydroxyapatite nanoparticle [47). The typical procedure for medicinal plant-mediated synthesis of nanocomposites is as per the following depicted in Fig. 1. The aqueous solution of mmol concentration of salts such as chlorides, nitrates, sulphates of corresponding metal or metal oxide nanoparticles is mixed with 10–20 ml of prepared medicinal plant leaf extract. If the dried extract is used, then an appropriate concentration of aqueous solution of extract is prepared and mixed with the salt solution which is the precursor of the metal and metal oxide nanoparticles. The salts should be chosen such that which on reduction gives its corresponding metal nanoparticles or converted into metal oxide nanoparticles. This mixture was continuously stirred at temperature 60–70 °C in a round bottom flask which is wrapped with aluminium foil to prevent unwanted photochemical reactions. The pH of the mixture was adjusted using 0.1 N HCl and 0.1 N NaOH solution. The stirring is continued until the colour of the mixture changes which indicates the formation of

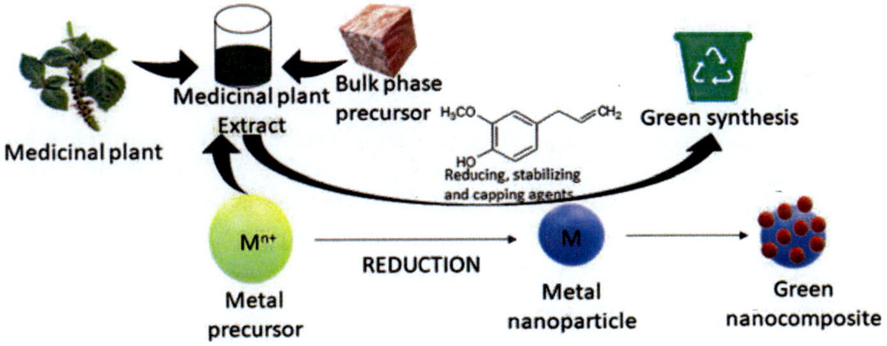

Fig. 1 Schematic representation of medicinal plant-mediated green nanocomposite synthesis. *Source* Author

metal and metal oxide nanoparticles. The time period for the formation of nanoparticles depends on the nature of reaction, metal and metal oxide nanoparticles. The reaction rate can be improved by using microwave radiation which completes the reaction in few minutes. The metal and metal oxide nanoparticles formed were separated from the mixture by centrifugation, and the solid product obtained is suspended and washed in distilled water to eliminate the biomass residue. The pure nanoparticles were dried in oven at 60–70 °C for 48 h and collected and used for future synthesis of nanocomposites.

2.3.3 Optimized Conditions

Several parameters affect the reaction rate for the formation of nanoparticles, size and shape of nanoparticles and uniform distribution of nanoparticles synthesized by the green synthesis method. The crucial parameters which decide the above-mentioned factors are concentration of precursor of metal nanoparticle, concentration of medicinal plant extract, temperature, pH, reaction time, etc.

Concentration of Precursor

The concentration of precursor of metal nanoparticle influences the reaction time and yield of the product nanoparticles synthesized. The increase in concentration of precursor decreases the reaction time and also changes the morphology of the nanoparticles synthesized. When the reducing effect of medicinal plant leaf extract is low, the higher concentration of precursors favours the formation of nanoparticles. Various concentrations of precursors used were reported by many researchers [48, 49].

Concentration of Medicinal Plant Extract

Different phyto constituents, for example, tannin, terpenoid, ketone, aldehyde, amide and so on are available in the concentrate which are essential for the reduction of precursor to metal nanoparticles. Many reported the effect of the various concentration of leaf extract on the size and shape of the nanoparticles formed [50, 51]. With increase in concentration of leaf extract, the nanoparticle produced possesses good biological activities.

Temperature

The temperature is a critical parameter in green synthesis of nanoparticles. The stability of medicinal plant extract depends on temperature, and an ambient temperature mostly room temperature is required. However, the synthesis of nanoparticles was performed at higher temperature by many researchers [51]. Microwave radiation can also be used as source of energy in preparing few nanoparticles [52].

pH

pH plays a crucial role in tailoring the size and shape of the nanoparticles synthesized by medicinal plant extract. The acidity and basicity nature of the extract affects the binding of metal and phytochemical sites which alter the shape and size of nanoparticles prepared. Low pH values of the medium result in larger-sized nanoparticles, and at higher pH values, smaller-sized nanoparticles are produced [53]. Similarly, the rate of the formation of the nanoparticles was higher at higher pH values [54].

Reaction Time

Reaction time of the green synthesis of nanoparticles mediated by medicinal plants also influences the size, shape and stability of the nanoparticles formed. The reaction time depends on the parameters such as temperature, concentration of medicinal plant extract, pH and concentration of precursor involved. Longer reaction time leads to the formation of larger-sized nanoparticles and uniform smaller-sized nanoparticle are formed with shorter reaction time. Thus, the size of the nanoparticles increases with increase in reaction time of the green synthesis [55].

2.3.4 Synthesis of Green Nanocomposites

The metal and metal oxide nanoparticles synthesized from green synthesis can be used in the preparation of green nanocomposites. These nanoparticles act as reinforcement phase which disperse in the matrix phase such as ceramics, metal oxides, polymers, graphene, graphene oxide and reduced graphene oxide. The generic method of synthesis of green nanocomposites is given in the following as shown in Fig. 2. Appropriate quantity of precursor of reinforcement phase and matrix phase is added to the 10–20 ml of medicinal plant leaf extract and continuous stirred at constant rpm speed at ambient temperature for few minutes to hours. The solid nanocomposite is separated from the extract by centrifugation, further washed with

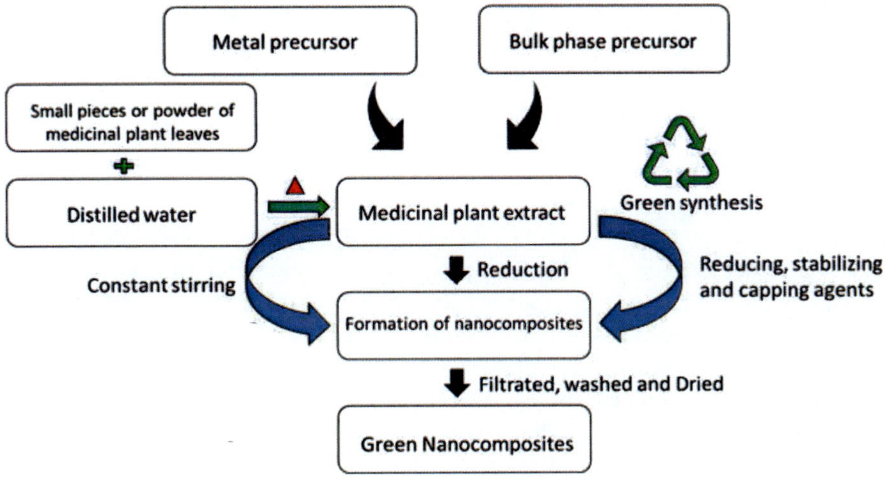

Fig. 2 Flow chart for green synthesis of nanocomposites. *Source* Author

water and or alcohol. The residue is dried in oven, collected and stored which can be used for specific applications. This synthetic method differs with the nature of reinforcement phase and matrix phase, temperature, pH, reaction time, etc. Recent development in the synthesis of such green nanocomposites attracts the attention of the researchers to adapt the green synthesis of nanocomposites. The green synthesis methods are easy and simple to perform with good yield, cost-effective, shorter reaction time and mainly environmentally safe. A few green composite syntheses to provide a better understanding of green synthesis of nanocomposites are summarized. Ag–ZnO nanoparticles were chosen as the reinforcement phase and reduced graphene oxide as matrix phase and prepared Ag–ZnO/rGO nanocomposite using leaf extract of *Stigmaphyllon ovatum* [56]. The nanoparticles are dispersed into the reduced graphene oxide by sonication. Ag-Au bimetallic nanocomposite was prepared by using *piper betal* leaf extract, and the mixture was irradiated using microwaves [57]. *Xanthium strumarium* seeds extract was used and synthesized graphene-metal nanocomposite by sonication method to disperse the metal nanoparticles on the graphene sheets [58]. *Achyranthes aspera* leaf extract was coated on the cotton fabrics and prepared Cu nanoparticle dispersed on the cotton fabrics [59].

3 Discussion

3.1 *Characterization of the Green Composites*

The synthesized green nanocomposites were characterized to access the surface morphology such as size, shape, chemical nature and distribution of the particle or fibre. Physico-chemical properties are noteworthy for behaviour, safety, biodispersion and viability of nanocomposites. Accordingly, nanocomposite characterization is important to assess the functional aspects of orchestrated nanocomposites. Figure 3 depicts the characterization of synthesized green nanocomposites by utilizing different techniques.

3.1.1 UV–VIS Spectroscopy

UV–VIS spectroscopy is an important and basic technique for characterization of nanoparticles. Certain nanocomposites possess optical properties which make them absorb specific frequencies of light. Synthesis of nanocomposites can be characterized by position of surface plasmon resonance (SPR) band detected at specific wavelength range unique for particular nanocomposite. The size of the nanocomposite can be tracked using these SPR bands as different particle size of nanocomposites exhibits SPR bands at different wavelengths. UV–VIS spectroscopy is speedy, straightforward, fundamental and explicit for different types of nanocomposites, needs only a concise timeframe for estimation. Sharp and narrow high intense peak at λ_{max}

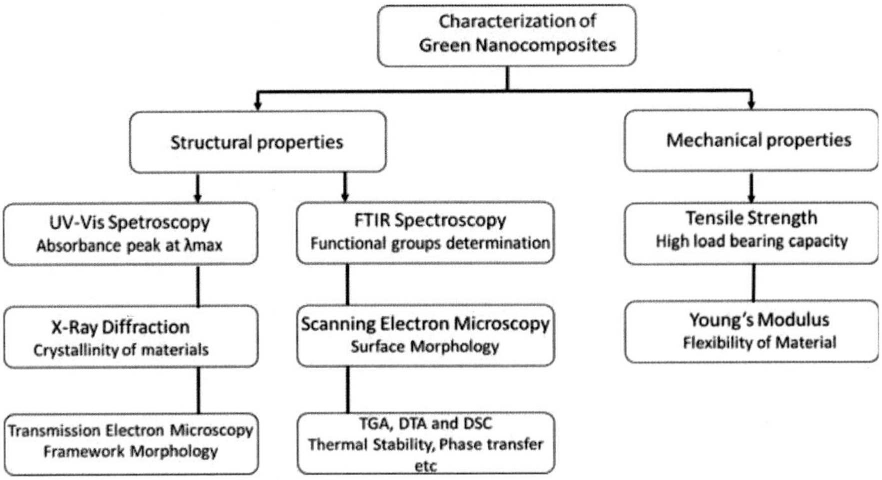

Fig. 3 Characterization techniques for analysis green nanocomposites. *Source* Author

is obtained at higher reaction time. The peak intensity increases with increase in reaction time indicating the completion of reaction and maximum yield of nanocomposites synthesized. On the other hand, peak broadening at λ_{max} appears when the initial concentration of metal precursor is increased. This may be accounted for aggregation of nanocomposites at higher concentration of metal precursor. At lower concentration of metal precursor, monodipersity of nanocomposites formation shows less peak broadening of SPR band. A similar trend in the peak broadening at λ_{max} takes place based on the pH of the extract. At low pH value of medicinal plant extract, the nanocomposite formation is hindered due to non-accessibility of functional group responsible for the reduction of metal precursor. However, at high pH of the extract, the peak broadening of SPR bands takes place indicating the higher yield of nanocomposites. The effect of temperature also influences the λ_{max} of SPR band in the UV–Vis spectra. At lower temperatures, there is practically no SPR band or very low intense SPR band appearing in the UV–Vis spectra. But at higher temperature (optimum temperature for the growth of nanocomposites), UV–Vis spectra show high intense sharp narrow peak at λ_{max} with a blue shift.

3.1.2 Fourier Transform InfraRed (FTIR) Spectroscopy

FTIR can give precision, reproducibility and moreover a perfect signal to noise proportion. By using FTIR spectroscopy, it becomes conceivable to distinguish little absorbance changes on the functional groups which performs qualification spectroscopy, where one could perceive the little units of amino acids to the entire protein. FTIR spectroscopy is frequently used to check whether biomolecules of the medicinal plant which are involved in the reduction process of nanocomposites. Additionally,

FTIR has also been confirmed the examination of nanoscaled materials, for instance, attestation of helpful molecules covalently joined onto silver, carbon nanotubes, graphene and gold nanoparticles or co-tasks occurring among in the midst of the reactant technique. A FTIR spectra of nanocomposites contain these characteristic peaks. The minor peak found at 3215 cm^{-1} represents the –NH stretch of primary and secondary amines or amides. The same peak may be due to the –OH stretch of alcohols and phenols; the peak at 3199 cm^{-1} is due to the –OH stretch of carboxylic acids; the smaller peak at 2920 cm^{-1} shows –CH stretch of alkanes and –OH stretch of carboxylic acids, 2258 cm^{-1} corresponding to –CN stretch of nitriles; 1644 cm^{-1} indicates –CC– stretch of alkenes and –NH bend of primary amines, 1390 cm^{-1} for CH_2 and CH_3 bending, CH_3 deformation and CH_2 rocking; 1114 cm^{-1} shows –CN stretch of aliphatic amines; 653 and 597 cm^{-1} are corresponding to C–Cl and C–Br stretch of alkyl halides and –CH bend of alkynes. FTIR reveals that carboxyl and amine groups may be involved in the reduction and stabilizing mechanism.

3.1.3 X-Ray Diffraction (XRD)

X-ray diffraction (XRD) is a most common analytical method which has been used for the assessment of both amorphous and crystalline structures, subjective recognizable of different compounds, estimating the level of crystallinity, quantitative goals of identifying species, molecule sizes, isomorphous substitution and so on. When X-ray beam light strikes any crystal, it prompts the diffraction pattern which reflects physico-chemical properties of crystal structures. In a powder diffraction method, diffracted patterns regularly begin from the powder sample and reflect its auxiliary physico-chemical highlights. Along these lines, XRD can look at the fundamental highlights of a broad assortment of materials, for instance, inorganic forces, superconductors, biomolecules, glasses, polymers and so forth. The amorphous or crystalline nature of nanocomposites can be determined from XRD patterns with the peaks at various 2θ values which corresponds to the lattice planes of the material. The average crystallite size of the nanocomposite crystallite can be estimated by the Scherrer equation. The peaks in the XRD pattern corresponding to different lattice planes can be assigned from the JCPDS file. For example, silver nanoparticle shows four intense peaks at 38.13°, 46.2°, 64.44° and 77.36° corresponding to the planes of (111), (200), (220) and (311), respectively. These lattice planes obtained were indexed to the face-centred crystal structure of silver nanoparticles (JCPDS file number 04-0783).

3.1.4 Scanning Electron Microscopy (SEM)

The surface morphology and shape of the nanocomposite can be characterized by scanning electron microscopy. Among various electron microscopy, SEM is a surface imaging strategy, totally prepared for determining different atom sizes, size dispersions, nanomaterials shapes and the surface morphology of the particles at the little scope and nanoscales. Using SEM, we can test the morphology of particles and get a micrograph from the picture by either by estimating physical examination of the particles, or by using specific programming. The SEM image of silver nanoparticles prepared by using medicinal plant leaf extract of *S. trilobatum* confirmed the development of silver nanostructures having spherical shape of crystallite sizes in the range 50–70 nm with uniform monodispersed without aggregation and well dispersed [60].

3.1.5 Energy Dispersive X-Ray (EDX) Analysis

Analysis through energy dispersive X-ray (EDX) spectrometers confirmed the existence of elemental constituents in the nanocomposites. The EDX spectrum shows the number of X-ray counts in the y-axis and energy in keV in x-axis. The EDX spectrum of silver nanoparticles shows an intense signal for the silver atoms in the nanoparticles at 3 keV and weak signal from "Cl" and "O" atoms which are attributed to the organic constituents of the medicinal plant. These weak signals are from the plant organic constituents. This confirms the presence of majority of silver nanoparticles in the system [60]. Any impurities present show an additional energy peak in the spectrum. The surface morphologies of the orchestrated copper cotton fabric nanocomposites by in situ synthetic technique were obtained by SEM investigation. The copper nanoparticles cotton fabric nanocomposites are shown as spherical particles of size of 95 nm. The presence of copper nanoparticles and cotton fabric nanocomposites shows two energy peaks which confirm their existence [61].

3.1.6 Transmission Electron Microscopy (TEM)

TEM is an important commonly used analytical technique which provides the basic framework morphology of nanomaterials. It is utilized to get quantitative information of particle and also grain size, size dispersion and morphology [60]. The electron beam is transmitted through the material which gives the structure of the surface to few nm depths inside the nanocomposite which provides a clear idea about the surface of the nanocomposites. Thus, the exactness of the particle size of the nanocomposites can be estimated by the TEM analysis.

3.1.7 TG-DTG and DSC Analysis

The thermal stability of nanocomposites was carried out by TG-DTG and DSC studies. From the thermogram, the moisture content, volatile products, decomposition reactions and by-products expel out from each step, the stability of the nanocomposites can be determined. Similarly, the DSC thermogram shows the crystalline nature and phase transfer involved in the nanocomposites.

3.1.8 Mechanical Properties

The mechanical properties of nanocomposites such as high load bearing capability, tensile stress (T. stress) and tensile strain (T. strain) were measured which indicate the strength and flexibility of nanocomposite material. Nowadays, metal nanocomposite material is used in versatile applications due to their high strength, smaller size and lightweight [62, 63].

3.2 Therapeutical Applications of Green Nanocomposites Synthesized from Medicinal Plants

The therapeutical applications of green nanocomposites synthesized from medicinal plant extract are numerous and few mentioned here around their remedial applications as antiviral, photosensitizer as well as radiosensitizer and anticancer operators. Figures 4 and 5 show the various applications and therapeutical use of green nanocomposites.

Silver nanocomposites as virucidal operators repress HIV-1, Tacaribe infection (TCRV), hepatitis B infection (HBV), recombinant respiratory syncytial infection (RSV), monkey pox infection, murine norovirus (MNV)-1 and flu A/H1N1 infections. Park et al. [56] created and assessed a novel micrometre-sized attractive half and half colloid (MHC) enhanced with green nanocomposites that could be utilized to inactivate viral pathogens ΦX174 and MNV with least possibility of likely discharge

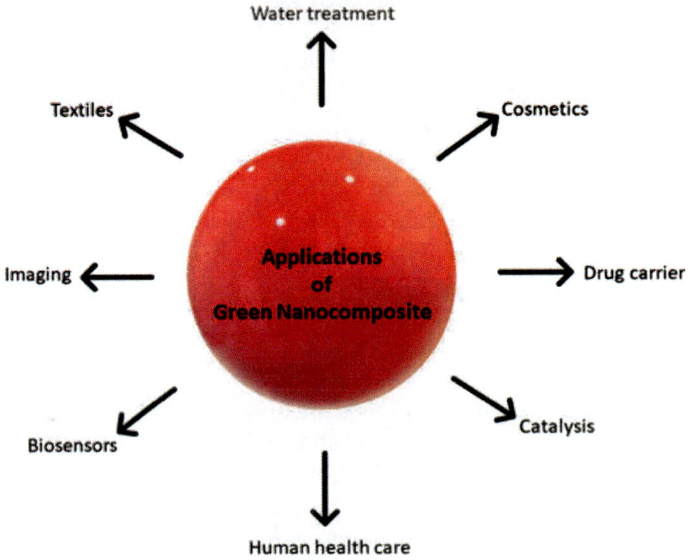

Fig. 4 Applications of Green Nanocomposites. *Source* Author

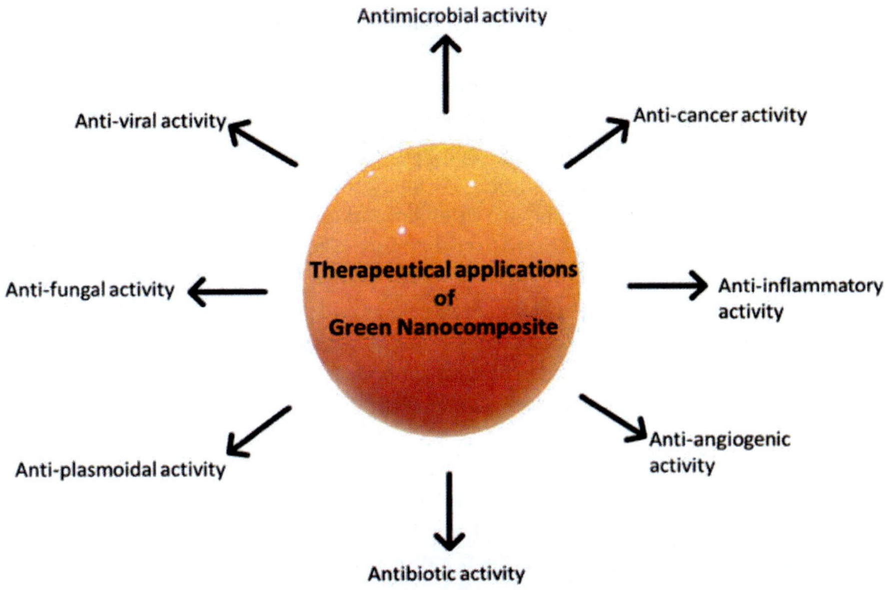

Fig. 5 Therapeutical applications of green nanocomposites. *Source* Author

into nature [64]. Xiang et al. [57] indicated that AgNPs have advantageous impacts in forestalling A/Human/Hubei/3/2005 (H3N2) flu infection disease both in vitro and in vivo [65]. Another investigation [66] shows that nanocomposites would be wise to antiviral movement (80–90% hindrance) against Herpes simplex infection (HSV)-1 and human parainfluenza infection (hPIV)-3 and were less cytotoxic to Vero cells. Moreover, it was discovered that these nanocomposites can restrain the replication of Vaccinia infection (VACV) [67].

Nanocomposites as photosensitizers or potentially radiosensitizers LSPR of nanoparticles empower the utilization of these nanocomposites in nonionizing radiation and ionizing radiation. In a report, it was shown that aptamer–Ag–Au shell–centre nanostructures have a high capacity to retain NIR illumination and can perform photothermal treatment of the A549 cells at a low light force thickness (0.20 W cm^{-2}) without wrecking the sound cells and the encompassing typical tissues [68]. Additionally, it was accounted for the graphene oxide@Ag-doxorubicin-DSPE-PEG2000-NGR (GO@Ag-DOX-NGR) demonstrated amazing chemophotothermal restorative viability, tumour-focusing on properties, NIR laser-controlled medication discharging capacities and X-beam imaging capacity in an in vivo murine tumour model [69]. Besides, it was uncovered that empty Au–Ag nanoshells (HGNS) indicated potential for photothermal treatment in light of their strength when PEGylated under laser enlightenment [70]. Likewise, Zhao et al. [71] revealed that Fe_3O_4/Ag/C225 (epidermal development factor receptor) joined with X-beam treatment could expand the affectability of human nasopharyngeal carcinoma cell lines (CNEs) to light.

3.2.1 Antimicrobial Activity of Green Nanocomposites

Medicinal plants-mediated orchestrated nanocomposites were productive in repressing the bacterial, yeast and dermatophytic human pathogens and especially Vasambu silver nanoparticles demonstrated great inhibitory action against Escherichia coli, Micrococcus sp., *Staphylococcus aureus*, *Corynebacterium diphtheriae* and *Candida albicans* though vilangam silver nano molecule indicated the great inhibitory action against *Pityrosporum ovale*. These natural blended nanocomposites are proficient than the artificially combined nanoparticles and comparable to that of synthetic triclosan. *Ocimum sanctum* and *Vitex negundo* were utilized against *S. aureus*, *Vibrio cholerae*, *Proteus vulgaris* and *Pseudomonas aeruginosa* and got the most extreme restraint at 150 μl of plant leaf nanoparticle [72]. Nanocomposite orchestrated utilizing *Euphorbia hirta* had a successful antifungal movement against *C. albicans*, *C. kefyr* and *A. niger* [73]. The antibacterial activity was exhibited utilizing electron microscopy with cell divider brought about the section of nanoparticles inside the bacterial cell and partner with bacterial cell segments by regulating tyrosine phosphorylation of putative peptides, and along these lines, it hinders microorganisms [74]. Since koduveli, vasambu, vilangam, Pavu and vettiver indicated high antimicrobial movement against bacterial human pathogen, these particles were read with anti-infection agents for its similar inhibitory action against *S. aureus* and *Micrococcus* sp., and with azole mixes against *C. albicans*. Every one of the five was demonstrating acceptable inhibitory movement against *S. aureus* while vasambu and vilangam not demonstrating movement against *Micrococcus* sp. If there should arise an occurrence of azole mixes, vasambu and vilangam demonstrated great inhibitory action against *C. albicans* even triclosan did not demonstrated inhibitory movement. Nanocomposite essentially synthesized by the incorporated molecule utilizing cell free filtrate of Lactobacillus acidophilus against *E. coli*, *P. aeruginosa* demonstrated bacterial strains which are protection from antitoxins are profoundly defenceless to silver nanoparticles [75]. Antibacterial test was finished utilizing papaya organic product removes interceded silver nanoparticles on same human pathogens, indicated exceptionally harmful against multidrug opposition microorganisms [76].

3.2.2 Antifungal Activity of Green Nanocomposites

At low concentration, vasambu and vilangam did not indicated any inhibitory action against contagious phytopathogens, for example, *Rhizactonia solani*, *Macrophomina phaseolina*, *Alternaria alternata*, *Fusarium oxysporum*, *Sclerotium rolfsii*, *Aspergillus niger*. However, at higher concentration the vasambu silver nano composite demonstrated better inhibitory action against *Fusarium oxysporum* whereas vilangam nanocomposite demonstrated a diminished inhibitory action against *Fusarium oxysporum*. Vasambu silver nanocomposites demonstrated marginally decreased inhibitory effect when contrasted with the accessible fungicide carbendazim (control). The silver nanocomposite combined with Argemone leaf extricate indicated inhibitory action against *Aspergillus flavus* [77]. Nanocomposites prepared by utilizing *E. hirta* indicated inhibitory action against *A. niger*, *A. flavus* and *A. fumigatus* [73].

3.2.3 Antiplasmoid Activity of Green Nanocomposites

Parasitic diseases (like jungle fever, leishmaniasis, tryanosomiasis) are one of the serious issues around the globe. Antiparasitic chemotherapy is the main decision of treatment for these parasitic diseases. The purpose behind this is these dis]eases do not inspire articulated resistant reaction; henceforth, successful inoculation may not be possible. Notwithstanding serious endeavours to control intestinal sickness, the malady keeps on being one of the best medical issues confronting Africa. Albeit various advances have been made towards understanding the illness, moderately scarcely any antimalarial drugs have been created in the last 30 years. Then, the customary drugs have been utilized to reward jungle fever for thousands of years. The plant is ordinarily utilized in customary medication in Kenya to cure malaria. Leaf concentrate of *Mentha piperita* (Lamiaceae) is excellent bioreductant for the union of silver and gold nanoparticles, and orchestrated nanoparticles are dynamic against clinically confined human pathogens, Staphylococcus aureus and Escherichia coli [78]. At any rate from ethnomedicinal use, there is some proof that the plant might be sheltered in people. *Cassia occidentalis* leaves, *Cryptolepis sanguinolenta* (*C. sanguinolenta*) root bark, *Euphorbia hirta* (*E. hirta*) entire plant, *Garcinia kola* stem bark and seed, *Morinda lucida* leaves and *Phyllanthus niruri* entire plant created over 60% hindrance of parasite development in vitro at a test focus g/mL. Concentrates from *E. hirta*, *C. sanguinolenta* and *Morinda morindoides* demonstrated noteworthy chemosuppression of parasitaemia in mice tainted with *Plasmodium berghei* at orally given dosages of 100–400 mg/kg per day [79]. The watched antiplasmodial activity might be identified with the nearness of terpenes, steroids, coumarins, flavonoids, phenolica acids, lignans, xanthones and anthraquonones [80].

3.2.4 Anticancer Activity of Green Nanocomposites

Cancer is one of the noticeable diseases which taint both developing the developed nation. In this manner, there is a prerequisite to grow new strategies which can lessen the fundamental reactions. The mechanism of action of green nanocomposites towards the drug delivery was depicted in Fig. 6 and found that customized cell demise was concentration dependent. Further, the synergistic impact on apoptosis utilizing uracil phosphoribosyltransferase cells and non-uracil phosphoribosyltransferase cells within the sight of fluorouracil was noticed [81]. In these conditions, it was seen that silver nanocompsites actuate apoptosis as well as sharpen malignant growth cells. The silver embedded attractive nanocomposites demonstrated noteworthy activity against bosom disease cells and coasting leukaemia cells [82]. Medicinal plant extracts mediated silver nanocomposites demonstrated a harmful impact

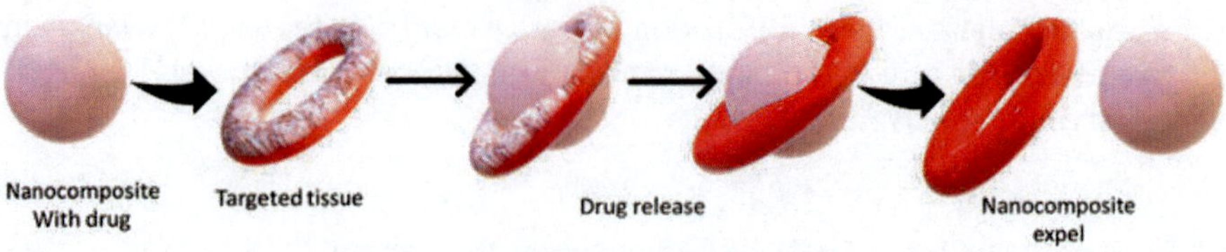

Fig. 6 Drug delivery of green nanocomposites. *Source* Author

on the human lung carcinoma cells (A549) which shows these nanocomposites could target cell-explicit poisonousness [83].

4 Conclusion

The chapter deals about the different types, composition, synthesis, properties, applications of green nanocomposites. Green nanocomposites synthesized by various chemical methods, physical methods, biological methods and common green synthetic procedure adopted with optimization parameters are discussed. The merits of medicinal plant-mediated green nanocomposite synthesis such as simple procedure, eco-friendly, high efficiency were analysed. The characterization techniques such as UV–Vis, FTIR, XRD, SEM, EDAX and TEM reveal the remarkable properties of the green nanocomposites synthesized using medicinal plant extract. Finally, the therapeutical applications of the various medicinal plant-mediated green nanocomposite synthesized are reviewed.

References

1. Zhang Q et al (2011) A systematic study of the synthesis of silver nanoplates: is aitrate a 'magic' reagent. J Am Chem Soc 133:18931–18939
2. Roldán MV et al (2013) Electrochemical method for Ag-PEG nanoparticles synthesis. J Nanopart. 2013:7
3. Sotiriou GA, Pratsinis SE (2010) Antibacterial activity of nanosilver ions and particles. Environ Sci Technol 44:5649–5654
4. Sotiriou GA et al (2011) Nanosilver on nanostructured silica: antibacterial activity and Ag surface area. Chem Eng J 170:547–554
5. Kholoud MM et al (2010) Synthesis and applications of silver nanoparticles. Arab J Chem 3:135–140
6. Tien D et al (2008) Discovery of ionic silver in silver nanoparticle suspension fabricated by arc discharge method. J Alloys Compounds 463:408–411
7. Kosmala A et al (2011) Synthesis of silver nano particles and fabrication of aqueous Ag inks for inkjet printing. Mater Chem Phys 129:1075–1080
8. Asanithi P et al (2012) Growth of silver nanoparticles by DC magnetron sputtering. J Nanomater 2012:8

9. Shivaji S et al (2011) Extracellular synthesis of antibacterial silver nanoparticles using psychrophilic bacteria. Process Biochem 46:1800–1807
10. Li G et al (2012) Fungus-mediated green synthesis of silver nanoparticles using Aspergillus terreus. Int J Mol Sci 13:466–476
11. Mourato A et al (2011) Biosynthesis of crystalline silver and gold nanoparticles by extremophilic yeasts. Bioinorg Chem Appl 2011:8
12. Iravani S (2011) Green synthesis of metal nanoparticles using plants. Green Chem 13:2638–2650
13. Suriyaraj SP, Ramadoss G, Chandraraj K, Selvakumar R (2019) One pot facile green synthesis of crystalline bio-ZrO_2 nanoparticles using Acinetobacter sp. KCSI1 under room temperature. Mater Sci Eng C 105:110021
14. Husseiny MI, El-Aziz MA, Badr Y, Mahmoud MA (2007) Biosynthesis of gold nanoparticles using *Pseudomonas aeruginosa*. Spectrochim Acta A 67:1003–1006
15. Yong P, Rowsen NA, Farr JPG, Harris IR, Macaskie LE (2002) Bioreduction and biocrystallization of palladium by *Desulfovibrio desulfuricans* NCIMB 8307. Biotechnol Bioeng 80:369–379
16. Golinska P, Wypij M, Ingle A.P, Gupt, I, Dahm H, Rai M (2014) Biogenic synthesis of metal nanoparticles from actinomycetes: biomedical applications and cytotoxicity. Appl Microbiol Biotechnol 98:8083–8097
17. Mahdieha M, Zolanvari AS, Azimeea M, Mahdieh (2012) Green biosynthesis of silver nanoparticles by *Spirulina platensis*. Scientia Irania F 19 (3):926–929
18. Khandel P, Shahi SK (2018) Mycogenic nanoparticles and their bio-prospective applications: current status and future challenges. J Nanostruct Chem 8:369–391
19. Priyabrata M, Satyajoti S, Deendayal M, Absar A, Islam khan M, Rajiv K, Murali S (2002) Extracellular synthesis of gold nanoparticles by the fungus *Fusarium oxysporum*. Chembiochem 3(5):461–464
20. Akhta MS, Panwar J, Yun YS (2013) Biogenic synthesis of metallic nanoparticles by plant extracts. ACS Sustain Chem Eng 1(6):591–602
21. Gopinath V, MubarakAli D, Priyadarshini S, Priyadharsshini NM, Thajuddin N, Velusamya P (2012) Biosynthesis of silver nanoparticles from *Tribulus terrestris* and its antimicrobial activity: a novel biological approach. Colloids Surfaces B Biointerfaces 96:69–74
22. Jeyaraj M, Rajesh M, Arun R, MubarakAli D, Sathishkumar G, Sivanandhan G, Dev GK, Manickavasagam M, Premkumar K, Thajuddin N, Ganapathi A (2013) An investigation on the cytotoxicity and caspase-mediated apoptotic effect of biologically synthesized silver nanoparticles using *Podophyllum hexandrum* on human cervical carcinoma cells. Colloids Surfaces B Biointerfaces 102:708–717
23. Priyadarshini S, Mainal A, Sonsudin F et al (2020) Biosynthesis of TiO_2 nanoparticles and their superior antibacterial effect against human nosocomial bacterial pathogens. Res Chem Intermed 46:1077–1089
24. Premkumar J, Sudhakar T, Dhakal A, Shrestha JB, Krishnakumar S, Balashanmugam P (2018) Synthesis of silver nanoparticles (AgNPs) from cinnamon against bacterial pathogens. Biocatal Agric Biotechnol 15:311–316
25. Hemanth Kumar NK, Andia JD, Manjunatha S, Murali M, Amruthesh KN, Jagannath S (2019) Antimitotic and DNA-binding potential of biosynthesized ZnO-NPs from leaf extract of *Justicia wynaadensis* (Nees) Heyne—a medicinal herb. Biocatal Agric Biotechnol 18:101024
26. Deepika S, Selvaraj CI, Roopan SM (2020) Screening bioactivities of *Caesalpinia pulcherrima* L. swartz and cytotoxicity of extract synthesized silver nanoparticles on HCT116 cell line. Mater Sci Eng C 106:110279
27. Chandra H, Patel D, Kumari P, Jangwan JS, Yadav S (2019) Phytomediated synthesis of zinc oxide nanoparticle of *Berberis aristata*: characterisation, antioxidant activity and antibacterial activity with special reference to urinary tract infection. Mater Sci Eng C 102:212–220
28. Francis S, Joseph S, Koshy EP, Mathew B (2017) Green synthesis and characterization of gold and silver nanoparticles using *Mussaendo glabrata* leaf extract and their environmental applications to dye degradation. Environ Sci Pollut Res Int 24:17347–17357

29. Devipriya D, Roopan SM (2017) *Cissus quadrangularis* mediated ecofriendly synthesis of copper oxide nanoparticles and its antifungal studies against *Aspergillus niger*, *Aspergillus flavus*. Mater Sci Eng C 80:38–44
30. Jamdagni P, Khatri P, Rana JS (2018) Green synthesis of zinc oxide nanoparticles using flower extract of Nyctanthes arbor-tristis and their antifungal activity. J King Saud Univ Sci 30(2):168–175
31. Farshchi HK, Azizi M, Jaafari MR, Nemati SH, Fotovat A (2018) Green synthesis of iron nanoparticles by rosemary extract and cytotoxicity effect evaluation on cancer cell lines. Biocatal Agric Biotechnol 16:54–62
32. Shah M, Fawcett D, Sharma S, Tripathy SK, Poinern GEJ (2015) Green synthesis of metallic nanoparticles via biological entities. Materials 8:7278–7308
33. Mittal AK, Chisti Y, Banerjee UC (2013) Synthesis of metallic nanoparticles using plants. Biotechnol Adv 31:346–356
34. Roy A, Bulut O, Some S, Mandal AK, Yilmaz MD (2019) Green synthesis of silver nanoparticles: biomolecule-nanoparticle organizations targeting antimicrobial activity. RSC Adv 9:2673–2702
35. Ranu BC, Dey R, Chatterjee T, Ahammed S (2012) Copper nanoparticle-catalyzed carbon carbon and carbon heteroatom bond formation with a greener perspective. Chemsuschem 1:22–44
36. Lin C, Tao K, Hua D, Ma Z, Zhou S (2013) Size effect of gold nanoparticles in catalytic reduction of p-nitrophenol with NaBH4. Molecules 18:12609–12620
37. Chen X, Mao SS (2007) Titanium dioxide nanomaterials: synthesis, properties, modifications, and applications. Chem Rev 107:2891–2959
38. Saha K, Agasti SS, Kim C, Li X, Rotello VM (2012) Gold nanoparticles in chemical and biological sensing. Chem Rev 112:2739–2779
39. Kołodziejczak-Radzimska A, Jesionowski T (2014) Zinc oxide—from synthesis to application: a review. Materials 7:2833–2881
40. Chen H-L, Zhu H-Y, Wang H, Dong L, Zhu J-J (2006) Sonochemical fabrication and characterization of ceria (CeO_2) nanowires. J Nanosci Nanotechnol 6(1):157–161
41. Sone BT, Manikandan E, Gurib-Fakim A, Maaza M (2015) Sm_2O_3 nanoparticles green synthesis via Callistemon viminalis' extract. J Alloys Compd 650:357–362
42. Diallo A, Beye AC, Doyle TB, Park E, Maaza M (2015) Green synthesis of Co_3O_4 nanoparticles via *Aspalathus linearis*: physical properties. Green Chem Lett Rev 8(3–4):30–36
43. Chandrasekhar M, Nagabhushana H, Sudheerkumar KH, Dhananjaya N, Sharma SC, Kavyashree D, Shivakumara C, Nagabhushana BM (2014) Comparison of structural and luminescence properties of Dy_2O_3 nanopowders synthesized by coprecipitation and green combustion routes. Mater Res Bull 55:237–245
44. Thovhogi N, Park E, Manikandan E, Maaza M, Gurib-Fakim A (2016) Physical properties of CdO nanoparticles synthesized by green chemistry via Hibiscus sabdariffa flower extract. J Alloys Compd 655:314–320
45. Thema FT, Manikandan E, Gurib-Fakim A, Maaza M (2016) Single phase bunsenite NiO nanoparticles green synthesis by *Agathosma betulina* natural extract. J Alloys Compd 657:655–661
46. Maensiri S, Laokul P, Klinkaewnarong J, Phokha S, Promarak V, Seraphin S (2008) Indium oxide (In_2O_3) nanoparticles using Aloe vera plant extract: synthesis and optical properties. J Optoelectron Adv Mater 10:161–165
47. Klinkaewnarong J, Swatsitang E, Masingboon C, Seraphin S, Maensiri S (2010) Synthesis and characterization of nanocrystalline hap powders prepared by using Aloe vera plant extracted solution. Curr Appl Phys 10(2):521–525
48. Marslin G, Dias AC, Selvakesavan RK, Gregory F, Sarmento B (2015) Antimicrobial activity of cream incorporated with silver nanoparticles biosynthesized from Withania somnifera. Int J Nanomed 10:5955–5963
49. Dhand V, Soumya L, Bharadwaj S, Chakra S, Bhatt D, Sreedhar B (2016) Green synthesis of silver nanoparticles using Coffea arabica seed extract and its antibacterial activity. Mater Sci Eng 58:36–43

50. Rashid MI, Mujawar LH, Rehan ZA, Qari H, Zeb J, Almeelbi T, Ismail IM (2016) One-step synthesis of silver nanoparticles using *Phoenix dactylifera* leaves extract and their enhanced bactericidal activity. J Mo. Liq 223:1114–1122
51. Rao K, Aziz S, Roome T, Razzak A, Sikandar B, Jamali KS, Imran M, Jabri T, Shah MR (2018) Gum acacia stabilized silver nanoparticles based nano-cargo for enhanced antiarthritic potentials of hesperidin in adjuvant induced arthritic rats. Artif Cells Nanomed Biotechnol 46:597–607
52. Jayaprakash N, Vijaya JJ, Kaviyarasu K, Kombaiah K, Kennedy LJ, Ramalingam RJ et al (2017) Green synthesis of Ag nanoparticles using Tamarind fruit extract for the antibacterial studies. J Photochem Photobiol B Biol 169:178–185
53. Muthu K, Priya S (2017) Green synthesis, characterization and catalytic activity of silver nanoparticles using *Cassia auriculata* flower extract separated fraction. Spectrochim Acta Part A Mol Biomol Spectrosc 179:66–72
54. Khalil MM, Ismail EH, El-Baghdady KZ, Mohamed D (2014) Green synthesis of silver nanoparticles using olive leaf extract and its antibacterial activity. Arab J Chem 7:1131–1139
55. Sathyavathi R, Krishna MB, Rao SV, Saritha R, Rao DN (2010) Biosynthesis of silver nanoparticles using *Coriandrum sativum* leaf extract and their application in nonlinear optics. Adv Sci Lett 3(2):138–143
56. Elias EE, Damian CO, Lei W, Lou C, Zhao Z (2019) Synthesis of nanostructured ZnO, AgZnO and the composites with reduced graphene oxide (rGO-AgZnO) using leaf extract of Stigmaphyllon ovatum. J Environ Chem Eng 7:103190
57. Arunkumar L, Sangappa K, Shasidhar (2019) Synthesis, characterization and antibacterial study of Ag-Au bimetallic nanocomposite by bioreduction using piper betal leaf extract. Heliyon 5:e02794
58. Jing W, Bipin C et al (2017) Biosynthesis of graphene-metal nanocomposites using plant extract and their biological activities. J Chem Technol Biotechnol 92:1428–1435
59. Jaswanth S, Umamahesh M, Arundathi M (2020) In situ green synthesis of antibacterial copper nanocomposite cotton fabrics using *Achyranthes aspera* leaf extract. J Appl Pharm Sci 10(5):104–109
60. Vanaja M, Paulkumar K, Gnanajobitha G, Rajeshkumar S, Malarkodi C, Annadurai G (2014) Herbal plant synthesis of antibacterial silver nanoparticles by *Solanum trilobatum* and its characterization. Hindawi Publish Corp Int J Metals 2014:692461
61. Sadanand V, Feng TH, Rajulu AV, Satyanarayana B (2017) Preparation and properties of low-cost cotton nanocomposite fabrics with in situgenerated copper nanoparticles by simple hydrothermal method. Int J Polymer Anal Char 22:587–594
62. Gouda M, Hebeish A (2010) Preparation and evaluation of CuO/chitosan nanocomposite for antibacterial finishing cotton fabric. J Indus Textiles 39:203–214
63. Li R, He M, Li T, Zhang L (2015) Preparation and properties of cellulose/silver nanocomposite fibers. Carbohydr Poly 115:269–275
64. Park S et al (2014) Antiviral properties of silver nanoparticles on a magnetic hybrid colloid. Appl Environ Microbiol 80:2343–2350
65. Xiang D et al (2013) Inhibition of A/Human/Hubei/3/2005 (H_3N_2) influenza virus infection by silver nanoparticles in vitro and in vivo. Int J Nanomed 8:4103–4114
66. Gaikwad S et al (2013) Antiviral activity of mycosynthesized silver nanoparticles against herpes simplex virus and human parainfluenza virus type 3. Int J Nanomed 8:4303–4314
67. Trefry JC, Wooley DP (2013) Silver nanoparticles inhibit vaccinia virus infection by preventing viral entry through a macropinocytosis-dependent mechanism. J Biomed Nanotechnol 9:1624–1635
68. Wu P et al (2013) High specific detection and near-infrared photothermal therapy of lung cancer cells with high SERS active aptamer–silver–gold shell–core nanostructures. Analyst 138:6501–6510
69. Shi J et al (2014) A tumor-targeting near-infrared laser-triggered drug delivery system based on GO@Ag nanoparticles for chemo-photothermal therapy and X-ray imaging. Biomaterials 35:5847–5861

70. Goodman AM et al (2014) The surprising in vivo instability of near-IR-absorbing hollow Au–Ag nanoshells. ACS Nano 8:3222–3231
71. Zhao D et al (2012) A novel multifunctional nanocomposite C225-conjugated Fe3O4/Ag enhances the sensitivity of nasopharyngeal carcinoma cells to radiotherapy. Acta Biochim Biophys Sin 44:678–684
72. Prabhu N, Divya TR, YamunaGowri K, Ayisha SS, Joseph PD (2010) Silver phytho nanoparticles and their antibacterial efficacy. Digest J Nanomater Biostruct 5:185–189
73. David E, Elumalai EK, Prasad TNVKV, Venkata K, Nagajyothi PC (2010) Green synthesis of silver nanoparticle using *Euphorbia hirta* L and their antifungal activities. Arch Appl Sci Res 2(6):76–81
74. Shrivastava S, Bera T, Roy A, Gajendra S, Ramachandrarao P, Dash D (2007) Characterization of enhanced antibacterial effects of novel silver nanoparticles. Nanotechnology 18:9
75. Cathrine R, Raghunathan R, Prasannakumar K (2010) Biosynthesis of silver nano particles using *L. acidophilus* (Probiotic bacteria) and its applications. Int J Nanotechnol Appl 4:217–222
76. Jain D, Kumar DS, Kachhwaha S, Kothari SL (2009) Synthesis of plant mediated silver nanoparticles using papaya fruit extract and evaluation of their antimicrobial activities. Digest J Nanomater Biostruct 4(3):557–563
77. Singh A, Jain D, Upadhyay MK, Khandelwal VHN (2010) Green synthesis of silver nanoparticles using *Argemone mexicana* leaf extracts and evaluation of their antimicrobial activities. Digest J Nanomater Biostruct 5:483–489
78. MubarakAli D, Thajuddin N, Jeganathan K, Gunasekaran M (2011) Plant extract mediated synthesis of silver and gold nanoparticles and its antibacterial activity against clinically isolated pathogens. Colloids Surf B Biointerfaces 85(2):360–365
79. Tona L, Cimanga RK, Mesia K, Musuamba CT, De Bruyne T, Apers S et al (1999) Antimalarial activity of 20 crude extracts from nine African medicinal plants used in Kinshasa, Congo. J Ethnopharmacol 68(1–3):193–203
80. Tona L, Cimanga RK, Mesia K, Musuamba CT, De Bruyne T, Apers S et al (2004) In vitro antiplasmodial activity of extracts and fractions from seven medicinal plants used in the Democratic Republic of Congo. J Ethnopharmacol 93(1):27–32
81. Gopinath P, Gogoi SK, Chattopadhyay A et al (2008) Implications of silver nanoparticle induced cell apoptosis for in vitro gene therapy. Nanotechnology 19(7):075104
82. Jun BH, Noh MS, Kim J et al (2010) Multifunctional silver-embedded magnetic nanoparticles as SERS nanoprobes and their applications. Small 6:119–125
83. Gurunathan S, Jeong JK, Han JW et al (2015) Multidimensional effects of biologically synthesized silver nanoparticles in Heli cobacter pylori, Helicobacter felis, and human lung (L132) and lung carcinoma A549 cells. Nanoscale Res Lett 10:10–17

Chapter 21
Green Approaches to Prepare Polymeric Composites for Wastewater Treatment

Durga Yadav, Priyanka, and Joydeep Dutta

1 Introduction

Earth is the only planet where we can find life because of the existence of water and atmosphere, which are utterly necessary for our survival. It is so crucial that one cannot even imagine life without water [1]. Further, it is equally important for the survival of plant as well as animal kingdom. Therefore, it is our sole responsibility to save water. Due to not only extensive use of water for industrial as well as domestic purposes but also due to deliberate negligence of people to save freshwater, the water table is gradually decreasing, which is a clear indication of water scarcity. Additionally, improper maintenance of industrial wastewater pushes us forward to confront the same issue, which in turn, is also responsible for causing environmental pollution. So at the outset, if we do not take any preventive measure, then one day we all have to face lots of challenges that would eventually destroy the whole kingdom [2]. Water pollution along with drought, inefficiency, and an exploding population are also contributing to freshwater crisis, threatening the sources we solely rely on for drinking water and other critical needs. Industries are the huge sources for water pollution. For instance, industries including paper, food processing, textile, leather tanning, cosmetics, plastics, rubber, printing, and dye manufacturing industries discharge huge volume of wastewater. Table 1 gives information about the major pollutants released from different types of industrial wastewater.

Thus, various contaminants ranging from macrosized garbages to ultra-fine chemicals (heavy metals ions like copper, mercury, lead, nickel, chromium, cadmium, different types of dyes) are ended up into rivers, lakes, streams, groundwater, and eventually into the oceans [14]. If this continues, then it would be difficult to curb

D. Yadav · Priyanka · J. Dutta (✉)
Department of Chemistry, Amity School of Applied Sciences, Amity University Haryana, Panchgaon, Near IMT Manesar, Gurugram, Haryana 122413, India
e-mail: dutta_joy@yahoo.co.in

S. Thomas and P. Balakrishnan (eds.), *Green Composites*,
Materials Horizons: From Nature to Nanomaterials,
https://doi.org/10.1007/978-981-15-9643-8_21

Table 1 Composition of various industrial wastewater in terms of major pollutants

S. No.	Various wastewater generating industries	Major components of effluents/industrial wastewater	References
1	Sugar mill	Sugar, alcohol, and heavy metals	[3]
2	Paper and pulp	Chlorinated complex products, dibenzo-p-dioxin, and benzofuran	[4]
3	Aluminum	Heavy metals (Cd, Mn, Pb, and Zn)	[5]
4	Chemical manufacturers	Ammonia, urea, heavy metals	[6]
5	Glass manufacturers	Chloride, fluoride, heavy metals (Fe, Cr, Cd, and Ni)	[7]
6	Leather products	Toluene, benzene, acids, alkalis, chromium salts, solvents, sulfides, dyes, heavy metals.	[8]
7	Plastic products	Polyolefins and poly(ethylene terephthalate) (PET)	[9]
8	Petrochemical	Polycyclic and aromatic hydrocarbons, phenols, metal derivatives, surface-active substances, sulfides, naphthenic acids, and other chemicals	[10]
9	Textiles	Dyes and pigments, the presence of detergents, and surfactants	[11]
10	Thermal power plant	Coal dust, CO_2, fly ash, chlorinated water, heavy metal residues (Hg, Bi, As, Cr, Cu, Pb)	[12]
11	Construction	Ignitable paint waste, strong acids, and bases	[13]
12	Printing	Heavy metals and its solution, waste inks, solvents, ink sludge	[5, 7]
13	Metal manufacturing	Paint waste, heavy metals, sludges, and cyanide	[7]

water pollution altogether. Various treatment methods such as chemical precipitation, ion exchange, electrochemical reaction, membrane filtration and reverse osmosis, oxidation process, biological treatment, and adsorption have been employed for removal of pollutants from wastewater but all the methods, especially the chemical treatment method has its own merits and demerits [15]. It is worth noting that recyclability of industrial wastewater can also cut down the water scarcity to a great extent but on the other hand, the involvement of various chemical processes to treat wastewater is too responsible for causing environmental pollution, further, leading to climate change. At the same time, all these techniques barring adsorption suffer from high operation cost and are responsible for the elimination of metallic sludge that causes hazard to environment. Thus, protecting the environment from this industrial wastewater, reusing, and recycling it is of utmost importance, which in turn, also renders us to devise cost-effective means to remove pollutants from the industrial wastewater [16]. From the last few decades, the use of polymer composites for wastewater treatment has received a considerable interest of researchers because of their high performance, versatile characteristics, low density, stiffness, and high strength [17]. Further, polymer composites as effective adsorbents are being widely used for wastewater treatment. Adsorption using sorbent is one of the most popular methods and has advantages over other methods in terms of removing toxic metals ion and dyes because it is eco-friendly, cost-effective, and efficient [18]. Polymeric composites can be defined as the heterogeneous system consisting of at least two materials, which have physically and chemically distinct phase and are separated by interfaces. These phases are termed as reinforcement (dispersed or discontinuous phase) and matrix (continuous phase). The reinforcement phase usually composed of fiber embedded within an organic polymer matrix of composites. The various types of fibers can be categorized into several heads, namely organic fibers (carbon, graphite, polymer, aramid), inorganic (glass), natural fibers (plant-based and animal-based), and synthetic fibers [19]. The matrices of composites are majorly classified as non-biodegradable and biodegradable matrices. The main problem associated with polymeric composites is their disposal, because they contain toxic ingredients which are difficult to recycle. They neither be degraded nor be decomposed on their own and pose a major threat to our ecological system. Hence, this strongly alarming issue of environmental pollution has induced a tremendous impetus to developing environmentally benign green composites through green approaches since the past few decades. There are many advantages of using bio-based green composites [20]. Needless to say, the green composites are quite eco-friendly, more efficient and reliable, and also promising alternatives to traditional composites for environmental sustainability. Therefore, here, we shall exclusively focus on development of various polymers-based green composites with a special emphasis on chitosan, an environmentally benign biomaterial due to its excellent and remarkable properties such as biodegradability, non-toxicity, metal binding ability, flocculation activity for effectively treatment of wastewater for its further use. Moreover, it is hoped that readers will get an opportunity to learn about the development of new chitosan-based green composites (CGC) and their subsequent applications for wastewater treatment.

2 Green Composites—Promising and Sustainable Alternatives for Wastewater Treatment

Before we go to details about the preparation of polymer-based green composites for wastewater treatment, it is quite essential to know what green composites are! Principally, a green composite is a combination of at least two or more than two constituents, one of which is mandatorily derived from a renewable resource and hence, is eco-friendly in nature. This definition applies to both the reinforcement and the matrix used in fabricating a composite. Green composites usually consist of a biodegradable polymer as a matrix and natural fibers as reinforcing materials [21]. Thus, results in the formation of an interface, which is used to bind them together. Biopolymers are chiefly obtained from either plant-based or animal-based renewable resources including poly(lactic acid) (PLA), polyhydroxybutyrates (PHB), starch, cellulose, chitin, and chitosan to name a few [22]. Natural fibers are more advantageous over synthetic fibers because of their good stiffness, low cost, less density, environmentally friendly, renewable, reasonable mechanical properties, high disposability, and carbon dioxide neutral life cycle [23].

Further, attempts are being made to use byproducts obtained from various food crops, which can be alternative sources for natural fibers such as corn husk [24], rice husk [25], wheat straw [26], ahankari, soy stalk [27], sunflower stalk, and bagasse [28].

Depending on the use of different reinforcement and polymer materials, green composites can be categorized into three main heads:

1. Totally renewable composites—when both reinforcement material and matrix are from renewable resources.
2. Partly renewable composites—when synthetic matrix is reinforced with natural biopolymers
3. Partly renewable composites—when natural matrix is reinforced with synthetic materials.

Moreover, renewable green composites containing natural fibers and biopolymers are completely biodegradable. On the other hand, partly renewable green composites containing synthetic polymer as a reinforcing agent and natural polymer as a matrix and vice versa are partially biodegradable. Therefore, partially degradable green composites cannot be called completely environmentally benign as one of the components gets degraded and other one does not rather remain in the environment.

Apart from defining the category of green composites, the processability of green composites, especially polymeric composites also plays a vital role, which ensures not only about an appropriate processing technique and equipment used under optimized processing parameters but also about the quality, functionality, cost-effectiveness of the product.

The processability of green composites is governed by the following factors:

1. Interphase in the composite—Generally, natural fibers are treated with weak acids containing functional groups such as –COOH, –CHO, and –OH in order to attach the latter ones on their surfaces as potential sites for binding with polymer matrices. This phenomenon prevents hydrophilic natural fibers from getting wet because of the presence of hydrophobic polymer matrices.
2. Hygroscopic nature of natural fibers—Adequate drying of natural fibers is needed before processing of their composites because hygroscopic nature of natural fibers may cause imperfection in morphology after molding.
3. Low strength—The processability of natural fibers is hindered because of their low mechanical strength. So, this problem can be overcome using filament winding, tape layup, or using pultrusion methods.

4. Use of high-temperature matrices—Processing of low-temperature natural fibers is not feasible with high-temperature matrices such as polyetherimide (PEI), poly(ether ether ketone) PEEK, and polyimide (PI).
5. Service environment—Service environment having either high moisture content or high temperature is not conducive for natural fiber-reinforced polymer composites.

Commonly used processes in preparation of green composites mainly include lay up method, injection molding, autoclave bonding, resin transfer molding, and filament winding [29].

With regard to environmental sustainability, green composites are more advantageous over traditional polymeric composites for wastewater treatment because the former ones are significantly less hazardous than the latter ones. In other words, traditional polymeric composites are more difficult to be degraded even after their disposal because the matrices obtained from non-renewable petroleum-based resources are non-biodegradable, whereas green composites are recyclable, reusable, and also fantastically tuneable with the environment. Besides these, green composites are endowed with some additional features such as better resistance to high temperature, antibacterial activity, and chelating properties. Thus, green composites are very promising materials to be used as adsorbents for removing the different kinds of pollutants from industrial wastewater.

3 Various Polymers-Based Green Composites for Wastewater Treatment

Due to the rapid growth of industrialization, a huge amount of hazardous chemical wastes is continuously generated from various industries, namely paper, textile, fertilizer or petrochemical industries, electroplating plants, and tannery industries, etc., and subsequently, the wastes are disposed into various water bodies. Consequently, the water bodies are getting contaminated due to improper disposal, which in turn, further creates water crisis. Therefore, the treatment of wastewater released from industries have become very challenging due to the lack of effective materials for the removal of toxic dyes, heavy metal ions, micro-pollutants, etc. These pollutants exhibit harmful effects on living species in water as well as on the human being [30]. Hence, treating these pollutants is the foremost step to develop polymer-based green composites. Further, to purify the contaminated water, various methods including chemical oxidation, photocatalytic degradation, biological treatment, and adsorption have been explored [31]. Of them, adsorption method has been considered to be an effective and environment-friendly method due to its ease of operation and cost-effectiveness. Some of the already reported adsorbents such as carbon nanotubes (CNT)/activated carbon fiber, hydrolyzed nanosilica incorporated polyacrylamide-g-xanthan gum, and polyacrylonitrile (PAN)/zeolite were made from activated carbons, zeolites, clays, grapheme oxide, silica gel, and activated alumina, which generally require high-cost production and lots of energy, and also on the other hand, results in the formation of byproducts during preparation. Furthermore, mostly adsorbents need high adsorption capacity toward certain type of pollutants, but these are not applicable to serve all the purposes. Recently, polymer-based green composites have received much attention of the researchers to treating wastewater as a consequence of their ease of synthesis, effective porosity, tunable morphology, good mechanical property, non-toxicity, insolubility in water, reversibility of ion (favorably cation), and sorption/desorption capability [32, 33]. Both synthetic and natural polymers are used for the fabrication of green composites. As natural polymers are obtained from renewable resources, so they can easily degrade within a proper time period. On the other hand, some synthetic polymers can also degrade to a certain extent provided if they are chemically bonded to natural compounds [34]. Notably, the "green composites" cannot be fully eco-compatible and recyclable, if the temperature exceeds 200 °C during the recycling process because the main properties get drastically altered under this condition. For this reason, many of the researchers have been focusing to develop 100% eco-sustainable and "green" composites since the last several years by replacing non-biodegradable polymer matrices with biodegradable ones. Many of the polymers such as alginate, carrageenan, gelatin, cellulose, chitin/chitosan, poly(lactic acid), polysulfone, poly(acrylic acid), and so on are used as reinforced matrices in the form of membranes, sponges, hydrogels, beads, films, and adsorbents to develop polymer-based green composites for wastewater treatment [35, 36]. In a study, Thakur et al. [37] reported the preparation of gelatin-Zr(IV) phosphate nanocomposite (GT/ZPNC) using sol-gel method for the treatment of aqueous environment containing methylene blue and fast green toxic dyes through photocatalytic degradation. In another attempt, Benhouria et al. [38] fabricated various beads such as bentonite-alginate beads, activated carbon-alginate beads, and activated carbon-bentonite-alginate beads using a very simple method to remove methylene blue (MB) with the adsorption capacity of 756.97 mg/g at 30 °C and reused six times.

Polymer-based nanocomposites (PNCs) also play prominent role in either separation or degradation of toxic effluents for water treatment [38]. Natural low-cost PNCs give sufficient active sites onto adsorbents to remove the toxic dyes from various industrial wastewater. In this regard, metal nanoparticles combined with the natural polymer matrix to develop advanced materials for effective water treatment by optimizing their properties, e.g., hydrophilicity, hydrophobicity, porosity, mechanical strength, and dispersibility. In addition, many more polymers-based green composites such as polyhydroxybutyrate-g-CNTs [39], zinc oxide/polyaniline nanocomposites [40], poly(methyl methacrylate) (PMMA)/TiO_2 nanotubes composites [41], poly(vinyl alcohol)/tetraethyl orthosilicate/aminopropyltriethoxysilane (PVA/TEOS/APTES) nanofiber membrane [42], TiO_2/poly(acrylamide-styrene sodium sulfonate) (TiO_2/(PAAm–SSS)) [43], and ammonium molybdophosphate-polyacrylonitrile (AMP–PAN) [44] were reported by several groups of researchers to remove toxic effluents from the polluted water. A few of them is summarized in Table 2.

Table 2 Various synthesis parameters of individual polymer-based green composites and their specific use for wastewater treatment

S. No.	Green composite (s)	Form (s)	Method of preparation	Temperature (°C)	pH	Application (s)	Numbers of cycles to reuse	Average removal efficiency per cycle (%)	References
1	Sodium alginate/gelatin-based ZnS-nanocomposite	Hydrogel	Swelling–shrinking process	30	5	Used as effective adsorbent for the removal of biebrich scarlet dye	4	97.37	[45]
				40	9	Used to remove crystal violet (CV) dyes from textile industry and also used in water pollution control sector	3	94.45	
2	Polysulfone/wood flour/nanoclay	Films	Solution casting	20	9	For the removal of methylene blue (MB) dyes from aqueous solutions	3	95.72	[46]
3	Carboxymethyl cellulose/k-carrageenan/activated montmorillonite composite	Beads	Precipitation	25–45	10	Same as polysulfone/wood flour/nanoclay	5	92	[47]
4	Citric acid (CA) crosslinked β-cyclodextrin (CD) polymer	Adsorbents	Esterification	25	6.5	To remove the bisphenol A (BPA) as the model pollutants	5	90	[48]
					11	Used specifically to remove cationic dyes such as MB		92	

(continued)

Table 2 (continued)

S. No.	Green composite (s)	Form (s)	Method of preparation	Temperature (°C)	pH	Application (s)	Numbers of cycles to reuse	Average removal efficiency per cycle (%)	References
					4	Used specifically to remove anionic dyes such as methyl orange (MO)		88	
5	κ-carrageenan and nano-silver chloride	Hydrogels	Stirring	25	6	It can be more effectively used than sodium alginate/gelatin-based ZnS-nanocomposite for the removal of CV as a cationic dye from wastewater	9	98	[49]
6	Gelatin/TiO_2/polyethyleneimine (PEI) (GTP)	Aerogels	Mixing	25	5.1–6	Can be applied for the removal of cationic dyes such as MB	5	85	[50]
					6.8	For the removal of anionic dyes such as congo red (CR)			
					4.62	Efficiently used for the removal of Cu(II) ions from aqueous solution	5	86.47	

(continued)

Table 2 (continued)

S. No.	Green composite (s)	Form (s)	Method of preparation	Temperature (°C)	pH	Application (s)	Numbers of cycles to reuse	Average removal efficiency per cycle (%)	References
7	Gelatin/β-cyclodextrin composite	Fibers	By electrospinning	20–60	8	Fiber adsorbents exhibit excellent adsorption capacity for removal of basic fuchsin, gentian violet, brilliant blue R and malachite green (MG) dyes from wastewater	9	73	[51]
8	Poly(acrylic acid-co-acrylamide)/attapulgite composite	–	Simple mixing	30	4.5–9.8	Low-cost adsorption agent used to remove MV dye from industrial water	4	95.6	[52]
9	Cross-linked graft copolymers of cellulose	Cross-linker	Bleaching method	30	7	Good adsorbent used to remove cationic dyes such as MG	5	48.33	[53]
				25	7	For the removal of cationic dyes as CV		50.17	
				25	2.2	Used to remove CR		77.7	
				30	6	Efficiently used to remove heavy metal ions such as Ni(II) ions		72.6	

(continued)

Table 2 (continued)

S. No.	Green composite (s)	Form (s)	Method of preparation	Temperature (°C)	pH	Application (s)	Numbers of cycles to reuse	Average removal efficiency per cycle (%)	References
				30	6	Also used to remove toxic Cu(II) metal ions from industrial wastewater		67.6	
10	Cellulose with hyperbranched polyethylenimine (hPEI)	Powder	Freeze drying	30	5	Finds application to remove anionic dyes such as CR	–	–	[54]
					8.6	Finds specific use to remove cationic dyes such as basic yellow 28 (BY28) from various pollutants of industrial wastewater			

4 Why Chitosan in Wastewater Treatment?

Although various kinds of materials have already been reported for the treatment of wastewater, however, an aggressive research is still going on toward the development of cheaper and effective technologies in terms of preparing polymer composites that cannot only remove the contaminants efficiently but also can improve the quality of treated effluents [55]. As various synthetic non-biodegradable polymers are used for the said purposes so in terms of environmental sustainability, the prepared polymer composites are no longer desirable because of their hazardous effects on environment when disposed. Therefore, a considerable attention has been paid to green, renewable, sustainable, and environmentally friendly materials for treatment of wastewater. From the literature survey, it has been found that biopolymers such as agar, alginate, chitin, chitosan, starch, and their derivatives have been used to prepare different adsorbents for the removal of a wide range of pollutants including heavy metal ions and both cationic and anionic dyes from wastewater [56, 57]. These polysaccharides are low cost, biodegradable, naturally abundant, and very effective in removing pollutants from wastewater. In addition, these are also extremely useful in terms of sustainable environment [58]. Among them, chitosan is the one and only polysaccharide, which is polycation in nature and that is why it has received lots of attention as effective biosorbent due to its efficiently chelating as well as metal binding ability, flocculating ability, biodegradability, non-toxicity, low cost, abundant availability on earth next to cellulose, and processability that makes it an ideal candidate for wastewater treatment applications. Moreover, the presence of free amino and hydroxyl functional groups can serve as active sites for adsorption of different types of inorganic and organic pollutants from wastewater [59]. Besides such exciting and remarkable properties, there are some limitations of using chitosan for various purposes because of its non-porosity, pH sensitivity, poor mechanical properties, insolubility in conventional organic solvents as well as in aqueous media [60]. To overcome these limitations, chitosan can easily be chemically modified using its free amino as well as hydroxyl functional groups. Modification of chitosan not only helps improve its physicochemical properties but also facilitates introducing special properties for specific applications. Furthermore, it can be easily processed into different forms such as nanofibers, nanoparticles, beads, gels, sponges films, hydrogels, and nanocomposites. A glimpse of various chitosan-based derivatives used for wastewater treatment is depicted in Fig. 1.

Chitosan has been extensively used as adsorbent for wastewater treatment [61]. Sharma et al. [62, 63] demonstrated the preparation of chitosan cross-linked-poly(alginic acid) nanohydrogel to remove Cr(VI) ions through adsorption from aqueous solution. In another study, Sharma et al. [62] synthesized sodium dodecyl sulfate-iron silicophosphate (SDS-FeSP) nanocomposites using co-precipitation method. These nanocomposites showed highly selective nature of adsorption toward specific metal ions, i.e., Zn^{+2} and Mg^{+2} ions. Further, chitosan has been combined with a wide range of various natural as well as synthetic polymers to fabricate chitosan-based green composites including α-Fe_2O_3 impregnated

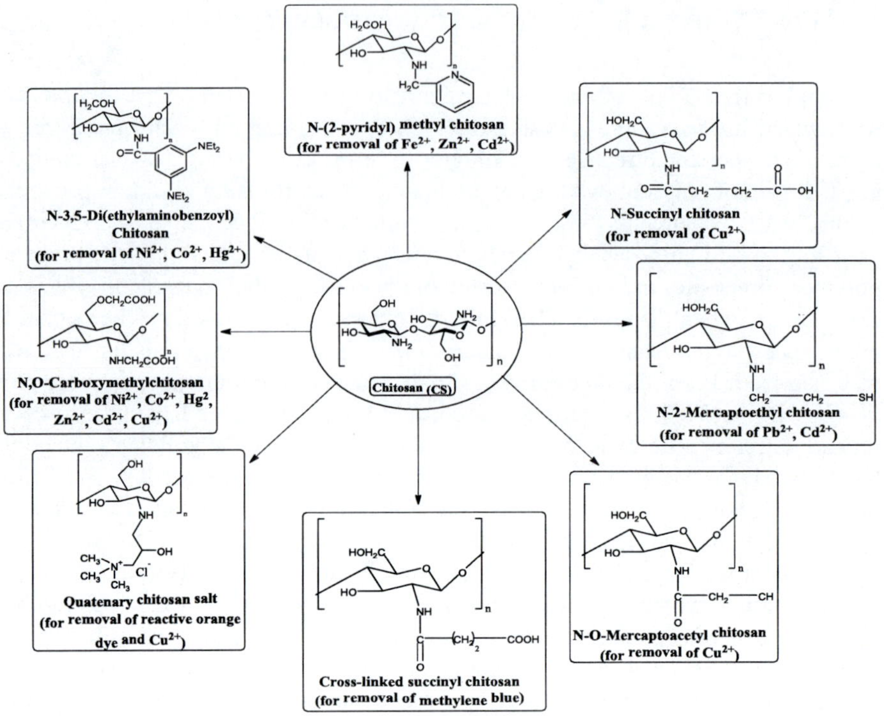

Fig. 1 Chemical modification of chitosan for its effective use in wastewater treatment (*Source* Author)

chitosan beads with As(III) as imprinted ions [63], thiourea-chitosan coating on the surface of magnetite (Fe_3O_4) [64], chitosan-polyaniline/ZnO hybrid composite [65], and polymethylmethacrylate-chitosan composites [66] to name a few. However, an attempt has been made to establish a comprehensive pictorial representation portrayed in Fig. 2 to show the importance of using chitosan for fabricating various green composites for wastewater treatment. In the following section, preparation of various chitosan-based green composites is discussed along with their applications as adsorbents for wastewater treatment.

5 Preparation of Chitosan-Based Green Composites (CGCs)

Depending on the logic behind preparing green composites as already discussed in Sect. 2, chitosan has been combined with various polymers of interest to prepare CGCs for their subsequent applications in wastewater treatment.

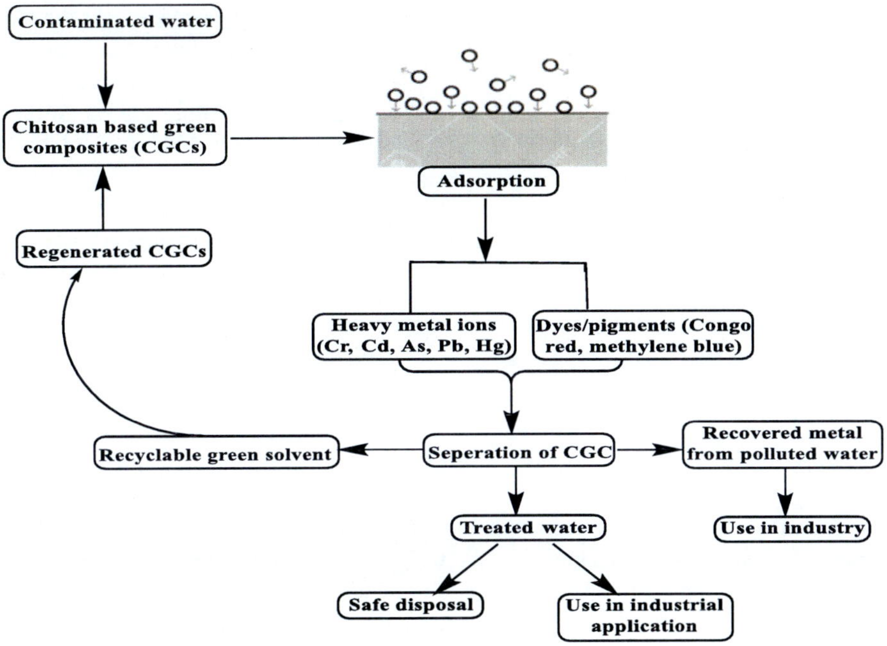

Fig. 2 A possible benefit of recycling of chitosan-based green composites used in wastewater treatment (*Source* Author)

5.1 Chitosan–Cellulose Acetate Composites

Chitosan–cellulose acetate composites are biodegradable, eco-friendly, and highly investigated biosorbent composites. These are usually used for the removal of metal ions from wastewater. In a study, Ghaee et al. [67] demonstrated the preparation of chitosan-cellulose acetate composite nanofiltration membrane for wastewater treatment. As chitosan (CS) membrane itself does not possess good mechanical strength, that is why cellulose acetate (CA) is combined with it to enhance an overall mechanical strength of the membrane because nanofiltration is basically a pressure-driven process. The entire preparation process can be elaborated as follows: initially, a solution of CA was prepared by dissolving varying amounts of CA in *N*-methyl pyrrolidone (NMP) containing poly(ethylene glycol) followed by casting the solution after deaeration onto a flat surface so as to obtain CA membrane. A 200-μm-slot applicator was used to maintain a proper thickness of the membrane. On the other hand, 0.5% (w/v) CS solution was prepared by dissolving CS into 10% (v/v) acetic acid solution and filtered. Then, the as-prepared CA membrane was dipped into eight consecutive solutions, which were segregated into two different parts, i.e., the first four parts were made of various concentrations of ethanol, namely 25, 50, 75, and 100%, in water as a co-solvent and the second four parts were made of various concentrations of n-hexane, namely 25, 50, 75, and 100%, using ethanol as a co-solvent for solvent exchange before the membrane was subjected to drying. After that the CA membrane was dipped into the CS solution for 3 min, and dried at room temperature. Cross-linking was then performed by placing the dry membrane into a 0.25% (w/w) glutaraldehyde aqueous solution for 30 min at 25 °C. Next, the membrane was thoroughly washed with distilled water to remove the unreacted glutaraldehyde residues, and finally, the membrane was dried using a filter paper. It was reported that under 506.5 kPa applied pressure, the said composite membrane efficiently rejected 81.03% copper from a common effluent treatment plant of wastewater.

5.2 *Chitosan-Calcium Alginate Composites*

Alginates can be used as stabilizers, viscosity modifiers, and gel-forming, film forming, water-binding agents [68]. Further, it is polyanionic in nature, whereas chitosan is polycationic in nature. Therefore, very easily both of them can be combined together to prepare chitosan–alginate composites. Vijaya et al. [69] demonstrated the preparation of chitosan–calcium alginate composites shown in Fig. 3 for the removal of Ni^{2+} ions from aqueous solution. In this work, the properties of calcium alginate and chitosan were greatly modified through the preparation of green chitosan–alginate composites. Briefly, calcium alginate beads were prepared by dispensing sodium alginate solution through the tip of a pipette into 2% calcium chloride solution. Then, these as-prepared beads were dropped into 4% chitosan gel, which was prepared by dissolving 4 g of chitosan in 2% acetic acid solution followed by stirring for 12 h to obtain chitosan-coated calcium alginate (CCCA). Next, the obtained chitosan coated beads were placed into 0.1 M NaOH solution for 4 h followed by washing with distilled water and drying for further application. It was reported that the maximum monolayer Ni^{2+} adsorption capacity calculated through the Langmuir adsorption isotherm was about 222.2 mg g^{-1}. In another work, Wan and Ngah [70] conducted a comparative study between chitosan–glutaraldehyde and chitosan–alginate beads setting various parameters such as agitation time, pH, adsorbent dosage, and initial concentration of each for the adsorption of Cu^{2+} ions from the aqueous solutions. Here, they reported the preparation of chitosan–alginate composite slightly in a different way. Succinctly, both the solutions of chitosan flakes and alginic acid were dissolved in acetic acid and distilled water, respectively. Then each of these solutions was aged overnight. Next, chitosan solution was further heated at 60 °C and stirred for about 30 min before the alginic acid solution was added to it. After that each of these two solutions was mixed together for another 20 min at 60 °C. The resulting solution was then poured into a 0.50 M NaOH solution so as to obtain chitosan–alginate beads, which were then filtered and rinsed with distilled water to remove the residual sodium hydroxide. Finally, the chitosan–alginate beads were dried and sieved. The chitosan–alginate beads were found to have a relatively better fit to nonlinear Langmuir isotherm than the chitosan–glutaraldehyde beads and also the adsorption capacity of the former ones was found to be 67.66 mg g^{-1}. However, from the above studies carried out by different groups of researchers, it is

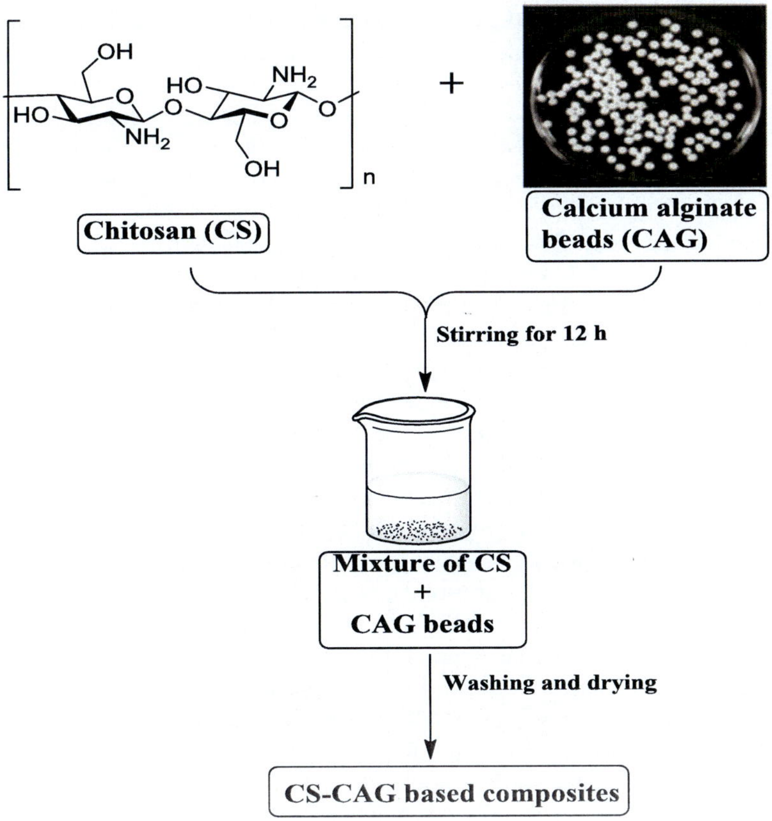

Fig. 3 A schematic representation for the preparation of chitosan-calcium alginate composites (*Source* Author)

quite obvious that chitosan–alginate composite is more effective in case of removing Ni^{2+} compared to Cu^{2+} from aqueous solution.

5.3 Chitosan-Clay Composites

5.3.1 Chitosan-Activated Clay Composites

Usually, activated clay shows effective adsorption properties toward organic dyes. Therefore, in a study, Chang and Juang [71] used this activated clay to combine with chitosan to prepare a composite with improved mechanical properties, which not only enhanced the adsorption properties but also prevented the dissolution of chitosan beads in highly acidic and alkaline medium [72, 73]. Briefly, 1 g of chitosan flakes was dissolved in 100 mL of 1 mol/L of acetic acid followed by adding 1 g of activated clay into it, and then, the solution mixture was agitated for 24 h using a disperser. Next, the resulting viscous solution was deaerated by placing it into a vacuum dryer for 3 h. After that this deaerated solution was sprayed dropwise through a nozzle into a known volume of deionized water containing 15% of NaOH and 95% of ethanol and allowed to leave for 24 h. Finally, the obtained chitosan beads were washed with deionized water until the pH of the rinsing solution reached neutral values. The activated clay chitosan composites were reported to be used as adsorbents for the removal of some common contaminants such as humic acid, tannic acid, basic dye methylene blue, and reactive dye 222 (RR222) [74–79] from water. The results revealed that the said composites more actively adsorbed the contaminants as compared to individual chitosan beads and activated clay as well.

In another study, Gecol et al. [80] prepared biosorbent chitosan coated montmorillonite clay for removing tungsten species from aqueous solution. The experimental results revealed that the biosorbent effectively adsorbed tungsten at optimal pH 4 in the range between 68.2–93.8%, 66.7–94.2%, and 53.6–93.7% from the simulated natural water containing varied amount of H_2CO_3, H_4SiO_4, and SO_4^{2-}, respectively. Nevertheless, Alcântara et al. [81] showed that the incorporation of various clays as nanofillers into the biopolymer matrices to developing various biopolymer–clay-based nanocomposites in the form of films including chitosan–clay bionanocomposite film drastically improved all the properties such as mechanical strength, water resistance, biodegradability, biocompatibility, and reduced water absorption capacity, which make the bionanocomposite films essential candidates for their diverse applications ranging from bioplastics to adsorbents. Further, they also proved that the individual biopolymer such as chitosan, alginate, and starch in the form of film was incapable of retaining heavy metals such as Cu^{2+} and Pb^{2+} because of their rapid dissolution in water and this phenomenon was in agreement with the result already reported by Chang and Juang [71]. Therefore, addition of fibrous clays as nanofillers to polysaccharide matrices not only strengthens the mechanical integrity of the bionanocomposites but also makes the bionanocomposites more suitable and cost-effective adsorbents for treating contaminated water.

5.3.2 Chitosan–Bentonite Composites

The role of clay as an adsorbent is quite known and thus is used frequently to remove unwanted chemical substances from aqueous solutions. On the other hand, chitosan is extensively used in biological applications as a consequence of its biocompatibility and biodegradability. Therefore, researchers are trying hard to find out new adsorbents by combining both of clay and chitosan together to prepare novel composites for the removal of different types of dyes and heavy metal ions from the aqueous solutions. Many of the researchers have reported the use of modified bentonite for said purposes. It is mainly composed of montmorillonite with a composition of Cao, MgO, SiO_2, Fe_2O_3, Al_2O_3, Na_2O, K_2O [82, 83]. In addition, bentonite is a 2:1 type of aluminosilicate, in which a unit layer of Al^{3+} octahedral sheet is sandwiched between two SiO^{4+} octahedral sheet [84]. Recently, Abukhadra et al. [85] developed a chitosan–bentonite composite supported with Co_3O_4 using a green approach. Figure 4 shows a probable pathway for the preparation of the same. The composite is an eco-friendly adsorbent can be applied for the removal of acidic dye such as congo red and inorganic species such as Cr^{4+} from decontaminated water. Experimental results revealed that the maximum adsorption for congo red and Cr^{4+} were found to be about 303 mg g^{-1} and 250 mg g^{-1}, respectively. Pseudo-second order and Langmuir equilibrium model was used to describe the system. The prepared composite showed high reusability and was effectively reused for decontamination

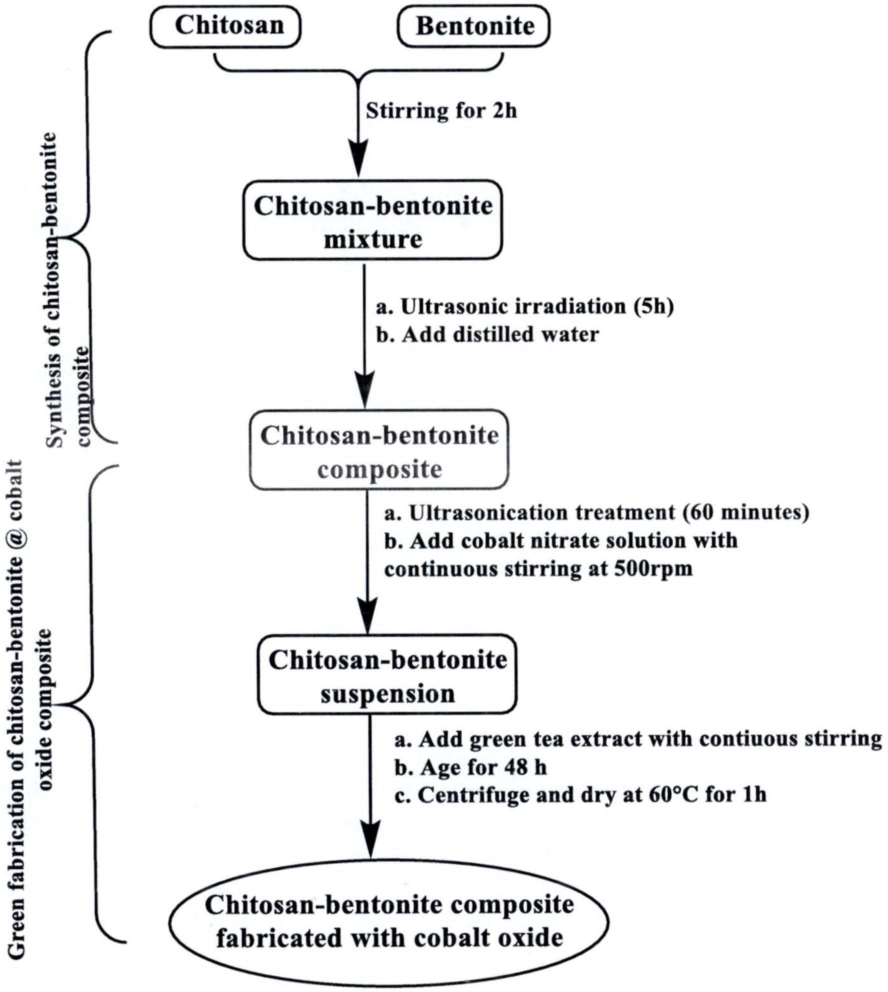

Fig. 4 A probable pathway for the preparation of chitosan–bentonite composite and its green fabrication with cobalt oxide (*Source* Author)

of congo red and Cr^{4+} for six runs. In another study, Liu et al. [86] reported the preparation of chitosan–bentonite composite, which was an efficient adsorbent for the removal of contaminants from wastewater because bentonite possesses good adsorption properties. Further, Wan et al. [87] developed a cross-linked chitosanbentonite composite to adsorb tartrazine dye. They reported that the maximum adsorption of tartrazine was found to be 2941 mg g^{-1}, which was best described by pseudo-second order and Langmuir isotherm model. Further, in a study, Anirudhan et al. [88] reported the development of a cross-linked methacrylic acid-g-chitosan–bentonite composite through graft copolymerization using *N*,*N′*-methylenebisacrylamide as a cross-linking agent for the removal of Th^{4+} from seawater and also showed that the maximum adsorption capacity of the composite was found to be about 110.5 mg g^{-1} at 30 °C. Moreover, in another study, Anirudhan and Rijith demonstrated the preparation of carboxyl terminated poly(methacrylic acid)-g-chitosan–bentonite composite for the removal of Ur^{6+} from water. They reported that the maximum adsorption was found to be 117.2 mg g^{-1} under specified condition such as at pH 5.5 and 30 °C [89]. In addition, different forms of modified bentonite encompassing chitosan modified bentonite, hexadecyl trimethyl ammonium bromide modified bentonite have also been comparatively studied in terms of adsorbing dye species from contaminated water.

5.3.3 Chitosan–Clinoptilolite Composites

Considering public health and ecosystem stability, it is quite essential to treat radioactive wastewater containing radiocobalt, i.e., ^{60}Co before its safe discharge because of its exposure to human causes deadly illness such as cancer. For treating such water, many researchers are working aggressively to prepare novel polymer composites for effective elimination of radiocobalt. In a study, Zhao et al. [90] demonstrated the use of clinoptilolite, a natural zeolite to prepare chitosan–clinoptilolite composite (CS-CLP) because of its outstanding mechanical and sorption properties, which in turn, also facilitates increasing not only the mechanical strength of chitosan but also that of the overall composite. In brief, equal amounts of chitosan and clinoptilolite were mixed together in a known volume of 5% (v/v) acetic acid solution under continuous stirring for 2 h followed by gradually adding epichlorohydrin, a cross-linker and vigorously stirring for 1 h so as to obtain CS-CLP hydrogel. Then the composite hydrogel was serially treated with sodium tripolyphosphate and sodium hydroxide solutions in order to form composite beads. Finally, the beads were dried and powdered to the desired size. A schematic diagram for the preparation of CS-CLP composites is shown in Fig. 5. The composite beads were used for the removal of Co (II) ions from nuclear wastewater, and later they conducted a comparative study of the CS-CLP composite beads with other adsorbents such as raw bentonite, formaldehyde modified bentonite [91], TiO_2/eggshell suspensions, natural halloysite nanotubes [92], titanium dioxide (TiO_2) [93], MWCNTs-g-VP [94], Mg_2Al-layered double hydroxide [95], and wastes containing boron in terms of maximum sorption capacity (q_{max}). The results revealed the sorption capacity of the CS-CLP composite

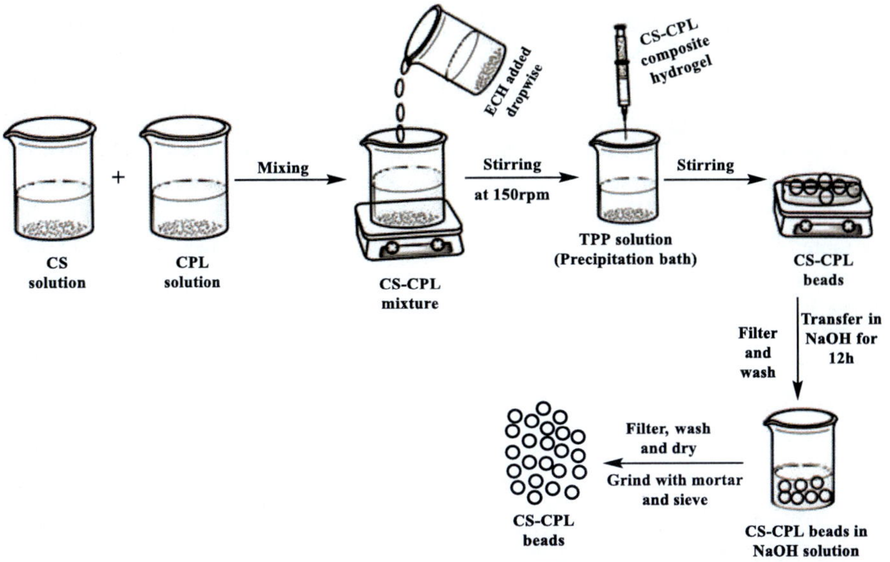

Fig. 5 A scheme for synthesis of chitosan–clinoptilolite (CS-CPL)-based composites beads (*Source* Author)

was superior than all the other said adsorbents barring the wastes containing boron under similar experimental conditions. Therefore, the CS-CLP composite may be the promising candidate for effective treatment of nuclear wastewater.

In another study, Maria et al. [96] conducted a comparative study for the evaluation of the adsorption capacities of CS-CLP composites for three environmentally harmful toxic divalent metal ions, namely Co^{2+}, Ni^{2+}, and Cu^{2+}, from aqueous solution. In this work, they analyzed the effects of initial metal ion concentration, initial pH value of the solution, contact time, and temperature on the adsorption capacity of the composite. The results revealed that the optimum adsorption of the composites took place at pH 5 and contact time required to achieve equilibrium was 24 h. Further, they also reported that the adsorption capacities of the CS-CLP composites for Co^{2+}, Ni^{2+}, and Cu^{2+} contained in aqueous solutions were 11.32 mmol g^{-1}, 7.94 mmol g^{-1}, and 4.209 mmol g^{-1}, respectively.

5.4 Hydroxyapatite–Chitosan Composites

The improper disposal of colored wastewater from the textile industries near water bodies such as water streams, and river poses a great threat to the environment. Some of the chemicals present in this wastewater are carcinogenic and thus, are responsible for causing a potential human health risk. Various methods such as chemical precipitation, adsorption, cations-exchange, reverse osmosis, electrodialysis, electrochemical reduction have been employed to treat this wastewater generated from textile industries [97]. Keeping in mind that adsorption is one of the best and industrially accepted methods for its effective treatment of wastewater, Hamzah and Salleh [98] reported the preparation of hydroxyapatite–chitosan composite for removal of remazol blue dyes from textile industrial wastewater. A detailed description for the preparation of the said composite is illustrated in Fig. 6. Briefly, hydroxyapatite extracted from egg shells was mixed with chitosan solution in acetic acid and stirred for 8 h. The solid white precipitate obtained was filtered and dried at ambient temperature, i.e., 30 °C so as to obtain hydroxyapatite–chitosan composite adsorbent. It

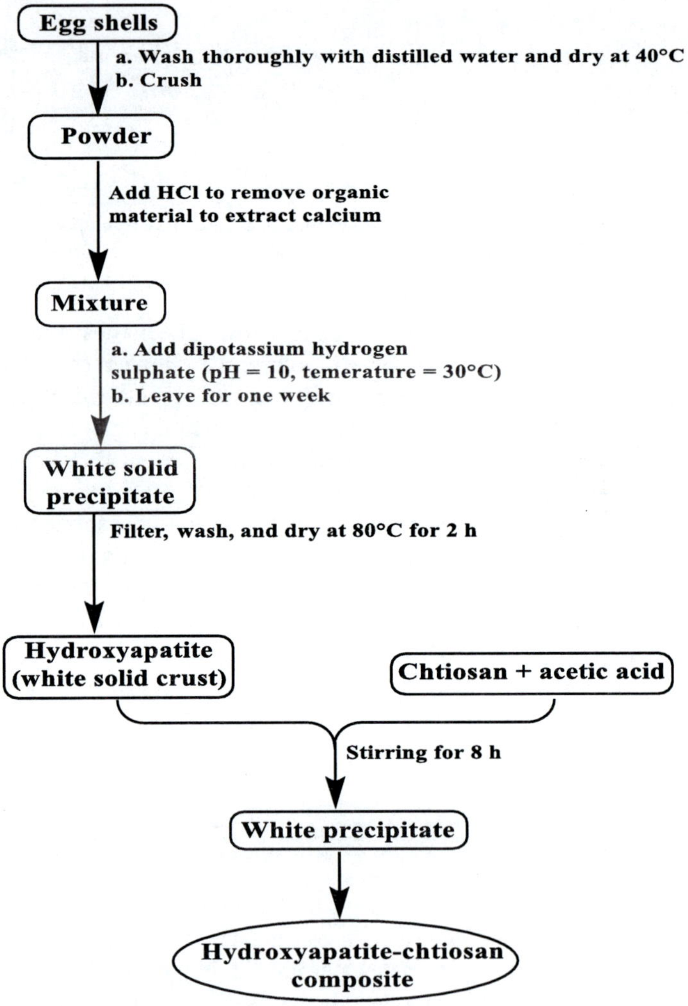

Fig. 6 Flowchart for the preparation of hydroxyapatite–chitosan composite (*Source* Author)

was found that the remazol blue dye removal efficiency of the prepared composite exclusively depended on pH. With increasing the pH of the solution percentage dye removal efficiency of the composite also increased. The optimal dye removal efficiency, i.e., 95% using the said composite was found at pH 6, and it might be due to the protonation of $-NH_2$ groups of chitosan in slightly acidic solution that yielded positively charged $-NH_3^+$ groups.

As we talk about chitosan-based green composites for wastewater treatment then from authors' perspective, keeping in mind the principle of green synthesis, a general scheme for the preparation of chitosan-based green composites is depicted in Fig. 7.

6 Removal of Heavy Metals from Industrial Wastewater Using CGCs

In the previous section, we mainly focused on the preparation of CGCs for their applications in wastewater treatment. Here, we shall discuss on other green composites based on chitosan and their uses for the removal of various metal ions found in various wastewater generated from different industries. Although there are various causes of environmental pollution but unfortunately, improper disposal of wastewater containing hazardous and harmful pollutants including heavy metals directly into the environment has also become one of the most alarming environmental issues that further aggregates the environmental stability [99, 100]. However, various toxic heavy metals such as Pb(II), Cd(II), Hg(II), and Cr(VI) generated by numerous industrial activities including mining industry, plating and smelting industry, fertilizers manufacturing, chemical industry, textile industry, pigments, and plastics to name a few are directly discharged into water bodies, which subsequently not only impact the human health but also affect the biological systems too [101]. Table 3 summarizes the tolerance limits of various toxic heavy metals contained in wastewater generated through various operations by different industries and their subsequent toxic effects on living organisms. It is worth mentioning that the natural water gets contaminated over a period of time due to mixing of heavy toxic metals, which directly affect the human health and also decreases the quality of other biological systems. Thus, the discharge of industrial wastewater with heavy metal contents should be performed according to the standard operating procedures prepared by a specific scientific body, which is duly approved by the government. In this way, the environmental pollution can be prevented to a great extent. So far various methods such as chemical precipitation, coagulation or flocculation, membrane filtration, electrochemical techniques, and ion exchange have been employed for the removal of toxic metal ions from wastewater [102]. Unfortunately, all these said methods are not effective for the removal of toxic metals at low concentrations because of the high operating costs, uses of large amounts of chemicals, and requirement of high amount of energy.

The elimination of heavy metals from wastewater by different methods is shown in Fig. 8. Due to low cost, adsorption is considered as an effective method to remove

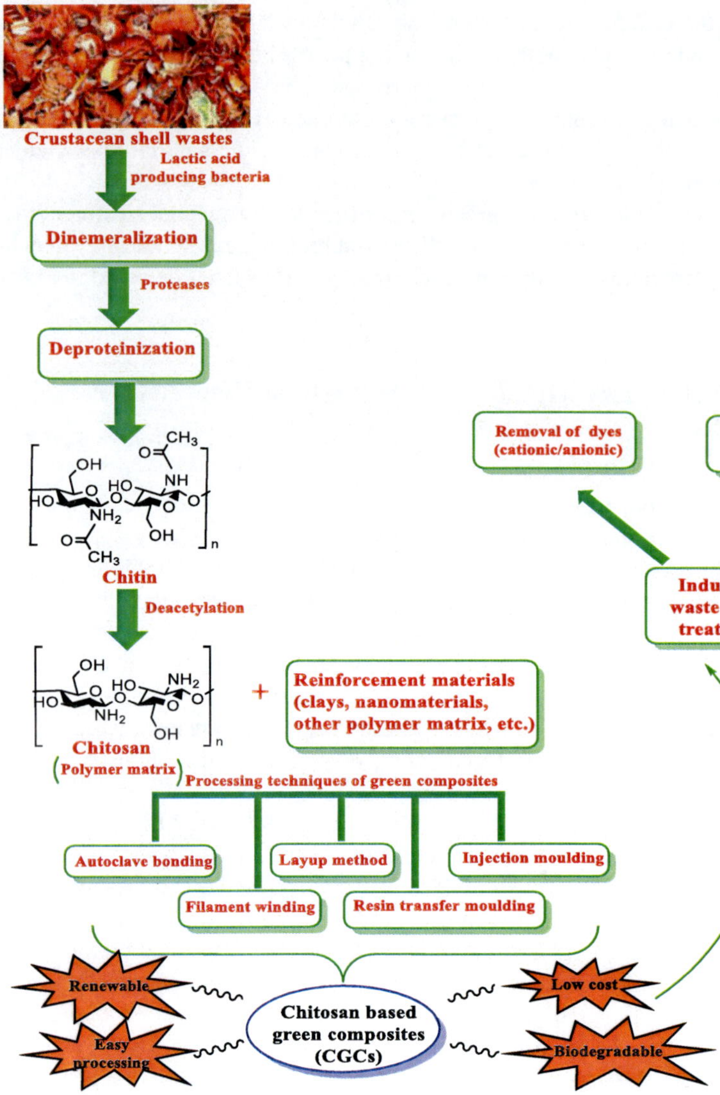

Fig. 7 A general approach to prepare CGCs and their applications in wastewater treatment (*Source* Author)

toxic metals from wastewater effluents at low and high concentrations. Further, the elimination of heavy metal ions from water effluents through adsorption follows all the principles of environmental sustainability. Moreover, adsorption method is associated with lots of advantages such as ease of flexibility and simplicity in designing, ease of operation, and economically viable to mention a few, which are also very crucial from environmental viewpoint. In addition, the adsorbents employed in this

Table 3 Influence of various toxic heavy metals released into the environment from different industrial sectors on living organisms

Source of industrial wastewater	Presence of metal	Tolerance limit	Effects on living organisms	References
Smelting and refining industries, thermal power plants	Arsenic (As)	25 Pbb	Causes skin cancer, visceral cancer, kidney failure, bladder damage, abdominal pain, and visceral cancer. Also decreases red and white blood cell production in the human body	[103, 104]
Lead acid battery manufacturing, metal plating and finishing, paints, ceramic and glass industries, ammunition, etc.	Lead (Pb)	0.05 ppm	Human brain and nervous system are greatly affected. Further, excessive amount of lead damages the circulatory system and kidney function. Moreover, it also leads to high blood pressure	[104]
Fertilizers, paints, chlorine-manufacturing, switchgears/batteries, and pharmaceuticals	Mercury (Hg)	0.001 ppm	Affects the human nervous system, digestive system, kidney function and circulatory function	[105]
Mining and processing of non-ferrous metal ores	Cadmium (Cd)	0.01 ppm	It is carcinogenic which causes itai-itai disease, kidney damage, and renal disorder	[106, 107]
Leather tanning, electroplating, and dyeing	Chromium (Cr)	0.05 ppm	Due to carcinogenic nature, it also causes cancer, kidney, and liver damage	[107]
Mining, paints and pigments, smelting, petroleum refinering, fertilizers, paper and pulp industries	Copper (Cu)	0.01 ppm	Having direct influence on the human blood coagulation and also responsible for causing melena hypertension, Wilson's disease, gastrointestinal distress, and insomnia	[107]
Electroplating and batteries industries	Nickel (Ni)	2 μg/L	Greatly affects lung embolism, asthma, liver, and kidney damage with chronic bronchitis	[108]
Metallurgical, steel manufacturing, heavy machinery, food and medicine industries	Iron (Fe)	1.0 ppm	Reduces the growth of plants. In addition, over exposure to iron has noticeable effects on human health such as tissue disorder	[109]
Metal polishing and metallurgical industries	Zinc (Zn)	0.5 ppm	Causes hypertension, lethargy, neurological disorder, and edema of lungs. Excessive zinc in drinking water also affects the internal organs in living organisms	[2]

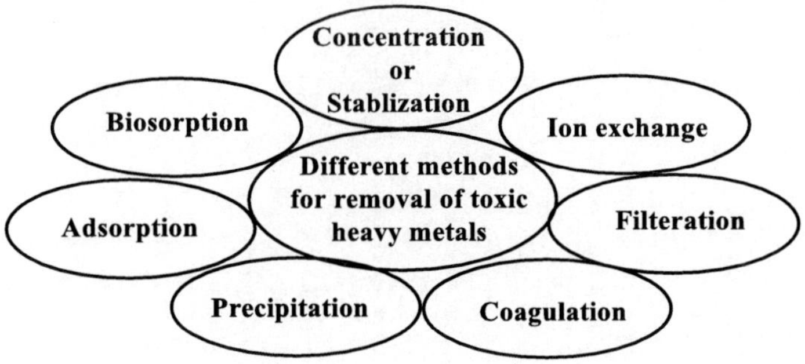

Fig. 8 Various methods for the removal of heavy toxic metals from industrial wastewater (*Source* Author)

method can be reused again through desorption method and thus, maintains the low operation cost and high efficiency [110]. Therefore, various polymers-based green composites in the form of mats, membranes, beads, films, hydrogels, etc., possessing superior properties have already been reported for the removal of heavy toxic metals from industrial wastewater streams. Among them, also CGCs have been proved to be effective adsorbents for the removal of not only precious metal ions such as Pd(II), Pt(IV); heavy metal oxyanions including Cr(VI), As(III), As(V)), but also radionuclides such as Co(II), Sr(II), Cs(I) apart from already mentioned heavy metal ions.

It has been known that addition of magnetic particles during preparation of polymer-based green composites is more advantageous for water purification because of its magnetic properties that not only enhance the adsorption rate but also increase the efficiencies of the composites for removal of various pollutants from wastewater. In a study, Liu et al. [111] reported the preparation of recyclable chitosan-based magnetic nanocomposites to eliminate heavy metal ions. The interaction between chitosan and metal ions in adsorption method is reversible, and it is said the metal ions can be removed from chitosan by increasing pH of the solution by adding deionized water under the influence of ultrasonic radiation. Usually, the interaction between adsorbate and adsorbent are carried out both through physical and chemical processes. But if the process takes place through chemical bonding, then it is difficult to describe its chemicals from adsorbent surface [112]. Furthermore, adsorption mechanism is associated with single or mixed mechanisms like metal chelation, electrostatic interactions, and ion pair formations. However, it also depends on some parameters such as pH, ion strength, and temperature of solution, metal ions, and the chitosan-based adsorbents. Many studies have been focused on modified chitosan for increasing adsorption capacity for the removal of various pollutants in the form of metal ions. To improve the adsorption capacity and selectivity, chitosan can be modified using physical or chemical methods. Due to the easy processability, chitosan can be molded into different forms, namely beads, sponge, membranes, fibers, flakes, films, gels, etc. Recently developed various chitosan-based green composites for the removal of heavy metal ions from wastewater and their adsorption capacities are summarized in Table 4.

Tang et al. [128] synthesized chitin/cellulose-based microporous membranes with good efficiency for the removal of heavy metal ions, namely mercury (Hg), copper (Cu), and lead (Pb), and they also reported that the chitin/cellulose-based microporous membranes could be easily regenerated for further use. In this, the adsorption of metal ions on chitin/cellulose-based membranes took place via metal chelation and ion exchange followed by green pathway for the removal of hazardous pollutants from wastewater.

In another study, Habiba et al. [117] demonstrated the preparation of chitosan/PVA/zeolite electrospun composites, for the removal of Cr(VI), Fe(III), and Ni(II) at medium concentration. They also showed that the adsorption capacity of the electrospun nanofibers remaind unaltered even after five consecutive recycling. Therefore, the composites could be reused many times for water treatment.

On the other hand, many of the researchers have reported the use of hydroxyapatite for the preparation of chitosan-based green composites as a consequence of its availability, de-fluroridation capacity, and cost-effectiveness for the removal of F^- from aqueous solution [129]. Further, lots of work have been performed to modify the nature of sorbents and make them viable for the applications for the removal of other pollutants from wastewater [126, 130–134]. Sundaram et al. [134] developed a bioinorganic composite of nano-hydroxyapatite and chitosan as sorbent for de-flouridation purpose. From the results, it was found that the adsorption capacity of the prepared sorbent was 1.560 mg g^{-1}, whereas the same for nano-hydroxyapatite was 1.296 mg g^{-1}.

In another study, Kousalya et al. [130] reported the development of another bioorganic composites of nano-hydroxyapatite using chitin and chitosan individually through a precipitation method for the removal of Cr^{6+} from aqueous solution. The results showed that the adsorption capacities of both the composites toward Cr^{6+} removal were found to be higher than that of nano-hydroxyapatite itself. The adsorption capacities for chitin-based and chitosan-based composites were reported to be 2.845 and 3.450 mg g^{-1}, respectively, which distinctively more than that of nano-hydroxyapatite, i.e., 2.720 mg g^{-1}.

In an attempt, Samandari et al. [126] demonstrated the preparation of another low-cost adsorbent composite based on chitosan and Fe-substituted hydroxyapatite in the form of beads and used for removal of basic dyes such as methylene blue and Pb^{2+} from wastewater. They showed that the prepared composite was temperature-dependent and the adsorption capacity of the composite increased with increasing temperature of the system. The adsorption capacities of the composite were reported to be 999.4 mg g^{-1} and 999.9 mg g^{-1} for Pb^{2+} and methylene blue, respectively, at 70 °C. In addition, Langmuir, Freundlich, and Temkin isotherm were used to describe the adsorption kinetics. Among these isotherm models, Langmuir model was reported to be most suitable method for describing the adsorption that strongly followed pseudo-second order kinetics.

In another attempt, Salah et al. [133] reported a comparative study of adsorption behavior between hydroxyapatite nanorods and nano-hydroxyapatite chitosan

Table 4 Use of chitosan-based green composites as adsorbents for removal of heavy metal ions from wastewater

S. No.	Adsorbent	Form	Adsorbate	Q_{max} (mg g^{-1})	pH	T (°C)	References
1	Chitosan/cotton fibers	–	Cr^{6+}	–	3	–	[113]
			Cu^{2+}	–	5		
2	Chitosan-activated carbon	Powder	Cu^{2+}	74.35	5–6	20	[114]
3	Chitosan/alginate	Nanoparticles	Cr^{4+}	–	–	–	[115]
4	Modified chitosan magnetic composite adsorbent (CS-CTA-MCM)	Powder	Cr^{4+}	171	2	25	[116]
5	Chitosan/clinoptilolite	Beads	Cu^{2+}	11.32	5	25	[96]
			Co^{2+}	7.94			
			Ni^{2+}	4.209			
6	Chitosan/poly(vinyl-alcohol)/zeolite	Nanofiber	Cr^{6+}	0.17	–	–	[117]
			Fe^{3+}	0.11			
			Ni^{2+}	0.03			
7	Chitosan–montmorillonite	Particles	Pb^{2+}	49.33	5	25–50	[118]
			Cu^{2+}	25.28			
			Cd^{2+}	20.61			
8	Chitosan-g-polyaniline	Powder	Pb^{2+}	16.07	6	50	[119]
			Cd^{2+}	14.33			
9	Chitosan/fly ash	Particles	Cd^{2+}	87.72	8	35	[120]
10	Chitosan-PVA	Beads	Co^{2+}	14.39	6	30	[121]
11	Quaternized chitosan-coated bentonite		Cr^{4+}	66.6	–	–	[122]
12	Chitosan/$CaCO_3$-silane	Nanocomposite	Cu^{2+}	33.90	–	–	[123]
13	Chitosan/Ag hydroxyapatite nanocomposite	Beads	Cu^{2+}	40.11	–	–	[124]

(continued)

Table 4 (continued)

S. No.	Adsorbent	Form	Adsorbate	Q_{max} (mg g^{-1})	pH	*T* (°C)	References
14	Xanthane-modified chitosan/poly(*N*-isopropyl acrylamide)	Hydrogel	Cu^{2+}	115.1	–	–	[125]
			Pb^{2+}	172.0			
			Ni^{2+}	66.9			
15	Chitosan/Fe-hydroxyapatite nanocomposite	Beads	Pb^{2+}	1837	–	–	[126]
			Ni^{2+}	1324			
16	Magnetic hydroxyapatite/chitosan		Ni^{2+}	112.36	–	–	[127]

composite toward removal of Cd^{2+} from aqueous solution. They showed that the adsorption capacities of the nanorods and the composite were 92 and 122 mg g^{-1}, respectively. In addition, a similar experiment was later performed by Mohammad et al. [131] for the removal of Pb^{2+} from aqueous solution. The respective adsorption capacities of nano-hydroxyapatite and chitosan-based composite for the removal of the latter metal ion were 180 and 190 mg g^{-1}. It again proved that the chitosan and nano-hydroxyapatite-based composite as an adsorbent was far superior than nano-hydroxyapatite itself.

7 Removal of Dyes from Industrial Wastewater Using CGCs

The water polluted by dyes causes a serious environmental as well as health problems for living organisms. The consumption of colored substances in various industries including chemicals, textiles, pulp and paper, metallurgy, leather, paint and coatings industry, food, packaging, pharmacy, and plastics to mention a few is gradually increasing to meet the aesthetic needs of the people. As such there is no adequate information on methods to be effectively used for the removal of dye effluents from wastewater. That is why, it is not a wise decision to depend on a single technique [135]. Approximately, 700,000 tons of synthetic dyes are produced globally per year, whereas in India, it is about to 80,000 tons and 5–10% of dyes are discharged in wastewater [136]. Among all the industries, the textile industries are accountable for producing about 54% of total waster water containing dyes throughout the world. Therefore, water quality is greatly influenced due to the presence of even very small amount of colored dyes and thus increases the number of environmental issues. Dyes are not easily degraded into the environment using standard biological treatments [137]. Due to their charge difference, the dyes are categorized into three heads such as anionic dyes (reactive, direct, and acid dyes), cationic dyes (basic dyes), and non-ionic dyes (dispersed dyes). Thus, the removal of dyes from wastewater is the foremost step to reduce water pollution and also to bring back the original quality of natural water. Dye removal can be processed by several physical methods encompassing adsorption, ion exchange, filtration, coagulation, and also by chemical methods such as Fenton's reagent, ozonization, photocatalysis, and biological methods including aerobic and anaerobic degradation. But due to some sort of limitations such as time consumption, biological methods cannot degrade problematic dyes, and thus, adsorbents used for such purposes cannot be reused during processing [138]. On the other hand, the coagulation method is also responsible for causing water pollution due to growth of colloids in wastewater. Furthermore, if reactive reagents such as chlorine required during chemical oxidation remain in excessive amount in water after treating, then it will also render water to be harmful for living organisms. Nevertheless, Fenton's reagent, ozonization, and photocatalysis methods are also not applicable due to high-cost operation [139].

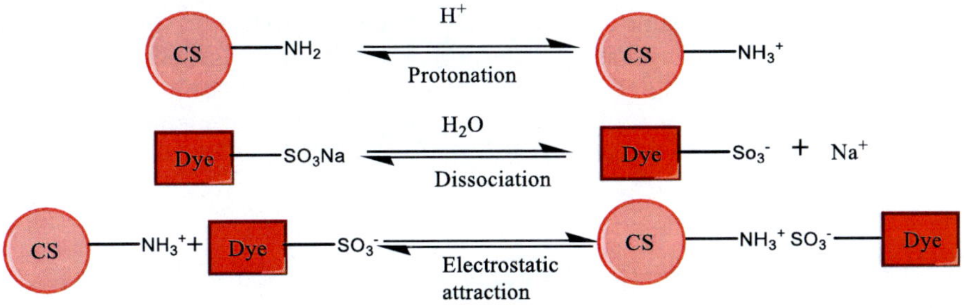

Fig. 9 An electrostatic interaction between chitosan and anionic dye during the adsorption (*Source* Author)

Therefore, there is a dire need to develop inexpensive and effective polymer-based green composites for treating industrial wastewater, which would consume small amount of chemicals and relatively less energy. Owing to their cost-effectiveness, easy availability, and eco-friendly nature, biopolymers such as chitin-/chitosan-based green composites are being developed to effectively treat industrial wastewater containing various dyes using adsorption method. In addition, there is an ample scope of chemical modifications of chitin/chitosan due to the presence of functional groups such as amino groups as well as hydroxyl groups [140]. Therefore, chitosan has received a great deal of attention of researchers owing to its outstanding features, particularly its chelating ability, which can be accentuated to develop CGCs as adsorbents for the removal of toxic dyes. Among various dyes, the anionic dyes are mostly studied because they contain negative charges and get adsorbed into chitosan carrying positive charge through electrostatic interaction. This phenomenon is portrayed in Fig. 9. However, the adsorption efficiency of CGC is greatly influenced not only by the physicochemical characteristics of chitosan such as crystallinity and degree of deacetylation of chitosan but also the pH of dye solution. In addition, some other limitations such as low mechanical strength and solubility are also associated with chitosan. The adsorption capacity, mechanical strength, and acid stability of chitosan can be improved by modifying it through chemical and physical methods. Modification of chitosan using physical method results in the formation of porous network structure with an added advantage of reusability of the adsorbent to be prepared but sometimes, it lowers the adsorption capacity of the said adsorbent due to shielding of active binding sites, which decelerate dye adsorption process. Therefore, chemical modification of chitosan is favoured in order to achieve high dye removal efficiency through its hydroxyl as well as amino functional groups. It is noteworthy to mention that the utilization of CGC-based adsorbents for the removal of cationic dyes has been studied in a very limited way because chitosan possess low affinity toward cationic dyes. It is attributed to the protonation of both chitosan and cationic dyes under acidic conditions, and thus, the electrostatic repulsion between them restricts the adsorption process. To overcome this issue, various CGC have been synthesized with other natural biopolymers, nanoclay, activated carbon, graphene, metal oxide nanoparticles, and its derivatives with better adsorption efficiency, stability in acids, and enhanced physicochemical properties for removal of toxic dyes from industrial wastewater. With the incorporation of above-mentioned materials, both the mechanical strength and the solubility of chitosan are greatly enhanced. Table 5 summarizes dye removal efficiencies of various CGCs from industrial wastewater under optimum conditions.

8 Conclusions and Future Directions

There is no doubt about it that due to rapid growth of industrialization and urbanization, the necessity of water is tremendously increasing. As already we have known that the entire world is depending on 1% of freshwater of the total water present in the biosphere, so it is very crucial for each one of us to save water for the survival of all living organisms else one day we must have to face a great problem of water scarcity. On the other hand, huge amounts of wastewater generated from various industries are neither properly treated nor always recycled for further operations, rather directly or indirectly disposed into the environment, which in turn, leads to environmental pollution. Although various polymer-based composites are already employed to treat wastewater, due to the involvement of various unwanted chemical processes as well as other chemicals used to fabricate traditional composites are not environmentally friendly. Further, it is to be noted that some polymeric composites neither get degraded nor decomposed naturally and pose a great threat to the environment. In addition, the water bodies such as ponds, lakes, and rivers are greatly polluted with the exposure of this untreated industrial wastewater, which in turn, not only disturbing the whole ecosystems but also causing several diseases due to the presence of toxic elements, compounds, etc., within it. Considering all these issues, researchers are aggressively working toward fabricating green composites for the effective treatment of industrial water because these are eco-friendly, efficient, reliable, and alternative to traditional composites for environmental sustainability. To address the above-mentioned issues, in this book chapter, preparations of various polymers-based green composites for the removal of various toxic metals and dyes from industrial wastewater have been discussed. In addition, a special focus has been made on green composites based on chitosan because of its biocompatibility, biodegradability, non-toxicity, chelating ability, and flocculation activity to name a few for the said purposes. Further, it has been shown that metal as well as dye removal efficiencies of the chitosan-based composites through adsorption depends on the chemical nature of materials and physicochemical experimental conditions such as temperature, pH, initial metal ion concentration of the system. Moreover, it has also been shown that chitosan-based green composites can be reused as adsorbents several times through recycling even after getting exhausted for the said purposes. Hence, due to this added advantage, chitosan-based green composites could be the most promising alternatives to conventional synthetic adsorbents in terms of removing heavy metal ions and dyes from wastewater.

Table 5 Removal of different dyes from industrial wastewater using CGCs

S. No.	Chitosan-based green composites	Form	Name of dyes	Type of dyes	Temperature (°C)	pH	Adsorption capacity (mg g^{-1})	References
1	Chitosan/bentonite	Hybrid form	Amaranth red (AR)	Anionic	25	2	362.1	[141]
			Methylene blue (MB)	Cationic		10	496.5	
2	Chitosan/cellulose	Hydrogels	Congo red (CR)	Anionic	30	3–11	381.7	[142]
3	Chitosan–montmorillonite	Beads	Remazol blue (RB)	Anionic	25	2–8	100–1600	[143]
4	Ethylenediamine-modified magnetic chitosan	Nanoparticles	Acid orange 7	Anionic	25	4	3.47	[144]
			Acid orange 10	Anionic		3	2.25	
5	Glutamine modified chitosan magnetic composite	Microspheres	Acid green	Anionic	25	7	179–652.41	[145]
6	Bentonite/chitosan@cobalt oxide	Nanoparticles	Congo red (CR)	Anionic	27	7	303	[85]
7	Coir pith activated carbon (CPAC), chitosan, and sodium dodecyl sulfate (SDS)	Beads	Malachite green (MG)	Cationic	50	4	6.30	[146]
8	Graphene oxide/chitosan	Sponge	Methylene blue (MB)	Cationic	45	4–8	275.5	[147]

(continued)

Table 5 (continued)

S. No.	Chitosan-based green composites	Form	Name of dyes	Type of dyes	Temperature (°C)	pH	Adsorption capacity (mg g^{-1})	References
9	Chitosan–montmorillonite composite		Nitrazine yellow	Azo dyes (anionic as well as cationic)	30	5	144.4	[148, 149]
			Eriochrome black T	Cationic	25	3	520	
10	Chitosan-g-poly(acrylamide)	Powder	Remazol violet	Azo	30	4	1428	[68]
			Procion yellow	Anionic			1000	

In view of the future of chitosan-based green composites, there are lots of prospects in terms of environmentally friendliness and sustainability. Further, several times these composites can be used with the same efficiency as earlier, which in turn, will not only reduce the manufacturing cost of the composite but also offer us a newer and greener environment. On contrary, the conventional treatment methods used for wastewater require high input cost and are also not reliable. Yet we need to depend to a certain extent on traditional polymeric composites for wastewater treatment unless we are able to largely produce polymer-based green composites, especially chitosan-based ones. It is worth mentioning that the most important challenges lying in the synthesis of green composites are selection of suitable materials, adoption of more accurate processing conditions and parameters, etc. To solve these problems, various possibilities can be explored by not only chemically modifying chitosan and but also subsequently using the chemically modified chitosan for the fabrication of green composites for wastewater treatment. Further, a lot of research is required to produce more efficient, cost-effective chitosan-based green composites to completely substitute the conventional adsorbents. Apart from laboratory research on chitosan and its derivatives for synthesis of green composites, pilot plant is needed to set up in order to manufacture chitosan-based green composites in large scale for their successful commercialization in terms of treating industrial wastewater. However, a future research should mainly focus on scalability and commercial availability of chitosan-based green composites to overcome the current limitations to save the environment from water pollution, which is greatly caused by the direct and indirect disposal of wastewater.

References

1. Mohad UAF, Abu HH, Takriff MS et al (2017) A review of the potentials, challenges and current status of microalgae biomass applications in industrial wastewater treatment. J Water Process Eng 20:8–21
2. William J, Cosgrove DP, Loucks (2015) Water management: current and future challenges and research directions. Water Resour Res 51:4823–4839
3. Agale MC, Patel NG, Patil AG (2013) Impact of sugar industry effluents on the quality of groundwater from Dahiwad Village, District Dhule (M.S). Arch Appl Sci Res 5:58–60
4. Pokhrel D, Viraraghvan T (2004) Treatment of pulp and paper mill wastewater—a review. Sci Total Environ 333:37–58
5. Tariq M, Ali M, Shah Z (2006) Characteristics of industrial effluents and their possible impacts on quality of underground water. Soil Environ 25:64–69
6. Yadav A, Neraliya S, Gopesh A (2007) Acute toxicity levels and ethological responses of Channa striatus to fertilizer industrial wastewater. J Environ Biol 28:159–162
7. Singh U, Singh S, Tiwari RK et al (2018) Water pollution due to discharge of industrial effluents with special reference to Uttar Pradesh, India—a review. Inter Arch Appl Sci Technol 9:111–121
8. Islam BI, Musa AE, Ibrahim EH et al (2014) Evaluation and characterization of tannery wastewater. J Forest Prod Indus 3:141–150
9. Santos A, Teixeira BAN, Agnelli JAM et al (2005) Characterization of effluents through a typical plastic recycling process: an evaluation of cleaning performance and environmental pollution. Resour Conserv Recycl 45:159–171

10. Suleimanov RA (1995) Conditions of waste fluid accumulation at petrochemical and processing enterprises and prevention of their harm to water bodies. Meditsina Trudai Promyshlennaia Ekologiia 12:31–36
11. Manikandan P, Palanisamy PN, Baskar R et al (2015) Physicochemical analysis of textile industrial effluents from Tripur city, TN, India. Int J Adv Res Sci Eng 4:93–104
12. Haneef F (2016) Quantitative assessment of heavy metals in coal-fired power plant waste water. Matter Int J Sci Technol 2:135–149
13. Bharagava RN, Saxena G, Chowdhary P (2017) Constructed wetlands: an emerging phytotechnology for degradation and detoxification of industrial wastewaters. In: Bharagava RN (ed) Environmental pollutants and their bioremediation approaches, 1st edn., pp 397–426
14. Mohan S, Gandhimathi R (2009) Removal of heavy metal ions from municipal solid waste leachate using coal fly ash as an adsorbent. J Hazard Mater 169:351–359
15. Rajasulochana P, Preethy V (2016) Comparison on efficiency of various techniques in treatment of waste and sewage water—a comprehensive review. Resour Efficient Technol 2:175–184
16. Feng X, Chu KH (2004) Cost optimization of industrial wastewater reuse systems. Process Saf Environ 82:249–255
17. Mohammed L, Ansari MNM, Pua G et al (2015) A review on natural fiber reinforced polymer composite and its applications. Int J Polym Sci 2015:1–15
18. Girijappa YGT, Rangappa SM, Parameswaranpillai J et al (2019) Natural fibers as sustainable and renewable resource for development of eco-friendly composites: a comprehensive review. Front Mater 6:1–14
19. Fu SY, Sun Z, Huang P et al (2019) Some of basic aspects of polymer nanocomposites: a critical review. Nano Mater Sci 1:2–30
20. Owusu PA, Asumadu-Sarkoodie S (2016) A review of renewable energy sources, sustainability issues and climate change mitigation. Cogent Eng 3:1–14
21. Satyanarayana KG, Arizaga GGC, Wypych F (2009) Biodegradable composites based on lignocellulosic fibers: an overview. Progress Polym Sci 34:982–1021
22. Bajpai PK, Singh I, Madaan J (2014) Development and characterization of PLA based green composites: a review. J Thermoplast Compos Mater 27:52–81
23. Kim JH, Lee SB, Kim SJ et al (2002) Rapid temperature/pH response of porous alginate-g-poly (*N*-isopropylacrylamide) hydrogels. Polym (Guildf) 43:7549–7558
24. Reddy N, Yang Y (2005) Properties and potential applications of natural cellulose fibres from corn husks. Green Chem 7:190–195
25. Fávaro SL, Lopes MS, de Carvalho Vieira, Neto AG et al (2010) Chemical, morphological, and mechanical analysis of rice husk/post-consumer polyethylene composites. Compos Part A Appl Sci Manuf 41:154–160
26. Pfister DP, Larock RC (2010) Green composites from a conjugated linseed oil-based resin and wheat straw. Appl Sci Manuf 41:1279–1288
27. Ahankari SS, Mohanty AK, Misra M (2011) Mechanical behaviour of agro-residue reinforced poly(3-hydroxybutyrate-co-3-hydroxyvalerate), (PHBV) green composites: a comparison with traditional polypropylene composites. Compos Sci Technol 71:653–657
28. Ashori A, Nourbakhsh A (2010) Bio-based composites from waste agricultural residues. J Waste Manag 3:680–684
29. Kandpal BC, Chaurasia R, Khurana V (2015) Recent advances in green composites—a review. Glob J Pharmacol 267–271
30. Zhang YR, Shen SL, Wang SQ et al (2014) A dual function magnetic nanomaterial modified with lysine for removal of organic dyes from water solution. Chem Eng J 239:250–256
31. Priya BSK, Shanker U, Gupta B et al (2018) RSM-CCD optimized in air synthesis of photocatalytic nanocomposite: application in removal-degradation of toxic brilliant blue. React Funct Polym 131:107–122
32. Ghaedi M, Karimi F, Barazesh B (2013) Removal of reactive orange 12 from aqueous solutions by adsorption on tin sulfide nanoparticle loaded on activated carbon. J Ind Eng Chem 19:756–763

33. Daraei P, Madaeni SS, Salehi E et al (2013) Novel thin film composite membrane fabricated by mixed matrix nanoclay/chitosan on PVDF microfiltration support: preparation, characterization and performance in dye removal. J Membr Sci 436:97–108
34. Karim Z, Mathew AP, Mouzon J et al (2014) Nanoporous membranes with cellulose nanocrystals as functional entity in chitosan: removal of dyes from water. Carbohydr Polym 112:668–676
35. Luo S, Netravali AN (1999) Interfacial and mechanical properties of environment-friendly "green" composites made from pineapple fibers and poly(hydroxybutyrate-co-valerate) resin. J Mater Sci 34:3709–3719
36. Yu L, Dean K, Li L (2006) Polymer blends and composites from renewable resources. Prog Polym Sci 31:576–602
37. Thakur M, Sharma G, Ahamad T (2017) Efficient photocatalytic degradation of toxic dyes from aqueous environment using gelatin-Zr (IV) phosphate nanocomposite and its antimicrobial activity. Colloids Surf B Biointerfaces 157:456–463
38. Benhouria, A, Islam MA, Zaghouane-Boudiaf (2015) Calcium alginate–bentonite–activated carbon composite beads as highly effective adsorbent for methylene blue. Chem Eng J 270:621–630
39. Sharma G, Thakur B, Naushad M et al (2018) Applications of nanocomposite hydrogels for biomedical engineering and environmental protection. Environ Chem Lett 16:113–146
40. Bankole MT, Abdulkareem AS, Mohammed IA et al (2019) Selected heavy metals removal from electroplating waste-water by purified and polyhydroxylbutyrate functionalized carbon nanotubes adsorbents. Sci Rep 9:1–19
41. Shukla SK, Singh NB, Rastog RP (2013) Efficient ammonia sensing over zinc oxide/polyaniline nanocomposite. Ind J Eng Mater Sci 20:319–324
42. Cantarella M, Sanz R, Buccheri MA (2016) PMMA/TiO_2 nanotubes composites for photocatalytic removal of organic compounds and bacteria from water. Mater Sci Semicond Process 42:58–61
43. Keshtkar AR, Irani M, Moosavian MA (2013) Removal of uranium (VI) from aqueous solutions by adsorption using a novel electrospun PVA/TEOS/APTES hybrid nanofiber membrane: comparison with casting PVA/TEOS/APTES hybrid membrane. J Radioanal Nucl Chem 295:563–571
44. Borai EH, Breky MME, Sayed MS (2015) Synthesis, characterization and application of titanium oxide nanocomposites for removal of radioactive cesium, cobalt and europium ions. J Colloid Interface Sci 450:17–25
45. Park Y, Lee YC, Shin et al (2010) Removal of cobalt, strontium and cesium from radioactive laundry wastewater by ammonium molybdophosphate–polyacrylonitrile (AMP–PAN). Chem Eng J 162:685–695
46. Priya Sharma, Kaith AK et al (2019) RSM-CCD optimized sodium alginate/gelatin based ZnS-nanocomposite hydrogel for the effective removal of biebrich scarlet and crystal violet dyes. Int J Biol Macromol 129:214–226
47. Alosaimi AM, Hussein MA, Abdelaal MY et al (2019) Polysulfone/wood flour/organoclay hybrid nanocomposites as efficient eco-friendly materials. Compos Interfaces. https://doi.org/10.1080/09276440.2019.1692615
48. Liu C, Omer AM, Kun OX (2018) Adsorptive removal of cationic methylene blue dye using carboxymethyl cellulose/k-carrageenan/activated montmorillonite composite beads: Isotherm and kinetic studies. Int J Biol Macromol 106:823–833
49. Zhou Y, Hu Y, Huang W et al (2018) A novel amphoteric β-cyclodextrin-based adsorbent for simultaneous removal of cationic/anionic dyes and bisphenol A. Chem Eng J 341:47–57
50. Dargahi M, Ghasemzadeh H, Bakhtiary A (2018) Highly efficient absorption of cationic dyes by nano composite hydrogels based on κ-carrageenan and nano silver chloride. Carbohydr Polym 181:587–595
51. Jiang J, Zhang Q, Zhan X et al (2019) A multifunctional gelatin-based aerogel with superior pollutants adsorption, oil/water separation and photocatalytic properties. Chem Eng J 358:1539–1551

52. Chen Y, Ma Y, Lu W (2018) Environmentally friendly gelatin/β-cyclodextrin composite fiber adsorbents for the efficient removal of dyes from wastewater. Molecules 23:1–17
53. Wang Y, Zeng L, Ren X et al (2010) Removal of methyl violet from aqueous solutions using poly (acrylic acid-co-acrylamide)/attapulgite composite. J Environ Sci 22:7–14
54. Kumar R, Sharma RK, Singh AP (2018) Removal of organic dyes and metal ions by cross-linked graft copolymers of cellulose obtained from the agricultural residue. J Environ Chem Eng 6:6037–6048
55. Zhu W, Liu L, Liao Q et al (2016) Functionalization of cellulose with hyper branched polyethylenimine for selective dye adsorption and separation. Cell 6:3785–3797
56. Barakat MA (2011) New trends in removing heavy metals from industrial wastewater. Arab J Chem 4:361–377
57. Salam OEA, Reiad NA, Shafei EI et al (2011) A study of the removal characteristics of heavy metals from wastewater by low-cost adsorbents. J Adv Res 2:297–303
58. Gupta KC, Ravikumar MNV (2000) Preparation characterization and release profiles of pH sensitive chitosan beads. Polym Int 49:141–146
59. Bolto BA (1995) Soluble polymers in water purification. Prog Polym Sci 20:987–1041
60. Muzzarelli RAA, Boudrant J, Meyer D (2012) Current views on fungal chitin/chitosan, human chitinases, food preservation, glucans, pectins and inulin: a tribute to Henri Braconnot, precursor of the carbohydrate polymers science, on the chitin bicentennial. Carbohydr Polym 87:995–1012
61. Sudha PN (2010) In: Kim SK (ed) Chitin, chitosan, oligosaccharides and their derivatives, 1st edn. CRC Press, Boca Raton, pp 561–585
62. Sharma G, Naushad M, Al-Muhtaseb AH et al (2017) Fabrication and characterization of chitosan-crosslinked-poly (alginic acid) nanohydrogel for adsorptive removal of Cr(VI) metal ion from aqueous medium. Int J Biol Macromol 95:484–493
63. Sharma G, Thakur B, Naushad M (2017) Fabrication and characterization of sodium dodecyl sulphate@ ironsilicophosphate nanocomposite: ion exchange properties and selectivity for binary metal ions. Mater Chem Phys 193:129–139
64. Claude B, Vironâ-Lamy C, Haupt K (2010) Synthesis of a molecularly imprinted polymer for the solid phase extraction of betulin and betulinic acid from plane bark. Phytochem Analysis 2:180–185
65. Fan L, Luo C, Lv Z et al (2011) Removal of Ag + from water environment using a novel magnetic thiourea-chitosan imprinted Ag^+. J Hazard Mater 194:193–201
66. Kannusamy P, Sivalingam T (2013) Synthesis of porous chitosan-polyaniline/ZnO hybrid composite and application for removal of reactive orange 16 dye. Colloid Surf B 108:229–238
67. Ghaee A, Niassar MS, Barzin J et al (2015) Preparation of chitosan/cellulose acetate composite nanofiltration membrane for wastewater treatment. Desalin Water Treat 57:14453–14460
68. Singh V, Sharma AK, Tripathi DN et al (2009) Poly(methylmethacrylate) grafted chitosan: an efficient adsorbent for anionic azo dyes. J Hazard Mater 161:955–966
69. Vijaya Y, Popuri SR, Boddu VM et al (2008) Modified chitosan and calcium alginate biopolymer sorbents for removal of nickel (II) through adsorption. Carbohydr Polym 72(2):261–271
70. Wan WS, Ngah FS (2008) Adsorption of Cu(II) ions in aqueous solution using chitosan beads, chitosan–GLA beads and chitosan–alginate beads. Chem Eng J 143:62–72
71. Chang MY, Juang RS (2004) Adsorption of tannic acid, humic acid and dyes from water using the composite of chitosan and activated clay. J Colloid Interface Sci 278:18–25
72. Guibal E, Charrier MJ, Saucedo I et al (1995) Enhancement of metal ion sorption performances of chitosan: effect of the structure on the diffusion properties. Langmuir 11:591–598
73. Rorrer GL, Hsien TY, Way JD (1993) Synthesis of porous-magnetic chitosan beads for removal of cadmium ions from wastewater. Ind Eng Chem Res 32:2170–2178
74. Hsieh CT, Teng H (2000) Influence of mesopore volume and adsorbate size on adsorption capacities of activated carbons in aqueous solutions. Carbon 38:863–869
75. Dentel SK, Jamrah AI, Sparks DL (1998) Sorption and cosorption of 1,2,4-trichlorobenzene and tannic acid by organo-clays. Water Res 32:3689

76. Jekel MR (1991) Particle stability in the presence of pre-ozonated humic acids. J Water Srt Aqua 40:18–24
77. Netpradita S, Thiravetyanb P, Towprayoona S (2003) Application of 'waste' metal hydroxide sludge for adsorption of azo reactive dyes. Water Res 37:763–772
78. Panswad T, Luangdilok W (2000) Decolorization of reactive dyes with different molecular structures under different environmental conditions. Water Res 34:4177–4184
79. Wu FC, Tseng RL, Juang RS (2001) Kinetic modeling of liquid-phase adsorption of reactive dyes and metal ions on chitosan. Water Res 35:613–618
80. Gecol H, Ergican E, Miakatsindila P (2005) Biosorbent for tungsten species removal from water: effects of co-occurring inorganic species. J Colloid Interface Sci 292:344–353
81. Alcântara ACS, Darder M, Aranda P et al (2004) Polysaccharide-fibrous clay bionanocomposites. Appl Clay Sci 96:2–8
82. Holzer L, Münch B, Rizzi M et al (2010) 3D-microstructure analysis of hydrated bentonite with cryo-stabilized pore water. Appl Clay Sci 47:330–342
83. Li Q, Yue QY, Sun HJ et al (2010) A comparative study on the properties, mechanism and process designs for the adsorption of non-ionic or anionic dyes onto cationic-polymer/bentonite. J Environ Manag 91:1601–1611
84. Wei JM, Zhu RL, Zhu JX et al (2009) Simultaneous sorption of crystal violet and 2-naphthol to bentonite with different CECs. J Hazard Mater 166:195–199
85. Abukhadra MRA, Alyaa B, Belal M (2019) Green fabrication of bentonite/chitosan@cobalt oxide composite (BE/CH@Co) of enhanced adsorption and advanced oxidation removal of Congo red dye and Cr (VI) from water. Int J Biol Macromol 126:402–413
86. Liu Q, Yang B, Zhang L et al (2015) Adsorption of an anionic azo dye by cross-linked chitosan/bentonite composite. Int J Biol Macromol 72:1129–1135
87. Wan NWS, Teong LC, Hanafiah MAKM (2011) Adsorption of dyes and heavy metal ions by chitosan composites: a review. Carbohydr Polym 83:1446–1456
88. Anirudhan TR, Rijith S, Tharun AR (2010) Adsorptive removal of thorium(IV) from aqueous solutions using poly(methacrylic acid)-grafted chitosan/bentonite composite matrix: process design and equilibrium studies. Colloid Surf A 368:13–22
89. Anirudhan TS, Rijith S, Nair RVR (2012) A highly efficient carboxyl-terminated hybrid adsorbent composite matrix for the adsorption of uranium(VI) and thorium(IV) from aqueous solutions and nuclear industry effluents. Desalin Water Treat 38:79–89
90. Zhao B, Zhang X, Dou C (2015) Adsorption property of methyl orange by chitosan coated on quartz sand in batch mode. Desalin Water Treat 55:1598–1608
91. Omar H, Arida H, Daifullah A (2009) Adsorption of 60Co radionuclides from aqueous solution by raw and modified bentonite. Appl Clay Sci 44:21–26
92. Li J, Wen F, Pan LS et al (2013) Removal of radiocobalt ions from aqueous solutions by natural halloysite nanotubes. J Radioanal Nucl Chem 295:431–438
93. Ren XM, Yang ST, Tan XL et al (2012) Investigation of radionuclide $^{60}Co(II)$ binding to TiO_2 by batch technique, surface complexation model and DFT calculations. Sci China-Chem 55:1752–1759
94. Zhao DL, Wang Y, Xuan H et al (2013) Removal of radiocobalt from aqueous solution by Mg2Al layered double hydroxide. J Radioanal Nucl Chem 295:1251–1259
95. Olgun A, Atar N (2011) Removal of copper and cobalt from aqueous solution onto waste containing boron impurity. Chem Eng J 167:140–147
96. Dinu MV, Dragan ES (2010) Evaluation of Cu^{2+}, Co^{2+} and Ni^{2+} ions removal from aqueous solution using a novel chitosan/clinoptilolite composite: kinetics and isotherms. Chem Eng J 160:157–163
97. Yaseen DA, Scholz M (2019) Textile dye wastewater characteristics and constituents of synthetic effluents: a critical review. Int J Environ Sci Technol 16:1193–1226
98. Hamzah S, Salleh MFM (2014) Hydroxyapatite/chitosan biocomposite for remazol blue dyes removal. Appl Mech Mater 695:106–109
99. Gautam RK, Sharma SK, Mahiya S et al (2015) Contamination of heavy metals in aquatic media: transport, toxicity and technologies for remediation, pp 1–24

100. Zhang L, Zhao B, Xu G et al (2018) Characterizing fluvial heavy metal pollutions under different rainfall conditions: implication for aquatic environment protection. Sci Total Environ 635:1495–1506
101. Ghasemi M, Naushad M, Ghasemi N et al (2014) Adsorption of Pb(II) from aqueous solution using new adsorbents prepared from agricultural waste: adsorption isotherm and kinetic studies. J Ind Eng Chem 20:2193–2199
102. Abdolali A, Ngo HH, Guo W et al (2016) A breakthrough biosorbent in removing heavy metals: equilibrium, kinetic, thermodynamic and mechanism analyses in a lab-scale study. Sci Total Environ 542:603–611
103. Wuana RA, Okieimen FE (2011) Heavy metals in contaminated soils: a review of sources, chemistry, risks and best available strategies for remediation. ISRN Ecol 2011:1–20
104. Sharma B, Singh S, Siddiqi NJ (2014) Biomedical implications of heavy metals induced imbalances in redox systems. Biomed Res Int 2014:1–26
105. Lauwerys R, Roels H, Buchet JP et al (1983) The influence of orally administered vitamin C or zinc on the absorption of and the biological response to lead. J Occup Environ Med 25:668–678
106. Wang Y, Fang J, Leonard SS et al (2004) Cadmium inhibits the electron transfer chain and induces reactive oxygen species. Free Radical Biol Med 36:1434–1443
107. Jaishankar M, Tseten T, Anbalagan N et al (2014) Toxicity, mechanism and health effects of some heavy metals. Interdisc Toxicol 7:60–72
108. Sharma RK, Agarwal M (2005) Biological effects of heavy metals: an overview. J Environ Biol 26:301–313
109. Abbaspoue N, Hurrell R, Kelishadi R (2014) Review on iron and its importance for human health. J Res Med Sci 19:164–174
110. Chiban M, Zerbet M, Carja G et al (2012) Application of low-cost adsorbents for arsenic removal: a review. J Environ Chem Ecotoxicol 4:91–102
111. Liu XW, Hu QY, Fang Z et al (2009) Magnetic chitosan nanocomposites: a useful recyclable took for heavy metal ion removal. Langmuir 25:3–8
112. Hickner MA, Herring AM, Coughlin A (2014) Anion exchange membranes: current status and moving forward. J Polym Sci Part B Polym Phys 51:1727–1735
113. Ferrero F, Tonetti C, Periolatto M (2014) Adsorption of chromate and cupric ions onto chitosan-coated cotton gauze. Carbohydr Polym 110:367–373
114. Liu Q, Yang B, Zhang L et al (2014) Simultaneous adsorption of aniline and Cu^{2+} from aqueous solution using activated carbon/chitosan composite. Desalin Water Treat 55:410–419
115. Gokila S, Gomathi T, Sudha PN et al (2017) Removal of the heavy metal ion chromium(VI) using chitosan and alginate nanocomposites. Int J Biol Macromol 104:1459–1468
116. Li K, Li P, Cai J (2016) Efficient adsorption of both methyl orange and chromium fro their aqueous mixture using a quaternary ammonium salt modified chitosan magnetic composite adsorbent. Chemosphere 154:310–318
117. Habiba U, Afifi AM, Salleh A (2017) Chitosan/(polyvinyl alcohol)/zeolite electrospun composite nanofibrous membrane for adsorption of Cr^{6+}, Fe^{3+} and Ni^{2+}. J Hazard Mater 322:182–194
118. Hu C, Zhu P, Cai M et al (2017) Comparative adsorption of Pb(II), Cu(II) and Cd(II) on chitosan saturated montmorillonite: kinetic, thermodynamic and equilibrium studies. Appl Clay Sci 143:320–326
119. Karthik R, Meenakshi S (2015) Removal of Pb(II) and Cd(II) ions from aqueous solution using polyaniline grafted chitosan. Chem Eng J 263:168–177
120. Pandey S, Tiwari S (2015) Facile approach to synthesize chitosan based composite-characterization and cadmium(II) ion adsorption studies. Carbohydr Polym 134:646–656
121. Zhao Y, Zhao X, Deng J et al (2015) Utilization of chitosan–clinoptilolite composite for the removal of radiocobalt from aqueous solution: kinetics and thermodynamics. J Radioanal Nucl Chem 695:106–109
122. Zhang L, Pan H, Hua HR et al (2016) Crosslinked quaternized chitosan/bentonite composite for the removal of amino black 10B from aqueous solutions. Int J Biol Macromol 93:217–225

123. Mallakpour S, Khadem E (2018) Chitosan/CaCO3-silane nanocomposites: synthesis, characterization, in vitro bioactivity and Cu(II) adsorption properties. Int J Biol Macromol 114:149–160
124. Li L, Iqbal J, Zhu Y et al (2018) Chitosan/Ag-hydroxyapatite nanocomposite beads as a potential adsorbent for the efficient removal of toxic aquatic pollutant. Int J Biol Macromol 120:1752–1759
125. Wu S, Wang F, Yaun H et al (2018) Fabrication of xanthate-modified chitosan/poly (*N*-isopropylacrylamide) composite hydrogel for the selective adsorption of Cu (II), Pb (II) and Ni (II) metal ions. Chem Eng Res Des 139:197–210
126. Samandari SS, Samandari SS, Nezafati N et al (2014) Efficient removal of lead (II) ions and methylene blue from aqueous solution using chitosan/Fe-hydroxyapatite nanocomposite beads. J Environ Manag 146:481–490
127. Le VT, Doan VD, Nguyen DD et al (2018) A novel cross-linked magnetic hydroxyapatite/chitosan composite: preparation, characterization, and application for Ni(II) ion removal from aqueous solution. Water Air Soil Pollut 229:101–114
128. Tang H, Chang C, Zhang L (2011) Efficient adsorption of Hg2+ ions on chitin/cellulose composite membranes prepared via environmentally friendly pathway. Chem Eng J 173:689–697
129. Sundaram CS, Viswanathan N, Meenakshi S (2007) Defluoridation chemistry of synthetic hydroxyapatite at nano scale: equilibrium and kinetic studies. J Hazard Mater 155:206–215
130. Kousalya GN, Rajiv Gandhi M, Meenakshi S (2010) Removal of toxic Cr(VI) ions from aqueous solution using nano-hydroxyapatite-based chitin and chitosan hybrid composites. Adsorp Sci Technol 28:49–64
131. Mohammad AM, Salah Eldin TA, Hassan MA et al (2015) Efficient treatment of lead-containing wastewater by hydroxyapatite/chitosan nanostructures. Arabian J Chem 10:683–690
132. Pandi K, Viswanathan N (2015) Synthesis and applications of eco-magnetic nano-hydroxyapatite chitosan composite for enhanced fluoride sorption. Carbohydr Polym 134:732–739
133. Salah TA, Mohammad AM, Hassan MA et al (2014) Development of nano-hydroxyapatite/chitosan composite for cadmium ions removal in wastewater treatment. J Taiwan Inst Chem Eng 45:1571–1577
134. Sundaram CS, Viswanathan N, Meenakshi S (2008) Uptake of fluoride by nano-hydroxyapatite/ chitosan, a bioinorganic composite. Bio Res Technol 99:8226–8230
135. Katheresan V, Kansedo J, Lau SY (2018) Efficiency of various recent wastewater dye removal methods: a review. J Environ Chem Eng 6:4676–4697
136. Piątkowski M, Janus Ł, Radwan-Pragłowska J et al (2017) Microwave-enhanced synthesis of biodegradable multifunctional chitosan hydrogels for wastewater treatment. Express Polym Lett 11:809–819
137. Wang JL, Xu LJ (2012) Advanced oxidation processes for wastewater treatment: formation of hydroxyl radical and application. Crit Rev Environ Sci Technol 42:251–325
138. Kandisa RV, Saibaba NVR, Shaik KB (2016) Dye removal by adsorption, a review. J Bioremediat Biodegrad 7:371–374
139. Pirkarami A, Olya ME (2017) Removal of dye from industrial waste- water with an emphasis on improving economic efficiency and degradation mechanism. J Saudi Chem Soc 21:S179–S186
140. Fan L, Luo C, Sun M et al (2012) Preparation of novel magnetic chitosan/ graphene oxide composite as effective adsorbents toward methylene blue. Bioresour Technol 114:703–706
141. Dotto GL, Rodrigues FK, Tanab EH et al (2016) Development of chitosan/bentonite hybrid composite to remove hazardous anionic and cationic dyes from colored effluents. J Environ Chem Eng 4:3230–3239
142. Wang Y, Wang H, Peng H et al (2018) Dye adsorption from aqueous solution by cellulose/chitosan composite: equilibrium, kinetics, and thermodynamics. Fibre Polym 19:340–349

143. Pereira FAR, Sousa KS, Cavalcanti GRS (2017) Green biosorbents based on chitosan-montmorillonite beads for anionic dye removal. J Environ Chem Eng 5:3309–3318
144. Zhou L, Jin J, Liu Z et al (2011) Adsorption of acid dyes from aqueous solutions by the ethylenediamine-modified magnetic chitosan nanoparticles. J Hazard Mater 185:1045–1052
145. Tao X, Li K, Yan H et al (2016) Simultaneous removal of acid green 25 and mercury ions from aqueous solutions using glutamine modified chitosan magnetic composite microspheres. Environ Pollut 209:21–29
146. Arumugam TK, Krishnamoorthy P, Rajagopalan NR et al (2019) Removal of malachite green from aqueous solutions using a modified chitosan composite. Int J Biol Macromol 128:655–664
147. Qi C, Zhao L, Lin Y et al (2018) Graphene oxide/chitosan sponge as a novel filtering material for the removal of dye from water. J Colloid Interface Sci 517:18–27
148. Oladipo AA, Gazi M, Yilmaz E (2015) Single and binary adsorption of azo and anthraquinone dyes by chitosan-based hydrogel: selectivity factor and Box-Behnken process design. Chem Eng Res Des 104:264–279
149. Ngwabebhoh FA, Erdem A, Yildiz Y et al (2016) Synergistic removal of Cu(II) and nitrazine yellow dye using an eco-friendly chitosan-montmorillonite hydrogel: optimization by response surface methodology. J Appl Polym Sci 133:1–14

Chapter 22
Molecular Imprinted Nanocomposites for Green Chemistry

Monireh Bakhshpour, Sevgi Aslıyüce, Neslihan Idil, Bo Mattiasson, and Adil Denizli

1 Nanocomposites

In the middle of an exciting research area, toward a technological future, a new attempt was made by scientists who came together from different disciplines to form 'Nanotechnology' (Greek; nanos = dwarf) [1] exploiting the nanotechnology. Improved properties are obtained by making some changes in the basic structure of material via reducing the dimensions of the material. Nano-sized materials could be prepared by changing their biological and physical properties as well as their chemical properties [2]. The basic principle relies on the atoms, which are the building blocks of material. It offers advantages for developing special materials by arranging the components. In this way, cost will decrease with the increase in quality. The purpose of these materials in the field of technology is low cost, high performance to make products and make great progress in the industry [3]. The other well-defined features of nanocomposites are mechanical, electrical, barrier properties and thermal stability.

Nanocomposites are a class of composites. These reinforcing materials have dimensions in nanometer size in the range of 1–1000 nm, composed of ultra-fine inorganic particles. Nanocomposites have superior properties over conventional composites due to their nanometer size characteristics, maximizing interface adhesion [4]. Nanocomposites enable the interaction with the increasing surface area. In conventional composite materials, volumetric interaction is less due to micro size particles.

M. Bakhshpour · S. Aslıyüce · A. Denizli (✉)
Biochemistry Division, Department of Chemistry, Hacettepe University, Ankara, Turkey
e-mail: denizli@hacettepe.edu.tr

N. Idil
Biotechnology Division, Department of Biology, Hacettepe University, Ankara, Turkey

B. Mattiasson
Department of Biotechnology, Lund University, Lund, Sweden

Indienz AB, Annebergs Gård, Billeberga, Lund, Sweden

S. Thomas and P. Balakrishnan (eds.), *Green Composites*,
Materials Horizons: From Nature to Nanomaterials,
https://doi.org/10.1007/978-981-15-9643-8_22

Nanomaterials have a very strong and high elastic modulus. They are also quite resistant to degradation [5]. They could be twisted, straightened, they can curl into small circles, so as a result of various stretches, they can stay unbroken. On the other hand, due to the ion-exchange clay in the structure of nanocomposite materials, their electrical conductivity will be high. When compared with the feature of conventional polymers and barrier of nanocomposite materials containing polymer–clay refers to the transition of matter through the polymer, it is easier to pass the matter through conventional polymers. However, this process has been found to face obstacles in nanocomposite material because of silicate layers. Above the glass transition temperature, the polymer chain bends, and its geometry is distorted. But in nanocomposite materials, glass transition temperature is higher than for pure polymers. For this reason, nanocomposites are more stable even at higher temperatures [6].

2 Green Chemistry

Green chemistry known as sustainable chemistry is referred to as the generation of chemical products and processes aiming for the reduction of environmental impact of the chemical processes. This approach relies on the application of materials which are non-toxic to living organisms and the environment. The principles of this technology are based on the life cycle of a chemical product including its synthesis, processing and ultimate disposal [7].

Using polymeric composites from renewable sources as a solution to the environmental challenges especially originating from plastic wastes has advantages over synthetic resources. From the point of this view, it offers some alternatives for maintaining the sustainable improvement of economically and ecologically desirable strategies. Green composites have presently gained great attention because of developing materials from biodegradable polymers, preserving fossil-based raw materials, obtaining fully biodegradability and decreasing the amount of carbon dioxide emission into the atmosphere. On the other hand, the implementation of agricultural resources including both waste and products performed to establish green resources is another reason green composite have received magnifical research interests. It is expected that applications of green composites enable the increase of speed of production and recycling with the increase of environmental compatibility [4].

Biodegradation refers to the degradation of the polymer in its natural environment, including alterations in the chemical, mechanical and structural properties, and conversion to other environmentally beneficial compounds. Natural-based polymers such as starch, lignin, cellulose acetate, poly-lactic acid (PLA), polyhydroxylakanoates (PHA), polyhydroxylbutyrate (PHB) and some synthetic sources (aliphatic and aromatic polyesters, polyvinyl alcohol, modified polyolefins) can be grouped as soluble biopolymers. Synthetic polymers cannot be renewed and do not fully comply with the concept of renewability and degradability. This is per se a challenge that today attracts a lot of interest. Degradation of petro-based polymers is important research. Both bio-based and petro-based polymeric structures play crucial roles in the construction of green composites [4].

3 Molecular Imprinting in Green Chemistry

Sorbents prepared by molecular imprinting can be reused in many applications which have effective properties in green chemistry so as to minimize keeping the solid waste generation. This is important for environmental protection because MIPs are highly stable against harsh conditions such as organic and inorganic solvents, high pressures and ambient temperatures. In the view of the process for producing MIP sorbents, sample preparation must comply with green chemistry principles. These principles can be explained by following directions[8]:

1. Analytical data have to be obtained on site.
2. Reagents need to be reduced during synthesis.
3. Miniature methods are superior over conventional techniques.
4. Waste needs to be minimized.
5. Multiple analyte analysis, which can also be translated into multi-template MIPs, is preferred over single analyte or single-template MIPs.
6. The usage of toxic chemicals should be prevented.

These principles can be combined with a molecular imprinting approach. Even though many attempts have been made in the preparation of synthetic methodology in recent years, molecular imprinting includes a high amount of solvents and various preparation steps [8].

Alternative solvents have been preferred in the preparation of MIPs. Besides that, cost-effectiveness and rapid synthesis of MIPs are the other important features. MIPs can be produced by creating a specific interaction between the template molecule and the functional monomers in the presence of a crosslinking agent and solvent. The resultant recognition regions are formed by removal of the template molecules whereby the recognition region is complementary to this template in size, shape and chemical functionality. The schematic presentation of molecular imprinting is shown in Fig. 1. Therefore, the specific binding sites were formed for the selective recognition of target molecules from any sample matrix. The interactions can be obtained by different mechanisms such as weak non-covalent hydrogen bonding interactions, ion-pairing, hydrophobic or dipolar interactions. Non-covalent, covalent and semi-covalent interactions can take place. Bulk, epitope, microcontact, precipitation, surface, suspension, gelation are the imprinting techniques frequently preferred for the development of MIPs. The synthesized material can be observed in different sizes changing from nano- to micron-size and prepared in various forms such as membranes, fibers, core–shell particles [9].

In recent years, MIPs have received great interest in many fields such as solid phase extraction, chromatographic separations, sensors. MIPs are well-established rigid materials having several superior properties over natural affinity molecules.

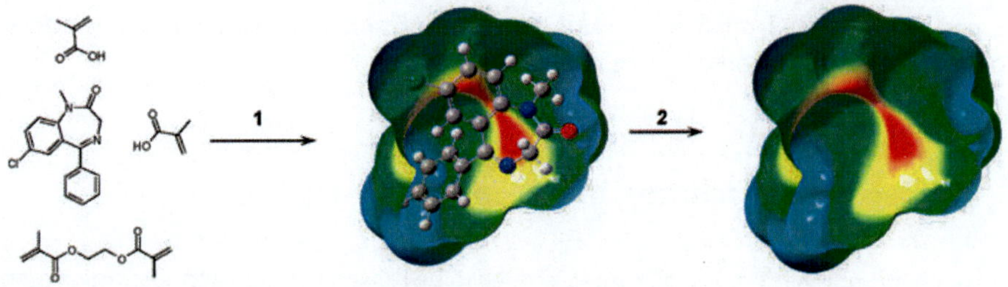

Fig. 1 Schematic presentation of the molecular imprinting of diazepam (a drug used to treat anxiety). Specific binding sites are generated using methacrylic acid as a functional monomer. After polymerization (step 1) and removal of the template (step 2), binding sites containing carboxyl groups are left in the polymer matrix [8].

The other property that has to be highlighted in terms of green chemistry is their reusability [10].

The MIP optimization process has a crucial role to obtain effective polymeric structures. Solvents are one of the most important components of the polymerization process. Water is an indispensable alternative, but using it leads to create robust interactions with the template and/or the monomers. It affects the stabilization of the complex and is involved in the production of the recognition regions. Application of green solvents can meet some requirements and many of the handicaps of conventional MIP designing platforms.

Bulk imprinting is based on the addition of template microorganism as a whole to the pre-polymerization mixture. After the polymerization step, the removal of template molecules from the polymeric structure is required. Therefore, recognition cavities with the shape and size complementary were formed in the polymer matrix. But conventional bulk imprinting approach is more preferable for the applications of small molecules [9, 11].

Surface imprinting is today a viable strategy providing the formation of thin polymeric films having template imprinted regions. Large biomolecules and even microorganisms could be readily imprinted with the advantage of efficient binding and removal steps. In this approach, the template fingerprint is produced by the stamping method on the transducer, whereby a more stable sensing affinity surface was generated. It is noteworthy to highlight that surface imprinting ensures an applicable route to detect microorganisms and macromolecules efficiently [9, 12].

Microcontact imprinting has been presented as an alternative strategy to form well-defined polymeric MIPs for relatively large molecules. A polymer layer was synthesized between the surface of the sensor and a template stamp and in a polymerization process. Microcontact imprinting has gained great interest since the printing structure is never wholly entrapped in the matrix along with both enabling the removal of the template and rebinding the target molecule [9, 12].

Epitope imprinting relies on the imprinting of a particular component of the molecule. The most significant feature has to be indicated to facilitate the formation of specific interactions using specific parts of the microorganisms. Therefore, definitely decreased non-specific binding resulted in obtaining improved affinity [12].

In green chemistry, it was indicated that the MIP applications are based on the use of less toxic molecules. Deep eutectic solvents can be used due to environmental friendly properties. It is noteworthy to mention that such eutectic solvents are shown to have superior features over them from commonly preferred ionic liquids. Their biodegradability, non-toxicity and low-cost characteristics provide several advantages in line with green technology [8].

Ionic liquids have been explored as both environmentally friendly reagents standing in front of porogenic solvents in molecular imprinting. The solvent remains trapped in the MIP, especially in bulk type. It is evaporated during drying and therefore causing atmospheric fouling. [8]. Since then, the potential of some ionic liquids as greener replacements to organic solvents has been applied [13].

4 Green Nanocomposite Materials

4.1 Cellulose-Based Nanocomposite

Cellulose nanofibers are emerging attractive biomaterials with high biocompatibility, surface/volume ratio, tensile strength, low-density, low-cost, non-abrasive, non-toxic, combustible and biodegradable properties. Cellulose is found as a most abundant biopolymer in nature. The cellulose nanofibers can improve the barrier and mechanical properties of the nanocomposite materials. Cellulose nanofibers can be mostly obtained simply via non-plant resources (bacteria), plant resources (wood, cotton, etc.), or polymerization procedures for many wide applications.

Bacterial cellulose is synthesized by certain species of bacteria via *Acetobacter xylinum* under special culturing conditions [14]. These bacterial cellulose nanofibers due to their structural and morphological properties with good mechanical strength and biodegradability have shown unique potential as native form or with various modifications using polymers, micro/nanoparticles and small molecules for several application areas. The thickness of bacterial celluloses is around 5–500 nm in diameter, and they are two forms of nanofibers, nano-whiskers and microfibril [15]. Here, the applicability and efficiency of the bacterial cellulose nanofibers and their composites focusing on the desired characteristics of bacterial cellulose are discussed in the fields of biomedical. The unique properties of nanofiber composite qualify it to be used for biomedical applications. The biocompatibility character of nanofiber makes it suitable to function appropriately to procreate the desired clinical outcome in the human body [16], without causing adverse effects. The nanocomposite cellulose is a versatile choice for biomedical applications. The use of versatile and adaptive nanocomposite cellulose-based polymers has recently been increased for recognition materials in the clinical area. The Web-like structure and high surface area of nanofibers have drawn attention to use as alternative nanocomposite materials with molecular imprinting technology. These unique properties made nanocomposite nanofibers as green alternative materials for using depletion, purification, separation of proteins and in drug delivery systems [16, 17].

Here, we discuss various kinds of polymeric nanocomposite nanofibers in studies for protein recognition. In a study, a novel fabrication technique with nanocomposite nanofibers was used for recognition of hemoglobin imprinted into the polymeric part of the nanofiber. The polymeric nanocomposite was synthesized by surface imprinting method using amino acid-based *N*-methacryloyl-(L)-histidinemethylester (MAH) functional monomers. The nanocomposite imprinted nanofibers were used for selective binding of hemoglobin in aqueous solution samples and hemolysate.

The characterization studies of these nanofibers were obtained by Fourier transform infrared spectrophotometer (FT-IR), scanning electron microscopy (SEM), micro-computed tomography (μCT) and atomic force microscopy (AFM) that is shown in Fig. 2. According to the results, these promising nanofibers can be used for purification of hemoglobin with significant selectivity, high adsorption capacity and reusability [18].

In the other study, a thin imprinted nanocomposite polymeric nanofiber for the removal of human serum albumin was obtained. The composite nanofiber materials were prepared using albumin as a template molecule and amino acid-based functional monomer for the selective depletion of albumin. The polymerization had occurred under constant stirring. As a magnification role of molecular imprinting, the selectivity of the nanofibers was carried out by studying potential binding of myoglobin and transferrin. Overall, the relative selectivity coefficients were reported 3.02 and 4.73 for albumin/myoglobin and albumin/transferrin, respectively. The photograph of two-dimensional Sodium Dodecyl Sulfate–Polyacrylamide Gel Electrophoresis (SDS-PAGE) of the depletion of albumin is shown in Fig. 3 [19].

In other studies, nanofibers were used to prepare lysozyme (Lyz) imprinted composite nanofibers. The surface imprinting strategy was used onto the surface of fibers matrix in the presence of Lyz as the template molecule. The obtained results showed that these imprinted composite nanofibers hold excellent potential for recognition of proteins such as Lyz with a significant selectivity [20]. Nanofibers as green materials also can be used as drug carriers [21]. For this aim, composite cellulose nanofibers were prepared to control antibiotic release. Molecular imprinting strategies were utilized for obtaining gentamicin imprinted cellulose nanofibers via in-situ graft polymerization in which methacrylic acid and *N*, *N*′-methylene bisacrylamide as a monomer and crosslinker were used, respectively. In this study, the delivery of gentamicin from nanofibers was investigated. The Korsmeyer–Peppas model was obtained as kinetic models.

Recently, Lin et al. prepared photo-responsive cellulose for the adsorption of pesticide residue. They used imprinting technology through surface-initiated atom transfer radical polymerization for preparation of selective recognition cavities onto the cellulose. The photoisomerization property of nanocomposite cellulose showed

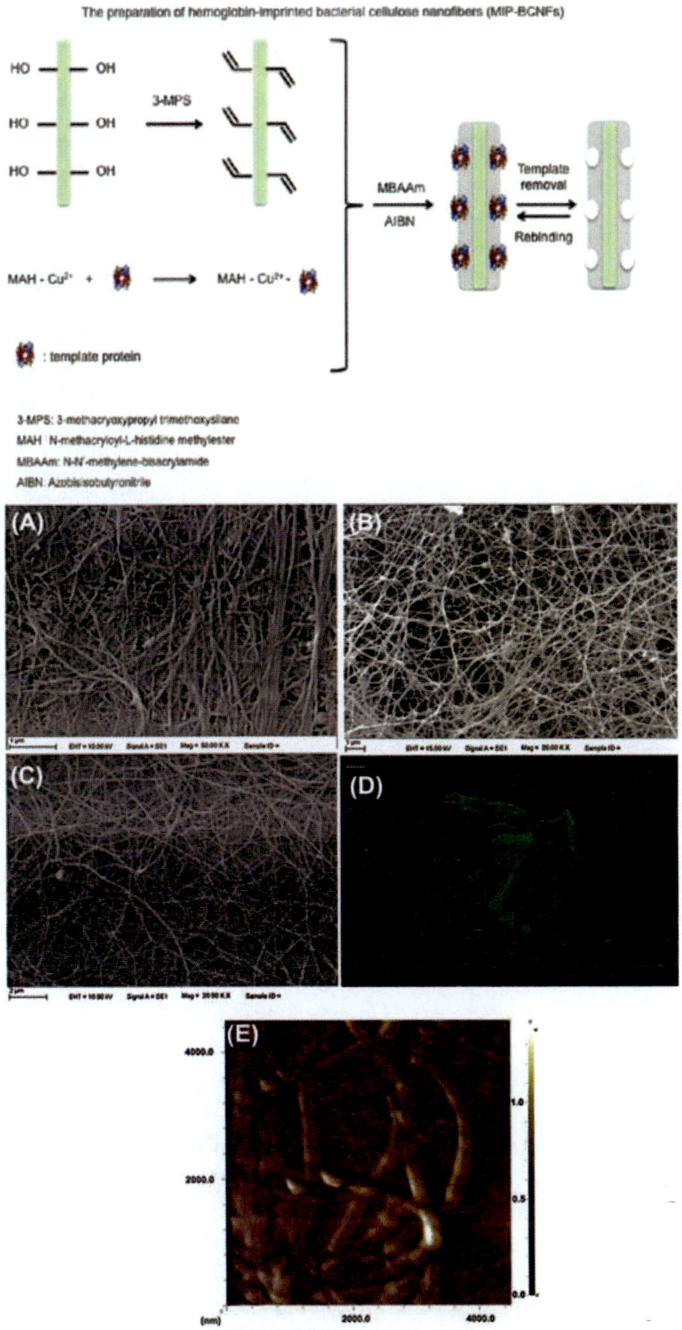

Fig. 2 Schematic preparation of hemoglobin imprinted bacterial cellulose. **a–c** SEM photographs of bacterial cellulose nanofibers (BCNFs), hemoglobin imprinted bacterial cellulose nanofibers (MIP1 BCNFs) and MIP2 BCNFs, and **d**, **e** three-dimensional μCT and AFM images of BCNFs, respectively [18].

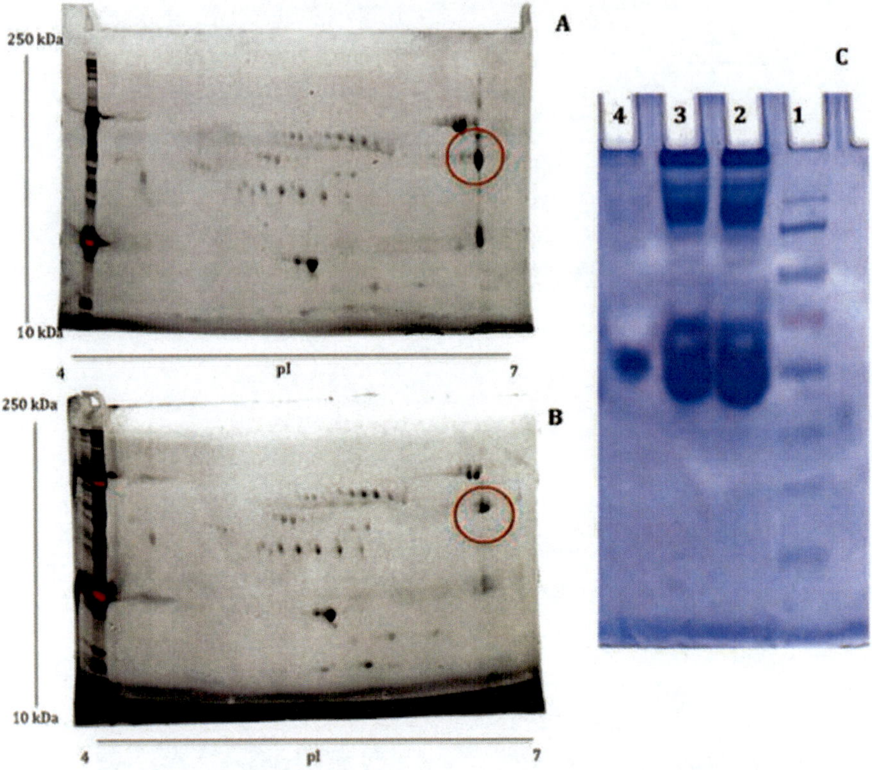

Fig. 3 2-D gel electrophoresis and SDS-PAGE analysis results of artificial human plasma before and after treatment with molecular imprinted bacterial cellulose nanofibers (MIP-cBCNFs) [19].

the excellent regeneration of pesticide residues. The schematic preparation of these photo-responsive materials is shown in Fig. 4 [22].

5 Polymeric-Based Nanocomposite

Besides some of the natural sources such as cellulose acetate, PHB and PHA, some synthetic sources can be used for the preparation of degradable polymeric green nanocomposite materials. Polyvinyl alcohol (PVA), aromatic polyesters and aliphatic polyesters are the important synthetic sources that have recently been used for this purpose. Although these polymers are not renewable and degradable like natural polymers, they can be applied to obtain green nanocomposite materials. Accordingly, these synthetic polymers are generally less preferred to use in biodegradable materials [23, 24]. In this chapter, some of these natural and synthetic polymers used in green

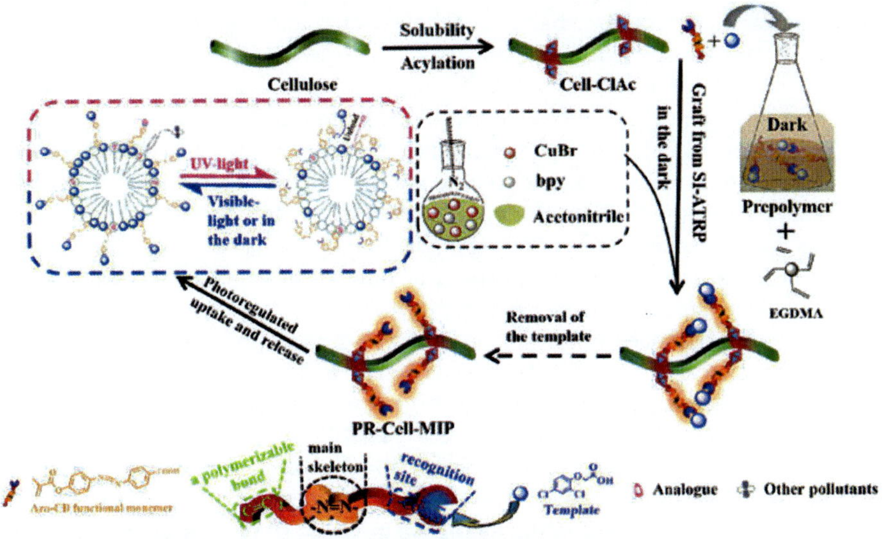

Fig. 4 Schematic preparation of these photo-responsive cellulose [22]

nanocomposites material are given as particular examples from the literature. MIP-based nanocomposites have been widely used to prepare robust, stable, biocompatible and cheap materials for the minimization of organic pollution in the environment and wastewater.

In a study, Bhatia and Kurian used Sorona polymers made from corn-derived 1,3-propanediol to indicate the effects of these nanocomposite polymers on cell survival using in vitro cell cultures. They showed the non-inflammatory and non-cytotoxic properties against cell lines in this study [25]. To this end, 3T3 fibroblast cells and J774 macrophage were used in the cytotoxicity and inflammatory experiments, respectively. The Sorona polymers have drawn attention as promising biomaterials used in plastics, fibers and films. Giavaresi and co-workers prepared soy-derived polymers to use them as bone fillers. They compared the characteristics of these green materials with a commercial p(D, L lactide–glycolide)-based bone graft. As a consequence, they indicated that there was no considerable difference in both ones. These results highlighted that the soy-derived polymers could be recommended as significant green materials [26].

Among the other polymeric materials, hydrogels obtained via freeze-thawing technology have attracted much attention as biomedical materials for, e.g., drug delivery, scaffolds and wound dressing materials. The advantage of this technology can be explained that the initiator and crosslinking are not needed during synthesis. The non-toxicity and biocompatibility features of hydrogels can be provided via the combination with chitosan, carbon nanotubes, etc. [27].

The history of the preparation of hydrogels began in the 1950s by Wichterle and Lim. They prepared the first hydrogels using hydroxyethyl methacrylate (HEMA) as a monomer and ethylene dimethacrylate (EDMA) as a crosslinker for the construction of contact lenses-based materials [28].

Wang et al. obtained iron oxide (Fe_3O_4) magnetic nanoparticles covered by estrone imprinted polymer. They controlled the size of nanoparticles during polymerization via semi-covalent imprinting technique. Firstly, they synthesized the estrone and silica monomer pre-complex (EstSi) via the reaction 3-(triethoxysilyl)propyl isocyanate with estrone. The aim of this process was to establish an interaction between the estrone and silica covering on the Fe_3O_4 core via a thermally reversible bond. Therefore, the prepared magnetic nanoparticles have been obtained to use in the biochemical area for separation of estrone [29].

In another study, one-pot green synthesis was designed by MIP strategy for the extraction and cleaning-up of hydrochlorothiazide in the urine samples. The sol–gel process was utilized, and Fe_3O_4 was used for obtaining magnetic property. On the other hand, aminopropyl trimethoxysilane (APTES), tetraethyl orthosilicate (TEOS) and hydrochlorothiazide were used in the imprinting process as functional monomer, crosslinker and template, respectively. During this process, a surfactant was particularly used to graft the silica imprinted nanoparticles. The nanoparticles were characterized by FT-IR, TEM, SEM and vibrating sample magnetometry (VSM). The prepared nanoparticles responded in the concentration range of 2.5–1000 μg/L template molecule. Also, the selectivity of prepared nanoparticles was successfully verified in a urine sample [30].

The titanium dioxide (TiO_2) nanoparticles have been used in the environmental area for the purpose of remediation of toxic materials. Therefore, the concentrations of organic pollutants have been minimized in wastewater by these photocatalyst nanoparticles. In a study, Ag_2S-based MIP-TiO_2 nanocomposite was developed via a sol–gel deposition technique. The prepared nanocomposite materials showed a great catalytic performance in comparing the anatase TiO_2. Degradation efficiency was reported as 92.22% for ethyl *p*-hydroxybenzoate. Also, the selectivity factor value was found to be 3.571 in comparison with phenol [31].

Bagheri et al. used the green strategy for the preparation of dual-template chitosan-based magnetic water-compatible biopolymer by molecular imprinting technique. Chitosan, as a multi-functional polymer, can be easily polymerized in mild conditions. They used these materials as sorbents for the determination of losartan and valsartan. In conclusion, they reported a green, bio- and water-compatible polymeric matrix for the selective extraction [32].

Mushrooms have been used as biotechnological products for the preparation of nanocomposites occasionally. Yildirim et al. prepared *Pleurotus ostreatus*-based chitosan nanocomposite for the removal of Reactive Orange 16 dye. They reported 65.5 mg/g adsorption capacity[33] .

In another study, instead of oxidation reaction, green chemistry was used for the preparation of iron nanoparticles. Lopez et al. developed nanoparticles from eucalyptus extracts in the green chemistry process. They used nanoparticles with molecular imprinting technology. These materials were prepared for the selective and specific analyte recognition. The adsorption capacities of green nanoparticles are higher than in chemical nanoparticles [34].

Supercritical carbon dioxide ($scCO_2$) which has been introduced as green solvent has advantages such as non-toxicity, low-cost, inertness, non-flammability and recyclability. Ye et al. prepared non-covalent MIP-based nanoparticles by using divinylbenzene and methacrylic acid. Propranolol was used as a target molecule. The size of nanoparticles was reported as 100 nm [35].

Xia et al. prepared 17-estradiol imprinted magnetic nanoparticles via ultrasonication-assisted synthesis. They used methacrylic acid (MAA) and ethylene glycol dimethacrylate (EGDMA) as monomer and crosslinker, respectively. They reported the size of the MIP-magnetic nanoparticles as 300 nm [36]. They showed a reasonable adsorption capacity compared to traditional approaches.

Gao et al. used a green synthesis of Mussel-inspired nanoparticles on carbon fiber surfaces. The low surface activity of carbon fibers is an excellent advantage that is used in many applications. They reported a novel bioinspired copolymerization of dopamine and poly(amidoamine) on the fiber surface (Fig. 5). They reported a novel promising strategy and green composite materials [37].

The recent advances in the field of both green synthesis and green applications were summarized by emphasizing the molecular imprinting technology. Table 1 shows representative examples of green chemistry strategies using particle-based molecularly imprinting.

Yan et al. prepared a three-dimensional magnetic polymer using carbon hybrid nanocomposite as the carrier in the presence of chlorogenic acid (CGA) as an imprinting template. Figure 6 shows the schematic preparation of the imprinted

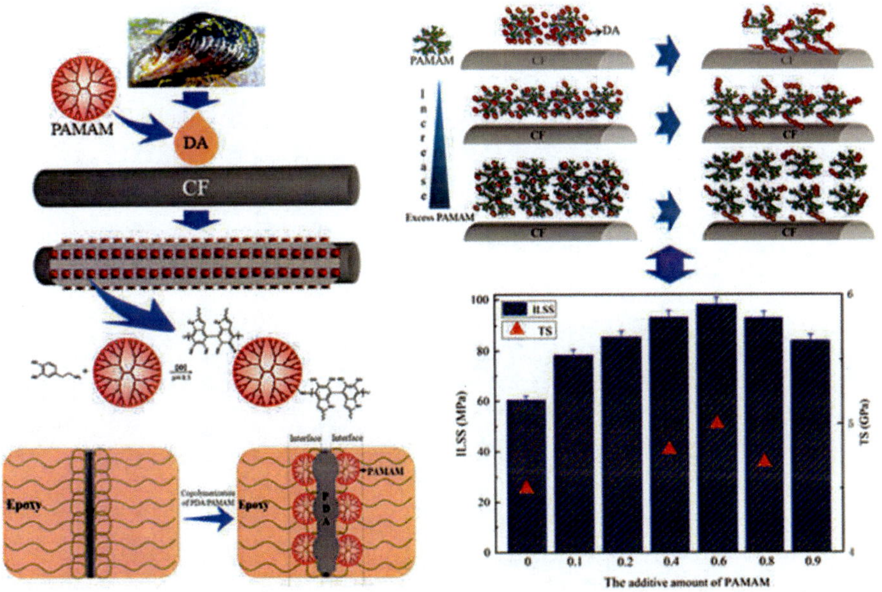

Fig. 5 Schematic preparation of bioinspired copolymerization of dopamine and poly(amidoamine) on the fiber surface [37].

Table 1 Representative examples of green chemistry strategies using particle-based molecularly imprinting

Materials	Monomers	Crosslinkers	Technique	Size of nanoparticles	Target molecules	References
Supercritical carbon dioxide	MAA	DVB	Non-covalent	100 nm	Propranolol	[35]
Supercritical carbon dioxide	MAA-4Vpy MMA	EGDMA	Non-covalent	250–300 nm	Aspirin and acetaminophen	[38]
Supercritical carbon dioxide	MAA	EGDMA	Non-covalent	300 nm	Bisphenol A	[39]
Supercritical carbon dioxide	MAA	EGDMA	Non-covalent	200 nm	Carbamazepine	[40]
Supercritical carbon dioxide	MAA/MMA	EGDMA	Non-covalent	150–200 nm	Gallic acid	[41]
Silica-coated Fe_3O_4 particles	NIPAM/AAM	MBAAM	–	12 nm	Lysozyme	[42]
Fe_3O_4 particles	HEMA/MAA	EGDMA	Non-covalent	18 nm	Pefloxacin mesylate	[43]
p(AN-co-DTCS) membrane	AN	DTCS	Photo-iniferter	500 nm	Theophylline	[44]
p(AN-co-DTCS) membrane	AN	DTCS	–	<10 nm	Lysozyme	[45]

*Methacrylic acid (MAA), Divinylbenzene (DVB), Methyl Methacrylate (MMA), 4-vinyl pyridine Methyl Methacrylate (4Vpy MMA), Ethylene glycol dimethacrylate (EGDMA), *N*,*N*′-Methylenebisacrylamide (MBAAM), *N*,*N*-dimethyl amino di thiocarbamoyl methyl styrene (DTCS), Acrylonitrile (AN), Hydroxyethyl methacrylate (HEMA), *N*-Isopropylacrylamide (NIPAM)

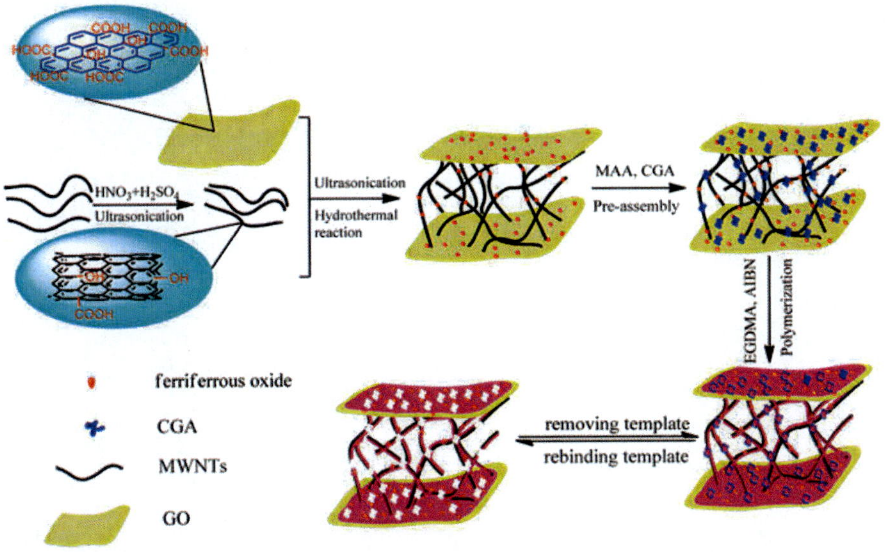

Fig. 6 Synthesis schematic diagram of 3D MMIPs [46].

magnetic polymer [46]. They characterized the three-dimensional MMIPs by SEM, FT-IR spectroscopy, thermogravimetric (TGA) analysis and UV spectrometry in detail. They showed that the imprinted layer was attached successfully on the surface of a three-dimensional magnetic carbon hybrid nanocomposite. The adsorption performance of the materials was studied. They showed the high adsorption capacity and fast adsorption toward CGA with a maximum adsorption capacity of 10.88 mg/g.

Huang et al. aimed to prepare molecularly imprinted nanoparticles for the detection of paracetamol during a green synthesis process [47]. They prepared novel polymeric nanoparticles at mild working condition, energy-saving and green (Fig. 7). They synthesized amphiphilic fluorescent and photo cross-linkable copolymers with carbazole groups. The obtained material can be coassembled with paracetamol as a template molecule and photoinitiator in aqueous solution. These nanoparticles were cross-linked by UV-irradiation to prepare hydrophilic and fluorescent MIP nanoparticles. The researchers showed 1.0 μM detection limit of the template in the concentration range of 4–1000 μM.

6 Carbon Nanotube-Based Nanocomposites

Nanotubes are made of carbon atoms, rounded or arranged in graphite-like layers. They can be synthesized with light elements such as boron, nitrogen and their metal complexes [48, 49]. Nanotubes can form crystalline fibers, which can be very long in

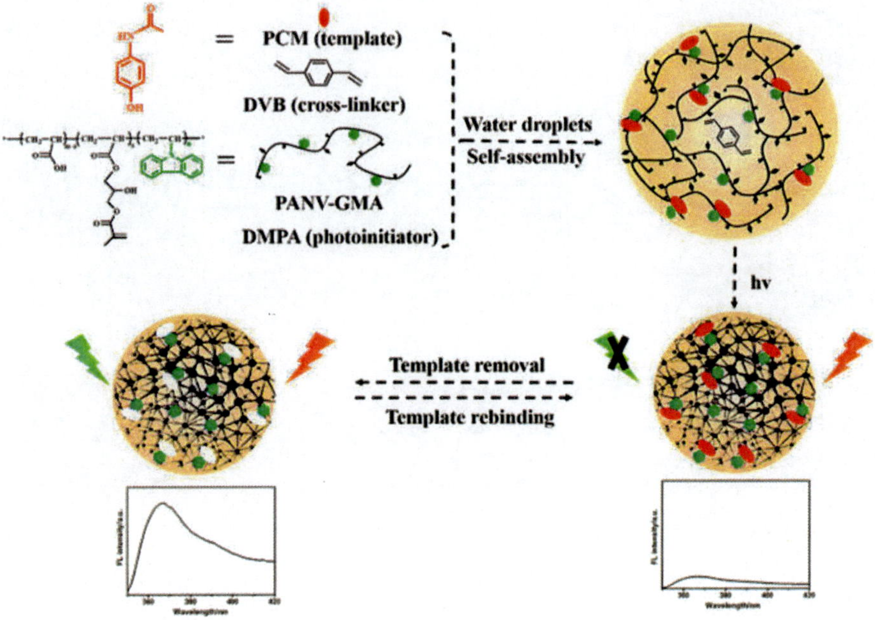

Fig. 7 Synthetic procedure of FMIP nanoparticles and its binding with PCM [47].

length compared to their diameter. These fiber structures are small enough to settle in the lung, and their toxic effects can pose many risks, such as cancer [50].

Carbon nanotubes (CNTs) are the most common research among all nanotubes. In recent years, their environmentally friendly and biocompatible properties have made them a priority. It was first reported to belong to the fullerene family [51]. There are two main types, single-walled carbon nanotubes (SWCNTs) and multi-walled carbon nanotubes (MWCNTs). The electrical conductivity of CNTs with high electron mobility has been used to develop electrochemical sensors from CNTs. Y orbital electrons delocalized in hexagonal atomic arrays contribute to this electrical conductivity. The arrangement of hexagonal atomic arrays in the shape of a honeycomb determines the electronic capacity of CNTs [52]. The surface of the CNTs can be functionalized by placing appropriate hydroxyl or carboxyl groups to the surface [53, 54]. CNTs are also chemically stable and suitable for enabling the combination with different materials and for the production of composite structures. However, defects can occur on the walls, and the small attachment angle of the holes makes the carbon atoms in the structure more prone to oxidation. Also, it is their disadvantage that they are not suspended in water and form aggregates. Thanks to their features, CNTs have been applied to solid phase extraction (SPE), adsorption, drug delivery, sensing and hydrogen storage devices [55–58] .

Molecular imprinting technique is used to modify the surface of CNTs with large surface area. Functional monomers and crosslinkers to be used according to the selected template molecule are effective to achieve high performance MIPs. As a functional monomer, aminopropyltriethoxysilane (APTES), phenyltrimethoxysilane (PTMOS), phenyltriethylorthosilicate (PTEOS), MAA, (MTMOS) are among the favorite ones. Tetraethylorthosilicate (TEOS) and EGDMA are also widely used as crosslinkers. Initiators are selected according to the polymerization method of molecularly imprinted CNTs to be synthesized [52]. Recent researches show that polymer composite-based CNTs promise higher performance than CNTs or polymer-based electrodes alone [59, 60]. The composite structures created with molecularly imprinted CNTs are mentioned with many materials.

One of the newest materials made with molecularly imprinted composite magnetic CNTs has been studied by Whang et al. In the study, lead ions were imprinted on thermo-sensitive surfaces of MWCNTs. Their thermo-sensitive feature was gained with *N*-isopropylacrylamide, and the optimization experiments were carried out by adsorption–desorption temperature change [61].

Sajini et al. prepared photo-sensitive composite polymers with MWCNTs on platinum electrodes. The polymeric structure was prepared for the enantiomeric analysis of L-phenylalanine benzyl ester. The researchers used 4-[(4-methacryloyloxy)phenylazo]benzoic acid by functionalizing the surface of the CNTs with vinyl. The desorption of the bound target molecule was carried out under UV light at 365 nm for 2 h. Cyclic voltammetry was applied for the electrochemical analysis of the binding potential of the L-phenylalanine benzyl ester against the D-form [62].

Anirudhan et al. modified the surface of MWCNTs with vinyl groups, using nitric acid oxidation and allylamine reaction for this process. Modified MWCNTs were synthesized for the imprinting of chlorpyrifos, and a composite polymeric matrix was prepared using methacrylic acid. Characterization studies were performed by FTIR, X-ray diffraction (XRD) and SEM analysis. Maximum adsorption capacity was found to be 38.14 mg/g at pH 7.0 buffer. When compared with the literature, resultant adsorption capacity value was higher than for many of the materials prepared for chlorpyrifos adsorption. This encouraging result can be attributed to the excess of the surface area of the material [63].

Qin et al. used epitope imprinting technology on the surface of magnetic CNTs to recognize cytochrome C [64]. They used peptide sequences, EGDMA and zinc acrylate to prepare MIPs. Metal chelation was applied to immobilize the epitope (Fig. 8). They showed specific recognition ability for cytochrome C with 11.7 imprinting factor and 780.0 mg/g maximum adsorption capacity. Also, they used the magnetic property of the material to be highly efficient and processed easily by the assistance of an external magnetic field. They showed a protocol to the detection of template protein in a complex sample via surface imprinting strategy and epitope imprinting approach.

Apart from covering the CNTs with different polymers when preparing composite materials, they can be also prepared by combining CNTs in other nanomaterials. The representative examples of these different complex polymers and target molecules are summarized in Table 2.

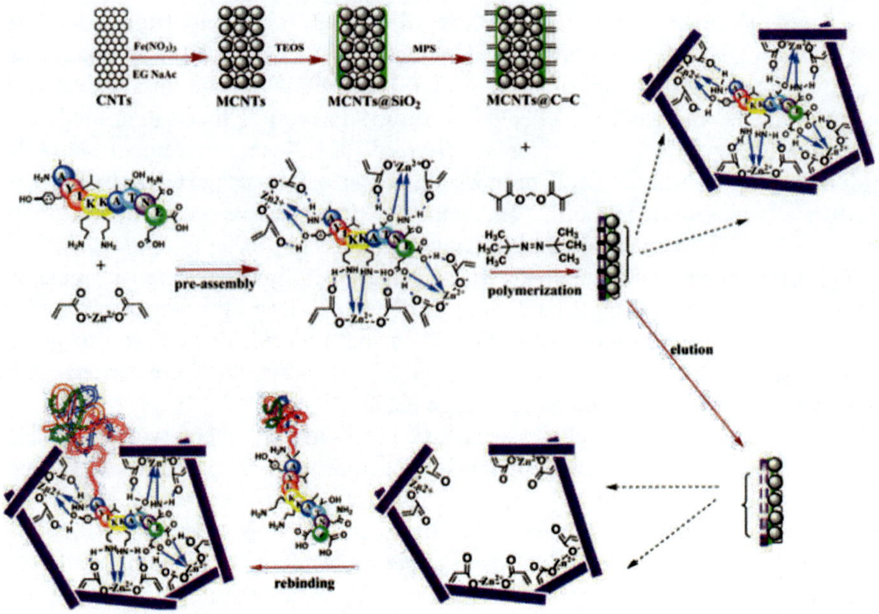

Fig. 8 Synthesis protocol of MCNTs@EMIP via surface imprinting and epitope imprinting [64].

7 Quantum Dots-Based Nanocomposite

Quantum dots (QDs) are nanometer-sized semiconductor crystals consisting of elements within the groups II to VI or III to V of the periodic table, which are called 'artificial atoms' because of their unique sizes [75]. On the absorption of light, nanocrystals produce excitons and electron hole recombination which leads to luminescence. QDs are often described as spheres, but they have a lattice structure similar to crystalline materials and bulk semiconductor materials. According to their sizes, they have thousands of atoms that can be found in each nanocrystal structure. They are suitable materials for molecular imprinting with a high surface area/volume ratio. Most of the QDs are synthesized as core/shell structures, which are made up of various metal complexes such as semiconductors, noble metals and magnetic transition metals. QDs have been widely used in bioimaging [76], drug delivery [77] and solar technology [78]. With their ability to measure even in small amount, they have been used as sensing materials to analyze the presence of many harmful and toxic components with molecularly imprinted QDs.

Although the preparation process of traditional QDs does not include green synthesis, QDs have an important role in many applications covering green strategies. Environmental pollutants such as pesticides, other endocrine disruptors and antibiotic residues could easily reach people through the water, soil and food chain, causing many health problems. Even the presence of traces in water and food, these

Table 2 Studies performed with different MIPs used to prepare composites with CNTs

Materials	Template molecules	Samples	Polymerization methods	Analysis technique	References
Au nanoparticle-chitosan	Tyramine	Yoghurt	Sol–gel process	Electrochemical sensor	[65]
SiO_2	Triclosan	Environmental water	Sol–gel process	HPLC	[66]
ZnO-chitosan	p-nitrophenol	Aqueous solution	Sol–gel process	Electrochemical impedance spectroscopy	[67]
Au nanoparticle	Hg(II)	Water	Electro polymerization	Electrochemical sensor	[68]
Au nanoparticle-chitosan	Chlortetracycline	Water	Free radical polymerization	Electrochemical sensor	[69]
SiO_2	Estron	Environmental water	Semi-covalent imprinting	HPLC	[70]
PbS nanoparticle-Fe_3O_4@Au-chitosan	Streptomycin	Aqueous solution	Sol–gel process	Electrochemical sensor	[71]
Cobalt nanoparticle-chitosan	Oxacillin	Blood serum	Sol–gel process	Electrochemical sensor	[72]
MXene	Fisetin	Cotinus coggygria leaves	Electro polymerization	Electrochemical sensor	[73]
SiO_2	Chlorpyrifos	Fruit vegetable	Electro polymerization	Electrochemical sensor	[74]

pollutants posing a great danger are very important to detect at their low concentrations. Therefore, HPLC [79], Raman spectroscopy [80], electrochemical sensors [81, 82], and gas chromatography-mass spectrometry [83] have been used in a number of complex and expensive techniques for the detection of trace amounts of pollutants. In recent years, fluorescence detection ability, unique electronic-optic properties and surface modification strategies of QDs have been applied in various studies as potential alternative analytical sensors to these traditional methods. When the performed studies in the literature are taken into consideration, green applications of molecularly imprinted QDs-based composites were produced. They are shown in Table 3.

7.1 Carbon Quantum Dots-Based Nanocomposite

Researchers have sought different sources of biocompatibility, considering the toxic properties of conventional QDs, environmental and cellular damage. Carbon quantum dots (CQDs) are produced as alternatives to semi-conductive QDs, which are partially toxic by the use of heavy metals in the synthesis phase. One of the new products of carbon nanomaterials, CQDs have been introduced as effective nanomaterials and they were obtained was discovered during the purification of SWCNTs by electrophoresis technique [104]. In recent years, CQDs have gained a lot of attention in terms of environmental friendliness and biocompatibility, and they are very preferable due to their stability, non-toxicity, simple synthesis and use of natural resources [105]. The synthesis of CQDs was performed using microwave and hydrothermal methods, and in addition, these techniques have gained importance in terms of green synthesis.

An interesting approach is that Ensaifi et al. used QDs for determining Promethazine hydrochloride (PrHy) by using orange juice as the carbon source. Imprinting process was carried out on the surface of the materials by sol–gel polymerization method. CQDs were prepared as approximately 5 nm but can be increased to about 60 nm in both MIP and NIP-CQDs. The optimization studies were performed in an aqueous solution. The applicability of these materials has been clarified in blood plasma, and the optical determination of PrHy was successfully done from plasma [106].

Hou et al. prepared selective and environmentally friendly CQDs for the detection of the tetracycline from milk, which is a broad-spectrum antibiotic and generally used in the treatment of livestock. Microwave is used for the synthesis of tetracycline printed composite materials. With this method, it was possible to simplify the experimental procedure and to save energy by producing quickly. Molecular imprinting was carried out on the surface of silica-coated CQDs, which were synthesized using citric acid. The limit of detection of the synthesized nanocomposites is specified as 5.48 nM, which is below the maximum value that is allowed to be found in the milk reported by many authorities [107]. In another study, tetracycline was used as a template by Yang and coworker synthesized N-doping CQDs. N-doping has been

Table 3 Green applications of molecularly imprinted QDs-based composites

Type of QDts	Size (nm)	Template	Sample	Sensing system	Limit of detection	References
Polypyrole/Graphene	0.7–1.2	Bisphenol A	Water	Electrochemical	0.04 μM	[84]
Mn-Dopped ZnS/Fe_3O_4	50	Pentachlorophenol	Water	Fluorescence	0.5 μmol/L	[85]
Polyindole/Graphene	353 ± 12	Dopamine	Serum Urine	Fluorescence	0.1 nM	[86]
Graphene	100	Ornidazole	Plasma	Luminescence	0.24 μM	[87]
Mn-Dopped ZnS	12.1	Pentachlorophenol	Water	Room-temperature Phosphorescence optosensor	86 nM	[88]
CdTe@CdS	457.4	Perfluorooctanoic acid	Water	Photoluminescence	25 nmol/L	[89]
Mn-Dopped ZnS	6–27	Nicosulfuron	Water	Fluorescence	1.1 nmol/L	[90]
CdTe	5	Phycocyanin	Water urine	Fluorescence	0.075 μM	[91]
CdTe	2–4	Sulfadiazine	Seawater	Fluorescence	0.7 μM	[92]
Mn-Dopped ZnS	10–30	4-Nitrophenol	Tap water	Fluorescence	76 nM	[93]
ZnS:Mn^{2+}	160	Diazinon	Water	Fluorescence	38.6 ng/mL	[94]
CdTe@SiO_2	5	p-nitrophenol	Tap and Lake water	Fluorescence	0.08 μmol/L	[95]
Mn-Dopped ZnS	36.3	Cyphenothrin	River water	Fluorescence	9.0 nmol/L	[96]
CdTe@SiO_2	53 ± 3	λ-cyhalothrin	Water	Fluorescence	–	[97]
CdTe		Tetracycline	Aqueous solution	Fluorescence	6.5 μM	[98]
Mn-dopped ZnS	4.2	Florfenicol	Meat	Optical	24 μmol/L	[99]
CdTe-SiO_2	52 ± 1.8	Deltamethrin	Fruit vegetable	Fluorescence	0.16 μg/mL	[100]

(continued)

Table 3 (continued)

Type of QDts	Size (nm)	Template	Sample	Sensing system	Limit of detection	References
CdTe	10	Parathion	Water	Fluorescence	0.218 μmol/L	[101]
$CdSe@SiO_2$	90	Pyrethroids	Water	Fluorescence	8.0 nM	[102]
CdSe/ZnS	Not reported	Carbaryl	Rice Cabbage	Fluorescence	1.47×10^{-7} mol/L	[103]

used to increase the fluorescent intensity of CQDs. The lower limit of tetracycline determination analyzed from fish by synchronous fluorescence spectroscopy was found as 9 μmol/L [108].

7.2 *Graphene Quantum Dots-Based Nanocomposite*

One of the interesting features of GQDs is that they have a low toxicity, easily functional carbon material and highly soluble in various solvents [109]. Although GQDs are points of carbon atoms like CQDs, there are differences in structure, synthesis methods, properties and applications. GQDs have a graphene lattice structure, while CQDs have a crystal or amorphous structure. GQDs consist of one or more layers of graphene and are below 20 nm, while CQDs are usually carbon nanoparticles below 10 nm [110]. GQDs have carboxylic acid regions on the edges, as in graphene, therefore they are water-soluble and can demonstrate an easy functionalization with organic–inorganic derivatives.

Zor et al. prepared a molecularly imprinted GQD-based composite photoluminescence sensor to analyze pesticides from seawater. In the first and second stages, GQDs were prepared using citric acid, and then, tributyltin was used as a template with pyrrole synthesis on their surface. At the last stage, GQD surface carboxylic groups are activated and encapsulated into magnetic silica particles [111].

Üzek et al. attached GQDs to the surface of bisphenol A imprinted polyvinyl alcohol-based nanoparticles by *N*-(3-dimethylaminopropyl)-*N'*-ethylcarbodiimide hydrochloride/*N*-hydroxysuccinimide conjugation method, which synthesized by hydrothermal pyrolysis from glutamic acid and aspartic acid. Experiments were carried out both with composite nanoparticles directly from the aqueous solution and with composite nanoparticles absorbed nitrocellulose papers. They used seawater as a real example and reported that the limit of detection is 4.2 ± 0.5 μg L^{-1} [112].

8 Conclusion

Nanocomposite materials are multi-phase solid materials that are used for embedding onto the polymeric matrix with reinforcing phases into other materials. The market production of these materials is increasing to support their global demands. Recently, the research has focused on the preparation of nature-friendly green nanocomposites polymeric materials. Therefore, these materials can be replaced instead of non-degradable polymers such as polyamides and polyolefins. The green nanocomposite materials are renewable and biodegradable. The significant attractions about these materials are that they are completely degradable in nature, biocompatible, environment friendly and useful in every way. Toward the end of their application, green nanocomposite materials can be readily degraded without any damage to Nature. In other words, the chemical structure of these biodegradable materials can change into other beneficial compounds by loss of structural properties [24]. To obtain useful green nanocomposite materials, green polymers should be preferred that are used as matrices [23, 113].

MIPs can easily be produced via synthetic methodologies. Evolving environmental awareness, by the way, significant regulations on chemical usage leads to the emergence of greener platforms. Avoiding pollution and minimizing wastes have been accomplished by the combination of molecularly imprinted green nanocomposites. In this concept, molecular imprinting has attracted tremendous interest in the green applications including novel alternative solvents.

References

1. Boholm M (2016) The use and meaning of nano in American English: towards a systematic description. Ampersand. https://doi.org/10.1016/j.amper.2016.10.001
2. Wypych F, Satyanarayana KG (2005) Functionalization of single layers and nanofibers: a new strategy to produce polymer nanocomposites with optimized properties. J Colloid Interface Sci. https://doi.org/10.1016/j.jcis.2004.12.028
3. Cannillo V et al (2006) Modeling of ceramic particles filled polymer-matrix nanocomposites. Compos Sci Technol. https://doi.org/10.1016/j.compscitech.2005.07.030
4. Adeosun SO et al (2012) Review of green polymer nanocomposites. J Miner Mater Charact Eng. https://doi.org/10.4236/jmmce.2012.114028
5. Nanocomposite science and technology. Mater Today (2003). https://doi.org/10.1016/s1369-7021(03)01139-8
6. Choi J et al (2011) The glass transition and thermoelastic behavior of epoxy-based nanocomposites: a molecular dynamics study. Polymer. https://doi.org/10.1016/j.polymer.2011.09.019
7. Viveiros R, Rebocho S, Casimiro T (2018) Green strategies for molecularly imprinted polymer development. Polymers. https://doi.org/10.3390/polym10030306
8. Madikizela LM, Ncube S, Chimuka L (2019) Green chemistry features in molecularly imprinted polymers preparation process, 1st edn. Comprehensive analytical chemistry, 1st edn. Elsevier B.V. https://doi.org/10.1016/bs.coac.2019.05.007
9. Idil N, Mattiasson B (2017) Imprinting of microorganisms for biosensor applications. Sensors (Switzerland). https://doi.org/10.3390/s17040708
10. Ye L, Mosbach K (2008) Molecular imprinting: synthetic materials as substitutes for biological antibodies and receptors. Chem Mater. https://doi.org/10.1021/cm703190w
11. Kempe M, Mosbach K (1995) Separation of amino acids, peptides and proteins on molecularly imprinted stationary phases. J Chromatogr A. https://doi.org/10.1016/0021-9673(94)00820-Y
12. Ertürk G, Mattiasson B (2017) Molecular imprinting techniques used for the preparation of biosensors. Sensors (Switzerland). https://doi.org/10.3390/s17020288
13. Wu X et al (2018) Recent advances in green reagents for molecularly imprinted polymers. RSC Adv. https://doi.org/10.1039/c7ra11047b
14. Klemm D et al (2005) Cellulose: fascinating biopolymer and sustainable raw material. Angew Chem Int Ed. https://doi.org/10.1002/anie.200460587
15. Azizi Samir MAS, Alloin F, Dufresne A (2005) Review of recent research into cellulosic whiskers, their properties and their application in nanocomposite field. Biomacromol. https://doi.org/10.1021/bm0493685
16. Tamahkar E, Kutsal T, Denizli A (2015) Surface imprinted bacterial cellulose nanofibers for cytochrome c purification. Process Biochem. https://doi.org/10.1016/j.procbio.2015.09.026

17. Bakhshpour M et al (2017a) Affinity binding of proteins to the modified bacterial cellulose nanofibers. J Chromatogr B Anal Technol Biomed Life Sci. https://doi.org/10.1016/j.jchromb.2017.03.021
18. Bakhshpour M et al. (2017b) Surface imprinted bacterial cellulose nanofibers for hemoglobin purification. Colloids Surf B Biointerfaces. https://doi.org/10.1016/j.colsurfb.2017.07.023
19. Göktürk I et al (2018) Protein depletion with bacterial cellulose nanofibers. J Chromatogr B Anal Technol Biomed Life Sci. https://doi.org/10.1016/j.jchromb.2018.08.030
20. Saylan Y, Tamahkar E, Denizli A (2017) Recognition of lysozyme using surface imprinted bacterial cellulose nanofibers. J Biomater Sci Polym Ed. https://doi.org/10.1080/09205063.2017.1364099
21. Tamahkar E, Bakhshpour M, Denizli A (2019) Molecularly imprinted composite bacterial cellulose nanofibers for antibiotic release. J Biomater Sci Polym Ed. https://doi.org/10.1080/09205063.2019.1580665
22. Lin C et al (2020) Fabrication of photo-responsive cellulose based intelligent imprinted material and selective adsorption on typical pesticide residue. Chem Eng J. https://doi.org/10.1016/j.cej.2020.124841
23. Kalia S, Kaith BS, Kaur I (2009) Pretreatments of natural fibers and their application as reinforcing material in polymer composites—a review. Polym Eng Sci. https://doi.org/10.1002/pen.21328
24. Jamshidian M et al (2010) Poly-lactic acid: production, applications, nanocomposites, and release studies. Compr Rev Food Sci Food Saf. https://doi.org/10.1111/j.1541-4337.2010.00126.x
25. Bhatia SK, Kurian JV (2008) Biological characterization of Sorona polymer from corn-derived 1,3-propanediol. Biotech Lett. https://doi.org/10.1007/s10529-007-9607-z
26. Giavaresi G et al (2010) Bone regeneration potential of a soybean-based filler: experimental study in a rabbit cancellous bone defects. J Mater Sci Mater Med. https://doi.org/10.1007/s10856-009-3870-6
27. Hassan CM, Peppas NA (2000) Structure and applications of poly(vinyl alcohol) hydrogels produced by conventional crosslinking or by freezing/thawing methods. Adv Polym Sci. https://doi.org/10.1007/3-540-46414-x_2
28. Wichterle O, Lím D (1960) Hydrophilic gels for biological use. Nature. https://doi.org/10.1038/185117a0
29. Wang X et al (2009) A molecularly imprinted polymer-coated nanocomposite of magnetic nanoparticles for estrone recognition. Talanta. https://doi.org/10.1016/j.talanta.2008.11.024
30. Arabi M, Ghaedi M, Ostovan A (2017) Development of a lower toxic approach based on green synthesis of water-compatible molecularly imprinted nanoparticles for the extraction of hydrochlorothiazide from human urine. ACS Sustain Chem Eng. https://doi.org/10.1021/acssuschemeng.6b02615
31. Liu X et al (2020) Photocatalytic degradation of wastewater by molecularly imprinted Ag2S-TiO_2 with high-selectively. Sci Rep. https://doi.org/10.1038/s41598-020-57925-8
32. Bagheri AR et al (2019) Dummy molecularly imprinted polymers based on a green synthesis strategy for magnetic solid-phase extraction of acrylamide in food samples. Talanta. https://doi.org/10.1016/j.talanta.2018.11.065
33. Yildirim A, Acay H, Baran F (2020) Synthesis and characterisation of mushroom-based nanocomposite and its efficiency on dye biosorption via antimicrobial activity. Int J Environ Anal Chem. https://doi.org/10.1080/03067319.2020.1739664
34. Sabín López A et al (2020) Synthesis of magnetic green nanoparticle—molecular imprinted polymers with emerging contaminants templates. J Environ Chem Eng. https://doi.org/10.1016/j.jece.2020.103889
35. Ye L et al (2006) Preparation of molecularly imprinted polymers in supercritical carbon dioxide. J Appl Polym Sci. https://doi.org/10.1002/app.24648
36. Xia X, Lai EPC, Ormeci B (2012) Ultrasonication-assisted synthesis of molecularly imprinted polymer-encapsulated magnetic nanoparticles for rapid and selective removal of 17β-estradiol from aqueous environment. Polym Eng Sci. https://doi.org/10.1002/pen.23126

37. Gao B et al (2019) Bioinspired modification via green synthesis of mussel-inspired nanoparticles on carbon fiber surface for advanced composite materials. ACS Sustain Chem Eng. https://doi.org/10.1021/acssuschemeng.8b03590
38. Yoon SD, Byun HS (2013) Molecularly imprinted polymers for selective separation of acetaminophen and aspirin by using supercritical fluid technology. Chem Eng J. https://doi.org/10.1016/j.cej.2013.04.052
39. Byun HS, Chun D (2017) Adsorption and separation properties of gallic acid imprinted polymers prepared using supercritical fluid technology. J Supercrit Fluids. https://doi.org/10.1016/j.supflu.2016.05.042
40. Lee JC, Kim CR, Byun HS (2014) Synthesis and adsorption properties of carbamazepine imprinted polymer by dispersion polymerization in supercritical carbon dioxide. Korean J Chem Eng. https://doi.org/10.1007/s11814-014-0178-0
41. Byun HS, Yang DS, Cho SH (2013) Synthesis and characterization of high selective molecularly imprinted polymers for bisphenol A and 2,4-dichlorophenoxyacetic acid by using supercritical fluid technology. Polymer. https://doi.org/10.1016/j.polymer.2012.11.079
42. Gai QQ et al (2010) Superparamagnetic lysozyme surface-imprinted polymer prepared by atom transfer radical polymerization and its application for protein separation. J Chromatogr A. https://doi.org/10.1016/j.chroma.2010.06.001
43. Liu Y et al (2012) Superparamagnetic surface molecularly imprinted nanoparticles for water-soluble pefloxacin mesylate prepared via surface initiated atom transfer radical polymerization and its application in egg sample analysis. J Chromatogr A. https://doi.org/10.1016/j.chroma.2012.01.045
44. Wang HY, Kobayashi T, Fujii N (1997) Surface molecular imprinting on photosensitive dithiocarbamoyl polyacrylonitrile membranes using photograft polymerization. J Chem Technol Biotechnol. https://doi.org/10.1002/(sici)1097-4660(199712)70:4%3c355::aid-jctb793%3e3.0.co;2-%23
45. Chen RR et al (2010) Novel surface-modified molecularly imprinted membrane prepared with iniferter for permselective separation of lysozyme. J Membr Sci. https://doi.org/10.1016/j.memsci.2010.07.026
46. Yan L et al (2016) Synthesis and application of novel 3D magnetic chlorogenic acid imprinted polymers based on a graphene-carbon nanotube composite. J Agric Food Chem. https://doi.org/10.1021/acs.jafc.6b00518
47. Huang J et al (2018) Green synthesis of water-compatible fluorescent molecularly imprinted polymeric nanoparticles for efficient detection of paracetamol. ACS Sustain Chem Eng. https://doi.org/10.1021/acssuschemeng.8b00823
48. Chopra N et al (1995) Boron nitride nanotubes. Science 269(5226):966–967
49. Zhang Y et al (1998) Coaxial nanocable: silicon carbide and silicon oxide sheathed with boron nitride and carbon. https://doi.org/10.1126/science.281.5379.973
50. Seaton A, Donaldson K (2005) Nanoscience, nanotoxicology, and the need to think small. Lancet 365(9463):923–924
51. Iijima S (1991) Helical microtubules of graphitic carbon. Nature (6348):56–58. https://doi.org/10.1038/354056a0
52. Dai H et al (2015) Synthesis and analytical applications of molecularly imprinted polymers on the surface of carbon nanotubes: a review. Microchim Acta 893–908. https://doi.org/10.1007/s00604-014-1376-5
53. Giordani S et al (2009) Organic functionalisation and characterisation of single-walled carbon nanotubes supramolecular chemistry of fullerenes and carbon nanotubes view project supramolecular chemistry view project organic functionalisation and characterisation of single-walled carbon nanotubes. https://doi.org/10.1039/b518111a
54. Bianco A, Kostarelos K, Prato M (2011) Making carbon nanotubes biocompatible and biodegradable. Chem Commun 47(37):10182–10188. https://doi.org/10.1039/c1cc13011k
55. Raffaelle RP et al (2005) Carbon nanotubes for power applications. Mater Sci Eng B Solid-State Mater Adv Technol 233–243. https://doi.org/10.1016/j.mseb.2004.09.034

56. Jiang L et al (2012) Adsorption behavior of pazufloxacin mesilate on amino-functionalized carbon nanotubes. J Nanosci Nanotechnol 12(9):7271–7279. https://doi.org/10.1166/jnn.2012.6562
57. Xiao D et al (2012) Magnetic carbon nanotubes: synthesis by a simple solvothermal process and application in magnetic targeted drug delivery system. J Nanopart Res 14(7). https://doi.org/10.1007/s11051-012-0984-4
58. Li L et al (2013) Interaction of carboxylated single-walled carbon nanotubes with bovine serum albumin. Spectrochim Acta Part A Mol Biomol Spectrosc 105:45–51. https://doi.org/10.1016/j.saa.2012.11.111
59. Liu A, Honma I, Zhou H (2007) Simultaneous voltammetric detection of dopamine and uric acid at their physiological level in the presence of ascorbic acid using poly(acrylic acid)-multiwalled carbon-nanotube composite-covered glassy-carbon electrode. Biosens Bioelectron 23(1):74–80. https://doi.org/10.1016/j.bios.2007.03.019
60. Umasankar Y, Chen SM (2008) Multi-walled carbon nanotubes with poly(methylene blue) composite film for the enhancement and separation of electroanalytical responses of catecholamine and ascorbic acid. Sens Actuators B Chem 130(2):739–749. https://doi.org/10.1016/j.snb.2007.10.040
61. Wang H et al (2020) Preparation of thermo-sensitive surface ion-imprinted polymers based on multi-walled carbon nanotube composites for selective adsorption of lead(II) ion. Colloids Surf A Physicochem Eng Aspects 585:124139. https://doi.org/10.1016/j.colsurfa.2019.124139
62. Sajini T, John S, Mathew B (2019) Tailoring of photo-responsive molecularly imprinted polymers on multiwalled carbon nanotube as an enantioselective sensor and sorbent for L-PABE. Compos Sci Technol 181:107676. https://doi.org/10.1016/j.compscitech.2019.06.003
63. Anirudhan TS, Alexander S (2013) Synthesis and characterization of vinyl-functionalized multiwalled carbon nanotubes based molecular imprinted polymer for the separation of chlorpyrifos from aqueous solutions. J Chem Technol Biotechnol 88(10):1847–1858. https://doi.org/10.1002/jctb.4039
64. Qin YP et al (2016) Preparation of high-efficiency Cytochrome c-imprinted polymer on the surface of magnetic carbon nanotubes by epitope approach via metal chelation and six-membered ring. ACS Appl Mater Interfaces. https://doi.org/10.1021/acsami.6b00794
65. Huang J et al (2011) A molecularly imprinted electrochemical sensor based on multiwalled carbon nanotube-gold nanoparticle composites and chitosan for the detection of tyramine. Food Res Int 44(1):276–281. https://doi.org/10.1016/j.foodres.2010.10.020
66. Gao R et al (2010) Synthesis and evaluation of molecularly imprinted core–shell carbon nanotubes for the determination of triclosan in environmental water samples. J Chromatogr A 1217(52):8095–8102
67. Hu Y et al (2012) Sensitive and selective imprinted electrochemical sensor for p-nitrophenol based on ZnO nanoparticles/carbon nanotubes doped chitosan film. Thin Solid 520(16):5314–5321
68. Fu XC et al (2012) Electropolymerized surface ion imprinting films on a gold nanoparticles/single-wall carbon nanotube nanohybrids modified glassy carbon electrode for electrochemical detection of trace mercury(II) in water. Anal Chim Acta 720:29–37. https://doi.org/10.1016/j.aca.2011.12.071
69. Lian W et al (2012) A molecularly imprinted sensor based on β-cyclodextrin incorporated multiwalled carbon nanotube and gold nanoparticles-polyamide amine dendrimer nanocomposites combining with water-soluble chitosan derivative for the detection of chlortetracycline. Food Control 26(2):620–627. https://doi.org/10.1016/j.foodcont.2012.02.023
70. Gao R et al (2011) Preparation and characterisation of core-shell CNTs@MIPs nanocomposites and selective removal of estrone from water samples. Talanta 83(3):757–764. https://doi.org/10.1016/j.talanta.2010.10.034
71. Hu Y et al (2012a) A sensitive and selective sensor-coated molecularly imprinted sol-gel film incorporating β-cyclodextrin-multi-walled carbon nanotubes and cobalt nanoparticles-chitosan for oxacillin determination. Surf Interface Anal 44(3):334–341. https://doi.org/10.1002/sia.3807

72. Hu Y et al (2012b) Selective and sensitive molecularly imprinted sol-gel film-based electrochemical sensor combining mercaptoacetic acid-modified PbS nanoparticles with Fe_3O_4@Au-multi-walled carbon nanotubes-chitosan. J Solid State Electrochem 16(3):857–867. https://doi.org/10.1007/s10008-011-1434-4
73. Ma X et al (2020) Hierarchical porous MXene/amino carbon nanotubes-based molecular imprinting sensor for highly sensitive and selective sensing of fisetin. Sens Actuators B Chem 309:127815. https://doi.org/10.1016/j.snb.2020.127815
74. Huang W et al (2020) A sensitive electrochemical sensor modified with multi-walled carbon nanotubes doped molecularly imprinted silica nanospheres for detecting chlorpyrifos. J Sep Sci 43(5):954–961. https://doi.org/10.1002/jssc.201901036
75. Chan WCW et al (2002) Luminescent quantum dots for multiplexed biological detection and imaging. Curr Opin Biotechnol 40–46. https://doi.org/10.1016/S0958-1669(02)00282-3
76. Wei J et al (2014) Dual functional carbon dots derived from cornflour via a simple one-pot hydrothermal route. Mater Lett. https://doi.org/10.1016/j.matlet.2014.02.090
77. Hua XW, Bao YW, Wu FG (2018) Fluorescent carbon quantum dots with intrinsic nucleolus-targeting capability for nucleolus imaging and enhanced cytosolic and nuclear drug delivery. ACS Appl Mater Interfaces. https://doi.org/10.1021/acsami.7b19549
78. Huang JJ et al (2014) An easy approach of preparing strongly luminescent carbon dots and their polymer based composites for enhancing solar cell efficiency. Carbon. https://doi.org/10.1016/j.carbon.2013.12.092
79. Watabe Y et al (2004) Determination of bisphenol A in environmental water at ultra-low level by high-performance liquid chromatography with an effective on-line pretreatment device. J Chromatogr A 1032(1–2):45–49. https://doi.org/10.1016/j.chroma.2003.11.079
80. Ren X et al (2018) Silver microspheres coated with a molecularly imprinted polymer as a SERS substrate for sensitive detection of bisphenol A. Microchim Acta 185(4):1–8. https://doi.org/10.1007/s00604-018-2772-z
81. Ensafi AA, Amini M, Rezaei B (2018) Molecularly imprinted electrochemical aptasensor for the attomolar detection of bisphenol A. Microchim Acta 185(5):1–7. https://doi.org/10.1007/s00604-018-2810-x
82. Goulart LA et al (2018) Synergic effect of silver nanoparticles and carbon nanotubes on the simultaneous voltammetric determination of hydroquinone, catechol, bisphenol A and phenol. Microchim Acta 185(1):1–9. https://doi.org/10.1007/s00604-017-2540-5
83. Kawaguchi M et al (2006) Liquid phase microextraction with in situ derivatization for measurement of bisphenol A in river water sample by gas chromatography-mass spectrometry. J Chromatogr A 1110(1–2):1–5. https://doi.org/10.1016/j.chroma.2006.01.061
84. Tan F et al (2016) An electrochemical sensor based on molecularly imprinted polypyrrole/graphene quantum dots composite for detection of bisphenol A in water samples. Sens Actuators B Chem 233:599–606. https://doi.org/10.1016/j.snb.2016.04.146
85. Yang M et al (2012) Magnetic nanoparticles and quantum dots co-loaded imprinted matrix for pentachlorophenol. J Hazard Mater 237–238:63–70. https://doi.org/10.1016/j.jhazmat.2012.07.064
86. Zhou X et al (2015) Facile synthesis of molecularly imprinted graphene quantum dots for the determination of dopamine with affinity-adjustable. ACS Appl Mater Interfaces 7(22):11741–11747. https://doi.org/10.1021/am5078478
87. Mehrzad-Samarin M, Faridbod F, Ganjali MR (2019) A luminescence nanosensor for ornidazole detection using graphene quantum dots entrapped in silica molecular imprinted polymer. Spectrochim Acta Part A Mol Biomol Spectrosc 206:430–436. https://doi.org/10.1016/j.saa.2018.08.026
88. Wang H et al (2020) Preparation of thermo-sensitive surface ion-imprinted polymers based on multi-walled carbon nanotube composites for selective adsorption of lead(II) ion. Colloids Surf A Physicochem Eng Aspects 585:124139. https://doi.org/10.1016/j.colsurfa.2019.124139
89. Zheng L et al (2019) Core-shell quantum dots coated with molecularly imprinted polymer for selective photoluminescence sensing of perfluorooctanoic acid. Talanta 194:1–6. https://doi.org/10.1016/j.talanta.2018.09.106

90. Ren X, Chen L (2017) Preparation of molecularly imprinted polymer coated quantum dots to detect nicosulfuron in water samples. Anal Bioanal Chem 407(26):8087–8095. https://doi.org/10.1007/s00216-015-8982-x
91. Wang X et al (2018) Quantum dots based imprinting fluorescent nanosensor for the selective and sensitive detection of phycocyanin: a general imprinting strategy toward proteins. Sens Actuators B Chem 255:268–274. https://doi.org/10.1016/j.snb.2017.08.068
92. Shi T et al (2019) Application of molecular imprinting polymer anchored on CdTe quantum dots for the detection of sulfadiazine in seawater. Mar Pollut Bull 146:591–597. https://doi.org/10.1016/j.marpolbul.2019.07.010
93. Liu J et al (2010) Preparation of surface imprinting polymer capped Mn-Doped ZnS quantum dots and their application for chemiluminescence detection of 4-nitrophenol in tap water. Anal Chem 82(17):7380–7386. https://doi.org/10.1021/ac101510b
94. Zhao Y et al (2012) Composite QDs@MIP nanospheres for specific recognition and direct fluorescent quantification of pesticides in aqueous media. Anal Chem 84(1):386–395. https://doi.org/10.1021/ac202735v
95. Wang Y et al (2015) A core-shell CdTe quantum dots molecularly imprinted polymer for recognizing and detecting p-nitrophenol based on computer simulation. RSC Adv 5(90):73424–73433. https://doi.org/10.1039/c5ra06889d
96. Ren X, Chen L (2014) Quantum dots coated with molecularly imprinted polymer as fluorescence probe for detection of cyphenothrin. Biosens Bioelectron. https://doi.org/10.1016/j.bios.2014.08.086
97. Wei X et al (2014) Synthesis of molecularly imprinted silica nanospheres embedded mercaptosuccinic acid-coated CdTe quantum dots for selective recognition of λ-cyhalothrin. J Lumin 153:326–332. https://doi.org/10.1016/j.jlumin.2014.03.055
98. Chao MR, Hu CW, Chen JL (2016) Glass substrates crosslinked with tetracycline-imprinted polymeric silicate and CdTe quantum dots as fluorescent sensors. Anal Chim Acta 925:61–69. https://doi.org/10.1016/j.aca.2016.04.037
99. Sadeghi S, Jahani M, Belador F (2016) The development of a new optical sensor based on the Mn doped ZnS quantum dots modified with the molecularly imprinted polymers for sensitive recognition of florfenicol. Spectrochim Acta Part A: Mol Biomol Spectrosc 159:83–89. https://doi.org/10.1016/j.saa.2016.01.043
100. Ge S et al (2011) Development of a novel deltamethrin sensor based on molecularly imprinted silica nanospheres embedded CdTe quantum dots. Spectrochim Acta Part A Mol Biomol Spectrosc 79(5):1704–1709. https://doi.org/10.1016/j.saa.2011.05.040
101. Tang J, Studies LX (2016) Development of a probe based on quantum dots embedded with molecularly imprinted polymers to detect parathion. Pol J Environ. https://doi.org/10.15244/pjoes/60888
102. Li H, Li Y, Cheng J (2010) Molecularly imprinted silica nanospheres embedded Cdse quantum dots for highly selective and sensitive optosensing of pyrethroids. Chem Mater 22(8):2451–2457. https://doi.org/10.1021/cm902856y
103. Zhang C et al (2015) Development of fluorescence sensing material based on CdSe/ZnS quantum dots and molecularly imprinted polymer for the detection of carbaryl in rice and Chinese cabbage. J Agric Food Chem 63(20):4966–4972. https://doi.org/10.1021/acs.jafc.5b01072
104. Xu X et al (2004) Electrophoretic analysis and purification of fluorescent single-walled carbon nanotube fragments. J Am Chem Soc. https://doi.org/10.1021/ja040082h
105. Rezaei B et al (2016) Application of modified carbon quantum dots/multiwall carbon nanotubes/pencil graphite electrode for electrochemical determination of dextromethorphan. IEEE Sens J 16(8):2219–2227. https://doi.org/10.1109/JSEN.2016.2517005
106. Ensafi AA, Nasr-Esfahani P, Rezaei B (2018) Synthesis of molecularly imprinted polymer on carbon quantum dots as an optical sensor for selective fluorescent determination of promethazine hydrochloride. Sens Actuators B Chem 257:889–896. https://doi.org/10.1016/j.snb.2017.11.050

107. Hou J et al (2016) Rapid microwave-assisted synthesis of molecularly imprinted polymers on carbon quantum dots for fluorescent sensing of tetracycline in milk. Talanta 146:34–40. https://doi.org/10.1016/j.talanta.2015.08.024
108. Yang J et al (2018) Detection of trace tetracycline in fish via synchronous fluorescence quenching with carbon quantum dots coated with molecularly imprinted silica. Spectrochim Acta Part A Mol Biomol Spectrosc 190:450–456. https://doi.org/10.1016/j.saa.2017.09.066
109. Zhu S et al (2011) Strongly green-photoluminescent graphene quantum dots for bioimaging applications. Chem Commun 47(24):6858–6860. https://doi.org/10.1039/c1cc11122a
110. Durán N et al (2016) Nanobiotechnology of carbon dots: a review. J Biomed Nanotechnol 1323–1347. https://doi.org/10.1166/jbn.2016.2225
111. Zor E et al (2015) Graphene quantum dots-based photoluminescent sensor: a multifunctional composite for pesticide detection. ACS Appl Mater Interfaces 7(36):20272–20279. https://doi.org/10.1021/acsami.5b05838
112. Üzek R et al (2019) A nitrocellulose paper strip for fluorometric determination of bisphenol A using molecularly imprinted nanoparticles. Microchim Acta 186(4):1–10. https://doi.org/10.1007/s00604-019-3323-y
113. Pandey JK et al (2010) Recent advances in the application of natural fiber based composites. Macromol Mater Eng. https://doi.org/10.1002/mame.201000095

Chapter 23
Review of Chemical Treatments of Natural Fibers: A Novel Plastination Approach

Reeghan Osmond, Kevin Golovin, and Abbas S. Milani

1 Introduction

Starting in the 1940s, following advancements in petrochemical technology, fiber-reinforced polymer (FRP) composites became a staple material in many industries because of their high strength and stiffness-to-weight ratios [1, 2]. In 2014, the US FRP market increased to a value of $8.2 billion [3]. The weight saved by using FRPs has improved fuel efficiency in aerospace, space, and automotive applications [1, 4]. They can also be found in buildings, sporting equipment, the marine industry, and wind turbine construction [1, 2, 5, 6].

The combination of the two distinct phases present in FRPs—fibers and polymer matrix—result in a combination of material properties that could not solely be achieved by either phase alone [7]. The fibers bear the majority of the load while the polymer matrix holds the fibers in place, distributes the load, and protects the fibers from the environment [7, 8]. Within the matrix, the fibers can come in different forms including chopped strands, chopped strand mats, continuous fiber-reinforced mats, and woven fabrics [8]. Depending on the fiber form that is used, a different fiber volume fraction is required in order to optimize mechanical properties (typically between 15 and 70%) [8]. Furthermore, as more fibers are oriented in the direction of loading, the mechanical properties of the FRP increase [8].

There are two main types of fibers: synthetic and natural. Synthetic fibers are not naturally occurring and include carbon, boron, aramid, and glass [7]. In particular, carbon fibers are created from polyacrylonitrile (PAN) after undergoing a process involving carbonization and graphitization which requires temperatures ranging from

R. Osmond · K. Golovin (✉) · A. S. Milani (✉)
School of Engineering, University of British Columbia, Kelowna, BC, Canada
e-mail: kevin.golovin@ubc.ca

A. S. Milani
e-mail: abbas.milani@ubc.ca

S. Thomas and P. Balakrishnan (eds.), *Green Composites*,
Materials Horizons: From Nature to Nanomaterials,
https://doi.org/10.1007/978-981-15-9643-8_23

1000 to 3000 °C [9]. These expensive, high-performance fibers are used extensively in the aerospace industry and make up more than 50% of the Boeing 787 Dreamliner's structure and 25% of the Airbus A380 [4, 10].

Although glass fibers have a lower tensile strength and modulus than carbon, they still exhibit good mechanical properties and are less expensive to produce [11]. One of the most common formulations of glass fiber is called E-glass which is made from alumina–borosilicate and other minerals [11]. The fibers are drawn through a platinum alloy bushing from the 1250 °C mineral melt, cooled, and then wound into roving [11].

Synthetic fibers require harsh processing conditions and may come from non-renewable resources. Conversely, natural fibers may be derived from plant stems, leaves, and fruit [12]. These plants absorb CO_2 and generate O_2 when growing [13]. The fibers are less abrasive and produce less toxic fumes during processing [13]. Natural fibers have been used for centuries in a variety of applications. Most commonly, flax (linen), jute, hemp, and cotton have been used to make clothing, ropes, and bags [14-17], while bamboo and types of wood have been used to make buildings [18, 19]. Only more recently, natural fiber-reinforced polymer (NFRP) composites have gained popularity because of their sustainable nature [16].

NFRPs have a comparable stiffness to glass fiber-reinforced polymer (GFRP) composites while maintaining a low cost [13]. Currently, they are being heavily used in the automotive industry for interior components such as door panels, seat backs, and engine covers [13, 20]. For example, in 2000, Audi released the A2 car whose door panels were made of flax, sisal, and polyurethane [20]. They have also been used to make toys, sporting equipment, and in electronic devices [13]. However, using NFRPs does come with a few main disadvantages, namely natural fibers do not bond well with the polymer matrix, have high moisture absorption, are susceptible to microbial attack, and are flammable [13, 21-24]. A widely used method for rectifying these problems is through chemical treatments. In this chapter, an overview of common and more recent developments in natural fiber chemical treatments will be discussed.

2 Background

Natural fibers are derived from different parts of plants. In Table 1, the properties of some natural fibers that are commonly used in NFRPs are tabulated. Notably, the fibers with higher mechanical properties come from the stems of different plants, and are known as bast fibers [16].

The bast fibers from ramie, jute, hemp, and flax plants all exhibit similar or higher Young's moduli in comparison with E-glass fiber. These fibers also have higher specific moduli because of their lower densities than E-glass fibers. In terms of mechanical properties, these fibers would likely be the best candidates for replacing E-glass fibers in high performance NFRPs.

Table 1 Properties of select natural fibers compared to E-glass fiber (reprinted from [12], Copyright (2012), with permission from Elsevier)

Fiber type	Density (g/cm^3)	Length (mm)	Diameter (μm)	Tensile strength (MPa)	Tensile modulus (GPa)	Specific modulus (approx)	Elongation (%)	Cellulose (wt%)	Hemi-cellulose (wt%)	Lignin (wt%)	Pectin (wt%)	Waxes (wt%)	Micro-fibrillar angle (°)	Moisture content (wt%)
E-glass	2.5–2.59	–	< 17	2000–3500	70–76	29	1.8–4.8	–	–	–	–	–	–	–
Abaca	1.5	–	–	400–980	6.2–20	9	1.0–10	56–63	20–25	7–13	1	3	–	5–10
Alfa	0.89	–	–	35	22	25	5.8	45.4	38.5	14.9	–	2	–	–
Bagasse	1.25	10–300	10–34	222–290	17–27.1	18	1.1	32–55.2	16.8	19–25.3	–	–	–	–
Bamboo	0.6–1.1	1.5–4	25–40	140–800	11–32	25	2.5–3.7	26–65	30	5–31	–	–	–	–
Banana	1.35	300–900	12–30	500	12	9	1.5–9	63–67.6	10–19	5	–	–	–	8.7–12
Coir	1.15–1.46	20–150	10–460	95–230	2.8–6	4	15–51.4	32–43.8	0.15–20	40–45	3–4	–	30–49	8
Cotton	1.5–1.6	10–60	10–45	287–800	5.5–12.6	6	3–10	82.7–90	5.7	<2	0–1	0.6	–	7.85–8.5
Curaua	1.4	35	7–10	87–1150	11.8–96	39	1.3–4.9	70.7–73.6	9.9	7.5–11.1	–	–	–	–
Flax	1.4–1.5	5–900	12–600	343–2000	27.6–103	45	1.2–3.3	62–72	18.6–20.6	2–5	2.3	1.5–1.7	5–10	8–12
Hemp	1.4–1.5	5–55	25–500	270–900	23.5–90	40	1–3.5	68–74.4	15–22.4	3.7–10	0.9	0.8	2–6.2	6.2–12
Henequen	1.2	–	–	430–570	10.1–16.3	11	3.7–5.9	60–77.6	4–28	8–13.1	–	0.5	–	–
Isora	1.2–1.3	–	–	500–600	–	–	5–6	74	–	23	–	1.09	–	–
Jute	1.3–1.49	1.5–120	20–200	320–800	8–78	30	1–1.8	59–71.5	13.6–20.4	11.8–13	0.2–0.4	0.5	8	12.5–13.7
Kenaf	1.4	–	–	223–930	14.5–53	24	1.5–2.7	31–72	20.3–21.5	8–19	3–5	–	–	–
Nettle	–	–	–	650	38	–	1.7	86	10	–	–	4	–	11–17
Oil palm	0.7–1.55	–	150–500	80–248	0.5–3.2	2	17–25	60–65	–	11–29	–	–	42–46	–
Piassava	1.4	–	–	134–143	1.07–4.59	2	7.8–21.9	28.6	25.8	45	–	–	–	–
PALF	0.8–1.6	900–1500	20–80	180–1627	1.44–82.5	35	1.6–14.5	70–83	–	5–12.7	–	–	14	11.8
Ramie	1.0–1.55	900–1200	20–80	400–1000	24.5–128	60	1.2–4.0	68.6–85	13–16.7	0.5–0.7	1.9	0.3	7.5	7.5–17
Sisal	1.33–1.5	900	8–200	363–700	9.0–38	17	2.0–7.0	60–78	10.0–14.2	8.0–14	10	2	10–22	10–22

Some common polymers used to make NFRPs include polypropylene, high-density polyethylene, polystyrene, epoxy, and polyester [25]. Since natural fibers are lighter than E-glass, a higher fiber content is needed to achieve a similar performance, and thus, less polymer matrix is needed to make NFRPs [26]. Most polymer production generates more harmful emissions and uses more energy than growing and processing natural fibers [26].

2.1 Structure and Chemical Composition

The inherent structure and chemical composition of natural fibers influence their properties. Natural fibers consist of macrofibrils that have multiple cell walls surrounding an open channel called the lumen [27, 28]. Each cell wall consists of a hemicellulose-lignin matrix reinforced with semi-crystalline cellulose microfibrils oriented at different angles [27]. There is one primary cell wall in each microfibril and three secondary cell walls [28]. The middle secondary cell wall is the thickest and most significant for determining mechanical properties [27]. Figure 1 shows an illustration of a typical natural fiber structure.

The amount of each chemical present in the fiber is dependent on how the plant was grown and from which part of the plant the fibers were taken [27]. Cellulose is hydrophilic, highly crystalline (up to 80% crystallinity), and contains repeating units of glucose [27-29]. Better mechanical properties can be achieved with higher cellulose content, along with a lower microfibril angle, and a higher degree of polymerization [27]. In Fig. 2, the chemical structure of cellulose is shown. Hemicellulose

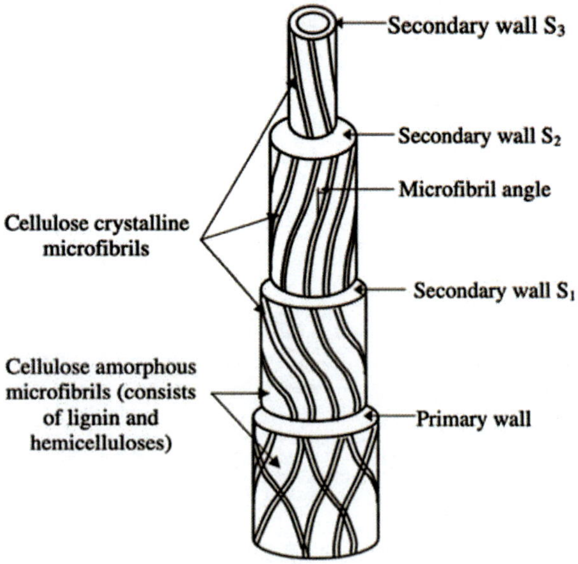

Fig. 1 Typical structure of a natural fiber (reprinted from [30], Copyright (2012), with permission from Elsevier)

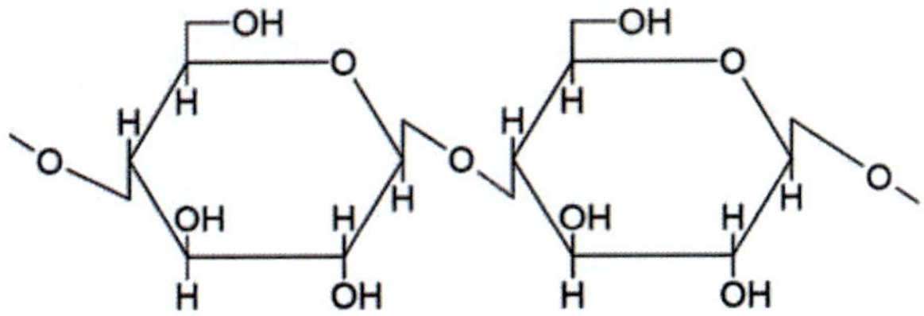

Fig. 2 Typical chemical structures of cellulose

is composed of highly branched polysaccharides that hydrogen bond to the cellulose [28]. The hydrophobic lignin in the matrix acts as a stiffener and protective layer that fills the gaps between the hemicellulose and cellulose [28]. Lignin is an amorphous polymer consisting of aromatic phenyl-propane [28]. Pectin, waxes, and water-soluble molecules are also present [28]. Pectin consists of polygalacturon acid and gives the fibers their flexibility [27].

2.2 *Interfacial Adhesion*

Natural fibers are polar, hydrophilic materials, whereas the polymer matrix used in NFRPs tends to be hydrophobic. Since the fiber and matrix are not very compatible, the interfacial adhesion between them is weak. This can affect mechanical properties such as toughness, tensile strength, and flexural strength because weak interfacial adhesion leads to poor stress transfer [13, 24]. Weak interfacial adhesion leads to NFRP failure between the fibers and the matrix [31]. The fiber can cleanly peel away from the matrix. However, when interfacial adhesion is strong, the matrix will fail first leaving residual polymer on the fibers. A strong interface can also reduce moisture absorption as there are less voids present for water to accumulate. The interfacial adhesion between the fibers and matrix can be improved through different mechanisms including mechanical interlocking, chemical bonding, and inter-diffusion bonding [13].

2.3 *Moisture Absorption Properties*

Hemicellulose is the main component in natural fibers that is responsible for moisture absorption [24]. The hydroxyl groups in the amorphous phase of cellulose are also readily available to uptake moisture, whereas only some of the hydroxyl groups are available in its crystalline form [27]. The amount of moisture absorbed by the fibers depends on the humidity of the surrounding air [27].

The polymer matrix used in NFRPs can also absorb moisture from the air and when immersed in water [24]. The water can then be transported to the fibers through either diffusion or capillarity, by flowing through imperfections like cracks [24]. Fiber swelling damages the surrounding matrix and allows more water to reach the fibers.

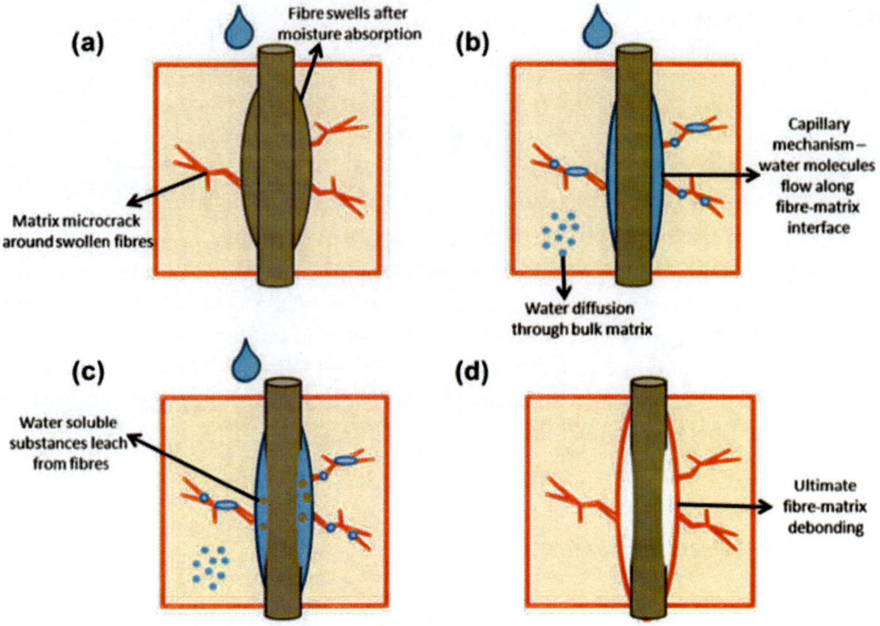

Fig. 3 Diagram depicting the effects of water on a NFRP fiber-matrix interface (reprinted from [24], Copyright (2013), with permission from Elsevier)

Water-soluble substances in the fibers are then removed from the fibers and matrix via capillaries along the fiber-matrix interface. This results in fiber-matrix debonding as depicted in Fig. 3. As the fiber volume fraction in the NFRP increases, so does the amount of moisture that it absorbs. After extended exposure to moisture, natural fibers are susceptible to microbial attack [24].

2.4 *Microbial Attack*

Natural fibers can be degraded by microorganisms (bacteria and fungi) that mainly produce cellulolytic (hydrolyzes cellulose) enzymes [32]. The enzymes work to release glucose from the cellulose, so that it can be used by the microorganisms as an energy source. Two different enzymes (exoglucanase or (1 → 4)-β-D-glucan cellobiohydrolase, and endoglucanase or endo-(1 → 4)-β-D-glucan 4-glucanohydrolase) attack either the crystalline or amorphous cellulose of fibers, breaking them down into smaller molecules. Subsequently, another two enzymes (β-D-glucosidase or β-D-glucoside glucohydrolase, and (1 → 4)-β-D-glucan 4-glucohydrolase) further break down these molecules into glucose. However, the wax protecting the fibers must first be removed before the hydrolytic breakdown of the cellulose can occur. This can be done by other types of enzymes. Hemicellulose and pectin are easier to breakdown than cellulose, so the presence of these substances would allow faster degradations.

Fungi can start degrading natural fibers at lower levels of moisture than bacteria. In particular, *Penicillium* and *Aspergillus* require much less moisture to grow than other types of fungi. After they are established, they can accumulate more moisture, which can promote the growth of other types of fungi. Bacteria usually require that the natural fibers remain saturated with water before they can grow.

2.5 *Flammability*

During the thermal degradation of natural fibers, water is desorbed, leading to the formation of dehydrocellulose and levoglucosan which decompose into char, gas, tar, and flammable and non-flammable volatiles [33]. Cellulose decomposes between 260 and 350 °C, hemicellulose decomposes between 200 and 260 °C, and lignin decomposes between 160 and 400 °C. Natural fibers with higher cellulose content are more flammable because they form more combustible substances as they thermally degrade. More cellulose crystallinity results in a higher flammability but also a higher ignition temperature. Crystalline cellulose has a higher activation energy, but it also produces more combustible substances. Furthermore, a high degree of polymerization, in addition to microfibular orientation, results in less flammable fibers. NFRPs that have a better interfacial adhesion between the fibers and the matrix are less flammable. Flame retardants can be added to the polymer, and the fibers can be treated to reduce flammability.

3 Chemical Treatments

Most natural fiber chemical treatments have been used to improve the interfacial adhesion and reduce the flammability of NFRPs. Plastination, however, attempts to address the moisture absorption challenges associated with natural fibers. Outlined in this section are a few commonly used chemical treatment processes along with the new plastination technique.

3.1 *Alkaline*

The alkaline treatment process removes lignin and wax from the fibers and disrupts the hydrogen bonding [34]. This leads to an increased surface roughness which results in better mechanical interlocking and more exposed bonding sites between the matrix and fibers. Typically, NaOH is used as the solution.

Van de Weyenberg et al. [35] studied alkaline-treated flax fibers reinforced with epoxy. They soaked loose slivers of flax in different concentrations of NaOH for 20 min. Then, the fibers were washed with water and acidified water. Afterward, the fibers were dried in an oven for eight hours. They also treated continuous flax fiber by following a similar process, using different NaOH concentrations and bathing times. Their results showed that the alkaline treatment caused the fibers to swell, lowering their mechanical properties. However, when they were added to the epoxy matrix, the longitudinal and transverse tensile strength of the composite significantly increased. This was due to the improved interfacial adhesion between the fibers and the matrix. They also noted that the importance of properly rinsing the fibers to ensure residual NaOH is removed because it can cause excess swelling leading to porosity in the composite.

Similar results were found when sugar palm reinforced epoxy [36] and kenaf-reinforced epoxy [37] were studied. Longer treatments were used in these studies: 1–48 h with different concentrations of NaOH. The mechanical properties of the fibers improved, and it was noted that the hydrophilic nature of the fibers was reduced after alkaline treatment.

A study [38] of alkaline-treated alfa-reinforced polyester found a ~60% improvement in flexural strength and modulus after a 24 h treatment.

3.2 Acetylation

In acetylation, the hydroxyl groups in the natural fibers are reacted with acetyl groups to make the fibers hydrophobic [39]. The general process involves soaking in an NaOH solution, washing, and then soaking in acetic anhydride with sulfuric acid [40]. Others have reported slightly different processes which all involved an acetic anhydride bath with another chemical added, such as a catalyst when wood is being acetylated [39, 41, 42].

Albano et al. [40] and Rana et al. [41] studied the thermal effects of using an acetylation treatment on sisal fibers. They found that treated fibers were much more thermally stable than untreated fibers. Hung et al. [42] found that weathered bamboo NFRPs had much higher flexural properties and mildew resistance when they were treated using acetylation. As well, Bledzki et al. [39] found that flax fiber-reinforced polypropylene had higher tensile and flexural strength. They also saw that the acetylated fibers absorbed 50% less moisture than untreated fibers.

3.3 Maleated Coupling Agents

Maleated coupling agents were originally used to improve the interfacial adhesion in glass fiber-reinforced polymer composites [43]. Now, they are also used to improve the compatibility between natural fibers and hydrophobic matrices. Two common maleated coupling agents are maleic anhydride-grafted polypropylene (MAPP) and maleic anhydride-grafted polyethylene (MAPE). In both cases, the anhydride groups of the coupling agent interact with the hydroxyl groups of the natural fibers.

Kaewkuk et al. [44] studied the effects of adding MAPP to sisal fiber-reinforced polymer composites. They also used alkaline and heat treatment for comparison. It was found that the composites containing MAPP resulted in the best mechanical properties. They also had the lowest moisture absorption. In this study, the polypropylene and fibers were combined using a mixer and injection moulded. Appropriate amounts of MAPP were simply added to the fiber-polypropylene mixture.

Other studies have been done where the fibers first underwent an alkaline treatment and MAPP was used afterward when making the composite [45, 46]. These studies showed an improvement in stiffness, strength, and impact properties in the NFRPs. However, the NFRPs with MAPP were found to be less thermally stable than polypropylene alone.

3.4 Plastination: A New Approach

Plastination is a new treatment process borrowed from anatomy in the preservation of human tissues, which involves replacing water present in the natural fibers with a polymer—typically silicone rubber [47]. In essence, plastination creates a composite of its own, where the replacement polymer acts as the matrix. It has been shown to improve the mechanical properties of bamboo [47]. Plastination also partially hindered the moisture absorption of the bamboo [47].

As it exhibits a low thermal conductivity (0.1511–0.1678 Wm^{-1} K^{-1}), plastination using silicone can help reduce the flammability of natural fibers [48]. At elevated temperatures, silicone generates a silica residue which acts as an insulator and traps volatiles within [49]. This reduces the amount of volatiles available, slowing the burning process [49]. There have also been many studies where antibiotic and antifungal molecules and particles have been added to silicone [50]. Thus, it could be possible to use such additives within the silicone during plastination to protect the natural fibers against microbial attack.

Although silicone is the most commonly used polymer in the plastination of human remains, epoxy, polyester, polyurethane, and acrylic have also been used [51]. These different polymers and others could also be used to plastinate natural fibers depending on which properties are desired. For instance, to improve interfacial bonding between the NFRP fibers and matrix, a compatible polymer could be used during the plastination process (i.e,. matching the hydrophobicity of the plastination polymer and the matrix).

3.4.1 History

Originally, plastination was developed by Dr. von Hagens in 1977 to preserve human remains for studying and teaching anatomy [51, 52]. Instead of specimens prepared using solutions of carcinogenetic chemicals, plastinated samples are non-odorous, dry, durable, and non-toxic [51-53]. Human hearts, intestines, lungs, brains, limbs, and whole bodies are examples of remains which have been plastinated [51, 52]. Dr. von Hagens went on to create the Body Worlds exhibition, starting in 1997 [54]. The first exhibition was in Mannheim, Germany, and has since been shown around the world [54, 55]. They have been used to show the anatomical structures of different body parts along with demonstrating things like the effect of smoking on the lungs [54]. In more recent exhibits, the bodies have been displayed more dynamically to emulate the living—making it easier for exhibit visitors to empathize with the specimens [54].

Since its inception, plastination has also been used to preserve animal remains, fungi, and most recently natural fibers [47, 52, 56]. In 2020, Dhir et al. [47] published a paper where they characterized plastinated bamboo. In the subsections that follow, a brief review of different plastination methods will be discussed, starting with silicone rubber plastination, the method previously used for natural fibers.

3.4.2 Silicone Rubber Plastination

There are two well-established methods for plastinating with silicone: the standard S10 and room temperature techniques. The standard S10 plastination technique, devised by Dr. von Hagens [52], involves four main steps: fixation, dehydration and defatting, forced impregnation, and curing.

Fixation is performed to stop the initial decay of the specimen. Next, the water and fats are removed from the specimen during the dehydration and defatting step. As the fat and water cannot be replaced directly by a polymer, acetone or ethanol are used as solvents. Stepwise ethanol dehydration is used when the specimen was embalmed because it can remove long alcohol chains. This method causes a considerable amount of shrinkage, and the ethanol needs to be replaced by a low boiling point, high vapor pressure solvent like acetone or methylene chloride afterwards. However, it is completed at room temperature making it ideal for samples used in histological examinations (no ice crystals are formed). The ethanol also removes the water and fat simultaneously. In contrast, the standard dehydration process utilizes freeze substitution in acetone. This causes only minimal shrinkage but can generate ice crystals within the specimen since it is conducted at −25 °C. This can harm the cellular structure of the specimen, making it hard to examine up-close. To remove the fats, another acetone bath is required at room temperature. Freeze substitution is commonly used to preserve tissues and cells in applications other than plastination [57]. Normally, a solvent plus a fixative are substituted for the ice [57].

During the forced impregnation step, the solvent in the specimen is replaced with a silicone [52]. The specimen is placed in a vacuum chamber at −25 °C and the pressure is gradually lowered until the solvent is completely boiled away. The process will be complete when 2–15 mm Hg is reached, depending on the solvent used. Lower viscosity polymers are ideal for plastination because they have a faster impregnation speed. On its own, vacuum impregnation has been used to impregnate woods with polymers and chemicals like methyl methacrylate, styrene, and nano-zinc oxide to reduce biological degradation [58-60]. Fruits and vegetables are also commonly vacuum impregnated with sugars, salts, enzymes, and nutrients [61].

Finally, the specimen is removed from the polymer bath and placed into another chamber for polymer curing [52]. The crosslinking curing agent is evaporated and saturates the atmosphere within the chamber. As it comes into contact with the specimen, the polymer on the outside is cured. To allow the curing agent to fully penetrate the specimen, the sample is removed from the chamber and bagged for multiple weeks. The entire standard S10 process can take up to 12 weeks to complete. Figure 4 shows a plastinated human heart.

A three-part silicone is used during the standard S10 plastination process, where only two parts are added to the vacuum chamber for impregnation. The first part, the base silicone (hydroxy-terminated polydimethylsiloxane), is called BIODUR S10, and the second part, the catalyst/chain extender (mainly dibutyltindilaurate), is called BIODUR S3 (Fig. 5a, b) [52, 63, 64]. The S3 initiates curing through polymerization. To ensure that the mixture does not become too viscous, forced impregnation needs to be completed at freezing temperatures. This also allows the excess silicone to be used for plastinating more specimens in the future. The third part, BIODUR S6, is used to crosslink the silicone and contains tetraethoxysilane (Fig. 5c).

Later, the standard S10 plastination technique was altered to allow for room temperature impregnation [65]. This method was initially developed by Daniel Corcoran of Dow Corning Corporation in 1998, using similar chemicals to those of the standard S10 plastination technique [63, 66]. Most of the steps in this process

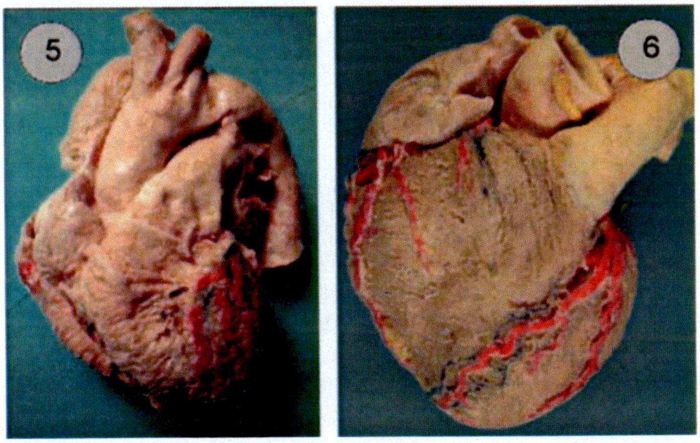

Fig. 4 Human hearts plastinated using the standard S10 technique with colouring (reprinted from [62], Copyright (2006), with permission from Elsevier)

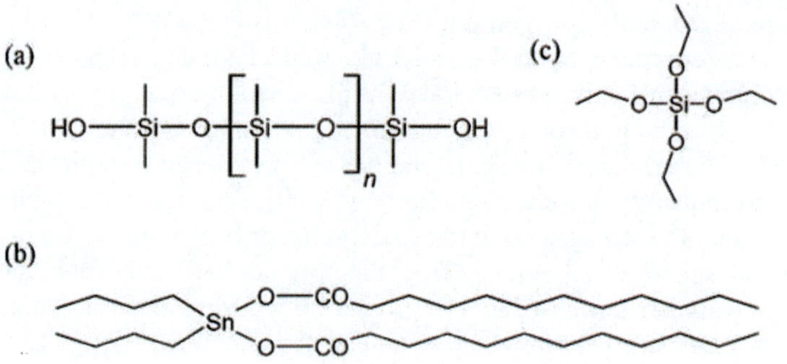

Fig. 5 Chemical structures of hydroxy-terminated polydimethylsiloxane (**a**), dibutyltindilaurate (**b**), and tetraethoxysilane (**c**)

remained the same, except the three-part silicone was mixed in a different order which changed the curing process. The same chemicals used in the standard S10 plastination technique can be used in the room temperature method [65]. In this instance, the S10 and S6 (base silicone and crosslinker) are first mixed together and used during forced impregnation. This mixture is more stable at room temperature because it does not include the catalyst. Then, the S3 catalyst is sprayed onto the specimen instead of using a gas curing chamber. The specimen is then wrapped in plastic, so that the S3 can penetrate the specimen and cure the silicone. This process is 65% faster than the standard S10 method [63]. However, specimens produced using the standard S10 method are more flexible and elastic. In general, mainly only the outer surface of room temperature specimens become cured.

Dr. Henry developed another version of the room temperature technique involving a four-part curing silicone [67]. This method follows the same general steps as outlined above where the cross-linker (NCSVI) and base silicone (NCSX) are first combined for impregnation. However, this is then followed by separate applications of the chain extender (NCSV) and catalyst (NCSIII). The specimen is placed in a closed chamber where the chain extender is vaporized similar to the standard S10 plastination technique. Finally, the catalyst is sprayed onto the specimen, and it is wrapped in plastic.

Another room temperature method was developed in China which uses a different silicone called the Su-Yi Chinese silicone [68]. The dehydration was conducted with acetone at room temperature. This was done in a series of graded acetone solutions, starting with 50% acetone and ending with 100%. The defatting and dehydration were completed simultaneously without much shrinkage. The Su-Yi Chinese silicone was used during forced impregnation which included two additional steps. Pre-and post-impregnation steps involved submerging the dehydrated specimens in the Su-Yi Chinese silicone without any vacuum applied to equilibrate. Vacuum pressure was gradually applied to the specimen during forced impregnation. Finally, the specimen was cured by spreading a mixture of the old silicone plus hardener onto the specimen.

Plastination has also been used to preserve fungi [56]. The methods used in this study were similar to the standard S10 plastination technique and the original room temperature method. However, instead of using the BIODUR chemicals, NCSX, NCSVI, and NCSIII were used in place of S10, S6, and S3, respectively. NCSV, the chain extender, was not used. Results from both methods were comparable, but the room temperature method was faster and had a glossier finish when there was too much NCSIII applied.

3.4.3 Plastination with Polymers Other Than Silicone Rubber

After silicone, the most commonly used polymers in plastination include epoxy and polyester. When using these polymers, the same general four steps in the process are followed [52]. However, the forced impregnation and curing steps differ from the standard S10 technique, as these methods are used to create specimen slices for examination under a microscope. Polyurethane and acrylic have been used to make transparent specimens [51, 69]. Commercially available polymer mixtures have also been used in plastination [69-71].

The standard epoxy technique is used for body, organ, and extremity slices, and it is called the E-12 technique [52]. Before plastination begins, the slices are cut to about 2.5 mm in thickness. After acetone dehydration, BIODUR E-12 (epoxy) and BIODUR E-1 (hardener) are mixed together to be used in forced impregnation. Rather than taking weeks to impregnate, the E-12 method only takes about one day. The slices are then placed between two glass plates and cured at room temperature or 50 °C. At 70 °C, human skin is damaged within one second of exposure, so high curing temperatures should not be used [72].

The standard polyester technique, the P 35 technique, is used for brain slices [52]. The slices are first cut to 4–6 mm thick and are then dehydrated in acetone. The defatting step is skipped to ensure that the slices remain opaque. Next, forced impregnation with BIODUR P 35 (polyester copolymer) and BIODUR A 9 (peroxide hardener) is completed at room temperature and in a darkened chamber in only one day. The vacuum pressure should not go lower than 10 mm Hg to ensure that the styrene in the mixture is not extracted. The polymer cures using UVA light at room temperature and a post-cure at 50 °C. During the curing process, the slices are kept between two double-walled glass plates.

3.4.4 Plastinating Natural Fibers

In Dhir et al.'s [47] study, culms and fiber bundles of *Inversa Bambusoide* (a species of bamboo) were plastinated using the original silicone room temperature technique. However, the fixation and defatting steps were omitted because natural fibers do not contain many lipids and do not decay as quickly as human remains. The density, moisture content, tensile strength, and tensile stiffness of the plastinated bamboo were measured and compared to that of virgin (non-plastinated) bamboo. The properties

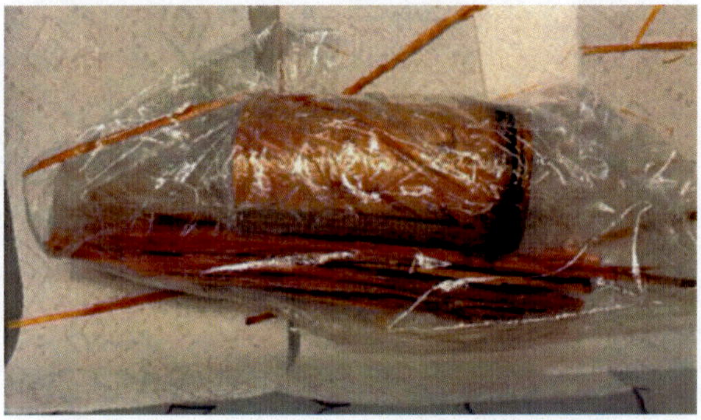

Fig. 6 Bamboo sprayed with S3 and wrapped in plastic [47]

of conditioned, plastinated, and virgin bamboo were also studied. Conditioning was defined as submersion in tap water (20 ± 2 °C) for four days. Figure 6 shows a sample of bamboo in the process of curing after forced impregnation.

After conditioning, the density of both the plastinated and virgin bamboo increased, meaning that they both absorbed water. However, the plastinated bamboo had a lesser increase in density. It was assumed that the high water vapor permeability of silicone was the reason for the water uptake of the plastinated samples. The moisture behaviour was also investigated, and it was found that plastinated bamboo contained 13–14% less moisture than virgin bamboo, before and after conditioning.

In terms of mechanical properties, the plastinated bamboo had a 70.8% higher tensile strength than virgin bamboo and a 22.9% increase in tensile stiffness. After conditioning, both plastinated and virgin bamboo exhibited a large decrease in mechanical properties due to the extra water absorption. However, conditioned, plastinated bamboo still showed a 73.3% higher tensile strength than conditioned virgin bamboo and a 55.3% higher tensile stiffness. These results were promising because of the large improvements observed in the mechanical properties of plastinated bamboo, both before and after conditioning. However, the plastinated samples did absorb water during conditioning which led to degraded mechanical properties. Even though the plastinated bamboo had some moisture present before conditioning, it was less than that of virgin bamboo. Thus, it appeared that plastinated bamboo has a lower affinity for moisture.

4 Outlook and Conclusions

Research is still needed to evaluate the microbial degradation, flame retardancy, and interfacial properties of plastinated bamboo along with other types of natural fibers. Plastination can also be a lengthy process. The Dhir et al. plastination process [47] took at least five days to complete. Since the room temperature plastination method was used, there may also have also been residual, uncured silicone present within the bamboo. Moreover, some ice crystals likely formed during acetone dehydration, damaging the bamboo's structure. This could have affected the mechanical properties of the samples before and after conditioning.

Although some plastination techniques used to preserve human remains utilize polymers that cure at 50 °C, higher temperatures should not be used to ensure that the specimens are not damaged. However, considering that natural fibers are stable up 160 °C, it is possible to use a heat-curing polymer during plastination. Furthermore, the forced impregnation step in natural fiber plastination could be done faster. As natural fibers are not transparent, UVA curing does not seem feasible. However, there is potential in using a polymer other than silicone to change the interfacial bonding properties to match that of the matrix. Perhaps plastination with the same polymer as the matrix could reduce the overall NFRP processing time.

NFRPs are a good replacement candidate for some synthetic FRPs because of their good mechanical and environmental properties. However, to achieve these attractive mechanical properties, generally chemical treatments are required. They are used to combat microbes, reduce flammability, improve interfacial bonding, and reduce moisture absorption. There are many different chemical treatments that try to address these challenges. In particular, alkaline treatments and maleated coupling agents are commonly used to improve interfacial bonding. Acetylation is used to address all of these challenges, and plastination has the potential to do this as well. There have already been results reported on the greatly improved mechanical properties and slight reduction in moisture absorption in plastinated bamboo. However, there is still research left to be conducted on optimizing plastination to further improve moisture absorption properties, process speed, and addressing the other challenges.

Acknowledgements The authors would like to acknowledge the stimulating discussions and support of Dr. Grant Bogyo and Mr. Ron Ryde from Ryde Holdings Ltd., as well as valuable assistance of colleagues at the Composites Research Network.

References

1. Bakis CE et al (2002) Fiber-reinforced polymer composites for construction—state-of-the-art review. J Compos Constr 6(2):73–87. https://doi.org/10.1061/(ASCE)1090-0268(2002)6:2(73)
2. Manjunatha CM, Taylor AC, Kinloch AJ, Sprenger S (2010) The tensile fatigue behaviour of a silica nanoparticle-modified glass fibre reinforced epoxy composite. Compos Sci Technol 70(1):193–199. https://doi.org/10.1016/j.compscitech.2009.10.012
3. Naqvi SR, Prabhakara HM, Bramer EA, Dierkes W, Akkerman R, Brem G (2018) A critical review on recycling of end-of-life carbon fibre/glass fibre reinforced composites waste using pyrolysis towards a circular economy. Resour Conserv Recycl 136(April):118–129. https://doi.org/10.1016/j.resconrec.2018.04.013
4. Yang Y, Boom R, Irion B, van Heerden DJ, Kuiper P, de Wit H (2012) Recycling of composite materials. Chem Eng Process Process Intensif 51:53–68. https://doi.org/10.1016/j.cep.2011.09.007
5. Roberts T (2007) Rapid growth forecast for carbon fibre market. Reinf Plast 51(2):10–13. https://doi.org/10.1016/S0034-3617(07)70051-6

6. Jawaid M, Abdul Khalil HPS (2011) Cellulosic/synthetic fibre reinforced polymer hybrid composites: a review. Carbohydr Polym 86(1):1–18. https://doi.org/10.1016/j.carbpol.2011.04.043.
7. Mallick PK (2007) Fiber-reinforced composites: materials, manufacturing, and design, 3rd edn. Taylor & Francis Group, LLC, Boca Raton
8. Edwards KL (1998) An overview of the technology of fibre-reinforced plastics for design purposes. Mater Des 19(1–2):1–10. https://doi.org/10.1016/s0261-3069(98)00007-7
9. Rahaman MSA, Ismail AF, Mustafa A (2007) A review of heat treatment on polyacrylonitrile fiber. Polym Degrad Stab 92(8):1421–1432. https://doi.org/10.1016/j.polymdegradstab.2007.03.023
10. Tang CS, Zimmerman JD, Nelson JI (2009) Managing new product development and supply chain risks: the Boeing 787 case. Supply Chain Forum an Int J 10(2):74–86. https://doi.org/10.1080/16258312.2009.11517219
11. Bingham PA, Wallenberger FT (2010) Fiberglass and glass technology: energy-friendly compositions and applications, 1st edn. Springer Science+Business Media, LLC, New York
12. Dittenber DB, Gangarao HVS (2012) Critical review of recent publications on use of natural composites in infrastructure. Compos Part A Appl Sci Manuf 43(8):1419–1429. https://doi.org/10.1016/j.compositesa.2011.11.019
13. Pickering KL, Efendy MGA, Le TM (2016) A review of recent developments in natural fibre composites and their mechanical performance. Compos Part A Appl Sci Manuf 83:98–112. https://doi.org/10.1016/j.compositesa.2015.08.038
14. Goyal A, Sharma V, Upadhyay N, Gill S, Sihag M (2014) Flax and flaxseed oil: an ancient medicine & modern functional food. J Food Sci Technol 51(9):1633–1653. https://doi.org/10.1007/s13197-013-1247-9
15. Riello G (2013) Cotton: the fabric that made the modern world. Cambridge University Press, New York
16. Mwaikambo L (2006) Review of the history, properties and application of plant fibres. Afr J Sci Technol 7(2):121
17. Webber CLI, Bledsoe VK (2002) Kenaf yield components and plant composition. Trends new crop new uses, pp 348–357
18. Sharma B, van der Vegte A (2020) Engineered bamboo for structural applications. In: Nonconventional and vernacular construction materials, 2nd edn. Woodhead Publishing, Sawston, pp 597–623
19. Ramage MH et al (2017) The wood from the trees: the use of timber in construction. Renew Sustain Energy Rev 68(October 2016):333–359. https://doi.org/10.1016/j.rser.2016.09.107
20. Koronis G, Silva A, Fontul M (2013) Green composites: a review of adequate materials for automotive applications. Compos Part B Eng 44(1):120–127. https://doi.org/10.1016/j.compositesb.2012.07.004
21. Faruk O, Bledzki AK, Fink HP, Sain M (2014) Progress report on natural fiber reinforced composites. Macromol Mater Eng 299(1):9–26. https://doi.org/10.1002/mame.201300008
22. Hill C, Hughes M (2010) Natural fibre reinforced composites opportunities and challenges. J Biobased Mater Bioenergy 4(2):148–158. https://doi.org/10.1166/jbmb.2010.1079
23. Tak Lau K, Yan Hung P, Zhu MH, Hui D (2018) Properties of natural fibre composites for structural engineering applications. Compos Part B Eng 136(November 2017):222–233. https://doi.org/10.1016/j.compositesb.2017.10.038.
24. Azwa ZN, Yousif BF, Manalo AC, Karunasena W (2013) A review on the degradability of polymeric composites based on natural fibres. Mater Des 47:424–442. https://doi.org/10.1016/j.matdes.2012.11.025
25. Ku H, Wang H, Pattarachaiyakoop N, Trada M (2011) A review on the tensile properties of natural fiber reinforced polymer composites. Compos Part B Eng 42(4):856–873. https://doi.org/10.1016/j.compositesb.2011.01.010
26. Joshi SV, Drzal LT, Mohanty AK, Arora S (2004) Are natural fiber composites environmentally superior to glass fiber reinforced composites? Compos Part A Appl Sci Manuf 35(3):371–376. https://doi.org/10.1016/j.compositesa.2003.09.016

27. Thomas S, Paul SA, Pothan LA, Deepa B (2011) Natural fibres: structure, properties and applications. In Cellulose fibers: bio- and nano-polymer composites. Springer, Berlin, Heidelberg, pp 3–42
28. Ho MP et al (2012) Critical factors on manufacturing processes of natural fibre composites. Compos Part B Eng 43(8):3549–3562. https://doi.org/10.1016/j.compositesb.2011.10.001
29. Bledzki AK, Gassan J (1999) Composites reinforced with cellulose based fibres. Prog Polym Sci 24:221–274. https://doi.org/10.1016/S0079-6700(98)00018-5
30. Kabir MM, Wang H, Lau KT, Cardona F (2012) Chemical treatments on plant-based natural fibre reinforced polymer composites: an overview. Compos Part B Eng 43(7):2883–2892. https://doi.org/10.1016/j.compositesb.2012.04.053
31. Chen P, Lu C, Yu Q, Gao Y, Li J, Li X (2006) Influence of fiber wettability on the interfacial adhesion of continuous fiber-reinforced PPESK composite. J Appl Polym Sci 102(3):2544–2551. https://doi.org/10.1002/app.24681
32. Szostak-Kotowa J (2004) Biodeterioration of textiles. Int Biodeterior Biodegrad 53(3):165–170. https://doi.org/10.1016/S0964-8305(03)00090-8
33. Chapple S, Anandjiwala R (2010) Flammability of natural fiber-reinforced composites and strategies for fire retardancy: a review. J Thermoplast Compos Mater 23(6):871–893. https://doi.org/10.1177/0892705709356338
34. Li X, Tabil LG, Panigrahi S (2007) Chemical treatments of natural fiber for use in natural fiber-reinforced composites: a review. J Polym Environ 15(1):25–33. https://doi.org/10.1007/s10924-006-0042-3
35. Van de Weyenberg I, Chi Truong T, Vangrimde B, Verpoest I (2006) Improving the properties of UD flax fibre reinforced composites by applying an alkaline fibre treatment. Compos Part A Appl Sci Manuf 37(9):1368–1376. https://doi.org/10.1016/j.compositesa.2005.08.016.
36. Bachtiar D, Sapuan SM, Hamdan MM (2008) The effect of alkaline treatment on tensile properties of sugar palm fibre reinforced epoxy composites. Mater Des 29(7):1285–1290. https://doi.org/10.1016/j.matdes.2007.09.006
37. Fiore V, Di Bella G, Valenza A (2015) The effect of alkaline treatment on mechanical properties of kenaf fibers and their epoxy composites. Compos Part B Eng 68:14–21. https://doi.org/10.1016/j.compositesb.2014.08.025
38. Rokbi M, Osmani H, Imad A, Benseddiq N (2011) Effect of chemical treatment on flexure properties of natural fiber-reinforced polyester composite. Procedia Eng 10:2092–2097. https://doi.org/10.1016/j.proeng.2011.04.346
39. Bledzki AK, Mamun AA, Lucka-Gabor M, Gutowski VS (2008) The effects of acetylation on properties of flax fibre and its polypropylene composites. Express Polym Lett 2(6):413–422. https://doi.org/10.3144/expresspolymlett.2008.50
40. Albano C, González J, Ichazo M, Kaiser D (1999) Thermal stability of blends of polyolefins and sisal fiber. Polym Degrad Stab 66(2):179–190. https://doi.org/10.1016/S0141-3910(99)00064-6
41. Rana AK, Basak RK, Mitra BC, Lawther M, Banerjee AN (1997) Studies of acetylation, of jute using simplified procedure and its characterization. J Appl Polym Sci 64(8):1517–1523. https://doi.org/10.1002/(SICI)1097-4628(19970523)64:8%3c1517::AID-APP9%3e3.0.CO;2-K
42. Hung KC, Chen YL, Wu JH (2012) Natural weathering properties of acetylated bamboo plastic composites. Polym Degrad Stab 97(9):1680–1685. https://doi.org/10.1016/j.polymdegradstab.2012.06.016
43. Keener TJ, Stuart RK, Brown TK (2004) Maleated coupling agents for natural fibre composites. Compos Part A Appl Sci Manuf 35(3):357–362. https://doi.org/10.1016/j.compositesa.2003.09.014
44. Kaewkuk S, Sutapun W, Jarukumjorn K (2013) Effects of interfacial modification and fiber content on physical properties of sisal fiber/polypropylene composites. Compos Part B Eng 45(1):544–549. https://doi.org/10.1016/j.compositesb.2012.07.036
45. Beckermann GW, Pickering KL (2008) Engineering and evaluation of hemp fibre reinforced polypropylene composites: fibre treatment and matrix modification. Compos Part A Appl Sci Manuf 39(6):979–988. https://doi.org/10.1016/j.compositesa.2008.03.010

46. El-Sabbagh A (2014) Effect of coupling agent on natural fibre in natural fibre/polypropylene composites on mechanical and thermal behaviour. Compos Part B Eng 57:126–135. https://doi.org/10.1016/j.compositesb.2013.09.047
47. Dhir DK, Rashidi A, Bogyo G, Ryde R, Pakpour S, Milani AS (2020) Environmental durability enhancement of natural fibres using plastination: a feasibility investigation on bamboo. Molecules 25(3):1–14. https://doi.org/10.3390/molecules25030474
48. Kuo ACM (1999) Poly(dimethylsiloxane). In: Polymer data handbook. Oxford University Press, Inc., pp 411–435.
49. Hamdani S, Longuet C, Perrin D, Lopez-cuesta JM, Ganachaud F (2009) Flame retardancy of silicone-based materials. Polym Degrad Stab 94(4):465–495. https://doi.org/10.1016/j.polymdegradstab.2008.11.019
50. Muñoz-Bonilla A, Fernández-García M (2012) Polymeric materials with antimicrobial activity. Prog Polym Sci 37(2):281–339. https://doi.org/10.1016/j.progpolymsci.2011.08.005
51. von Hagens G (1979) Impregnation of soft biological specimens with thermosetting resins and elastomers. Anat Rec 194(2):247–255. https://doi.org/10.1002/ar.1091940206
52. von Hagens G, Tiedemann K, Kriz W (1987) The current potential of plastination. Anat Embryol (Berl) 175:411–421
53. Douglass C, Glover R (2003) Plastination: preservation technology enhances biology teaching. Am Biol Teach 65(7):503–510. https://doi.org/10.1662/0002-7685(2003)065[0503:pptebt]2.0.co;2
54. Pashaei S (2010) A brief review on the history, methods and applications of plastination. Int J Morphol 28(4):1075–1079. https://doi.org/10.4067/S0717-95022010000400014
55. Walter T (2004) Plastination for display: a new way to dispose of the dead. J R Anthropol Inst 10(3):603–627. https://doi.org/10.1111/j.1467-9655.2004.00204.x
56. Looney B, Henry RW (2014) Fruitbody worlds, plastination of muhsrooms. Fungi 7(1):45–49
57. Feder N, Sidman RL (1958) Methods and principles of fixation by freeze-substitution. J Biophys Biochem Cytol 4(5):593–600. https://doi.org/10.1083/jcb.4.5.593
58. Mathias LJ, Lee S, Wright JR, Warren SC (1991) Improvement of wood properties by impregnation with multifunctional monomers. J Appl Polym Sci 42(1):55–67. https://doi.org/10.1002/app.1991.070420107
59. Husain MM, Khan MA, Ali MA, Ali KMI, Mustafa AI (1996) Impregnation mode in wood plastic composite. Radiat Phys Chem 48(6):781–786
60. Clausen CA, Green F, Kartal SN (2010) Weatherability and leach resistance of wood impregnated with nano-zinc oxide. Nanoscale Res Lett 5(9):1464–1467. https://doi.org/10.1007/s11671-010-9662-6
61. Zhao Y, Xie J (2004) Practical applications of vacuum impregnation in fruit and vegetable processing. Trends Food Sci Technol 15(9):434–451. https://doi.org/10.1016/j.tifs.2004.01.008
62. Steinke H, Spanel-Borowski K (2006) Coloured plastinates. Ann Anat 188(2):177–182. https://doi.org/10.1016/j.aanat.2005.10.001
63. Starchik D, Henry RW (2015) Comparison of cold and room temperature silicone plastination techniques using tissue core samples and a variety of plastinates. J Plast 27(2):13–19
64. Chaynes P, Mingotaud AF (2004) Analysis of commercial plastination agents. Surg Radiol Anat 26(3):235–238. https://doi.org/10.1007/s00276-003-0216-9
65. von Hagens G (1986) Plastination folder. Biodur Products GmbH, Heidelberg
66. Raoof A (2007) Silicone plastination of biological tissue: room-temperature technique DowTM/corcoran technique ande products. J Int Soc Plast
67. Henry RW (2007) Silicone plastination of biological tissue: room-temperature technique North Carolina technique and products. J Int Soc Plast pp 26–30
68. Tianzhong Z, Jingren L, Kerming Z (1998) Plastination at room temperature. J Plast 13(2):21–25
69. Steinke H, Pfeiffer S, Spanel-Borowski K (2002) A new plastination technique for head slices containing brain. Ann Anat 184(4):353–358. https://doi.org/10.1016/S0940-9602(02)80055-3
70. Ottone NE, del Sol M, Fuentes R (2016) Report on a sheet plastination technique using commercial epoxy resin. Int J Morphol 34(3):1039–1043. https://doi.org/10.4067/s0717-95022016000300036

71. Sakamoto Y, Miyake Y, Kanahara K, Kajita H, Ueki H (2006) Chemically reactivated plastination with Shin-Etsu silicone KE-108. J Int Soc Plast 21:11–16
72. Moritz AR, Henriques FC (1947) Studies of thermal injury: II. The relative importance of time and surface temperature in the causation of scutaneous burns. Am J Pathol 23(5):695–720